Advances in Experimental Medicine and Biology

Volume 1045

Editorial Board
IRUN R. COHEN, *The Weizmann Institute of Science, Rehovot, Israel*
ABEL LAJTHA, *N.S.Kline Institute for Psychiatric Research, Orangeburg, NY, USA*
JOHN D. LAMBRIS, *University of Pennsylvania Philadelphia, PA, USA*
RODOLFO PAOLETTI, *University of Milan, Milan, Italy*
NIMA REZAEI, *Children's Medical Center, Tehran University of Medical Sciences, Tehran, Iran*

More information about this series at http://www.springer.com/series/5584

Yasushi Kawaguchi • Yasuko Mori
Hiroshi Kimura
Editors

Human Herpesviruses

Editors
Yasushi Kawaguchi
Division of Molecular Virology
Department of Microbiology and
Immunology
The Institute of Medical Science
The University of Tokyo
Minato-ku, Tokyo, Japan

Department of Infectious Disease
Control, International Research Center
for Infectious Diseases
The Institute of Medical Science
The University of Tokyo
Minato-ku, Tokyo, Japan

Research Center for Asian Infectious
Diseases
The Institute of Medical Science
The University of Tokyo
Minato-ku, Tokyo, Japan

Yasuko Mori
Division of Clinical Virology
Kobe University Graduate School
of Medicine
Kobe, Hyogo, Japan

Hiroshi Kimura
Department of Virology
Nagoya University Graduation School
of Medicine
Nagoya, Aichi, Japan

ISSN 0065-2598   ISSN 2214-8019 (electronic)
Advances in Experimental Medicine and Biology
ISBN 978-981-10-7229-1   ISBN 978-981-10-7230-7 (eBook)
https://doi.org/10.1007/978-981-10-7230-7

Library of Congress Control Number: 2018941559

© Springer Nature Singapore Pte Ltd. 2018
Chapter 22: This is a U.S. government work and its text is not subject to copyright protection in the United States; however, its text may be subject to foreign copyright protection 2018
This work is subject to copyright. All rights are reserved by the Publisher, whether the whole or part of the material is concerned, specifically the rights of translation, reprinting, reuse of illustrations, recitation, broadcasting, reproduction on microfilms or in any other physical way, and transmission or information storage and retrieval, electronic adaptation, computer software, or by similar or dissimilar methodology now known or hereafter developed.
The use of general descriptive names, registered names, trademarks, service marks, etc. in this publication does not imply, even in the absence of a specific statement, that such names are exempt from the relevant protective laws and regulations and therefore free for general use.
The publisher, the authors and the editors are safe to assume that the advice and information in this book are believed to be true and accurate at the date of publication. Neither the publisher nor the authors or the editors give a warranty, express or implied, with respect to the material contained herein or for any errors or omissions that may have been made. The publisher remains neutral with regard to jurisdictional claims in published maps and institutional affiliations.

Printed on acid-free paper

This Springer imprint is published by the registered company Springer Nature Singapore Pte Ltd.
The registered company address is: 152 Beach Road, #21-01/04 Gateway East, Singapore 189721, Singapore

Preface

Herpesviridae encompasses a large family of double-stranded DNA viruses with unique biological features, enabling them to establish latency after primary infection prior to reactivation later in life. Following the discovery of different human herpesvirus subspecies, an abundance of research has accumulated in both basic research and clinical medicine that tie their infective process to human disease. The significance of herpesvirus infection is increasing not only in clinical fields but also in biological aspects.

A vast majority of biological features are still masked in mystery. Additionally, strategies for treatment and prevention have not yet been established against most human herpesvirus species. To date, nine human herpesvirus species are known, and each can cause a variety of diseases during the primary infection and reactivation stages.

In this book, experts introduce and review several topics on each human herpesviruses. The book divides herpesviridae into three subfamilies: alphaherpesviruses, betaherpesviruses, and gammaherpesviruses. Each subfamily is unique in its specific biological and clinical characteristics. One of the most important features of this book is that it covers aspects of basic research and clinical medicines. The most current research on herpesviridae is outlined in this book and is sure to attract a wide range of readers.

This book would not be possible without all of the authors who contributed their time and effort, for which I am grateful. Additionally, the writing of this book would not have been feasible without the guidance and expertise of the Herpesvirus Study Group in Japan. I would like to thank all the current and former members of the group.

Nagoya, Japan Hiroshi Kimura

Contents

Part I   Alphaherpesviruses

1 **The Role of HSV Glycoproteins in Mediating Cell Entry**........... 3
Jun Arii and Yasushi Kawaguchi

2 **Virus Assembly and Egress of HSV**............................ 23
Colin Crump

3 **Us3 Protein Kinase Encoded by HSV: The Precise Function and Mechanism on Viral Life Cycle**............................ 45
Akihisa Kato and Yasushi Kawaguchi

4 **Oncolytic Virotherapy by HSV**................................ 63
Daisuke Watanabe and Fumi Goshima

5 **Neurological Disorders Associated with Human Alphaherpesviruses**... 85
Jun-ichi Kawada

6 **Antiviral Drugs Against Alphaherpesvirus**..................... 103
Kimiyasu Shiraki

7 **Vaccine Development for Varicella-Zoster Virus**................ 123
Tomohiko Sadaoka and Yasuko Mori

Part II   Betaherpesviruses

8 **Glycoproteins of HHV-6A and HHV-6B**.......................... 145
Huamin Tang and Yasuko Mori

9 **Betaherpesvirus Virion Assembly and Egress**................... 167
William L. Close, Ashley N. Anderson, and Philip E. Pellett

10 **Chromosomal Integration by Human Herpesviruses 6A and 6B**.... 209
Louis Flamand

| 11 | **Structural Aspects of Betaherpesvirus-Encoded Proteins** 227
Mitsuhiro Nishimura and Yasuko Mori |
|---|---|
| 12 | **Betaherpesvirus Complications and Management During Hematopoietic Stem Cell Transplantation**. 251
Tetsushi Yoshikawa |
| 13 | **Vaccine Development for Cytomegalovirus** 271
Naoki Inoue, Mao Abe, Ryo Kobayashi, and Souichi Yamada |

Part III Gammaherpesviruses

| 14 | **KSHV Genome Replication and Maintenance in Latency** 299
Keiji Ueda |
|---|---|
| 15 | **Signal Transduction Pathways Associated with KSHV-Related Tumors** 321
Tadashi Watanabe, Atsuko Sugimoto, Kohei Hosokawa, and Masahiro Fujimuro |
| 16 | **Pathological Features of Kaposi's Sarcoma-Associated Herpesvirus Infection** 357
Harutaka Katano |
| 17 | **EBV-Encoded Latent Genes**. 377
Teru Kanda |
| 18 | **Encyclopedia of EBV-Encoded Lytic Genes: An Update** 395
Takayuki Murata |
| 19 | **Animal Models of Human Gammaherpesvirus Infections** 413
Shigeyoshi Fujiwara |
| 20 | **Gastritis-Infection-Cancer Sequence of Epstein-Barr Virus-Associated Gastric Cancer** 437
Masashi Fukayama, Akiko Kunita, and Atsushi Kaneda |
| 21 | **EBV in T-/NK-Cell Tumorigenesis** 459
Hiroshi Kimura |
| 22 | **Vaccine Development for Epstein-Barr Virus** 477
Jeffrey I. Cohen |

Index. 495

Part I
Alphaherpesviruses

Chapter 1
The Role of HSV Glycoproteins in Mediating Cell Entry

Jun Arii and Yasushi Kawaguchi

Abstract The successful entry of herpes simplex virus (HSV) into a cell is a complex process requiring the interaction of several surface viral glycoproteins with host cell receptors. These viral glycoproteins are currently thought to work sequentially to trigger fusogenic activity, but the process is complicated by the fact that each glycoprotein is known to interact with a range of target cell surface receptor molecules. The glycoproteins concerned are gB, gD, and gH/gL, with at least four host cell receptor molecules known to bind to gB and gD alone. Redundancy among gD receptors is also evident and binding to both the gB and gD receptors simultaneously is known to be required for successful membrane fusion. Receptor type and tissue distribution are commonly considered to define the extent of viral tropism and thus the magnitude of pathogenesis. Viral entry receptors are therefore attractive pharmaceutical target molecules for the prevention and/or treatment of viral infections. However, the large number of HSV glycoprotein receptors makes a comprehensive understanding of HSV pathogenesis in vivo difficult. Here we summarize our current understanding of the various HSV glycoprotein cell surface receptors, define their redundancy and binding specificity, and discuss the significance of these interactions for viral pathogenesis.

Keywords HSV · Glycoprotein · Entry · Receptor · Pathogenesis · Tropism

---

J. Arii (✉) · Y. Kawaguchi
Division of Molecular Virology, Department of Microbiology and Immunology, The Institute of Medical Science, The University of Tokyo, Minato-ku, Tokyo, Japan

Department of Infectious Disease Control, International Research Center for Infectious Diseases, The Institute of Medical Science, The University of Tokyo, Minato-ku, Tokyo, Japan

Research Center for Asian Infectious Diseases, The Institute of Medical Science, The University of Tokyo, Minato-ku, Tokyo, Japan
e-mail: jun-arii@ims.u-tokyo.ac.jp

1.1 Introduction

In order to replicate, enveloped viruses must be able to fuse with the membrane of a living cell and deliver their genetic material into its cytoplasm. The process of membrane fusion is initiated by binding of a virus to an appropriate receptor on the cell surface and is mediated by a virus-encoded membrane fusion protein (fusogen). Herpesviruses are enveloped double-stranded DNA viruses that establish lifelong latent infections in their natural hosts (Pellet and Roizman 2013). The *Herpesviridae* family is subdivided into the *Alphaherpesvirinae*, *Betaherpesvirinae*, and *Gammaherpesvirinae* subfamilies, based on their molecular and biological properties (Pellet and Roizman 2013). Herpes simplex virus 1 and -2 (HSV-1 and-2) are prototypes of the alphaherpesvirus subfamily and are among the most common pathogenic agents in humans, causing a variety of conditions such as mucocutaneous disease, keratitis, skin disease, and encephalitis (Roizman et al. 2013). After primary infection at a peripheral site, the virus establishes a lifelong latency in sensory neurons from which it periodically reactivates to cause lesions at or near the primary infection site (Roizman et al. 2013). Entry of herpesvirus is a complex process in which fusion alone requires at least three conserved proteins, namely, glycoprotein B (gB), glycoprotein H (gH), and glycoprotein L (gL). Fusion also requires additional non-conserved glycoproteins specific to individual herpesviruses (Roizman et al. 2013). The envelope glycoprotein, glycoprotein D (gD), is specific to alphaherpesviruses and is essential for cell entry in the majority of them (Ligas and Johnson 1988).

Since initiation of the viral life cycle depends entirely upon a successful entry step, entry receptors are effectively the keys that are able to unlock the cell membranes of target cells and thus determine viral tropism in vivo. Studies on viral entry receptors can also reveal the molecular mechanism of their pathogenesis and suggest interventions that may interfere with this process. HSV has a characteristic neurotropism but may also cause disease in epithelial tissues (Roizman et al. 2013). Curiously, this tropism is not reflected in tissue culture systems; HSV replicates well in cell lines derived from neural or epithelial origins but also replicates in others including fibroblast and endothelial cell lines. The molecular mechanism underlying this in vitro observation is not well studied but clearly demonstrates the importance of in vivo studies in order to fully understand natural HSV pathogenesis. Here, we summarize recent papers describing newly identified HSV receptors and the roles of the wide range of receptors involved in HSV pathogenesis.

1.2 The Orchestration of Herpesvirus Membrane Fusion

Successful HSV fusion requires a multitude of viral glycoproteins (Turner et al. 1998) and cellular receptors to interact in a sequential process (Eisenberg et al. 2012) (Fig. 1.1). Firstly, gD must bind to its receptor and undergo a conformational change (Krummenacher et al. 2005; Lazear et al. 2008). Secondly, this activated

1 The Role of HSV Glycoproteins in Mediating Cell Entry

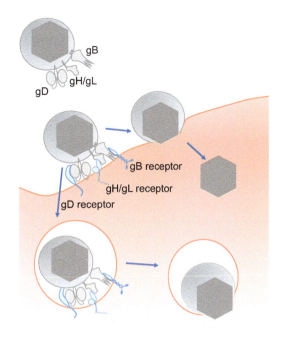

Fig. 1.1 Illustration of the steps involved during HSV entry. Four glycoproteins gB, gD, and the gH/gL complex on the viral envelope are essential for entry and fusion. These are known to each have a specific receptor. Fusion between the viral envelope and the cell can occur either at the plasma membrane or at the endosomal membrane

Table 1.1 Glycoproteins essential for HSV entry

Glycoprotein	Conservation	Function	Cellular binding partner
gD	*Alphaherpesvirinae*	Receptor recognition	HVEM, nectins, 3-OS HS
gH/gL	*Herpesviridae*	Activator of gB	Integrins
gB	*Herpesviridae*	Fusogen	PILRα, MAG, NMHC-IIs

form of gD must facilitate conversion of gH/gL into a form that is able to interact with gB (Avitabile et al. 2007; Chowdary et al. 2010; Atanasiu et al. 2007, 2016, 2010). Thirdly, the activated gB must bring the two membranes together to allow insertion of fusion loops into the cell membrane (Heldwein et al. 2006; Zeev-Ben-Mordehai et al. 2016). This leads to fusion of the two membranes and delivery of the capsid into the cell. Although the molecular mechanisms by which gD activates gH/gL, or gH/gL activates gB, are not known, it is clear that gB is the only HSV glycoprotein that is capable of becoming a fusogen (Heldwein et al. 2006; Chowdary et al. 2010) (Table 1.1).

1.3 Two HSV Entry Pathways

The process by which enveloped virus enters a cell can be divided into two distinct pathways (Fig. 1.1):

1. The virus enters the cell by direct fusion of the viral envelope and the plasma membrane.

2. The viral envelope fuses with an endosomal membrane and virions are transported into endosomes via endocytosis (Nicola 2016).

Endocytosis-mediated entry usually requires a low pH for fusion, whereas this is not required for fusion with the plasma membrane. HSV is known to employ multiple entry pathways, including low pH-dependent and -independent routes, that are determined by the cell type involved (Nicola 2016). Vero cells and human neurons are thought to allow HSV entry by fusion at the plasma membrane (Lycke et al. 1988; Wittels and Spear 1991), whereas HSV enters HeLa, CHO cells expressing gD receptors, and epithelial cells via endocytosis (Nicola et al. 2003, 2005). Curiously, the same set of viral proteins are required for both routes (Nicola and Straus 2004), and other envelope proteins have been shown to be nonessential for low-pH entry (Komala Sari et al. 2013). The molecular characteristics that determine the mode of viral entry mechanism are not known.

1.4 Receptor Recognition of gD

Although it is not conserved among *Herpesviridae*, gD is essential for HSV entry. Successful HSV entry depends upon the binding of gD to one of its many cell surface receptors, which include HVEM (herpesvirus entry mediator), a member of the tumor necrosis factor receptor family (Montgomery et al. 1996), nectin-1 or nectin-2, cell adhesion molecules belonging to the immunoglobulin (Ig) superfamily (Warner et al. 1998; Geraghty et al. 1998), and specific modifications of heparan sulfate (3-O-S HS) catalyzed by particular isoforms of 3-O-sulfotransferase (Shukla et al. 1999). CHO cells are resistant to HSV infection, but expression of these molecules on the cell surface makes CHO cells susceptible to HSV infection (Montgomery et al. 1996; Geraghty et al. 1998). However, gD receptors bind gD independently and do not act as co-receptors. Among these receptors, nectin-1 and HVEM are used for entry by all the clinical strains of HSV-1 and HSV-2 tested, regardless of their origin (Montgomery et al. 1996). In contrast, nectin-2 is a receptor for some mutant forms of HSV-1 and HSV-2 (Warner et al. 1998), whereas 3-O-S HS is only used by HSV-1 (Shukla et al. 1999). Nectin-1 is overexpressed in neurons and nectin-1 antibody (but not HVEM antibody) blocks HSV entry into neurons (Richart et al. 2003; Simpson et al. 2005). In contrast, HVEM is overexpressed in lymphoid cells which are not targets for productive HSV infection in vivo. Thus gD-HVEM interactions were initially considered important for modulating the host's immune response, despite the fact that epithelial cells and fibroblasts express both HVEM and nectin-1 (Montgomery et al. 1996).

1.5 The Role of the gH/gL Complex

The gH/gL complex is highly conserved among herpesviruses and is an absolute requirement for viral entry, although the precise role of gH/gL in fusion remains uncertain. Crystal structures of gH/gL complexes do not resemble any other known viral fusogen structures (Chowdary et al. 2010). Formation of a gB-gH-gL complex is critical for fusion and inhibited by neutralizing antibody to gH, suggesting that the gH/gL complex activates gB through direct binding (Chowdary et al. 2010). In addition, gB interacts with gH through its cytoplasmic domains, and mutation in the cytoplasmic domain of gH reduces fusion activity suggesting that gH also regulates gB fusion activity through direct interaction via its cytoplasmic tail (Silverman and Heldwein 2013; Rogalin and Heldwein 2015).

In addition to its role as an activator of gB, gH is known to bind cellular receptors such as integrins. Pioneering experiments have shown that gH binds to αvβ3 integrin via its RGD motif (Parry et al. 2005) but that the RGD sequence in gH is not required for HSV entry (Galdiero et al. 1997). The role of integrins in the fusion step is postulated to be via gH and αvβ3 integrin interaction triggering intracellular signals which facilitate capsid transport (Cheshenko et al. 2014).

1.6 gB as a Class III Fusogen

Of the four essential HSV glycoproteins, only gB is considered to be a fusogen (Heldwein et al. 2006), indeed gB has been identified specifically as a class III fusogen (Roche et al. 2006; Heldwein et al. 2006; Kadlec et al. 2008). The gB ectodomain architecture is similar to the post-fusion structure of vesicular stomatitis virus (VSV) G protein and the baculovirus gp64 protein. The sequences of these proteins are not conserved, but domains of class III fusogens share much conformational similarity (Kadlec et al. 2008; Roche et al. 2006; Heldwein et al. 2006). Curiously, there is a fundamental difference between HSV gB and the other two fusogens; conformational changes in VSV G and gp64, but not HSV gB, are pH-driven and reversible (Kadlec et al. 2008; Roche et al. 2006). Cellular entry of VSV and baculovirus depends upon endocytosis, where low endosomal pH induces conformational changes in VSV G and gp64, thus exposing their fusion sequences (Backovic and Jardetzky 2009). In the case of HSV entry, for which low pH is not strictly necessary, it is still unknown how gB is activated. The most convincing hypothesis is that the fusogenic activity of gB is triggered by the gH/gL complex, although precise details at the molecular level are unclear.

1.7 PILRα Associates with gB and Mediates HSV-1 Entry

In addition to the activation by gH/gL, gB must bind to specific receptors in order for HSV to achieve fusion. It is well known that although gB associates with cell-surface heparan sulfate (Herold et al. 1994), it is not essential for membrane fusion but promotes viral adsorption on the cell surface (Banfield et al. 1995; Laquerre et al. 1998). The soluble form of gB binds to heparan sulfate-deficient cells and blocks HSV infection of some cell lines (Bender et al. 2005), indicating that molecules other than heparan sulfate mediate gB-associated HSV infection. The first HSV gB receptor to be identified was the paired immunoglobulin-like type 2 receptor (PILR)α (Satoh et al. 2008). PILRα belongs to a group of paired receptors that consists of both highly homologous activating and inhibitory receptors that are widely involved in the regulation of immune responses. PILRα is mainly expressed on myeloid cells such as monocytes, macrophages, and dendritic cells (Satoh et al. 2008; Fournier et al. 2000). PILRα-transfected CHO cells become permissive to HSV-1 infection in the same way as is observed in those transfected with gD receptors. Endogenous PILRα is also functional, and CD14-positive monocytes, which express both PILRα and HVEM, are susceptible to HSV-1 infection. This infection is blocked by either anti-PILRα or anti-HVEM antibodies. In contrast, HSV-1 does not infect the CD14-negative population, which expresses HVEM but which does not express PILRα. These results suggest that HSV entry requires both gD and gB receptor interactions and that although CHO cells express both gD and gB receptors, it is at levels that are too low to mediate effective HSV-1 infection.

HSV is thought to enter cells either via fusion of the virion envelope with the host cell plasma membrane or via endocytosis, depending on the cell type (Nicola 2016). HSV uptake into wild-type CHO cells, and in those transduced with the gD receptor, is known to be mediated by endocytosis. However, HSV-1 entry into CHO cells expressing PILRα occurs by direct fusion at the plasma membrane (Arii et al. 2009). Thus, expression of PILRα on CHO cells results in an alternative HSV entry pathway. Similarly, later reports demonstrate that signals from the gH receptors $\alpha v \beta 3$, $\alpha v \beta 6$, and $\alpha v \beta 8$-integrins change the mode of HSV entry from direct fusion at the plasma membrane to endocytosis (Gianni et al. 2010, 2013). Thus, cell surface receptors appear to determine which of the two entry routes is utilized.

It is widely accepted that gB, gD, gH, and gL are necessary and sufficient for HSV entry (Turner et al. 1998), but the contribution of other glycoproteins is also likely to be important. For example, gC is not essential for HSV replication but is responsible for adsorption to cell-surface heparan sulfate (Herold et al. 1991). The HSV gK/UL20 protein complex is incorporated into virion envelopes and plays a significant role in secondary envelopment (Baines et al. 1991; Jayachandra et al. 1997; Hutchinson and Johnson 1995). These proteins are also known to interact with gB and are responsible for the syncytium phenotype (Chouljenko et al. 2009, 2010). The mutant virus, carrying a 37-amino acid deletion at the gK amino terminus, is unable to enter CHO-PILRα cells but is more efficient at entering CHO cells expressing HVEM and nectin-1 than the wild-type virus (Chowdhury et al. 2013).

Although the precise mechanism is not known, mutant gK/UL20 may be able to associate more effectively with the gB-PILRα protein complex and thus regulate membrane fusion more efficiently.

The interaction between gB and PILRα has been studied in more detail. PILRα has a related activating receptor, PILRβ. Although they are structurally similar, activating PILRβ does not promote HSV infection. Differential protein sequence analysis indicates that the tryptophan residue at position 139 in the Ig-like V-type domain of PILRα is important for the binding interaction with gB and subsequent membrane fusion (Fan and Longnecker 2010). PILRα recognizes a wide variety of O-glycosylated mucins and related proteins and regulates a broad range of immune responses. The gB sialylated O-glycans at T53 and T480 have been shown to be important for binding PILRα and for facilitating viral entry (Wang et al. 2009; Arii et al. 2010b). The crystal structure of PILRα in its complex with a sialylated O-linked sugar T antigen (sTn) shows that PILRα undergoes large conformational changes in order to simultaneously recognize both the sTn O-glycan and the compact peptide structure constrained by proline residues (Kuroki et al. 2014).

1.8 MAG and NMHC-II Are Alternative gB Receptors

Because PILRα is expressed on restricted lineages, identification of PILRα as a gB receptor led to the prediction that other gB receptors must exist. The most promising candidates were proteins with structural similarity to PILRα. Sialic acid-binding Ig-like lectins (Siglecs) showed relatively high homology with human PILRα. Of these, Siglec-4 (also called myelin-associated glycoprotein; MAG) associated with HSV-1 gB and promoted HSV-1 infection (Suenaga et al. 2010). However, expression of MAG is restricted to glial cells. In addition, the recombinant HSV-1 expressing mutant gB, which is unable to interact with PILRα, could gain entry and replicate in both epithelial and neural cell lines as well as in the wild-type HSV-1 (Arii et al. 2010b). These results suggest that HSV-1 gB interacts with other types of unrelated receptor that are similar to gD, e.g., HVEM and nectin-1.

Another gB receptor, non-muscle myosin heavy chain IIA (NMHC-IIA), was identified as a cellular protein that coprecipitates with gB from PILRα-negative fibroblasts but not from a PILRα-positive macrophage cell line (Arii et al. 2010a). Human promyelocytic HL60 cells express NMHC-IIA at low levels and are relatively resistant to HSV-1 infection, but expression of NMHC-IIA on HL60 cells significantly increases susceptibility to HSV-1 infection. Anti-NMHC-IIA serum not only inhibited HSV-1 infection in HL60 cells expressing NMHC-IIA but also in epithelial cell lines such as Vero, HaCaT, HCE-T, and NCI-H292. It is well known that NMHC-IIA mainly functions in the cytoplasm and not on the cell surface although, interestingly, cell surface expression of NMHC-IIA was markedly and rapidly induced during the initiation of HSV-1 entry. This relocalization of NMHC-IIA during HSV infection depends on myosin light chain kinase (MLCK), which regulates non-muscle myosin (NM)-II by phosphorylation (Arii et al. 2010a).

The observation that PILRα changed the HSV entry route from endocytosis to direct fusion at the plasma membrane suggests that the nature or expression levels of gB receptors affects HSV entry mode. However, NMHC-IIA antibody blocked HSV-1 infection both in Vero cells (that depend on direct fusion at the plasma membrane) and CHO-K1 cells overexpressing nectin-1 (that depend on endocytosis) (Arii et al. 2010a). Thus NMHC-IIA contributes to both entry routes but does not influence entry route selection.

In vitro most cells express NMHC-IIA, but some cell lines don't express NMHC-IIA at all. In vivo NMHC-IIA is expressed highly in epithelial tissue but not in neural tissue, and thus its tissue-specific expression doesn't completely explain the observed HSV tropism. Mammalian cells express three genetically distinct isoforms of NMHC-II (designated NMHC-IIA, NMHC-IIB, and NMHC-IIC) (Vicente-Manzanares et al. 2009). The three NMHC-II isoforms are highly conserved and have both common and unique properties (Vicente-Manzanares et al. 2009). Most human tissues express different ratios of the NMHC-II isoforms (Vicente-Manzanares et al. 2009) with NMHC-IIB predominating in neuronal tissue. As might be predicted, NMHC-IIB associates with HSV-1 gB and mediates HSV-1 entry (Arii et al. 2015). Thus NMHC-IIs appear to be present on all cells susceptible to HSV-1 infection in vitro and in vivo.

1.9 NMHC-II and Other Pathogens

Accumulating evidence suggests that NMHC-II is also used by other pathogens for cell entry. Epstein-Barr virus (EBV) is a ubiquitous gamma herpesvirus that causes B-cell lymphomas and nasopharyngeal carcinoma (NPC). Recently a subset of spherical nasopharyngeal epithelial cells (NPECs) were reported to be efficiently infected by EBV (Xiong et al. 2015). EBV entry into NPECs depends on NMHC-IIA and is blocked by antisera to NMHC-IIA. Importantly, the authors also show that NMHC-IIA aggregates on the surface of the sphere-like NPECs that are susceptible to EBV infection but not on that of another subset of NPEC that are not susceptible. NMHC-IIA is also important for the entry of viruses that are not in the *Herpesviridae* family, e.g., severe fever with thrombocytopenia syndrome virus (SFTSV), a member of the *Bunyaviridae* family, and porcine reproductive and respiratory syndrome virus (PRRSV), a member of the *Arteriviridae* family (Gao et al. 2016; Sun et al. 2014). Intriguingly, upregulation of cell surface NMHC-IIA, as seen during HSV-1 entry, was also observed during SFTFV and PRRS entry.

It is well known that different families of DNA and RNA viruses use the same proteins as entry receptors. Cell surface proteins that are suitable for use as entry receptors are possibly rare, and thus each virus might use the same molecules depending on their individual tropism. Although cell surface expression of NMHC-IIA is limited, its stability, avidity to the viral proteins, and broad distribution in vivo are possibly advantageous. As described above, NMHC-IIA is expressed on the cell surface before viral infection in the case of EBV entry into NPECs. Herpesvirus ancestors might have originally infected these types of cells and then broadened their target cell range by increasing the cell surface expression of NMHC-IIA.

1.10 Attempts to Demonstrate the Significance of gD Receptors In Vivo Using Knockout Mice

Entry receptor characteristics were originally studied in cultured cells, especially by expressing receptors in normally nonpermissive cells. These studies were unable to reveal the significance or contribution of each receptor in viral pathogenesis without further in vivo experiments. Mice infected with HSV show manifestations of disease that are similar to those found in humans, i.e., encephalitis, keratitis, or genital disease. Since the murine HSV receptors are known to be paralogs of their human equivalents, mice are excellent models for the study of HSV entry requirements. To reveal the significance of each receptor in vivo, an attractive approach is to knockout each receptor in mice and then infect these mice with wild-type HSV. An initial study using gD receptor, nectin-1 and HVEM knockout mice with vaginal HSV infections, clearly differentiates the roles of these receptors in vivo, highlighting the large redundancy between them (Taylor et al. 2007). HSV-2 infection of the vaginal epithelium occurred in the absence of either HVEM or nectin-1 but was virtually undetectable when both receptors were absent. This observation indicates that either HVEM or nectin-1 is necessary for HSV-2 infection of vaginal epithelium (Fig. 1.2, Table 1.2). In agreement with this result, human epithelial cells and fibroblasts were reported to express both HVEM and nectin-1 (Wang et al. 2015; Simpson et al. 2005). Although the precise mechanism is not known, it has been reported that in mice, HSV predominantly uses nectin-1 rather than HVEM, both in vitro (Manoj et al. 2004; Simpson et al. 2005) and in vivo (Taylor et al. 2007). Nectin-1 knockout reduced efficiency of vaginal epithelium infection and viral spread to the nervous system, whereas HVEM knockout did not.

The significance of nectin-1 during HSV-2 infection has been clearly demonstrated using the cranial model which shows that nectin-1 is indispensable for HSV-2 infection of neurons in the brain and for the development of encephalitis (Kopp et al. 2009). In this model, however, HVEM also has a role in the infection of non-parenchymal cells of the brain, although this infection has no consequences in terms of disease progression. These results are in agreement with in vitro studies (Richart et al. 2003; Simpson et al. 2005) and in vivo studies in which nectin-1 was the predominant gD receptor in neurons (Fig. 1.2, Table 1.2).

The importance of nectin-1 in the murine model is not restricted to its role in neurons. Nectin-1, but not HVEM, is also indispensable for HSV-2 corneal infection (Karaba et al. 2012). Interestingly, both HVEM and nectin-1 are required for HSV-1 corneal infection (Karaba et al. 2011) (Fig. 1.2). Although HVEM and nectin-1 mediate HSV-1 and HSV-2 infection equally well in vitro, these two serotypes

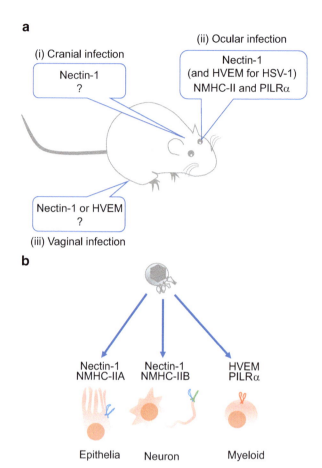

Fig. 1.2 In vivo experiments reveal the significance of the gB and gD receptors. (**a**) Summary of in vivo experiments. (i) Nectin-1 is essential for cranial infection. Either HVEM or PILRα is dispensable for HSV infection in the brain. (ii) In ocular infection, nectin-1 is required for both HSV-1 and HSV-2 pathogenesis. HVEM alone is required for HSV-1 pathogenesis. Both PILRα and NMHC-II contribute to HSV-1 pathogenesis through ocular infection. (iii) Either nectin-1 or HVEM is required for infection via the intravaginal route. (**b**) Predicted HSV receptor usage in various cell types

Table 1.2 HSV receptor requirements in vivo

Glycoprotein	Receptor	Distribution	Requirement in vivo
gD	HVEM	Ubiquitous except neural cells (highly expressed on lymphoid and myeloid cells)	Epithelium
	Nectin-1	Ubiquitous (highly expressed on neural cells)	CNS, epithelium
	Nectin-2	Ubiquitous	?
	3-OS Hs	Ubiquitous	?
gB	PILRα	Myeloid cells	Peripheral site
	MAG	Glial cells	?
	NMHC-IIA	Ubiquitous except neural cells	Peripheral site
	NMHC-IIB	Neural cells	?

possibly have different receptor requirements in vivo. It is well known that HSV-1 and HSV-2 infections are transmitted via different routes and involves different areas of the body, e.g., viral infections of the eye are usually caused by HSV-1 (Roizman et al. 2013). The molecular mechanisms behind these differences in pathogenesis are not yet known, but an attractive hypothesis is that the difference depends on HVEM. However, there is still a great deal of overlap between the epidemiology and clinical manifestations of these two viruses (Roizman et al. 2013).

As described above, either nectin-1 or HVEM is necessary for HSV-1 entry into cells in vitro. The fact that HSV-1 requires both nectin-1 and HVEM for infection of murine cornea in vivo suggests that two types of cells are important for HSV-1 corneal infection: one expressing nectin-1 (such as epithelia and/or neurons) and the other expressing HVEM. Human peripheral monocytes have been reported to be susceptible to HSV-1 infection in both an HVEM and a PILRα-dependent manner (Satoh et al. 2008), but, intriguingly, HSV-2 is unable to use PILRα as an entry receptor (Arii et al. 2009). Murine cornea is not well studied, but monocytes might be required to provide the correct receptors for productive HSV-1 infection.

1.11 Mutant Viruses Confirm the Importance of Nectin-1 In Vivo

Studies on knockout mice have provided many insights into the molecular basis of HSV infection, but the method has some limitations. The genetic background of most knockout mice is C57BL/6, which is known to be resistant to lethal HSV-1 infection (Lopez 1975). Although the receptor usage of HSV-1 and -2 during murine ocular infection is clearly different, it is difficult to know whether their distinct tropism might result in different clinical symptoms in humans. Moreover, if a receptor plays a critical role in the regulation of viral infection in vivo, other than viral entry, it would be difficult to determine whether an experimental result using knockout mice might be due solely to an effect on viral entry or to an effect on some other, as yet unknown, function of the receptor. For example, nectins are known to play a role in a variety of cell-cell junctions and cell-cell contacts (Mandai et al. 2015), which could contribute to viral dissemination efficiency in vivo. Also, HVEM modulates activation of lymphocytes by binding to B and T lymphocyte attenuator (BTLA) protein or TNF superfamily member 14 (LIGHT) protein (Mandai et al. 2015). Therefore, viral manipulation analyses should be used in animal models to investigate the role of viral entry receptors in vivo, without modifying any host cell functions, and may complement data derived from studies using knockout mice and vice versa.

In vitro analysis shows that different amino acids of gD are required for HVEM or nectin-1-dependent fusion (Spear et al. 2006). Crystal structures of an ectodomain of gD bound to ectodomains of HVEM and nectin-1 confirm that each receptor binds a distinct region of gD. The HVEM-binding site is limited to the N-terminal hairpin of gD (residues 1–32), whereas gD binds to nectin-1 using residues from the C- and N-terminal extensions (residues 35–38 and residues 219–221) (Carfi et al. 2001; Di Giovine et al. 2011; Lu et al. 2014). Mutant viruses can be created that are deficient in their recognition of nectin-1, but not of HVEM, and vice versa and are called HVEM-1-restricted or nectin-1-restricted mutants. Their genomes have been engineered to express gD R222N/F223I or A3C/Y38C, respectively (Uchida et al. 2009). A cell line which expresses both nectin-1 and HVEM was susceptible to the nectin-1-restricted mutant, but not to the HVEM-restricted mutant, suggesting, again, that nectin-1 is the dominant entry receptor. These recombinant viruses offer the benefit that they can be tested in animal models without the physiological defects inherent in knockout mice. However, these recombinant HSV-1 mutants were constructed from the K26GFP strain, which is avirulent in mice, and have not been studied in vivo.

Similarly, a D215G/R222N/F223I HSV-2 gD mutant, which is able to use HVEM but not nectin-1 as its receptor, couldn't infect neuronal cells but could infect HVEM-positive epithelial cells (Wang et al. 2012). Also, HVEM-restricted HSV-2 could not infect sensory ganglia in mice after intramuscular inoculation. These results suggest that HSV-2 uses both nectin-1 and HVEM, but that nectin-1 is predominantly used in neurons, in agreement with the knockout mice experiments.

1.12 Significance of gB Receptors In Vivo

The history of the gB receptor is short compared to that of the gD receptor, but the role of gB receptors in vivo has been analyzed. The first reported gB receptor, PILRα, is expressed mainly in immune system cells such as macrophages and dendritic cells (Fournier et al. 2000; Satoh et al. 2008). There is no doubt that these cells restrict HSV pathogenesis (Cheng et al. 2000; Kodukula et al. 1999) and that HSV infects myeloid cells and modulates them (Bosnjak et al. 2005; Melchjorsen et al. 2006). However, the importance of HSV entry into these cells for pathogenesis is not fully understood.

Because sialylated O-glycans on gB T53 and T480 are required for binding to PILRα, the recombinant HSV-1 carrying gB-T53A/T480A is defective in PILRα-dependent viral entry (Arii et al. 2010b). In contrast, the mutations have no effect on viral entry, viral attachment to heparan sulfate or viral replication in PILRα-negative cells. The gB-T53A/T480A mutations also had no effect on neurovirulence in mouse models but reduced viral replication, pathogenesis, and neuroinvasiveness following corneal infection (Arii et al. 2010b). These results suggest that PILRα-dependent viral entry is not important in the CNS but that it is important in peripheral sites (Fig. 1.2, Table 1.2). Because both HVEM- and PILRα-dependent entry is required for HSV-1 pathogenesis in the murine cornea, HSV-1 infection of HVEM- and PILRα-positive cells, such as monocytes, might facilitate viral pathogenesis in the cornea by attenuating the host immune system.

Despite the similarity between PILRα and MAG, the gB-T53A/T480A mutant can enter cells via MAG-dependent entry (Arii et al. 2010b). Thus, the significance of MAG, which is restricted to glial cells in vivo, is not yet certain. PILRα and MAG knockout mice are available, but it is difficult to evaluate the significance of these receptors by using them. PILRα knockout mice are not able to regulate recruitment of neutrophils and are thus highly susceptible to lipopolysaccharide (LPS)-induced endotoxic shock (Wang et al. 2013). It is therefore important to demonstrate that if PILRα knockout mice can resist HSV-1 corneal infection, an activated immune system due to the PILRα knockout is not hindering the role of the entry receptor. Similarly, MAG knockout induces axon degeneration (Pan et al. 2005), which might indirectly affect HSV-1 neuroinvasiveness.

Since NMHC-IIA and -IIB are important for development, knockout of NMHC-IIA or NMHC-IIB is lethal (Tullio et al. 1997; Conti et al. 2004). In addition, the specific gB residues required for interaction with NMHC-II are not yet known. Nevertheless, the importance of NMHC-II in HSV-1 pathogenesis has been investigated using drugs. During HSV-1 entry NMHC-II redistributes to the cell surface via phosphorylation of NM-II by MLCK. A specific inhibitor of MLCK, ML-7 is able to disturb both NMHC-II relocalization and HSV-1 entry in vitro (Arii et al. 2010a). In the murine corneal model, treatment of mouse eyes with ML-7 before HSV-1 inoculation reduced viral replication, herpes stromal keratitis severity, and mortality rate (Arii et al. 2010a). These results indicate that regulation of NMHC-II redistribution to the cell surface during initiation of HSV infection is required for efficient HSV pathogenesis in vitro and in vivo (Fig. 1.2, Table 1.2). However, these experiments were unable to illustrate the role of NMHC-II, if any, in mouse brains. Viral manipulation analyses of NMHC-IIs in animal models, which have already been performed in the case of HVEM and PILRα, are thus important to further understand its role in HSV pathogenesis.

1.13 Concluding Remarks

For some considerable time, only one gD receptor was thought to be required for HSV entry. Now, identification of gB receptors and in vivo analyses of HSV receptors clearly reveal the requirements and redundancy of multiple receptors. The predominant role of nectin-1 appears to be to contribute the characteristic in vivo neurotropism of HSV, while studies on gB receptors reveal the unexpected role of immune cells in HSV pathogenesis. However, many aspects of HSV pathogenesis still remain to be illuminated. These include the molecular mechanisms behind the preference for nectin-1 compared to the other gD receptors in vivo (and even in some cell lines in vitro), and the decision between which of the two available modes of cellular entry is used (direct fusion at the plasma membrane or endocytosis). The most promising answers to these questions may lie in the existence of new classes of HSV receptors, or the existence of as yet unknown host factors, that are central to the mediation of cell entry.

References

Arii J, Uema M, Morimoto T, Sagara H, Akashi H, Ono E, Arase H, Kawaguchi Y (2009) Entry of herpes simplex virus 1 and other alphaherpesviruses via the paired immunoglobulin-like type 2 receptor alpha. J Virol 83(9):4520–4527. https://doi.org/10.1128/JVI.02601-08

Arii J, Goto H, Suenaga T, Oyama M, Kozuka-Hata H, Imai T, Minowa A, Akashi H, Arase H, Kawaoka Y, Kawaguchi Y (2010a) Non-muscle myosin IIA is a functional entry receptor for herpes simplex virus-1. Nature 467(7317):859–862. https://doi.org/10.1038/nature09420

Arii J, Wang J, Morimoto T, Suenaga T, Akashi H, Arase H, Kawaguchi Y (2010b) A single-amino-acid substitution in herpes simplex virus 1 envelope glycoprotein B at a site required for binding to the paired immunoglobulin-like type 2 receptor alpha (PILRalpha) abrogates PILRalpha-dependent viral entry and reduces pathogenesis. J Virol 84(20):10773–10783. https://doi.org/10.1128/JVI.01166-10

Arii J, Hirohata Y, Kato A, Kawaguchi Y (2015) Nonmuscle myosin heavy chain IIb mediates herpes simplex virus 1 entry. J Virol 89(3):1879–1888. https://doi.org/10.1128/JVI.03079-14

Atanasiu D, Whitbeck JC, Cairns TM, Reilly B, Cohen GH, Eisenberg RJ (2007) Bimolecular complementation reveals that glycoproteins gB and gH/gL of herpes simplex virus interact with each other during cell fusion. Proc Natl Acad Sci U S A 104(47):18718–18723. https://doi.org/10.1073/pnas.0707452104

Atanasiu D, Saw WT, Cohen GH, Eisenberg RJ (2010) Cascade of events governing cell-cell fusion induced by herpes simplex virus glycoproteins gD, gH/gL, and gB. J Virol 84(23):12292–12299. https://doi.org/10.1128/JVI.01700-10

Atanasiu D, Saw WT, Eisenberg RJ, Cohen GH (2016) Regulation of HSV glycoprotein induced cascade of events governing cell-cell fusion. J Virol 90:10535. https://doi.org/10.1128/JVI.01501-16

Avitabile E, Forghieri C, Campadelli-Fiume G (2007) Complexes between herpes simplex virus glycoproteins gD, gB, and gH detected in cells by complementation of split enhanced green fluorescent protein. J Virol 81(20):11532–11537. https://doi.org/10.1128/JVI.01343-07

Backovic M, Jardetzky TS (2009) Class III viral membrane fusion proteins. Curr Opin Struct Biol 19(2):189–196. https://doi.org/10.1016/j.sbi.2009.02.012

Baines JD, Ward PL, Campadelli-Fiume G, Roizman B (1991) The UL20 gene of herpes simplex virus 1 encodes a function necessary for viral egress. J Virol 65(12):6414–6424

Banfield BW, Leduc Y, Esford L, Schubert K, Tufaro F (1995) Sequential isolation of proteoglycan synthesis mutants by using herpes simplex virus as a selective agent: evidence for a proteoglycan-independent virus entry pathway. J Virol 69(6):3290–3298

Bender FC, Whitbeck JC, Lou H, Cohen GH, Eisenberg RJ (2005) Herpes simplex virus glycoprotein B binds to cell surfaces independently of heparan sulfate and blocks virus entry. J Virol 79(18):11588–11597. https://doi.org/10.1128/JVI.79.18.11588-11597.2005

Bosnjak L, Miranda-Saksena M, Koelle DM, Boadle RA, Jones CA, Cunningham AL (2005) Herpes simplex virus infection of human dendritic cells induces apoptosis and allows cross-presentation via uninfected dendritic cells. J Immunol 174(4):2220–2227

Carfi A, Willis SH, Whitbeck JC, Krummenacher C, Cohen GH, Eisenberg RJ, Wiley DC (2001) Herpes simplex virus glycoprotein D bound to the human receptor HveA. Mol Cell 8(1):169–179

Cheng H, Tumpey TM, Staats HF, van Rooijen N, Oakes JE, Lausch RN (2000) Role of macrophages in restricting herpes simplex virus type 1 growth after ocular infection. Invest Ophthalmol Vis Sci 41(6):1402–1409

Cheshenko N, Trepanier JB, Gonzalez PA, Eugenin EA, Jacobs WR Jr, Herold BC (2014) Herpes simplex virus type 2 glycoprotein H interacts with integrin alphavbeta3 to facilitate viral entry and calcium signaling in human genital tract epithelial cells. J Virol 88(17):10026–10038. https://doi.org/10.1128/JVI.00725-14

Chouljenko VN, Iyer AV, Chowdhury S, Chouljenko DV, Kousoulas KG (2009) The amino terminus of herpes simplex virus type 1 glycoprotein K (gK) modulates gB-mediated virus-induced cell fusion and virion egress. J Virol 83(23):12301–12313. https://doi.org/10.1128/JVI.01329-09

Chouljenko VN, Iyer AV, Chowdhury S, Kim J, Kousoulas KG (2010) The herpes simplex virus type 1 UL20 protein and the amino terminus of glycoprotein K (gK) physically interact with gB. J Virol 84(17):8596–8606. https://doi.org/10.1128/JVI.00298-10

Chowdary TK, Cairns TM, Atanasiu D, Cohen GH, Eisenberg RJ, Heldwein EE (2010) Crystal structure of the conserved herpesvirus fusion regulator complex gH-gL. Nat Struct Mol Biol 17(7):882–888. https://doi.org/10.1038/nsmb.1837

Chowdhury S, Chouljenko VN, Naderi M, Kousoulas KG (2013) The amino terminus of herpes simplex virus 1 glycoprotein K is required for virion entry via the paired immunoglobulin-like type-2 receptor alpha. J Virol 87(6):3305–3313. https://doi.org/10.1128/JVI.02982-12

Conti MA, Even-Ram S, Liu C, Yamada KM, Adelstein RS (2004) Defects in cell adhesion and the visceral endoderm following ablation of nonmuscle myosin heavy chain II-A in mice. J Biol Chem 279(40):41263–41266. https://doi.org/10.1074/jbc.C400352200

Di Giovine P, Settembre EC, Bhargava AK, Luftig MA, Lou H, Cohen GH, Eisenberg RJ, Krummenacher C, Carfi A (2011) Structure of herpes simplex virus glycoprotein D bound to the human receptor nectin-1. PLoS Pathog 7(9):e1002277. https://doi.org/10.1371/journal.ppat.1002277

Eisenberg RJ, Atanasiu D, Cairns TM, Gallagher JR, Krummenacher C, Cohen GH (2012) Herpes virus fusion and entry: a story with many characters. Virus 4(5):800–832. https://doi.org/10.3390/v4050800

Fan Q, Longnecker R (2010) The Ig-like v-type domain of paired Ig-like type 2 receptor alpha is critical for herpes simplex virus type 1-mediated membrane fusion. J Virol 84(17):8664–8672. https://doi.org/10.1128/JVI.01039-10

Fournier N, Chalus L, Durand I, Garcia E, Pin JJ, Churakova T, Patel S, Zlot C, Gorman D, Zurawski S, Abrams J, Bates EE, Garrone P (2000) FDF03, a novel inhibitory receptor of the immunoglobulin superfamily, is expressed by human dendritic and myeloid cells. J Immunol 165(3):1197–1209

Galdiero M, Whiteley A, Bruun B, Bell S, Minson T, Browne H (1997) Site-directed and linker insertion mutagenesis of herpes simplex virus type 1 glycoprotein H. J Virol 71(3):2163–2170

Gao J, Xiao S, Xiao Y, Wang X, Zhang C, Zhao Q, Nan Y, Huang B, Liu H, Liu N, Lv J, Du T, Sun Y, Mu Y, Wang G, Syed SF, Zhang G, Hiscox JA, Goodfellow I, Zhou EM (2016) MYH9 is an essential factor for porcine reproductive and respiratory syndrome virus infection. Sci Rep 6:25120. https://doi.org/10.1038/srep25120

Geraghty RJ, Krummenacher C, Cohen GH, Eisenberg RJ, Spear PG (1998) Entry of alphaherpesviruses mediated by poliovirus receptor-related protein 1 and poliovirus receptor. Science 280(5369):1618–1620

Gianni T, Gatta V, Campadelli-Fiume G (2010) {alpha}V{beta}3-integrin routes herpes simplex virus to an entry pathway dependent on cholesterol-rich lipid rafts and dynamin2. Proc Natl Acad Sci U S A 107(51):22260–22265. https://doi.org/10.1073/pnas.1014923108

Gianni T, Salvioli S, Chesnokova LS, Hutt-Fletcher LM, Campadelli-Fiume G (2013) alphavbeta6- and alphavbeta8-integrins serve as interchangeable receptors for HSV gH/gL to promote endocytosis and activation of membrane fusion. PLoS Pathog 9(12):e1003806. https://doi.org/10.1371/journal.ppat.1003806

Heldwein EE, Lou H, Bender FC, Cohen GH, Eisenberg RJ, Harrison SC (2006) Crystal structure of glycoprotein B from herpes simplex virus 1. Science 313(5784):217–220. https://doi.org/10.1126/science.1126548

Herold BC, WuDunn D, Soltys N, Spear PG (1991) Glycoprotein C of herpes simplex virus type 1 plays a principal role in the adsorption of virus to cells and in infectivity. J Virol 65(3):1090–1098

Herold BC, Visalli RJ, Susmarski N, Brandt CR, Spear PG (1994) Glycoprotein C-independent binding of herpes simplex virus to cells requires cell surface heparan sulphate and glycoprotein B. J Gen Virol 75(Pt 6):1211–1222. https://doi.org/10.1099/0022-1317-75-6-1211

Hutchinson L, Johnson DC (1995) Herpes simplex virus glycoprotein K promotes egress of virus particles. J Virol 69(9):5401–5413

Jayachandra S, Baghian A, Kousoulas KG (1997) Herpes simplex virus type 1 glycoprotein K is not essential for infectious virus production in actively replicating cells but is required for efficient envelopment and translocation of infectious virions from the cytoplasm to the extracellular space. J Virol 71(7):5012–5024

Kadlec J, Loureiro S, Abrescia NG, Stuart DI, Jones IM (2008) The postfusion structure of baculovirus gp64 supports a unified view of viral fusion machines. Nat Struct Mol Biol 15(10):1024–1030. https://doi.org/10.1038/nsmb.1484

Karaba AH, Kopp SJ, Longnecker R (2011) Herpesvirus entry mediator and nectin-1 mediate herpes simplex virus 1 infection of the murine cornea. J Virol 85(19):10041–10047. https://doi.org/10.1128/JVI.05445-11

Karaba AH, Kopp SJ, Longnecker R (2012) Herpesvirus entry mediator is a serotype specific determinant of pathogenesis in ocular herpes. Proc Natl Acad Sci U S A 109(50):20649–20654. https://doi.org/10.1073/pnas.1216967109

Kodukula P, Liu T, Rooijen NV, Jager MJ, Hendricks RL (1999) Macrophage control of herpes simplex virus type 1 replication in the peripheral nervous system. J Immunol 162(5):2895–2905

Komala Sari T, Pritchard SM, Cunha CW, Wudiri GA, Laws EI, Aguilar HC, Taus NS, Nicola AV (2013) Contributions of herpes simplex virus 1 envelope proteins to entry by endocytosis. J Virol 87(24):13922–13926. https://doi.org/10.1128/JVI.02500-13

Kopp SJ, Banisadr G, Glajch K, Maurer UE, Grunewald K, Miller RJ, Osten P, Spear PG (2009) Infection of neurons and encephalitis after intracranial inoculation of herpes simplex virus requires the entry receptor nectin-1. Proc Natl Acad Sci U S A 106(42):17916–17920. https://doi.org/10.1073/pnas.0908892106

Krummenacher C, Supekar VM, Whitbeck JC, Lazear E, Connolly SA, Eisenberg RJ, Cohen GH, Wiley DC, Carfi A (2005) Structure of unliganded HSV gD reveals a mechanism for receptor-mediated activation of virus entry. EMBO J 24(23):4144–4153. https://doi.org/10.1038/sj.emboj.7600875

Kuroki K, Wang J, Ose T, Yamaguchi M, Tabata S, Maita N, Nakamura S, Kajikawa M, Kogure A, Satoh T, Arase H, Maenaka K (2014) Structural basis for simultaneous recognition of an O-glycan and its attached peptide of mucin family by immune receptor PILRalpha. Proc Natl Acad Sci U S A 111(24):8877–8882. https://doi.org/10.1073/pnas.1324105111

Laquerre S, Argnani R, Anderson DB, Zucchini S, Manservigi R, Glorioso JC (1998) Heparan sulfate proteoglycan binding by herpes simplex virus type 1 glycoproteins B and C, which differ in their contributions to virus attachment, penetration, and cell-to-cell spread. J Virol 72(7):6119–6130

Lazear E, Carfi A, Whitbeck JC, Cairns TM, Krummenacher C, Cohen GH, Eisenberg RJ (2008) Engineered disulfide bonds in herpes simplex virus type 1 gD separate receptor binding from fusion initiation and viral entry. J Virol 82(2):700–709. https://doi.org/10.1128/JVI.02192-07

Ligas MW, Johnson DC (1988) A herpes simplex virus mutant in which glycoprotein D sequences are replaced by beta-galactosidase sequences binds to but is unable to penetrate into cells. J Virol 62(5):1486–1494

Lopez C (1975) Genetics of natural resistance to herpesvirus infections in mice. Nature 258(5531):152–153

Lu G, Zhang N, Qi J, Li Y, Chen Z, Zheng C, Gao GF, Yan J (2014) Crystal structure of herpes simplex virus 2 gD bound to nectin-1 reveals a conserved mode of receptor recognition. J Virol 88(23):13678–13688. https://doi.org/10.1128/JVI.01906-14

Lycke E, Hamark B, Johansson M, Krotochwil A, Lycke J, Svennerholm B (1988) Herpes simplex virus infection of the human sensory neuron. An electron microscopy study. Arch Virol 101(1–2):87–104

Mandai K, Rikitake Y, Mori M, Takai Y (2015) Nectins and nectin-like molecules in development and disease. Curr Top Dev Biol 112:197–231. https://doi.org/10.1016/bs.ctdb.2014.11.019

Manoj S, Jogger CR, Myscofski D, Yoon M, Spear PG (2004) Mutations in herpes simplex virus glycoprotein D that prevent cell entry via nectins and alter cell tropism. Proc Natl Acad Sci U S A 101(34):12414–12421. https://doi.org/10.1073/pnas.0404211101

Melchjorsen J, Siren J, Julkunen I, Paludan SR, Matikainen S (2006) Induction of cytokine expression by herpes simplex virus in human monocyte-derived macrophages and dendritic cells is dependent on virus replication and is counteracted by ICP27 targeting NF-kappaB and IRF-3. J Gen Virol 87(Pt 5):1099–1108. https://doi.org/10.1099/vir.0.81541-0

Montgomery RI, Warner MS, Lum BJ, Spear PG (1996) Herpes simplex virus-1 entry into cells mediated by a novel member of the TNF/NGF receptor family. Cell 87(3):427–436

Nicola AV (2016) Herpesvirus entry into host cells mediated by endosomal low pH. Traffic 17(9):965–975. https://doi.org/10.1111/tra.12408

Nicola AV, Straus SE (2004) Cellular and viral requirements for rapid endocytic entry of herpes simplex virus. J Virol 78(14):7508–7517. https://doi.org/10.1128/JVI.78.14.7508-7517.2004

Nicola AV, McEvoy AM, Straus SE (2003) Roles for endocytosis and low pH in herpes simplex virus entry into HeLa and Chinese hamster ovary cells. J Virol 77(9):5324–5332

Nicola AV, Hou J, Major EO, Straus SE (2005) Herpes simplex virus type 1 enters human epidermal keratinocytes, but not neurons, via a pH-dependent endocytic pathway. J Virol 79(12):7609–7616. https://doi.org/10.1128/JVI.79.12.7609-7616.2005

Pan B, Fromholt SE, Hess EJ, Crawford TO, Griffin JW, Sheikh KA, Schnaar RL (2005) Myelin-associated glycoprotein and complementary axonal ligands, gangliosides, mediate axon stability in the CNS and PNS: neuropathology and behavioral deficits in single- and double-null mice. Exp Neurol 195(1):208–217. https://doi.org/10.1016/j.expneurol.2005.04.017

Parry C, Bell S, Minson T, Browne H (2005) Herpes simplex virus type 1 glycoprotein H binds to alphavbeta3 integrins. J Gen Virol 86(Pt 1):7–10. https://doi.org/10.1099/vir.0.80567-0

Pellet PE, Roizman B (2013) Herpesviridae. In: Knipe DM, Howley PM, Cohen JI et al (eds) Fields virology, 6th edn. Lippincott-Williams &Wilkins, Philadelphia, pp 1802–1822

Richart SM, Simpson SA, Krummenacher C, Whitbeck JC, Pizer LI, Cohen GH, Eisenberg RJ, Wilcox CL (2003) Entry of herpes simplex virus type 1 into primary sensory neurons in vitro is mediated by Nectin-1/HveC. J Virol 77(5):3307–3311

Roche S, Bressanelli S, Rey FA, Gaudin Y (2006) Crystal structure of the low-pH form of the vesicular stomatitis virus glycoprotein G. Science 313(5784):187–191. https://doi.org/10.1126/science.1127683

Rogalin HB, Heldwein EE (2015) Interplay between the herpes simplex virus 1 gB Cytodomain and the gH Cytotail during cell-cell fusion. J Virol 89(24):12262–12272. https://doi.org/10.1128/JVI.02391-15

Roizman B, Knipe DM, Whitley RJ (2013) Herpes simplex viruses. In: Knipe DM, Howley PM, Cohen JI et al (eds) Fields virology, 6th edn. Lippincott-Williams &Wilkins, Philadelphia, pp 1823–1897

Satoh T, Arii J, Suenaga T, Wang J, Kogure A, Uehori J, Arase N, Shiratori I, Tanaka S, Kawaguchi Y, Spear PG, Lanier LL, Arase H (2008) PILRalpha is a herpes simplex virus-1 entry coreceptor that associates with glycoprotein B. Cell 132(6):935–944. https://doi.org/10.1016/j.cell.2008.01.043

Shukla D, Liu J, Blaiklock P, Shworak NW, Bai X, Esko JD, Cohen GH, Eisenberg RJ, Rosenberg RD, Spear PG (1999) A novel role for 3-O-sulfated heparan sulfate in herpes simplex virus 1 entry. Cell 99(1):13–22

Silverman JL, Heldwein EE (2013) Mutations in the cytoplasmic tail of herpes simplex virus 1 gH reduce the fusogenicity of gB in transfected cells. J Virol 87(18):10139–10147. https://doi.org/10.1128/JVI.01760-13

Simpson SA, Manchak MD, Hager EJ, Krummenacher C, Whitbeck JC, Levin MJ, Freed CR, Wilcox CL, Cohen GH, Eisenberg RJ, Pizer LI (2005) Nectin-1/HveC mediates herpes simplex virus type 1 entry into primary human sensory neurons and fibroblasts. J Neurovirol 11(2):208–218. https://doi.org/10.1080/13550280590924214

Spear PG, Manoj S, Yoon M, Jogger CR, Zago A, Myscofski D (2006) Different receptors binding to distinct interfaces on herpes simplex virus gD can trigger events leading to cell fusion and viral entry. Virology 344(1):17–24. https://doi.org/10.1016/j.virol.2005.09.016

Suenaga T, Satoh T, Somboonthum P, Kawaguchi Y, Mori Y, Arase H (2010) Myelin-associated glycoprotein mediates membrane fusion and entry of neurotropic herpesviruses. Proc Natl Acad Sci U S A 107(2):866–871. https://doi.org/10.1073/pnas.0913351107

Sun Y, Qi Y, Liu C, Gao W, Chen P, Fu L, Peng B, Wang H, Jing Z, Zhong G, Li W (2014) Nonmuscle myosin heavy chain IIA is a critical factor contributing to the efficiency of early infection of severe fever with thrombocytopenia syndrome virus. J Virol 88(1):237–248. https://doi.org/10.1128/JVI.02141-13

Taylor JM, Lin E, Susmarski N, Yoon M, Zago A, Ware CF, Pfeffer K, Miyoshi J, Takai Y, Spear PG (2007) Alternative entry receptors for herpes simplex virus and their roles in disease. Cell Host Microbe 2(1):19–28. https://doi.org/10.1016/j.chom.2007.06.005

Tullio AN, Accili D, Ferrans VJ, Yu ZX, Takeda K, Grinberg A, Westphal H, Preston YA, Adelstein RS (1997) Nonmuscle myosin II-B is required for normal development of the mouse heart. Proc Natl Acad Sci U S A 94(23):12407–12412

Turner A, Bruun B, Minson T, Browne H (1998) Glycoproteins gB, gD, and gHgL of herpes simplex virus type 1 are necessary and sufficient to mediate membrane fusion in a Cos cell transfection system. J Virol 72(1):873–875

Uchida H, Shah WA, Ozuer A, Frampton AR Jr, Goins WF, Grandi P, Cohen JB, Glorioso JC (2009) Generation of herpesvirus entry mediator (HVEM)-restricted herpes simplex virus type 1 mutant viruses: resistance of HVEM-expressing cells and identification of mutations that rescue nectin-1 recognition. J Virol 83(7):2951–2961. https://doi.org/10.1128/JVI.01449-08

Vicente-Manzanares M, Ma X, Adelstein RS, Horwitz AR (2009) Non-muscle myosin II takes centre stage in cell adhesion and migration. Nat Rev Mol Cell Biol 10(11):778–790

Wang J, Fan Q, Satoh T, Arii J, Lanier LL, Spear PG, Kawaguchi Y, Arase H (2009) Binding of herpes simplex virus glycoprotein B (gB) to paired immunoglobulin-like type 2 receptor alpha depends on specific sialylated O-linked glycans on gB. J Virol 83(24):13042–13045. https://doi.org/10.1128/JVI.00792-09

Wang K, Kappel JD, Canders C, Davila WF, Sayre D, Chavez M, Pesnicak L, Cohen JI (2012) A herpes simplex virus 2 glycoprotein D mutant generated by bacterial artificial chromosome mutagenesis is severely impaired for infecting neuronal cells and infects only Vero cells expressing exogenous HVEM. J Virol 86(23):12891–12902. https://doi.org/10.1128/JVI.01055-12

Wang J, Shiratori I, Uehori J, Ikawa M, Arase H (2013) Neutrophil infiltration during inflammation is regulated by PILRalpha via modulation of integrin activation. Nat Immunol 14(1):34–40. https://doi.org/10.1038/ni.2456

Wang K, Goodman KN, Li DY, Raffeld M, Chavez M, Cohen JI (2015) A Herpes Simplex Virus 2 (HSV-2) gD mutant impaired for neural tropism is superior to an HSV-2 gD subunit vaccine to protect animals from challenge with HSV-2. J Virol 90(1):562–574. https://doi.org/10.1128/JVI.01845-15

Warner MS, Geraghty RJ, Martinez WM, Montgomery RI, Whitbeck JC, Xu R, Eisenberg RJ, Cohen GH, Spear PG (1998) A cell surface protein with herpesvirus entry activity (HveB) confers susceptibility to infection by mutants of herpes simplex virus type 1, herpes simplex virus type 2, and pseudorabies virus. Virology 246(1):179–189

Wittels M, Spear PG (1991) Penetration of cells by herpes simplex virus does not require a low pH-dependent endocytic pathway. Virus Res 18(2–3):271–290

Xiong D, Du Y, Wang HB, Zhao B, Zhang H, Li Y, Hu LJ, Cao JY, Zhong Q, Liu WL, Li MZ, Zhu XF, Tsao SW, Hutt-Fletcher LM, Song E, Zeng YX, Kieff E, Zeng MS (2015) Nonmuscle myosin heavy chain IIA mediates Epstein-Barr virus infection of nasopharyngeal epithelial cells. Proc Natl Acad Sci U S A 112(35):11036–11041. https://doi.org/10.1073/pnas.1513359112

Zeev-Ben-Mordehai T, Vasishtan D, Hernandez Duran A, Vollmer B, White P, Prasad Pandurangan A, Siebert CA, Topf M, Grunewald K (2016) Two distinct trimeric conformations of natively membrane-anchored full-length herpes simplex virus 1 glycoprotein B. Proc Natl Acad Sci U S A 113(15):4176–4181. https://doi.org/10.1073/pnas.1523234113

Chapter 2
Virus Assembly and Egress of HSV

Colin Crump

Abstract The assembly and egress of herpes simplex virus (HSV) is a complicated multistage process that involves several different cellular compartments and the activity of many viral and cellular proteins. The process begins in the nucleus, with capsid assembly followed by genome packaging into the preformed capsids. The DNA-filled capsids (nucleocapsids) then exit the nucleus by a process of envelopment at the inner nuclear membrane followed by fusion with the outer nuclear membrane. In the cytoplasm nucleocapsids associate with tegument proteins, which form a complicated protein network that links the nucleocapsid to the cytoplasmic domains of viral envelope proteins. Nucleocapsids and associated tegument then undergo secondary envelopment at intracellular membranes originating from late secretory pathway and endosomal compartments. This leads to assembled virions in the lumen of large cytoplasmic vesicles, which are then transported to the cell periphery to fuse with the plasma membrane and release virus particles from the cell. The details of this multifaceted process are described in this chapter.

Keywords HSV · Herpes simplex virus · Virus assembly · Virus egress

2.1 Capsid Assembly and Genome Packaging

The first stage of forming new HSV particles is the assembly of the icosahedral capsid. Like all herpesviruses, the HSV capsid is an approximately 125 nm diameter icosahedron with $T = 16$ symmetry (Schrag et al. 1989), composed of 162 capsomers connected by 320 triplexes (2 copies of VP23 and 1 copy of VP19C) (Newcomb et al. 1993; Okoye et al. 2006). The 162 capsomers include 150 hexons (6 copies of VP5), which make up the edges and faces of the icosahedron, and 11 of the 12 vertices are pentons (5 copies of VP5). The twelfth vertex is the portal complex, a dodecamer of pUL6 arranged in a ring structure, through which the genome is packaged during assembly and released during entry (Newcomb et al. 2001). In

C. Crump (✉)
Department of Pathology, University of Cambridge, Cambridge, UK
e-mail: cmc56@cam.ac.uk

addition, 900 copies of the small capsid protein VP26 decorate the outer surface of the capsid, with one copy of VP26 on the tip of each VP5 in the 150 hexons. Furthermore, five copies of a heterodimer composed of pUL17 and pUL25, termed the capsid vertex-specific component (CVSC), associate with each penton. The CVSC is thought to be important for both capsid stability and association with the tegument (Thurlow et al. 2006; Toropova et al. 2011; Trus et al. 2007). One final viral protein component of the capsid is VP24, the protease that processes the scaffold during DNA encapsidation (Sheaffer et al. 2000; Stevenson et al. 1997).

Capsids initially assemble as a procapsid around a scaffold complex composed of ~1900 subunits of two related proteins: pUL26.5 that contains the scaffold core domain and pUL26 that has the viral protease (VP24) fused to the N-terminus of the scaffold core domain via a linker. Approximately 90% of the scaffold is composed of pUL26.5 and 10% is pUL26 (Aksyuk et al. 2015). Both pUL26.5 and pUL26 interact with the major capsid protein VP5 via their identical C-termini. The assembly is thought to initiate with the portal complex associated with scaffold proteins (Newcomb et al. 2005), followed by progressive addition of scaffold-bound VP5 together with preformed triplexes, which produces spherical procapsids containing a single portal complex (Newcomb et al. 1996, 2003; Spencer et al. 1998). Viral DNA, primarily in concatemeric form after synthesis by rolling-circle replication, is packaged into preformed procapsids via the pUL6 portal and requires the action of the terminase complex (pUL15-pUL28-pUL33) (Heming et al. 2014). The terminase complex interacts with packaging signals (pac sequences) in the terminal repeat region at the free end of newly synthesised viral DNA and drives ATP-dependent translocation of the viral genome into the procapsid. Once DNA packaging is complete, the terminase complex cleaves the concatemeric viral DNA at the next terminal repeat to separate the packaged single genome length of DNA from the rest of the concatemer (Tong and Stow 2010). As viral DNA begins to be packaged into the procapsid, the protease domain of pUL26 scaffold protein is activated, causing its autocatalytic release from the N-terminus of pUL26 to become VP24, the free protease protein. VP24 cleaves both pUL26 and pUL26.5 near their C-termini releasing the core scaffold domains, termed VP21 and VP22a, respectively, from their bound VP5. The majority of the cleaved scaffold protein products are released from the capsid providing the space for the viral genome, although at least some of the VP24 protease domain is retained inside the capsid (McClelland et al. 2002; Sheaffer et al. 2000). This whole process leads to large structural changes resulting in the rearrangement of the spherical procapsid into the stable icosahedral capsid containing the viral genome (often termed C-capsids or nucleocapsids) (Heymann et al. 2003; Roos et al. 2009). Two other forms of stable icosahedral capsids that lack DNA, A-capsids and B-capsids, are also produced during this process; A-capsids contain little or no scaffold protein, whereas B-capsids retain an inner shell of processed scaffold (Cardone et al. 2012a). Both A-capsids and B-capsids are thought to be dead-end products that result from defective or abortive DNA encapsidation, and these DNA-less capsids rarely exit the nucleus.

2.2 Nuclear Egress

Upon completion of capsid assembly and genome packaging, the resulting nucleocapsids need to escape the confines of the nuclear envelope. The nuclear envelope is a formidable barrier, being composed of two phospholipid bilayers – the inner nuclear membrane facing the nucleoplasm and the outer nuclear membrane facing the cytoplasm. There are numerous pores within the nuclear envelope, where the inner nuclear membrane and outer nuclear membrane fuse, which are filled by nuclear pore complexes that tightly regulate the transport of cargo between the cytoplasm and nucleus. The size exclusion of these pores is typically around 39 nm diameter or less (Pante and Kann 2002) and is thus too small to accommodate herpesvirus capsids, which are ~125 nm. Instead, transport of nucleocapsids into the cytoplasm is achieved by budding of nucleocapsids at the inner nuclear membrane to form primary enveloped particles (also termed perinuclear virions) within the perinuclear space, followed by fusion of primary enveloped particles with the outer nuclear membrane to release nucleocapsids into the cytoplasm. While many details of this process are still unclear, much progress has been made recently in understanding this unusual mode of intracellular transport.

To gain access to the inner nuclear membrane, the underlying nuclear lamina must be penetrated. The nuclear lamina is a dense mesh of intermediate filament-type proteins (lamins) and associated proteins, which interacts with chromatin and aids the structural integrity of the nucleus. Local disruption of the nuclear lamina to enable nucleocapsid access to the inner nuclear membrane is facilitated through phosphorylation of lamins and associated proteins by viral and cellular kinases, including pUS3 and PKC isoforms (Bjerke and Roller 2006; Leach and Roller 2010).

The budding of nucleocapsids at the inner nuclear membrane is driven by the nuclear egress complex (NEC), a heterodimer of pUL31 and a tail-anchored membrane protein pUL34. The NEC recruits PKC isoforms (Park and Baines 2006) and is itself a target for phosphorylation by pUS3 (Kato et al. 2005), which facilitates the correct localisation of the NEC to the nuclear membrane (Reynolds et al. 2001). The other HSV protein kinase, pUL13, can also regulate the localisation of the NEC, either by phosphorylation of pUS3 or by a pUS3-independent mechanism (Kato et al. 2006).

The recruitment of capsids to the inner nuclear membrane involves the interaction of the NEC with pUL25, part of the heterodimeric CVSC present on the vertices of capsids. DNA-filled capsids (C-capsids/nucleocapsids) have higher levels of occupancy of the CVSC on their vertices than either A- or B-capsids (Newcomb et al. 2006; Sheaffer et al. 2001), providing a mechanism by which DNA-filled capsids can be selected for nuclear export (O'Hara et al. 2010; Yang and Baines 2011).

Recent structural studies have begun to uncover the molecular mechanisms by which the NEC mediates primary envelopment of herpesvirus capsids. The NEC oligomerises on the inner nuclear membrane to form a hexagonal scaffold that coats the inner surface of the budding membrane and links the membrane to the nucleocapsid (Bigalke and Heldwein 2017; Zeev-Ben-Mordehai et al. 2015). The NEC has an intrinsic activity to deform membranes and cause membrane scission, suggesting the NEC alone is sufficient for forming perinuclear enveloped virus particles (Hagen et al. 2015). However, other viral proteins may be involved in regulating nuclear egress, including pUL16 (Gao et al. 2017), pUL21 (Le Sage et al. 2013), pUL47 (Liu et al. 2014) and pUS3 (Reynolds et al. 2002), as well as the nonstructural proteins pUL24 (Lymberopoulos et al. 2011) and ICP22 (Maruzuru et al. 2014). One protein with a well-established, although enigmatic, role in nuclear egress is the viral kinase pUS3. Deletion of the US3 gene or introduction of an inactivating mutation in the kinase domain of pUS3 results in the accumulation of primary enveloped virions in the perinuclear space, often observed as bulges protruding into the nucleoplasm termed herniations (Reynolds et al. 2002; Ryckman and Roller 2004). This suggests a role for pUS3 kinase activity in regulating the fusion of primary enveloped virions with the outer nuclear membrane or dissociation of the nucleocapsid from the NEC (Newcomb et al. 2017). However, this appears to be a more facilitatory, non-essential function because US3 deletion viruses are viable and nucleocapsids are still able to gain access to the cytoplasm, undergo secondary envelopment and release infectious virions from cells. It is possible that pUL13, or host kinases, can at least partially compensate for loss of pUS3 function.

The precise composition of perinuclear virions is unknown, with much uncertainty about the presence of various tegument and envelope proteins. Regarding tegument proteins, immuno-electron microscopy studies have suggested that both pUS3 and VP16 are present in perinuclear virions (Naldinho-Souto et al. 2006; Reynolds et al. 2002), and proteomics analysis of partially purified perinuclear virions identified pUL49 as a component of these particles (Padula et al. 2009). However, for many of the viral proteins proposed to regulate nuclear egress, it is unclear if they become components of primary enveloped particles or indeed whether their roles in nucleocapsid transport across the nuclear envelope are direct or indirect. Recent cryo-electron microscopy studies of primary enveloped virions have also demonstrated limited space between the NEC and nucleocapsid, suggesting few tegument proteins are likely to be incorporated to a significant level during nuclear egress and the majority of tegument is acquired in the cytoplasm (Newcomb et al. 2017).

Whether there are any viral membrane proteins that are specifically incorporated into primary enveloped virions and what functional roles they play in nuclear egress is also unclear. Several viral membrane proteins localise to the nuclear envelope in infected cells and thus could be incorporated into perinuclear virions, including gB, gD, gH and gM (Baines et al. 2007; Farnsworth et al. 2007b; Wills et al. 2009). Given the need for perinuclear virions to fuse with the outer nuclear membrane, the presence of the viral entry proteins gB, gD and gH could indicate a potential role of these proteins in the outer nuclear membrane fusion event. Indeed, some evidence suggests gB is involved in HSV-1 nuclear egress, possibly in a redundant manner with gH, and that this activity of gB in nuclear egress is regulated by pUS3 (Farnsworth et al. 2007b; Wisner et al. 2009; Wright et al. 2009). However, gB deletion viruses still efficiently assemble and release virions from infected cells, albeit lacking gB, suggesting any role of this glycoprotein in nuclear egress is facilitatory rather than essential (Farnsworth et al. 2007b). Furthermore, in pseudorabies virus, a related alphaherpesvirus, deletion of gB and gH does not affect nuclear egress (Klupp et al. 2008).

Following de-envelopment, nucleocapsids detach from the NEC leaving the NEC proteins behind in the outer nuclear membrane in a process that may partially rely on pUS3 kinase activity (Newcomb et al. 2017; Reynolds et al. 2002). Once released, the nucleocapsid must recruit then tegument and undergo secondary envelopment to form an infectious virion.

2.3 Tegument Assembly

The tegument is a complex proteinaceous layer that connects the nucleocapsid to the viral envelope, which in HSV contains up to 24 different viral proteins (Table 2.1). As well as performing a structural role within the virion, the tegument is also a reservoir for proteins that modulate host cell function, such as the ubiquitin ligase ICP0 and the virion host shut-off protein (Vhs/pUL41), which are important for antagonising antiviral host responses (Boutell and Everett 2013; Smiley 2004). Unlike the icosahedral herpesvirus capsid, the structure of the tegument is rather poorly defined; the lack of symmetry within the tegument prevents high-resolution single-particle structural analysis by cryo-electron microscopy. Tegument proteins are often broadly subdivided into 'inner' and 'outer' tegument, with inner tegument proteins more tightly associated with the nucleocapsid and outer tegument proteins weakly associated with the nucleocapsid and/or associated with the inner surface of the envelope. These definitions mainly come from biochemical experiments investigating how labile the association of tegument proteins with nucleocapsids is to increasing salt concentration, following disruption of the viral enveloped with detergent. Therefore, such designations do not necessarily provide information of the structural organisation of these proteins within virions. Recently, some details of tegument organisation have begun to be uncovered by modern techniques in fluorescence microscopy analysis of single virus particles (Bohannon et al. 2013; Laine et al. 2015).

The tegument protein that is most tightly associated to nucleocapsids is pUL36 (also termed VP1/2), the C-terminal domain of which has been shown to interact with pUL25, part of the heterodimeric CVSC present on nucleocapsid pentons (Coller et al. 2007). Single-particle analysis of nucleocapsids obtained from purified virions that have been stripped of their envelope and all tegument proteins except pUL36 identified extra density protruding from capsid vertices, suggesting that at least part of pUL36, most likely its C-terminus, is the one tegument protein that does display some icosahedral symmetry (Cardone et al. 2012b). More recently, it has been shown that the presence of pUL36 is necessary for the CVSC to form suggesting that pUL36 residues may contribute to the observed CVSC density in cryo-EM reconstructions of nucleocapsids or that pUL36 is required to stabilise the structure of the pUL17 and pUL25 heterodimer (Fan et al. 2015).

Table 2.1 HSV structural proteins

	Gene	Protein name	Amino acids[a]	Mass (kDa)[a]	Description
Capsid	UL6	pUL6	676	74.1	Portal protein
	UL17	pUL17	703	74.6	CVSC protein
	UL18	VP23	318	34.3	Triplex protein
	UL19	VP5	1374	149.1	Major capsid protein, hexon and penton
	UL25	pUL25	580	62.7	CVSC protein
	UL26	VP24	247	26.6	Protease (N-terminal domain of pUL26)
	UL35	VP26	112	12.1	Binds hexon VP5 tip
	UL38	VP19C	465	50.3	Triplex protein
Tegument	RL1	ICP34.5	248	26.2	Neurovirulence factor
	RL2	ICP0	775	78.5	E3 ubiquitin ligase
	RS1	ICP4	1298	132.8	Essential gene; viral transcription factor
	UL7	pUL7	296	33.1	
	UL11	pUL11	96	10.5	Myristoylated and palmitoylated
	UL13	pUL13	518	57.2	Serine/threonine-protein kinase
	UL14	pUL14	219	23.9	
	UL16	pUL16	373	40.4	
	UL21	pUL21	535	57.6	
	UL23	TK	376	41.0	Thymidine kinase
	UL36	VP1/2	3112	333.6	Essential gene
	UL37	pUL37	1123	120.6	Essential gene
	UL41	Vhs	489	54.9	Endoribonuclease
	UL46	VP11/12	718	78.2	
	UL47	VP13/14	693	73.8	
	UL48	VP16	490	54.3	Essential gene; transcriptional activator of IE genes
	UL49	VP22	301	32.3	
	UL50	dUTPase	371	39.1	Deoxyuridine 5′-triphosphate nucleotidohydrolase
	UL51	pUL51	244	25.5	Palmitoylated
	UL55	pUL55	186	20.5	
	US2	pUS2	291	32.5	
	US3	pUS3	481	52.8	Serine/threonine-protein kinase
	US10	pUS10	306	33.5	
	US11[b]	pUS11	161	17.8	
Envelope	UL1	gL	224	24.9	Heterodimer with gH; essential for entry
	UL10	gM	473	51.4	Multiple transmembrane domains; forms complex with gN

(continued)

Table 2.1 (continued)

Gene	Protein name	Amino acids[a]	Mass (kDa)[a]	Description
UL20	pUL20	222	24.2	Multiple transmembrane domains; forms complex with gK
UL22	gH	838	90.4	Heterodimer with gL; essential for entry
UL27	gB	904	100.3	Essential for entry
UL43[b]	pUL43	417	42.9	Multiple transmembrane domains
UL44	gC	511	55.0	Heparan sulphate binding
UL45	pUL45	172	18.2	
UL49.5[b]	gN	91	9.2	Forms complex with gM
UL53	gK	338	37.6	Multiple transmembrane domains; forms complex with pUL20
UL56	pUL56	197	21.2	
US4	gG	238	25.2	Chemokine binding protein
US5[b]	gJ	92	9.6	
US6	gD	394	43.3	Essential for entry
US7	gI	390	41.4	Forms complex with gE; Fc receptor
US8	gE	550	59.1	Forms complex with gI; Fc receptor
US9	pUS9	90	10.0	

[a]Values from UniProt for HSV-1 strain 17; not including co-translational or post-translational modifications
[b]Unclear if present in virus particle for HSV (Loret et al. 2008)

As well as being tightly associated with nucleocapsids, pUL36 is one of the few tegument proteins, along with pUL37 and VP16, that are essential for HSV assembly; loss of functional pUL36 leads to accumulation of non-enveloped nucleocapsids in the cytoplasm, suggesting a failure of tegument to associate with nucleocapsids in the absence of pUL36 (Desai 2000; Roberts et al. 2009). Furthermore, pUL36 is the largest protein in herpesviruses, being >3000 amino acids in HSV, and has been shown to interact with the two other essential assembly proteins pUL37 and VP16 (Mijatov et al. 2007; Svobodova et al. 2012). This has led to the logical suggestion that pUL36 could serve as a platform or central organiser for subsequent assembly of the rest of the tegument. In addition to pUL36, pUL37 and pUS3 are usually classified as inner tegument proteins, whereas most of the other tegument proteins are generally considered outer tegument proteins. However, the association properties of most tegument proteins with nucleocapsids remain undefined, with investigations hampered due to low copy numbers within virions as well as a lack of suitable detection reagents for many tegument proteins.

The location where the tegument first begins to associate with nucleocapsids during virion assembly is unclear, with disagreement in the literature regarding the association of tegument proteins with capsids in the nucleus. Given the tight association of the inner tegument pUL36 with nucleocapsids and the importance of pUL36 for association of pUL37 and VP16 with nucleocapsids and subsequent virion assembly, it is reasonable to suggest pUL36 will be one of the first tegument proteins to interact with nucleocapsids during virion morphogenesis. However, there is evidence both for and against pUL36 associating with capsids in the nucleus (Bucks et al. 2007; Fan et al. 2015; Henaff et al. 2013; Radtke et al. 2010). Furthermore, it doesn't appear that pUL36 has an important role in the nuclear egress stage of the assembly pathway because deletion of UL36 does not prevent transport of nucleocapsids from the nucleus into the cytoplasm (Desai 2000; Roberts et al. 2009). The observations that some tegument proteins are important for efficient nuclear egress, as described above, could suggest association of these tegument proteins with nucleocapsids before or during primary envelopment, and indeed pUS3 has been detected in primary enveloped virions (Henaff et al. 2013; Reynolds et al. 2002). However, it is also possible tegument proteins could function in a regulatory manner during nuclear egress but not physically associate with nucleocapsids at this stage and then become incorporated into assembling virions by interaction with nucleocapsid and/or other tegument or envelope proteins in the cytoplasm. An example of the complexities in interpreting when tegument proteins become incorporated into assembling virions is the major tegument protein VP16. It is well established that VP16 is imported into the nucleus for its role during immediate-early viral gene expression (Campbell et al. 1984), VP16 has been observed in perinuclear virions by immuno-EM studies (Naldinho-Souto et al. 2006) and VP16 interacts with the inner tegument protein pUL36 (Svobodova et al. 2012), suggesting potential association with capsids in the nucleus. However, VP16-negative capsids can be readily observed in the cytoplasm of infected cells, VP16 deletion inhibits virion assembly but does not appear to affect nuclear egress, and VP16 interacts with the cytoplasmic domain of gH as well as several other 'outer' tegument proteins (pUL41, pUL46, pUL47 and VP22), suggesting VP16 incorporation into virions occurs in the cytoplasm (Elliott et al. 1995; Gross et al. 2003; Mossman et al. 2000; Smibert et al. 1994; Svobodova et al. 2012; Vittone et al. 2005). It should also be born in mind that the tegument assembly process could be somewhat flexible, whereby a few copies of some tegument proteins can associate with nucleocapsids in the nucleus and be carried across the nuclear envelope, but then further copies of these tegument proteins assemble onto nucleocapsids in the cytoplasm. Alternatively, nuclear-localised tegument proteins may transiently associate with nucleocapsid prior to or during nuclear egress and then dissociate once the nucleocapsid reaches the cytoplasm before reacquisition later during virion assembly. Future development of imaging technologies that allow direct observation of the dynamics of virion assembly at the single-particle level will hopefully shed light on this topic.

Regardless of whether individual tegument proteins can or do associate with nucleocapsids before they exit the nucleus, it is clear that the majority of the tegument assembles in the cytoplasm. To form the complex tegument layer, a network of protein-protein interactions with significant redundancy is thought to occur, including tegument-tegument, tegument-capsid and tegument-envelope interactions (Lee et al. 2008; Vittone et al. 2005). The inner tegument proteins pUL36, pUL37 and pUS3 are presumably recruited to nucleocapsids before the outer tegument proteins, many of which may assemble during the secondary envelopment process by virtue of interacting with the cytoplasmic domains of viral envelope proteins. Several tegument proteins have been shown to be important for efficient virion assembly, and so they may facilitate the formation of the complex tegument layer, although as mentioned above only three, pUL36, pUL37 and VP16 appear to be essential for virion assembly, suggesting reasonable flexibility within the assembly process that can compensate for the loss of one or more 'non-essential' components.

VP16, pUL47 and pUL49 are the most prevalent proteins in the tegument, with copy numbers estimated to be ca. 500–1500 per virion for each of these proteins (Clarke et al. 2007; Newcomb et al. 2012). These proteins may be central organisers of the tegument structure: VP16 has been shown to interact with pUL41 (Vhs), pUL46, pUL47 and pUL49, as well as pUL36 and the cytoplasmic domain of gH (Elliott et al. 1995; Gross et al. 2003; Smibert et al. 1994; Svobodova et al. 2012; Vittone et al. 2005); pUL47 also interacts with pUL17 providing another link between the tegument and nucleocapsids (Scholtes et al. 2010); pUL49 also interacts with pUL16, ICP0 and the cytoplasmic domains of gD, gE and gM (Farnsworth et al. 2007a; Maringer et al. 2012; Starkey et al. 2014).

In addition to the essential pUL36 and pU37, there are six other tegument proteins that are conserved throughout the herpesvirus family, and while not 'essential' these tegument proteins are also important for virion assembly. Firstly, there is pUL11, pUL16 and pUL21, which have been shown to form a tripartite complex that associates with membranes via the lipid anchors present on pUL11 and through interaction with the cytoplasmic domain of gE (Han et al. 2012). There is also evidence that both pUL16 and pUL21 interact with capsids, suggesting the pUL11-pUL16-pUL21 complex can directly connect the envelope with the nucleocapsid (de Wind et al. 1992; Meckes and Wills 2007). Secondly, there are pUL7, pUL14 and pUL51 which may also form a complex. pUL7 and pUL51 have recently been shown to form a complex through direct protein-protein interaction, and pUL14 has also been shown to interact with pUL51 (Albecka et al. 2017; Oda et al. 2016; Roller and Fetters 2015). As yet it is unclear if a tripartite complex of pUL7-pUL14-pUL51 forms or if there are independent pUL7-pUL51 and pUL14-pUL51 complexes. Deletion of each of these three proteins leads to defects in cytoplasmic virion assembly, and the stability of pUL7 and pUL51 relies on each other (Albecka et al. 2017; Oda et al. 2016). Similar to pUL11, pUL51 is associated with membranes via a lipid anchor providing additional links between the tegument and envelope. Therefore, it appears there are at least two independent protein complexes that can link the envelope to the underlying tegument and the nucleocapsid that are conserved throughout the herpesvirus family.

While many interactions between tegument proteins have been identified, it is important to note that such an extensive and seemingly redundant network of interactions between these proteins makes it problematic to investigate the precise roles of individual components or interactions during virion assembly. It is often difficult to know whether identified interactions are direct or indirect and whether they occur within the virion structure or during other, non-assembly activities of these multifunctional virus proteins. Elucidating the molecular details of tegument protein structures, both in isolation and in complex with each other, will be needed to shed further light on these complex assembly events.

2.4 Secondary Envelopment

The final stage of assembling mature HSV particles is secondary envelopment, sometimes referred to as final envelopment, the process by which nucleocapsids with a full complement of tegument proteins are encased within a lipid bilayer containing all the viral envelope proteins. This occurs at cytoplasmic membranes resulting in HSV particles inside the lumen of cytoplasmic compartments. To orchestrate this process, a complex series of interactions must occur between viral capsid, tegument and envelope proteins, as well as with cellular membranes and host proteins involved in regulating membrane modulation and transport.

The cellular compartment from which the final envelope of HSV is derived is somewhat unclear. It often stated that the trans-Golgi network (TGN), a major sorting station for membrane proteins of the secretory and endocytic pathways, is the site of HSV secondary envelopment. Evidence for this comes from the observation of co-localisation of several viral envelope glycoproteins with markers of the TGN (Beitia Ortiz de Zarate et al. 2004; Crump et al. 2004; Foster et al. 2004b; McMillan and Johnson 2001). Furthermore, it has been shown that appending localisation signals from cellular TGN resident proteins onto HSV-1 glycoproteins leads to incorporation into virions (Whiteley et al. 1999), and inhibitors that perturb Golgi/TGN function, such as brefeldin A, block HSV-1 assembly (Cheung et al. 1991; Koyama and Uchida 1994), supporting the notion of envelopment of virions by TGN membrane. However, there is also evidence for endosomes being the sites of HSV-1 secondary envelopment. Electron microscopy studies have clearly shown the presence of endocytic tracers within membrane compartments that are wrapping cytoplasmic nucleocapsids (Hollinshead et al. 2012). Furthermore, blocking the function of specific Rab GTPases that are associated with endosomal trafficking inhibits HSV-1 assembly (Johns et al. 2011; Zenner et al. 2011), and inhibition of endocytosis prevents the incorporation of viral glycoproteins into virions (Albecka et al. 2016). These data point to endosomal compartments being a major source of membrane during HSV secondary envelopment. However, because of the highly dynamic and fluid nature of the secretory and endocytic pathways in cells, it can be difficult to accurately interpret cellular membrane compartment identity, particularly in infected cells where the cytopathic effects of HSV infection are known to perturb membrane traffic and cellular organelle structure (Henaff et al. 2012). Furthermore, the majority of cellular proteins that are defined as markers of particular organelles frequently undergo rapid transport between various secretory and endocytic compartments, and so in the context of HSV infection defining the identity of a membrane by the presence of such cellular markers can be fraught with difficulties. It must also be considered that the secondary envelopment of HSV could involve membranes that originate from more than one cellular compartment, including TGN and endosomes. Given the size of HSV particles, several vesicles containing viral envelope proteins may need to fuse to provide sufficient membrane in order to wrap the large nucleocapsid/tegument complex that makes up the internal structure of the virus particle.

Despite the uncertainties about secondary envelopment compartment identity, there is little doubt that this occurs at cellular membranes that are derived from late secretory pathway (e.g. the TGN) and/or endosomal pathway compartments, and not at early secretory pathway membranes such as the endoplasmic reticulum (ER) or Golgi apparatus. Artificially targeting viral envelope proteins to the ER prevents their incorporation into virions (Browne et al. 1996; Whiteley et al. 1999), and blocking ER-Golgi transport inhibits virus assembly and leads to accumulation of viral envelope proteins in the ER (Zenner et al. 2011).

As with the rest of the virion structure, the envelope of HSV is also highly complex, containing up to 16 different viral transmembrane proteins (Table 2.1). Therefore, all of these different membrane proteins need to be accumulated in the appropriate cellular compartment(s) so that they can be incorporated into mature virions. Due to the highly dynamic and interconnected nature of the late secretory and endocytic pathways, cellular membrane proteins are often localised to discrete compartments through an active retrieval mechanism, whereby proteins that leave the 'home' compartment are recycled back through vesicle transport. Therefore, it is of little surprise that many viral membrane proteins have adopted a similar mechanism by encoding targeting motifs in their cytoplasmic domains that mimic equivalent cellular motifs for packaging into transport vesicles. For example, tyrosine-based targeting motifs in glycoprotein B (gB) and glycoprotein E (gE) have been shown to direct endocytosis and intracellular targeting of these viral envelope proteins (Alconada et al. 1999; Beitia Ortiz de Zarate et al. 2004). However, not all HSV envelope proteins possess recognised targeting motifs, including the essential entry proteins gD and gH-gL. These envelope proteins appear to rely on interaction with other HSV proteins including gM and the gK-pUL20 complex to mediate their endocytosis and correct localisation to intracellular assembly compartments during infection (Crump et al. 2004; Lau and Crump 2015; Ren et al. 2012).

To help drive efficient virion assembly, several interactions between the tegument and envelope are thought to occur in a co-operative and partially redundant manner. The redundant nature of these envelope-tegument interactions is demonstrated by the observations that single deletion of gB, gD or gE results in little or no attenuation of secondary envelopment, whereas the combined deletion of gB and gD or gD and gE causes a dramatic inhibition of secondary envelopment leading to the accumulation of large aggregates of non-enveloped cytoplasmic capsids (Farnsworth et al. 2003; Johnson et al. 2011). Similar observations have been made for gM and the lipid-anchored tegument protein pUL11, with only minor defects in assembly caused by the loss of either protein individually, whereas deletion of both gM and pUL11 causes a profound inhibition of secondary envelopment (Leege et al. 2009). These observations support the notion that numerous interactions between envelope proteins and the underlying tegument occur to facilitate secondary envelopment, although loss of some of these interactions can be tolerated by the virus.

Two envelope proteins that appear to be particularly important for secondary envelopment are gK and pUL20. These multifunctional membrane proteins form a complex, and their correct subcellular localisation is reliant on each other (Foster et al. 2004b). Loss of either gK or pUL20 leads to significant defects in secondary envelopment (Foster et al. 2004a; Jayachandra et al. 1997; Melancon et al. 2004). The gK-pUL20 complex has been shown to interact with the pUL37 tegument protein as well as being important for the subcellular localisation of gD and gH-gL suggesting important functions of the gK-pUL20 complex during secondary envelopment in organising viral envelope proteins and mediating interactions with the tegument (Jambunathan et al. 2014; Lau and Crump 2015).

Consistent with the idea of a complex and redundant series of tegument-glycoprotein interactions helping to drive secondary envelopment, several tegument proteins have been reported to interact with the cytoplasmic domains of envelope proteins; VP16 with gH (Gross et al. 2003); VP22 with gD, gE and gM (Farnsworth et al. 2007a; Maringer et al. 2012); pUL11 with gD and gE (Farnsworth et al. 2007a; Han et al. 2011); and pUL37 with gK (Jambunathan et al. 2014). In fact, envelope-tegument interactions appear to be sufficient to drive secondary envelopment because this process can occur in the absence of nucleocapsids, forming the so-called light particles that contain most, if not all, tegument and envelope proteins but lack a nucleocapsid (Rixon et al. 1992; Szilagyi and Cunningham 1991).

The final stage of secondary envelopment is membrane scission to separate the virus from the host cell membrane, giving rise to a virion contained within the lumen of a vesicle. Similar to many enveloped viruses, HSV utilises the membrane scission activity of the cellular endosomal sorting complex required for transport (ESCRT) machinery for this final stage of assembly (Crump et al. 2007; Pawliczek and Crump 2009). The ESCRT machinery is a series of protein complexes that are normally involved in remodelling host membranes with the same topology as virus budding, for example, budding of material from the cytoplasm into the lumen of multivesicular bodies (Henne et al. 2011). Currently it is unclear how HSV recruits and regulates the ESCRT machinery at sites of secondary envelopment. Other viruses are known to recruit ESCRT complexes via their matrix proteins (Votteler and Sundquist 2013), and so tegument protein(s) appear the most likely candidates to directly or indirectly interact with ESCRT proteins. Indeed HSV-1 pUL36 has been shown to interact with the ESCRT-I subunit TSG101 (Calistri et al. 2015). However, TSG101 does not appear to be essential for HSV-1 assembly as siRNA depletion of this cellular protein does not inhibit virus replication (Pawliczek and Crump 2009). Given the complexity and redundancy within viral protein interactions during secondary envelopment, recruitment of the ESCRT machinery could be mediated by several viral proteins in a similarly redundant fashion, providing more than one mechanism to engage this important cellular machinery.

2.5 Virion Transport and Release

The completion of secondary envelopment results in virions being contained within large intracellular vesicles. To release viruses to the extracellular environment, these virion-containing vesicles need to be transported to the cell periphery and fuse with the plasma membrane. Currently there is little understanding of which cellular pathways are used during virus release and how they are controlled, although undoubtedly many host regulators of secretion are involved.

Given their size and the distances involved, it is likely that most if not all virion-containing vesicles require transport along microtubules via the activity of plus-end-directed kinesin motor proteins to travel from sites of secondary envelopment to the cell periphery. This will be particularly important upon reactivation of HSV from latency due to the long distances between the cell body and axon termini in neurons. There is some disagreement in the literature about the nature of virus particles that are transported along the axons during virus egress. Two extreme models are (1) HSV particles that undergo secondary envelopment in the cell body of neurons and are then transported along the axons as mature virions in large transport vesicles and (2) nucleocapsids and membrane compartments containing envelope proteins that are transported separately along the axons and then secondary envelopment occurs at axon termini (Cunningham et al. 2013; Kratchmarov et al. 2012; Taylor and Enquist 2015). There is evidence for both these models, although it is also possible that a combination of different types of transport occurs, with fully assembled virions in transport vesicles, partially enveloped virions and membraneless nucleocapsids all capable of being transported along the axons. Despite these uncertainties, there can be little doubt that kinesin motors must be recruited and activated to enable efficient virus egress. Kinesin-1, also known as conventional kinesin, has been shown to interact with the envelope protein pUS9 and with the tegument protein pUS11, giving two potential mechanisms for recruitment and regulation of this microtubule motor (Diefenbach et al. 2015, 2002). Interestingly pseudorabies virus (PRV) appears to recruit kinesin-3 via the combined activities of pUS9 and the gE-gI heterodimer (Kratchmarov et al. 2013). In HSV-1, deletion of pUS9 and gE has also been shown to inhibit egress of virions from the cell body into the axons in neuronal cells, lending weight to the hypothesis that pUS9 and gE-gI promote microtubule-based transport during virus egress, although defects in secondary envelopment are also observed in the absence of pUS9 and gE (DuRaine et al. 2017). Furthermore, the lack of pUS9 does not appear to affect HSV egress in epithelial cells, suggesting different mechanisms are required for virion transport to the cell periphery in neuronal and non-neuronal cells.

The gE-gI complex is well established as being important for cell-to-cell spread of HSV in epithelial cells. Loss of gE-gI function causes a substantial decrease in the spread of infection between cells in monolayer cultures, which is thought to be due to reduced targeting of virus secretion to cell junctions (Dingwell et al. 1994; Dingwell and Johnson 1998; Johnson et al. 2001). The targeting of virus secretion to cell junctions is thought to rely on the activity of specific sorting signals in the cytoplasmic domains of both gE and gI (Farnsworth and Johnson 2006; McMillan and Johnson 2001). However, whether this is related to kinesin motor activity or interactions with some other cellular transport regulators is unclear.

Upon reaching the cell periphery, the large virion-containing vesicles will encounter cortical actin, a dense mesh of cytoskeleton underlying the plasma membrane. This could pose a significant barrier to secretion of HSV from infected cells, by restricting access of the large virion-containing transport vesicles to the plasma membrane. A similar problem is faced by large cellular secretory vesicles, such as melanosomes, and is overcome through the action actin motors such as Myosin Va. HSV may well utilise a similar mechanism; Myosin Va has been shown to be important for efficient HSV-1 secretion from epithelial cells (Roberts and Baines 2010), although the details of how this actin-based motor is recruited or regulated by the virus is unclear.

The final stage of HSV egress is the fusion of the virion-containing transport vesicle with the plasma membrane to release the newly synthesised virus from the producing cell. Such vesicle fusion events within cells rely on the action of several proteins including Rab GTPases, membrane tethering complexes and SNARE fusion proteins. There are many different pathways for secretion of proteins from cells that are regulated by different subsets of membrane traffic regulators, and it is currently unclear which cellular Rab, tether or SNARE proteins are involved in HSV secretion. Some Rab proteins, namely, Rab3A, Rab6A, Rab8A and Rab11A, as well as the SNARE protein SNAP-25, have been shown to localise to vesicles containing virus particles or viral tegument and glycoproteins (Hogue et al. 2014; Miranda-Saksena et al. 2009), although whether they have any functional role in HSV egress is unknown.

It is conceivable that secretion of the newly formed viruses is controlled entirely by endogenous cellular factors that are normally recruited to the membranes that make up the secondary envelopment compartments. However, it seems more likely that these processes are co-opted and controlled by the virus. Viral proteins that recruit or regulate the transport, docking and fusion of virion-containing vesicles would need to be present on the cytoplasm-facing domain of such vesicles. The most obvious candidates for recruiting the necessary host cell factors are viral envelope proteins and membrane-associated tegument proteins that could remain on the vesicle membrane, in addition to being incorporated into virions. Both pUS9 and gE-gI are good candidates to be present on the vesicle membrane as this would position their cytoplasmic domains to be available for recruiting and regulating host trafficking proteins such as kinesins. The tegument protein pUL51, which is membrane-associated due to a lipid anchor modification, has been shown to be important for virus cell-to-cell spread and so is also a good candidate for being retained on virion-containing exocytic vesicle membranes (Albecka et al. 2017; Roller et al. 2014). However, which specific subset of viral proteins remain on the cytoplasmic surface of the vesicle membranes, how they are retained or recruited and how they function to control transport, docking and fusion of these large virion-containing vesicles with the plasma membrane remain to be discovered.

2.6 Summary

As described in this chapter, the assembly and egress of herpesviruses is a complex multistage process that requires the co-ordinated activities of numerous viral and cellular factors. While we know many details about the structure and assembly of the capsid and the components of mature virions, there is still much to discover. Future research to analyse the process of HSV assembly at the single-particle level in real time and to uncover the detailed mechanisms of host factor involvement at specific stages of virus morphogenesis will be important to shed new light on this fascinating subject.

References

Aksyuk AA, Newcomb WW, Cheng N, Winkler DC, Fontana J, Heymann JB, Steven AC (2015) Subassemblies and asymmetry in assembly of herpes simplex virus procapsid. MBio 6(5):e01525–e01515. https://doi.org/10.1128/mBio.01525-15

Albecka A, Laine RF, Janssen AF, Kaminski CF, Crump CM (2016) HSV-1 glycoproteins are delivered to virus assembly sites through dynamin-dependent endocytosis. Traffic 17(1):21–39. https://doi.org/10.1111/tra.12340

Albecka A, Owen DJ, Ivanova L, Brun J, Liman R, Davies L, Ahmed MF, Colaco S, Hollinshead M, Graham SC, Crump CM (2017) Dual function of the pUL7-pUL51 tegument protein complex in herpes simplex virus 1 infection. J Virol 91(2). https://doi.org/10.1128/JVI.02196-16

Alconada A, Bauer U, Sodeik B, Hoflack B (1999) Intracellular traffic of herpes simplex virus glycoprotein gE: characterization of the sorting signals required for its trans-Golgi network localization. J Virol 73(1):377–387

Baines JD, Wills E, Jacob RJ, Pennington J, Roizman B (2007) Glycoprotein M of herpes simplex virus 1 is incorporated into virions during budding at the inner nuclear membrane. J Virol 81(2):800–812. https://doi.org/10.1128/JVI.01756-06

Beitia Ortiz de Zarate I, Kaelin K, Rozenberg F (2004) Effects of mutations in the cytoplasmic domain of herpes simplex virus type 1 glycoprotein B on intracellular transport and infectivity. J Virol 78(3):1540–1551

Bigalke JM, Heldwein EE (2017) Have NEC coat, will travel: structural basis of membrane budding during nuclear Egress in herpesviruses. Adv Virus Res 97:107–141. https://doi.org/10.1016/bs.aivir.2016.07.002

Bjerke SL, Roller RJ (2006) Roles for herpes simplex virus type 1 UL34 and US3 proteins in disrupting the nuclear lamina during herpes simplex virus type 1 egress. Virology 347(2):261–276. https://doi.org/10.1016/j.virol.2005.11.053

Bohannon KP, Jun Y, Gross SP, Smith GA (2013) Differential protein partitioning within the herpesvirus tegument and envelope underlies a complex and variable virion architecture. Proc Natl Acad Sci U S A 110(17):E1613–E1620. https://doi.org/10.1073/pnas.1221896110

Boutell C, Everett RD (2013) Regulation of alphaherpesvirus infections by the ICP0 family of proteins. J Gen Virol 94(Pt 3):465–481. https://doi.org/10.1099/vir.0.048900-0

Browne H, Bell S, Minson T, Wilson DW (1996) An endoplasmic reticulum-retained herpes simplex virus glycoprotein H is absent from secreted virions: evidence for reenvelopment during egress. J Virol 70(7):4311–4316

Bucks MA, O'Regan KJ, Murphy MA, Wills JW, Courtney RJ (2007) Herpes simplex virus type 1 tegument proteins VP1/2 and UL37 are associated with intranuclear capsids. Virology 361(2):316–324. S0042-6822(06)00880-4 [pii] https://doi.org/10.1016/j.virol.2006.11.031

Calistri A, Munegato D, Toffoletto M, Celestino M, Franchin E, Comin A, Sartori E, Salata C, Parolin C, Palu G (2015) Functional interaction between the ESCRT-I component TSG101 and the HSV-1 tegument ubiquitin specific protease. J Cell Physiol 230(8):1794–1806. https://doi.org/10.1002/jcp.24890

Campbell ME, Palfreyman JW, Preston CM (1984) Identification of herpes simplex virus DNA sequences which encode a trans-acting polypeptide responsible for stimulation of immediate early transcription. J Mol Biol 180(1):1–19

Cardone G, Heymann JB, Cheng N, Trus BL, Steven AC (2012a) Procapsid assembly, maturation, nuclear exit: dynamic steps in the production of infectious herpesvirions. Adv Exp Med Biol 726:423–439. https://doi.org/10.1007/978-1-4614-0980-9_19

Cardone G, Newcomb WW, Cheng N, Wingfield PT, Trus BL, Brown JC, Steven AC (2012b) The UL36 tegument protein of herpes simplex virus 1 has a composite binding site at the capsid vertices. J Virol 86(8):4058–4064. https://doi.org/10.1128/JVI.00012-12

Cheung P, Banfield BW, Tufaro F (1991) Brefeldin A arrests the maturation and egress of herpes simplex virus particles during infection. J Virol 65(4):1893–1904

Clarke RW, Monnier N, Li H, Zhou D, Browne H, Klenerman D (2007) Two-color fluorescence analysis of individual virions determines the distribution of the copy number of proteins in herpes simplex virus particles. Biophys J 93(4):1329–1337. https://doi.org/10.1529/biophysj.107.106351

Coller KE, Lee JI, Ueda A, Smith GA (2007) The capsid and tegument of the alphaherpesviruses are linked by an interaction between the UL25 and VP1/2 proteins. J Virol 81 (21):11790–11797. JVI.01113-07 [pii] https://doi.org/10.1128/JVI.01113-07

Crump CM, Bruun B, Bell S, Pomeranz LE, Minson T, Browne HM (2004) Alphaherpesvirus glycoprotein M causes the relocalization of plasma membrane proteins. J Gen Virol 85(Pt 12):3517–3527. https://doi.org/10.1099/vir.0.80361-0

Crump CM, Yates C, Minson T (2007) Herpes simplex virus type 1 cytoplasmic envelopment requires functional Vps4. J Virol 81 (14):7380–7387. JVI.00222-07 [pii] https://doi.org/10.1128/JVI.00222-07

Cunningham A, Miranda-Saksena M, Diefenbach R, Johnson D (2013) Letter in response to: making the case: married versus separate models of alphaherpes virus anterograde transport in axons. Rev Med Virol 23(6):414–418. https://doi.org/10.1002/rmv.1760

de Wind N, Wagenaar F, Pol J, Kimman T, Berns A (1992) The pseudorabies virus homology of the herpes simplex virus UL21 gene product is a capsid protein which is involved in capsid maturation. J Virol 66(12):7096–7103

Desai PJ (2000) A null mutation in the UL36 gene of herpes simplex virus type 1 results in accumulation of unenveloped DNA-filled capsids in the cytoplasm of infected cells. J Virol 74(24):11608–11618

Diefenbach RJ, Miranda-Saksena M, Diefenbach E, Holland DJ, Boadle RA, Armati PJ, Cunningham AL (2002) Herpes simplex virus tegument protein US11 interacts with conventional kinesin heavy chain. J Virol 76(7):3282–3291

Diefenbach RJ, Davis A, Miranda-Saksena M, Fernandez MA, Kelly BJ, Jones CA, LaVail JH, Xue J, Lai J, Cunningham AL (2015) The basic domain of herpes simplex virus 1 pUS9 recruits Kinesin-1 to facilitate egress from neurons. J Virol 90(4):2102–2111. https://doi.org/10.1128/JVI.03041-15

Dingwell KS, Johnson DC (1998) The herpes simplex virus gE-gI complex facilitates cell-to-cell spread and binds to components of cell junctions. J Virol 72(11):8933–8942

Dingwell KS, Brunetti CR, Hendricks RL, Tang Q, Tang M, Rainbow AJ, Johnson DC (1994) Herpes simplex virus glycoproteins E and I facilitate cell-to-cell spread in vivo and across junctions of cultured cells. J Virol 68(2):834–845

DuRaine G, Wisner TW, Howard P, Williams M, Johnson DC (2017) Herpes simplex virus gE/gI and US9 promote both envelopment and sorting of virus particles in the cytoplasm of neurons, two processes that precede anterograde transport in axons. J Virol 91(11). https://doi.org/10.1128/JVI.00050-17

Elliott G, Mouzakitis G, O'Hare P (1995) VP16 interacts via its activation domain with VP22, a tegument protein of herpes simplex virus, and is relocated to a novel macromolecular assembly in coexpressing cells. J Virol 69(12):7932–7941

Fan WH, Roberts AP, McElwee M, Bhella D, Rixon FJ, Lauder R (2015) The large tegument protein pUL36 is essential for formation of the capsid vertex-specific component at the capsid-tegument interface of herpes simplex virus 1. J Virol 89(3):1502–1511. https://doi.org/10.1128/JVI.02887-14

Farnsworth A, Johnson DC (2006) Herpes simplex virus gE/gI must accumulate in the trans-Golgi network at early times and then redistribute to cell junctions to promote cell-cell spread. J Virol 80(7):3167–3179. https://doi.org/10.1128/JVI.80.7.3167-3179.2006

Farnsworth A, Goldsmith K, Johnson DC (2003) Herpes simplex virus glycoproteins gD and gE/gI serve essential but redundant functions during acquisition of the virion envelope in the cytoplasm. J Virol 77(15):8481–8494

Farnsworth A, Wisner TW, Johnson DC (2007a) Cytoplasmic residues of herpes simplex virus glycoprotein gE required for secondary envelopment and binding of tegument proteins VP22 and UL11 to gE and gD. J Virol 81 (1):319–331. JVI.01842-06 [pii] https://doi.org/10.1128/JVI.01842-06

Farnsworth A, Wisner TW, Webb M, Roller R, Cohen G, Eisenberg R, Johnson DC (2007b) Herpes simplex virus glycoproteins gB and gH function in fusion between the virion envelope and the outer nuclear membrane. Proc Natl Acad Sci U S A 104(24):10187–10192. https://doi.org/10.1073/pnas.0703790104

Foster TP, Melancon JM, Baines JD, Kousoulas KG (2004a) The herpes simplex virus type 1 UL20 protein modulates membrane fusion events during cytoplasmic virion morphogenesis and virus-induced cell fusion. J Virol 78(10):5347–5357

Foster TP, Melancon JM, Olivier TL, Kousoulas KG (2004b) Herpes simplex virus type 1 glycoprotein K and the UL20 protein are interdependent for intracellular trafficking and trans-Golgi network localization. J Virol 78(23):13262–13277. https://doi.org/10.1128/JVI.78.23.13262-13277.2004

Gao J, Hay TJM, Banfield BW (2017) The product of the herpes simplex virus 2 UL16 gene is critical for the egress of capsids from the nuclei of infected cells. J Virol 91(10). https://doi.org/10.1128/JVI.00350-17

Gross ST, Harley CA, Wilson DW (2003) The cytoplasmic tail of herpes simplex virus glycoprotein H binds to the tegument protein VP16 in vitro and in vivo. Virology 317(1):1–12

Hagen C, Dent KC, Zeev-Ben-Mordehai T, Grange M, Bosse JB, Whittle C, Klupp BG, Siebert CA, Vasishtan D, Bauerlein FJ, Cheleski J, Werner S, Guttmann P, Rehbein S, Henzler K, Demmerle J, Adler B, Koszinowski U, Schermelleh L, Schneider G, Enquist LW, Plitzko JM, Mettenleiter TC, Grunewald K (2015) Structural basis of vesicle formation at the inner nuclear membrane. Cell 163(7):1692–1701. https://doi.org/10.1016/j.cell.2015.11.029

Han J, Chadha P, Meckes DG Jr, Baird NL, Wills JW (2011) Interaction and interdependent packaging of tegument protein UL11 and glycoprotein e of herpes simplex virus. J Virol 85(18):9437–9446. https://doi.org/10.1128/JVI.05207-11

Han J, Chadha P, Starkey JL, Wills JW (2012) Function of glycoprotein E of herpes simplex virus requires coordinated assembly of three tegument proteins on its cytoplasmic tail. Proc Natl Acad Sci U S A 109(48):19798–19803. https://doi.org/10.1073/pnas.1212900109

Heming JD, Huffman JB, Jones LM, Homa FL (2014) Isolation and characterization of the herpes simplex virus 1 terminase complex. J Virol 88(1):225–236. https://doi.org/10.1128/JVI.02632-13

Henaff D, Radtke K, Lippe R (2012) Herpesviruses exploit several host compartments for envelopment. Traffic 13(11):1443–1449. https://doi.org/10.1111/j.1600-0854.2012.01399.x

Henaff D, Remillard-Labrosse G, Loret S, Lippe R (2013) Analysis of the early steps of herpes simplex virus 1 capsid tegumentation. J Virol 87(9):4895–4906. https://doi.org/10.1128/JVI.03292-12

Henne WM, Buchkovich NJ, Emr SD (2011) The ESCRT pathway. Dev Cell 21(1):77–91. https://doi.org/10.1016/j.devcel.2011.05.015

Heymann JB, Cheng N, Newcomb WW, Trus BL, Brown JC, Steven AC (2003) Dynamics of herpes simplex virus capsid maturation visualized by time-lapse cryo-electron microscopy. Nat Struct Biol 10(5):334–341. https://doi.org/10.1038/nsb922

Hogue IB, Bosse JB, Hu JR, Thiberge SY, Enquist LW (2014) Cellular mechanisms of alpha herpesvirus egress: live cell fluorescence microscopy of pseudorabies virus exocytosis. PLoS Pathog 10(12):e1004535. https://doi.org/10.1371/journal.ppat.1004535

Hollinshead M, Johns HL, Sayers CL, Gonzalez-Lopez C, Smith GL, Elliott G (2012) Endocytic tubules regulated by Rab GTPases 5 and 11 are used for envelopment of herpes simplex virus. EMBO J 31(21):4204–4220. https://doi.org/10.1038/emboj.2012.262

Jambunathan N, Chouljenko D, Desai P, Charles AS, Subramanian R, Chouljenko VN, Kousoulas KG (2014) Herpes simplex virus 1 protein UL37 interacts with viral glycoprotein gK and membrane protein UL20 and functions in cytoplasmic virion envelopment. J Virol 88(11):5927–5935. https://doi.org/10.1128/JVI.00278-14

Jayachandra S, Baghian A, Kousoulas KG (1997) Herpes simplex virus type 1 glycoprotein K is not essential for infectious virus production in actively replicating cells but is required for efficient envelopment and translocation of infectious virions from the cytoplasm to the extracellular space. J Virol 71(7):5012–5024

Johns HL, Gonzalez-Lopez C, Sayers C, Hollinshead M, Elliott G (2011) A role for human Rab6 in Herpes Simplex Virus Morphogenesis. In: 36th international Herpesvirus workshop, Gdansk, Poland

Johnson DC, Webb M, Wisner TW, Brunetti C (2001) Herpes simplex virus gE/gI sorts nascent virions to epithelial cell junctions, promoting virus spread. J Virol 75(2):821–833. https://doi.org/10.1128/JVI.75.2.821-833.2001

Johnson DC, Wisner TW, Wright CC (2011) Herpes simplex virus glycoproteins gB and gD function in a redundant fashion to promote secondary envelopment. J Virol 85(10):4910–4926. https://doi.org/10.1128/JVI.00011-11

Kato A, Yamamoto M, Ohno T, Kodaira H, Nishiyama Y, Kawaguchi Y (2005) Identification of proteins phosphorylated directly by the Us3 protein kinase encoded by herpes simplex virus 1. J Virol 79(14):9325–9331. https://doi.org/10.1128/JVI.79.14.9325-9331.2005

Kato A, Yamamoto M, Ohno T, Tanaka M, Sata T, Nishiyama Y, Kawaguchi Y (2006) Herpes simplex virus 1-encoded protein kinase UL13 phosphorylates viral Us3 protein kinase and regulates nuclear localization of viral envelopment factors UL34 and UL31. J Virol 80(3):1476–1486. https://doi.org/10.1128/JVI.80.3.1476-1486.2006

Klupp B, Altenschmidt J, Granzow H, Fuchs W, Mettenleiter TC (2008) Glycoproteins required for entry are not necessary for egress of pseudorabies virus. J Virol 82(13):6299–6309. https://doi.org/10.1128/JVI.00386-08

Koyama AH, Uchida T (1994) Inhibition by Brefeldin A of the envelopment of nucleocapsids in herpes simplex virus type 1-infected Vero cells. Arch Virol 135(3–4):305–317

Kratchmarov R, Taylor MP, Enquist LW (2012) Making the case: married versus separate models of alphaherpes virus anterograde transport in axons. Rev Med Virol 22(6):378–391. https://doi.org/10.1002/rmv.1724

Kratchmarov R, Kramer T, Greco TM, Taylor MP, Ch'ng TH, Cristea IM, Enquist LW (2013) Glycoproteins gE and gI are required for efficient KIF1A-dependent anterograde axonal transport of alphaherpesvirus particles in neurons. J Virol 87(17):9431–9440. https://doi.org/10.1128/JVI.01317-13

Laine RF, Albecka A, van de Linde S, Rees EJ, Crump CM, Kaminski CF (2015) Structural analysis of herpes simplex virus by optical super-resolution imaging. Nat Commun 6:5980. https://doi.org/10.1038/ncomms6980

Lau SY, Crump CM (2015) HSV-1 gM and the gK/pUL20 complex are important for the localization of gD and gH/L to viral assembly sites. Virus 7(3):915–938. https://doi.org/10.3390/v7030915

Le Sage V, Jung M, Alter JD, Wills EG, Johnston SM, Kawaguchi Y, Baines JD, Banfield BW (2013) The herpes simplex virus 2 UL21 protein is essential for virus propagation. J Virol 87(10):5904–5915. https://doi.org/10.1128/JVI.03489-12

Leach NR, Roller RJ (2010) Significance of host cell kinases in herpes simplex virus type 1 egress and lamin-associated protein disassembly from the nuclear lamina. Virology 406(1):127–137. https://doi.org/10.1016/j.virol.2010.07.002

Lee JH, Vittone V, Diefenbach E, Cunningham AL, Diefenbach RJ (2008) Identification of structural protein-protein interactions of herpes simplex virus type 1. Virology 378 (2):347–354. S0042-6822(08)00390-5 [pii] https://doi.org/10.1016/j.virol.2008.05.035

Leege T, Fuchs W, Granzow H, Kopp M, Klupp BG, Mettenleiter TC (2009) Effects of simultaneous deletion of pUL11 and glycoprotein M on virion maturation of herpes simplex virus type 1. J Virol 83(2):896–907. https://doi.org/10.1128/JVI.01842-08

Liu Z, Kato A, Shindo K, Noda T, Sagara H, Kawaoka Y, Arii J, Kawaguchi Y (2014) Herpes simplex virus 1 UL47 interacts with viral nuclear egress factors UL31, UL34, and Us3 and regulates viral nuclear egress. J Virol 88(9):4657–4667. https://doi.org/10.1128/JVI.00137-14

Loret S, Guay G, Lippe R (2008) Comprehensive characterization of extracellular herpes simplex virus type 1 virions. J Virol 82 (17):8605–8618. JVI.00904-08 [pii] https://doi.org/10.1128/JVI.00904-08

Lymberopoulos MH, Bourget A, Ben Abdeljelil N, Pearson A (2011) Involvement of the UL24 protein in herpes simplex virus 1-induced dispersal of B23 and in nuclear egress. Virology 412(2):341–348. https://doi.org/10.1016/j.virol.2011.01.016

Maringer K, Stylianou J, Elliott G (2012) A network of protein interactions around the herpes simplex virus tegument protein VP22. J Virol 86(23):12971–12982. https://doi.org/10.1128/JVI.01913-12

Maruzuru Y, Shindo K, Liu Z, Oyama M, Kozuka-Hata H, Arii J, Kato A, Kawaguchi Y (2014) Role of herpes simplex virus 1 immediate early protein ICP22 in viral nuclear egress. J Virol 88(13):7445–7454. https://doi.org/10.1128/JVI.01057-14

McClelland DA, Aitken JD, Bhella D, McNab D, Mitchell J, Kelly SM, Price NC, Rixon FJ (2002) pH reduction as a trigger for dissociation of herpes simplex virus type 1 scaffolds. J Virol 76(15):7407–7417

McMillan TN, Johnson DC (2001) Cytoplasmic domain of herpes simplex virus gE causes accumulation in the trans-Golgi network, a site of virus envelopment and sorting of virions to cell junctions. J Virol 75(4):1928–1940. https://doi.org/10.1128/JVI.75.4.1928-1940.2001

Melancon JM, Foster TP, Kousoulas KG (2004) Genetic analysis of the herpes simplex virus type 1 UL20 protein domains involved in cytoplasmic virion envelopment and virus-induced cell fusion. J Virol 78(14):7329–7343. https://doi.org/10.1128/JVI.78.14.7329-7343.2004

Mijatov B, Cunningham AL, Diefenbach RJ (2007) Residues F593 and E596 of HSV-1 tegument protein pUL36 (VP1/2) mediate binding of tegument protein pUL37. Virology 368(1):26–31. https://doi.org/10.1016/j.virol.2007.07.005

Miranda-Saksena M, Boadle RA, Aggarwal A, Tijono B, Rixon FJ, Diefenbach RJ, Cunningham AL (2009) Herpes simplex virus utilizes the large secretory vesicle pathway for anterograde transport of tegument and envelope proteins and for viral exocytosis from growth cones of human fetal axons. J Virol 83 (7):3187–3199. JVI.01579-08 [pii] https://doi.org/10.1128/JVI.01579-08

Mossman KL, Sherburne R, Lavery C, Duncan J, Smiley JR (2000) Evidence that herpes simplex virus VP16 is required for viral egress downstream of the initial envelopment event. J Virol 74(14):6287–6299

Naldinho-Souto R, Browne H, Minson T (2006) Herpes simplex virus tegument protein VP16 is a component of primary enveloped virions. J Virol 80(5):2582–2584. https://doi.org/10.1128/JVI.80.5.2582-2584.2006

Newcomb WW, Trus BL, Booy FP, Steven AC, Wall JS, Brown JC (1993) Structure of the herpes simplex virus capsid. Molecular composition of the pentons and the triplexes. J Mol Biol 232(2):499–511. https://doi.org/10.1006/jmbi.1993.1406

Newcomb WW, Homa FL, Thomsen DR, Booy FP, Trus BL, Steven AC, Spencer JV, Brown JC (1996) Assembly of the herpes simplex virus capsid: characterization of intermediates observed during cell-free capsid formation. J Mol Biol 263(3):432–446. https://doi.org/10.1006/jmbi.1996.0587

Newcomb WW, Juhas RM, Thomsen DR, Homa FL, Burch AD, Weller SK, Brown JC (2001) The UL6 gene product forms the portal for entry of DNA into the herpes simplex virus capsid. J Virol 75(22):10923–10932. https://doi.org/10.1128/JVI.75.22.10923-10932.2001

Newcomb WW, Thomsen DR, Homa FL, Brown JC (2003) Assembly of the herpes simplex virus capsid: identification of soluble scaffold-portal complexes and their role in formation of portal-containing capsids. J Virol 77(18):9862–9871

Newcomb WW, Homa FL, Brown JC (2005) Involvement of the portal at an early step in herpes simplex virus capsid assembly. J Virol 79(16):10540–10546. https://doi.org/10.1128/JVI.79.16.10540-10546.2005

Newcomb WW, Homa FL, Brown JC (2006) Herpes simplex virus capsid structure: DNA packaging protein UL25 is located on the external surface of the capsid near the vertices. J Virol 80(13):6286–6294. https://doi.org/10.1128/JVI.02648-05

Newcomb WW, Jones LM, Dee A, Chaudhry F, Brown JC (2012) Role of a reducing environment in disassembly of the herpesvirus tegument. Virology 431(1–2):71–79. https://doi.org/10.1016/j.virol.2012.05.017

Newcomb WW, Fontana J, Winkler DC, Cheng N, Heymann JB, Steven AC (2017) The primary enveloped virion of herpes simplex virus 1: its role in nuclear egress. MBio 8(3). https://doi.org/10.1128/mBio.00825-17

Oda S, Arii J, Koyanagi N, Kato A, Kawaguchi Y (2016) The interaction between herpes simplex virus 1 tegument proteins UL51 and UL14 and its role in Virion morphogenesis. J Virol 90(19):8754–8767. https://doi.org/10.1128/JVI.01258-16

O'Hara M, Rixon FJ, Stow ND, Murray J, Murphy M, Preston VG (2010) Mutational analysis of the herpes simplex virus type 1 UL25 DNA packaging protein reveals regions that are important after the viral DNA has been packaged. J Virol 84(9):4252–4263. https://doi.org/10.1128/JVI.02442-09

Okoye ME, Sexton GL, Huang E, McCaffery JM, Desai P (2006) Functional analysis of the triplex proteins (VP19C and VP23) of herpes simplex virus type 1. J Virol 80(2):929–940. https://doi.org/10.1128/JVI.80.2.929-940.2006

Padula ME, Sydnor ML, Wilson DW (2009) Isolation and preliminary characterization of herpes simplex virus 1 primary enveloped virions from the perinuclear space. J Virol 83(10):4757–4765. https://doi.org/10.1128/JVI.01927-08

Pante N, Kann M (2002) Nuclear pore complex is able to transport macromolecules with diameters of about 39 nm. Mol Biol Cell 13(2):425–434. https://doi.org/10.1091/mbc.01-06-0308

Park R, Baines JD (2006) Herpes simplex virus type 1 infection induces activation and recruitment of protein kinase C to the nuclear membrane and increased phosphorylation of lamin B. J Virol 80(1):494–504. https://doi.org/10.1128/JVI.80.1.494-504.2006

Pawliczek T, Crump CM (2009) Herpes simplex virus type 1 production requires a functional ESCRT-III complex but is independent of TSG101 and ALIX expression. J Virol 83 (21):11254–11264. JVI.00574-09 [pii] https://doi.org/10.1128/JVI.00574-09

Radtke K, Kieneke D, Wolfstein A, Michael K, Steffen W, Scholz T, Karger A, Sodeik B (2010) Plus- and minus-end directed microtubule motors bind simultaneously to herpes simplex virus capsids using different inner tegument structures. PLoS Pathog 6(7):e1000991. https://doi.org/10.1371/journal.ppat.1000991

Ren Y, Bell S, Zenner HL, Lau SY, Crump CM (2012) Glycoprotein M is important for the efficient incorporation of glycoprotein H-L into herpes simplex virus type 1 particles. J Gen Virol 93(Pt 2):319–329. https://doi.org/10.1099/vir.0.035444-0

Reynolds AE, Ryckman BJ, Baines JD, Zhou Y, Liang L, Roller RJ (2001) U(L)31 and U(L)34 proteins of herpes simplex virus type 1 form a complex that accumulates at the nuclear rim and is required for envelopment of nucleocapsids. J Virol 75(18):8803–8817

Reynolds AE, Wills EG, Roller RJ, Ryckman BJ, Baines JD (2002) Ultrastructural localization of the herpes simplex virus type 1 UL31, UL34, and US3 proteins suggests specific roles in primary envelopment and egress of nucleocapsids. J Virol 76(17):8939–8952

Rixon FJ, Addison C, McLauchlan J (1992) Assembly of enveloped tegument structures (L particles) can occur independently of virion maturation in herpes simplex virus type 1-infected cells. J Gen Virol 73(Pt 2):277–284. https://doi.org/10.1099/0022-1317-73-2-277

Roberts KL, Baines JD (2010) Myosin Va enhances secretion of herpes simplex virus 1 virions and cell surface expression of viral glycoproteins. J Virol. JVI.00732-10 [pii] https://doi.org/10.1128/JVI.00732-10

Roberts AP, Abaitua F, O'Hare P, McNab D, Rixon FJ, Pasdeloup D (2009) Differing roles of inner tegument proteins pUL36 and pUL37 during entry of herpes simplex virus type 1. J Virol 83(1):105–116. JVI.01032-08 [pii] https://doi.org/10.1128/JVI.01032-08

Roller RJ, Fetters R (2015) The herpes simplex virus 1 UL51 protein interacts with the UL7 protein and plays a role in its recruitment into the virion. J Virol 89(6):3112–3122. https://doi.org/10.1128/JVI.02799-14

Roller RJ, Haugo AC, Yang K, Baines JD (2014) The herpes simplex virus 1 UL51 gene product has cell type-specific functions in cell-to-cell spread. J Virol 88(8):4058–4068. https://doi.org/10.1128/JVI.03707-13

Roos WH, Radtke K, Kniesmeijer E, Geertsema H, Sodeik B, Wuite GJ (2009) Scaffold expulsion and genome packaging trigger stabilization of herpes simplex virus capsids. Proc Natl Acad Sci U S A 106(24):9673–9678. https://doi.org/10.1073/pnas.0901514106

Ryckman BJ, Roller RJ (2004) Herpes simplex virus type 1 primary envelopment: UL34 protein modification and the US3-UL34 catalytic relationship. J Virol 78(1):399–412

Scholtes LD, Yang K, Li LX, Baines JD (2010) The capsid protein encoded by U(L)17 of herpes simplex virus 1 interacts with tegument protein VP13/14. J Virol 84(15):7642–7650. https://doi.org/10.1128/JVI.00277-10

Schrag JD, Prasad BV, Rixon FJ, Chiu W (1989) Three-dimensional structure of the HSV1 nucleocapsid. Cell 56(4):651–660

Sheaffer AK, Newcomb WW, Brown JC, Gao M, Weller SK, Tenney DJ (2000) Evidence for controlled incorporation of herpes simplex virus type 1 UL26 protease into capsids. J Virol 74(15):6838–6848

Sheaffer AK, Newcomb WW, Gao M, Yu D, Weller SK, Brown JC, Tenney DJ (2001) Herpes simplex virus DNA cleavage and packaging proteins associate with the procapsid prior to its maturation. J Virol 75(2):687–698. https://doi.org/10.1128/JVI.75.2.687-698.2001

Smibert CA, Popova B, Xiao P, Capone JP, Smiley JR (1994) Herpes simplex virus VP16 forms a complex with the virion host shutoff protein vhs. J Virol 68(4):2339–2346

Smiley JR (2004) Herpes simplex virus virion host shutoff protein: immune evasion mediated by a viral RNase? J Virol 78(3):1063–1068

Spencer JV, Newcomb WW, Thomsen DR, Homa FL, Brown JC (1998) Assembly of the herpes simplex virus capsid: preformed triplexes bind to the nascent capsid. J Virol 72(5):3944–3951

Starkey JL, Han J, Chadha P, Marsh JA, Wills JW (2014) Elucidation of the block to herpes simplex virus egress in the absence of tegument protein UL16 reveals a novel interaction with VP22. J Virol 88(1):110–119. https://doi.org/10.1128/JVI.02555-13

Stevenson AJ, Morrison EE, Chaudhari R, Yang CC, Meredith DM (1997) Processing and intracellular localization of the herpes simplex virus type 1 proteinase. J Gen Virol 78(Pt 3):671–675. https://doi.org/10.1099/0022-1317-78-3-671

Svobodova S, Bell S, Crump CM (2012) Analysis of the interaction between the essential herpes simplex virus 1 tegument proteins VP16 and VP1/2. J Virol 86 (1):473–483. JVI.05981-11 [pii] https://doi.org/10.1128/JVI.05981-11

Szilagyi JF, Cunningham C (1991) Identification and characterization of a novel non-infectious herpes simplex virus-related particle. J Gen Virol 72(Pt 3):661–668

Taylor MP, Enquist LW (2015) Axonal spread of neuroinvasive viral infections. Trends Microbiol 23(5):283–288. https://doi.org/10.1016/j.tim.2015.01.002

Thurlow JK, Murphy M, Stow ND, Preston VG (2006) Herpes simplex virus type 1 DNA-packaging protein UL17 is required for efficient binding of UL25 to capsids. J Virol 80(5):2118–2126. https://doi.org/10.1128/JVI.80.5.2118-2126.2006

Tong L, Stow ND (2010) Analysis of herpes simplex virus type 1 DNA packaging signal mutations in the context of the viral genome. J Virol 84(1):321–329. https://doi.org/10.1128/JVI.01489-09

Toropova K, Huffman JB, Homa FL, Conway JF (2011) The herpes simplex virus 1 UL17 protein is the second constituent of the capsid vertex-specific component required for DNA packaging and retention. J Virol 85(15):7513–7522. https://doi.org/10.1128/JVI.00837-11

Trus BL, Newcomb WW, Cheng N, Cardone G, Marekov L, Homa FL, Brown JC, Steven AC (2007) Allosteric signaling and a nuclear exit strategy: binding of UL25/UL17 heterodimers to DNA-filled HSV-1 capsids. Mol Cell 26(4):479–489. https://doi.org/10.1016/j.molcel.2007.04.010

Vittone V, Diefenbach E, Triffett D, Douglas MW, Cunningham AL, Diefenbach RJ (2005) Determination of interactions between tegument proteins of herpes simplex virus type 1. J Virol 79 (15):9566–9571. 79/15/9566 [pii] https://doi.org/10.1128/JVI.79.15.9566-9571.2005

Votteler J, Sundquist WI (2013) Virus budding and the ESCRT pathway. Cell Host Microbe 14(3):232–241. https://doi.org/10.1016/j.chom.2013.08.012

Whiteley A, Bruun B, Minson T, Browne H (1999) Effects of targeting herpes simplex virus type 1 gD to the endoplasmic reticulum and trans-Golgi network. J Virol 73(11):9515–9520

Meckes DG, Jr., Wills JW (2007) Dynamic interactions of the UL16 tegument protein with the capsid of herpes simplex virus. J Virol 81 (23):13028–13036. doi:https://doi.org/10.1128/JVI.01306-07

Wills E, Mou F, Baines JD (2009) The U(L)31 and U(L)34 gene products of herpes simplex virus 1 are required for optimal localization of viral glycoproteins D and M to the inner nuclear membranes of infected cells. J Virol 83(10):4800–4809. https://doi.org/10.1128/JVI.02431-08

Wisner TW, Wright CC, Kato A, Kawaguchi Y, Mou F, Baines JD, Roller RJ, Johnson DC (2009) Herpesvirus gB-induced fusion between the virion envelope and outer nuclear membrane during virus egress is regulated by the viral US3 kinase. J Virol 83(7):3115–3126. https://doi.org/10.1128/JVI.01462-08

Wright CC, Wisner TW, Hannah BP, Eisenberg RJ, Cohen GH, Johnson DC (2009) Fusion between perinuclear virions and the outer nuclear membrane requires the fusogenic activity of herpes simplex virus gB. J Virol 83(22):11847–11856. https://doi.org/10.1128/JVI.01397-09

Yang K, Baines JD (2011) Selection of HSV capsids for envelopment involves interaction between capsid surface components pUL31, pUL17, and pUL25. Proc Natl Acad Sci U S A 108(34):14276–14281. https://doi.org/10.1073/pnas.1108564108

Zeev-Ben-Mordehai T, Weberruss M, Lorenz M, Cheleski J, Hellberg T, Whittle C, El Omari K, Vasishtan D, Dent KC, Harlos K, Franzke K, Hagen C, Klupp BG, Antonin W, Mettenleiter TC, Grunewald K (2015) Crystal structure of the herpesvirus nuclear egress complex provides insights into inner nuclear membrane remodeling. Cell Rep 13(12):2645–2652. https://doi.org/10.1016/j.celrep.2015.11.008

Zenner HL, Yoshimura S, Barr FA, Crump CM (2011) Analysis of Rab GTPase-activating proteins indicates that Rab1a/b and Rab43 are important for herpes simplex virus 1 secondary envelopment. J Virol 85 (16):8012–8021. JVI.00500-11 [pii] https://doi.org/10.1128/JVI.00500-11

Chapter 3
Us3 Protein Kinase Encoded by HSV: The Precise Function and Mechanism on Viral Life Cycle

Akihisa Kato and Yasushi Kawaguchi

Abstract All members of the *Alphaherpesvirinae* subfamily encode a serine/threonine kinase, designated Us3, which is not conserved in the other subfamilies. Us3 is a significant virulence factor for herpes simplex virus type 1 (HSV-1), which is one of the best-characterized members of the *Alphaherpesvirinae* family. Accumulating evidence indicates that HSV-1 Us3 is a multifunctional protein that plays various roles in the viral life cycle by phosphorylating a number of viral and cellular substrates. Therefore, the identification of Us3 substrates is directly connected to understanding Us3 functions and mechanisms. To date, more than 23 phosphorylation events upregulated by HSV-1 Us3 have been reported. However, few of these have been shown to be both physiological substrates of Us3 in infected cells and directly linked with Us3 functions in infected cells. In this chapter, we summarize the 12 physiological substrates of Us3 and the Us3-mediated functions. Furthermore, based on the identified phosphorylation sites of Us3 or Us3 homolog physiological substrates, we reverified consensus phosphorylation target sequences on the physiological substrates of Us3 and Us3 homologs in vitro and in infected cells. This information might aid the further identification of novel Us3 substrates and as yet unidentified Us3 functions.

A. Kato (✉)
Division of Molecular Virology, Department of Microbiology and Immunology, The Institute of Medical Science, The University of Tokyo, Tokyo, Japan

Division of Viral Infection, Department of Infectious Disease Control, International Research Center for Infectious Diseases, The Institute of Medical Science, The University of Tokyo, Tokyo, Japan
e-mail: akihisak@ims.u-tokyo.ac.jp

Y. Kawaguchi
Division of Molecular Virology, Department of Microbiology and Immunology, The Institute of Medical Science, The University of Tokyo, Minato-ku, Tokyo, Japan

Department of Infectious Disease Control, International Research Center for Infectious Diseases, The Institute of Medical Science, The University of Tokyo, Minato-ku, Tokyo, Japan

Research Center for Asian Infectious Diseases, The Institute of Medical Science, The University of Tokyo, Minato-ku, Tokyo, Japan

Keywords HSV · Us3 · Phosphorylation · Protein kinase · Consensus phosphorylation target sequence · LOGO algorithm

3.1 Introduction

The reversible phosphorylation of cellular proteins mediated by protein kinases and phosphatases is universally utilized by eukaryotes and prokaryotes and is one of the most intensively studied posttranslational modifications (Sugiyama and Ishihama 2016). Protein kinases mainly catalyze the transfer of γ-phosphate from ATP to the serine, threonine, and tyrosine residues of target proteins but rarely to the histidine, aspartic acid, lysine, and arginine residues (Sugiyama and Ishihama 2016). Phosphorylation changes the higher-order structure of proteins resulting in a functional change of the target protein, such as switching on/off of enzymatic activity, subcellular localization and molecular recognition ability (Knighton et al. 1991; Johnson et al. 1996; Nolen et al. 2004).

Many viruses have evolved mechanisms to utilize the phosphorylation system for the regulation of their own viral proteins and to establish a cellular environment for efficient viral replication and virulence (Terry et al. 2012; Arii et al. 2010; Arend et al. 2017). Phosphorylation events in cells infected with herpesviruses is of particular interest because, unlike most other viruses, herpesviruses encode virus-specific protein kinase(s) (Kawaguchi and Kato 2003; Jacob et al. 2011). In this chapter, we summarize the current understanding of Us3 protein kinases encoded by herpes simplex virus type 1, one of the best-characterized members of the *Alphaherpesvirinae* family.

3.2 Overview of HSV-1 Us3 Protein Kinase

In the mid-1980s, it was reported that the protein kinase activity of infected cells was markedly elevated by herpes simplex virus type 1 (HSV-1) or pseudorabies virus (PRV) infection (Purves et al. 1986; Katan et al. 1985). This activity was designated as virus-induced protein kinase (ViPK), which was purified from wild-type PRV-infected cells (Katan et al. 1985). Based on the unique biochemical property of ViPK that functions at a high optimal KCl concentration, it was predicted that herpesviruses may encode a viral specific protein kinase (Purves et al. 1986). In 1986, substrate specificity analysis using a peptide library showed that $R_nX(S/T)YY$ was the consensus target sequence of ViPK encoded by HSV-1 and PRV, where n is ≥ 2; X can be Arg, Ala, Val, Pro, or Ser; and Y can be any amino acid except an acidic residue (Purves et al. 1986; Leader 1993; Leader et al. 1991). Furthermore, a sequencing study of the HSV-1 genome revealed that the amino acid sequence of the HSV-1 Us3 gene contains motifs shared by host cellular protein kinases

(McGeoch and Davison 1986). In the following year, a specific antibody against Us3 was generated, and Us3 was identified as a ViPK. In 1988, Us3 was reported to be a significant virulence factor for HSV-1 infection in mice (Frame et al. 1987). Interestingly, all members of the *Alphaherpesvirinae* subfamily encode Us3 homologs that are not conserved in the other subfamilies (Deruelle and Favoreel 2011). To date, deletion of the Us3 homolog genes from HSV-1, HSV-2, varicella zoster virus (VZV), PRV, Marek's disease virus 1 (MDV-1), or bovine herpesvirus 1 or 5 (BHV-1 or BHV-5) is known to impair cell-type-dependent viral replication in cell cultures (Deruelle and Favoreel 2011). Notably, the protein kinase activity of HSV-1 Us3 was shown to be critical for viral virulence in mouse peripheral sites (e.g., eyes and vagina) and the central nervous system (CNS) of mice following intracranial and peripheral infection, respectively (Morimoto et al. 2009; Sagou et al. 2009).

HSV-1 Us3 can protect infected cells from apoptosis (Yu and He 2016; Leopardi et al. 1997; Munger and Roizman 2001; Wang et al. 2011; Benetti et al. 2003), promote the vesicle-mediated nucleocytoplasmic transport of nucleocapsids through nuclear membranes (Mou et al. 2007; Ryckman and Roller 2004; Reynolds et al. 2002; Johnson and Baines 2011), control infected cell morphology or microtubule networks (Naghavi et al. 2013; Munger and Roizman 2001; Kato et al. 2008), escape from host immune systems (Sloan et al. 2003, Sen et al. 2013; Wang et al. 2013; Rao et al. 2011), promote gene expression by blocking histone deacetylation (Walters et al. 2010; Poon et al. 2003, 2006), stimulate mRNA translation (Chuluunbaatar et al. 2010), regulate the intracellular trafficking of viral and cellular proteins, and upregulate viral enzymes (Kato et al. 2008, 2009, 2011, 2014c) in infected cells (Fig. 3.1). These observations suggest that Us3 is a multifunctional protein that plays various roles in the viral life cycle by phosphorylating a number of viral substrates. In agreement with this hypothesis, it was reported that HSV-1 Us3 is a promiscuous protein kinase that might phosphorylate more substrates than originally predicted (Mou et al. 2007). Therefore, the identification of Us3 substrates is directly connected to an understanding of Us3 function and mechanisms. In general, identification of the physiological substrate of a viral kinase and its phosphorylation sites requires the demonstration that the substrate and its phosphorylation sites are specifically and directly phosphorylated by the kinase in vitro and that the phosphorylation of the substrate is altered in cells infected with mutant viruses lacking the protein kinase activity and/or blocking phosphorylation site(s). As shown in Tables 3.1 and 3.2, more than 23 phosphorylation events upregulated by HSV-1 Us3 have been reported. Of note, nine of these proteins are physiological substrates of Us3 in infected cells and are directly linked with Us3 functions in infected cells. These include UL34, UL31, gB, UL47, vdUTPase, Us3 itself, kinesin family member 3A (KIF3A), tuberous sclerosis complex 2 (TSC2), and interferon regulatory factor 3 (IRF3). In addition, three host proteins, bcl-2-associated agonist of cell death (BAD), protein kinase A (PKA), and lamin A/C, were shown to be the physiological substrates of Us3 in vitro and in infected cells although the phosphorylation sites remain unknown. In the next section, we will introduce the 12 physiological substrates and Us3-mediated functions.

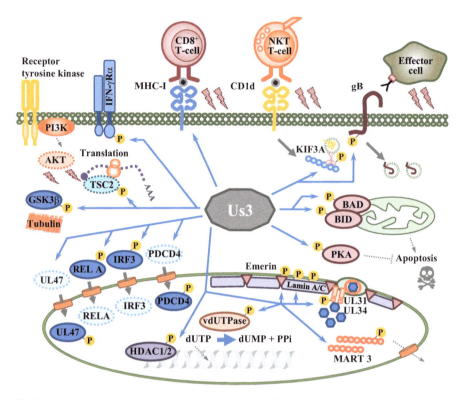

Fig. 3.1 Summary of the major functions associated with HSV-1 Us3 protein kinase. Abbreviations: AKT, protein kinase B; BAD, bcl-2-associated agonist of cell death; BID, BH3-interacting domain death agonist; CD1d, major histocompatibility complex class I-like antigen-presenting molecule; GSK3β, glycogen synthase kinase 3β; KIF3A, kinesin-like protein; HDAC1/2, histone deacetylase 1 and 2; IFN-γRα, interferon gamma receptor 1; IRF3, interferon regulatory factor 3; MATR3, matrin 3; MHC-I, major histocompatibility complex class I; PDCD4, programmed cell death protein 4; PKA, protein kinase A; PI3K, phosphatidylinositol 4,5-bisphosphate 3-kinase; RELA, transcription factor p65; TSC2, tuberous sclerosis complex 2

Table 3.1 Viral proteins phosphorylated or modified by HSV-1 Us3

Protein name	Direct phosphorylation by purified Us3 in vitro	Phosphorylation or modification in infected cells	Function regulated by phosphorylation	Viral virulence upregulated by phosphorylation
Us3	Detected	Detected	Activation of its protein kinase activity	HSK[a]
UL34	Detected	Detected	Promotion of viral replication only at early infection	Unknown
UL31	Detected	Detected	Promotion of nuclear egress	Unknown
gB	Detected	Detected	Downregulation of its cell surface expression by promoting endocytosis	HSK[a]
UL47	Detected	Detected	Promotion of nuclear localization	HSK[a]
vdUTPase	Detected	Detected	Maintenance of viral genome integrity by activating dUTPase activity	Neurovirulence[b]
Us8A	Not detected	Detected	Promotion of viral replication in trigeminal ganglia	Neuroinvasiveness[c]
Us9	Detected	Unknown	Unknown	Unknown
ICP22	Detected	Detected	Unknown	Unknown

HSK herpes stromal keratitis
[a]HSK indicates herpes stromal keratitis in the eyes of mice following ocular infection
[b]Neurovirulence indicates viral virulence in the CNS of mice following intracranial infection
[c]Neuroinvasiveness indicates the viral ability to invade from the peripheral site to the CNS of mice following ocular infection

Table 3.2 Host proteins phosphorylated or modified by HSV-1 Us3

Protein name	Direct phosphorylation by purified Us3 in vitro	Phosphorylation of modification in infected cells	Regulated function
KIF3A	Detected	Detected	Suppression of NKT cell function by modulating intracellular trafficking of CD1d
BAD	Detected	Detected	Blocking apoptosis
PKA	Detected	Detected	Blocking apoptosis
PDCD4	Unknown	Detected	Blocking apoptosis
BID	Detected	Unknown	Blocking apoptosis
TSC2	Detected	Detected	Stimulation of mRNA translation
GSK3β	Unknown	Detected	Microtubules rearrangement by inactivating GSK3β activity
HDAC2	Not detected [a]	Detected	Promotion of gene expression
IFN-γRα	Unknown	Detected	Accumulation of IFN-γRα dependent gene transcripts
IRF3	Detected	Detected	Suppression of type I interferon signaling
p65/RelA	Unknown	Detected	Suppression of IL-8 production
Lamin A/C	Detected	Detected	Promotion of nuclear egress by elevating its solubility
Emerin	Unknown	Detected	Unknown
Matrin3	Not detected[a]	Detected	Promotion of its nuclear retention

[a]Not detected indicates that purified Us3 or Us3 homologs did not phosphorylate the protein in vitro

3.3 Protein Phosphorylation or Modification Mediated by Us3

3.3.1 Us3 Autophosphorylates at Ser-147

The Us3 autophosphorylation site was first identified by bioinformatic analysis. Subsequent in vitro kinase assays and experiments using an antibody that specifically reacted with phosphorylated Ser-147 on Us3 confirmed that Us3 at Ser-147 was specifically autophosphorylated by Us3 itself both in vitro and in infected cells (Kato et al. 2008; Sagou et al. 2009). Us3 protein phosphorylated at Ser-147 purified from infected cells displayed higher kinase activity than Us3 not phosphorylated at Ser-147 (Sagou et al. 2009). Although only a small fraction (~6%) of total Us3 protein is autophosphorylated at Ser-147 in infected cells, autophosphorylation was required for the proper localization of Us3 and the ability of Us3 to induce wild-type cytopathic effects in infected cells (Sagou et al. 2009). Furthermore, Us3 autophosphorylation at Ser-147 promoted the development of herpes stromal keratitis (HSK) disease and viral replication in the eyes of mice following ocular infection (Sagou et al. 2009). In contrast, autophosphorylation had no effect on viral virulence in the CNS of mice following intracranial infection (Sagou et al. 2009). These observations suggest that the protein kinase activity of Us3 is tightly regulated by autophosphorylation at Ser-147 in infected cells and that this regulation partly has a critical role in viral replication and virulence in vivo.

3.3.2 Us3 Phosphorylates UL31 at Ser-11, Ser-24, Ser-26, Ser-27, Ser-40, and/or Ser-43

HSV-1 UL31 is a nuclear matrix-associated phosphoprotein localized within the nuclear membrane that interacts with HSV-1 UL34 to form a heterodimeric complex termed the nuclear egress complex (NEC) (Bigalke and Heldwein 2017; Johnson and Baines 2011). The NEC is essential for nuclear egress, in which nucleocapsids bud through the inner nuclear membrane (INM) (primary envelopment) and the enveloped nucleocapsids then fuse with the outer nuclear membrane (ONM) (de-envelopment) (Hagen et al. 2015; Bigalke et al. 2014). UL31 escorts nucleocapsids to inner nuclear envelope sites, the sites of primary envelopment (Funk et al. 2015). UL31 produced by cells infected with a Us3 null mutant virus migrated in a denaturing gel faster than those produced by cells infected with wild-type HSV-1, suggesting that Us3 mediates the phosphorylation of UL31 in infected cells (Poon and Roizman 2005; Kato et al. 2005). Consistent with this hypothesis, it was reported that Us3 phosphorylates UL31 Ser-11, Ser-24, Ser-26, Ser-27, Ser-40, and/or Ser-43 both in vitro and in infected cells (Mou et al. 2009). The null mutation in Us3 mislocalized both UL34 and UL31 into punctate structures at the nuclear rim and produced nuclear membrane invaginations containing multiple primary enveloped virions (Reynolds et al. 2001, 2002). Importantly, alanine substitutions at the six Us3 phosphorylation sites of UL31 mislocalized both UL34 and UL31 into punctate structures at the nuclear rim, and a kinase-dead mutation in Us3 accumulated enveloped virus particles at the interior perimeter of the nucleus (Mou et al. 2009). Furthermore, a phosphomimetic mutation in the six Ser of UL31 restored the wild-type localization of UL31 and UL34 in the absence of Us3 protein kinase activity (Mou et al. 2009). These observations suggest that UL31 is the major physiological substrate of Us3 responsible for regulating the localization of both UL31 and UL34 at the nuclear rim.

3.3.3 Us3 Phosphorylates UL34 at Thr-195 and Ser-198

HSV-1 UL34 is a type II membrane protein that forms the NEC with UL31 as described above (Bigalke and Heldwein 2017; Johnson and Baines 2011). In 1991, Purves et al. reported that Us3 mediated the phosphorylation of UL34 at Thr-195 and/or Ser-198 in infected cells (Purves et al. 1991). In agreement with this report, purified Us3 directly and specifically phosphorylated purified UL34 at Thr-195 and Ser-198 in vitro (Kato et al. 2005). However, alanine mutations in UL34 at Thr-195 and Ser-198 had little effect on viral replication and on the localization of UL34 and UL31 and viral nuclear egress in infected cells (Ryckman and Roller 2004). Thus, although UL34 was the first Us3 substrate identified, the significance of its phosphorylation by Us3 remains poorly understood.

3.3.4 Us3 Phosphorylates gB at Thr-887

The Us3 phosphorylation site Thr-887 in the cytoplasmic tail of glycoprotein B (gB) was identified using the same procedures used to identify the initial Us3 autophosphorylation site as described above (Kato et al. 2009). Interestingly, the Us3 phosphorylation site of gB is located near the endocytosis motif. An alanine mutation in this phosphorylation site downregulated the cell surface expression of gB in infected cells by promoting the endocytosis of gB, whereas a phosphomimetic mutation of gB at Thr-877 restored the wild-type phenotype (Kato et al. 2009; Imai et al. 2011). It was reported that the cell surface expression of gB is an important target of antibody-dependent cellular cytotoxic and is considered to promote viral cell-to-cell spread (Wisner and Johnson 2004; Kohl et al. 1990). From these evidences, we hypothesize that the cell surface expression of gB is strictly regulated for efficient viral replication, especially in the presence of the host immune system. In agreement with this hypothesis, blocking the Us3 phosphorylation of gB at Thr-887 significantly reduced the development of HSK disease and viral replication in the eyes of mice following ocular infection (Imai et al. 2010). Furthermore, a recombinant virus carrying both a null mutation in gH and a T887A mutation in gB caused the aberrant accumulation of primary enveloped virions in membranous vesicle structures adjacent to the nuclear membrane in infected cells (Wisner et al. 2009). These observations suggest that the Us3 phosphorylation of gB at Thr-887 plays a critical role in viral replication and pathogenic manifestations in the peripheral sites of mice by regulating the cell surface expression of gB and de-envelopment fusion during the nuclear egress of nucleocapsids.

3.3.5 Us3 Phosphorylates UL47 at Ser-77

Bioinformatic analysis predicted that UL47 has four putative Us3 phosphorylation sites (Kato et al. 2011). Interestingly, Us3 regulated the correct nuclear localization of UL47 in infected cells, and three of the four putative Us3 phosphorylation sites (Ser-77, Ser-88, and Thr-685) in UL47 were located close to the protein nuclear localization signal (NLS) (codons 50 to 68) and nuclear export signal (NES) (codons 658 to 667) (Kato et al. 2011). Mapping analysis by Us3 in an in vitro kinase assay showed that the Ser-77 of UL47 is a Us3 phosphorylation site of UL47 and that an alanine mutation in this site (S77A) reduced the development of HSK disease and viral replication in the eyes of mice following ocular infection as in the case of the Us3 phosphorylation of gB and Us3 itself (Kato et al. 2011). Notably, it was reported that the phosphorylation of a protein near its NLS is a key and common mechanism by which the transport of NLS-containing proteins into the nucleus can be regulated. This suggests that the Us3 phosphorylation of UL47 at Ser-77 regulates the nuclear localization of UL47 in infected cells (Kato et al. 2011). In agreement with this, a S77A mutation in UL47 impaired the nuclear localization of UL47 in infected cells, whereas a phosphomimetic mutation in this site restored wild-type nuclear localization. Furthermore, UL47 formed a complex(es) with UL34, UL31, and/or Us3 (all critical for viral nuclear egress) and promoted nuclear egress of the nucleocapsids (Liu et al. 2014). These observations suggest that the Us3 phosphorylation of UL47 at Ser-77 facilitates viral nuclear egress by promoting the nuclear localization of UL47 in infected cells.

3.3.6 Us3 Phosphorylates Viral dUTPase (vdUTPase) at Ser-187

A large-scale phosphoproteomic analysis of titanium dioxide affinity chromatography-enriched phosphopeptides from HSV-1-infected cells using high-accuracy mass spectrometry showed that the Ser-187 of HSV-1-encoded dUTPase (vdUTPase) is a phosphorylation site (Kato et al. 2014c). Amino acid sequences around vdUTPase Ser-187 resemble the consensus sequence of the Us3 phosphorylation site. Based on this information, it was shown that Us3 phosphorylates vdUTPase at Ser-187 both in vitro and in infected cells (Kato et al. 2014c). This phosphorylation upregulates the activity of vdUTPase to hydrolyze dUTP to dUMP and pyrophosphate in infected cells (Kato et al. 2014c). Because DNA polymerase readily misincorporates dUTP into replicating DNA causing point mutations and strand breakage, dUTP hydrolysis by dUTPase is required for accurate DNA replication (Vertessy and Toth 2009). It has long been assumed that viruses encode a dUTPase to compensate for low cellular dUTPase activity if present in their host cells, for example, in resting and differentiated cells such as neurons and macrophages where cellular dUTPase activity was reported to be low. In agreement with this hypothesis, sufficient dUTPase activity is required for efficient HSV-1 replication, and the upregulation of vdUTPase activity by Us3 phosphorylation requires viral genome integrity by compensating for low cellular dUTPase activity in the CNS of mice following intracranial infection (Kato et al. 2014a, 2015). To the best of our knowledge, HSV proteins critical for viral virulence in the CNS are, in almost all cases, also involved in viral pathogenicity at peripheral sites. Surprisingly, the Us3 phosphorylation of vdUTPase at Ser-187 was critical for viral virulence in the CNS, but not for pathogenic effects in the eyes and vaginas of mice following ocular and vaginal infection, respectively (Kato et al. 2014b). Therefore, vdUTPase is a Us3 substrate responsible for HSV-1 virulence in the CNS, and this phosphorylation is a specific mechanism involved in HSV-1 virulence in the CNS.

3.3.7 Us3 Phosphorylates KIF3A at Ser-687

Us3 downregulates the cell surface expression of the major histocompatibility complex class I-like antigen-presenting molecule (CD1d) by suppressing its endocytic recycling (Rao et al. 2011; Xiong et al. 2015). CD1d molecules on antigen-presenting cells present antigenic lipids to the T-cell receptor on natural killer T-cells, resulting in the rapid production of cytokines and cytotoxic proteins (Getz and Reardon 2017). Mature CD1d molecules constantly recycle between the cell surface and intracellular endosomal compartment, presumably to survey intracellular lipid antigens (Getz and Reardon 2017). Xiong et al. found that (i) the type II kinesin motor protein KIF3A is critical for the cell surface expression of CD1d, (ii) Us3 phosphorylates KIF3A at Ser-687 both in vitro and in infected cells, and (iii) the Us3-mediated downregulation of CD1d is completely absent in cells transfected with a green fluorescent protein-tagged KIF3A S687A mutant (Xiong et al. 2015). These observations suggest that the Us3 phosphorylation of KIF3A at Ser-687 inhibits natural killer T-cell function by modulating the intracellular trafficking of CD1d in infected cells.

3.3.8 Us3 Phosphorylates BAD

Us3 suppresses apoptosis induced by sorbitol, proapoptotic cellular proteins or replication-incompetent mutant HSV-1 (Leopardi et al. 1997) (Munger and Roizman 2001; Yu and He 2016). In vitro, Us3 directly phosphorylates human Bcl-2-associated agonist of cell death (BAD) (Kato et al. 2005; Cartier et al. 2003), which is a proapoptotic member of the Bcl-2 gene family, and the overexpression of Us3 mediates a posttranslational modification of mouse BAD and blocks its cleavage, which activated apoptosis (Munger and Roizman 2001). The phosphorylation of mouse BAD at Ser-112, Ser-136, and Ser-155 (human BAD Ser-75, 99 and 118) by cellular protein kinases, such as protein kinase A (PKA) and protein kinase B (Akt) and/or p90 ribosomal S6 kinase (RSK), inhibited the apoptotic activity of BAD (Niemi and MacKeigan 2013). Amino acid sequences around the three phosphorylation sites of BAD are similar to the consensus sequence recognized by Us3. These observations suggest that Us3 phosphorylates BAD at Ser-112, Ser-136, and/or Ser-155 to inhibit apoptosis. However, Us3 also inhibited apoptosis induced by the overexpression of mutant mouse BAD in which the Ser-112, Ser-136, and Ser-115 of BAD were substituted with alanines (Benetti et al. 2003). In addition, Us3 blocked apoptosis induced by the overexpression of BH3-interacting domain death agonist (BID), a factor parallel to BAD in the apoptotic pathway, and apoptosis regulator BAX, a factor downstream of BAD in the apoptotic pathway (Ogg et al. 2004). Therefore, it is reasonable to conclude that Us3 inhibits both BAD and other factor(s) downstream of BAD to block apoptosis.

3.3.9 Us3 Phosphorylates PKA

The phosphorylation target site specificity of HSV-1 Us3 is similar to that of PKA or Akt (Benetti and Roizman 2004; Chuluunbaatar et al. 2010), and some antibodies that recognize the phosphorylated substrate sequences of PKA react with Us3 phosphorylation sites (Kato et al. 2008, 2009, 2011; Xiong et al. 2015). Interestingly, Benetti et al. reported that (1) purified Us3 directly phosphorylated peptides containing the PKA regulatory type IIα subunit (PRKAR2A) sequence in vitro, (2) Us3 mediated a posttranslational modification of PRKAR2A in infected cells, and (3) activation of PKA by forskolin blocked apoptosis induced by a replication-incompetent mutant HSV-1 or by BAD independent of Us3 in infected cells (Benetti and Roizman 2004). These results suggest that a major determinant of the antiapoptotic activity of Us3 is involved in the phosphorylation of PKA substrates by either or both enzymes. Furthermore, PKA directly phosphorylated the Us3 phosphorylation sites of gB and Us3 itself in vitro although the protein kinase activity of Us3 is responsible for their phosphorylation in infected cells (Kato et al. 2008, 2009). Therefore, it would be interesting to analyze whether Us3 regulates some of the physiological substrates of PKA in infected cells.

3.3.10 Us3 Phosphorylates TSC2 at Ser-939

Viruses, as obligate intracellular parasites, dynamically modulate de novo protein synthesis by hijacking the host translational machinery. Mammalian target of rapamycin complex 1 (mTORC1) inactivated the translational repressor activity of eukaryotic translation initiation factor 4E-binding protein 1 (4E-BP1) by phosphorylating 4E-BP1 at both Thr-37 and Thr-46 (Gingras et al. 1999; Huang and Manning 2009). In addition, Akt neutralized the tuberin (TSC2)-mediated negative regulation of mTORC1 by phosphorylating TSC2 at Ser-929 and Thr-1462 (Inoki et al. 2002; Huang and Manning 2009). HSV-1 infection induced the phosphorylation and degradation of 4E-BP1 to stimulate cap-dependent translation independent of Akt in infected cells (Chuluunbaatar et al. 2010). Interestingly, (1) Us3 phosphorylated TSC2 at Ser-929 and Thr-1462 both in vitro and in infected cells, (2) the overexpression of a TSC2 mutant carrying S929A and T1462A double mutations prevented the degradation of 4E-BP1 in HSV-1-infected cells, and (3) the knockdown of TSC2 by siRNA rescued the reduction of viral replication caused by the Us3 deletion (Chuluunbaatar et al. 2010). These results suggest that Us3 activates mTORC1 via the neutralization of TSC2-negative regulators in infected cells. However, Us3 suppressed the activation of Akt by blocking the phosphorylation of Akt at Ser-473 in infected cells (Benetti and Roizman 2006; Eaton et al. 2014). Therefore, we hypothesized that Us3 inhibits Akt activation and hijacks downstream Akt targets such as the Akt/TSC2/mTORC1 signaling axis in infected cells. This hypothesis is supported by the observation that Us3 controls the stable formation of microtubules in infected cells by upregulating the phosphorylation of glycogen synthase kinase 3β (GSK3β) at Ser-21, which is a phosphorylation site usually mediated by Akt (Naghavi et al. 2013).

3.3.11 Us3 Phosphorylates Lamin A/C at Multiple Sites

As described above, the nucleocapsids of herpesviruses assemble in the nucleoplasm and bud through the INM (primary envelopment), and then the enveloped nucleocapsids fuse with the ONM (de-envelopment) (Johnson and Baines 2011). To access the INM, nucleocapsids must bypass the nuclear lamina, a dense meshwork of type V microfilaments categorized as either A-type (lamin A/C) or B-type (lamin B1, B2) (Harr et al. 2015). Us3 directly phosphorylates lamin A/C at multiple sites both in vitro and in infected cells (Mou et al. 2007). Interestingly, Us3 elevated the solubility of lamin A/C from an endogenous pre-existing nuclear lamina in vitro (Mou et al. 2007). These observations suggest that the Us3 phosphorylation of lamin A/C contributes to the efficient transport of nucleocapsids to the INM of infected cells. Furthermore, in infected cells, Us3 also mediates the phosphorylation of emerin (Morris et al. 2007; Leach et al. 2007), a member of the LEM domain class of INM proteins that binds to a number of nuclear components including lamins, barrier-to-autointegration factor (BAF), as well as F-actin, and is believed to be involved in maintaining nuclear integrity (Harr et al. 2015). Biochemical extraction experiments and immunofluorescence assays showed that the association of emerin with the INM was significantly reduced during HSV-1 infection (Morris et al. 2007). Currently, it is not known whether emerin is a physiological substrate of Us3 or whether the phosphorylation of emerin induced by HSV-1 infection is involved in its association with the INM. However, the phosphorylation of emerin at Ser-175 was reported to inhibit its binding to BAF in a *Xenopus* egg cell-free system (Hirano et al. 2005). Therefore, it would be interesting to identify the Us3 phosphorylation sites of lamin A/C or emerin and investigate the effect of phosphorylation on the transport efficiency of nucleocapsids to the INM in infected cells.

3.3.12 Us3 Phosphorylates IRF3 at Ser-175

A Us3 null mutant virus was reported to be more sensitive to IFN-α compared with wild-type virus, suggesting Us3 mediates resistance to IFN-α treatment (Deruelle and Favoreel 2011). Wang et al. reported that (1) the overexpression of Us3 inhibited IFN-β production by Sendai virus (SeV) infection and the overexpression of constitutive active IRF3 and that (2) Us3 phosphorylated and interacted with IFR3 in HSV-1-infected cells (Wang et al. 2013). Furthermore, dimers of activated IRF3 translocated to the nucleus from the cytoplasm and elevated the transcription of interferons α and β, as well as other interferon-induced genes (Wang et al. 2013). As expected, the overexpression of Us3 in part canceled SeV-induced dimerization and the nuclear translocation of IRF3 (Wang et al. 2013). Furthermore, alanine replacement at the Ser-175 of IRF3 prevented Us3 phosphorylation and the transcriptional activity of IRF3 in cells transfected with Us3 (Wang et al. 2013). These observations suggest that Us3 modulates type-1 interferon signaling through the phosphorylation of IRF3 at Ser-175 in infected cells. Interestingly, Epstein-Barr virus encoding BGLF4 protein kinase phosphorylated IRF-3 at Ser-123, Ser-175, and Thr-180 and inhibited the transcriptional activity of IRF3 (Wang et al. 2009). There is no evidence showing that the Ser-175 of IRF3 is phosphorylated except in herpesvirus uninfected cells. Therefore, herpesviruses have evolved unique strategies to overcome IRF3-mediated antiviral responses.

3.4 Future Perspectives

In this section, we reverified the consensus phosphorylation target sequence of Us3 using the LOGO algorithm (http://weblogo.threeplusone.com/) based on phosphorylation sites of "the physiological substrates of Us3 and VZV encoding Us3 homologs, ORF66" reported by five other groups (Fig. 3.2a). As shown in Fig. 3.2b, the physiological substrate of Us3 or Us3 homologs tends to contain 3.25 basic residues on the N-terminal side and 0.92 acidic residues or a tyrosine residue on the C-terminal side. Although peptides with a +1 and/or +2 acidic residue have been reported to not be optimal phosphorylation targets (Purves et al. 1986; Leader 1993; Leader et al. 1991), an acidic residue on the C-terminal side of Us3 target residues does not exert a negative influence on the physiological substrates of Us3. This information may aid the further identification of novel Us3 substrates and our understanding of unidentified Us3 functions in the near future.

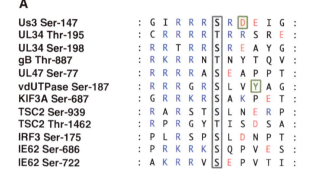
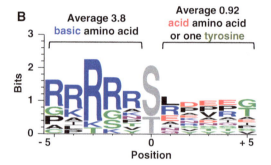

Fig. 3.2 (a) Amino acid sequence around the phosphorylation sites of the physiological substrates of Us3 and VZV encoding Us3 homologs. Phosphorylation sites are surrounded by gray squares. Red color indicates basic amino acids. Blue color indicates acid amino acids. (b) Consensus sequence images of the physiological substrates of Us3 and VZV encoding Us3 homologs generated by WebLogo (Crooks et al. 2004)

References

Arend KC, Lenarcic EM, Vincent HA, Rashid N, Lazear E, McDonald IM, Gilbert TS, East MP, Herring LE, Johnson GL, Graves LM, Moorman NJ (2017) Kinome profiling identifies drug-gable targets for novel Human Cytomegalovirus (HCMV) antivirals. Mol Cell Proteomics 16(4 suppl 1):S263–S276. https://doi.org/10.1074/mcp.M116.065375

Arii J, Goto H, Suenaga T, Oyama M, Kozuka-Hata H, Imai T, Minowa A, Akashi H, Arase H, Kawaoka Y, Kawaguchi Y (2010) Non-muscle myosin IIA is a functional entry receptor for herpes simplex virus-1. Nature 467(7317):859–862. https://doi.org/10.1038/nature09420

Benetti L, Roizman B (2004) Herpes simplex virus protein kinase US3 activates and functionally overlaps protein kinase A to block apoptosis. Proc Natl Acad Sci U S A 101(25):9411–9416. https://doi.org/10.1073/pnas.0403160101

Benetti L, Roizman B (2006) Protein kinase B/Akt is present in activated form throughout the entire replicative cycle of deltaU(S)3 mutant virus but only at early times after infection with wild-type herpes simplex virus 1. J Virol 80(7):3341–3348. https://doi.org/10.1128/JVI.80.7.3341-3348.2006

Benetti L, Munger J, Roizman B (2003) The herpes simplex virus 1 US3 protein kinase blocks caspase-dependent double cleavage and activation of the proapoptotic protein BAD. J Virol 77(11):6567–6573

Bigalke JM, Heldwein EE (2017) Have NEC coat, will travel: structural basis of membrane budding during nuclear egress in herpesviruses. Adv Virus Res 97:107–141. https://doi.org/10.1016/bs.aivir.2016.07.002

Bigalke JM, Heuser T, Nicastro D, Heldwein EE (2014) Membrane deformation and scission by the HSV-1 nuclear egress complex. Nat Commun 5:4131. https://doi.org/10.1038/ncomms5131

Cartier A, Komai T, Masucci MG (2003) The Us3 protein kinase of herpes simplex virus 1 blocks apoptosis and induces phosphorylation of the Bcl-2 family member Bad. Exp Cell Res 291(1):242–250

Chuluunbaatar U, Roller R, Feldman ME, Brown S, Shokat KM, Mohr I (2010) Constitutive mTORC1 activation by a herpesvirus Akt surrogate stimulates mRNA translation and viral replication. Genes Dev 24(23):2627–2639. https://doi.org/10.1101/gad.1978310

Crooks GE, Hon G, Chandonia JM, Brenner SE (2004) WebLogo: a sequence logo generator. Genome Res 14(6):1188–1190. https://doi.org/10.1101/gr.849004

Deruelle MJ, Favoreel HW (2011) Keep it in the subfamily: the conserved alphaherpesvirus US3 protein kinase. J Gen Virol 92(Pt 1):18–30. https://doi.org/10.1099/vir.0.025593-0

Eaton HE, Saffran HA, Wu FW, Quach K, Smiley JR (2014) Herpes simplex virus protein kinases US3 and UL13 modulate VP11/12 phosphorylation, virion packaging, and phosphatidylinositol 3-kinase/Akt signaling activity. J Virol 88(13):7379–7388. https://doi.org/10.1128/JVI.00712-14

Frame MC, Purves FC, McGeoch DJ, Marsden HS, Leader DP (1987) Identification of the herpes simplex virus protein kinase as the product of viral gene US3. J Gen Virol 68(Pt 10):2699–2704. https://doi.org/10.1099/0022-1317-68-10-2699

Funk C, Ott M, Raschbichler V, Nagel CH, Binz A, Sodeik B, Bauerfeind R, Bailer SM (2015) The herpes simplex virus protein pUL31 escorts nucleocapsids to sites of nuclear egress, a process coordinated by its N-terminal domain. PLoS Pathog 11(6):e1004957. https://doi.org/10.1371/journal.ppat.1004957

Getz GS, Reardon CA (2017) Natural killer T cells in atherosclerosis. Nat Rev Cardiol 14(5):304–314. https://doi.org/10.1038/nrcardio.2017.2

Gingras AC, Gygi SP, Raught B, Polakiewicz RD, Abraham RT, Hoekstra MF, Aebersold R, Sonenberg N (1999) Regulation of 4E-BP1 phosphorylation: a novel two-step mechanism. Genes Dev 13(11):1422–1437

Hagen C, Dent KC, Zeev-Ben-Mordehai T, Grange M, Bosse JB, Whittle C, Klupp BG, Siebert CA, Vasishtan D, Bauerlein FJ, Cheleski J, Werner S, Guttmann P, Rehbein S, Henzler K, Demmerle J, Adler B, Koszinowski U, Schermelleh L, Schneider G, Enquist LW, Plitzko JM, Mettenleiter TC, Grunewald K (2015) Structural basis of vesicle formation at the inner nuclear membrane. Cell 163(7):1692–1701. https://doi.org/10.1016/j.cell.2015.11.029

Harr JC, Luperchio TR, Wong X, Cohen E, Wheelan SJ, Reddy KL (2015) Directed targeting of chromatin to the nuclear lamina is mediated by chromatin state and A-type lamins. J Cell Biol 208(1):33–52. https://doi.org/10.1083/jcb.201405110

Hirano Y, Segawa M, Ouchi FS, Yamakawa Y, Furukawa K, Takeyasu K, Horigome T (2005) Dissociation of emerin from barrier-to-autointegration factor is regulated through mitotic phosphorylation of emerin in a xenopus egg cell-free system. J Biol Chem 280(48):39925–39933. https://doi.org/10.1074/jbc.M503214200

Huang J, Manning BD (2009) A complex interplay between Akt, TSC2 and the two mTOR complexes. Biochem Soc Trans 37(Pt 1):217–222. https://doi.org/10.1042/BST0370217

Imai T, Sagou K, Arii J, Kawaguchi Y (2010) Effects of phosphorylation of herpes simplex virus 1 envelope glycoprotein B by Us3 kinase in vivo and in vitro. J Virol 84(1):153–162. https://doi.org/10.1128/JVI.01447-09

Imai T, Arii J, Minowa A, Kakimoto A, Koyanagi N, Kato A, Kawaguchi Y (2011) Role of the herpes simplex virus 1 Us3 kinase phosphorylation site and endocytosis motifs in the intracellular transport and neurovirulence of envelope glycoprotein B. J Virol 85(10):5003–5015. https://doi.org/10.1128/JVI.02314-10

Inoki K, Li Y, Zhu T, Wu J, Guan KL (2002) TSC2 is phosphorylated and inhibited by Akt and suppresses mTOR signalling. Nat Cell Biol 4(9):648–657. https://doi.org/10.1038/ncb839

Jacob T, Van den Broeke C, Favoreel HW (2011) Viral serine/threonine protein kinases. J Virol 85(3):1158–1173. https://doi.org/10.1128/JVI.01369-10

Johnson DC, Baines JD (2011) Herpesviruses remodel host membranes for virus egress. Nat Rev Microbiol 9(5):382–394. https://doi.org/10.1038/nrmicro2559

Johnson LN, Noble ME, Owen DJ (1996) Active and inactive protein kinases: structural basis for regulation. Cell 85(2):149–158

Katan M, Stevely WS, Leader DP (1985) Partial purification and characterization of a new phosphoprotein kinase from cells infected with pseudorabies virus. Eur J Biochem 152(1):57–65

Kato A, Yamamoto M, Ohno T, Kodaira H, Nishiyama Y, Kawaguchi Y (2005) Identification of proteins phosphorylated directly by the Us3 protein kinase encoded by herpes simplex virus 1. J Virol 79(14):9325–9331. https://doi.org/10.1128/JVI.79.14.9325-9331.2005

Kato A, Tanaka M, Yamamoto M, Asai R, Sata T, Nishiyama Y, Kawaguchi Y (2008) Identification of a physiological phosphorylation site of the herpes simplex virus 1-encoded protein kinase Us3 which regulates its optimal catalytic activity in vitro and influences its function in infected cells. J Virol 82(13):6172–6189. https://doi.org/10.1128/JVI.00044-08

Kato A, Arii J, Shiratori I, Akashi H, Arase H, Kawaguchi Y (2009) Herpes simplex virus 1 protein kinase Us3 phosphorylates viral envelope glycoprotein B and regulates its expression on the cell surface. J Virol 83(1):250–261. https://doi.org/10.1128/JVI.01451-08

Kato A, Liu Z, Minowa A, Imai T, Tanaka M, Sugimoto K, Nishiyama Y, Arii J, Kawaguchi Y (2011) Herpes simplex virus 1 protein kinase Us3 and major tegument protein UL47 reciprocally regulate their subcellular localization in infected cells. J Virol 85(18):9599–9613. https://doi.org/10.1128/JVI.00845-11

Kato A, Hirohata Y, Arii J, Kawaguchi Y (2014a) Phosphorylation of herpes simplex virus 1 dUTPase upregulated viral dUTPase activity to compensate for low cellular dUTPase activity for efficient viral replication. J Virol 88(14):7776–7785. https://doi.org/10.1128/JVI.00603-14

Kato A, Shindo K, Maruzuru Y, Kawaguchi Y (2014b) Phosphorylation of a herpes simplex virus 1 dUTPase by a viral protein kinase, Us3, dictates viral pathogenicity in the central nervous system but not at the periphery. J Virol 88(5):2775–2785. https://doi.org/10.1128/JVI.03300-13

Kato A, Tsuda S, Liu Z, Kozuka-Hata H, Oyama M, Kawaguchi Y (2014c) Herpes simplex virus 1 protein kinase Us3 phosphorylates viral dUTPase and regulates its catalytic activity in infected cells. J Virol 88(1):655–666. https://doi.org/10.1128/JVI.02710-13

Kato A, Arii J, Koyanagi Y, Kawaguchi Y (2015) Phosphorylation of herpes simplex virus 1 dUTPase regulates viral virulence and genome integrity by compensating for low cellular dUTPase activity in the central nervous system. J Virol 89(1):241–248. https://doi.org/10.1128/JVI.02497-14

Kawaguchi Y, Kato K (2003) Protein kinases conserved in herpesviruses potentially share a function mimicking the cellular protein kinase cdc2. Rev Med Virol 13(5):331–340. https://doi.org/10.1002/rmv.402

Knighton DR, Zheng JH, Ten Eyck LF, Ashford VA, Xuong NH, Taylor SS, Sowadski JM (1991) Crystal structure of the catalytic subunit of cyclic adenosine monophosphate-dependent protein kinase. Science 253(5018):407–414

Kohl S, Strynadka NC, Hodges RS, Pereira L (1990) Analysis of the role of antibody-dependent cellular cytotoxic antibody activity in murine neonatal herpes simplex virus infection with antibodies to synthetic peptides of glycoprotein D and monoclonal antibodies to glycoprotein B. J Clin Invest 86(1):273–278. https://doi.org/10.1172/JCI114695

Leach N, Bjerke SL, Christensen DK, Bouchard JM, Mou F, Park R, Baines J, Haraguchi T, Roller RJ (2007) Emerin is hyperphosphorylated and redistributed in herpes simplex virus type 1-infected cells in a manner dependent on both UL34 and US3. J Virol 81(19):10792–10803. https://doi.org/10.1128/JVI.00196-07

Leader DP (1993) Viral protein kinases and protein phosphatases. Pharmacol Ther 59(3):343–389

Leader DP, Deana AD, Marchiori F, Purves FC, Pinna LA (1991) Further definition of the substrate specificity of the alpha-herpesvirus protein kinase and comparison with protein kinases A and C. Biochim Biophys Acta 1091(3):426–431

Leopardi R, Van Sant C, Roizman B (1997) The herpes simplex virus 1 protein kinase US3 is required for protection from apoptosis induced by the virus. Proc Natl Acad Sci U S A 94(15):7891–7896

Liu Z, Kato A, Shindo K, Noda T, Sagara H, Kawaoka Y, Arii J, Kawaguchi Y (2014) Herpes simplex virus 1 UL47 interacts with viral nuclear egress factors UL31, UL34, and Us3 and regulates viral nuclear egress. J Virol 88(9):4657–4667. https://doi.org/10.1128/JVI.00137-14

McGeoch DJ, Davison AJ (1986) Alphaherpesviruses possess a gene homologous to the protein kinase gene family of eukaryotes and retroviruses. Nucleic Acids Res 14(4):1765–1777

Morimoto T, Arii J, Tanaka M, Sata T, Akashi H, Yamada M, Nishiyama Y, Uema M, Kawaguchi Y (2009) Differences in the regulatory and functional effects of the Us3 protein kinase activities of herpes simplex virus 1 and 2. J Virol 83(22):11624–11634. https://doi.org/10.1128/JVI.00993-09

Morris JB, Hofemeister H, O'Hare P (2007) Herpes simplex virus infection induces phosphorylation and delocalization of emerin, a key inner nuclear membrane protein. J Virol 81(9):4429–4437. https://doi.org/10.1128/JVI.02354-06

Mou F, Forest T, Baines JD (2007) US3 of herpes simplex virus type 1 encodes a promiscuous protein kinase that phosphorylates and alters localization of lamin A/C in infected cells. J Virol 81(12):6459–6470. https://doi.org/10.1128/JVI.00380-07

Mou F, Wills E, Baines JD (2009) Phosphorylation of the U(L)31 protein of herpes simplex virus 1 by the U(S)3-encoded kinase regulates localization of the nuclear envelopment complex and egress of nucleocapsids. J Virol 83(10):5181–5191. https://doi.org/10.1128/JVI.00090-09

Munger J, Roizman B (2001) The US3 protein kinase of herpes simplex virus 1 mediates the post-translational modification of BAD and prevents BAD-induced programmed cell death in the absence of other viral proteins. Proc Natl Acad Sci U S A 98(18):10410–10415. https://doi.org/10.1073/pnas.181344498

Naghavi MH, Gundersen GG, Walsh D (2013) Plus-end tracking proteins, CLASPs, and a viral Akt mimic regulate herpesvirus-induced stable microtubule formation and virus spread. Proc Natl Acad Sci U S A 110(45):18268–18273. https://doi.org/10.1073/pnas.1310760110

Niemi NM, MacKeigan JP (2013) Mitochondrial phosphorylation in apoptosis: flipping the death switch. Antioxid Redox Signal 19(6):572–582. https://doi.org/10.1089/ars.2012.4982

Nolen B, Taylor S, Ghosh G (2004) Regulation of protein kinases; controlling activity through activation segment conformation. Mol Cell 15(5):661–675. https://doi.org/10.1016/j.molcel.2004.08.024

Ogg PD, McDonell PJ, Ryckman BJ, Knudson CM, Roller RJ (2004) The HSV-1 Us3 protein kinase is sufficient to block apoptosis induced by overexpression of a variety of Bcl-2 family members. Virology 319(2):212–224. https://doi.org/10.1016/j.virol.2003.10.019

Poon AP, Roizman B (2005) Herpes simplex virus 1 ICP22 regulates the accumulation of a shorter mRNA and of a truncated US3 protein kinase that exhibits altered functions. J Virol 79(13):8470–8479. https://doi.org/10.1128/JVI.79.13.8470-8479.2005

Poon AP, Liang Y, Roizman B (2003) Herpes simplex virus 1 gene expression is accelerated by inhibitors of histone deacetylases in rabbit skin cells infected with a mutant carrying a cDNA copy of the infected-cell protein no. 0. J Virol 77(23):12671–12678

Poon AP, Gu H, Roizman B (2006) ICP0 and the US3 protein kinase of herpes simplex virus 1 independently block histone deacetylation to enable gene expression. Proc Natl Acad Sci U S A 103(26):9993–9998. https://doi.org/10.1073/pnas.0604142103

Purves FC, Deana AD, Marchiori F, Leader DP, Pinna LA (1986) The substrate specificity of the protein kinase induced in cells infected with herpesviruses: studies with synthetic substrates [corrected] indicate structural requirements distinct from other protein kinases. Biochim Biophys Acta 889(2):208–215

Purves FC, Spector D, Roizman B (1991) The herpes simplex virus 1 protein kinase encoded by the US3 gene mediates posttranslational modification of the phosphoprotein encoded by the UL34 gene. J Virol 65(11):5757–5764

Rao P, Pham HT, Kulkarni A, Yang Y, Liu X, Knipe DM, Cresswell P, Yuan W (2011) Herpes simplex virus 1 glycoprotein B and US3 collaborate to inhibit CD1d antigen presentation and NKT cell function. J Virol 85(16):8093–8104. https://doi.org/10.1128/JVI.02689-10

Reynolds AE, Ryckman BJ, Baines JD, Zhou Y, Liang L, Roller RJ (2001) U(L)31 and U(L)34 proteins of herpes simplex virus type 1 form a complex that accumulates at the nuclear rim and is required for envelopment of nucleocapsids. J Virol 75(18):8803–8817

Reynolds AE, Wills EG, Roller RJ, Ryckman BJ, Baines JD (2002) Ultrastructural localization of the herpes simplex virus type 1 UL31, UL34, and US3 proteins suggests specific roles in primary envelopment and egress of nucleocapsids. J Virol 76(17):8939–8952

Ryckman BJ, Roller RJ (2004) Herpes simplex virus type 1 primary envelopment: UL34 protein modification and the US3-UL34 catalytic relationship. J Virol 78(1):399–412

Sagou K, Imai T, Sagara H, Uema M, Kawaguchi Y (2009) Regulation of the catalytic activity of herpes simplex virus 1 protein kinase Us3 by autophosphorylation and its role in pathogenesis. J Virol 83(11):5773–5783. https://doi.org/10.1128/JVI.00103-09

Sen J, Liu X, Roller R, Knipe DM (2013) Herpes simplex virus US3 tegument protein inhibits Toll-like receptor 2 signaling at or before TRAF6 ubiquitination. Virology 439(2):65–73. https://doi.org/10.1016/j.virol.2013.01.026

Sloan DD, Zahariadis G, Posavad CM, Pate NT, Kussick SJ, Jerome KR (2003) CTL are inactivated by herpes simplex virus-infected cells expressing a viral protein kinase. J Immunol 171(12):6733–6741

Sugiyama N, Ishihama Y (2016) Large-scale profiling of protein kinases for cellular signaling studies by mass spectrometry and other techniques. J Pharm Biomed Anal 130:264–272. https://doi.org/10.1016/j.jpba.2016.05.046

Terry LJ, Vastag L, Rabinowitz JD, Shenk T (2012) Human kinome profiling identifies a requirement for AMP-activated protein kinase during human cytomegalovirus infection. Proc Natl Acad Sci U S A 109(8):3071–3076. https://doi.org/10.1073/pnas.1200494109

Vertessy BG, Toth J (2009) Keeping uracil out of DNA: physiological role, structure and catalytic mechanism of dUTPases. Acc Chem Res 42(1):97–106. https://doi.org/10.1021/ar800114w

Walters MS, Kinchington PR, Banfield BW, Silverstein S (2010) Hyperphosphorylation of histone deacetylase 2 by alphaherpesvirus US3 kinases. J Virol 84(19):9666–9676. https://doi.org/10.1128/JVI.00981-10

Wang JT, Doong SL, Teng SC, Lee CP, Tsai CH, Chen MR (2009) Epstein-Barr virus BGLF4 kinase suppresses the interferon regulatory factor 3 signaling pathway. J Virol 83(4):1856–1869. https://doi.org/10.1128/JVI.01099-08

Wang X, Patenode C, Roizman B (2011) US3 protein kinase of HSV-1 cycles between the cytoplasm and nucleus and interacts with programmed cell death protein 4 (PDCD4) to block apoptosis. Proc Natl Acad Sci U S A 108(35):14632–14636. https://doi.org/10.1073/pnas.1111942108

Wang S, Wang K, Lin R, Zheng C (2013) Herpes simplex virus 1 serine/threonine kinase US3 hyperphosphorylates IRF3 and inhibits beta interferon production. J Virol 87(23):12814–12827. https://doi.org/10.1128/JVI.02355-13

Wisner TW, Johnson DC (2004) Redistribution of cellular and herpes simplex virus proteins from the trans-golgi network to cell junctions without enveloped capsids. J Virol 78(21):11519–11535. https://doi.org/10.1128/JVI.78.21.11519-11535.2004

Wisner TW, Wright CC, Kato A, Kawaguchi Y, Mou F, Baines JD, Roller RJ, Johnson DC (2009) Herpesvirus gB-induced fusion between the virion envelope and outer nuclear membrane during virus egress is regulated by the viral US3 kinase. J Virol 83(7):3115–3126. https://doi.org/10.1128/JVI.01462-08

Xiong R, Rao P, Kim S, Li M, Wen X, Yuan W (2015) Herpes simplex virus 1 US3 phosphorylates cellular KIF3A to downregulate CD1d expression. J Virol 89(13):6646–6655. https://doi.org/10.1128/JVI.00214-15

Yu X, He S (2016) The interplay between human herpes simplex virus infection and the apoptosis and necroptosis cell death pathways. Virol J 13:77. https://doi.org/10.1186/s12985-016-0528-0

Chapter 4
Oncolytic Virotherapy by HSV

Daisuke Watanabe and Fumi Goshima

Abstract Oncolytic virotherapy is a kind of antitumor therapy using viruses with natural or engineered tumor-selective replication to intentionally infect and kill tumor cells. An early clinical trial has been performed in the 1950s using wild-type and non-engineered in vitro-passaged virus strains and vaccine strains (first generation oncolytic viruses). Because of the advances in biotechnology and virology, the field of virotherapy has rapidly evolved over the past two decades and innovative recombinant selectivity-enhanced viruses (second generation oncolytic viruses). Nowadays, therapeutic transgene-delivering "armed" oncolytic viruses (third generation oncolytic viruses) have been engineered using many kinds of viruses. In this chapter, the history, mechanisms, rationality, and advantages of oncolytic virotherapy by herpes simplex virus (HSV) are mentioned. Past and ongoing clinical trials by oncolytic HSVs (G207, HSV1716, NV1020, HF10, Talimogene laherparepvec (T-VEC, OncoVEX^{GM-CSF})) are also summarized. Finally, the way of enhancement of oncolytic virotherapy by gene modification or combination therapy with radiation, chemotherapy, or immune checkpoint inhibitors are discussed.

Keywords Herpes simplex virus · HSV · Oncolytic virotherapy · G207 · HSV1716 · NV1020 · HF10 · Talimogene laherparepvec · T-VEC · OncoVEX^{GM-CSF}

4.1 Introduction

Cancer is the second worldwide cause of death, exceeded only by cardiovascular diseases (Pérez-Herrero and Fernández-Medarde 2015). For local and nonmetastatic cancers, surgery is the most effective and valuable treatment but is inefficient when the cancer has spread throughout the body. For advanced and metastatic

---

D. Watanabe (✉)
Department of Dermatology, Aichi Medical University School of Medicine, Nagakute, Japan
e-mail: dwatanab@aichi-med-u.ac.jp

F. Goshima
Department of Virology, Graduate School of Medicine, Nagoya University, Nagoya, Japan

cancers, systemic cytotoxic chemotherapy and/or radiation therapy has been used, but in some cancers such as malignant melanoma, these therapies are ineffective. Recent advances in understanding of cancer biology and immunology have spurred the development of numerous targeted therapies including molecular-targeted therapies or immunotherapies. In particular, a class of immune modulatory drugs targeting the immune checkpoint pathways like anti-PD-1 antibody (nivolumab) has demonstrated remarkable durable remissions in a part of advanced malignant melanoma patients (Tang et al. 2016). However, these agents cause many systemic adverse reactions, such as pneumonitis, colitis, and autoimmune diseases (Dossett et al. 2015; Spain et al. 2016). The high prices of these drugs are also problematic.

Besides these agents, another immunotherapy with tumor destruction using oncolytic virus (oncolytic virotherapy) has been studied for several decades and showed significant progress in recent years. This review focuses on the progress of oncolytic virotherapy, especially by using herpes simplex virus (HSV) for malignant internal tumors and brain tumors based on the mechanisms and clinical development. We also discuss the attempts for enhancing the effectiveness of oncolytic virotherapy by gene modification or combination therapies.

4.2 What Is Oncolytic Virotherapy?

Oncolytic virotherapy is a kind of antitumor therapy using viruses with natural or engineered tumor-selective replication to intentionally infect and kill tumor cells. The phenomenon that tumor regression following naturally acquired virus infections has been known over 100 years before. In 1904, a patient with chronic myelogenous leukemia had a dramatic decrease in white blood cells during a "flu-like" illness (Dock 1904). In 1912, a woman with cervical carcinoma responded to repeated rabies vaccinations (DePace 1912). An early clinical trial has been performed in the 1950s using wild-type and non-engineered in vitro-passaged virus strains and vaccine strains (first generation oncolytic viruses). For example, a clinical trial was performed using 30 patients with cervical cancer treated with different adenovirus serotypes (Huebner et al. 1956). More than 50% of patients showed a marked to moderate local tumor response; no systemic responses were reported, but the prolongation of survival was not significantly significant.

Because of the advances in biotechnology and virology, the field of virotherapy has rapidly evolved over the past two decades and innovative recombinant selectivity-enhanced viruses (second generation oncolytic viruses). In 1991, Martuza et al. first reported oncolytic virotherapy against mice glioma (malignant brain tumor) model using genetically engineered HSV. In that report, intraneoplastic inoculation of a thymidine kinase-negative mutant of herpes simplex virus-1 (dlsptk) prolonged survival of nude mice with intracranial U87 gliomas (Martuza et al. 1991). Nowadays, therapeutic transgene-delivering "armed" oncolytic viruses (third generation oncolytic viruses) have been engineered using many kinds of viruses. Table 4.1 shows the viruses studying as oncolytic virus today.

Table 4.1 Candidates of oncolytic virus

DNA viruses	Herpes simplex virus 1	Herpesviridae
	Human adenovirus 5	Adenoviridae
	Vaccinia virus	Poxviridae
	Myxoma virus	
RNA viruses	Echovirus(type I)	Picornaviridae
	Coxsackie virus	
	Poliovirus	
	Measles virus	Paramyxoviridae
	Newcastle disease virus	
	Mumps virus	
	Vesicular stomatitis virus	Rhabdoviridae
	Reovirus	Reoviridae
	Influenza virus	Myxoviridae

4.3 Mechanisms of Oncolytic Virotherapy

Mainly, there are two kinds of mechanisms for oncolytic viruses to kill cancer cells (Sze et al. 2013). The first and direct oncolytic effects are caused by viral infection itself and tumor cell lysis. By infection into tumor cells, viruses can produce viral proteins that are antigenic. After the lysis of infected cell, new virions are released and will infect neighboring cancer cells.

By releasing tumor antigens and triggering an immune response by infection, viruses can act as immunomodulators or tumor vaccines. By viral infection, inflammation occurs, and innate and acquired immune cells including cytotoxic T lymphocytes, natural killer cells, dendritic cells, and phagocytic cells will eliminate cancer cells. In addition, development of memory against tumor antigens will begin by these immune responses and will act on distant metastases as well (Fig. 4.1).

4.4 HSV as Oncolytic Virus

Herpes simplex virus type 1 (HSV-1) and type 2 (HSV-2) are important human pathogens that cause a variety of skin diseases from recurrent herpes labialis, herpes genitalis, and Kaposi's varicelliform eruption (eczema herpeticum) to life-threatening diseases such as herpes encephalitis and neonatal herpes (Nishiyama 2004). Especially among immunocompromised patients, the virus can be systemically disseminated and cause fatal infection (Witt et al. 2009). HSV was the first of the human herpes viruses to be discovered and has been the most intensively studied. Since the publication of the complete genomic DNA sequence of HSV-1 in 1988 (McGeoch et al. 1988), a number of studies have focused on elucidating the

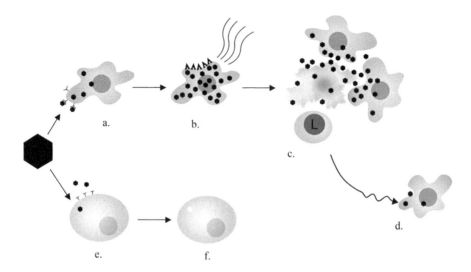

Fig. 4.1 An oncolytic virus, represented by black hexagons, attacks a cancerous cell (*a*). As a result of a natural tropism for the cell type or a specificity for a tumor-related cell surface antigen or receptor, the virus enters the malignant cell more readily than it would enter into a normal cell (*e*), which may not exhibit the same receptivity to infection. Upon infecting the malignant cell, the permissive nature of the cell that allows malignant genetic material to propagate also allows unchecked replication of the virus (*b*). An infection of a normal cell (*f*) is abortive as a result of the cell's ability to recognize and to destroy abnormal genetic material. The infected cell (*b*) will produce viral proteins that are antigenic and may alert the immune system, and, if the virus is genetically armed, the cell may produce cytokines and other signaling chemicals to activate the immune system. The infected cancer cell is eventually overwhelmed by the viral infection and lyses, releasing new viral particles locally to infect neighboring malignant cells (*c*). Lysis releases new viral and tumor- related antigens, which may be recognized and attacked by the immune system, represented here by a lymphocyte (L). Lysis-related viremia may result in infection of distant metastases (*d*), transforming a locoregional effect into a systemic effect. Activation of the immune system with elevation of systemic cytokine levels and activated leukocytes further enhances the systemic effect. Parts of the immune system may also develop memory and learn to recognize tumor antigens, potentially providing a more durable defense against residual and recurrent disease. (Adapted from Sze et al. 2013)

So far, many types of viruses, including HSV, adenoviruses, and adeno-associated viruses, have been engineered and evaluated for their potential as therapeutic agents in the treatment of malignant neoplasm

roles of individual HSV genes in viral replication and pathogenicity. These studies also identified the HSV gene products involved in the regulation of gene expression, interaction with the host cell, and evasion from the host immune system. The resulting depth of knowledge of HSV has allowed the development of potential therapeutic agents and vectors for several applications in human diseases.

4.4.1 Structure, Natural History, and Gene Function of HSV

HSV is an enveloped, double-stranded linear DNA virus about 100 nm in diameter. The viral particle is an icosahedral capsid containing the viral DNA with a genome of 152 kb encoding over 74 distinct genes. The capsid is surrounded by an amorphous layer known as the tegument, which contains viral structural and regulatory proteins, and external envelope containing numerous glycoproteins (Fig. 4.2a). Following primary infection from skin or mucosa, HSV enters into nerve endings and is then transported to the dorsal root and trigeminal ganglia where the virus establishes latent infection. The virus is reactivated by stimuli such as UV irradiation, mental or physical stress, or menstruation and causes symptomatic or asymptomatic recurrent infection (Mori and Nishiyama 2005).

The HSV genome consists of two long structures of unique sequences (designated long (UL) and short (US)), both of which are flanked by a pair of inverted repeat regions (TRL–IRL and IRS–TRS). There is a single copy of the "a" sequence, which contains the specific signals for packaging of viral DNA into capsids (Taylor et al. 2002), at each terminus and one at the junction between IRL and IRS (Fig. 4.2b).

HSV genes are classified into three groups by the regulation of their expression: immediate early (IE), early (E), and late (L). The IE gene products regulate gene

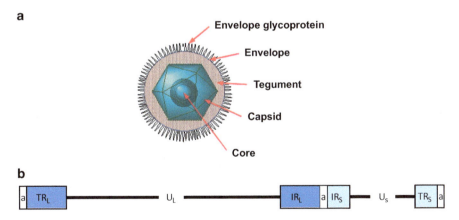

Fig. 4.2 (a) Structure of HSV. HSV is an enveloped, double-stranded linear DNA virus which diameter is about 100 nm. The virus particle comprehends an icosahedral capsid, which contains the viral DNA with a genome of 152 kb encoding over 74 distinct genes. Around the capsid, there is an amorphous layer known as the tegument, containing viral structural and regulatory proteins, surrounded by external envelope glycoproteins (b) HSV genome. HSV has genome of 152 kb encoding over 74 distinct genes. The HSV genome consists of two long structures of unique sequences (designated long (UL) and short (Us)), both of which are flanked by a pair of inverted repeat regions (TRL-IRL and IRs-TRs). There is a single copy of the "a" sequence at each terminus and one at the junction between IRL and IRs

transcription and include the US12 gene product, ICP47 which is responsible for decreasing MHC class I expression in infected cells via inhibition of the transporter associated with antigen presentation (TAP). The E gene products promote viral DNA synthesis in either viral DNA replication or in nucleic acid metabolism. Most of the L gene products are virion components such as capsid proteins, tegument proteins, and envelope glycoproteins (Nishiyama 1996).

HSV genes can also be divided into two groups according to whether or not they are essential for viral replication in cultured cells. Approximately half of the genes are essential genes that are necessary for viral replication and which encode capsid proteins, viral DNA replication proteins, viral DNA cleavage/packaging proteins, and some envelope glycoproteins. HSV is unable to replicate with even a single dysfunction in an essential gene. The remaining genes have been shown to be dispensable for replication. These accessory genes encode enzymes involved in nucleic acid metabolism, regulatory proteins required for efficient viral replication, proteins for protecting the virus and infected cells from the host immunity, and genes with other undetermined functions. Although the accessory genes are not necessary for viral replication in cell culture, the expression of these genes enables the virus to replicate effectively in a variety of cell types under different conditions, resulting in the replication and survival of HSV in humans (Mori and Nishiyama 2006). The functions of HSV gene products, their expression phase, and whether they are essential or accessory genes are summarized in Table 4.2.

4.4.2 Rationality of Using HSV for Oncolytic Virotherapy

HSV offers a number of advantages as an oncolytic agent.

1. Unlike many other viruses that only bind to a single receptor, HSV has four cellar receptors and has broad host range that allows the virus to infect and replicate almost all cell lines. As a result, oncolytic viruses derived from HSV can be applied therapeutically to many different types of tumors. In addition, in contrast to other oncolytic viruses, the property might protect against the rapid development of resistance to virotherapy using HSV.
2. HSV can infect both in replicating and non-replicating cells such as neuronal cells. This property enables oncolytic HSV to applying brain tumors such as glioblastoma.
3. HSV has the potential for incorporating a large size of foreign DNA. It is useful when making therapeutic transgene-delivering "armed" oncolytic viruses.
4. Undesired infection or toxicity from the virus replication can be controlled by effective anti-herpetic agents such as acyclovir and famciclovir.
5. Compared to adenoviruses, lytic infection by HSV usually kills target cells much more rapidly and effectively. For example, HSV can form visible plaques in cultured cells in 2 days, in contrast to 7 to 9 days for an adenovirus. In in vitro

Table 4.2 Functions of herpes simplex virus gene products

Gene	Essential(E)/ dispensable(D)	Times of expression (IE/E/L)	Gene products and functions
Regulation of gene expression			
RL2	D	IE	ICP0: promiscuous transactivator with E3 ubiquitin ligase domains
UL54	E	IE	ICP27: regulation of gene expression at posttranscriptional level
RS1	E	IE	ICP4: major regulatory protein
US1	D	IE	ICP22: regulatory protein that enhances the expression of late genes
Nucleic acid metabolism			
UL2	D	E	Uracil DNA glycosidase
UL23	D	E	Thymidine kinase, selective activation of aciclovir and ganciclovir
UL39	D	E	Ribonucleotide reductase large subunit with protein kinase activity
UL40	D	E	Ribonucleotide reductase small subunit
UL50	D	E	Deoxyuridine triphosphatase
DNA replication			
UL5	E	E	DNA helicase, a component of DNA primase/helicase complex
UL8	E	E	A component of DNA primase/helicase complex
UL9	E	E	Replication origin-binding protein
UL29	E	E	ICP8: single-strand DNA binding protein
UL30	E	E	DNA polymerase catalytic subunit
UL42	E	E	DNA polymerase accessory subunit
UL52	E	E	DNA primase, a component of DNA primase/helicase complex
DNA cleavage/packaging			
UL6	E	L	Associated with capsids, a subunit of the portal complex
UL15	E	L	DNA terminase activity
UL25	E	L	Associated with capsids, seals capsids after DNA packaging
UL32	E	L	Not associated with capsids, required for correct localization of capsids
UL33	E	L	Not associated with capsids, interact with UL14
Protein kinase			
UL13	D	E	Protein kinase
US3	D	L	Protein kinase with antiapoptotic activity
Capsid protein			
UL18	E	L	VP23: forms triplex with VP19c

(continued)

Table 4.2 (continued)

Gene	Essential(E)/ dispensable(D)	Times of expression (IE/E/L)	Gene products and functions
UL19	E	L	VP5: major capsid protein
UL26	E	L	VP24 and VP21 are the products of the self-cleavage of UL26
UL26.5	E	L	VP22a: scaffolding protein
UL35	D	L	VP26
UL38	E	L	VP19c: a component of the intercapsomeric triplex
Tegument protein			
UL36	E	L	ICP1/2: involved in both uncoating and egress
UL41	D	L	Vhs: virion host shutoff protein, causes the degradation of mRNA
UL46	D	L	VP11/12: interacts with UL48
UL47	D	L	VP13/14: enhances immediate early gene expression
UL48	E	L	VP16: stimulating immediate early gene expression
UL49	E	L	VP22: intercellular trafficking activity
Envelope glycoprotein			
UL1	E	L	gL: forms a complex with gH, involved in entry, egress, and cell-to-cell spread
UL22	E	L	gH: forms a complex with gH, involved in entry, egress, and cell-to-cell spread
UL27	E	L	gB: required for entry
UL44	D	L	gC: involved in adsorption, C3b-binding activity
UL53	D	L	gK: involved in egress
US6	E	L	gD
US7	D	L	gI: forms a complex with gE, cell-to-cell spread
US8	D	L	gE: Fc receptor activity, forms a complex with gI, cell-to-cell spread
Others			
RL1	D	L	$\gamma_1 34.5$: requires protein synthesis by binding to protein phosphatase 1
RL3	D	–	LAT: latency associated transcript
US12	D	IE	ICP47: TAP-binding protein, involved in MHC class I downregulation

studies, it has also shown that HSV can kill almost 100% of cultured cancer cells at a multiplicity of infection (MOI) of 0.01 (Fu and Zhang 2002).

6. HSV can infect many kinds of animals. Due to the similarity in the viral pathogenicity in mice, guinea pigs, and monkeys, to that in humans, preclinical studies of oncolytic HSV can be performed relatively easily by using these animal models.

7. The determination of complete open reading frames and identification of disease-related viral genes (Marconi et al. 2008; Todo 2008).
8. The risk of introducing an insertional mutation during HSV oncolytic therapy appears minimal because HSVs rarely integrate into cellular DNA. While strong immunogenicity and cell toxicity induced by HSV infection are major disadvantages for developing gene delivery vectors using HSV, they are beneficial when developing vaccine vectors or anticancer agent by HSV recombination.

4.4.3 Important Genes for Making Effective Oncolytic HSVs

4.4.3.1 Immediate Early Genes

ICP0 (infected cell polypeptide 0) is the RL2 gene product. It belongs to the immediate early proteins (IE) that it required for effective initiation of viral lytic infection and reactivation from latent infection of HSV (Bringhurst and Schaffer 2006). ICP0 is a 775-amino acid really interesting new gene (RING)-finger-containing protein that possesses E3 ubiquitin ligase activity, which is required for ICP0 to activate HSV-1 gene expression; disrupt nuclear domain (ND) 10 structures; mediate the degradation of cellular proteins including cdc34, Sp100, and PML; and evade the host cell's intrinsic and innate antiviral defenses. This protein degradation may create a favorable microenvironment for viral replication (Boehmer and Nimonkar 2003; Lilley et al. 2005). ICP0 also prevents cellular rRNA degradation (Sobol and Mossman 2006). It has been reported that ICP0 mutation impairs viral replication in normal cells. On the other hand, the ICP0 mutant KM100 virus exhibits an oncolytic effect on tumor cells, causing tumor regression and increased survival in experimental breast cancer models in mice (Hummel et al. 2005).

ICP4 is also a regulator of viral transcription that is required for productive infection. Since viral genes are transcribed by cellular RNA polymerase II (RNA pol II), ICP4 must interact with components of the pol II machinery to regulate viral gene expression. It has been shown previously that ICP4 interacts with TATA box-binding protein (TBP), TFIIB, and the TBP-associated factor 1 (TAF1) in vitro (Zabierowski and Deluca 2008).

NV1066, which has only one of two originally present copies of both ICP0 and ICP4, has an antitumor effect against breast cancer, pleural cancer, bladder cancer, and esophageal cancer (Mullerad et al. 2005; Stiles et al. 2006a; b). Interestingly, this virus not only destroys tumor cells but also induces apoptosis in uninfected cells via the cellular bystander pathway. This apoptosis hinders viral spread from cell to cell. Previous studies showed that pharmaceutically inhibiting apoptosis can improve the oncolytic viral proliferation and the antitumor effect (Stanziale et al. 2004).

ICP47 also belongs to IE proteins. It inhibits the transporter associated with antigen presentation (TAP), decreasing MHC class I expression and preventing infected cells from presenting viral to CD8+ cells (Hill et al. 1995). The lack of ICP47 increases MHC class I expression, which might induce an enhanced antitumor immune response. The bovine herpesvirus 1 (BHV-1) TAP-inhibitor (UL49.5)-expressing oncolytic virus showed superior efficacy treating bladder and breast cancer in murine preclinical models that was dependent upon a CD8+ T-cell response. In addition to treating directly injected, subcutaneous tumors, UL49.5-oncolytic virotherapy reduced untreated, contralateral subcutaneous tumor size and naturally occurring metastasis (Pourchet et al. 2016).

4.4.3.2 Early and Late Genes

Ribonucleotide reductase (RR) catalyzes the reduction of ribonucleotides to deoxyribonucleotides. As a result, it provides sufficient precursors for the de novo synthesis of DNA. Because HSV has its own RR, replication of the virus is independent of the host cell cycle. By inactivating the viral RR gene, viral replication is completely under the control of host cell dividing conditions. RR-deficient HSV-1 such as hrR3 was expected to exhibit selective oncolytic effects and increase their potency when combined with radiation; however, complementary toxicity was seen between radiation and hrR3, without evidence of viral replication (Spear et al. 2000).

The γ34.5 gene product (ICP34.5) enables the virus to replicate in neurons and spread within the brain. When this gene is deleted, HSV-1 cannot complete a lytic infection in neurons and thus cannot cause encephalitis (Kanai et al. 2012). Attenuated viruses that are mutated in the γ34.5 may be useful for malignant tumors in the central nervous system.

4.5 How to Make Oncolytic HSVs?

There are several established techniques for generating recombinant HSV. Traditionally, recombinant HSV mutants have been generated by homologous recombination between purified HSV DNA and a recombination plasmid in co-transfected cells (Bataille and Epstein 1995). An alternative procedure is the transfection of cells with overlapping cosmids containing appropriate insertions or deletions. Expression of genes contained in cosmids leads, through recombination, to the construction of full-length viral genome (Kong et al. 1999). In these methods, there are several problems such as the inefficiency of recombination and the need to screen or select plaques for the correct recombinant. This has hampered the development of new recombinant HSV vectors.

Recently, novel recombinant technique using bacterial artificial chromosome (BAC) has enabled the cloning of the whole HSV genome as a BAC plasmid and its subsequent manipulation in E. coli (Stavropoulos and Strathdee 1998). BAC cloning requires the insertion of mini F plasmid sequences and antibiotic resistance genes into the viral genome. The total length of these BAC backbone sequences is usually greater than 6 kb. Insertion of BAC sequences into the wild-type HSV genome (152 kb) increases the genome length to approximately 158 kb, leaving insufficient space for the insertion of additional sequences. To avoid deleterious effects of the BAC sequences, including growth defects and potential transmission between bacteria and man, some herpes virus BAC clones have been constructed with loxP site-flanked BAC sequences that can be removed by Cre recombinase (Tanaka et al. 2003). One potential disadvantage of the BAC system is the potential for higher rates of error in DNA replication in bacteria than eukaryotic cells, whether or not this will prove to be a problem is not yet known.

4.6 Oncolytic Viruses Derived from HSV-1 that Have Reached Clinical Testing

There have been several clinical trials with HSV-1 mutants as oncolytic agents using gene deletion described before. These mutants have been applied for the treatment of malignant brain tumors or malignant melanoma and other solid tumors (Table 4.3).

4.6.1 G207

G207 has been credited as the first oncolytic virus generated by genetic engineering technology. The virus was constructed from HSV-1 by deleting both copies of the $\gamma 34.5$ gene and an insertional mutation in the ICP6 gene (Mineta et al. 1995). First, the safety and efficacy of G207 were demonstrated in preclinical animal models (Mineta et al. 1995, Sundaresan et al. 2000. Todo et al. 2000). Then G207 was tested in the treatment of malignant glioma in a phase I clinical trial. Up to 3×10^9 plaque-forming units (pfu) of virus was injected into tumors of 21 patients and revealed that the virus was well tolerated (Markert et al. 2009). Because of a lack of convincing evidence of clinical efficacy, G207 clinical development has not yet reached the phase II stage of testing.

Table 4.3 Studies of oncolytic HSV in clinical trials

Virus	Mutation	Tumor type	Phase
G207	UL39 $^-$, $\gamma_1 34.5$ $^-$	Glioma	I
HSV1716	$\gamma_1 34.5^-$	Melanoma, hepatocellular carcinoma, glioblastoma, mesothelioma, neuroblastoma	I/II
NV1020	UL24 $^-$, UL56 $^-$	Liver metastasis of colon cancer	I
T-VEC	UL39 $^-$, $\gamma_1 34.5^-$, US12$^-$, GM-CSF$^+$	Melanoma, head and neck cancer, and pancreatic cancer	III
HF10	UL43$^-$, UL49.5$^-$, UL55$^-$, UL56$^-$, LAT$^-$	Recurrent breast cancer, melanoma, and pancreatic cancer	I/II

4.6.2 HSV1716

HSV1716 is a spontaneous mutant of a replication selective HSV-1 that bears a deletion of 759 bp in each copy of the γ34.5 gene. Two separate phase I clinical trials have evaluated the safety of HSV1716 with high-grade glioma (HGG) and with stage IV melanoma (Harrow et al. 2004; MacKie et al. 2001). In the HGG study, HSV1716 DNA was detected by PCR at the sites of inoculation. In several patients, an immune response to the virus was detected. Although it remains unclear whether the immune response to the virus contributes to the eradication of cancer cells infected by the virus, a significant increase in long-term survival following surgery was also observed. In the melanoma trial, immunohistochemical staining of injected nodules revealed that virus replication was confined to tumor cells and had no toxicity in patients.

4.6.3 NV1020

NV1020 has deletions of both UL56 genes and one copy of the γ34.5 gene. In preclinical studies, NV1020 was evaluated as oncolytic agents in several solid tumors outside the brain (Advani et al. 1999; Cozzi et al. 2001; Ebright et al. 2002). In these studies, the virus showed effectiveness in treating several tumors in both mouse and rat models. Then NV1020 was evaluated against liver metastases from colorectal cancer. Followed by chemotherapy, NV1020 was tested in 12 patients with colorectal cancer hepatic metastases. It was well tolerated when NV1020 was given through hepatic arterial infusion. Reported side reactions were mainly transient febrile reactions and transient lymphopenia. Over half of the treated patients showed partial responses or stable disease, indicating therapeutic efficacy (Geevarghese et al. 2010; Kemeny et al. 2006; Sze et al. 2012).

4.6.4 Talimogene Laherparepvec (T-VEC)

T-VEC, formerly known as OncoVEX^{GM-CSF}, has deletion of the genesγ34.5 and US12 which encodes ICP47 and contains the gene encoding human granulocyte macrophage colony-stimulating factor (GM-CSF) (Liu et al. 2003). Gene modification of this virus was intended to increase the lytic activity of the virus (over deletion of the ICP47 gene) and to potentiate the ability of virotherapy to induce antitumor immunity (deletion of ICP47 combined with insertion of GM-CSF). Preclinical studies of OncoVEX^{GM-CSF} showed that this virus can effectively reduced injected tumors and also induced antitumor immunity that could protect animals against tumor rechallenge (Liu et al. 2003). The safety of OncoVEX^{GM-CSF} was evaluated in a phase I study in patients with metastatic breast, head/neck and gastrointestinal cancers, and malignant melanoma. Overall, intralesional administration of the virus was well tolerated by patients (Hu et al. 2006). In phase II study, OncoVEX^{GM-CSF} was injected in patients with metastatic melanoma (Senzer et al. 2009). The overall response rate was 26%. Surprisingly, all responding patients showed regressions of both injected and noninjected lesions (Fig. 4.3). An increase in CD8+ T-cells and a reduction in CD4+FoxP3+ regulatory T-cells were detected in biopsy samples of regressing lesions (Kaufman et al. 2010).

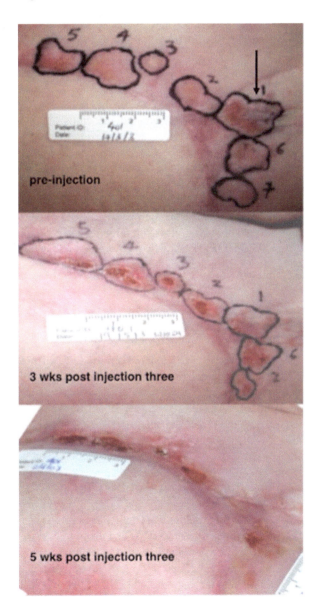

Fig. 4.3 Breast cancer patient treated with T-VEC. The injection was made into tumor 1 (arrow). Tumor regressions were observed both in injected and in noninjected lesions, consistent with both direct oncolytic and immune-mediated antitumor effects. (Adapted from Hu et al. 2006)

A randomized phase III trial was performed in 291 patients with unresected stage IIIB to IV melanoma, with 127 patients receiving subcutaneous GM-CSF as the control arm (OPTiM; NCT00769704) (Andtbacka et al. 2015). T-VEC was administered at a concentration of 10^8 plaque-forming units (pfu)/mL injected into 1 or more skin or subcutaneous tumors on days 1 and 15 of each 28-day cycle for up to 12 months, while GM-CSF was administered at a dose of 125 µg/m2/day subcutaneously for 14 consecutive days followed by 14 days of rest, in 28-day treatment cycles for up to 12 months. At the primary analysis, 290 deaths had occurred (T-VEC, n = 189; GM-CSF, n = 101). T-VEC treatment produced a significant improvement in (1) durable response rate (TVEC 16% vs. GM-CSF control arm 2%), (2) objective response rate (26% vs. 6%), and (3) complete response rate (11% vs. 1%). The difference of the median overall survival rate, a secondary end point of this trial, between T-VEC and GM-CSF treatment groups was 4.4 months. The most common adverse events with T-VEC were fatigue, chills, and pyrexia, but the only grade 3 or 4 treatment-related adverse event, occurring in over 2% of patients, was cellulitis (T-VEC, n = 6; GM-CSF, n = 1). There were no fatal treatment-related adverse events. Median overall survival (OS) was 23.3 months for the T-VEC arm versus 18.9 months for the GM-CSF arm (hazard ratio, 0.79; P = 0.051), but the difference in OS became significant (P = 0.049) by the time of drug application (Andtbacka et al. 2015). This phase III trial was the first to prove that local intralesional injections with an oncolytic virus can not only suppress the growth of injected tumors, and in 2015, the US Food and Drug Administration (FDA) approved T-VEC as a first oncolytic HSV for the treatment of advanced inoperable malignant melanoma.

4.6.5 HF10

HSV-1 mutant strain HF10, derived from an in vitro-passaged laboratory strain of HSV-1, is an alternative candidate for an oncolytic HSV. Previous studies have shown that HF10 does not cause any neurological symptoms in mice when inoculated into the peripheral tissues and organs due to its inability to invade the central nervous system (Nishiyama et al. 1991). The HF10 genome has a deletion of 3832 bp to the right of the UL and UL/IRL junction. Sequences from 6025 to 8319 bp have also been deleted from the TRL, and 6027 bp of DNA has been inserted in an inverted orientation. Sequence analysis revealed that HF10 lacks the expression of functional UL43, UL49.5, UL55, UL56, and LAT (Ushijima et al. 2007). Although the detailed mechanisms of the HF10 phenotype are not clear, the lack of the UL56 gene and LAT may play an important role. UL56 associates with the kinesin motor protein KIF1A, and the absence of UL56 reduces the neuroinvasiveness of HSV without affecting viral replication in vitro (Koshizuka et al. 2005). The LAT promoter region is also known to be associated with neurovirulence (Jones et al. 2005). The mechanisms of HF10 in tumor selectivity are also unknown, but the differences in the IFN pathway between normal cells and cancer cells may be involved (Nawa et al. 2008).

The loss of HF10 neuroinvasiveness and its high potency of replication in tumor cells contribute to the usefulness of HF10 as an oncolytic virotherapy for non-brain malignancy. HF10 therapy exhibited striking antitumor efficacy of peritoneally disseminated internal malignancies of immunocompetent mice models (Kimata et al. 2003; Kohno et al. 2005; Teshigahara et al. 2004; Watanabe et al. 2008). In a BALB/c mouse model of disseminated peritoneal colon carcinoma, 100% of intraperitoneally HF10-treated mice survived without remarkable side effects (Takakuwa et al. 2003). HF10 virotherapy using a mouse melanoma model was also studied. In the intraperitoneal melanoma model, all mice survived when given intraperitoneal injections of HF10 compared to none of the control mice (Fig. 4.4a, b). In the subcutaneous melanoma model, intratumoral inoculation of HF10 showed not only tumor growth inhibition at the injected site but also the induction of systemic antitumor immune responses in mice (Watanabe et al. 2008).

Several clinical trials have been done with HF10 thus (Nakao et al. 2004; Kimata et al. 2006; Fujimoto et al. 2006). Six patients with recurrent breast cancer who were treated with HF10 showed no serious adverse effects, and distinct tumor regression was observed in all patients (Kimata et al. 2006). Another clinical trial was carried out in three patients with advanced head and neck squamous cell carcinoma (Fujimoto et al. 2006). Although no significant tumor regression was found after injection with HF10 at a low dose, pathological examination revealed extensive tumor cell death and fibrosis, with marked infiltration of CD4+ or CD8+ T-cells. Moreover, a number of HSV antigen-positive cells were detected within the tumor even at 2 weeks postinjection. These studies suggest that HF10 is safe and effective for oncolytic virotherapy. Currently, phase I/II trial of HF10 in patients with solid cutaneous tumors, including melanomas, has been completed (NCT01017185) in the USA.

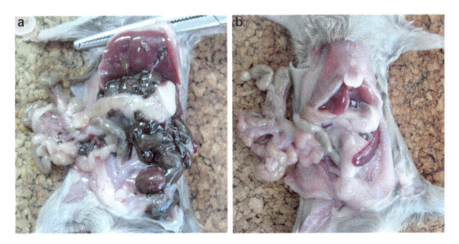

Fig. 4.4 Tumor growth reduction by HF10 in a murine intraperitoneal melanoma model. DBA/2 mice were injected intraperitoneally with 1×10^5 clone M3 cells and then were injected with PBS (control) or 1×10^7 pfu of HF10 at days 6,7, and 8. Representative clinical pictures of control (**a**) and HF10-treated (**b**) mice at day 14

4.7 Enhancement of Oncolytic Virotherapy by Gene Modification

4.7.1 Receptor Retargeted Mutants

HSV has several receptors to enter into host cells. For instance, glycoproteins gC and gB binds to heparan sulfate and glycoprotein gD binds to herpesvirus entry mediator (HVEM) (Salameh et al. 2012). There have been many attempts to make effective infection of oncolytic HSVs into cancer cells with limiting cell specificity by altering the receptors of the viruses.

IL-13 is the ligand of the IL-13 receptor 2α, expressed in glioblastoma and high-grade astrocytoma (Sengupta et al. 2014). IL-13 insertion mutants into gC or gD has been constructed as oncolytic HSVs against these tumors (Zhou et al. 2002; Zhou and Roizman 2006).

Another example is HER-2. HER-2 is a member of the EGFR (epidermal growth factor receptor) family. This protein is overexpressed in breast and ovarian cancers, gastric carcinomas, glioblastomas, and so on (Jackson et al. 2013). R-LM249 was created by replacing gD dispensable region with the sequence for the single-chain antibody trastuzumab, which targets human epidermal HER-2 (Menotti et al. 2008). In preclinical study, a therapeutic effect of R-LM249 against a murine model of HER2 glioblastoma has been reproted (Gambini et al. 2012). R-LM249 also showed a therapeutic effect against peritoneal and brain metastases of ovarian and breast cancers by intraperitoneal injections (Nanni et al. 2013).

4.7.2 Modified (Armed and Targeted) Oncolytic HSV

To enhance antitumor responses of oncolytic HSVs, many studies have been conducted. One strategy of this is to generate insert immunostimulatory genes into oncolytic HSVs. Numerous immune-stimulating genes have been inserted into various oncolytic HSVs including IL-2, IL-12, IL-15, IL-18, tumor necrosis factor alpha, CD80 (B7.1), and GM-CSF like T-VEC (Nakashima and Chiocca 2014). These genes have many functions to activate, proliferate, differentiate, and maturate innate and acquired immune cells important for antitumor responses such as macrophage, dendritic cells, natural killer cells, cytotoxic T-cells, helper T-cells, and B cells.

Another way is to generate HSVs expressing therapeutic genes, including those that can activate prodrugs. There are many reports of oncolytic HSVs that have been modified to code for enzymes that catalyze prodrugs into active substrates, for example, HSV1yCD codes for the yeast cytosine deaminase (CD) enzyme that converts the nontoxic 5- uorocytosine into 5-FU (Nakamura et al. 2001).

4.7.3 Oncolytic HSVs as Amplicon Vector

Amplicon vectors are HSV-1 particles that carry a concatemeric form of a DNA plasmid, named the amplicon plasmid, instead of the viral genome. An amplicon plasmid has one origin of replication (generally *ori-S*) and one packaging signal (pac or *a*) from HSV-1, in addition to the transgenic sequences of interest. The vector has identical structure to wild-type HSV-1, so it has same immunological and host-range as wild-type virus.

HF10 has been investigated as a helper virus. An HF10-packaged mouse GM-CSF-expressing amplicon (mGM-CSF amplicon) was used to infect subcutaneously inoculated murine colorectal tumor cells (CT26 cells), and the antitumor effects were compared to tumors treated only with HF10. When mice subcutaneously inoculated with CT26 cells were intratumorally injected with HF10 or mGM-CSF amplicon, greater tumor regression and prolonged survival was seen in mGM-CSF amplicon-treated animals (Kohno et al. 2007). This amplicon system might be one of the good tools to used for tailor-made therapy.

4.8 Combination Therapy

4.8.1 Combination with Radiation

There is a report about intratumoral HSV G207 injection to glioma patients prior to a single palliative fraction of radiotherapy (Markert et al. 2014). The combination therapy showed some synergistic activity. Combination with chemoradiotherapy was also considered. Combined chemoradiotherapy with cisplatin and intratumoral injection of T-VEC for stage III/IV head and neck cancer patients showed 93% of complete response (CR) (Harrington et al. 2010).

4.8.2 Combination with Chemotherapy

Combination therapy with oncolytic HSV and chemotherapy was first evaluated with HSV1716 and four standard chemotherapeutic drugs: methotrexate, cisplatin, mitomycin C, and doxorubicin (Toyoizumi et al. 1999). Since then, there have been many studies reporting the increased efficacy of oncolytic HSV in combination with a many kinds of existing and potentially new anticancer drugs including cyclophosphamide, docetaxol, etoposide, 5- uorouracil (5-FU), and so on. In our laboratory, combination therapy with HF10 and chemotherapy has been studied (Braidwood et al. 2013).

For instance, we have shown enhanced antitumoral activity was shown in murine colorectal cancer model by HF10 inoculation following GEM treatment even in the distal tumor (Esaki et al. 2013). In murine subcutaneous melanoma model, intratumoral HF10 inoculation significantly inhibited tumor growth. When mice were treated with HF10 and dacarbazine (DTIC), the combination therapy induced a robust systemic antitumor immune response and prolonged survival. IFN-γ secretion from splenocytes of the HF10-DTIC combination therapy group showed more IFN-γ secretion than did the other groups (Tanaka et al. unpublished data).

4.8.3 Combination with Immune Checkpoint Inhibitors

Because oncolytic virotherapy has an aspect of cancer immunotherapy, combination therapy with oncolytic virotherapy and with immune checkpoint inhibitors is promising.

A phase Ib study of T-VEC and the anti-CTLA-4 antibody ipilimumab were administrated to patients with untreated, advanced cutaneous melanoma. The overall response rate by immune-related response criteria was 50% (Puzanov et al. 2016). The result was higher than would be expected from ipilimumab alone (10%). High regression rates were observed. The phase I study of combination therapy with pembrolizumab (NCT02263508) is ongoing. With regard to HF10, a phase II study of combination treatment with ipilimumab in patients with unresectable or metastatic melanoma is ongoing both in USA (NCT02272855) and in Japan.

4.9 Conclusion

In summary, HSV has many advantages for cancer therapy, and significant progress has been made in generating more effective oncolytic HSVs. Although much more work is required to better understand the efficacy and safety issues of these oncolytic HSVs before clinical use, the results from extensive preclinical and clinical trials have clearly demonstrated the potential of HSV recombinants for oncolytic viruses. Moreover, combination therapy with oncolytic HSVs and conventional chemotherapy radiotherapy and immune checkpoint inhibitors will expand the potential of oncolytic virotherapy.

References

Advani SJ, Chung SM, Yan SY et al (1999) Replication-competent, nonneuroinvasive genetically engineered herpes virus is highly effective in the treatment of therapy-resistant experimental human tumors. Cancer Res 59:2055–2058

Andtbacka RH, Kaufman HL, Collichio F et al (2015) Talimogene laherparepvec improves durable response rate in patients with advanced melanoma. J Clin Oncol 33:2780–2788

Bataille D, Epstein AL (1995) Herpes simplex virus type 1 replication and recombination. Biochimie 77:787–795

Boehmer PE, Nimonkar AV (2003) Herpes virus replication. IUBMB Life 55:3–22

Braidwood L, Graham SV, Graham A et al (2013) Oncolytic herpes viruses, chemotherapeutics, and other cancer drugs. Oncolytic Virother 2:57–74

Bringhurst RM, Schaffer PA (2006) Cellular stress rather than stage of the cell cycle enhances the replication and plating efficiencies of herpes simplex virus type 1 ICP0- viruses. J Virol 80:4528–4537

Cozzi PJ, Malhotra S, Mcauliffe P et al (2001) Intravesical oncolytic viral therapy using attenuated, replication-competent herpes simplex viruses G207 and Nv1020 is effective in the treatment of bladder cancer in an orthotopic syngeneic model. FASEB J 15:1306–1308

DePace NG (1912) Sulla Scomparsa di un enorme cancro begetante del callo dell'utero senza cura chirurgica [Italian]. Ginecol 9:82

Dock G (1904) Influence of complicating diseases upon leukemia. Am J Med Sci 127:563–592

Dossett LA, Kudchadkar RR, Zager JS (2015) BRAF and MEK inhibition in melanoma. Expert Opin Drug Saf 14:559–570

Ebright MI, Zager JS, Malhotra S et al (2002) Replication-competent herpes virus NV1020 as direct treatment of pleural cancer in a rat model. J Thorac Cardiovasc Surg 124:123–129

Esaki S, Goshima F, Kimura H et al (2013) Enhanced antitumoral activity of oncolytic herpes simplex virus with gemcitabine using colorectal tumor models. Int J Cancer 132:1592–1601

Fu X, Zhang X (2002) Potent systemic antitumor activity from an oncolytic herpes simplex virus of syncytial phenotype. Cancer Res 62:2306–2312

Fujimoto Y, Mizuno T, al SS (2006) Intratumoral injection of herpes simplex virus HF10 in recurrent head and neck squamous cell carcinoma. Acta Otolaryngol 126:1115–1117

Gambini E, Reisoli E, Appolloni I et al (2012) Replication-competent herpes simplex virus retargeted to HER2 as therapy for high-grade glioma. Mol Ther 20:994–1001

Geevarghese SK, Geller DA, De Haan HA et al (2010) Phase I/II study of oncolytic herpes simplex virus NV1020 in patients with extensively pretreated refractory colorectal cancer metastatic to the liver. Hum Gene Ther 21:1119–1128

Harrington KJ, Hingorani M, Tanay MA et al (2010) Phase I/II study of oncolytic HSV GM-CSF in combination with radiotherapy and cisplatin in untreated stage III/IV squamous cell cancer of the head and neck. Clin Cancer Res 16:4005–4015

Harrow S, Papanastassiou V, Harland J et al (2004) HSV1716 injection into the brain adjacent to tumour following surgical resection of high-grade glioma: safety data and long-term survival. Gene Ther 11:1648–1658

Hill A, Jugovic P, York I et al (1995) Herpes simplex virus turns off the TAP to evade host immunity. Nature 375:411–415

Hu JCC, Coffin RS, Davis CJ (2006) A phase I study of OncoVEX^{GM-CSF}, a second-generation oncolytic herpes simplex virus expressing granulocyte macrophage colony-stimulating factor. Clin Cancer Res 12:6737–6747

Huebner RJ, Rowe WP, Schatten WE et al (1956) Studies on the use of viruses in the treatment of carcinoma of the cervix. Cancer 9:1211–1218

Hummel JL, Safroneeva E, Mossman KL (2005) The role of ICP0-Null HSV-1 and interferon signaling defects in the effective treatment of breast adenocarcinoma. Mol Ther 12:1101–1110

Jackson C, Browell D, Gautrey H et al (2013) Clinical significance of HER-2 splice variants in breast cancer progression and drug resistance. Int J Cell Biol 2013:973584

Jones C, Inman M, Peng W et al (2005) The herpes simplex virus type 1 locus that encodes the latency-associated transcript enhances the frequency of encephalitis in male BALB/c mice. J Virol 79:14465–14469

Kanai R, Zaupa C, Sgubin D et al (2012) Effect of γ34.5 deletions on oncolytic herpes simplex virus activity in brain tumors. J Virol 86:4420–4431

Kaufman HL, Kim DW, DeRaffele G et al (2010) Local and distant immunity induced by intralesional vaccination with an oncolytic herpes virus encoding GM-CSF in patients with stage IIIc and IV melanoma. Ann Surg Oncol 17:718–730

Kemeny N, Brown K, Covey A et al (2006) Phase I, open-label, dose-escalating study of a genetically engineered herpes simplex virus, NV1020, in subjects with metastatic colorectal carcinoma to the liver. Hum Gene Ther 17:1214–1224

Kimata H, Takakuwa H, Goshima F et al (2003) Effective treatment of disseminated peritoneal colon cancer with new replication-competent herpes simplex viruses. Hepato-Gastroenterology 50:961–966

Kimata H, Imai T, Kikumori T et al (2006) Pilot study of oncolytic viral therapy using mutant herpes simplex virus (HF10) against recurrent metastatic breast cancer. Ann Surg Oncol 13:1078–1084

Kohno S, Luo C, Goshima F et al (2005) Herpes simplex virus type 1 mutant HF10 oncolytic viral therapy for bladder cancer. Urology 66:1116–1121

Kohno SI, Luo C, Nawa A et al (2007) Oncolytic virotherapy with an HSV amplicon vector expressing granulocyte-macrophage colony-stimulating factor using the replication-competent HSV type 1 mutant HF10 as a helper virus. Cancer Gene Ther 14:918–926

Kong Y, Yang T, Geller AI (1999) An efficient in vivo recombination cloning procedure for modifying and combining HSV-1 cosmids. J Virol Methods 80:129–136

Koshizuka T, Kawaguchi Y, Nishiyama Y (2005) Herpes simplex virus type 2 mem- brane protein UL56 associates with the kinesin motor protein KIF1A. J Gen Virol 86:527–533

Lilley CE, Carson CT, Muotri AR et al (2005) DNA repair proteins affect the lifecycle of herpes simplex virus 1. Proc Natl Acad Sci U S A 102:5844–5849

Liu BL, Robinson M, Han ZQ et al (2003) ICP34.5 deleted herpes simplex virus with enhanced oncolytic, immune stimulating, and anti-tumour properties. Gene Ther 10:292–303

MacKie RM, Stewart B, Brown SM (2001) Intralesional injection of herpes simplex virus 1716 in metastatic melanoma. Lancet 357:525–526

Markert JM, Liechty PG, Wang W et al (2009) Phase Ib trial of mutant herpes simplex virus G207 inoculated pre-and post-tumor resection for recurrent GBM. Mol Ther 17:199–207

Markert JM, Razdan SN, Kuo HC et al (2014) A phase 1 trial of oncolytic HSV-1, G207, given in combination with radiation for recurrent GBM demonstrates safety and radiographic responses. Mol Ther 22:1048–1055

Martuza RL, Malick A, Markert JM et al (1991) Experimental therapy of human glioma by means of a genetically engineered virus mutant. Science 252:854–885

Marconi P, Argnani R, Berto E et al (2008) HSV as a vector in vaccine development and gene therapy. Hum Vaccin 4:91–105

McGeoch DJ, Dalrymple MA, Davison AJ et al (1988) The complete DNA sequence of the long unique region in the genome of herpes simplex virus type 1. J Gen Virol 69:1531–1574

Menotti L, Cerretani A, Hengel H et al (2008) Construction of a fully retargeted herpes simplex virus 1 recombinant capable of entering cells solely via human epidermal growth factor receptor 2. J Virol 82(20):10153–10161

Mineta T, Rabkin SD, Yazaki T et al (1995) Attenuated multi-mutated herpes simplex virus-1 for the treatment of malignant gliomas. Nat Med 1:938–943

Mori I, Nishiyama Y (2005) Herpes simplex virus and varicella-zoster virus: why do these human alphaherpesviruses behave so differently from one another? Rev Med Virol 15:393–406

Mori I, Nishiyama Y (2006) Accessory genes define the relationship between the herpes simplex virus and its host. Microbes Infect 8:2556–2562

Mullerad M, Bochner BH, Adusumilli PS et al (2005) Herpes simplex virus based gene therapy enhances the efficacy of mitomycin C for the treatment of human bladder transitional cell carcinoma. J Urol 174:741–746

Nakamura H, Mullen JT, Chandrasekhar S et al (2001) Multimodality therapy with a replication-conditional herpes simplex virus 1 mutant that expresses yeast cytosine deaminase for intratumoral conversion of 5-fluorocytosine to 5-fluorouracil. Cancer Res 61:5447–5452

Nakao A, Kimata H, al IT (2004) Intratumoral injection of herpes simplex virus HF10 in recurrent breast cancer. Ann Oncol 15:988–989

Nakashima H, Chiocca EA (2014) Modification of HSV-1 to an oncolytic virus. Methods Mol Biol 1144:117–127

Nanni P, Gatta V, Menotti L et al (2013) Preclinical therapy of disseminated HER-2+ ovarian and breast carcinomas with a HER-2-retargeted oncolytic herpesvirus. PLoS Pathog 9:e1003155

Nawa A, Luo C, Zhang L et al (2008) Non- engineered, naturally oncolytic herpes simplex virus HSV1 HF-10: applications for cancer gene therapy. Curr Gene Ther 8:208–221

Nishiyama Y (1996) Herpesvirus genes: molecular basis of viral replication and pathogenicity. Nagoya J Med Sci 59:107–119

Nishiyama Y (2004) Herpes simplex virus gene products: the accessories reflect her lifestyle well. Rev Med Virol 14:33–46

Nishiyama Y, Kimura H, Daikoku T (1991) Complementary lethal invasion of the central nervous system by nonneuroinvasive herpes simplex virus types 1 and 2. J Virol 65:4520–4524

Pérez-Herrero E, Fernández-Medarde A (2015) Advanced targeted therapies in cancer: drug nanocarriers, the future of chemotherapy. Eur J Pharm Biopharm 93:52–79

Pourchet A, Fuhrmann SR, Pilones KA et al (2016) CD8(+) T-cell immune evasion enables oncolytic virus immunotherapy. EBioMedicine 5:59–67

Puzanov I, Milhem MM, Minor D et al (2016) Talimogene laherparepvec in combination with ipilimumab in previously untreated, unresectable stage IIIB-IV melanoma. J Clin Oncol 34:2619–2626

Salameh S, Sheth U, Shukla D (2012) Early events in herpes simplex virus lifecycle with implications for an infection of lifetime. Open Virol J 6:1–6

Sengupta S, Thaci B, Crawford AC et al (2014) Interleukin-13 receptor alpha 2-targeted glioblastoma immunotherapy. Biomed Res Int 2014:952128

Senzer NN, Kaufman HL, Amatruda T et al (2009) Phase II clinical trial of a granulocyte-macrophage colony-stimulating factor-encoding, second-generation oncolytic herpesvirus in patients with unresectable metastatic melanoma. J Clin Oncol 27:5763–5771

Sobol PT, Mossman KL (2006) ICP0 prevents RNase L-independent rRNA cleavage in herpes simplex virus type 1-infected cells. J Virol 80:218–225

Spain L, Diem S, Larkin J (2016) Management of toxicities of immune checkpoint inhibitors. Cancer Treat Rev 44:51–60

Spear MA, Sun F, Eling DJ et al (2000) Cytotoxicity, apoptosis, and viral replication in tumor cells treated with oncolytic ribonucleotide reductase-defective herpes simplex type 1 virus (hrR3) combined with ionizing radiation. Cancer Gene Ther 7:1051–1059

Stanziale SF, Petrowsky H, Adusumilli PS et al (2004) Infection with oncolytic herpes simplex virus-1 induces apoptosis in neighboring human cancer cells: a potential target to increase anticancer activity. Clin Cancer Res 10:3225–3232

Stavropoulos TA, Strathdee CA (1998) An enhanced packaging system for helper- dependent herpes simplex virus vectors. J Virol 72:7137–7143

Stiles BM, Adusumilli PS, Bhargava A et al (2006a) Minimally invasive localization of oncolytic herpes simplex viral therapy of metastatic pleural cancer. Cancer Gene Ther 13:53–64

Stiles BM, Adusumilli PS, Stanziale SF et al (2006b) Estrogen enhances the efficacy of an oncolytic HSV-1 mutant in the treatment of estrogen receptor-positive breast cancer. Int J Oncol 28:1429–1439

Sundaresan P, Hunter WD, Martuza RL et al (2000) Attenuated, replication-competent herpes simplex virus type 1 mutant G207: safety evaluation in mice. J Virol 74:3832–3841

Sze DY, Iagaru AH, Gambhir SS et al (2012) Response to intra-arterial oncolytic virotherapy with the herpes virus NV1020 evaluated by [18F]fluorodeoxyglucose positron emission tomography and computed tomography. Hum Gene Ther 23:91–97

Sze DY, Reid TR, Rose SC (2013) Oncolytic virotherapy. J Vasc Interv Radiol 24:1115–1122

Takakuwa H, Goshima F, Nozawa N et al (2003) Oncolytic viral therapy using a spontaneously generated herpes simplex virus type 1 variant for disseminated peritoneal tumor in immunocompetent mice. Arch Virol 148:813–825

Tanaka M, Kagawa H, Yamanashi Y et al (2003) Construction of an excisable bacterial artificial chromosome containing a full-length infectious clone of herpes simplex virus type 1: viruses reconstituted from the clone exhibit wild-type properties in vitro and in vivo. J Virol 77:1382–1391

Tang T, Eldabaje R, Yang L (2016) Current status of biological therapies for the treatment of metastatic melanoma. Anticancer Res 36:3229–3241

Taylor TJ, Brockman MA, McNamee EE et al (2002) Herpes simplex virus. Front Biosci 7:d752–d764

Teshigahara O, Goshima F, Takao K et al (2004) Oncolytic viral therapy for breast cancer with herpes simplex virus type 1 mutant HF 10. J Surg Oncol 85:42–47

Todo T (2008) Oncolytic virus therapy using genetically engineered herpes simplex viruses. Front Biosci 200813:2060–2064

Todo T, Feigenbaum F, Rabkin SD et al (2000) Viral shedding and biodistribution of G207, a multimutated, conditionally replicating herpes simplex virus type 1, after intracerebral inoculation in aotus. Mol Ther 2:588–595

Toyoizumi T, Mick R, Abbas AE et al (1999) Combined therapy with chemotherapeutic agents and herpes simplex virus type 1 ICP34.5 mutant (HSV-1716) in human non- small cell lung cancer. Hum Gene Ther 10:3013–3029

Ushijima Y, Luo C, Goshima F et al (2007) Determination and analysis of the DNA sequence of highly attenuated herpes simplex virus type 1 mutant HF10, a potential oncolytic virus. Microbes Infect 9:142–149

Watanabe D, Goshima F, Mori I et al (2008) Oncolytic virotherapy for malignant melanoma with herpes simplex virus type 1 mutant HF10. J Dermatol Sci 50:185–196

Witt MN, Braun GS, Ihrler S et al (2009) Occurrence of HSV-1-induced pneumonitis in patients under standard immunosuppressive therapy for rheumatic, vasculitic, and connective tissue disease. BMC Pulm Med 18:22

Zabierowski SE, Deluca NA (2008) Stabilized binding of TBP to the TATA box of herpes simplex virus type 1 early (tk) and late (gC) promoters by TFIIA and ICP4. J Virol 82:3546–3554

Zhou G, Roizman B (2006) Construction and properties of a herpes simplex virus 1 designed to enter cells solely via the IL-13alpha2 receptor. Proc Natl Acad Sci U S A 103:5508–5513

Zhou G, Ye GJ, al DW (2002) Engineered herpes simplex virus 1 is dependent on IL13Ralpha 2 receptor for cell entry and independent of glycoprotein D receptor interaction. Proc Natl Acad Sci U S A 99:15124–15129

Chapter 5
Neurological Disorders Associated with Human Alphaherpesviruses

Jun-ichi Kawada

Abstract Herpes simplex virus (HSV) encephalitis is the most common cause of sporadic fatal encephalitis worldwide, and central nervous system (CNS) involvement is observed in approximately one-third of neonatal HSV infections. In recent years, single-gene inborn errors of innate immunity have been shown to be associated with susceptibility to HSV encephalitis. Temporal lobe abnormalities revealed by magnetic resonance imaging—the most sensitive imaging method for HSV encephalitis—are considered strong evidence for the disease. Detection of HSV DNA in the cerebrospinal fluid by polymerase chain reaction (PCR) is the gold standard for the diagnosis of HSV encephalitis and neonatal meningoencephalitis. Intravenous acyclovir for 14–21 days is the standard treatment in HSV encephalitis. Neurological outcomes in neonates are improved by intravenous high-dose acyclovir for 21 days followed by oral acyclovir suppressive therapy for 6 months.

Varicella-zoster virus (VZV) causes a wide range of CNS manifestations. VZV encephalitis typically occurs after primary infection, and reactivation of VZV may cause encephalitis. On the other hand, VZV infection of cerebral arteries produces vasculopathy, which can manifest as ischemic stroke. Vasculopathy can occur after primary infection or reactivation of VZV. PCR detection of VZV DNA in the cerebrospinal fluid can be used for the diagnosis of encephalitis or vasculopathy. Although there are no controlled treatment trials to assess VZV treatments of encephalitis or vasculopathy, intravenous acyclovir is a common treatment.

Keywords Herpes simplex virus · HSV · Herpes simplex encephalitis, HSE · Varicella-zoster virus, VZV · VZV encephalitis · VZV vasculopathy

---

J. Kawada (✉)
Department of Pediatrics, Nagoya University Graduate School of Medicine, Nagoya, Japan
e-mail: kawadaj@med.nagoya-u.ac.jp

Although hundreds of viruses are known as causative agents of central nervous system (CNS) infection, only a limited subset is responsible for most cases in which a specific cause is identified. Among *Alphaherpesvirinae*, herpes simplex virus (HSV)-1, HSV-2, and varicella-zoster virus (VZV) are commonly identified viruses causing encephalitis. Early recognition is important because acyclovir (ACV) treatment reduces morbidity and morbidity. This section summarizes the pathogenesis, clinical manifestation, diagnosis, and treatment of CNS infections caused by HSV and VZV.

5.1 HSV Types 1 and 2

5.1.1 Herpes Simplex Encephalitis

5.1.1.1 Epidemiology

HSV encephalitis (HSE) is the most commonly identified cause of sporadic fatal encephalitis in the United States and other industrialized nations (Whitley 1990, 2006; Huppatz et al. 2009; Granerod et al. 2010). HSE accounts for approximately 10–20% of all cases of viral encephalitis (Levitz 1998). In a nationwide retrospective study in Sweden, the incidence of confirmed HSE cases was 2.2 per million people annually (Hjalmarsson et al. 2007). HSE occurs sporadically throughout the year, and the age-specific incidence is bimodal, with peaks in the young and in the elderly. HSV-1 causes more than 90% of HSE cases in adults, while HSV-2 infection typically causes aseptic meningitis. On the other hand, HSV-2 is a common cause of acute generalized encephalitis in neonates (Corey et al. 1988).

5.1.1.2 Pathogenesis

The pathogenesis of HSE remains elusive. Histopathologically, replicating HSV includes the ballooning of infected cells and the appearance of chromatin within the nuclei of cells followed by degeneration of the nuclei. Furthermore, an influx of mononuclear cells can be detected in infected tissue. HSE results in acute inflammation, congestion, and/or hemorrhage most prominently in the temporal lobes, usually occurring asymmetrically in adults and more diffusely in newborns (Whitley 2006; Whitley et al. 2007).

Although primary and recurrent HSV infections can lead to HSE, the route of access of HSV to the CNS remains controversial. Approximately one-third of HSE cases are considered to be a consequence of immediate CNS invasion via the trigeminal nerve or olfactory tract following an episode of primary HSV-1 infection of the oropharynx (Levitz 1998). Most patients with primary infection are younger than 18 years of age. Additionally, HSE can be caused by CNS invasion after an episode of recurrent HSV infection, which is believed to represent viral reactivation with subsequent spread. Reactivation of latent HSV in situ within the CNS is another hypothetical mechanism for HSE pathogenesis (Whitley 2006). Various animal models have been developed to mimic the possible routes of infection. Some models have demonstrated that intranasal inoculation of HSV produces focal lesions localized to the temporal lobe, similar to those observed in human cases of HSE (Hudson et al. 1991). Another murine model revealed that inoculation of HSV into tooth pulp leads to an encephalitis primarily affecting the temporal cortex and limbic system (Barnett et al. 1994). Both direct virus-mediated and indirect immune-mediated mechanisms are thought to play a role in CNS damage, providing one explanation for why HSE is not more common among immunocompromised hosts despite recurrent mucocutaneous infections (Piret and Boivin 2015).

Immune control of HSV requires components from both innate and adaptive immune responses, including type I interferon (IFN), NK cells, and cytotoxic T-cells (Egan et al. 2013). Recent studies revealed that single-gene inborn errors of innate immunity are associated with susceptibility to specific infections. Toll-like receptors (TLRs) play crucial roles in the innate immune response, as TLRs 2, 3, and 9 recognize HSV (Kurt-Jones et al. 2004; Ashkar et al. 2004; Lund et al. 2003). Viral glycoproteins on the HSV particles are first sensed by TRL 2 at the cell surface. After entry, viral genomic DNA is detected by endosomal TLR 9 and other cytosolic double-strand (ds)DNA sensors. Replication of the viral genome leads to accumulation of intermediate dsRNAs, which are then sensed by endosomal TLR 3. TLR 3 is widely distributed throughout the CNS, where it may prevent the spread of HSV through the generation of IFNs (Kielian 2009). Seminal studies have demonstrated that mutations in genes encoding components in the TRL 3-mediated IFN-α/IFN-β pathway confer susceptibility to HSE. Mutations in *UNC93B1*, *TLR3*, *TRAF3*, *TRIF*, *TBK1*, *STAT1*, *NEMO*, and *IRF3* genes have been identified in patients with HSE (Casrouge et al. 2006; Zhang et al. 2007; Perez de Diego et al. 2010; Sancho-Shimizu et al. 2011; Herman et al. 2012; Dupuis et al. 2003; Audry et al. 2011; Andersen et al. 2015). A common theme among the identified genetic defects is that they lead to reduced IFN responses in cell culture after HSV-1 infection or stimulation through the TLR 3 pathway. While most of the identified mutations are inherited by an autosomal dominant mechanism, several cases with autosomal recessive mutations have also been identified. These findings may help explain why HSV becomes neuroinvasive to cause HSE in a small minority of individuals.

5.1.1.3 Clinical Features

Symptoms and Signs

Focal neurological findings are sometimes seen in the acute phase of HSE, while initial symptoms are frequently nonspecific. Furthermore, similar symptoms can occur in other viral and bacterial infections of the CNS, with headache and fever presenting in ~80 and 90% of cases, respectively (Raschilas et al. 2002). Other common features include altered levels of consciousness, focal or generalized seizures, cranial nerve deficits, hemiparesis, dysphasia, aphasia, and ataxia (Raschilas et al.

2002). Although several seizures at the onset of illness are not uncommon, status epilepticus is rare. Later in the clinical course, patients may have diminished comprehension, paraphasic spontaneous speech, impaired memory, and loss of emotional control (Solomon et al. 2012). In pediatric patients with HSE, nonneurologic complications such as fever, fatigue, and vomiting are dominant, and the frequency of neurological symptoms at disease onset is lower than in adults (Schleede et al. 2013). Atypical presentations may occur in immunocompromised patients.

Laboratory Findings

Examination of the cerebrospinal fluid (CSF) shows a lymphocytic pleocytosis of 10–1000 white blood cells per μL in most patients with HSE (Nahmias et al. 1982). However, a normal CSF profile can occur early in the course of the disease; this observation has been made in immunocompetent hosts and in a case series of patients taking tumor necrosis factor inhibitors (Bradford et al. 2009). Furthermore, one retrospective multicenter review of pediatric patients with HSE found that 13% of children with HSE had normal CSF profiles (Schleede et al. 2013). HSE is often hemorrhagic, and red blood cells and xanthochromia can be detected in the CSF. The presence of red cells in the CSF is associated with a worse prognosis (Poissy et al. 2012).

Several studies detected N-methyl-D-aspartate receptor (NMDAR) antibodies in the serum and/or CSF in 20–30% patients with HSE (Pruss et al. 2012; Westman et al. 2016). Furthermore, NMDAR antibody synthesis was associated with relapse of HSE (Armangue et al. 2014). It is possible that the virus-induced destruction of neurons initiates a primary autoimmune response against NMDAR. Alternatively, CNS inflammation in the course of HSE may lead to immunological activation, resulting in a polyspecific B-cell activation (Pruss et al. 2012). While the clinical significance of NMDAR antibodies in patients with HSE remains unclear, one study suggested the association between NMDAR and impaired neurological recovery (Westman et al. 2016).

Imaging Studies and Electroencephalogram

Temporal lobe abnormalities detected by brain imaging are considered strong evidence of HSE. Typical computed tomography (CT) and magnetic resonance imaging (MRI) findings of HSE are shown in Fig. 5.1. Temporal lobe lesions are predominantly unilateral and may have a mass effect (Levitz 1998). MRI is significantly more sensitive than CT, especially in the early course of the disease (Domingues et al. 1998). MRI should be performed on all patients with suspected HSE. Approximately, 90% of patients with HSE have MRI abnormalities involving the temporal lobe within 48 h, and several studies have shown that MRI with diffusion-weighted imaging may be helpful in the early diagnosis of HSE (Domingues et al. 1998; McCabe et al. 2003; Heiner and Demaerel 2003). HSE is the most commonly identified cause of temporal lobe encephalitis. However, temporal lobe involvement on MRI can be observed in other infectious and

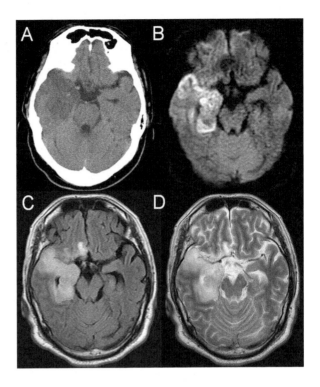

Fig. 5.1 CT and MRI findings in an adult patient with HSE. CT and MRI were taken during the acute stage of the illness. CT image of the brain shows low density area in the right temporal lobe (**a**). MRI with diffusion-weighted imaging (**b**), FLAIR (**c**), and T2 (**d**) show abnormal signals in the right temporal lobe

noninfectious etiologies including tuberculosis, VZV infection, malignancies, and vascular diseases (Chow et al. 2015). Other modalities such as MR spectroscopy, single-photon emission CT, or brain fluorodeoxyglucose positron-emission tomography can be used, but such tools are not yet sufficiently informative nor widely available for routine use in patients with HSE (Solomon et al. 2012).

Electroencephalogram (EEG) recordings may show abnormalities early in the course of the disease, demonstrating diffuse slow focal abnormalities in the temporal regions or periodic lateralizing epileptiform discharges (PLEDs). EEG abnormalities involving temporal lobes are seen in approximately 75% of patients with HSE (Domingues et al. 1997). Although many EEG findings in patients with HSE are nonspecific, such technology can be helpful in distinguishing whether abnormal behavior is due to a primary psychiatric disease or encephalitis. Additionally, the EEG is useful to identify nonconvulsive or subtle motor seizures, which may occur in patients with HSE (Solomon et al. 2012).

5.1.1.4 Diagnosis

Previously, diagnosis of HSE was dependent on brain biopsy, with identification of HSV in tissues by cell culture or immunohistochemical staining. Although brain biopsy has high sensitivity and specificity, it requires an invasive procedure, and the results may not be available for several days. However, several studies have shown that detection of HSV DNA in the CSF by polymerase chain reaction (PCR) has overall sensitivity and specificity of >95% for diagnosis of HSE compared to brain biopsy (Aurelius et al. 1991; Lakeman and Whitley 1995). Therefore, detection of HSV DNA in the CSF by PCR has become the gold standard for the diagnosis of HSE. PCR results show positivity early in the course of the illness and remain positive during the first week of therapy. However, false-negative results have been reported, most notably in CSF samples obtained within 72 h of illness onset. Therefore, caution should be used in stopping ACV therapy in patients with strongly suspected HSE on the sole basis of a single negative CSF PCR test obtained within 72 h of symptom onset, unless a suitable alternative diagnosis has been established (De Tiege et al. 2003; Elbers et al. 2007). Furthermore, there is no standardized assay of HSV PCR, and assay sensitivity may vary among laboratories. PCR assays with high sensitivity such as nested or real-time PCR should be applied for diagnosis of HSE because the CSF may contain small amounts of HSV DNA in some patients (Kawada et al. 2004b; Schloss et al. 2009). Real-time PCR is suitable for monitoring HSV load in CSF during ACV treatment. However, the HSV viral load is not associated with disease outcome (Poissy et al. 2012).

5.1.1.5 Treatment

ACV is a nucleoside analogue with strong antiviral activity against HSV and VZV. Because HSE is the most commonly identified cause of viral encephalitis and can be treated with ACV, it has become general practice to initiate ACV treatment once the CSF and/or imaging findings suggest viral encephalitis, without waiting for confirmation of HSV by PCR. In immunocompetent patients, presumptive ACV treatment might be safely discontinued if a negative HSV PCR result is obtained after 72 h following onset of neurological symptoms, with unaltered consciousness, normal MRI, and a CSF white cell count of less than 5 cells/mm^2 (Solomon et al. 2012). On the other hand, pediatric studies have shown that the use of presumptive ACV treatment for all patients with encephalopathy, without regard to the likely diagnosis, can be harmful (Kneen et al. 2010; Gaensbauer et al. 2014). Although ACV is generally considered to be a safe drug, serious side effects have been reported including renal impairment, hepatitis, and bone marrow failure.

Intravenous ACV (10 mg/kg three times a day) was shown to reduce mortality and morbidity of HSE in randomized trials in the1980s (Skoldenberg et al. 1984; Whitley et al. 1986). However, outcome is often poor, especially in patients with advanced age, reduced coma score, or delays of more than 48 h between hospital admission and initiation of treatment (Raschilas et al. 2002). The duration of treatment in the original trials of ACV for HSE was 10 days; however, occasional cases of HSE relapse with ACV-sensitive HSV were reported subsequently (VanLandingham et al. 1988; Dennett et al. 1996). Ongoing immune-mediated and inflammatory reactions to the infection can be the major pathogenic process of relapse, while there is evidence for continuing viral replication in some cases

(Yamada et al. 2003). Therefore, 14–21 days of intravenous ACV has become the standard treatment in confirmed HSE cases, although subsequent relapse can occur. Some studies advocate reevaluating HSV PCR in the CSF at 14–21 days and continuing treatment until a negative result is obtained (Solomon et al. 2012).

Oral ACV therapy is not recommended for the treatment of HSE because such delivery does not achieve adequate levels in the CSF. However, valaciclovir (its valine ester) has good bioavailability and is converted to ACV after absorption. One study evaluating the pharmacokinetics of orally administered valaciclovir in patients with HSE found that patients achieved and maintained therapeutic concentrations of valaciclovir in the CSF (Pouplin et al. 2011). Valaciclovir may have a role in ongoing treatments, particularly in patients with detectable HSV in the CSF after 2–3 weeks, as low-level viral replication in the CSF may contribute to relapse or progressive neurological morbidity. However, one clinical trial revealed that an additional 3-month course of oral valaciclovir therapy after standard treatment with intravenous ACV did not improve the outcome as measured by neuropsychological testing (Gnann et al. 2015).

Retrospective studies have suggested some benefits connected with the addition of corticosteroids to ACV treatment. Since the advent of ACV, corticosteroids have often been used, especially in patients with marked cerebral edema, brain shift, or increased intracranial pressure (Kamei et al. 2005; Ramos-Estebanez et al. 2014). The role of corticosteroids remains controversial because they could facilitate viral replication, in theory. On the other hand, in addition to exhibiting direct HSV-mediated cytolysis, HSE is also characterized by acute and persistent intrathecal inflammatory responses and possibly by autoimmune phenomena (Pruss et al. 2012; Armangue et al. 2014). Corticosteroid administration under a specialist's supervision may have a role in treating patients with HSE; however, further studies to define the role of immune responses and explore the therapeutic potential of adjunctive therapy with anti-inflammatory and immunomodulating agents are necessary (Martinez-Torres et al. 2008).

5.1.2 Neonatal HSV Meningoencephalitis

5.1.2.1 Epidemiology

Neonatal infection with HSV occurs in ~3–10 per 100,000 live births in the United States and other countries (Corey and Wald 2009; Jones et al. 2014). Most cases of neonatal herpes result from maternal infection and transmission, usually during passage through the contaminated infected birth canal of a mother with asymptomatic genital herpes. The risk for infection is higher in infants born to mothers with primary genital infection (30–50%) than those with recurrent genital infection (<3%) (Corey and Wald 2009).

Neonatal HSV infection is classified into three main categories: localized skin, eye, and mouth (SEM); CNS disease (also called meningoencephalitis); and disseminated disease. Meningoencephalitis accounts for approximately one-third of neonatal cases with HSV, while CNS involvement may be observed in SEM or disseminated disease (Kimberlin 2007). While HSV-2 has historically been the predominant serotype causing genital herpes and neonatal herpes, HSV-1 is increasingly being identified as causing more cases of genital herpes and neonatal herpes in the United States and some European countries (Pinninti and Kimberlin 2014).

5.1.2.2 Pathogenesis

Neonatal HSV meningoencephalitis may occur as a result of localized retrograde spread from the nasopharynx and olfactory nerves to the brain or through hematogenous spread. Compared to adults, neonates are particularly susceptible to poor neurological outcomes of meningoencephalitis from HSV. While explanations for this increased susceptibility to HSV infection in newborns remain elusive, differences in skin barrier function or aspects of HSV immunity compared with adults may contribute to their increased susceptibility (Kawada et al. 2004a; Kollmann et al. 2012; Gantt and Muller 2013).

5.1.2.3 Clinical Features

Infants with neonatal HSV meningoencephalitis typically present at 5–11 days of life with clinical findings suggestive of bacterial meningitis, including irritability, lethargy, poor feeding, poor tone, and seizures. Approximately 60–70% of these patients have skin lesions at some point during the course of the illness, with most such disease manifestations occurring within the first month of life (Pinninti and Kimberlin 2014).

5.1.2.4 Diagnosis

Pleocytosis is usually present in affected neonates, and detection of HSV DNA in the CSF by PCR is the most sensitive laboratory test for confirming the diagnosis (Kimura et al. 1991; Kimberlin et al. 1996). However, false-negative results have also been reported in CSF specimens obtained early in the course of illness and in samples obtained several days into ACV therapy (Kimberlin et al. 1996). HSV DNA is detected in the CSF of ~25% of affected neonates with apparent localization to the SEM and in more than 50% of neonates with disseminated HSV disease. Furthermore, patients with meningoencephalitis exhibit higher HSV DNA loads than other groups (Kimura et al. 2002), and HSV DNA loads are higher in neonates with meningoencephalitis than in older patients with HSE (Ando et al. 1993; Kawada et al. 2004b). These differences may be due to the permissiveness of immature brain cells for replication or to an immature immune response.

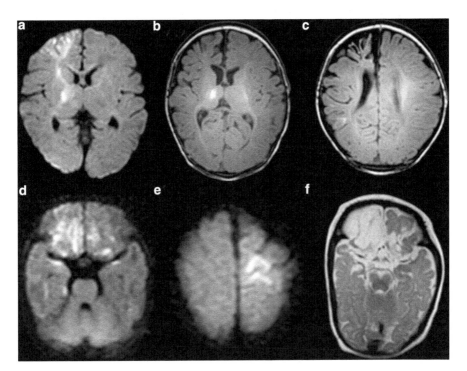

Fig. 5.2 Characteristic distribution of brain lesions in neonatal HSV meningoencephalitis
The right frontal watershed lesion and right thalamic lesion (upper panels: **a~c**)
Diffusion-weighted images taken 3 days after disease onset (**a**). FLAIR images exhibit the right thalamic lesion (**b**) and the right frontal area 3 months after disease onset (**c**)
The bilateral inferior frontal lesions and left perirolandic cortical lesion (lower panels: **d~f**)
Diffusion-weighted imaging taken 5 days after disease onset (**d, e**) and on T2-weighted MRI taken at 3 months old (**f**)
Courtesy of Hiroyuki Kidokoro, MD, PhD, Nagoya University Hospital

In neonates with HSV meningoencephalitis, brain imaging with MRI is recommended to determine the location and extent of brain involvement. One retrospective review of neonatal HSV meningoencephalitis showed that neurodevelopmental sequelae correlated with MRI abnormalities (Bajaj et al. 2014). The MRI findings can be variable, and diffusion-weighted imaging can reveal changes not visible on conventional MRI (Dhawan et al. 2006). Several days to a week into the illness, MRI studies may show parenchymal brain edema or abnormal attenuation, hemorrhage, or destructive lesions. In addition to the classic temporal lobe lesions, imaging abnormalities may be multifocal or limited to the brainstem or cerebellum (Vossough et al. 2008). Variable MRI findings of neonatal HSV meningoencephalitis are shown in Fig. 5.2.

5.1.2.5 Treatment

Parenteral ACV is recommended for all categories of neonatal HSV infection. With the advent of antiviral therapy, 1-year mortality rates for neonates with HSV meningoencephalitis declined from 50% to 4% (Whitley et al. 1980; Kimberlin et al. 2001). However, many survivors suffer substantial neurologic sequelae. The current recommendation of ACV therapy for neonatal HSV meningoencephalitis is 20 mg/kg three times a day, given intravenously for a minimum of 21 days. The benefits of this dose compared with a lower dose of ACV (10 mg/kg three times a day) were established in an open-label study in which neonates with meningoencephalitis or disseminated HSV were treated with ACV (20 mg/kg three times a day) for 21 days (Whitley et al. 1991; Kimberlin et al. 2001). As HSV DNA detection in the CSF at or after completion of ACV therapy is associated with poor outcomes, HSV PCR should be repeated in the CSF of all neonates with HSV meningoencephalitis to confirm a negative PCR result (Kimberlin et al. 1996).

Following parenteral ACV treatment, oral ACV suppressive therapy for 6 months improves neurodevelopmental outcomes in infants with HSV meningoencephalitis and prevents skin recurrences in infants with various disease categories of neonatal HSV infection (Kimberlin et al. 2011). Thus, infants surviving neonatal HSV infection are recommended to receive oral ACV suppression at 300 mg/m^2/dose, administered three times daily for 6 months. However, the effectiveness of long-term suppression with oral ACV in reducing the risk of CNS recurrence after neonatal HSV meningoencephalitis is unknown. Furthermore, CNS recurrences during or after oral suppression have been reported (Fonseca-Aten et al. 2005; Kato et al. 2015).

5.2 Varicella-Zoster Virus

Primary infection with varicella-zoster virus (VZV) causes varicella (chickenpox), after which the virus becomes latent in the cranial nerve and dorsal root ganglia. Reactivation of the latent virus produces shingles (herpes zoster), mainly in elderly or immunosuppressed patients. VZV causes a wide range of CNS manifestations including encephalitis, meningitis, Ramsay Hunt syndrome, cerebellitis, myelitis, and vasculopathy. The CNS manifestation caused by VZV can occur as a result of both primary and reactivated disease. In rare cases, CNS diseases caused by reactivation of the VZV vaccine virus can occur. Compared to the wild-type VZV-associated CNS diseases that most often cause encephalitis in children, the VZV vaccine strain is more often associated with meningitis (Pahud et al. 2011).

5.2.1 VZV Encephalitis

In studies involving all age groups, VZV has been reported to be the second most infectious cause of encephalitis after HSV; however, the incidence among children has declined considerably with routine vaccination (Granerod et al. 2010; Britton et al. 2016). In cases with primary infection (varicella), VZV encephalitis usually follows the rash at 2–8 days. However, it occasionally occurs before the rash or even in patients with no rash. Reactivation of VZV may also cause encephalitis, especially in elderly or immunocompromised patients. The onset is typically insidious, and there may be no rash.

The histopathologic findings in brain tissue from fatal cases of VZV encephalitis can include vasculitis of large and small vessels, demyelination, axonal damage, and neuronal degeneration; however, histological abnormalities are often minimal (Barnes and Whitley 1986; Amlie-Lefond et al. 1995). It is possible that inflammatory or immune-mediated pathogenesis might be the primary cause of encephalitis. VZV may spread to the CNS centripetally (toward the spinal cord or brain) or centrifugally (toward the vessels of the brain) by hematological or transaxonal transport. It has been suggested that VZV encephalitis might be primarily a vasculopathic disease and that symptoms of brain involvement may be derived not directly from the viral effect but, secondarily, from the productive virus infection within large and small cerebral arteries.

The most frequent acute symptoms in patients with VZV encephalitis are fever and altered mental status. Compared to HSE, focal neurological signs and seizures are less frequently observed (Pollak et al. 2012). Encephalitic symptoms resolve rapidly in some patients. In most patients with VZV encephalitis, mild lymphocytic pleocytosis is found in the CSF, and PCR detection of VZV DNA is useful for diagnosis, especially in patients without rash. High VZV DNA loads in the CSF are correlated with the severity of CNS disease, and quantitative PCR might be useful in monitoring the viral response during antiviral therapy (Aberle et al. 2005). However, VZV DNA can be detected in the CSF from varicella-zoster or herpes zoster cases without neurological symptoms (Persson et al. 2009).

Although there is no clinical trial of ACV treatment for VZV encephalitis, intravenous ACV is commonly used in accordance with the treatment of severe varicella (Kneen et al. 2012). Compared to patients with HSE, those with VZV encephalitis might need higher doses of ACV because VZV is less sensitive to ACV than HSV.

5.2.2 VZV Vasculopathy

VZV infection of cerebral arteries produces vasculopathy, manifesting in ischemic or hemorrhagic stroke. Vasculopathy can occur after primary infection or reactivation of VZV. Varicella is an important risk factor for childhood ischemic stroke, and it is estimated that ~1 in 15,000 cases of varicella are associated with subsequent stroke; most occur within 12 months of infection (Askalan et al. 2001). Several studies reported an increased risk of stroke after a zoster attack. A nationwide retrospective cohort study in Taiwan determined that the adjusted hazard ratios for risk of stroke after zoster and zoster ophthalmicus during the 1-year follow-up period were 1.31 and 4.28, respectively (Kang et al. 2009). VZV vasculopathy is more common in immunocompromised individuals, such as patients with HIV infection (Berkefeld et al. 2000).

The likely mechanism by which VZV causes stroke is by direct infection of cerebral arteries, resulting in pathological changes including thrombosis, necrosis, dissection, and aneurysm formation (Kleinschmidt-DeMasters and Gilden 2001). Several pathological and virological analyses in autopsy studies have revealed VZV DNA or antigen within the walls of cerebral arteries (Eidelberg et al. 1986; Gilden et al. 1996). Additionally, transient protein S deficiency caused by autoantibodies against phospholipids and coagulation proteins during or after varicella showed association with stroke events (Manco-Johnson et al. 1996).

One or several large or small cerebral arteries may be affected by VZV vasculopathy, and clinical presentations vary widely including headache, mental status changes, ataxia, visual loss, and hemiplegia (Grahn and Studahl 2015). Some patients may present with symptoms consistent with encephalitis followed by a focal deficit. A positive PCR finding for VZV DNA in the CSF is diagnostic, whereas a negative PCR result does not exclude the diagnosis of VZV vasculopathy because VZV DNA can be detected only during the first 2 weeks of the disease (Nagel et al. 2008). On the other hand, anti-VZV IgG antibody in the CSF is detectable during the second week after infection and is a more sensitive indicator of VZV vasculopathy than detection of VZV DNA (Nagel et al. 2008). To confirm the intrathecal synthesis of VZV IgG, the serum/CSF ratio of antibody should be calculated.

Although there are no controlled treatment trials to assess the treatment of VZV vasculopathy, intravenous ACV (10–15 mg/kg three times daily) is a commonly used treatment based on expert opinion or a descriptive case series (Nagel et al. 2008). As histologic specimens often demonstrate an inflammatory response in infected cerebral arteries, adding corticosteroids to ACV might be beneficial. However, prospective studies are necessary to determine the optimal dose, duration of ACV treatment, and benefit of concurrent steroid therapy in a controlled setting.

References

Aberle SW, Aberle JH, Steininger C, Puchhammer-Stockl E (2005) Quantitative real time PCR detection of Varicella-zoster virus DNA in cerebrospinal fluid in patients with neurological disease. Med Microbiol Immunol 194(1–2):7–12. https://doi.org/10.1007/s00430-003-0202-1

Amlie-Lefond C, Kleinschmidt-DeMasters BK, Mahalingam R, Davis LE, Gilden DH (1995) The vasculopathy of varicella-zoster virus encephalitis. Ann Neurol 37(6):784–790. https://doi.org/10.1002/ana.410370612

Andersen LL, Mork N, Reinert LS, Kofod-Olsen E, Narita R, Jorgensen SE, Skipper KA, Honing K, Gad HH, Ostergaard L, Orntoft TF, Hornung V, Paludan SR, Mikkelsen JG, Fujita T, Christiansen M, Hartmann R, Mogensen TH (2015) Functional IRF3 deficiency in a patient with herpes simplex encephalitis. J Exp Med 212(9):1371–1379. https://doi.org/10.1084/jem.20142274

Ando Y, Kimura H, Miwata H, Kudo T, Shibata M, Morishima T (1993) Quantitative analysis of herpes simplex virus DNA in cerebrospinal fluid of children with herpes simplex encephalitis. J Med Virol 41(2):170–173

Armangue T, Leypoldt F, Malaga I, Raspall-Chaure M, Marti I, Nichter C, Pugh J, Vicente-Rasoamalala M, Lafuente-Hidalgo M, Macaya A, Ke M, Titulaer MJ, Hoftberger R, Sheriff H, Glaser C, Dalmau J (2014) Herpes simplex virus encephalitis is a trigger of brain autoimmunity. Ann Neurol 75(2):317–323. https://doi.org/10.1002/ana.24083

Ashkar AA, Yao XD, Gill N, Sajic D, Patrick AJ, Rosenthal KL (2004) Toll-like receptor (TLR)-3, but not TLR4, agonist protects against genital herpes infection in the absence of inflammation seen with CpG DNA. J Infect Dis 190(10):1841–1849. https://doi.org/10.1086/425079

Askalan R, Laughlin S, Mayank S, Chan A, MacGregor D, Andrew M, Curtis R, Meaney B, deVeber G (2001) Chickenpox and stroke in childhood: a study of frequency and causation. Stroke 32(6):1257–1262

Audry M, Ciancanelli M, Yang K, Cobat A, Chang HH, Sancho-Shimizu V, Lorenzo L, Niehues T, Reichenbach J, Li XX, Israel A, Abel L, Casanova JL, Zhang SY, Jouanguy E, Puel A (2011) NEMO is a key component of NF-kappaB- and IRF-3-dependent TLR3-mediated immunity to herpes simplex virus. J Allergy Clin Immunol 128(3):610–617, e611–614. https://doi.org/10.1016/j.jaci.2011.04.059

Aurelius E, Johansson B, Skoldenberg B, Staland A, Forsgren M (1991) Rapid diagnosis of herpes simplex encephalitis by nested polymerase chain reaction assay of cerebrospinal fluid. Lancet 337(8735):189–192

Bajaj M, Mody S, Natarajan G (2014) Clinical and neuroimaging findings in neonatal herpes simplex virus infection. J Pediatr 165(2):404–407, e401. https://doi.org/10.1016/j.jpeds.2014.04.046

Barnes DW, Whitley RJ (1986) CNS diseases associated with varicella zoster virus and herpes simplex virus infection. Pathogenesis and current therapy. Neurol Clin 4(1):265–283

Barnett EM, Jacobsen G, Evans G, Cassell M, Perlman S (1994) Herpes simplex encephalitis in the temporal cortex and limbic system after trigeminal nerve inoculation. J Infect Dis 169(4):782–786

Berkefeld J, Enzensberger W, Lanfermann H (2000) MRI in human immunodeficiency virus-associated cerebral vasculitis. Neuroradiology 42(7):526–528

Bradford RD, Pettit AC, Wright PW, Mulligan MJ, Moreland LW, McLain DA, Gnann JW, Bloch KC (2009) Herpes simplex encephalitis during treatment with tumor necrosis factor-alpha inhibitors. Clin Infect Dis 49(6):924–927. https://doi.org/10.1086/605498

Britton PN, Khoury L, Booy R, Wood N, Jones CA (2016) Encephalitis in Australian children: contemporary trends in hospitalisation. Arch Dis Child 101(1):51–56. https://doi.org/10.1136/archdischild-2015-308468

Casrouge A, Zhang SY, Eidenschenk C, Jouanguy E, Puel A, Yang K, Alcais A, Picard C, Mahfoufi N, Nicolas N, Lorenzo L, Plancoulaine S, Senechal B, Geissmann F, Tabeta K, Hoebe K, Du X, Miller RL, Heron B, Mignot C, de Villemeur TB, Lebon P, Dulac O, Rozenberg F, Beutler B,

Tardieu M, Abel L, Casanova JL (2006) Herpes simplex virus encephalitis in human UNC-93B deficiency. Science 314(5797):308–312. https://doi.org/10.1126/science.1128346

Chow FC, Glaser CA, Sheriff H, Xia D, Messenger S, Whitley R, Venkatesan A (2015) Use of clinical and neuroimaging characteristics to distinguish temporal lobe herpes simplex encephalitis from its mimics. Clin Infect Dis 60(9):1377–1383. https://doi.org/10.1093/cid/civ051

Corey L, Wald A (2009) Maternal and neonatal herpes simplex virus infections. N Engl J Med 361(14):1376–1385. https://doi.org/10.1056/NEJMra0807633

Corey L, Whitley RJ, Stone EF, Mohan K (1988) Difference between herpes simplex virus type 1 and type 2 neonatal encephalitis in neurological outcome. Lancet 1(8575–6):1–4

De Tiege X, Heron B, Lebon P, Ponsot G, Rozenberg F (2003) Limits of early diagnosis of herpes simplex encephalitis in children: a retrospective study of 38 cases. Clin Infect Dis 36(10):1335–1339. https://doi.org/10.1086/374839

Dennett C, Klapper PE, Cleator GM (1996) Polymerase chain reaction in the investigation of "relapse" following herpes simplex encephalitis. J Med Virol 48(2):129–132. https://doi.org/10.1002/(SICI)1096-9071(199602)48:2<129::AID-JMV2>3.0.CO;2-B

Dhawan A, Kecskes Z, Jyoti R, Kent AL (2006) Early diffusion-weighted magnetic resonance imaging findings in neonatal herpes encephalitis. J Paediatr Child Health 42(12):824–826. https://doi.org/10.1111/j.1440-1754.2006.00986.x

Domingues RB, Tsanaclis AM, Pannuti CS, Mayo MS, Lakeman FD (1997) Evaluation of the range of clinical presentations of herpes simplex encephalitis by using polymerase chain reaction assay of cerebrospinal fluid samples. Clin Infect Dis 25(1):86–91

Domingues RB, Fink MC, Tsanaclis AM, de Castro CC, Cerri GG, Mayo MS, Lakeman FD (1998) Diagnosis of herpes simplex encephalitis by magnetic resonance imaging and polymerase chain reaction assay of cerebrospinal fluid. J Neurol Sci 157(2):148–153

Dupuis S, Jouanguy E, Al-Hajjar S, Fieschi C, Al-Mohsen IZ, Al-Jumaah S, Yang K, Chapgier A, Eidenschenk C, Eid P, Al Ghonaium A, Tufenkeji H, Frayha H, Al-Gazlan S, Al-Rayes H, Schreiber RD, Gresser I, Casanova JL (2003) Impaired response to interferon-alpha/beta and lethal viral disease in human STAT1 deficiency. Nat Genet 33(3):388–391. https://doi.org/10.1038/ng1097

Egan KP, Wu S, Wigdahl B, Jennings SR (2013) Immunological control of herpes simplex virus infections. J Neurovirol 19(4):328–345. https://doi.org/10.1007/s13365-013-0189-3

Eidelberg D, Sotrel A, Horoupian DS, Neumann PE, Pumarola-Sune T, Price RW (1986) Thrombotic cerebral vasculopathy associated with herpes zoster. Ann Neurol 19(1):7–14. https://doi.org/10.1002/ana.410190103

Elbers JM, Bitnun A, Richardson SE, Ford-Jones EL, Tellier R, Wald RM, Petric M, Kolski H, Heurter H, MacGregor D (2007) A 12-year prospective study of childhood herpes simplex encephalitis: is there a broader spectrum of disease? Pediatrics 119(2):e399–e407. https://doi.org/10.1542/peds.2006-1494

Fonseca-Aten M, Messina AF, Jafri HS, Sanchez PJ (2005) Herpes simplex virus encephalitis during suppressive therapy with acyclovir in a premature infant. Pediatrics 115(3):804–809. https://doi.org/10.1542/peds.2004-0777

Gaensbauer JT, Birkholz M, Pfannenstein K, Todd JK (2014) Herpes PCR testing and empiric acyclovir use beyond the neonatal period. Pediatrics 134(3):e651–e656. https://doi.org/10.1542/peds.2014-0294

Gantt S, Muller WJ (2013) The immunologic basis for severe neonatal herpes disease and potential strategies for therapeutic intervention. Clin Dev Immunol 2013:369172. https://doi.org/10.1155/2013/369172

Gilden DH, Kleinschmidt-DeMasters BK, Wellish M, Hedley-Whyte ET, Rentier B, Mahalingam R (1996) Varicella zoster virus, a cause of waxing and waning vasculitis: the New England Journal of Medicine case 5-1995 revisited. Neurology 47(6):1441–1446

Gnann JW Jr, Skoldenberg B, Hart J, Aurelius E, Schliamser S, Studahl M, Eriksson BM, Hanley D, Aoki F, Jackson AC, Griffiths P, Miedzinski L, Hanfelt-Goade D, Hinthorn D, Ahlm C, Aksamit A, Cruz-Flores S, Dale I, Cloud G, Jester P, Whitley RJ, National Institute of Allergy and Infectious Diseases Collaborative Antiviral Study Group (2015) Herpes simplex encephalitis: lack of clinical benefit of long-term valacyclovir therapy. Clin Infect Dis 61(5):683–691. https://doi.org/10.1093/cid/civ369

Grahn A, Studahl M (2015) Varicella-zoster virus infections of the central nervous system – prognosis, diagnostics and treatment. J Infect 71(3):281–293. https://doi.org/10.1016/j.jinf.2015.06.004

Granerod J, Ambrose HE, Davies NW, Clewley JP, Walsh AL, Morgan D, Cunningham R, Zuckerman M, Mutton KJ, Solomon T, Ward KN, Lunn MP, Irani SR, Vincent A, Brown DW, Crowcroft NS, Group UKHPAAoES (2010) Causes of encephalitis and differences in their clinical presentations in England: a multicentre, population-based prospective study. Lancet Infect Dis 10(12):835–844. https://doi.org/10.1016/S1473-3099(10)70222-X

Heiner L, Demaerel P (2003) Diffusion-weighted MR imaging findings in a patient with herpes simplex encephalitis. Eur J Radiol 45(3):195–198

Herman M, Ciancanelli M, Ou YH, Lorenzo L, Klaudel-Dreszler M, Pauwels E, Sancho-Shimizu V, Perez de Diego R, Abhyankar A, Israelsson E, Guo Y, Cardon A, Rozenberg F, Lebon P, Tardieu M, Heropolitanska-Pliszka E, Chaussabel D, White MA, Abel L, Zhang SY, Casanova JL (2012) Heterozygous TBK1 mutations impair TLR3 immunity and underlie herpes simplex encephalitis of childhood. J Exp Med 209(9):1567–1582. https://doi.org/10.1084/jem.20111316

Hjalmarsson A, Blomqvist P, Skoldenberg B (2007) Herpes simplex encephalitis in Sweden, 1990–2001: incidence, morbidity, and mortality. Clin Infect Dis 45(7):875–880. https://doi.org/10.1086/521262

Hudson SJ, Dix RD, Streilein JW (1991) Induction of encephalitis in SJL mice by intranasal infection with herpes simplex virus type 1: a possible model of herpes simplex encephalitis in humans. J Infect Dis 163(4):720–727

Huppatz C, Durrheim DN, Levi C, Dalton C, Williams D, Clements MS, Kelly PM (2009) Etiology of encephalitis in Australia, 1990–2007. Emerg Infect Dis 15(9):1359–1365. https://doi.org/10.3201/eid1509.081540

Jones CA, Raynes-Greenow C, Isaacs D, Neonatal HSVSI, Contributors to the Australian Paediatric Surveillance U (2014) Population-based surveillance of neonatal herpes simplex virus infection in Australia, 1997–2011. Clin Infect Dis 59(4):525–531. https://doi.org/10.1093/cid/ciu381

Kamei S, Sekizawa T, Shiota H, Mizutani T, Itoyama Y, Takasu T, Morishima T, Hirayanagi K (2005) Evaluation of combination therapy using aciclovir and corticosteroid in adult patients with herpes simplex virus encephalitis. J Neurol Neurosurg Psychiatry 76(11):1544–1549. https://doi.org/10.1136/jnnp.2004.049676

Kang JH, Ho JD, Chen YH, Lin HC (2009) Increased risk of stroke after a herpes zoster attack: a population-based follow-up study. Stroke 40(11):3443–3448. https://doi.org/10.1161/STROKEAHA.109.562017

Kato K, Hara S, Kawada J, Ito Y (2015) Recurrent neonatal herpes simplex virus infection with central nervous system disease after completion of a 6-month course of suppressive therapy: case report. J Infect Chemother 21(12):879–881. https://doi.org/10.1016/j.jiac.2015.08.005

Kawada J, Kimura H, Ito Y, Ando Y, Tanaka-Kitajima N, Hayakawa M, Nunoi H, Endo F, Morishima T (2004a) Evaluation of systemic inflammatory responses in neonates with herpes simplex virus infection. J Infect Dis 190(3):494–498. https://doi.org/10.1086/422325

Kawada J, Kimura H, Ito Y, Hoshino Y, Tanaka-Kitajima N, Ando Y, Futamura M, Morishima T (2004b) Comparison of real-time and nested PCR assays for detection of herpes simplex virus DNA. Microbiol Immunol 48(5):411–415

Kielian T (2009) Overview of toll-like receptors in the CNS. Curr Top Microbiol Immunol 336:1–14. https://doi.org/10.1007/978-3-642-00549-7_1

Kimberlin DW (2007) Herpes simplex virus infections of the newborn. Semin Perinatol 31(1):19–25. https://doi.org/10.1053/j.semperi.2007.01.003

Kimberlin DW, Lakeman FD, Arvin AM, Prober CG, Corey L, Powell DA, Burchett SK, Jacobs RF, Starr SE, Whitley RJ (1996) Application of the polymerase chain reaction to the diagnosis and management of neonatal herpes simplex virus disease. National Institute of Allergy and Infectious Diseases Collaborative Antiviral Study Group. J Infect Dis 174(6):1162–1167

Kimberlin DW, Lin CY, Jacobs RF, Powell DA, Corey L, Gruber WC, Rathore M, Bradley JS, Diaz PS, Kumar M, Arvin AM, Gutierrez K, Shelton M, Weiner LB, Sleasman JW, de Sierra TM, Weller S, Soong SJ, Kiell J, Lakeman FD, Whitley RJ, National Institute of Allergy and Infectious Diseases Collaborative Antiviral Study Group (2001) Safety and efficacy of high-dose intravenous acyclovir in the management of neonatal herpes simplex virus infections. Pediatrics 108(2):230–238

Kimberlin DW, Whitley RJ, Wan W, Powell DA, Storch G, Ahmed A, Palmer A, Sanchez PJ, Jacobs RF, Bradley JS, Robinson JL, Shelton M, Dennehy PH, Leach C, Rathore M, Abughali N, Wright P, Frenkel LM, Brady RC, Van Dyke R, Weiner LB, Guzman-Cottrill J, CA MC, Griffin J, Jester P, Parker M, Lakeman FD, Kuo H, Lee CH, Cloud GA, National Institute of Allergy and Infectious Diseases Collaborative Antiviral Study Group (2011) Oral acyclovir suppression and neurodevelopment after neonatal herpes. N Engl J Med 365(14):1284–1292. https://doi.org/10.1056/NEJMoa1003509

Kimura H, Futamura M, Kito H, Ando T, Goto M, Kuzushima K, Shibata M, Morishima T (1991) Detection of viral DNA in neonatal herpes simplex virus infections: frequent and prolonged presence in serum and cerebrospinal fluid. J Infect Dis 164(2):289–293

Kimura H, Ito Y, Futamura M, Ando Y, Yabuta Y, Hoshino Y, Nishiyama Y, Morishima T (2002) Quantitation of viral load in neonatal herpes simplex virus infection and comparison between type 1 and type 2. J Med Virol 67(3):349–353. https://doi.org/10.1002/jmv.10084

Kleinschmidt-DeMasters BK, Gilden DH (2001) Varicella-Zoster virus infections of the nervous system: clinical and pathologic correlates. Arch Pathol Lab Med 125(6):770–780. https://doi.org/10.1043/0003-9985(2001)125<0770:VZVIOT>2.0.CO;2

Kneen R, Jakka S, Mithyantha R, Riordan A, Solomon T (2010) The management of infants and children treated with aciclovir for suspected viral encephalitis. Arch Dis Child 95(2):100–106. https://doi.org/10.1136/adc.2008.144998

Kneen R, Michael BD, Menson E, Mehta B, Easton A, Hemingway C, Klapper PE, Vincent A, Lim M, Carrol E, Solomon T, National Encephalitis Guidelines D, Stakeholder G (2012) Management of suspected viral encephalitis in children – Association of British Neurologists and British Paediatric Allergy, Immunology and Infection Group national guidelines. J Infect 64(5):449–477. https://doi.org/10.1016/j.jinf.2011.11.013

Kollmann TR, Levy O, Montgomery RR, Goriely S (2012) Innate immune function by Toll-like receptors: distinct responses in newborns and the elderly. Immunity 37(5):771–783. https://doi.org/10.1016/j.immuni.2012.10.014

Kurt-Jones EA, Chan M, Zhou S, Wang J, Reed G, Bronson R, Arnold MM, Knipe DM, Finberg RW (2004) Herpes simplex virus 1 interaction with Toll-like receptor 2 contributes to lethal encephalitis. Proc Natl Acad Sci U S A 101(5):1315–1320. https://doi.org/10.1073/pnas.0308057100

Lakeman FD, Whitley RJ (1995) Diagnosis of herpes simplex encephalitis: application of polymerase chain reaction to cerebrospinal fluid from brain-biopsied patients and correlation with disease. National Institute of Allergy and Infectious Diseases Collaborative Antiviral Study Group. J Infect Dis 171(4):857–863

Levitz RE (1998) Herpes simplex encephalitis: a review. Heart Lung 27(3):209–212

Lund J, Sato A, Akira S, Medzhitov R, Iwasaki A (2003) Toll-like receptor 9-mediated recognition of Herpes simplex virus-2 by plasmacytoid dendritic cells. J Exp Med 198(3):513–520. https://doi.org/10.1084/jem.20030162

Manco-Johnson MJ, Nuss R, Key N, Moertel C, Jacobson L, Meech S, Weinberg A, Lefkowitz J (1996) Lupus anticoagulant and protein S deficiency in children with postvaricella purpura fulminans or thrombosis. J Pediatr 128(3):319–323

Martinez-Torres F, Menon S, Pritsch M, Victor N, Jenetzky E, Jensen K, Schielke E, Schmutzhard E, de Gans J, Chung CH, Luntz S, Hacke W, Meyding-Lamade U, Investigators G (2008) Protocol for German trial of Acyclovir and corticosteroids in Herpes-simplex-virus-encephalitis (GACHE): a multicenter, multinational, randomized, double-blind, placebo-controlled German, Austrian and Dutch trial [ISRCTN45122933]. BMC Neurol 8:40. https://doi.org/10.1186/1471-2377-8-40

McCabe K, Tyler K, Tanabe J (2003) Diffusion-weighted MRI abnormalities as a clue to the diagnosis of herpes simplex encephalitis. Neurology 61(7):1015–1016

Nagel MA, Cohrs RJ, Mahalingam R, Wellish MC, Forghani B, Schiller A, Safdieh JE, Kamenkovich E, Ostrow LW, Levy M, Greenberg B, Russman AN, Katzan I, Gardner CJ, Hausler M, Nau R, Saraya T, Wada H, Goto H, de Martino M, Ueno M, Brown WD, Terborg C, Gilden DH (2008) The varicella zoster virus vasculopathies: clinical, CSF, imaging, and virologic features. Neurology 70(11):853–860. https://doi.org/10.1212/01.wnl.0000304747.38502.e8

Nahmias AJ, Whitley RJ, Visintine AN, Takei Y, Alford CA Jr (1982) Herpes simplex virus encephalitis: laboratory evaluations and their diagnostic significance. J Infect Dis 145(6):829–836

Pahud BA, Glaser CA, Dekker CL, Arvin AM, Schmid DS (2011) Varicella zoster disease of the central nervous system: epidemiological, clinical, and laboratory features 10 years after the introduction of the varicella vaccine. J Infect Dis 203(3):316–323. https://doi.org/10.1093/infdis/jiq066

Perez de Diego R, Sancho-Shimizu V, Lorenzo L, Puel A, Plancoulaine S, Picard C, Herman M, Cardon A, Durandy A, Bustamante J, Vallabhapurapu S, Bravo J, Warnatz K, Chaix Y, Cascarrigny F, Lebon P, Rozenberg F, Karin M, Tardieu M, Al-Muhsen S, Jouanguy E, Zhang SY, Abel L, Casanova JL (2010) Human TRAF3 adaptor molecule deficiency leads to impaired Toll-like receptor 3 response and susceptibility to herpes simplex encephalitis. Immunity 33(3):400–411. https://doi.org/10.1016/j.immuni.2010.08.014

Persson A, Bergstrom T, Lindh M, Namvar L, Studahl M (2009) Varicella-zoster virus CNS disease—Viral load, clinical manifestations and sequels. J Clin Virol 46(3):249–253. https://doi.org/10.1016/j.jcv.2009.07.014

Pinninti SG, Kimberlin DW (2014) Management of neonatal herpes simplex virus infection and exposure. Arch Dis Child Fetal Neonatal Ed 99(3):F240–F244. https://doi.org/10.1136/archdischild-2013-303762

Piret J, Boivin G (2015) Innate immune response during herpes simplex virus encephalitis and development of immunomodulatory strategies. Rev Med Virol 25(5):300–319. https://doi.org/10.1002/rmv.1848

Poissy J, Champenois K, Dewilde A, Melliez H, Georges H, Senneville E, Yazdanpanah Y (2012) Impact of Herpes simplex virus load and red blood cells in cerebrospinal fluid upon herpes simplex meningo-encephalitis outcome. BMC Infect Dis 12:356. https://doi.org/10.1186/1471-2334-12-356

Pollak L, Dovrat S, Book M, Mendelson E, Weinberger M (2012) Varicella zoster vs. herpes simplex meningoencephalitis in the PCR era. A single center study. J Neurol Sci 314(1–2):29–36. https://doi.org/10.1016/j.jns.2011.11.004

Pouplin T, Pouplin JN, Van Toi P, Lindegardh N, Rogier van Doorn H, Hien TT, Farrar J, Torok ME, Chau TT (2011) Valacyclovir for herpes simplex encephalitis. Antimicrob Agents Chemother 55(7):3624–3626. https://doi.org/10.1128/AAC.01023-10

Pruss H, Finke C, Holtje M, Hofmann J, Klingbeil C, Probst C, Borowski K, Ahnert-Hilger G, Harms L, Schwab JM, Ploner CJ, Komorowski L, Stoecker W, Dalmau J, Wandinger KP (2012) N-methyl-D-aspartate receptor antibodies in herpes simplex encephalitis. Ann Neurol 72(6):902–911. https://doi.org/10.1002/ana.23689

Ramos-Estebanez C, Lizarraga KJ, Merenda A (2014) A systematic review on the role of adjunctive corticosteroids in herpes simplex virus encephalitis: is timing critical for safety and efficacy? Antivir Ther 19(2):133–139. https://doi.org/10.3851/IMP2683

Raschilas F, Wolff M, Delatour F, Chaffaut C, De Broucker T, Chevret S, Lebon P, Canton P, Rozenberg F (2002) Outcome of and prognostic factors for herpes simplex encephalitis in adult patients: results of a multicenter study. Clin Infect Dis 35(3):254–260. https://doi.org/10.1086/341405

Sancho-Shimizu V, Perez de Diego R, Lorenzo L, Halwani R, Alangari A, Israelsson E, Fabrega S, Cardon A, Maluenda J, Tatematsu M, Mahvelati F, Herman M, Ciancanelli M, Guo Y, AlSum

Z, Alkhamis N, Al-Makadma AS, Ghadiri A, Boucherit S, Plancoulaine S, Picard C, Rozenberg F, Tardieu M, Lebon P, Jouanguy E, Rezaei N, Seya T, Matsumoto M, Chaussabel D, Puel A, Zhang SY, Abel L, Al-Muhsen S, Casanova JL (2011) Herpes simplex encephalitis in children with autosomal recessive and dominant TRIF deficiency. J Clin Invest 121(12):4889–4902. https://doi.org/10.1172/JCI59259

Schleede L, Bueter W, Baumgartner-Sigl S, Opladen T, Weigt-Usinger K, Stephan S, Smitka M, Leiz S, Kaiser O, Kraus V, van Baalen A, Skopnik H, Hartmann H, Rostasy K, Lucke T, Schara U, Hausler M (2013) Pediatric herpes simplex virus encephalitis: a retrospective multicenter experience. J Child Neurol 28(3):321–331. https://doi.org/10.1177/0883073812471428

Schloss L, Falk KI, Skoog E, Brytting M, Linde A, Aurelius E (2009) Monitoring of herpes simplex virus DNA types 1 and 2 viral load in cerebrospinal fluid by real-time PCR in patients with herpes simplex encephalitis. J Med Virol 81(8):1432–1437. https://doi.org/10.1002/jmv.21563

Skoldenberg B, Forsgren M, Alestig K, Bergstrom T, Burman L, Dahlqvist E, Forkman A, Fryden A, Lovgren K, Norlin K et al (1984) Acyclovir versus vidarabine in herpes simplex encephalitis. Randomised multicentre study in consecutive Swedish patients. Lancet 2(8405):707–711

Solomon T, Michael BD, Smith PE, Sanderson F, Davies NW, Hart IJ, Holland M, Easton A, Buckley C, Kneen R, Beeching NJ, National Encephalitis Guidelines D, Stakeholder G (2012) Management of suspected viral encephalitis in adults—Association of British Neurologists and British Infection Association National Guidelines. J Infect 64(4):347–373. https://doi.org/10.1016/j.jinf.2011.11.014

VanLandingham KE, Marsteller HB, Ross GW, Hayden FG (1988) Relapse of herpes simplex encephalitis after conventional acyclovir therapy. JAMA 259(7):1051–1053

Vossough A, Zimmerman RA, Bilaniuk LT, Schwartz EM (2008) Imaging findings of neonatal herpes simplex virus type 2 encephalitis. Neuroradiology 50(4):355–366. https://doi.org/10.1007/s00234-007-0349-3

Westman G, Studahl M, Ahlm C, Eriksson BM, Persson B, Ronnelid J, Schliamser S, Aurelius E (2016) N-methyl-d-aspartate receptor autoimmunity affects cognitive performance in herpes simplex encephalitis. Clin Microbiol Infect 22(11):934–940. https://doi.org/10.1016/j.cmi.2016.07.028

Whitley R, Arvin A, Prober C, Burchett S, Corey L, Powell D, Plotkin S, Starr S, Alford C, Connor J et al (1991) A controlled trial comparing vidarabine with acyclovir in neonatal herpes simplex virus infection. Infectious Diseases Collaborative Antiviral Study Group. N Engl J Med 324(7):444–449. https://doi.org/10.1056/NEJM199102143240703

Whitley R, Kimberlin DW, Prober CG (2007) Pathogenesis and disease. In: Arvin A, Campadelli-Fiume G, Mocarski E et al (eds) Human herpesviruses: biology, therapy, and immunoprophylaxis. Cambridge University Press, Cambridge

Whitley RJ (1990) Viral encephalitis. N Engl J Med 323(4):242–250. https://doi.org/10.1056/NEJM199007263230406

Whitley RJ (2006) Herpes simplex encephalitis: adolescents and adults. Antivir Res 71(2–3):141–148. https://doi.org/10.1016/j.antiviral.2006.04.002

Whitley RJ, Alford CA, Hirsch MS, Schooley RT, Luby JP, Aoki FY, Hanley D, Nahmias AJ, Soong SJ (1986) Vidarabine versus acyclovir therapy in herpes simplex encephalitis. N Engl J Med 314(3):144–149. https://doi.org/10.1056/NEJM198601163140303

Whitley RJ, Nahmias AJ, Soong SJ, Galasso GG, Fleming CL, Alford CA (1980) Vidarabine therapy of neonatal herpes simplex virus infection. Pediatrics 66(4):495–501

Yamada S, Kameyama T, Nagaya S, Hashizume Y, Yoshida M (2003) Relapsing herpes simplex encephalitis: pathological confirmation of viral reactivation. J Neurol Neurosurg Psychiatry 74(2):262–264

Zhang SY, Jouanguy E, Ugolini S, Smahi A, Elain G, Romero P, Segal D, Sancho-Shimizu V, Lorenzo L, Puel A, Picard C, Chapgier A, Plancoulaine S, Titeux M, Cognet C, von Bernuth H, Ku CL, Casrouge A, Zhang XX, Barreiro L, Leonard J, Hamilton C, Lebon P, Heron B, Vallee L, Quintana-Murci L, Hovnanian A, Rozenberg F, Vivier E, Geissmann F, Tardieu M, Abel L, Casanova JL (2007) TLR3 deficiency in patients with herpes simplex encephalitis. Science 317(5844):1522–1527. https://doi.org/10.1126/science.1139522

Chapter 6
Antiviral Drugs Against Alphaherpesvirus

Kimiyasu Shiraki

Abstract The discovery of acyclovir and penciclovir has led to the development of a successful systemic therapy for treating herpes simplex virus infection and varicella-zoster virus infection, and the orally available prodrugs, valacyclovir and famciclovir, have improved antiviral treatment compliance. Acyclovir and penciclovir are phosphorylated by viral thymidine kinase and are incorporated into the DNA chain by viral DNA polymerase, resulting in chain termination. Helicase-primase plays an initial step in DNA synthesis to separate the double strand into two single strands (replication fork) and is a new target of antiviral therapy. The helicase-primase inhibitors (HPIs) pritelivir and amenamevir have novel mechanisms of action, drug resistance properties, pharmacokinetic characteristics, and clinical efficacy for treating genital herpes. The clinical study of amenamevir in herpes zoster has been completed, and amenamevir has been submitted for approval for treating herpes zoster in Japan. The clinical use of HPIs will be the beginning of a new era of anti-herpes therapy.

Keywords Acyclovir · Prodrug · Valacyclovir · Famciclovir · Antivirals · Helicase-primase · Amenamevir · Chain termination · Resistance

6.1 Introduction

Dr. Elion, a Nobel laureate, has pioneered an anti-herpetic drug, acyclovir (ACV), capable for systemic administration in herpes simplex virus (HSV) and varicella-zoster (VZV) infection with a wide safety margin and a very high therapeutic index (Elion 1989; Elion et al. 1977), leading to the development and situation of current anti-herpes medicine treatment. Various anti-influenza virus drugs have been developed such as neuraminidase inhibitors (Von Itzstein et al. 1993), RNA polymerase inhibitor (favipiravir) (Furuta et al. 2002), a proton pump inhibitor (amantadine)

K. Shiraki (✉)
Department of Virology, University of Toyama, Toyama, Japan
e-mail: kshiraki@med.u-toyama.ac.jp

(Jackson et al. 1963), and cap-dependent endonuclease inhibitor (S-033188) (Koszalka et al. 2017). By contrast, for the last 40 years, antiherpetic drugs have been limited to inhibitors of viral DNA synthesis because ACV and penciclovir (PCV) and prodrugs, valacyclovir and famciclovir, are satisfactory for the treatment and prevention of apparent HSV and VZV diseases. In addition to current antiherpetic therapy inhibiting DNA synthesis through DNA polymerase (DNApol), it has taken time to develop new antiherpetic drugs with different mechanism of action, and novel helicase-primase (HP) inhibitors (HPIs) of HSV and VZV have been developed and will expand new anti-herpes drug therapy.

DNA synthesis inhibitors affecting DNApol are categorized into five groups by their mechanism of action as shown in Fig. 6.1. The first group inhibits DNA polymerization by blocking the incorporation of normal deoxyribonucleoside triphosphates (dNTPs), but the inhibitor itself is not incorporated into DNA. Foscarnet, vidarabine, and sorivudine belong to this group.

The second group functions through chain termination by inhibiting chain elongation that stops at the incorporated site due to the lack of the 3′OH group in the deoxyribose part for the binding to the next base. ACV is phosphorylated by viral thymidine kinase (TK), and its triphosphate form is incorporated into the DNA chain. Next, the chain can no longer incorporate dNTPs, resulting in the inhibition of DNA synthesis because ACV lacking 3′OH cannot form the phosphodiester bonding with the 5′OH- of the next dNTP. Valacyclovir is a prodrug of ACV with improved oral absorption and compliance.

In the third group, the incorporation of the agents into the DNA chain does not stop at the incorporated site but terminates several bases ahead of the newly incorporated DNA chain. PCV, its prodrug famciclovir, and ganciclovir (GCV) possess 3′OH- groups for the phosphodiester bonding with the 5′OH- of normal dNTPs and stop after chain elongation of several bases possibly due to the unstable structure.

Brivudin, idoxuridine, and ribavirin are incorporated into the RNA or DNA strand, respectively, and the complementary strand synthesis to this incorporated strand generates many mismatches, resulting in the production of nonfunctional proteins leading to replication-incompetent virus production (lethal mutagenesis).

Recent advancement in antiherpetic drugs has resulted in the development of novel viral HPIs, and clinical studies on genital herpes using two HPIs, ASP2151 (amenamevir) and BAY 57-1293 (pritelivir), have been successfully conducted (Tyring et al. 2012; Wald et al. 2014, 2016). HPIs inhibit the initial stage of DNA replication before DNApol functions and have no effect on viral DNApol activity. Although the modes of DNA synthesis inhibition are different and specific to each antiviral agent, they exhibit similar efficacy by inhibiting viral DNA synthesis in vitro and in vivo. One of helicase-primase inhibitors, amenamevir, has recently completed its clinical trial on herpes zoster using once-daily dose and been approved and used for the treatment of herpes zoster in Japan. Thus, the clinical application of HPIs suggests the possibility that they may assume an important position similar to ACV in HSV and VZV therapy.

1. Native nucleosides

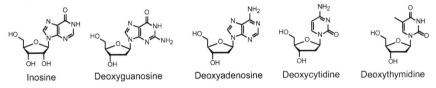

Inosine   Deoxyguanosine   Deoxyadenosine   Deoxycytidine   Deoxythymidine

2. DNA polymerase (DPase) inhibitors without incorporation

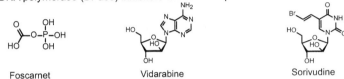

Foscarnet   Vidarabine   Sorivudine

3. Chain terminators at the incorporation site

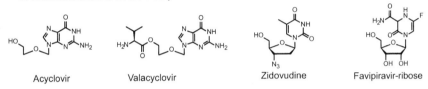

Acyclovir   Valacyclovir   Zidovudine   Favipiravir-ribose

4. Chain terminators after incorporation and elongation of several bases

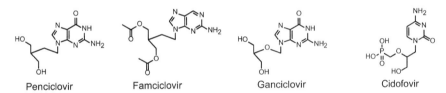

Penciclovir   Famciclovir   Ganciclovir   Cidofovir

5. Incorporation into viral DNA with replication-incompetent virus production

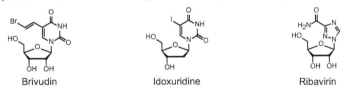

Brivudin   Idoxuridine   Ribavirin

Fig. 6.1 Antiviral compounds acting on viral DNA polymerase and their categories. (*1*) Native nucleosides are inosine, deoxyadenosine, deoxyguanosine, deoxycytidine, and deoxythymidine. (*2*) DNA polymerase (DNApol) inhibitors without incorporation are foscarnet, vidarabine, and sorivudine. (*3*) Chain terminators at the incorporation site are acyclovir and valacyclovir; an anti-human immunodeficiency virus drug, zidovudine; and an anti-influenza drug, favipiravir. (*4*) Chain terminators after incorporation and elongation of several bases are penciclovir, famciclovir, ganciclovir, and cidofovir. (*5*) Drugs that incorporate into viral DNA with replication-incompetent virus production are brivudin, idoxuridine, and ribavirin

6.2 Deoxyribonucleotide (dNTP) Synthesis in HSV- and VZV-Infected Cells

Purine is synthesized from amino acid to inosine monophosphate (IMP) and then to adenosine monophosphate (AMP) and guanosine monophosphate (GMP), and cytosine and uridine monophosphates are synthesized from amino acids as shown in Fig. 6.2. Ribonucleotide monophosphates (rNMPs) synthesized de novo are the ribose form (RNA type), and subsequently their triphosphate forms are the substrates for RNA. On the other hand, as the substrates for DNA, ribose forms of ribonucleotide diphosphate (rNDP) should be converted to the deoxyribose forms of dNDPs by ribonucleotide reductase (RR). The diphosphate form of dNTPs, except for that of uridine, is processed to the triphosphate form (dNTP) as the substrates of DNA. Uridine is not a substrate of DNA, and its diphosphate form, dUDP, becomes the monophosphate form dUMP. dUMP is a substrate for thymidylate synthase (TS) and is converted into thymidine monophosphate (dTMP) successively to thymidine triphosphate (dTTP) for DNA. TS is a key enzyme for TMP synthesis in a de novo pathway and requires folic acid as the coenzyme for C1 unit (methyl residue) transfer to produce TMP. Therefore, TS is a target anticancer drug of the chemotherapy agent 5-fluorouracil (5-FU); an immunosuppressant, methotrexate; and the anti-VZV agent sorivudine monophosphate. Thus, blocking the TS pathway results in a severe outcome of cancer cells or immunoregulatory cells depending on the importance of this pathway for DNA synthesis. TK is a key enzyme in the salvage pathway, which recycles nucleosides, to supply TMP, independent of the pathway from dUMP to TMP through TS. DNA polymerase (DNApol) is the final step for DNA synthesis using synthesized dNTPs through RR, TS, and TK.

Thus, RR, TS, and TK are important enzymes for dNTP synthesis and their supply to herpes virus DNA synthesis. HSV encodes RR and TK, while VZV encodes RR, TS, and TK. HSV and VZV can replicate in cells that do not synthesize DNA at the time of infection, but these cells synthesize proteins through mRNA synthesis. Thus, most of the cells synthesize RNA without DNA synthesis; therefore, rNTPs are abundant in the cells. HSV and VZV infection induce viral RR in these cells, and rNDPs are converted to dNDPs by viral RR. In addition to viral TK, dNMPs, including TMP, are supplied for viral DNA synthesis via the salvage pathway; thus, cells infected with HSV and VZV are supported for viral DNA synthesis by supplying dNMPs in infected cells through viral RR from rNDP and TK from thymidine. VZV encodes TS in addition to RR and TK, thus supporting viral DNA synthesis. Thus, viral RR, TK, and TS play important roles for the replication of HSV and VZV in infected cells. Cytomegalovirus (CMV) encodes UL97 that phosphorylates ganciclovir instead of TK (Littler et al. 1992; Sullivan et al. 1992), but RR (large subunit) lacks many catalytic residues (Patrone et al. 2003). Thus, CMV has a different dNTP supply pathway for DNA synthesis with HSV and VZV.

On the other hand, host cells possess RR, TK, and TS for their replication, and they are dependent on the cell cycle or cellular activity. Cellular TK phosphorylates thymidine, while cellular TK does not phosphorylate ACV and PCV; therefore,

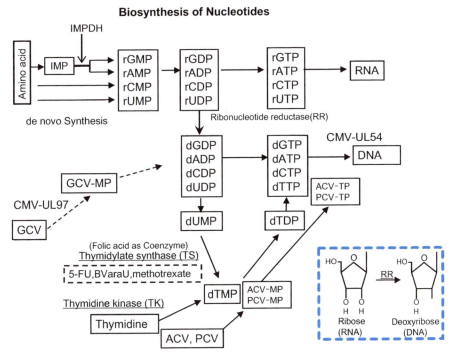

Fig. 6.2 Biosynthesis of nucleotides. Purine and pyrimidine are synthesized de novo from amino acids as ribose form nucleotides and inosine monophosphate (IMP) that are modified by IMP dehydrogenase to adenosine monophosphate (rAMP) and guanosine monophosphate (rGMP). Next, nucleotide monophosphate (rNMP) is phosphorylated to triphosphate forms (rNTP), and these become the substrate for RNA. The ribose form of nucleotide diphosphate (rNDP) is converted to the 2′-deoxyribose form (dNDP) by cellular or viral ribonucleotide reductase (RR) as shown in the lower box. When viral RR is induced by HSV and VZV infection, dNDPs are synthesized in the early phase of infection and are supplied for viral DNA synthesis to facilitate and activate viral DNA synthesis even in cells that do not actively synthesize cellular DNA. Thymidine is an important substrate of DNA and is supplied in two ways—from uridine monophosphate (UMP) to thymidine monophosphate (TMP) by thymidylate synthase (TS) (de novo pathway) and from the systemic circulation by thymidine kinase (TK) (salvage pathway). The important role of TS in thymidine biosynthesis can be easily understood by blocking this pathway with the anticancer drug 5-fluorouracil (5-FU) and immunosuppressant methotrexate. Sorivudine (BVaraU) is an anti-VZV agent, and its monophosphate form inhibits TS. Because sorivudine itself is phosphorylated and its monophosphate blocks TMP formation by the inhibition of TS activity, sorivudine shows potent anti-VZV action at the low concentration by the reduction of the competing TMP supply on viral DNA polymerase (DNApol). Acyclovir (ACV) and penciclovir (PCV) are phosphorylated by viral TK and are further phosphorylated to the triphosphate form by cellular enzymes. ACV-TP and PCV-TP are incorporated into viral DNA by viral DNApol, resulting in chain termination. Concerning the anti-CMV drug ganciclovir, it is phosphorylated by CMV-UL97 and incorporated into viral DNA by CMV-UL54 DNA polymerase

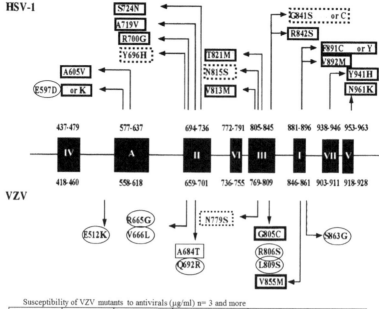

Fig. 6.3 Viral DNA polymerase mutations of ACV-resistant mutants in the HSV-1 and VZV DNA polymerase gene (Kamiyama et al. 2001) Filled boxes show the conserved regions I–VII of the HSV-1 DNA polymerase gene. The reported mutation sites of HSV and VZV DNA polymerase mutants are summarized, and these sites are substrate recognition sites for ACV, foscarnet, vidarabine, and aphidicolin

The lower table shows the susceptibility of VZV V855M, G805C, and N779S mutants to ACV, foscarnet/phosphonoacetic acid, vidarabine, and aphidicolin. These three mutants indicate the recognition sites of antiviral drugs between the foscarnet-arabinose moiety of the vidarabine group and aphidicolin group

▭ indicates ACV and foscarnet/phosphonoacetic acid-resistant mutants

┈┈ indicates ACV-resistant but foscarnet/phosphonoacetic acid-hypersensitive mutants

◯ indicates foscarnet/phosphonoacetic acid-resistant mutants

immunocompetent patients (Daikoku et al. 2016; Englund et al. 1990; Honda et al. 2001; Okuda et al. 2004; Reyes et al. 2003; Stranska et al. 2005). ACV treatment of infected cells increased the frequency of guanosine homopolymeric (G-string)-string mutation in the TK gene of HSV and VZV (Daikoku et al. 2016; Ida et al. 1999; Sasadeusz et al. 1997), while PCV treatment induced TK mutation quite rarer in VZV-infected cells than ACV treatment (Ida et al. 1999). This contrasting action between ACV and PCV is due to the mode of chain termination and the proofreading activity of herpesvirus DNApol. Herpesvirus DNApol has alkaline deoxyribonuclease (DNase) activity, which functions as the proofreading activity of DNApol.

RNA-dependent RNA polymerase or Taq DNA polymerase does not demonstrate proofreading activity. The presence of proofreading activity results in higher fidelity (1 in 10^6) of DNA polymerase than the lower fidelity (1 in 10^4) of RNA polymerase of RNA viruses or Taq DNA polymerase (Drake 1993). Thus, RNA viruses generate mutations more frequently than herpesviruses, and the proofreading activity is important in maintaining the fidelity of the genome during herpesvirus replication. Incorporation of ACV and proofreading activity induce mutation in the G-string parts as follows. When ACV is incorporated into the DNA chain at the terminus, the misincorporated ACV is removed by proofreading DNase and replaced with dGTP. These frequent correction cycles of the incorporation and removal of ACV are repeated at the G-string parts, and these G-string parts become the hot spots of mutation by the misincorporation of ACV, resulting in the deletion, addition, or substitution of nucleotides in the G-string parts of the TK gene. This type of mutation may occur in the whole genome, but some in the essential genes become fatal to the virus by the loss of function. The TK gene is nonessential and a target of ACV resistance, and the TK mutants are selected in the presence of ACV and are visualized. Although subclinical, this process was detected in the clinical isolates from patients with genital herpes treated with ACV. ACV treatment induces G-string mutations in the virus population in the genital lesions, and these mutants become latent, reactivate, and appear in the genital lesions in the patients (Daikoku et al. 2016). There is no problem as current ACV therapy, but such a change is subclinically occurring.

While PCV is incorporated into DNA but allows to the elongation of several bases, proofreading does not occur, and, subsequently, the G-string parts are not the hot spots of mutation, resulting in a quite lower mutation rate than that of ACV. Thus, mutants isolated in PCV treatment are rare.

6.6 Sorivudine

Sorivudine is phosphorylated to the diphosphate form by TK of HSV-1 and VZV, and the inhibition of TS activity by sorivudine monophosphate caused VZV to be quite susceptible to sorivudine. The TMP supply from UMP is blocked through inhibiting TS, and this increases the ratio of antiviral sorivudine monophosphate per TMP in VZV-infected cells (Cohen and Seidel 1993; De Clercq 2005; Kawai et al. 1993; Machida et al. 1982; Yokota et al. 1989). The IC_{50} of sorivudine is extremely low at 0.0035 μM, and sorivudine showed potent anti-VZV activity and better efficacy than acyclovir. Sorivudine 40 mg/day showed significantly more efficacy than ACV 4 g/day in the treatment of herpes zoster in HIV-infected adults (Bodsworth et al. 1997). Sorivudine was licensed for herpes zoster in Japan in 1993.

Bromovinyluracil of the sorivudine metabolite irreversibly binds and inhibits dihydropyrimidine dehydrogenase activity, and this enzyme is important as the degrading enzyme of 5-fluorouracil (5FU), an anticancer drug. When sorivudine was used in patients with cancer treated with 5-FU, the 5-FU concentration in the blood was increased by interfering with the 5-FU catabolism by inhibition of the 5-FU-degrading enzyme by bromovinyluracil. The increased 5-FU caused severe hematopoietic toxicity of 5-FU in the reduction of leukocytes and platelets causing 15 patient deaths with 5-FU and sorivudine treatment. If not combined with 5-FU, sorivudine is an excellent anti-VZV drug.

6.7 Brivudin

(E)-5-(2-Bromovinyl)-2′-deoxyuridine (BVDU/brivudin) is phosphorylated by viral TK, inhibits viral DNA synthesis, and shows strong activity toward VZV at lower concentrations than ACV (De Clercq 2004; De Clercq et al. 1982). A double-blind survey study was conducted on 608 herpes zoster patients treated with 1×125 mg oral brivudin ($n = 309$) or 5×800 mg ACV ($n = 299$), both for 7 days, during two prospective, randomized clinical herpes zoster trials. The survey was aimed to evaluate the outcome of the two treatment regimens in postherpetic neuralgia (PHN). The incidence of PHN, defined as zoster-associated pain occurring or persisting after rash healing, was significantly lower in brivudin recipients (32.7%) than in ACV recipients (43.5%, $P = 0.006$) (Wassilew et al. 2003). Brivudin is used for the treatment of herpes zoster in adult patients and may reduce the incidence of PHN. The use of brivudin with 5-FU requires caution, because like sorivudine, it can enhance the hematopoietic toxicity of 5-FU.

6.8 DNA Polymerase Inhibitors

The viral DNA polymerase inhibitors foscarnet and rarely vidarabine in intravenous preparations are used for drug-resistant HSV and VZV infection. Current treatment with ACV or PCV is satisfactory in HSV and VZV infection because immunocompromised patients, especially those with human immunodeficiency virus infection, are well controlled, and immunocompromised patients who need prolonged antiviral treatment are limited. Foscarnet and vidarabine do not require phosphorylation for their antiviral action and are used for TK-deficient HSV and VZV. Vidarabine inhibits viral DNA synthesis at concentrations below those required to inhibit host cell DNA synthesis (Shipman Jr et al. 1976) and may have multiple sites of action within an infected cell (Kamiyama et al. 2001; Suzuki et al. 2006). It is phosphorylated to its active triphosphate form by cellular kinases (Schwartz et al. 1984). Thus, vidarabine can inhibit TK-deficient mutants of HSV and VZV that are resistant to ACV, and the active site of vidarabine on DNApol is different from that of ACV but similar to that of foscarnet (Kamiyama et al. 2001; Larder and Darby 1986; Miwa et al. 2005; Shiraki et al. 1990). Vidarabine is less efficient with more adverse events than ACV (Whitley et al. 1986). Therefore, the clinical use of vidarabine is limited.

Foscarnet is a pyrophosphate analog that is released from dNTPs on DNApol; thus, foscarnet directly acts on DNApol (Kern et al. 1981; Ostrander and Cheng 1980). The inhibitory action of foscarnet depends on the ratio of the numbers of DNApol and foscarnet molecules; therefore, foscarnet is not influenced by the supply of deoxyribonucleotides, such as ACV or PCV (Yajima et al. 2017). Foscarnet is available as an intravenous preparation and requires attention regarding the electrolyte balance in the blood and renal function after its administration. Foscarnet is used for ACV- or ganciclovir-resistant virus.

The mutation sites of HSV and VZV DNApol are shown in Fig. 6.3. There are two recognition groups of ACV-resistant HSV and VZV in their DNApols: the foscarnet-arabinose moiety of the vidarabine group versus the aphidicolin group as described in the section of ACV-/PCV-resistant mutants. ACV-resistant mutants with foscarnet-vidarabine resistance (VZV G805C, V855M) are more sensitive to aphidicolin than the wild-type parent virus, and those with foscarnet-vidarabine hypersensitivity are more resistant to aphidicolin. ACV-resistant mutants with foscarnet-arabinose (vidarabine) hypersensitivity (VZV N779S) are more resistant to ACV than those with foscarnet-arabinose (vidarabine) resistance (Kamiyama et al. 2001).

Cidofovir has a phosphorylated form and does not require initial phosphorylation by TK or CMV-UL97 (Cundy 1999; Safrin et al. 1997). Cidofovir is an injectable antiviral medication used for the treatment of CMV retinitis in individuals with AIDS. Cidofovir-diphosphate inhibits viral DNApol of herpesviruses, orthopoxviruses, adenoviruses, polyomaviruses, and papillomaviruses (Beadle et al. 2002). A prodrug form of cidofovir, brincidofovir (CMX001), is orally available, and clinical studies are under way including Ebolavirus infection (Dunning et al. 2016).

6.9 Helicase-Primase in DNA Synthesis

Double-stranded DNA needs to become separated into two single strands (replication fork) before DNA synthesis, and their complementary strands are synthesized from each DNA strand to make two new double-stranded DNA molecules in the process of DNA replication (Fig. 6.4). Helicase-primase is responsible for both unwinding viral DNA at the replication fork, separating double-stranded DNA into two single strands, and synthesizing RNA primers (Okazaki fragment) in the lagging strand for DNA synthesis. DNApol starts complementary DNA synthesis from these two strands. This HP enzyme complex consists of three proteins—a helicase, a primase, and cofactor subunits—which are well conserved among *Herpesviridae* viruses and are called UL5 (helicase, VZVORF55), UL52 (primase, VZVORF6), and UL8 (cofactor, VZVORF52), respectively. UL5 unwinds duplex DNA ahead of the fork and separates the double strand into two single strands. UL52 lays down RNA primers that the two-subunit DNA polymerase (UL30/UL42) extends. The helicase-primase complex possesses multi-enzymatic activities, including DNA-dependent ATPase and helicase localized in the helicase subunit and primase in the

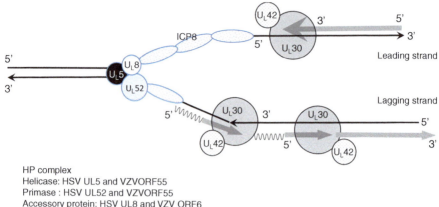

Fig. 6.4 Mechanism of DNA synthesis and viral helicase-primase (HP) complex (modified from Boehmer and Lehman 1997). Figure shows the role of the helicase-primase complex (UL5, UL8, UL52 of HSV and ORF55, ORF6, ORF52 of VZV), DNA polymerase complex (UL42, UL30 of HSVDNA polymerase), and ICP8 single-stranded DNA-binding protein of HSV. HSV UL5 and VZVORF55 (helicase) unwind double-stranded DNA and separate double strands into two single strands, making the replication fork. HSV UL52 and VZVORF55 (primase) synthesize RNA primers (Okazaki fragments) for DNA synthesis. DNA polymerase and its accessory protein (UL42) bind to each single strand and synthesize complementary DNA to each strand. The single-stranded DNA-binding protein, ICP8, binds to single-stranded template DNA. The arrows indicate the direction of movement of the DNA replication proteins (Shiraki 2017)

primase subunit; all of these enzymatic activities are needed for the helicase-primase complex to function in viral DNA replication. HP is quite different from topoisomerases to wind or relax double-stranded DNA. Therefore, HP is an important enzyme for DNA synthesis and a conserved enzyme from *E. coli* to *Homo sapiens*, and HSV, VZV, and CMV have their own HP as an essential gene product in their replication.

6.10 Helicase-Primase Inhibitor (HPI)

HPI inhibits the single-stranded, DNA-dependent ATPase, helicase, and primase activities by binding to the helicase-primase complex (Biswas et al. 2014; Chono et al. 2010, 2012; James et al. 2015). There are three classes of herpesvirus HPIs – thiazole urea, BAY 57-1293 (pritelivir) (Kleymann et al. 2002), 2-aminothiazolylphenyl derivatives, BILS 179 BS (Crute et al. 2002; Spector et al. 1998), and oxadiazolylphenyl type ASP2151 (amenamevir) (Fig. 6.5) (Chono et al. 2010). Interestingly, the former two classes of HPIs, BAY 57-1293 and BILS 179 BS,

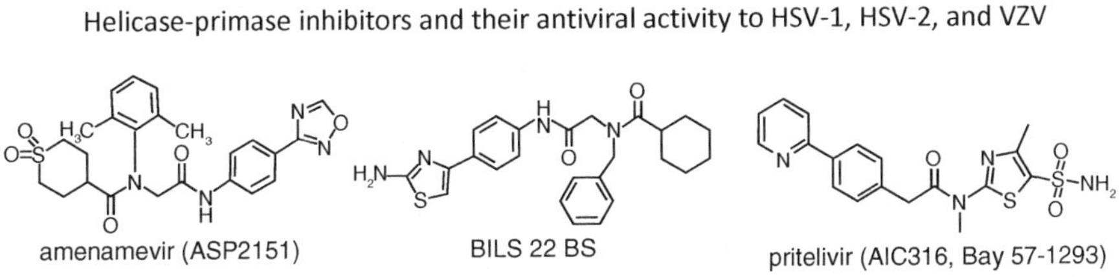

Helicase-primase inhibitors and their antiviral activity to HSV-1, HSV-2, and VZV

amenamevir (ASP2151)   BILS 22 BS   pritelivir (AIC316, Bay 57-1293)

Fig. 6.5 Structure and antiviral activity of helicase-primase inhibitors. The structures of three helicase-primase inhibitors (HPI) and their spectrum of antiviral activity are shown in the figure and table, respectively. Three HPIs of amenamevir—oxadiazolylphenyl type (ASP2151) (Chono et al. 2010); BILS 179 BS, 2-amino-thiazolylphenyl derivatives (Crute et al. 2002; Spector et al. 1998); and pritelivir, thiazole urea (BAY 57-1293) (Kleymann et al. 2002)—were developed and exhibited lower EC_{50} concentrations than those of acyclovir, and amenamevir and pritelivir have been evaluated for their clinical efficacy. Interestingly, amenamevir possesses anti-VZV activity and is based on the anti-VZV activity. A clinical study on amenamevir for herpes zoster has been completed and submitted for approval of the licensure of herpes zoster in Japan

inhibit HSV-1 and HSV-2 but not VZV, and amenamevir possesses antiviral activity not only against HSV-1 and HSV-2 but also against VZV. HPIs require low concentrations to inhibit viral growth of HSV-1 and HSV-2 at the effective concentrations for 50% plaque reduction (EC_{50}s) of 0.014 to 0.060 µM and 0.023 to 0.046 µM, respectively, and the anti-VZV activity of amenamevir (0.038–0.10 µM) is more potent against all strains tested than against ACV (1.3-5.9 µM) (Chono et al. 2010).

These HPIs are virus specific with low cytotoxicity in vitro, are orally available and effective against HSV infection, and are well tolerated in mice. As the target molecules are different from ACV, PCV, foscarnet, and vidarabine, their mechanism of action and antiviral and pharmacokinetic profiles are unique to HPIs. Foscarnet inhibits viral DNA polymerase activity through direct binding to the pyrophosphate binding site (James and Prichard 2014). Both HPIs and foscarnet directly inhibit viral enzymes. On the other hand, ACV is phosphorylated by HSV TK and cellular kinase, which convert it to the active form, ACV-TP, and then incorporate it into the elongating viral DNA strand, resulting in chain termination (James and Prichard 2014). Because ACV-TP competes with dGTP to inhibit viral DNA synthesis, the ratio of ACV-TP to dGTP directly affects the effectiveness of ACV in virus replication. Thus, it is reasonable that amenamevir and foscarnet, which directly inhibit the viral enzyme, suppressed viral replication more effectively than ACV after viral DNA synthesis becomes active (Yajima et al. 2017). ACV-TP competes with dGTP on viral DNApol in infected cells, and anti-HSV activity is attenuated when dGTP supply becomes abundant after infection. The EC_{50}s of ACV to HSV-1 was approximately 1.5 µM up 5 h after infection and increased after 7.5 h to 12.4 µM at 12.5 h after infection as shown in Fig. 6.6a. In contrast, HPIs target the enzyme and not the nucleoside analog, and anti-HSV activity of amenamevir is not influenced regardless of the time course after infection. Thus, the susceptibility to ACV increased to 8 to 11 times that of initial EC_{50} at 12 h after infection when active DNA synthesis is progressing by the supply of dGTP, while the susceptibility of HSV-1 to foscarnet

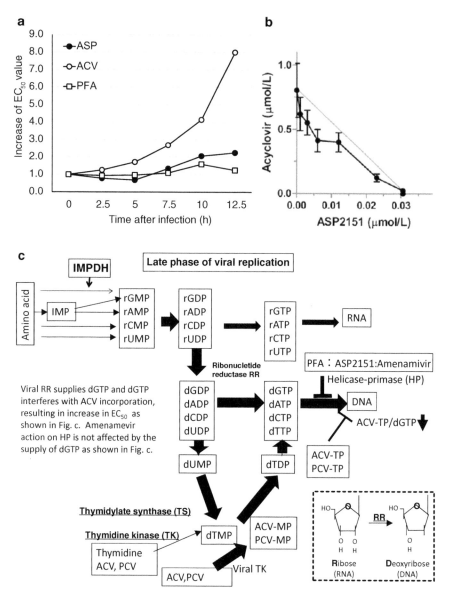

Fig. 6.6 Synergism of amenamevir (ASP2151) with ACV against the VZV Kawaguchi strain analyzed by isobologram (Chono et al. 2013). (**a**) Time course of the susceptibility (EC$_{50}$) changes in infected cells to ACV, ASP2151, and foscarnet (PFA) every 2.5 h after infection (Yajima et al. 2017). The increase in EC$_{50}$ values is expressed as the ratio of those at 0 h. The susceptibility of HSV-1 to ACV increased more than 7.5 h after infection, but ASP2151 and PFA were not influenced. The effects of the increase of dGTP for viral DNA synthesis in the late phase of infected cells reduced the antiviral activity of ACV, but ASP2151 and PFA were not affected by viral DNA synthesis and its related cellular factors. (**b**) The solid straight line (gray) indicates the theoretical additive antiviral activity in combination with ASP2151 and ACV (Chono et al. 2013).

and amenamevir was not influenced by the replication cycle. The antiviral activity that is not affected by the replication cycle is a great advantage of HPIs over current anti-herpetic drugs and allows the once-daily dose for human use.

6.11 HPI-Resistant Viruses

HPI-resistant virus and ACV-resistant virus in the wild-type virus stock were compared in HSV-1 and HSV-2. The ACV-resistant virus was found in one in 10^3 to 10^4 plaque-forming units (PFU) (Chono et al. 2012), and this value was consistent with that in the other reports (Coen et al. 1982; Daikoku et al. 2016; Parris and Harrington 1982). HPI-resistant virus was found in one in 10^6 to 10^7 PFU (Chono et al. 2012); thus, the HPI-resistant virus is rare. Any mutation to lose function in the TK gene leads to the ACV-resistant virus; by contrast, HPI mutants should preserve the function of HP in the restricted amino acid change in UL5 or UL52 (Chono et al. 2010, 2012). Thus, HPI-resistant mutants are quite rarer than ACV-resistant mutants.

The HPI-resistant viruses have been isolated, and their mutation sites have been analyzed. HP is an essential gene product, and these mutations maintain their basic function for replication but avoid the interaction of HPIs by changing the structures of helicase and primase. Sequencing analyses revealed several single-base-pair substitutions resulting in amino acid changes in the helicase and primase of amenamevir-resistant HSV mutants (Chono et al. 2010). Amino acid alterations in the helicase subunit were commonly clustered near helicase motif IV in the UL5 helicase gene of both HSV-1 and HSV-2 among HPIs, while the primase subunit substitution associated with reduced susceptibility was found only in amenamevir-resistant HSV-1 mutants. Interestingly, we found that R367H with S364G substitution in the UL52 primase gene (double mutation) enhanced the resistance to amenamevir compared with S364G substitution alone (single mutation). HPI-resistant HSV mutants show susceptibility to ACV and attenuated growth capability in vitro and pathogenicity than the parent virus in HSV-infected mice (Chono et al. 2012). Mutations in either helicase or primase of HP complex against amenamevir might confer defects in viral replication and pathogenicity.

Fig. 6.6 (continued) Each point (EC_{50}) is shown as the mean ± standard error from four independent experiments. Significant synergism was observed by the combination of ASP2151 and ACV ($P = 0.0005$), and a low concentration of ASP2151 showed strong synergism with ACV (**c**) Metabolic pathway in the stage of viral DNA synthesis when ribonucleotides are efficiently converted to deoxyribonucleotides by viral RR for viral DNA synthesis (Shiraki 2017). dGTP is massively supplied for DNA synthesis at about 60,000 and 90,000 dGTPs per one DNA molecule of VZV and HSV, respectively, and this supply of dGTP reduces the incorporation of ACV-TP into viral DNA with the competition of ACV-TP with the massive supply of dGTP, resulting in attenuation of the inhibition of viral DNA synthesis by ACV. This results in the increased EC_{50} value of ACV as shown in Fig. (**a**). PFA and ASP2151 directly act on DNApol and HP, respectively, and inhibit viral DNA synthesis without any influence by the supply of dGTP. Amenamevir efficiently inhibits viral growth in the early and late phases of infection, in contrast to ACV, as indicated in Fig. (**c**)

Synergism of amenamevir was observed with ACV, PCV, and vidarabine in HSV-2 and VZV. Synergism of amenamevir with ACV was observed at all concentrations in vitro by isobologram analysis, and amenamevir achieved synergism at its low concentrations in HSV-1, HSV-2, and VZV (Fig. 6.6b) (Chono et al. 2013). The combination of amenamevir and valacyclovir in oral administration showed significant synergistic activity in HSV-infected mice. This synergistic activity of amenamevir and ACV or PCV indicates the maximization of anti-herpetic therapy, possible reduction in increased toxicity by increasing the dose of ACV or PCV, and possible reduction in the generation of resistant virus in the prolonged treatment of chronic infection in immunocompromised patients. Combination therapy may be a useful approach to treat herpes infections suspected to be caused by nucleoside analog drug-resistant virus variants and represents more effective therapeutic options than monotherapy, particularly for severe disease conditions, such as herpes encephalitis or patients with immunosuppression.

The pharmacokinetic profile of HPIs suggests the oral 1-day dose can attain the concentration exhibiting antiherpetic activity for the entire day, and this excellent property exceeds that of valacyclovir and famciclovir in maintaining the antiviral level, when they are used in the suppressive therapy of genital herpes. This long-lasting antiviral status indicates that reactivation from the ganglia would be completely inhibited with subsequent viral shedding, leading to the sexual transmission of HSV. Thus, HPIs might stop genital lesions and viral shedding in healthy persons with genital herpes and subsequent sexual transmission of HSV; additionally, HPIs would have favorable characteristics as antiherpetic drugs in suppressive therapy.

Due to promising preclinical profiles on antiviral activity, safety, tolerability, and pharmacokinetics, HPIs, pritelivir and amenamevir, were selected as development candidates, and their clinical efficacies have been evaluated in two phase-2 clinical studies for patients with genital herpes (Tyring et al. 2012; Wald et al. 2014, 2016). The clinical study of pritelivir on the viral shedding of genital herpes comparing daily oral doses of 100 mg of pritelivir with 500 mg of valacyclovir showed better efficacy on genital lesions and viral shedding than valacyclovir. Genital lesions were present on 1.9% of days in the pritelivir group vs 3.9% in the valacyclovir group (RR, 0.40; 95% CI, 0.17-0.96; $P = 0.04$). The frequency of shedding episodes did not differ by group, with 1.3 per person-month for pritelivir and 1.6 per person-month for valacyclovir (RR, 0.80; 95% CI, 0.52 to 1.22; $P = 0.29$) (Wald et al. 2016). HSV shedding among placebo recipients was detected on 16.6% of days; shedding among pritelivir recipients was detected on 18.2% of days among those receiving 5 mg daily, 9.3% of days among those receiving 25 mg daily, 2.1% of days among those receiving 75 mg daily, and 5.3% of days among those receiving 400 mg weekly. The percentage of days with genital lesions was also significantly reduced, from 9.0% in the placebo group to 1.2% in both the group receiving 75 mg of pritelivir daily (relative risk, 0.13; 95% CI, 0.02 to 0.70) and group receiving 400 mg weekly (relative risk, 0.13; 95% CI, 0.03 to 0.52). Pritelivir reduced the rates of genital HSV shedding and days with lesions in a dose-dependent manner in otherwise healthy persons with genital herpes (Wald et al. 2014). One dose of valacyclovir does not maintain antiviral activity for the entire day but suppresses apparent reactivation with some breakthrough. By contrast, the excellent pharmacokinetic profile of HPIs with administration once a day can maintain the anti-HSV activity for the entire day, suggesting that HPIs would inhibit HSV reactivation, even viral shedding, in patients with genital herpes, as well as the transmission of HSV completely from a healthy person with genital herpes.

The clinical study of amenamevir on herpes zoster compares once-daily oral doses of amenamevir with three doses valacyclovir. Amenamevir has been approved as the first HPI drug in clinical use and successfully used for the treatment of herpes zoster in Japan.

6.12 Conclusion

This chapter has introduced the current anti-herpetic drugs and newly developed HPIs. HPI works at a low concentration in vitro, and the resistant virus is rarer than acyclovir and synergistic with ACV. HPIs have shown efficacy in genital herpes in a once-daily dose. The investigation of amenamevir has been completed a clinical study on herpes zoster, and the drug has been used for the treatment of herpes zoster in Japan. Thus, HPIs will be the next-generation drugs for HSV and VZV. Moreover, various anti-herpes virus drugs are under development (Disease NIOaaI Herpes Drugs in Development n.d.)

References

Beadle JR, Hartline C, Aldern KA et al (2002) Alkoxyalkyl esters of cidofovir and cyclic cidofovir exhibit multiple-log enhancement of antiviral activity against cytomegalovirus and herpesvirus replication in vitro. Antimicrob Agents Chemother 46:2381–2386

Biswas S, Sukla S, Field HJ (2014) Helicase-primase inhibitors for herpes simplex virus: looking to the future of non-nucleoside inhibitors for treating herpes virus infections. Future Med Chem 6:45–55

Bodsworth NJ, Boag F, Burdge D et al (1997) Evaluation of sorivudine (BV-araU) versus acyclovir in the treatment of acute localized herpes zoster in human immunodeficiency virus-infected adults. The Multinational Sorivudine Study Group. J Infect Dis 176:103–111

Boehmer PE, Lehman IR (1997) Herpes simplex virus DNA replication. Annu Rev Biochem 66:347–384

Chono K, Katsumata K, Kontani T et al (2010) ASP2151, a novel helicase-primase inhibitor, possesses antiviral activity against varicella-zoster virus and herpes simplex virus types 1 and 2. J Antimicrob Chemother 65:1733–1741

Chono K, Katsumata K, Kontani T et al (2012) Characterization of virus strains resistant to the herpes virus helicase-primase inhibitor ASP2151 (Amenamevir). Biochem Pharmacol 84:459–467

Chono K, Katsumata K, Suzuki H et al (2013) Synergistic activity of amenamevir (ASP2151) with nucleoside analogs against herpes simplex virus types 1 and 2 and varicella-zoster virus. Antivir Res 97:154–160

Coen DM, Schaffer PA, Furman PA et al (1982) Biochemical and genetic analysis of acyclovir-resistant mutants of herpes simplex virus type 1. Am J Med 73:351–360

Cohen JI, Seidel KE (1993) Generation of varicella-zoster virus (VZV) and viral mutants from cosmid DNAs: VZV thymidylate synthetase is not essential for replication in vitro. Proc Natl Acad Sci U S A 90:7376–7380

Crute JJ, Grygon CA, Hargrave KD et al (2002) Herpes simplex virus helicase-primase inhibitors are active in animal models of human disease. Nat Med 8:386–391

Cundy KC (1999) Clinical pharmacokinetics of the antiviral nucleotide analogues cidofovir and adefovir. Clin Pharmacokinet 36:127–143

Daikoku T, Tannai H, Honda M et al (2016) Subclinical generation of acyclovir-resistant herpes simplex virus with mutation of homopolymeric guanosine strings during acyclovir therapy. J Dermatol Sci 82:160–165

De Clercq E (2004) Discovery and development of BVDU (brivudin) as a therapeutic for the treatment of herpes zoster. Biochem Pharmacol 68:2301–2315

De Clercq E (2005) (E)-5-(2-bromovinyl)-2′-deoxyuridine (BVDU). Med Res Rev 25:1–20

De Clercq E, Descamps J, Ogata M et al (1982) In vitro susceptibility of varicella-zoster virus to E-5-(2-bromovinyl)-2′-deoxyuridine and related compounds. Antimicrob Agents Chemother 21:33–38

Disease NIOaaI Herpes Drugs in Development (n.d.) In: Division of AIDS Anti-HIV/OI/TB Therapeutics Database

Drake JW (1993) Rates of spontaneous mutation among RNA viruses. Proc Natl Acad Sci U S A 90:4171–4175

Dunning J, Kennedy SB, Antierens A et al (2016) Experimental treatment of Ebola virus disease with Brincidofovir. PLoS One 11:e0162199

Elion GB (1989) Nobel lecture in physiology or medicine—1988. The purine path to chemotherapy. In Vitro Cell Dev Biol 25:321–330

Elion GB, Furman PA, Fyfe JA et al (1977) Selectivity of action of an antiherpetic agent, 9-(2-hydroxyethoxymethyl) guanine. Proc Natl Acad Sci U S A 74:5716–5720

Englund JA, Zimmerman ME, Swierkosz EM et al (1990) Herpes simplex virus resistant to acyclovir. A study in a tertiary care center. Ann Intern Med 112:416–422

Furuta Y, Takahashi K, Fukuda Y et al (2002) In vitro and in vivo activities of anti-influenza virus compound T-705. Antimicrob Agents Chemother 46:977–981

Honda M, Okuda T, Hasegawa T et al (2001) Effect of long-term, low-dose acyclovir suppressive therapy on susceptibility to acyclovir and frequency of acyclovir resistance of herpes simplex virus type 2. Antivir Chem Chemother 12:233–239

Ida M, Kageyama S, Sato H et al (1999) Emergence of resistance to acyclovir and penciclovir in varicella-zoster virus and genetic analysis of acyclovir-resistant variants. Antivir Res 40:155–166

Jackson GG, Muldoon RL, Akers LW (1963) Serological evidence for prevention of influenzal infection in volunteers by an anti-influenzal drug adamantanamine hydrochloride. Antimicrob Agents Chemoter (Bethesda) 161:703–707

James SH, Prichard MN (2014) Current and future therapies for herpes simplex virus infections: mechanism of action and drug resistance. Curr Opin Virol 8:54–61

James SH, Larson KB, Acosta EP et al (2015) Helicase-primase as a target of new therapies for herpes simplex virus infections. Clin Pharmacol Ther 97:66–78

Kamiyama T, Kurokawa M, Shiraki K (2001) Characterization of the DNA polymerase gene of varicella-zoster viruses resistant to acyclovir. J Gen Virol 82:2761–2765

Kawai H, Yoshida I, Suzutani T (1993) Antiviral activity of 1-beta-D-arabinofuranosyl-E-5-(2-bromovinyl)uracil against thymidine kinase negative strains of varicella-zoster virus. Microbiol Immunol 37:877–882

Kern ER, Richards JT, Overall JC Jr et al (1981) A comparison of phosphonoacetic acid and phosphonoformic acid activity in genital herpes simplex virus type 1 and type 2 infections of mice. Antivir Res 1:225–235

Kleymann G, Fischer R, Betz UA et al (2002) New helicase-primase inhibitors as drug candidates for the treatment of herpes simplex disease. Nat Med 8:392–398

Koszalka P, Tilmanis D, Hurt AC (2017) Influenza antivirals currently in late-phase clinical trial. Influenza Other Respir Viruses 1(3):240–246

Larder BA, Darby G (1986) Susceptibility to other antiherpes drugs of pathogenic variants of herpes simplex virus selected for resistance to acyclovir. Antimicrob Agents Chemother 29:894–898

Littler E, Stuart AD, Chee MS (1992) Human cytomegalovirus UL97 open reading frame encodes a protein that phosphorylates the antiviral nucleoside analogue ganciclovir. Nature 358:160–162

Machida H, Kuninaka A, Yoshino H (1982) Inhibitory effects of antiherpesviral thymidine analogs against varicella-zoster virus. Antimicrob Agents Chemother 21:358–361

Miwa N, Kurosaki K, Yoshida Y et al (2005) Comparative efficacy of acyclovir and vidarabine on the replication of varicella-zoster virus. Antivir Res 65:49–55

Okuda T, Kurokawa M, Matsuo K et al (2004) Suppression of generation and replication of acyclovir-resistant herpes simplex virus by a sensitive virus. J Med Virol 72:112–120

Ostrander M, Cheng YC (1980) Properties of herpes simplex virus type 1 and type 2 DNA polymerase. Biochim Biophys Acta 609:232–245

Parris DS, Harrington JE (1982) Herpes simplex virus variants restraint to high concentrations of acyclovir exist in clinical isolates. Antimicrob Agents Chemother 22:71–77

Patrone M, Percivalle E, Secchi M et al (2003) The human cytomegalovirus UL45 gene product is a late, virion-associated protein and influences virus growth at low multiplicities of infection. J Gen Virol 84:3359–3370

Reyes M, Shaik NS, Graber JM et al (2003) Acyclovir-resistant genital herpes among persons attending sexually transmitted disease and human immunodeficiency virus clinics. Arch Intern Med 163:76–80

Safrin S, Cherrington J, Jaffe HS (1997) Clinical uses of cidofovir. Rev Med Virol 7:145–156

Sasadeusz JJ, Tufaro F, Safrin S et al (1997) Homopolymer mutational hot spots mediate herpes simplex virus resistance to acyclovir. J Virol 71:3872–3878

Schwartz PM, Novack J, Shipman C Jr et al (1984) Metabolism of arabinosyladenine in herpes simplex virus-infected and uninfected cells. Correlation with inhibition of DNA synthesis and role in antiviral selectivity. Biochem Pharmacol 33:2431–2438

Shipman C Jr, Smith SH, Carlson RH et al (1976) Antiviral activity of arabinosyladenine and arabinosylhypoxanthine in herpes simplex virus-infected KB cells: selective inhibition of viral deoxyribonucleic acid synthesis in synchronized suspension cultures. Antimicrob Agents Chemother 9:120–127

Shiraki K (2017) Helicase-primase inhibitor amenamevir for herpesvirus infection: towards practical application for treating herpes zoster. Drugs Today 53(11):573

Shiraki K, Namazue J, Okuno T, Yamanishi K, Takahashi M (1990) Novel sensitivity of acyclovir-resistant varicella-zoster virus to anti-herpetic drugs. Antivir Chem Chemother 1:373–375

Shiraki K, Ochiai H, Namazue J et al (1992) Comparison of antiviral assay methods using cell-free and cell-associated varicella-zoster virus. Antivir Res 18:209–214

Spector FC, Liang L, Giordano H et al (1998) Inhibition of herpes simplex virus replication by a 2-amino thiazole via interactions with the helicase component of the UL5-UL8-UL52 complex. J Virol 72:6979–6987

Stranska R, Schuurman R, Nienhuis E et al (2005) Survey of acyclovir-resistant herpes simplex virus in the Netherlands: prevalence and characterization. J Clin Virol 32:7–18

Sullivan V, Talarico CL, Stanat SC et al (1992) A protein kinase homologue controls phosphorylation of ganciclovir in human cytomegalovirus-infected cells. Nature 358:162–164

Suzuki M, Okuda T, Shiraki K (2006) Synergistic antiviral activity of acyclovir and vidarabine against herpes simplex virus types 1 and 2 and varicella-zoster virus. Antivir Res 72:157–161

Tyring S, Wald A, Zadeikis N et al (2012) ASP2151 for the treatment of genital herpes: a randomized, double-blind, placebo- and valacyclovir-controlled, dose-finding study. J Infect Dis 205:1100–1110

Vere Hodge RA, Cheng Y-C (1993) The mode of action of penciclovir. Antivir Chem Chemother 4:13–24

Von Itzstein M, Wu WY, Kok GB et al (1993) Rational design of potent sialidase-based inhibitors of influenza virus replication. Nature 363:418–423

Wald A, Corey L, Timmler B et al (2014) Helicase-primase inhibitor pritelivir for HSV-2 infection. N Engl J Med 370:201–210

Wald A, Timmler B, Magaret A et al (2016) Effect of pritelivir compared with valacyclovir on genital HSV-2 shedding in patients with frequent recurrences: a randomized clinical trial. JAMA 316:2495–2503

Wassilew SW, Wutzler P, Brivddin Herpes Zoster Study G (2003) Oral brivudin in comparison with acyclovir for herpes zoster: a survey study on postherpetic neuralgia. Antivir Res 59:57–60

Whitley RJ, Alford CA, Hirsch MS et al (1986) Vidarabine versus acyclovir therapy in herpes simplex encephalitis. N Engl J Med 314:144–149

Yajima M, Yamada H, Takemoto M et al (2017) Profile of anti-herpetic action of ASP2151 (amenamevir) as a helicase-primase inhibitor. Antivir Res 139:95–101

Yokota T, Konno K, Mori S et al (1989) Mechanism of selective inhibition of varicella zoster virus replication by 1-beta-D-arabinofuranosyl-E-5-(2-bromovinyl)uracil. Mol Pharmacol 36:312–316

Chapter 7
Vaccine Development for Varicella-Zoster Virus

Tomohiko Sadaoka and Yasuko Mori

Abstract Varicella-zoster virus (VZV) is the first and only human herpesvirus for which a licensed live attenuated vaccine, vOka, has been developed. vOka has highly safe and effective profiles; however, worldwide herd immunity against VZV has not yet been established and it is far from eradication. Despite the successful reduction in the burden of VZV-related illness by the introduction of the vaccine, some concerns about vOka critically prevent worldwide acceptance and establishment of herd immunity, and difficulties in addressing these criticisms often relate to its ill-defined mechanism of attenuation. Advances in scientific technologies have been applied in the VZV research field and have contributed toward uncovering the mechanism of vOka attenuation as well as VZV biology at the molecular level. A subunit vaccine targeting single VZV glycoprotein, rationally designed based on the virological and immunological research, has great potential to improve the strategy for eradication of VZV infection in combination with vOka.

Keywords Varicella-zoster virus · Live attenuated vaccine · Next-generation sequencing · In vitro latency system · VZV-specific cellular immunity · Herpes zoster subunit vaccine

7.1 Introduction

Varicella-zoster virus (VZV) is a ubiquitous alphaherpesvirus and highly communicable pathogen spreading by airborne transmission only among humans. VZV is a multiple cell-tropic virus mainly targeting T lymphocytes, epithelial cells, and neurons. Primary infection with VZV causes varicella (chickenpox), characterized by systemic vesicular rash accompanied by T-cell-associated viremia. During primary infection, VZV gains access to and establishes lifelong latency in ganglionic

---

T. Sadaoka (✉) · Y. Mori
Division of Clinical Virology, Center for Infectious Diseases, Kobe University Graduate School of Medicine, Kobe, Hyogo, Japan
e-mail: tomsada@crystal.kobe-u.ac.jp

neurons along the entire human neuraxis. Months to years later, when VZV-specific cellular immunity but not humoral immunity wanes, VZV can reactivate to cause herpes zoster (shingles), characterized by painful unilateral and dermatomal rash (Arvin and Gilden 2013).

In general, varicella is a self-limiting disease but can sometimes be followed by serious complications, including bacterial sepsis, pneumonia, hepatitis, encephalitis, and hemorrhage, and occasionally results in death. These serious complications and death are more prominent in infants, adults, and severely immunocompromised individuals (Arvin and Gilden 2013; Gershon et al. 2015). Congenital VZV infection is caused by maternal varicella suffered during the first 20 weeks of pregnancy and results in 2% severe embryopathy (Pastuszak et al. 1994).

The lifetime risk of herpes zoster (HZ) is estimated around 30% in infected individuals. Postherpetic neuralgia (PHN), the most common complication of HZ, is clinically defined as a pain persisting for at least 3 months after resolution of HZ rash, and risk of PHN increases with age. VZV meningitis, meningoencephalitis, meningoradiculitis, cerebellitis, myelopathy, and vasculopathy may develop after HZ but with lesser frequency than PHN. VZV reactivation causes serious ocular disorders including stromal keratitis, acute retinal necrosis, and progressive outer retinal necrosis. Neurological diseases other than PHN and ocular disorders can also happen without rash as seen in zoster *sine herpete*, a chronic radicular pain without rash caused by VZV reactivation (Arvin and Gilden 2013). Intensive studies by Gilden (1937–2016) and colleagues have identified that VZV vasculopathy and giant cell arthritis are strongly associated with productive VZV infection in cerebral and temporal arteries, respectively (Nagel et al. 2011, 2015), and most recently, the presence of VZV antigen in granulomatous arteritis of the aorta is also reported (Gilden et al. 2016).

VZV is the only human herpesvirus for which a live attenuated vaccine is licensed in several countries. Originally developed in Japan in 1974 (Takahashi et al. 1974) and known as strain vOka, the vaccine has been safely and effectively used to prevent both varicella and zoster including their complications. However, VZV-related diseases still occur all over the world. In some developing countries, VZV infection is less concern, if compared to other more serious infectious diseases, but the growing number of immunocompromised individuals in the world requires establishment of worldwide herd immunity against VZV. Even in developed countries where a routine varicella vaccination has been performed, there has been far from eradication of VZV infection, the common final object of infectious disease research. To achieve these, several distinct research strategies have been actively pursued in VZV research field.

In this chapter, we first summarize vOka biology to make clear what we do and do not know about vOka and then focus on recent (1) advances in VZV research uncovering vOka attenuation mechanism toward improving attenuated live VZV vaccine more safely and effectively, (2) findings regarding VZV-specific T-cell immunity, and (3) the significant improvement of HZ subunit vaccine.

7.2 Live Attenuated VZV Vaccine, vOka

7.2.1 Development of vOka

In 1974, Takahashi (1928–2013) and colleagues developed a live attenuated VZV vaccine (Takahashi et al. 1974). His enthusiasm and effort toward developing VZV vaccine and it's insightful history are well reviewed by Ozaki and Asano (the latter is one of the co-developers of vOka) which summarizes current clinical aspects of vOka as "varicella vaccine" in Japan (Ozaki and Asano 2016).

A clinical VZV isolate was derived from the vesicular fluid of a 3-year-old boy with typical varicella and named Oka after his family name. The Oka isolate was first expanded in human embryonic lung (HEL) cells at 37 °C, resulting in parental Oka strain (pOka) widely used as VZV wild-type strain. The pOka was serially cultured 11 times in HEL cells at 34 °C and 12 times in guinea pig embryonic fibroblasts (GPEF) at 37 °C by the concept of classic and empirical viral growth attenuation technique, semi-permissive culture. The resulting virus was further cultured three times in human diploid cells, WI-38, at 37 °C. Either virus before (six passages in GPEF) or after culturing in WI-38 cells was subcutaneously administrated to healthy children with no history of varicella and was proved to be safe with no clinical reactions and effective with compatible seroconversion rate and antibody titer against VZV. Importantly, after confirming its safety and efficacy in healthy children, the attenuated virus after a sixth passage in GPEF was subcutaneously vaccinated to 23 high-risk children of severe varicella with no history of varicella in a hospital to prevent varicella dissemination from a 3-year-old boy with nephrosis developing typical varicella symptom. Underlying diseases among 23 children included nephritis, nephrosis, enteritis, hepatitis, purulent meningitis, arthritis, asthma, myelitis, hemangioma, purpura, and ventricular septal defect, and 12 children received steroid therapy. After vaccination, no varicella dissemination was observed in the hospital except for two children with mild vesicular rash (Takahashi et al. 1974).

Based on the safety and efficacy profiles of the Oka attenuated virus on high-risk children as well as healthy children from the initial work (Takahashi et al. 1974) and several following clinical works mainly conducted in high-risk children (Hattori et al. 1976; Ozaki et al. 1978; Katsushima et al. 1982; Kamiya et al. 1984), the Oka strain was recognized "*to have the most desirable attribute of low virulence while inducing an adequate antibody response and protection against diseases*" by WHO (World Health Organization) (WHO experts committee on biological standardization 1994) and has been used in more than 80 countries to date. At present, a virus after a third passage in WI-38 cells is the Oka vaccine seed stock for vaccine Oka (vOka) preparations from three different providers: OkaVax from Biken for preventing both varicella and herpes zoster, VariVax and ZostaVax from Merck for varicella and herpes zoster, respectively, and VarilRix from GSK for varicella.

7.2.2 Clinical Aspects of vOka

A single dose of varicella vaccine was 80–85% effective in preventing disease of any severity and more than 95% effective in preventing severe varicella. The vaccination program reduced disease incidence by 57–90%, hospitalizations by 75–88%, deaths by 74%, and direct inpatient and outpatient medical expenditures by 74% as summarized in (Marin et al. 2008). A higher dose of vOka administration [1,000–3,000 plaque-forming units (PFU) for varicella vs. 19,400–60,000 PFU for HZ in general (but in Japan, 42,000–67,000 PFU) for both varicella and HZ] is approved for use in adults 50 years of age or older and is recommended for adults 60 years of age or older based on the data from the shingles prevention study. In this study, one subcutaneous administration reduced the burden of illness due to HZ by 61.1%, reduced the incidence of HZ by 51.3%, and reduced the incidence of PHN by 66.5% (Oxman et al. 2005).

Despite a dramatic decline in varicella disease by one-dose vaccination, continuing outbreaks of varicella had been reported among elementary school-aged population with high coverage rate. Therefore, in 2006, a two-dose varicella vaccine schedule was recommended by the Centers for Disease Control and Prevention (CDC) (Marin et al. 2007). Varicella within 42 days postvaccination, also known as breakthrough varicella, can occur up to 34.2% by one-dose vaccination (Takayama et al. 1997). Breakthrough varicella is normally milder than natural varicella, but some serious complications similar to those occurring in unvaccinated individuals have been reported even in healthy vaccines, albeit at less than 25% frequency (Chaves et al. 2008b), and occasionally resulted fatal in immunocompromised patients (Yoshikawa et al. 2016). The US two-dose vaccination program reduced odds of developing varicella by 95% and appears to have significantly reduced the number, size, and duration of outbreaks (Shapiro et al. 2011; Leung et al. 2015). However, the two-dose vaccination schedule has not yet been incorporated into the routine vaccination strategies for several countries.

The risk of HZ substantially decreased among vaccinated children aged <10 years, and its widespread use was postulated to reduce overall burden of HZ, while the incidence of HZ among 10–19-year-olds increased during the same period (Civen et al. 2009). The varicella vaccine is generally safe and well tolerated; however, both in healthy and immunocompromised individuals, vOka can establish latency as well as wild-type VZV and reactivate and cause HZ that is often indistinguishable from that caused by wild-type VZV (Chaves et al. 2008a; Galea et al. 2008; Goulleret et al. 2010; Weinmann et al. 2013). HZ associated with vOka is most common at the site of vaccine inoculation (Hardy et al. 1991), but vOka-related HZ only in the trigeminal nerve area in healthy 2-year-old girl without previous varicella history was reported in 2016, and this finding suggests that monitoring vOka-related HZ incidence rate is expected to elucidate many aspects of varicella vaccine safety (Iwasaki et al. 2016).

In 1965, Hope-Simpson hypothesized that both endogenous and exogenous exposure of VZV may be a boost to one's immunity and significantly delays the onset of HZ (Hope-Simpson 1965). Consistent with his hypothesis, exposure to varicella boosted immunity to HZ, and then mass varicella vaccination was postulated to reduce endemic HZ by reducing exposure to varicella (Brisson et al. 2002). The population-based long-term analysis using medical record (1945–1960 and 1980–2007 in Olmsted County, Minnesota, USA) revealed that the incidence of HZ has indeed increased >4-fold over the last six decades, but there is no change in the rate of increase before and after the introduction of varicella vaccination, and this increase is unlikely to be due to any predictable reasons including the introduction of varicella vaccination, antiviral therapies, or change in the prevalence of immunocompromised individuals (Kawai et al. 2016).

In general, vOka vaccination has proven to be safe and well tolerated among healthy individuals and even in some immunocompromised populations to prevent varicella and HZ. However, some concerns, as described above, cause the delay of licensure in some European countries and give the excuse to anti-vaccine movement, resulting in failure to establish herd immunity against VZV. At the same time, the molecular mechanism of vOka attenuation remains poorly understood and contributed to concern over the vaccine. Improving our understanding of attenuation could lead to worldwide acceptance of vOka. The mechanism(s) of attenuation is ill-defined, mainly due to the lack of a robust animal model which recapitulates whole viral life cycle in human. In any animal model established, the wild-type VZV infection does not sufficiently recapitulate any disease in humans, and hence neither the vOka virulence nor the effect of vOka vaccination can be adequately confirmed.

7.2.3 *Molecular Genetics of vOka Before the Next-Generation Sequencing Era*

In 2000, partial sequencing of vOka and pOka by Gomi et al. identified that multiple single nucleotide polymorphisms (SNPs) accumulated in ORF62 gene region of vOka, which is a duplicated gene (ORF62/ORF71) and encodes the immediate-early major transactivator protein, IE62. Interestingly, other transregulating coding genes, ORF4, ORF10, ORF61, and ORF63, as well as ORF14 (glycoprotein C), did not contain any SNP differences from pOka. Strikingly, not all SNPs within ORF62 gene accumulated on single vOka genome, suggesting vOka vaccine comprised heterogeneous (mixed) populations (Gomi et al. 2000). Subsequently, the comparison of the whole-genome sequences between vOka and pOka was completed by the same group. In the vOka genome, 42 SNPs were identified in 21 ORFs, in the tandem repeat regions, R1, R3, and R4, and in the OriS. Of 42 SNPs, 6 were in noncoding region, 16 were synonymous changes (no change to amino acid sequence), and 20 caused non-synonymous changes (change in the amino acid residue). Nearly 30% of SNPs (12/42) accumulated in ORF62, and 4 were synonymous and 8 were non-synonymous. And again not all genome contained the vOka-specific SNPs at all positions indicated that vOka vaccine is a mixture of genetically distinct haplotypes (Gomi et al. 2002).

ORF0, initially referred as ORFS/L, was discovered in 2000 and encodes a cytoplasmic protein expressing during lytic infection in vitro and in vivo. In vOka, a SNP (TGA to CGA) was observed in ORF0, and this eliminates a stop codon after a residue 129 in the wild-type VZV including pOka and extends vOka ORF0 protein by 92 amino acids (Kemble et al. 2000). In 2012, whole-genome sequencing of VZV strain Ellen was completed and identified the same SNP in ORF0 with vOka. The Ellen is known to have become highly attenuated after at least 90 passages in cultured cells. Additionally, the Ellen also shared two SNPs in ORF62 with vOka in the absence of other common SNPs. These findings sharing only three SNPs between two unrelated attenuated VZV strains suggested that these three SNPs are the determinants of vOka attenuation (Peters et al. 2012), and this claim seems to meet the rising concept that VZV attenuation requires a set of SNPs within two or three ORFs. However, this concept has only arisen from the negative results that SNPs in ORF62 had no responsibility for vOka attenuation (Zerboni et al. 2005a) and SNP in ORF0 alone had also no responsibility for vOka attenuation (Koshizuka et al. 2010) as described in Sect. 7.2.4. Thus far this concept is not yet supported by any molecular based genetic approaches to creating the recombinant virus which contains all vOka type SNPs within ORF0 and ORF62 in the pOka virus or contains all pOka type SNPs within ORF0 and ORF62 in the vOka virus.

The administration of live attenuated vOka following rash formation either by varicella or zoster provided unique opportunity to determine the candidate genotypes for remaining vOka virulence in vivo. The discrepancy has been reported among the following studies: one has reported that single rash vesicle contains single genotype but different vesicles in the same individual contain different genotype (Quinlivan et al. 2004), while others have found that the individual vesicles contain multiple genotypes (Loparev et al. 2007; Thiele et al. 2011); however, these studies provided the evidence that rashes are not caused by a single vOka haplotype. Comparison of the allele frequencies between rash-causing vOka and original vOka revealed that the wild-type alleles at four loci within vOka, which may survive during the Oka attenuation process through semi-permissive culture, were significantly more prevalent among rash-causing vOka viruses than original vOka viruses including nucleotide position at 560 in ORF0, 105,169 between ORF61 and ORF62, and 105,356 and 107,797 in ORF62. In other words, the vOka viruses harboring these four wild-type SNPs have been selected for and caused rashes, but these findings were not proven experimentally.

7.2.4 Molecular Approaches Uncovering vOka Attenuation Mechanism

VZV has no animal model which can recapitulate the whole viral life cycle observed in the natural host, human. Several attempts had been made, and the most successful animal model to analyzing VZV pathogenesis is the SCID-hu mouse model, in which human fetal tissues were xenografted in mice with severely combined immunodeficiency (SCID) and innate responses that modulate infectious process can be assessed independently of adaptive immunity, which is lacked in SCID mice. VZV-SCID-hu mouse model was first developed to analyze cell tropisms of VZV for human T lymphocytes and epidermal cells in SCID mice implanted with human fetal thymus/liver or fetal skin, respectively (Moffat et al. 1995). Since then, several impressive works regarding molecular mechanisms for VZV pathogenesis in vivo have been published to date as summarized by Zerboni et al. (2014).

Prior to uncovering the heterogenetic character of vOka, attenuation of vOka as well as strain Ellen in human skin was first confirmed using SCID-hu mice implanted with human fetal skin compared to the wild-type VZV (Moffat et al. 1998), while SCID-hu mice implanted with human fetal thymus/liver infected with vOka or the wild-type VZV resulted in slower replication of vOka than the wild-type VZV in T lymphocytes, but viral titers were comparable at by day 14 (Moffat et al. 1995). The SCID-hu mice implanted with human fetal dorsal root ganglia (DRG) have been also used to analyze VZV neurotropism during lytic infection. In this DRG, VZV acquires quiescent state after productive infection stage, but never reactivates, indicating that the true latency is not established in the model. During productive infection in the DRG, vOka exhibited the same pattern with pOka regarding short-time replication before acquiring quiescent state, indicating that vOka is not attenuated for neurotropism (Zerboni et al. 2005b), despite the existence of that acute lytic phase in neurons before establishment of latency has not been proved in human.

The attenuation in human skin and the lack of attenuation in human T-cell and neuron of vOka have been shown in SCID-hu mice, but the genetic basis of vOka attenuation has not yet been defined in SCID-hu mice. However, by creating chimeric viruses containing several combinations of pOka and vOka genome segment using the cosmid-based mutagenesis, SNP(s) laid between ORF30 and ORF55, not in ORF62/71, has been proven to contribute vOka attenuation in human skin-xenografted SCID-hu mice. The vOka cosmids were derived from a single haplotype of vOka mixture and contain only one of eight non-synonymous SNP in ORF62 (Zerboni et al. 2005a). Subsequently, it has proved that the SNP in ORF0 is not by itself sufficient for the attenuation of vOka by showing that the replacement of ORF0 SNP in pOka to vOka type did not alter the plaque size in cultured cell and vice versa using the BAC mutagenesis system (Koshizuka et al. 2010). These results in combination with the observations regarding the common SNPs between unrelated vOka and Ellen described in the Sect. 7.2.3 have gradually evoked the concept that vOka attenuation requires a set of SNPs within two or three ORFs, especially ORF0 and ORF62 combination. However, the vOka derived from the cosmid showed reduced growth in human skin-xenografted SCID-hu mice even though it contains only one vOka-type SNP within ORF62 (Zerboni et al. 2005a), and ORF0 in the vOka cosmid turned out to have the wild-type gene sequence later (Peters et al. 2012), again indicating that molecular mechanism of vOka attenuation should be confirmed by experimental data.

7.3 Molecular Genetics of vOka by Deep Sequencing Technology

Recent advances in sequencing technologies enable scientist to not only whole genomes but also to perform "deep" whole-genome sequencing, a process that allows the measuring of allele frequencies at every nucleotide positions from heterogeneous populations, reflecting the population diversity within a single sample. Deep sequencing of VariVax Merck vaccine batches has identified more than 100 new variant sites, and the allele frequencies of most, but not all, of these new SNPs were typically less than 10% (Victoria et al. 2010; Depledge et al. 2014).

Comparative analyses of vOka batches from VarilRix (GSK) and VariVax and ZostaVax (both from Merck) as well as the working seed stock from BIKEN identified 137 SNPs that were conserved in all tested batches as "core SNPs" for vOka. Of the 137 core SNPs, 53 were found in supposedly noncoding regions, 127 resulted in synonymous changes, and 57 encoded non-synonymous changes. In ORF62 gene, 28 SNPs accumulated and accounted for 20.4% of total SNPs. No other single ORF contained more than 4.38% of total core SNPs. In all three preparations, one is fixed (106,262, R958G in ORF62) and five are near fixation (>90%) [560, *130R in ORF0; 105,544, V1197A; 105,705, A1143; 107,252, S628G; and 108,111, P341 (all in ORF62)] among the core SNPs. Of these six core SNPs, four was shared with strain Ellen [560, *130R in ORF0; 106,262, R958G; 107,252, S628G; and 108,111, P341 (all in ORF62)] (Depledge et al. 2016). Another live attenuated varicella vaccine SuduVax was originally developed using the same methodology as vOka but based on the strain MAV/06, a wild-type VZV isolated in South Korea (Kim et al. 2011). SuduVax shares all six core SNPs with three vOka preparations, and their allele frequencies are above 99% in SuduVax, while the allele frequencies at other SNPs in SuduVax vary from those in vOka at several degrees (Jeon et al. 2016). Overall, these results increase the likelihood that some combination of these six SNPs in ORF0 and ORF62 are critically important for vOka attenuation. However, this still needs to be supported by experimental data for conclusive statement.

Deep sequencing analysis in combination with targeted enrichment methodology enabled direct sequencing of low amount of viral DNA with the information of allele frequencies at every nucleotide position without changing population diversity by avoiding any bias like culturing virus post-isolation, PCR amplification, or cloning steps (Depledge et al. 2011, 2014). By using the combined technologies, the allele frequencies of vOka from the rash vesicles in multiple individuals during vOka-associated varicella were compared with those of original vaccine, and it became evident that no one vOka haplotype is responsible for varicella pathogenesis and every vesicle contains different vOka haplotypes, confirming previous findings (Quinlivan et al. 2004; Loparev et al. 2007; Thiele et al. 2011). The population diversity, indicated by the allele frequencies, of the rash-forming vOka viruses was less than those of the vOka in vaccine, and the wild-type allele was significantly most favored at 12 loci among the rash-forming vOka viruses: 9 causing non-synonymous changes in 5 proteins (560, *130R in ORF0; 19,063, E199R in ORF13;

58,595, I593V in ORF31; 97,479, V495A in ORF55; 97,748, A585T in ORF55; 105,356, I1260VA I ORF62; 106,001, K1045E in ORF62; 107,599, V512A in ORF62; and 107,797, L446P in ORF62), 2 were between intergenic regions (102,002, c to t between ORF60 and ORF61; 105,169, a to g between ORF61 and ORF62), and 1 was located within ORF63 promoter region (109,237 a to g). This finding suggested that the selection of the rash-forming vOka viruses has occurred after vaccination and before establishment of latency, and these wild-type SNPs contribute to the skin rash formation, the indicative of less attenuated vOka haplotypes in skin tropism (Depledge et al. 2014).

Using the same combined methodology, the comparison of the population diversity between the varicella-causing vOka viruses soon after vaccination and the HZ-causing vOka viruses after certain period of latency was also performed (Depledge et al. 2014). Different from the comparison of the population diversity of the varicella-causing vOka viruses to the original vOka vaccine, there was no significant change between the varicella-causing vOka viruses and the HZ-causing vOka viruses. Thus, those vOka variants that persist after vaccination and cause varicella appeared equally able to establish latency and reactivate to cause HZ. In other words, the skin tropism of VZV cannot be segregated from the neurotropism, providing important data to the current approaches to advance vOka vaccine not to establish latency or reactivate from latency.

7.4 vOka Attenuation in In Vitro VZV Latency System

In 1965, Hope-Simpson hypothesized that HZ is a spontaneous manifestation of varicella infection based on several careful examinations by numerous researchers in the 1800s and 1900s and wrote *"Following the primary infection (chickenpox), virus becomes latent in sensory ganglia, where it can be reactivated from time to time (herpes zoster)"* (Hope-Simpson 1965). The site of VZV latency in human sensory ganglia was first proven by Gilden et al. in 1983 and further extended to the ganglia along the entire human neuraxis in 1990 (Gilden et al. 1983; Mahalingam et al. 1990). There have been only two consensuses about the mechanism of VZV latency and reactivation at molecular level. One is that the VZV latent genome in human trigeminal ganglia persists with fused termini likely in a circular (episomal) configuration (Clarke et al. 1995). The other is that VZV can gain access to neuron/ganglia by two possible routes during varicella; one is by retrograde axonal transport from the cutaneous lesions and the other is by hematogenous transfer from VZV-infected T-cells during viremia.

An inability to reactivate VZV in any reported model in vitro and in vivo casts doubt on several observations about VZV latency obtained from existing animal models and human cadaveric ganglia harboring VZV genome, especially regarding viral proteins or even viral transcripts during latency (Azarkh et al. 2010; Cohen 2010; Zerboni et al. 2012; Ouwendijk et al. 2012a, b). In latently infected human ganglia in vivo, (1) viral genome is maintained with fused termini, (2) viral transcription is severely restricted, and (3) the virus can reactivate. An in vitro model of VZV latency that satisfies all the three criteria was until very recently unavailable and thus prevented detailed studies that could potentially reveal the mechanism of VZV latency and reactivation (Gilden et al. 2015).

In 2015, Markus and colleagues reported two in vitro models of latency and reactivation of VZV using human embryonic stem cell (hESC)-derived neurons. In the first model, acyclovir (ACV) was essential to establish quiescent state which is not absolutely required in human. In the second model, human neurons derived from hESC were infected with cell-free recombinant pOka expressing GFP-fused ORF66 protein by selective retrograde axonal transport on a microfluidic device. ACV was not required for establishing latency in the second model. During latency, transcripts of immediate-early gene ORF63 and late gene ORF31 were readily detected in parallel with viral DNA by qPCR. By inhibition of PI3K signaling (LY294002) in combination with temperature shift from 37 °C to 34 °C, viral DNA and RNA increased and GFP expression became visible in a few cells (Markus et al. 2015).

Concurrently, Sadaoka and colleagues have developed an in vitro latency system using hESC-derived neurons which fulfill all the criteria (Sadaoka et al. 2016). In the system, human neurons were differentiated from neural stem cells derived from hESC on a microfluidic device and selectively infected from axon termini with pOka in a cell-free manner, resulting in the establishment of latency following controlled reactivation. Viral genome was maintained without detection of any tested viral transcripts including immediate-early genes ORF61, ORF62, and ORF63; early gene ORF16; and spliced late gene ORF42/45 by RT-qPCR up to 70 days postinfection. At 14 days postinfection, viral DNA configuration was circular, and reactivation could be induced by adding anti-NGF antibody following apparent viral DNA replication and mRNA expression. Distinct from the model by Markus and colleagues, reactivated virus could be passaged onto different cell type, the representative of full reactivation. Despite no detection of tested viral mRNAs during latency by RT-qPCR which is sensitive enough to detect less than ten copies per reaction, RNAseq analysis detected widespread transcriptions albeit low abundance and whether these transcripts are still functional or just a noise will need to be addressed by more advanced technologies like single cell-based transcriptional profiling. Among several compounds tested for reactivation, only anti-NGF antibody, but not the inhibitor of PI3K (LY294002), could reactivate the virus in vitro, and these were partially confirmed in human trigeminal ganglia removed within 24 hours after death by showing the ability of anti-NGF antibody and inability of LY294002 in induction of VZV DNA replication (Cohrs et al. 2016).

Using the in vitro latency system, latency and reactivation of vOka were compared to those of pOka (Sadaoka et al. 2016). In the system, vOka could establish latency at a similar rate to pOka. By treatment of anti-NGF antibody, vOka could reactivate from latency, but reactivation rate of vOka was significantly lesser than that of pOka, indicating that vOka is not attenuated for establishment of latency but is attenuated for reactivation. Deep sequencing analysis in combination with the targeted enrichment methodology has shown that the population diversity of vOka was maintained during latency, but its heterogeneity was apparently lost after reactivation, consistent with the findings in vivo that the population diversity of zoster-causing vOka is lesser than that of vOka vaccine (Depledge et al. 2014). However, no wild-type SNP which was suggested to be included in rash-forming vOka (Depledge et al. 2014) became dominant at any loci after reactivation of vOka in vitro except 100114 in ORF58 (K53N), which was not included in rash-forming vOka and was also dominant in latency population in vitro. The key for uncovering the attenuation mechanism of vOka by molecular genetic approach should be the understanding of the population diversity of pOka. For heterogeneity of vOka, molecular genetic approach has been mainly employed in vOka based on the pOka sequence information by traditional Sanger sequence technique containing no SNP information (Gomi et al. 2002). The analysis of the allele frequencies of pOka revealed that pOka also consisted of a heterogeneous population but to a much lesser extent than vOka. For instance, two alleles at single nucleotide in one ORF are equally detected in pOka population by deep sequencing analysis (Sadaoka et al. 2016), but one allele had not been reported in the pOka sequence information (Gomi et al. 2002), and this unreported allele was completely missing from any vOka preparations, indicating that the nucleotide which was thought to be identical between pOka and vOka has possibility to be strikingly different and to contribute in vOka attenuation.

7.5 VZV-Specific Cellular Immunity in Older Individuals and vOka Vaccination for HZ Prevention

The biggest risk factor for HZ is the progressive age-related decline in cell-mediated immunity to VZV, but not humoral immunity. The strong correlation of HZ incidence with aging has been linked to a decrease in the frequency of VZV-specific T-cells (Levin et al. 2003), while antibody response against VZV is extremely stable with half-lives of approximately 50 years (Amanna et al. 2007). In recipients of hematopoietic cell transplantation, reconstitution of CD4 T-cell immunity against VZV reduced the risk of HZ; on the other hand, VZV antibody is not sufficient to prevent its reactivation (Hata et al. 2002). Therefore, the objective of vaccination against HZ is to increase the frequency of VZV-specific long-lived memory T-cells poised to produce IFN-gamma, which is the most well-characterized factor in controlling HZ (Levin et al. 2008).

In general, several sequential events take place after vaccination or natural infection to generate antigen-specific T-cells. Initially, these T-cells exponentially expand and differentiate into effector cells. While most of these effector cells undergo apoptosis and are short-lived, a small subset survives this contraction phase to constitute memory T-cell precursor that finally differentiates into long-lived memory T-cells. Declining T-cell responsiveness to stimulation with age has been implied to relate to in particular reduced ability to proliferate due to telomeric erosion and expression of p16. In addition to initial expansion following contraction (T-cell frequency) and final differentiation (T-cell functionality), T-cell receptor (TCR) diversity is a defining hallmark of the antigen-reactive T-cell repertoire. With age, the diversity contracts two- to fivefold, but still the repertoire remains highly diverse. The administration of higher-dose vOka in elderly as HZ vaccine resulted in 63.9% prevention of HZ for 60–69-year-old ages but only 37.6% for 70 years and older (Oxman et al. 2005). This age-dependent decline of protective effect against HZ might be explained by the decrease of VZV-specific T-cells even after vOka vaccination. However, the stage at which generation of VZV-specific T-cells is critically impaired has not yet been identified.

To build predictive models of T-cell responses against vOka vaccination in older populations, 39 individuals between the age of 50 and 79 years including 9 monozygotic twin pairs were immunized with vOka (Qi et al. 2016b). The majority of activated VZV-specific T-cells measured by IFN-gamma-specific ELISPOT assay was of the CD4 T-cell subset, and their peak of frequencies was approximately tenfold higher between day 8 and 14 after vaccination and then declined to an average threefold higher by day 28 than before vaccination, while frequencies of global CD4 and CD8 populations did not change over the course of the vaccine response. By including identical twin pairs in the cohort, a genetic influence on vOka vaccine-induced responses could be also analyzed. While B-cell responses were more influenced by nonheritable factors which were evidenced on the observation that no difference on induction of VZV-specific antibodies between twins, twin pairs were more similar than unrelated individuals in the generation of memory T-cells. When effector cell differentiation and contraction were assessed separately, either phase of T-cell response was not significantly more similar between twins than between unrelated individuals. The expansion of VZV-specific T-cells was relatively independent of age. Conversely, the accelerated VZV-specific T-cell loss after the peak response was mainly a function of age, resulting in diminished generation of long-lived memory T-cells in older individuals. Transcriptome analysis in activated CD4 T-cells at the time of peak response to vOka vaccination identified the correlation between the contraction phase of the T-cell response and gene modulations related to cell cycle regulation and DNA repair. These pathways might be the target to reduce T-cell attrition and to improve the survival of expanded T effector cell populations and thereby increase the effectiveness of vOka vaccine, especially in older populations.

In another cohort study including three identical twin pairs and three unrelated individuals older than 50 years, the TCR diversity of VZV-specific CD4 T-cells before and after vOka vaccination was examined (Qi et al. 2016a). There were large differences in the repertoire richness of VZV-specific CD4 T-cells among individuals, even in identical twins. A genetic influence was seen for the sharing of individual TCR sequences from antigen-reactive cells but not for repertoire richness or the selection of dominant clones. By vOka vaccination, the repertoire in individuals was diversified by preferentially expanding infrequent T-cell clones unevenly, including recruiting new specificities from the naïve repertoire instead of further promoting the selection of dominant TCR sequencing. Viral replication might be required for sufficient clonal expansion of the low-frequency VZV-specific T-cells for protection against HZ; however, vOka seemed not to replicate efficiently after vaccination. Neither viral sequences in peripheral blood nor an inflammatory response in the days subsequent to vaccination was found in the study. The current vOka vaccination strategy with a single administration for HZ prevention has been built on the premise of refreshing a recall response and has not been optimized for rebuilding the memory compartment and for selecting for dominance of new T-cell clones, which may relate to incomplete protection against HZ. These suggest the importance of establishment of the rational vaccination strategy to generate VZV-specific memory T-cells in clonal sizes necessary for immune protection against HZ.

7.6 Herpes Zoster Subunit Vaccine Targeting Glycoprotein E

In 2015, a result from a phase 3 clinical trial for a subunit vaccine containing VZV glycoprotein E (gE) and the $AS01_B$ adjuvant system (called HZ/su, GlaxoSmithKline Biologicals) was reported. This phase 3 trial has been conducted in 18 countries, involved 15,411 participants, and evaluated the safety and efficacy of HZ/su in older adults ($\geq$50 years of age) (ZOE-50) (Lal et al. 2015).

VZV gE is the most abundant glycoprotein both in virions and infected cells and is a major target of the VZV-specific CD4 T-cell response (Arvin et al. 1986; Brunell et al. 1987; Grose 1990; Harper et al. 1990). The $AS01_B$ adjuvant system contains 3-O-desacyl-4′-monophosphoryl lipid A (MPL) and the saponin QS21 and promotes both strong CD4 T-cell and humoral immune responses to recombinant proteins by activating Toll-like receptor 4 and by increasing antigen uptake and retention by dendritic cells (Vandepapelière et al. 2008; Kester et al. 2009; Leroux-Roels et al. 2010; Coffman et al. 2010). In mice immunized with a recombinant gE protein purified from CHO-K1 cells with the same adjuvant system, this protein was highly immunogenic and induced abundant anti-gE neutralizing antibodies, but it could not induce any VZV-specific T-cell immune response by restimulation ex vivo. However, when mice were immunized with a DNA encoding the recombinant gE, the protein could proliferate T-cells after restimulation ex vivo (Jacquet et al. 2002), suggesting that the recombinant gE has a potential to induce efficient VZV gE-specific T-cell immunity only in the presence of gE-specific memory T-cell population by VZV natural infection or vOka vaccination during childhood. Consistent with this, the recombinant gE with $AS01_B$ induced gE- and VZV-specific CD4 T-cell responses as well as antibody responses in a VZV-primed mouse model. In any animal except human, VZV cannot efficiently replicate. In this priming model, however, mice could be primed by administration of vOka subcutaneously in the scruff of the neck with induction of VZV-specific cell-mediated immune response. Among tested combinations of the recombinant gE with several adjuvants, $AS01_B$ elicited higher frequencies of CD4 T-cells producing IFN-gamma and IL-2 and was superior to vOka in any immunological aspects including induction of anti-gE antibody (Dendouga et al. 2012).

During phase 1 and 2 clinical trials, the safety and immunogenicity of HZ/su was evaluated in comparison with vOka. Few grade 3 unsolicited adverse events and no severe adverse event were reported only in HZ/su-vaccinated older adults with just two cases of chills and one case of insomnia, but these symptoms lasted 1 day and resolved without treatment. Fatigue, myalgia, headache, and injection site pain were the most common solicited reactions for HZ/su group and occurred more frequently than for the vOka group. Overall HZ/su was well tolerated among all participants including younger adult group (18–30 years) and older adult group (50–70 years). Two doses of HZ/su induce significantly stronger gE- and VZV-specific CD4 T-cell and antibody responses than two doses of vOka, resulting in superior immunogenicity of HZ/su to vOka (Leroux-Roels et al. 2012).

In the phase 3 trial ZOE-50, HZ was confirmed in six participants in the HZ/su-vaccinated group (7698 participants) and in 210 participants in the placebo group (7713 participants), resulting in 97.2% of vaccine efficacy against HZ during a mean follow-up period of 3.2 years. In the vaccine group, solicited reports of injection-site and systemic reactions within 7 days of vaccination were more frequent. Solicited or unsolicited grade 3 symptoms were reported in 17.0% of vaccine recipients and 3.2% of placebo recipients (Lal et al. 2015).

Concurrently with ZOE-50, a second phase 3 clinical trial named ZOE-70 has been conducted at the same site to examine the safety and efficacy of HZ/su in adults 70 years of age or older (13,900 participants), the most important group given their higher risk of HZ. A mean follow-up period of ZOE-70 was 3.7 years. In the study, the efficacy against PHN was also assessed. Both assessments included pooled participants 70 years of age or older from ZOE-50 (2696 participants), and total participants were 16,596. Vaccine efficacy against HZ was 91.3% and against PHN was 88.8%. In ZOE-70 alone, there was no difference in vaccine efficacy against HZ between participants 70–79 years of age (90.0%) and 80 years of age or older (89.1%). Solicited reports of injection-site and systemic reactions within 7 days of vaccination were more frequent in HZ/su group (79.0%) than placebo group (29.5%), while the overall incidence of serious adverse events, potential immune-mediated diseases, and deaths was similar between two groups (Cunningham et al. 2016).

The longest follow-up study of safety and immunogenicity was from the phase 2 clinical trial conducted in Czech Republic, Germany, Sweden, and the Netherlands including 129 participants with two-dose administrations (60–84 years of age). Six years after vaccination, no vaccine-related serious adverse events were reported, and the gE-specific cell-mediated immune response was on average 3.8 times higher than prevaccination values with 25% decrease from the value at 3 years, while the anti-gE antibody concentration was 7.3 times higher than prevaccination values with 20% decrease from the 3 years value, indicating that HZ/su may have potential to provide long-term protection against HZ in older adults (Chlibek et al. 2016).

Overall, HZ/su dramatically reduced the risk of HZ and PHN among adults 50 years of age or older including 80 years of age or older without substantial safety concerns and has potential for long-term immunity to prevent HZ with an acceptable safety profile among older adults. At present, an immunological threshold has not been established for the protection against HZ, and further long-term clinical efficacy studies with a measurement of the frequencies of VZV-specific CD4 T-cells are essential.

7.7 Summary

With the development of live attenuated vOka as a highly safe and effective varicella vaccine, we have successfully reduced varicella diseases; however, we are still a long way from the eradication of VZV, and even herd immunity against VZV has not yet been established. Furthermore, the incidence of HZ has been increasing for reasons unknown. Of course, we have to take into consideration that the vOka has been administrated as HZ vaccine only for these 10 years and has not been widely used. Most of the concerns about administration of "live" vOka stem from its ill-defined mechanism for attenuation. Recently developed HZ/su might be particularly useful, once licensed, for worldwide distribution because it seems to be effective to prevent HZ and PHN and non-replicating subunit vaccine possibly available in immunocompromised individuals who cannot receive live vOka. Although HZ/su seems to be effective for preventing HZ, it requires preexistence of VZV-specific T-cell immunity to function as HZ vaccine and seems not be able to induce primary cellular immunity against VZV by itself, which can be established by vOka. Recent advances in research technology have gradually revealed the attenuation mechanism as well as the virulence of vOka at molecular level. These results will be helpful to improve live attenuated vOka as widely acceptable vaccine by enhancing its immunogenicity and reducing its virulence, especially to develop vaccines that do not establish (or at least reactivate) from latency in the near future.

Acknowledgment The authors thank Dr. Daniel P. Depledge for helpful discussions and critical reading of the manuscript. This work was supported in part by the Japan Herpesvirus Infections Forum, the Takeda Science Foundation, the Japan Foundation for Pediatric Research, and Japan Society for the Promotion of Science (JSPS KAKENHI JP17K008858).

References

Amanna IJ, Carlson NE, Slifka MK (2007) Duration of humoral immunity to common viral and vaccine antigens. N Engl J Med 357:1903–1915. https://doi.org/10.1056/NEJMoa066092

Arvin AM, Gilden D (2013) Varicella-zoster virus. In: Knipe DM, Howley PM (eds) Fields virology, 6th edn. Wolters Kluwer Health, Philadelphia, pp 2015–2057

Arvin AM, Kinney-Thomas E, Shriver K et al (1986) Immunity to varicella-zoster viral glycoproteins, gp I (gp 90/58) and gp III (gp 118), and to a nonglycosylated protein, p 170. J Immunol 137:1346–1351

Azarkh Y, Gilden DH, Cohrs RJ (2010) Molecular characterization of varicella zoster virus in latently infected human ganglia: physical state and abundance of VZV DNA, quantitation of viral transcripts and detection of VZV-specific proteins. Curr Top Microbiol Immunol 342:229–241. https://doi.org/10.1007/82_2009_2

Brisson M, Gay NJ, Edmunds WJ, Andrews NJ (2002) Exposure to varicella boosts immunity to herpes-zoster: implications for mass vaccination against chickenpox. Vaccine 20:2500–2507

Brunell PA, Novelli VM, Keller PM, Ellis RW (1987) Antibodies to the three major glycoproteins of varicella-zoster virus: search for the relevant host immune response. J Infect Dis 156:430–435

Chaves SS, Haber P, Walton K et al (2008a) Safety of varicella vaccine after licensure in the United States: experience from reports to the vaccine adverse event reporting system, 1995–2005. J Infect Dis 197(Suppl 2):S170–S177. https://doi.org/10.1086/522161

Chaves SS, Zhang J, Civen R et al (2008b) Varicella disease among vaccinated persons: clinical and epidemiological characteristics, 1997–2005. J Infect Dis 197(Suppl 2):S127–S131. https://doi.org/10.1086/522150

Chlibek R, Pauksens K, Rombo L et al (2016) Long-term immunogenicity and safety of an investigational herpes zoster subunit vaccine in older adults. Vaccine 34:863–868. https://doi.org/10.1016/j.vaccine.2015.09.073

Civen R, Chaves SS, Jumaan A et al (2009) The incidence and clinical characteristics of herpes zoster among children and adolescents after implementation of varicella vaccination. Pediatr Infect Dis J 28:954–959. https://doi.org/10.1097/INF.0b013e3181a90b16

Clarke P, Beer T, Cohrs RJ, Gilden DH (1995) Configuration of latent varicella-zoster virus DNA. J Virol 69:8151–8154

Coffman RL, Sher A, Seder RA (2010) Vaccine adjuvants: putting innate immunity to work. Immunity 33:492–503. https://doi.org/10.1016/j.immuni.2010.10.002

Cohen JI (2010) Rodent models of varicella-zoster virus neurotropism. Curr Top Microbiol Immunol 342:277–289. https://doi.org/10.1007/82_2010_11

Cohrs RJ, Badani H, Baird NL et al (2016) Induction of varicella zoster virus DNA replication in dissociated human trigeminal ganglia. J Neurovirol 23:1–6. https://doi.org/10.1007/s13365-016-0480-1

Cunningham AL, Lal H, Kovac M et al (2016) Efficacy of the herpes zoster subunit vaccine in adults 70 years of age or older. N Engl J Med 375:1019–1032. https://doi.org/10.1056/NEJMoa1603800

Dendouga N, Fochesato M, Lockman L et al (2012) Cell-mediated immune responses to a varicella-zoster virus glycoprotein E vaccine using both a TLR agonist and QS21 in mice. Vaccine 30:3126–3135. https://doi.org/10.1016/j.vaccine.2012.01.088

Depledge DP, Palser AL, Watson SJ et al (2011) Specific capture and whole-genome sequencing of viruses from clinical samples. PLoS One 6:e27805. https://doi.org/10.1371/journal.pone.0027805

Depledge DP, Kundu S, Jensen NJ et al (2014) Deep sequencing of viral genomes provides insight into the evolution and pathogenesis of varicella zoster virus and its vaccine in humans. Mol Biol Evol 31:397–409. https://doi.org/10.1093/molbev/mst210

Depledge DP, Yamanishi K, Gomi Y et al (2016) Deep sequencing of distinct preparations of the live attenuated Varicella-Zoster Virus vaccine reveals a conserved Core of attenuating single-nucleotide polymorphisms. J Virol 90:8698–8704. https://doi.org/10.1128/JVI.00998-16

Galea SA, Sweet A, Beninger P et al (2008) The safety profile of varicella vaccine: a 10-year review. J Infect Dis 197(Suppl 2):S165–S169. https://doi.org/10.1086/522125

Gershon AA, Breuer J, Cohen JI et al (2015) Varicella zoster virus infection. Nat Rev Dis Primers 1:15016. https://doi.org/10.1038/nrdp.2015.16

Gilden DH, Vafai A, Shtram Y et al (1983) Varicella-zoster virus DNA in human sensory ganglia. Nature 306:478–480

Gilden DH, Nagel M, Cohrs R et al (2015) Varicella Zoster Virus in the nervous system. F1000Res. https://doi.org/10.12688/f1000research.7153.1

Gilden DH, White T, Boyer PJ et al (2016) Varicella Zoster Virus infection in granulomatous arteritis of the Aorta. J Infect Dis 213:1866–1871. https://doi.org/10.1093/infdis/jiw101

Gomi Y, Imagawa T, Takahashi M, Yamanishi K (2000) Oka varicella vaccine is distinguishable from its parental virus in DNA sequence of open reading frame 62 and its transactivation activity. J Med Virol 61:497–503

Gomi Y, Sunamachi H, Mori Y et al (2002) Comparison of the complete DNA sequences of the Oka varicella vaccine and its parental virus. J Virol 76:11447–11459

Goulleret N, Mauvisseau E, Essevaz-Roulet M et al (2010) Safety profile of live varicella virus vaccine (Oka/Merck): five-year results of the European Varicella Zoster Virus Identification Program (EU VZVIP). Vaccine 28:5878–5882. https://doi.org/10.1016/j.vaccine.2010.06.056

Grose C (1990) Glycoproteins encoded by varicella-zoster virus: biosynthesis, phosphorylation, and intracellular trafficking. Annu Rev Microbiol 44:59–80. https://doi.org/10.1146/annurev.mi.44.100190.000423

Hardy I, Gershon AA, Steinberg SP, LaRussa P (1991) The incidence of zoster after immunization with live attenuated varicella vaccine. A study in children with leukemia. Varicella vaccine collaborative study group. N Engl J Med 325:1545–1550. https://doi.org/10.1056/NEJM199111283252204

Harper DR, Kangro HO, Heath RB (1990) Antibody responses in recipients of varicella vaccine assayed by immunoblotting. J Med Virol 30:61–67

Hata A, Asanuma H, Rinki M et al (2002) Use of an inactivated varicella vaccine in recipients of hematopoietic-cell transplants. N Engl J Med 347:26–34. https://doi.org/10.1056/NEJMoa013441

Hattori A, Ihara T, Iwasa T et al (1976) Use of live varicella vaccine in children with acute leukaemia or other malignancies. Lancet 2:210. https://doi.org/10.1016/S0140-6736(76)92397-7

Hope-Simpson RE (1965) The nature of herpes zoster: a long-term study and a new hypothesis. Proc R Soc Med 58:9–20

Iwasaki S, Motokura K, Honda Y et al (2016) Vaccine-strain herpes zoster found in the trigeminal nerve area in a healthy child: a case report. J Clin Virol 85:44–47. https://doi.org/10.1016/j.jcv.2016.10.022

Jacquet A, Haumont M, Massaer M et al (2002) Immunogenicity of a recombinant varicella-zoster virus gE-IE63 fusion protein, a putative vaccine candidate against primary infection and zoster reactivation. Vaccine 20:1593–1602

Jeon JS, Won YH, Kim IK et al (2016) Analysis of single nucleotide polymorphism among Varicella-Zoster Virus and identification of vaccine-specific sites. Virology 496:277–286. https://doi.org/10.1016/j.virol.2016.06.017

Kamiya H, Kato T, Isaji M et al (1984) Immunization of acute leukemic children with a live varicella vaccine (Oka strain). Biken J 27:99–102

Katsushima N, Yazaki N, Sakamoto M et al (1982) Application of a live varicella vaccine to hospitalized children and its follow-up study. Biken J 25:29–42

Kawai K, Yawn BP, Wollan P, Harpaz R (2016) Increasing incidence of herpes zoster over a 60-year period from a population-based study. Clin Infect Dis 63:221–226. https://doi.org/10.1093/cid/ciw296

Kemble GW, Annunziato PW, Lungu O et al (2000) Open reading frame S/L of varicella-zoster virus encodes a cytoplasmic protein expressed in infected cells. J Virol 74:11311–11321

Kester KE, Cummings JF, Ofori-Anyinam O et al (2009) Randomized, double-blind, phase 2a trial of falciparum malaria vaccines RTS,S/AS01B and RTS,S/AS02A in malaria-naive adults: safety, efficacy, and immunologic associates of protection. J Infect Dis 200:337–346. https://doi.org/10.1086/600120

Kim JI, Jung GS, Kim YY et al (2011) Sequencing and characterization of Varicella-zoster virus vaccine strain SuduVax. Virol J 8:547. https://doi.org/10.1186/1743-422X-8-547

Koshizuka T, Ota M, Yamanishi K, Mori Y (2010) Characterization of varicella-zoster virus-encoded ORF0 gene – comparison of parental and vaccine strains. Virology 405:280–288. https://doi.org/10.1016/j.virol.2010.06.016

Lal H, Cunningham AL, Chlibek R et al (2015) Efficacy of an adjuvanted herpes zoster subunit vaccine in older adults. N Engl J Med 372:2087–2096. https://doi.org/10.1056/NEJMoa1501184

Leroux-Roels I, Koutsoukos M, Clement F et al (2010) Strong and persistent CD4+ T-cell response in healthy adults immunized with a candidate HIV-1 vaccine containing gp120, Nef and Tat antigens formulated in three Adjuvant Systems. Vaccine 28:7016–7024. https://doi.org/10.1016/j.vaccine.2010.08.035

Leroux-Roels I, Leroux-Roels G, Clement F et al (2012) A phase 1/2 clinical trial evaluating safety and immunogenicity of a varicella zoster glycoprotein e subunit vaccine candidate in young and older adults. J Infect Dis 206:1280–1290. https://doi.org/10.1093/infdis/jis497

Leung J, Lopez AS, Blostein J et al (2015) Impact of the US two-dose Varicella vaccination program on the epidemiology of Varicella outbreaks: data from nine states, 2005–2012. Pediatr Infect Dis J 34:1105–1109. https://doi.org/10.1097/INF.0000000000000821

Levin MJ, Smith JG, Kaufhold RM et al (2003) Decline in varicella-zoster virus (VZV)-specific cell-mediated immunity with increasing age and boosting with a high-dose VZV vaccine. J Infect Dis 188:1336–1344. https://doi.org/10.1086/379048

Levin MJ, Oxman MN, Zhang JH et al (2008) Varicella-zoster virus-specific immune responses in elderly recipients of a herpes zoster vaccine. J Infect Dis 197:825–835. https://doi.org/10.1086/528696

Loparev VN, Rubtcova E, Seward JF et al (2007) DNA sequence variability in isolates recovered from patients with postvaccination rash or herpes zoster caused by Oka varicella vaccine. J Infect Dis 195:502–510. https://doi.org/10.1086/510532

Mahalingam R, Wellish M, Wolf W et al (1990) Latent varicella-zoster viral DNA in human trigeminal and thoracic ganglia. N Engl J Med 323:627–631. https://doi.org/10.1056/NEJM199009063231002

Marin M, Güris D, Chaves SS et al (2007) Prevention of varicella: recommendations of the Advisory Committee on Immunization Practices (ACIP). MMWR Recomm Rep 56:1–40

Marin M, Meissner HC, Seward JF (2008) Varicella prevention in the United States: a review of successes and challenges. Pediatrics 122:e744–e751. https://doi.org/10.1542/peds.2008-0567

Markus A, Lebenthal-Loinger I, Yang IH et al (2015) An in vitro model of latency and reactivation of Varicella Zoster Virus in human stem cell-derived neurons. PLoS Pathog 11:e1004885. https://doi.org/10.1371/journal.ppat.1004885

Moffat JF, Stein MD, Kaneshima H, Arvin AM (1995) Tropism of varicella-zoster virus for human CD4+ and CD8+ T lymphocytes and epidermal cells in SCID-hu mice. J Virol 69:5236–5242

Moffat JF, Zerboni L, Kinchington PR et al (1998) Attenuation of the vaccine Oka strain of varicella-zoster virus and role of glycoprotein C in alphaherpesvirus virulence demonstrated in the SCID-hu mouse. J Virol 72:965–974

Nagel MA, Traktinskiy I, Azarkh Y et al (2011) Varicella zoster virus vasculopathy: analysis of virus-infected arteries. Neurology 77:364–370. https://doi.org/10.1212/WNL.0b013e3182267bfa

Nagel MA, White T, Khmeleva N et al (2015) Analysis of varicella-zoster virus in temporal arteries biopsy positive and negative for Giant cell arteritis. JAMA Neurol 72:1281–1287. https://doi.org/10.1001/jamaneurol.2015.2101

Ouwendijk WJD, Choe A, Nagel MA et al (2012a) Restricted varicella-zoster virus transcription in human trigeminal ganglia obtained soon after death. J Virol 86:10203–10206. https://doi.org/10.1128/JVI.01331-12

Ouwendijk WJD, Flowerdew SE, Wick D et al (2012b) Immunohistochemical detection of intraneuronal VZV proteins in snap-frozen human ganglia is confounded by antibodies directed against blood group A1-associated antigens. J Neurovirol 18:172–180. https://doi.org/10.1007/s13365-012-0095-0

Oxman MN, Levin MJ, Johnson GR et al (2005) A vaccine to prevent herpes zoster and postherpetic neuralgia in older adults. N Engl J Med 352:2271–2284. https://doi.org/10.1056/NEJMoa051016

Ozaki T, Asano Y (2016) Development of varicella vaccine in Japan and future prospects. Vaccine 34:3427–3433. https://doi.org/10.1016/j.vaccine.2016.04.059

Ozaki T, Nagayoshi S, Morishima T et al (1978) Use of a live varicella vaccine for acute leukemic children shortly after exposure in a children's ward. Biken J 21:69–72

Pastuszak AL, Levy M, Schick B et al (1994) Outcome after maternal varicella infection in the first 20 weeks of pregnancy. N Engl J Med 330:901–905. https://doi.org/10.1056/NEJM199403313301305

Peters GA, Tyler SD, Carpenter JE et al (2012) The attenuated genotype of Varicella-Zoster Virus includes an ORF0 transitional stop codon mutation. J Virol 86:10695–10703. https://doi.org/10.1128/JVI.01067-12

Qi Q, Cavanagh MM, Le Saux S et al (2016a) Diversification of the antigen-specific T cell receptor repertoire after varicella zoster vaccination. Sci Transl Med 8:332ra46. https://doi.org/10.1126/scitranslmed.aaf1725

Qi Q, Cavanagh MM, Le Saux S et al (2016b) Defective T memory cell differentiation after Varicella Zoster vaccination in older individuals. PLoS Pathog 12:e1005892. https://doi.org/10.1371/journal.ppat.1005892

Quinlivan ML, Gershon AA, Steinberg SP, Breuer J (2004) Rashes occurring after immunization with a mixture of viruses in the Oka vaccine are derived from single clones of virus. J Infect Dis 190:793–796. https://doi.org/10.1086/423210

Sadaoka T, Depledge DP, Rajbhandari L et al (2016) In vitro system using human neurons demonstrates that varicella-zoster vaccine virus is impaired for reactivation, but not latency. Proc Natl Acad Sci U S A 113:E2403–E2412. https://doi.org/10.1073/pnas.1522575113

Shapiro ED, Vazquez M, Esposito D et al (2011) Effectiveness of 2 doses of varicella vaccine in children. J Infect Dis 203:312–315. https://doi.org/10.1093/infdis/jiq052

Takahashi M, Otsuka T, Okuno Y et al (1974) Live vaccine used to prevent the spread of varicella in children in hospital. Lancet 2:1288–1290

Takayama N, Minamitani M, Takayama M (1997) High incidence of breakthrough varicella observed in healthy Japanese children immunized with live attenuated varicella vaccine (Oka strain). Acta Paediatr Jpn 39:663–668

Thiele S, Borschewski A, Küchler J et al (2011) Molecular analysis of varicella vaccines and varicella-zoster virus from vaccine-related skin lesions. Clin Vaccine Immunol 18:1058–1066. https://doi.org/10.1128/CVI.05021-11

Vandepapelière P, Horsmans Y, Moris P et al (2008) Vaccine adjuvant systems containing monophosphoryl lipid A and QS21 induce strong and persistent humoral and T cell responses against hepatitis B surface antigen in healthy adult volunteers. Vaccine 26:1375–1386. https://doi.org/10.1016/j.vaccine.2007.12.038

Victoria JG, Wang C, Jones MS et al (2010) Viral nucleic acids in live-attenuated vaccines: detection of minority variants and an adventitious virus. J Virol 84:6033–6040. https://doi.org/10.1128/JVI.02690-09

Weinmann S, Chun C, Schmid DS et al (2013) Incidence and clinical characteristics of herpes zoster among children in the varicella vaccine era, 2005–2009. J Infect Dis 208:1859–1868. https://doi.org/10.1093/infdis/jit405

WHO experts committee on biological standardization (1994) Requirements for varicella vaccine (live)

Yoshikawa T, Ando Y, Nakagawa T, Gomi Y (2016) Safety profile of the varicella vaccine (Oka vaccine strain) based on reported cases from 2005 to 2015 in Japan. Vaccine 34:4943–4947. https://doi.org/10.1016/j.vaccine.2016.08.044

Zerboni L, Hinchliffe S, Sommer MH et al (2005a) Analysis of varicella zoster virus attenuation by evaluation of chimeric parent Oka/vaccine Oka recombinant viruses in skin xenografts in the SCIDhu mouse model. Virology 332:337–346. https://doi.org/10.1016/j.virol.2004.10.047

Zerboni L, Ku C-C, Jones CD et al (2005b) Varicella-zoster virus infection of human dorsal root ganglia in vivo. Proc Natl Acad Sci U S A 102:6490–6495. https://doi.org/10.1073/pnas.0501045102

Zerboni L, Sobel RA, Lai M et al (2012) Apparent expression of varicella-zoster virus proteins in latency resulting from reactivity of murine and rabbit antibodies with human blood group a determinants in sensory neurons. J Virol 86:578–583. https://doi.org/10.1128/JVI.05950-11

Zerboni L, Sen N, Oliver SL, Arvin AM (2014) Molecular mechanisms of varicella zoster virus pathogenesis. Nat Rev Microbiol 12:197–210. https://doi.org/10.1038/nrmicro3215

Part II
Betaherpesviruses

Chapter 8
Glycoproteins of HHV-6A and HHV-6B

Huamin Tang and Yasuko Mori

Abstract Recently, human herpesvirus 6A and 6B (HHV-6A and HHV-6B) were classified into distinct species. Although these two viruses share many similarities, cell tropism is one of their striking differences, which is partially because of the difference in their entry machinery. Many glycoproteins of HHV-6A/B have been identified and analyzed in detail, especially in their functions during entry process into host cells. Some of these glycoproteins were unique to HHV-6A/B. The cellular factors associated with these viral glycoproteins (or glycoprotein complex) were also identified in recent years. Detailed interaction analyses were also conducted, which could partially prove the difference of entry machinery in these two viruses. Although there are still issues that should be addressed, all the knowledges that have been earned in recent years could not only help us to understand these viruses' entry mechanism well but also would contribute to the development of the therapy and/or prophylaxis methods for HHV-6A/B-associated diseases.

Keywords HHV-6 · Glycoprotein · Tropism · Entry · CD46 · CD134

8.1 Introduction of HHV-6

8.1.1 General Characteristics of HHV-6

HHV-6, initially named as human B-lymphotropic virus (HBLV), was isolated from the peripheral blood mononuclear cells in patients with lymphoproliferative disorders by Salahuddin et al. in 1986 (Salahuddin et al. 1986). Latter studies showed that the virus has much more wide range of host cells and infects T-cell efficiently (Lusso et al. 1988, 1987). Based on the analysis of its biological and genetic

H. Tang (✉)
Department of Immunology, Nanjing Medical University, Nanjing, China
e-mail: htang@njmu.edu.cn

Y. Mori
Division of Clinical Virology, Center for Infectious Diseases, Kobe University Graduate School of Medicine, Kobe, Hyogo, Japan

properties, the virus had been officially classified as a member of *Herpesviridae* family, *Betaherpesvirinae* subfamily (Wyatt et al. 1990; Aubin et al. 1991; Campadelli-Fiume et al. 1993). In this subfamily, HHV-6, together with human herpesvirus 7 (HHV-7), belongs to beta-2 group, and the human cytomegalovirus (HCMV) was the prototype virus of beta-1 group (Roizmann et al. 1992).

After identification of HHV-6 first strain, GS strain, numerous other HHV-6 strains have been isolated (Schmiedel et al. 2016; Torrisi et al. 1999; Foa-Tomasi et al. 1992). Strikingly different from the strains of HCMV, recombination between HHV-6 strains rarely occurs (Kasolo et al. 1997). Viral genome sequence analysis shows these virus strains could be divided into two groups, initially named as variant A and B, now reclassified as HHV-6A and HHV-6B species (Ablashi et al. 2013). The prototype virus for HHV-6A is U1102 virus isolated from Uganda and characterized in the UK (Downing et al. 1987), and the prototype ones for group B are Z29 in the USA (isolated from AIDS patient in Zambia) (Lopez et al. 1988) and HST in Japan (isolated from the patient with exanthem subitum) (Yamanishi et al. 1988). The major species prevalent in the USA, Europe, and Japan is HHV-6B (Okuno et al. 1989; Saxinger et al. 1988). Interestingly, a study in South Africa (Zambia) showed similar prevalence of these two viruses (Kasolo et al. 1997). The reason for the variance of geographical distribution of these two viruses still needs to be addressed.

Most HHV-6 primary infection occurs within 2 years of childhood. HHV-6B is the causative agent for exanthem subitum during its primary infection (Okuno et al. 1989; Yamanishi et al. 1988). Lots of studies show the association of HHV-6 with other diseases like chronic fatigue syndrome (Chapenko et al. 2006), multiple sclerosis (Leibovitch and Jacobson 2014), drug rash with eosinophilia, systemic symptoms (Gentile et al. 2010), etc.; however, the cause relationship between HHV-6 infection and these diseases still needed to be elucidated. Strikingly different from other human herpesviruses, integration of HHV-6 full genome into host genome has been reported (Arbuckle et al. 2010); the mobilization and reactivation of HHV-6 virus from integration have also been reported (Prusty et al. 2013; Kaufer and Flamand 2014).

8.1.2 Virion Structure

HHV-6 shares the similar structure with other herpesviruses, consisting of four elements (an electron-opaque core containing linear viral DNA, an icosahedral capsid encapsulating the core, a proteinaceous layer surrounding the capsid, and a lipid bilayer envelop from host cells with viral proteins embedded in or associated it). The diameter of mature HHV-6 virion is about 160–220 nm, and the viral genome is about 160 kbps. Overall identity of viral genome sequence between HHV-6A and HHV-6B is about 90% with some low identical regions like U86-U100 (about 72%). The numbers of predictable viral open reading frames (ORF) in HHV-6A and HHV-6B genomes are different, 133 ORFs for HHV-6A (U1102 strain) and 119 for HHV-6B (Z29 strain or HST strain) (Isegawa et al. 1999; Gompels et al. 1995; Dominguez et al. 1999).

8.1.3 Life Cycle of HHV-6

Just like other human herpesviruses, HHV-6 infection starts with the attachment of virus particles to target cells. Specific virus-cell interaction begins with the binding of the HHV-6 ligand to its cellular receptor. Subsequently, HHV-6 envelope fuses with cellular membrane after endocytosis of the virion into cytoplasm, which is confirmed by the experiment of treatment of the target cells with chloroquine, a drug that disrupts the endocytic pathway, almost completely inhibited the viral infectivity (Cirone et al. 1992). Then, the capsid is released and transported to nuclear pore through cytoskeletal network (including dynein and dynactin components) (Dohner et al. 2002). Viral DNA is released into the nucleus, leaving the capsid outside. During lytic infection, once HHV-6 DNA enters the nucleus, the virus DNA is circularized and progeny viral DNA is synthesized by the method of "rolling circle." In productively infected cell, viral genes could be divided into immediate-early, early, and late genes based on their expression kinetics, while during its latent infection, only a set of virus genes are expressed (Schiewe et al. 1994; Mirandola et al. 1998; Kondo et al. 2002).

8.2 Viral Glycoproteins Contribute to Virus Entry

8.2.1 General Characteristics About Virus Entry

Viruses are obligate parasites of host cells, depending on the host machinery for their own replication. Thus, entry into host cells is an obligatory step for virus propagation. Enveloped virus entry process could be divided into several steps: attachment to cell surface, cellular receptor binding, fusion of the envelope with plasma membrane, and un-coating of viral genome. The envelope glycoprotein(s) play a key role during these steps.

Attachment of virus to the cell surface was considered as the first step for virus entry, which may lead to the accumulation of virus particles on the cell surface and therefore promote the specific binding of viral ligand to cellular receptor. There are glycosaminoglycan chains on cell surface proteoglycans which could be initial docking sites for pathogens. Among the glycosaminoglycan chains, heparan sulfate is particularly important, at least for viruses. Many herpesviruses have been reported to use the heparan sulfate for virus attachment, and viral glycoproteins binding to these molecules have also been reported, e.g., gB and gC in HSV-1 and HSV-2 (Williams and Straus 1997; Trybala et al. 2000; Tal-Singer et al. 1995; Herold et al. 1991) and gB in HHV-7 (Skrincosky et al. 2000; Secchiero et al. 1997). As to HHV-6, infection of the cells is only inhibited by heparin or heparan sulfate at high concentrations; thus, it is still unknown whether there is functional interaction of viral glycoprotein with heparan sulfate during HHV-6 entry (Conti et al. 2000).

Viral ligands are different from virus to virus and so are the viral receptors. Ligand-receptor interaction is the key step for virus entry, which also initiates the membrane fusion in enveloped viruses. How viral proteins participate in these steps is different among viruses (several typical models are listed below).

Model 1: in the case of many flaviviruses, e.g., dengue virus, the very similar membrane protein, envelope glycoprotein (E), participates in both receptor-binding and membrane fusion steps of the virus entry. Receptor binding of these flaviviruses to the host cells elicits the endocytosis of the incoming viruses into target cells. Fusion of viral envelope with the host membrane is initiated by low pH-induced conformational change of the E protein, followed by release of the genomic RNA into the cytoplasm (Stiasny et al. 2011; Harrison 2008).

Model 2: as to the viruses like influenza virus or human immunodeficiency virus (HIV), one precursor viral protein associated with receptor binding and membrane fusion is processed into two associated subunits. Hemagglutinin 1 and 2 (H1 and H2) for influenza virus and gp120/gp41 for virus (HIV), function both as receptor-binding and fusion protein. HIV-1 membrane fusion is triggered by receptor and co-receptor binding. HIV 160 kDa protein precursor is posttranslationally cleaved into envelope proteins gp120 (for receptor binding) and gp41 (for membrane fusion). gp120 and gp41 remain associated until gp120 binds to CD4 receptor on CD4+ T lymphocytes. The structural rearrangement in gp120 after its receptor binding enables further interaction of gp120 with cellular co-receptors (CCR5 and CXCR4) and leads to gp120 dissociation from gp41. The dissociation of gp120 from gp41 causes a conformational change of gp41, which leads the exposure of the fusion peptide in gp41 and subsequently the membrane fusion (Liu et al. 2008; Land and Braakman 2001; Hallenberger et al. 1992; Furuta et al. 1998; Doms and Moore 2000; Chan et al. 1997). Similar events occur during influenza virus entry. HA glycoprotein is cleaved into HA1 (for receptor binding) and HA2 (for membrane fusion) during virus maturation process. Binding of HA1 to the virus receptor (sialic acid on cellular glycoproteins, different from most other viruses using proteins as their cellular receptors) initiates the internalization of the virus and subsequently exposure of fusion peptide in HA2 (Steinhauer 1999; Skehel and Wiley 2000; Ivanovic et al. 2013).

Model 3: slightly different from the viruses in model 2, the viruses belonging to this category are engulfed, and then viral ligands are processed in endocytic vesicles, which initiate the receptor binding to their receptor. Ebola virus is the representative for these viruses. GP of Ebola virus is expressed as a single-chain precursor and posttranslationally processed into the disulfide-linked fragments, GP1 and GP2. After attachment and internalization of the virus, the receptor-binding domain (RBD) at the apex of GP1 is exposed by the cleavage of a glycan cap and a mucin-like domain (MLD) from it. The RBD of GP1 interacts with the endosomal receptor for Ebola virus, Niemann-Pick C1 (NPC1), and then the fusion loop in GP2 is subsequently exposed and initiates the viral-cellular membrane fusion (Saeed et al. 2010; Moller-Tank and Maury 2015; Cote et al. 2011; Chandran et al. 2005; Brecher et al. 2012).

Model 4: as to herpesviruses, receptor binding and membrane fusion are considered to be carried out by different viral proteins. Entry of HSV is mostly investigated among herpesvirus family. Binding of gD to HSV cellular receptor initiates the specific cell-virus interaction, and gH, gL, and gB are responsible for the virus fusion (Spear 2004; Atanasiu et al. 2007; Akhtar et al. 2008; Akhtar and Shukla 2009).

8.2.2 Cell Tropism of HHV-6

Although the original name for HHV-6 is human B-lymphotropic virus, B cells could only be infected by HHV-6 when they are immortalized by EBV. Latter studies showed that HHV-6 is a T-lymphotropic and neurotropic virus (Lusso et al. 1988; De Bolle et al. 2005). By now, no continuous cell line could be recommended for virus isolation. The primary isolation of HHV-6 from a human specimen usually requires co-cultivation with primary high susceptible cells consisting peripheral blood mononuclear cells (PBMs) or umbilical cord blood mononuclear cells (CBMCs). Routine culture of HHV-6 in laboratories has been adapted to use CD4+ T-cell lines. However, the usage of T-cell lines for HHV-6 propagation is different. HSB-2 and JJhan cells are for HHV-6A and MT4 and Molt3 for HHV-6B. The infection ability of these two viruses on particular T-cell lines is one of determinants for their classification into the two species, HHV-6A and HHV-6B (Ablashi et al. 1991).

Both HHV-6A and HHV-6B have high detection rate in central nerve system and could infect astrocytes and microglial and neuron cells (Chan et al. 2001; Cuomo et al. 2001). The association of these viruses with central nerve diseases has also been reported, like progressive multifocal leukoencephalopathy (Mock et al. 1999; Daibata et al. 2001), multiple sclerosis (Noseworthy et al. 2000; Challoner et al. 1995), etc. Besides T lymphocytes and nerve system cells, HHV-6 could also infect a broad range of cells, although the infectability may be different between HHV-6A and HHV-6B (Roush et al. 2001; Ozaki et al. 2001; Lusso et al. 1993; Luppi et al. 1995, 1999; Kakimoto et al. 2002; Ishikawa et al. 2002; He et al. 1996; Harma et al. 2003; Fox et al. 1990; Donati et al. 2003; Chen et al. 1994; Chan et al. 2001; Cermelli et al. 1996; Caruso et al. 2002).

In an ex vivo experiment, both HHV-6A and HHV-6B could productively infect human tonsil tissue fragments in the absence of exogenous stimulation. These two viruses could efficiently infect CD4+T lymphocytes expressing a non-naive phenotype; however, only HHV-6A efficiently infected CD8+ T-cells (Grivel et al. 2003). One reason for this cell tropism difference may come from the different cellular receptor usage of these two viruses. As to HHV-6A, its cellular receptor is a ubiquitous molecule, CD46, which is compatible with a broad cell tropism. However, productive HHV-6A infection is limited to a relatively small range of cell, which suggests there should be some other restriction factors acting beyond entry step of viral life cycle (Santoro et al. 1999). As to HHV-6B, CD134 has been identified as an entry receptor for this virus. However, CD134 is mainly expressed on activated

T-cells, which is not compatible with the target cell range of HHV-6B and implies unidentified cell receptor exits (Tang et al. 2013).

Laboratory culture of the viruses could affect the viral genomic composition, which may also influence cellular tropism of the virus. As reported in HHV-6B Z29 strain, expansion of repetitive sequences from the origin of lytic replication, terminal direct repeat (Dominguez et al. 1999; Gompels and Macaulay 1995; Stamey et al. 1995), has been observed, which is less dramatic as that in HCMV but may affect the virus tropism.

HHV-6 is a human-restricted virus. There are still no animal models effectively supporting HHV-6 infection, though much progress has been made. Intravenous injections of HHV-6 into marmosets resulted in different symptoms for HHV-6A and HHV-6B, with the neurological symptoms developed in the case of HHV-6A but not HHV-6B. However, intranasal infection resulted in completely asymptomatic and elicited limited, if any, antibody responses even in the case of HHV-6A (Leibovitch et al. 2013). In a study, human CD46 transgenic mouse model was used for HHV-6 infection, and viral protein production and syncytia development could only be confirmed in the case of HHV-6A, which may be consistent with different receptor usage for these two viruses (Reynaud et al. 2014). Humanized Rag2−/− γc−/− mouse (RAG-hu) model was also used for HHV-6 investigation, and HHV-6A DNA was detected only sporadically in plasma and blood cells, which is probably due to inefficient replication and establishment of latent infection of the virus in these mice (Tanner et al. 2013).

8.2.3 HHV-6 Entry Envelope Glycoproteins

Herpesviruses are enveloped, and their lipid bilayers are derived from preexisting host membrane, in the case of HHV-6 from trans-Golgi network (TNG) or post-TNG vacuoles, during virus budding. At the very location, viral glycoproteins are accumulated and finally incorporated into final virus envelope (Mori et al. 2008). At least, eight glycoproteins (gH, gL, gM, gN, gB, gO, gQ1, gQ2) have been reported to be expressed on HHV-6 envelope. The former five proteins are conserved in *Herpesviridae* family. gO is conserved in beta-herpesvirus subfamily, and gQ are only expressed in HHV-6 and HHV-7(Sadaoka et al. 2006; Mori et al. 2004).

8.2.3.1 HHV-6 gH/gL/gQ1/gQ2 Complex

Discovery of gQ Proteins and Its Complex

Although its name was designated later, gQ gene product was first described by Pfeiffer et al. in 1993. They generated several neutralizing antibodies for HHV-6 infection and found some of these antibodies recognized a peptide coded from a 624-bp genomic fragment from an HHV-6 strain GS genomic library constructed in the λgtll expression system. The antigen recognized by these antibodies was a component of gp82-gp105 complex (Pfeiffer et al. 1993). Later, the same group identified a 2.5K cDNA encoding a potential protein of 650 amino acids containing the very peptide they had identified in λgtll expression system. The originated mRNA for the cDNA was highly spliced, consisting of 12 exons (Pfeiffer et al. 1995). Later, a more abundant mRNA coded from this region in HHV-6 genome was identified by Mori group. They designate the gene product as gQ (abbreviation for glycoprotein Q) (Mori et al. 2003a). And this form of gQ forms a complex with gH and gL, which interacts with CD46, HHV-6A cellular receptor (Mori et al. 2003b).

Much more interestingly, in a northern blot analysis of the transcripts from gQ coding region in HHV-6A genome, Mori group identified a small transcript from this region, and a glycoprotein was coded from this transcript, which was also incorporated into the gH/gL/gQ complex. They designated the protein as gQ2 and the previous one as gQ1; the gH/gL/gQ complex comes to be a tetracomplex, gH/gL/gQ1/gQ2 complex. Although the gQ1 and gQ2 were first identified in HHV-6A-infected cells, the homologous proteins were identified in HHV-6B-infected cells, which were confirmed by corresponding antibodies for these proteins (Kawabata et al. 2011).

Functional Analysis of gH/gL/gQ1/gQ2 Complex

gQ1 and gH gene products were identified by the neutralization antibodies for these proteins, which indicates their role during HHV-6 entry. However, the detailed function of gH- and gQ-containing complex had not been elucidated before HHV-6 entry receptor, CD46, was identified. CD46 is one of the cell surface antigen downregulated during HHV-6 infection. Immunoprecipitation and pull-down experiments showed that CD46 interacts with gH-containing complex (Santoro et al. 1999, 2003). The gH-containing complex was finally defined as gH/gL/gQ1/gQ2 complex by Mori group (Akkapaiboon et al. 2004; Mori et al. 2003a, b). And the most important function of gH/gL/gQ1/gQ2 complex was eventually defined as the viral ligand for HHV-6 cellular receptor, CD46, during the virus entry into host cells.

Although CD46 was first reported as the entry receptor for both HHV-6A and HHV-6B, latter evidences showed an alternative receptor exists for HHV-6B. CD46 is a ubiquitous immunoregulatory receptor, expressed on all nucleated cells (Riley-Vargas et al. 2004; Liszewski et al. 2005), which is more compatible with relatively wider cell tropism of HHV-6A. HHV-6A induces the cell fusion when CHO (Chinese hamster ovary) cell expresses CD46 even in the absence of viral protein synthesis (FFWO), but this is not the case of HHV-6B (HST strain) (Pedersen et al. 2006; Mori et al. 2002). A monoclonal antibody for CD46 could block HHV-6A infection but not HHV-6B infection (unpublished data). Furthermore, in a co-immunoprecipitation assay, gH antibody could coprecipitate CD46 from HHV-6A-infected cells but not from HHV-6B-infected cells (Oyaizu et al. 2012). HHV-6B-specific receptor was finally identified as CD134, a member of TNFR superfamily. Interaction of CD134 with HHV-6B gH/gL/gQ1/gQ2 complex, but not HHV-6A homologous complex, was confirmed. Ectopic expression of CD134 in some HHV-6B-resistant cell lines converts these cells to be susceptible to HHV-6B infection (Tang et al. 2013). Thus, HHV-6A and HHV-6B take the usage of different cellular receptors for their host cell entry, which may partially explain their different cell tropism.

Detailed analysis of the interaction of HHV-6 cellular receptor and ligand has also been conducted. In physiological conditions, CD46 functions as a cofactor for serine protease factor I to inactivate C3b and C4b, components of convertases in complement pathway, which protects host cells from damage by complement factors (Riley-Vargas et al. 2004). However, recent studies show that CD46 is like a pathogen magnet and functions as the entry receptor for both viruses (measles virus (Kemper and Atkinson 2009), all species of B adenoviruses except 3 and 7(Gaggar et al. 2003), in addition to HHV-6A) and bacteria (*Streptococcus pyogenes* (Rezcallah et al. 2005), *Neisseria gonorrhoeae* (Kallstrom et al. 2001), and *Neisseria meningitides* (Kallstrom et al. 1997)). Different usage of CD46 domains for these pathogens' entry has also been analyzed. Extracellular region of CD46 contains four short consensus repeats (SCR) fold into a compact beta-barrel domain surrounded by flexible loops (Seya et al. 1990). For HHV-6A, at least SCR2 and SCR3 of CD46 are required (Mori et al. 2002; Greenstone et al. 2002). In the analysis using CD46 deletions, Mori group found that SCR4 was also required for HHV-6A infection and FFWO induced by HHV-6A (Mori et al. 2002). For measle virus entry, only SCR1 and SCR2 are required (Greenstone et al. 2002). Four different isoforms of CD46 derived from alternative splice has also been reported. Hansen et al. reported that although different expression patterns on T-cell lines have little influence on the sensitivity of these cells to HHV-6A infection, T-cell lines with the equal frequency of CD46 isoforms could be sensitive for a particular HHV-6B strain, PL1 infection (Hansen et al. 2017).

Relatively detailed functional analysis of HHV-6A gH/gL/gQ1/gQ2 complex shows that all four components of the complex were required for the complex maturation (transport of the complex from ER to Golgi) and its binding to its cellular receptor, CD46 (Tang et al. 2011). Partially exchange of the complex components with HHV-6B homologous proteins showed that HHV-6A gQ1 and gQ2 play the key role for the complex binding to CD46 (Jasirwan et al. 2014). Furthermore, analysis of an HHV-6A-specific neutralizing antibody for gQ1 showed that a short amino sequence in HHV-6A gQ1 (494-4987a.a) is important for HHV-6A gH/gL/gQ1/gQ2 complex function (Maeki et al. 2013).

Discovery of HHV-6B-specific receptor explained several different characters between HHV-6A and HHV-6B during their infection. CD46 had been initially considered as the receptor for both HHV-6A and HHV-6B. However, HHV-6B has relatively narrow host cell range compared with HHV-6A; furthermore, some anti-CD46 antibodies and soluble CD46 could block HHV-6A infection, but not HHV-6B infection. In the unique infection phenomenon of HHV-6, CD46 antibody or soluble CD46 could inhibit FFWO in the case of HHV-6A, but not HHV-6B (Mori et al. 2002). All these observations about HHV-6 led to the identification of HHV-6B-specific receptor, CD134. Compared with broad expression of CD46, CD134 is primarily expressed on the activated T-cell surface (Weinberg et al. 1996), which may be corresponding to HHV-6B relatively narrow host cell range and may explain the different T-cell sensitive to HHV-6A and HHV-6B infection.

Different from the requirement of all four components of HHV-6A gH/gL/gQ1/gQ2 complex for CD46 binding, HHV-6B gQ1 and gQ2 are required and sufficient for CD134 binding (Tang et al. 2014). Furthermore, an amino acid residue in HHV-6 gQ1 (E127), important for receptor binding, was identified by homologous-part-exchange analysis of HHV-6A and HHV-6B gQ1s (Tang and Mori 2015).

CD134, a type I transmembrane protein, is a co-stimulator for T-cell activation (Higgins et al. 1999). Ectodomain of CD134 contains four cysteine-rich domains (CRD) and a stalk domain (Marsters et al. 1992), in which CRD2 are required for HHV-6B complex binding. Furthermore, two amino acid residues (K79 and W86) in CRD2 were confirmed to play a key role for its interaction with HHV-6B complex, and the positive charge of K79 may correspond to the negative charge of E127 of HHV-6B gQ1 and contribute the conformational access for the interaction. More interestingly, when their homologous residues in murine CD134 (mCD134) were humanized (mCD134 D75K/Q82W), the mutant interacts with HHV-6B complex. As there is no suitable animal model for HHV-6B study, these findings may contribute to HHV-6 animal model development (Tang and Mori 2015; Tang et al. 2014).

8.2.3.2 HHV-6 gH/gL/gO Complex

gO glycoprotein is conserved in beta herpesviruses (Mori et al. 2004; Sadaoka et al. 2006). It forms a trimeric complex with gH and gL and incorporated into virion. Different forms of HHV-6 gO (120-130KDa gO gO-130K; 75-80KDa gO gO-80K) were expressed in HHV-6-infected cells. Only the gO-80K was incorporated into HHV-6 virion (Mori et al. 2004). Latter study showed that the C-terminal of gO-130K is cut off during its maturation process, although whether the cleavage is important for gO maturation and function is still unknown (Tang et al. 2015). Deletion of HCMV or MCMV gO results in severe growth deficit (Wille et al. 2010; Scrivano et al. 2010; Jiang et al. 2008); by contrast, gO-deficient HHV-6A virus shows no obvious growth defect during the infection of the virus in T-cells (Tang et al. 2015). Much detailed functional analysis of HCMV and MCMV gO has been reported. HCMV gH/gL/gO trimer is considered to be involved with the infection of HCMV into fibroblasts, and the gH/gL/pUL128L pentamer is required for the infection of endothelial, epithelial, and myeloid cells (Ryckman et al. 2006; Vanarsdall and Johnson 2012; Vanarsdall et al. 2011). Recently, HCMV gH/gL/gO was reported to be also required for infection of endothelial/epithelial cells (Zhou et al. 2015), and its receptor was confirmed to be platelet-derived growth factor-α (PDGFRα) (Kabanova et al. 2016), initially recognized as HCMV gB receptor (Soroceanu et al. 2008). The detailed gO function during HHV-6 infection is less known. HHV-6 gH/gL/gO does not bind to CD46 and may associate with an unknown cellular factor and contribute to HHV-6 infection in other type cells than T-cells.

8.2.3.3 HHV-6 gB

gB is a conserved glycoprotein in herpesvirus family. HHV-6 gB was identified by comparison of HHV-6 genome with other herpesvirus genomes, and it is coded from U39 region of HHV-6 genome (Foa-Tomasi et al. 1992). HHV-6 gB is a type I transmembrane protein with the length of 830 amino acids (about 112KDa). During its maturation process, gB is proteolytically cleaved into two subunits of 64 and 58KDa by host furin protease, which are covalently linked via a disulfide bond (Ellinger et al. 1993; Chou and Marousek 1992). Sequence divergence of HHV-6A and HHV-6B viruses has been reported. Virus-specific monoclonal antibodies for HHV-6A or HHV-6B have been raised for the differentiation of clinical isolates of these two viruses. Even in HHV-6B isolates, two subgroups, gB-B1 and gB-B2, exist based on the phylogenetic analysis (Campadelli-Fiume et al. 1993; Achour et al. 2008). Cytoplasmic tail domain (CTD) of gB in other herpesviruses has been demonstrated to be important for regulation of gB function, like enhancing, reducing, or even abolishing cell fusion (Foster et al. 2001; Beitia Ortiz de Zarate et al. 2007; Garcia et al. 2013). Deletion of the sequence coding for HHV-6A gB's CTD from HHV-6A genome results in the abolishment of the reconstitution of infectious virus from the mutant virus genome. More detailed study showed that gB without CTD diffusely distributed in cytoplasm compared that of wild-type HHV-6 gB accumulated in TGN, which indicated that gB CTD is required for efficient intracellular transport of gB (Mahmoud et al. 2016). Fusion of virus envelope with cell membrane is a key step for virus entry. gB, together with gH/gL, has been considered to be the core fusion machinery for herpesvirus entry. Tanaka Y et al. reported that HHV-6A gB and the gH/gL/gQ1/gQ2 complex are the minimum components required for the fusion induced by HHV-6A; furthermore, those proteins should be expressed in cis for the fusion (Tanaka et al. 2013).

8.2.3.4 HHV-6 gM/gN Complex

gM and gN are conserved in herpesvirus family. They form a complex and express on herpesvirus virion. It has been suggested that gM/gN complex has different function among the members of herpesvirus family. For most well-studied alphaherpesviruses, e.g., HSV and pseudorabies virus, gM is a nonessential gene for virus propagation (Dijkstra et al. 1996; Baines and Roizman 1991), although deletion of gM results in attenuation of these viruses' growth (MacLean et al. 1991; Browne et al. 2004; Baines and Roizman 1991). However, in the case of HCMV and HHV-6, gM is essential for the production of infectious virus (Kawabata et al. 2012; Hobom et al. 2000). gM/gN complex could facilitate the fusion step during virus entry into host cells (Koyano et al. 2003; Klupp et al. 2000; Kim et al. 2013; Crump et al. 2004) and also function during virus maturation and egress process (Lau and Crump 2015; Chouljenko et al. 2012). HHV-6 gM is coded from U72 and expressed as a type III transmembrane glycoprotein. Cytoplasmic tail of HHV-6 gM contains the motifs required for its intracellular trafficking, including YXXΦ conserved in all herpesvirus family (Owen and Evans 1998) and an acidic cluster probably required for TGN targeting (Voorhees et al. 1995). Efficiently intracellular traffic of gM also required the presence of gN, which is coded from HHV-6 U46 (Kawabata et al. 2012). Interestingly, HHV-6 gM/gN complex interacts with VAMP3 (vesicle-associated membrane protein 3), a v-SNARE (soluble N-ethylmaleimide-sensitive factor attachment protein receptor) protein that resides in recycling endosomes and endosome-derived transport vesicles. Interestingly, VAMP3 is also incorporated into HHV-6 virions. As v-SNARE could interact with target membranes (t-SNAREs) to form trans-SNARE complexes, it is still needed to be elucidated whether VAMP3 contributes to the membrane fusion step during HHV-6 infection (Kawabata et al. 2014).

8.3 Other HHV-6 Glycoproteins

Besides the viral glycoproteins described above, there are still some other glycoproteins (or predicted to be glycoproteins) expressed in HHV-6-infected cells. Whether they are all incorporated into HHV-6 virion, their full functions still need investigation.

8.3.1 U20

U20 is a conserved glycoprotein expressed in the beta-human herpesviruses of HHV-6 and HHV-7, but not in HCMV. During virus infection, host immune system would be activated and commit antiviral action. Activation of TNFR1 pathway is one of such antiviral activities. Expression of U20 is sufficient to inhibit proinflammatory and apoptotic TNFR1 signaling pathways, which may contribute to the immune evasion of the virus (Kofod-Olsen et al. 2012).

8.3.2 U21

U21 is a type I transmembrane protein conserved only in HHV-6 and HHV-7. Although U21s in HHV-6 and HHV-7 share only 30% identity, they function similarly to bind to and divert MHC I molecules to endolysomal compartment for degradation, which contribute to escape of these viruses from immune detection (May et al. 2010; Glosson and Hudson 2007).

8.3.3 U23

U23 gene is also the unique gene for Roseoloviruses. It is expressed at late stage during HHV-6 infection and is a nonessential for virus propagation. During HHV-6 infection, U23 is accumulation within TGN, but not incorporated into HHV-6 virions. Detailed functional analysis for HHV-6 U23 is still needed (Hayashi et al. 2014).

8.4 Host Glycoproteins in HHV-6 Virions

It has been reported that cellular glycoprotein could be incorporated into mature virions (Stegen et al. 2013; Shaw et al. 2008; Bechtel et al. 2005). CD63, a member of the tetraspanin family, was reported to be incorporated into HHV-6 virions. As CD63 is associated with intracellular vesicle transport, it may contribute to HHV-6 maturation and/or egress by multivesicular body (MVB) pathway (Mori et al. 2008). MHC class I molecule was also found in HHV-6 virions. It is still needed to be elucidated that whether shedding of these molecules from infected cells contributes to the immune evasion of these HHV-6-infected cells (Ota et al. 2014).

8.5 Summary and Future Perspectives

Herpesvirus infection is a very sophisticated process in target cells. Analysis of virus cellular tropism and identification of cellular receptors for HHV-6 infection has contributed much to our understanding of several steps of this process. Detailed analysis of receptor-ligand interaction may help us to develop HHV-6-specific therapy and prophylaxis methods, which are not available now.

To date, most studies of HHV-6 glycoproteins have being focusing on the entry steps of the virus. Nevertheless, functions of homologous HHV-6 glycoproteins in other herpesviruses have also been confirmed in virus egress and maturation process, as herpesvirus virions undergo envelopment, de-envelopment, and re-envelopment process during its nuclear and extracellular egress. Much excellent works would be needed for such analysis in HHV-6. Even as to the virus entry step, much more detailed subdivision (e.g., membrane closing, hemifusion, and complete fusion steps of viral-cellular membrane fusion) has been conducted in the case of other virus entries, such as HSV. What is the case of HHV-6 still needs to be found out. Structural function analysis has also being carried out for other viral glycoprotein studies (Zhang et al. 2011; Chandramouli et al. 2015), but not in HHV-6. Much more efforts are needed for such investigation.

Reference

Ablashi DV, Balachandran N, Josephs SF, Hung CL, Krueger GR, Kramarsky B, Salahuddin SZ, Gallo RC (1991) Genomic polymorphism, growth properties, and immunologic variations in human herpesvirus-6 isolates. Virology 184(2):545–552

Ablashi D, Agut H, Alvarez-Lafuente R, Clark DA, Dewhurst S, Diluca D, Flamand L, Frenkel N, Gallo R, Gompels UA, Hollsberg P, Jacobson S, Luppi M, Lusso P, Malnati M, Medveczky P, Mori Y, Pellett PE, Pritchett JC, Yamanishi K, Yoshikawa T (2013) Classification of HHV-6A and HHV-6B as distinct viruses. Arch Virol 159(5):863–870. https://doi.org/10.1007/s00705-013-1902-5

Achour A, Malet I, Le Gal F, Dehee A, Gautheret-Dejean A, Bonnafous P, Agut H (2008) Variability of gB and gH genes of human herpesvirus-6 among clinical specimens. J Med Virol 80(7):1211–1221. https://doi.org/10.1002/jmv.21205

Akhtar J, Shukla D (2009) Viral entry mechanisms: cellular and viral mediators of herpes simplex virus entry. FEBS J 276(24):7228–7236. https://doi.org/10.1111/j.1742-4658.2009.07402.x

Akhtar J, Tiwari V, Oh MJ, Kovacs M, Jani A, Kovacs SK, Valyi-Nagy T, Shukla D (2008) HVEM and nectin-1 are the major mediators of herpes simplex virus 1 (HSV-1) entry into human conjunctival epithelium. Invest Ophthalmol Vis Sci 49(9):4026–4035. https://doi.org/10.1167/iovs.08-1807

Akkapaiboon P, Mori Y, Sadaoka T, Yonemoto S, Yamanishi K (2004) Intracellular processing of human herpesvirus 6 glycoproteins Q1 and Q2 into tetrameric complexes expressed on the viral envelope. J Virol 78(15):7969–7983. https://doi.org/10.1128/JVI.78.15.7969-7983.2004

Arbuckle JH, Medveczky MM, Luka J, Hadley SH, Luegmayr A, Ablashi D, Lund TC, Tolar J, De Meirleir K, Montoya JG, Komaroff AL, Ambros PF, Medveczky PG (2010) The latent human herpesvirus-6A genome specifically integrates in telomeres of human chromosomes in vivo and in vitro. Proc Natl Acad Sci U S A 107(12):5563–5568. https://doi.org/10.1073/pnas.0913586107

Atanasiu D, Whitbeck JC, Cairns TM, Reilly B, Cohen GH, Eisenberg RJ (2007) Bimolecular complementation reveals that glycoproteins gB and gH/gL of herpes simplex virus interact with each other during cell fusion. Proc Natl Acad Sci U S A 104(47):18718–18723. https://doi.org/10.1073/pnas.0707452104

Aubin JT, Collandre H, Candotti D, Ingrand D, Rouzioux C, Burgard M, Richard S, Huraux JM, Agut H (1991) Several groups among human herpesvirus 6 strains can be distinguished by Southern blotting and polymerase chain reaction. J Clin Microbiol 29(2):367–372

Baines JD, Roizman B (1991) The open reading frames UL3, UL4, UL10, and UL16 are dispensable for the replication of herpes simplex virus 1 in cell culture. J Virol 65(2):938–944

Bechtel JT, Winant RC, Ganem D (2005) Host and viral proteins in the virion of Kaposi's sarcoma-associated herpesvirus. J Virol 79(8):4952–4964. https://doi.org/10.1128/JVI.79.8.4952-4964.2005

Beitia Ortiz de Zarate I, Cantero-Aguilar L, Longo M, Berlioz-Torrent C, Rozenberg F (2007) Contribution of endocytic motifs in the cytoplasmic tail of herpes simplex virus type 1 glycoprotein B to virus replication and cell-cell fusion. J Virol 81(24):13889–13903. https://doi.org/10.1128/JVI.01231-07

Brecher M, Schornberg KL, Delos SE, Fusco ML, Saphire EO, White JM (2012) Cathepsin cleavage potentiates the Ebola virus glycoprotein to undergo a subsequent fusion-relevant conformational change. J Virol 86(1):364–372. https://doi.org/10.1128/JVI.05708-11

Browne H, Bell S, Minson T (2004) Analysis of the requirement for glycoprotein m in herpes simplex virus type 1 morphogenesis. J Virol 78(2):1039–1041

Campadelli-Fiume G, Guerrini S, Liu X, Foa-Tomasi L (1993) Monoclonal antibodies to glycoprotein B differentiate human herpesvirus 6 into two clusters, variants A and B. J Gen Virol 74(Pt 10):2257–2262

Caruso A, Rotola A, Comar M, Favilli F, Galvan M, Tosetti M, Campello C, Caselli E, Alessandri G, Grassi M, Garrafa E, Cassai E, Di Luca D (2002) HHV-6 infects human aortic and heart microvascular endothelial cells, increasing their ability to secrete proinflammatory chemokines. J Med Virol 67(4):528–533. https://doi.org/10.1002/jmv.10133

Cermelli C, Concari M, Carubbi F, Fabio G, Sabbatini AM, Pecorari M, Pietrosemoli P, Meacci M, Guicciardi E, Carulli N, Portolani M (1996) Growth of human herpesvirus 6 in HEPG2 cells. Virus Res 45(2):75–85

Challoner PB, Smith KT, Parker JD, MacLeod DL, Coulter SN, Rose TM, Schultz ER, Bennett JL, Garber RL, Chang M et al (1995) Plaque-associated expression of human herpesvirus 6 in multiple sclerosis. Proc Natl Acad Sci U S A 92(16):7440–7444

Chan DC, Fass D, Berger JM, Kim PS (1997) Core structure of gp41 from the HIV envelope glycoprotein. Cell 89(2):263–273

Chan PK, Ng HK, Hui M, Cheng AF (2001) Prevalence and distribution of human herpesvirus 6 variants A and B in adult human brain. J Med Virol 64(1):42–46

Chandramouli S, Ciferri C, Nikitin PA, Calo S, Gerrein R, Balabanis K, Monroe J, Hebner C, Lilja AE, Settembre EC, Carfi A (2015) Structure of HCMV glycoprotein B in the postfusion conformation bound to a neutralizing human antibody. Nat Commun 6:8176. https://doi.org/10.1038/ncomms9176

Chandran K, Sullivan NJ, Felbor U, Whelan SP, Cunningham JM (2005) Endosomal proteolysis of the Ebola virus glycoprotein is necessary for infection. Science 308(5728):1643–1645. https://doi.org/10.1126/science.1110656

Chapenko S, Krumina A, Kozireva S, Nora Z, Sultanova A, Viksna L, Murovska M (2006) Activation of human herpesviruses 6 and 7 in patients with chronic fatigue syndrome. J Clin Virol 37(Suppl 1):S47–S51. https://doi.org/10.1016/S1386-6532(06)70011-7

Chen M, Popescu N, Woodworth C, Berneman Z, Corbellino M, Lusso P, Ablashi DV, DiPaolo JA (1994) Human herpesvirus 6 infects cervical epithelial cells and transactivates human papillomavirus gene expression. J Virol 68(2):1173–1178

Chou S, Marousek GI (1992) Homology of the envelope glycoprotein B of human herpesvirus-6 and cytomegalovirus. Virology 191(1):523–528

Chouljenko DV, Kim IJ, Chouljenko VN, Subramanian R, Walker JD, Kousoulas KG (2012) Functional hierarchy of herpes simplex virus 1 viral glycoproteins in cytoplasmic virion envelopment and egress. J Virol 86(8):4262–4270. https://doi.org/10.1128/JVI.06766-11

Cirone M, Zompetta C, Angeloni A, Ablashi DV, Salahuddin SZ, Pavan A, Torrisi MR, Frati L, Faggioni A (1992) Infection by human herpesvirus 6 (HHV-6) of human lymphoid T cells occurs through an endocytic pathway. AIDS Res Hum Retroviruses 8(12):2031–2037. https://doi.org/10.1089/aid.1992.8.2031

Conti C, Cirone M, Sgro R, Altieri F, Zompetta C, Faggioni A (2000) Early interactions of human herpesvirus 6 with lymphoid cells: role of membrane protein components and glycosaminoglycans in virus binding. J Med Virol 62(4):487–497

Cote M, Misasi J, Ren T, Bruchez A, Lee K, Filone CM, Hensley L, Li Q, Ory D, Chandran K, Cunningham J (2011) Small molecule inhibitors reveal Niemann-Pick C1 is essential for Ebola virus infection. Nature 477(7364):344–348. https://doi.org/10.1038/nature10380

Crump CM, Bruun B, Bell S, Pomeranz LE, Minson T, Browne HM (2004) Alphaherpesvirus glycoprotein M causes the relocalization of plasma membrane proteins. J Gen Virol 85(Pt 12):3517–3527. https://doi.org/10.1099/vir.0.80361-0

Cuomo L, Trivedi P, Cardillo MR, Gagliardi FM, Vecchione A, Caruso R, Calogero A, Frati L, Faggioni A, Ragona G (2001) Human herpesvirus 6 infection in neoplastic and normal brain tissue. J Med Virol 63(1):45–51

Daibata M, Hatakeyama N, Kamioka M, Nemoto Y, Hiroi M, Miyoshi I, Taguchi H (2001) Detection of human herpesvirus 6 and JC virus in progressive multifocal leukoencephalopathy complicating follicular lymphoma. Am J Hematol 67(3):200–205. https://doi.org/10.1002/ajh.1108

De Bolle L, Van Loon J, De Clercq E, Naesens L (2005) Quantitative analysis of human herpesvirus 6 cell tropism. J Med Virol 75(1):76–85. https://doi.org/10.1002/jmv.20240

Dijkstra JM, Visser N, Mettenleiter TC, Klupp BG (1996) Identification and characterization of pseudorabies virus glycoprotein gM as a nonessential virion component. J Virol 70(8):5684–5688

Dohner K, Wolfstein A, Prank U, Echeverri C, Dujardin D, Vallee R, Sodeik B (2002) Function of dynein and dynactin in herpes simplex virus capsid transport. Mol Biol Cell 13(8):2795–2809. https://doi.org/10.1091/mbc.01-07-0348

Dominguez G, Dambaugh TR, Stamey FR, Dewhurst S, Inoue N, Pellett PE (1999) Human herpesvirus 6B genome sequence: coding content and comparison with human herpesvirus 6A. J Virol 73(10):8040–8052

Doms RW, Moore JP (2000) HIV-1 membrane fusion: targets of opportunity. J Cell Biol 151(2):F9–14

Donati D, Akhyani N, Fogdell-Hahn A, Cermelli C, Cassiani-Ingoni R, Vortmeyer A, Heiss JD, Cogen P, Gaillard WD, Sato S, Theodore WH, Jacobson S (2003) Detection of human herpesvirus-6 in mesial temporal lobe epilepsy surgical brain resections. Neurology 61(10):1405–1411

Downing RG, Sewankambo N, Serwadda D, Honess R, Crawford D, Jarrett R, Griffin BE (1987) Isolation of human lymphotropic herpesviruses from Uganda. Lancet 2(8555):390

Ellinger K, Neipel F, Foa-Tomasi L, Campadelli-Fiume G, Fleckenstein B (1993) The glycoprotein B homologue of human herpesvirus 6. J Gen Virol 74(Pt 3):495–500. https://doi.org/10.1099/0022-1317-74-3-495

Foa-Tomasi L, Guerrini S, Huang T, Campadelli-Fiume G (1992) Characterization of human herpesvirus-6(U1102) and (GS) gp112 and identification of the Z29-specified homolog. Virology 191(1):511–516

Foster TP, Melancon JM, Kousoulas KG (2001) An alpha-helical domain within the carboxyl terminus of herpes simplex virus type 1 (HSV-1) glycoprotein B (gB) is associated with cell fusion and resistance to heparin inhibition of cell fusion. Virology 287(1):18–29. https://doi.org/10.1006/viro.2001.1004

Fox JD, Briggs M, Ward PA, Tedder RS (1990) Human herpesvirus 6 in salivary glands. Lancet 336(8715):590–593. doi:0140-6736(90)93392-3 [pii]

Furuta RA, Wild CT, Weng Y, Weiss CD (1998) Capture of an early fusion-active conformation of HIV-1 gp41. Nat Struct Biol 5(4):276–279

Gaggar A, Shayakhmetov DM, Lieber A (2003) CD46 is a cellular receptor for group B adenoviruses. Nat Med 9(11):1408–1412. https://doi.org/10.1038/nm952

Garcia NJ, Chen J, Longnecker R (2013) Modulation of Epstein-Barr virus glycoprotein B (gB) fusion activity by the gB cytoplasmic tail domain. MBio 4(1):e00571-00512. https://doi.org/10.1128/mBio.00571-12

Gentile I, Talamo M, Borgia G (2010) Is the drug-induced hypersensitivity syndrome (DIHS) due to human herpesvirus 6 infection or to allergy-mediated viral reactivation? Report of a case and literature review. BMC Infect Dis 10:49. doi: 1471-2334-10-49 [pii] 10.1186/1471-2334-10-49

Glosson NL, Hudson AW (2007) Human herpesvirus-6A and -6B encode viral immunoevasins that downregulate class I MHC molecules. Virology 365(1):125–135. https://doi.org/10.1016/j.virol.2007.03.048

Gompels UA, Macaulay HA (1995) Characterization of human telomeric repeat sequences from human herpesvirus 6 and relationship to replication. J Gen Virol 76(Pt 2):451–458. https://doi.org/10.1099/0022-1317-76-2-451

Gompels UA, Nicholas J, Lawrence G, Jones M, Thomson BJ, Martin ME, Efstathiou S, Craxton M, Macaulay HA (1995) The DNA sequence of human herpesvirus-6: structure, coding content, and genome evolution. Virology 209(1):29–51. https://doi.org/10.1006/viro.1995.1228

Greenstone HL, Santoro F, Lusso P, Berger EA (2002) Human Herpesvirus 6 and Measles Virus employ distinct CD46 domains for receptor function. J Biol Chem 277(42):39112–39118. https://doi.org/10.1074/jbc.M206488200

Grivel JC, Santoro F, Chen S, Faga G, Malnati MS, Ito Y, Margolis L, Lusso P (2003) Pathogenic effects of human herpesvirus 6 in human lymphoid tissue ex vivo. J Virol 77(15):8280–8289

Hallenberger S, Bosch V, Angliker H, Shaw E, Klenk HD, Garten W (1992) Inhibition of furin-mediated cleavage activation of HIV-1 glycoprotein gp160. Nature 360(6402):358–361. https://doi.org/10.1038/360358a0

Hansen AS, Bundgaard BB, Biltoft M, Rossen LS, Hollsberg P (2017) Divergent tropism of HHV-6AGS and HHV-6BPL1 in T cells expressing different CD46 isoform patterns. Virology 502:160–170. https://doi.org/10.1016/j.virol.2016.12.027

Harma M, Hockerstedt K, Lautenschlager I (2003) Human herpesvirus-6 and acute liver failure. Transplantation 76(3):536–539. https://doi.org/10.1097/01.TP.0000069233.13409.DF

Harrison SC (2008) The pH sensor for flavivirus membrane fusion. J Cell Biol 183(2):177–179. https://doi.org/10.1083/jcb.200809175

Hayashi M, Yoshida K, Tang H, Sadaoka T, Kawabata A, Jasirwan C, Mori Y (2014) Characterization of the human herpesvirus 6A U23 gene. Virology 450-451:98–105. https://doi.org/10.1016/j.virol.2013.12.004

He J, McCarthy M, Zhou Y, Chandran B, Wood C (1996) Infection of primary human fetal astrocytes by human herpesvirus 6. J Virol 70(2):1296–1300

Herold BC, WuDunn D, Soltys N, Spear PG (1991) Glycoprotein C of herpes simplex virus type 1 plays a principal role in the adsorption of virus to cells and in infectivity. J Virol 65(3):1090–1098

Higgins LM, McDonald SA, Whittle N, Crockett N, Shields JG, MacDonald TT (1999) Regulation of T cell activation in vitro and in vivo by targeting the OX40-OX40 ligand interaction: amelioration of ongoing inflammatory bowel disease with an OX40-IgG fusion protein, but not with an OX40 ligand-IgG fusion protein. J Immunol 162(1):486–493

Hobom U, Brune W, Messerle M, Hahn G, Koszinowski UH (2000) Fast screening procedures for random transposon libraries of cloned herpesvirus genomes: mutational analysis of human cytomegalovirus envelope glycoprotein genes. J Virol 74(17):7720–7729

Isegawa Y, Mukai T, Nakano K, Kagawa M, Chen J, Mori Y, Sunagawa T, Kawanishi K, Sashihara J, Hata A, Zou P, Kosuge H, Yamanishi K (1999) Comparison of the complete DNA sequences of human herpesvirus 6 variants A and B. J Virol 73(10):8053–8063

Ishikawa K, Hasegawa K, Naritomi T, Kanai N, Ogawa M, Kato Y, Kobayashi M, Torii N, Hayashi N (2002) Prevalence of herpesviridae and hepatitis virus sequences in the livers of patients with fulminant hepatitis of unknown etiology in Japan. J Gastroenterol 37(7):523–530. https://doi.org/10.1007/s005350200081

Ivanovic T, Choi JL, Whelan SP, van Oijen AM, Harrison SC (2013) Influenza-virus membrane fusion by cooperative fold-back of stochastically induced hemagglutinin intermediates. Elife 2:e00333. https://doi.org/10.7554/eLife.00333

Jasirwan C, Furusawa Y, Tang H, Maeki T, Mori Y (2014) Human herpesvirus-6A gQ1 and gQ2 are critical for human CD46 usage. Microbiol Immunol 58(1):22–30. https://doi.org/10.1111/1348-0421.12110

Jiang XJ, Adler B, Sampaio KL, Digel M, Jahn G, Ettischer N, Stierhof YD, Scrivano L, Koszinowski U, Mach M, Sinzger C (2008) UL74 of human cytomegalovirus contributes to virus release by promoting secondary envelopment of virions. J Virol 82(6):2802–2812. https://doi.org/10.1128/JVI.01550-07

Kabanova A, Marcandalli J, Zhou T, Bianchi S, Baxa U, Tsybovsky Y, Lilleri D, Silacci-Fregni C, Foglierini M, Fernandez-Rodriguez BM, Druz A, Zhang B, Geiger R, Pagani M, Sallusto F, Kwong PD, Corti D, Lanzavecchia A, Perez L (2016) Platelet-derived growth factor-alpha receptor is the cellular receptor for human cytomegalovirus gHgLgO trimer. Nat Microbiol 1(8):16082. https://doi.org/10.1038/nmicrobiol.2016.82

Kakimoto M, Hasegawa A, Fujita S, Yasukawa M (2002) Phenotypic and functional alterations of dendritic cells induced by human herpesvirus 6 infection. J Virol 76(20):10338–10345

Kallstrom H, Liszewski MK, Atkinson JP, Jonsson AB (1997) Membrane cofactor protein (MCP or CD46) is a cellular pilus receptor for pathogenic Neisseria. Mol Microbiol 25(4):639–647

Kallstrom H, Blackmer Gill D, Albiger B, Liszewski MK, Atkinson JP, Jonsson AB (2001) Attachment of Neisseria gonorrhoeae to the cellular pilus receptor CD46: identification of domains important for bacterial adherence. Cell Microbiol 3(3):133–143

Kasolo FC, Mpabalwani E, Gompels UA (1997) Infection with AIDS-related herpesviruses in human immunodeficiency virus-negative infants and endemic childhood Kaposi's sarcoma in Africa. J Gen Virol 78(Pt 4):847–855. https://doi.org/10.1099/0022-1317-78-4-847

Kaufer BB, Flamand L (2014) Chromosomally integrated HHV-6: impact on virus, cell and organismal biology. Curr Opin Virol 9:111–118. https://doi.org/10.1016/j.coviro.2014.09.010

Kawabata A, Oyaizu H, Maeki T, Tang H, Yamanishi K, Mori Y (2011) Analysis of a neutralizing antibody for human herpesvirus 6B reveals a role for glycoprotein Q1 in viral entry. J Virol 85(24):12962–12971. https://doi.org/10.1128/JVI.05622-11

Kawabata A, Jasirwan C, Yamanishi K, Mori Y (2012) Human herpesvirus 6 glycoprotein M is essential for virus growth and requires glycoprotein N for its maturation. Virology 429(1):21–28. https://doi.org/10.1016/j.virol.2012.03.027

Kawabata A, Serada S, Naka T, Mori Y (2014) Human herpesvirus 6 gM/gN complex interacts with v-SNARE in infected cells. J Gen Virol 95(Pt 12):2769–2777. https://doi.org/10.1099/vir.0.069336-0

Kemper C, Atkinson JP (2009) Measles virus and CD46. Curr Top Microbiol Immunol 329:31–57

Kim IJ, Chouljenko VN, Walker JD, Kousoulas KG (2013) Herpes simplex virus 1 glycoprotein M and the membrane-associated protein UL11 are required for virus-induced cell fusion and efficient virus entry. J Virol 87(14):8029–8037. https://doi.org/10.1128/JVI.01181-13

Klupp BG, Nixdorf R, Mettenleiter TC (2000) Pseudorabies virus glycoprotein M inhibits membrane fusion. J Virol 74(15):6760–6768

Kofod-Olsen E, Ross-Hansen K, Schleimann MH, Jensen DK, Moller JM, Bundgaard B, Mikkelsen JG, Hollsberg P (2012) U20 is responsible for human herpesvirus 6B inhibition of tumor necrosis factor receptor-dependent signaling and apoptosis. J Virol 86(21):11483–11492. https://doi.org/10.1128/JVI.00847-12

Kondo K, Shimada K, Sashihara J, Tanaka-Taya K, Yamanishi K (2002) Identification of human herpesvirus 6 latency-associated transcripts. J Virol 76(8):4145–4151

Koyano S, Mar EC, Stamey FR, Inoue N (2003) Glycoproteins M and N of human herpesvirus 8 form a complex and inhibit cell fusion. J Gen Virol 84(Pt 6):1485–1491. https://doi.org/10.1099/vir.0.18941-0

Land A, Braakman I (2001) Folding of the human immunodeficiency virus type 1 envelope glycoprotein in the endoplasmic reticulum. Biochimie 83(8):783–790

Lau SY, Crump CM (2015) HSV-1 gM and the gK/pUL20 complex are important for the localization of gD and gH/L to viral assembly sites. Viruses 7(3):915–938. https://doi.org/10.3390/v7030915

Leibovitch EC, Jacobson S (2014) Evidence linking HHV-6 with multiple sclerosis: an update. Curr Opin Virol 9:127–133. https://doi.org/10.1016/j.coviro.2014.09.016

Leibovitch E, Wohler JE, Cummings Macri SM, Motanic K, Harberts E, Gaitan MI, Maggi P, Ellis M, Westmoreland S, Silva A, Reich DS, Jacobson S (2013) Novel marmoset (Callithrix jacchus) model of human Herpesvirus 6A and 6B infections: immunologic, virologic and radiologic characterization. PLoS Pathog 9(1):e1003138. https://doi.org/10.1371/journal.ppat.1003138

Liszewski MK, Kemper C, Price JD, Atkinson JP (2005) Emerging roles and new functions of CD46. Springer Semin Immunopathol 27(3):345–358. https://doi.org/10.1007/s00281-005-0002-3

Liu J, Bartesaghi A, Borgnia MJ, Sapiro G, Subramaniam S (2008) Molecular architecture of native HIV-1 gp120 trimers. Nature 455(7209):109–113. https://doi.org/10.1038/nature07159

Lopez C, Pellett P, Stewart J, Goldsmith C, Sanderlin K, Black J, Warfield D, Feorino P (1988) Characteristics of human herpesvirus-6. J Infect Dis 157(6):1271–1273

Luppi M, Barozzi P, Maiorana A, Marasca R, Trovato R, Fano R, Ceccherini-Nelli L, Torelli G (1995) Human herpesvirus-6: a survey of presence and distribution of genomic sequences in normal brain and neuroglial tumors. J Med Virol 47(1):105–111

Luppi M, Barozzi P, Morris C, Maiorana A, Garber R, Bonacorsi G, Donelli A, Marasca R, Tabilio A, Torelli G (1999) Human herpesvirus 6 latently infects early bone marrow progenitors in vivo. J Virol 73(1):754–759

Lusso P, Salahuddin SZ, Ablashi DV, Gallo RC, Di Marzo Veronese F, Markham PD (1987) Diverse tropism of HBLV (human herpesvirus 6). Lancet 2(8561):743

Lusso P, Markham PD, Tschachler E, di Marzo Veronese F, Salahuddin SZ, Ablashi DV, Pahwa S, Krohn K, Gallo RC (1988) In vitro cellular tropism of human B-lymphotropic virus (human herpesvirus-6). J Exp Med 167(5):1659–1670

Lusso P, Malnati MS, Garzino-Demo A, Crowley RW, Long EO, Gallo RC (1993) Infection of natural killer cells by human herpesvirus 6. Nature 362(6419):458–462. https://doi.org/10.1038/362458a0

MacLean CA, Efstathiou S, Elliott ML, Jamieson FE, McGeoch DJ (1991) Investigation of herpes simplex virus type 1 genes encoding multiply inserted membrane proteins. J Gen Virol 72(Pt 4):897–906. https://doi.org/10.1099/0022-1317-72-4-897

Maeki T, Hayashi M, Kawabata A, Tang H, Yamanishi K, Mori Y (2013) Identification of the human herpesvirus 6A gQ1 domain essential for its functional conformation. J Virol 87(12):7054–7063. https://doi.org/10.1128/JVI.00611-13

Mahmoud NF, Jasirwan C, Kanemoto S, Wakata A, Wang B, Hata Y, Nagamata S, Kawabata A, Tang H, Mori Y (2016) Cytoplasmic tail domain of glycoprotein B is essential for HHV-6 infection. Virology 490:1–5. https://doi.org/10.1016/j.virol.2015.12.018

Marsters SA, Frutkin AD, Simpson NJ, Fendly BM, Ashkenazi A (1992) Identification of cysteine-rich domains of the type 1 tumor necrosis factor receptor involved in ligand binding. J Biol Chem 267(9):5747–5750

May NA, Glosson NL, Hudson AW (2010) Human herpesvirus 7 u21 downregulates classical and nonclassical class I major histocompatibility complex molecules from the cell surface. J Virol 84(8):3738–3751. https://doi.org/10.1128/JVI.01782-09

Mirandola P, Menegazzi P, Merighi S, Ravaioli T, Cassai E, Di Luca D (1998) Temporal mapping of transcripts in herpesvirus 6 variants. J Virol 72(5):3837–3844

Mock DJ, Powers JM, Goodman AD, Blumenthal SR, Ergin N, Baker JV, Mattson DH, Assouline JG, Bergey EJ, Chen B, Epstein LG, Blumberg BM (1999) Association of human herpesvirus 6 with the demyelinative lesions of progressive multifocal leukoencephalopathy. J Neurovirol 5(4):363–373

Moller-Tank S, Maury W (2015) Ebola virus entry: a curious and complex series of events. PLoS Pathog 11(4):e1004731. https://doi.org/10.1371/journal.ppat.1004731

Mori Y, Seya T, Huang HL, Akkapaiboon P, Dhepakson P, Yamanishi K (2002) Human herpesvirus 6 variant A but not variant B induces fusion from without in a variety of human cells through a human herpesvirus 6 entry receptor, CD46. J Virol 76(13):6750–6761

Mori Y, Akkapaiboon P, Yang X, Yamanishi K (2003a) The human herpesvirus 6 U100 gene product is the third component of the gH-gL glycoprotein complex on the viral envelope. J Virol 77(4):2452–2458

Mori Y, Yang X, Akkapaiboon P, Okuno T, Yamanishi K (2003b) Human herpesvirus 6 variant A glycoprotein H-glycoprotein L-glycoprotein Q complex associates with human CD46. J Virol 77(8):4992–4999

Mori Y, Akkapaiboon P, Yonemoto S, Koike M, Takemoto M, Sadaoka T, Sasamoto Y, Konishi S, Uchiyama Y, Yamanishi K (2004) Discovery of a second form of tripartite complex containing gH-gL of human herpesvirus 6 and observations on CD46. J Virol 78(9):4609–4616

Mori Y, Koike M, Moriishi E, Kawabata A, Tang H, Oyaizu H, Uchiyama Y, Yamanishi K (2008) Human herpesvirus-6 induces MVB formation, and virus egress occurs by an exosomal release pathway. Traffic 9(10):1728–1742. https://doi.org/10.1111/j.1600-0854.2008.00796.x

Noseworthy JH, Lucchinetti C, Rodriguez M, Weinshenker BG (2000) Multiple sclerosis. N Engl J Med 343(13):938–952. https://doi.org/10.1056/NEJM200009283431307

Okuno T, Takahashi K, Balachandra K, Shiraki K, Yamanishi K, Takahashi M, Baba K (1989) Seroepidemiology of human herpesvirus 6 infection in normal children and adults. J Clin Microbiol 27(4):651–653

Ota M, Serada S, Naka T, Mori Y (2014) MHC class I molecules are incorporated into human herpesvirus-6 viral particles and released into the extracellular environment. Microbiol Immunol 58(2):119–125. https://doi.org/10.1111/1348-0421.12121

Owen DJ, Evans PR (1998) A structural explanation for the recognition of tyrosine-based endocytotic signals. Science 282(5392):1327–1332

Oyaizu H, Tang H, Ota M, Takenaka N, Ozono K, Yamanishi K, Mori Y (2012) Complementation of the function of glycoprotein H of human herpesvirus 6 variant A by glycoprotein H of variant B in the virus life cycle. J Virol 86(16):8492–8498. https://doi.org/10.1128/JVI.00504-12

Ozaki Y, Tajiri H, Tanaka-Taya K, Mushiake S, Kimoto A, Yamanishi K, Okada S (2001) Frequent detection of the human herpesvirus 6-specific genomes in the livers of children with various liver diseases. J Clin Microbiol 39(6):2173–2177. https://doi.org/10.1128/JCM.39.6.2173-2177.2001

Pedersen SM, Oster B, Bundgaard B, Hollsberg P (2006) Induction of cell-cell fusion from without by human herpesvirus 6B. J Virol 80(19):9916–9920. https://doi.org/10.1128/JVI.02693-05

Pfeiffer B, Berneman ZN, Neipel F, Chang CK, Tirwatnapong S, Chandran B (1993) Identification and mapping of the gene encoding the glycoprotein complex gp82-gp105 of human herpesvirus 6 and mapping of the neutralizing epitope recognized by monoclonal antibodies. J Virol 67(8):4611–4620

Pfeiffer B, Thomson B, Chandran B (1995) Identification and characterization of a cDNA derived from multiple splicing that encodes envelope glycoprotein gp105 of human herpesvirus 6. J Virol 69(6):3490–3500

Prusty BK, Krohne G, Rudel T (2013) Reactivation of chromosomally integrated human herpesvirus-6 by telomeric circle formation. PLoS Genet 9(12):e1004033. https://doi.org/10.1371/journal.pgen.1004033

Reynaud JM, Jegou JF, Welsch JC, Horvat B (2014) Human herpesvirus 6A infection in CD46 transgenic mice: viral persistence in the brain and increased production of proinflammatory chemokines via Toll-like receptor 9. J Virol 88(10):5421–5436. https://doi.org/10.1128/JVI.03763-13

Rezcallah MS, Hodges K, Gill DB, Atkinson JP, Wang B, Cleary PP (2005) Engagement of CD46 and alpha5beta1 integrin by group A streptococci is required for efficient invasion of epithelial cells. Cell Microbiol 7(5):645–653. https://doi.org/10.1111/j.1462-5822.2004.00497.x

Riley-Vargas RC, Gill DB, Kemper C, Liszewski MK, Atkinson JP (2004) CD46: expanding beyond complement regulation. Trends Immunol 25(9):496–503. https://doi.org/10.1016/j.it.2004.07.004

Roizmann B, Desrosiers RC, Fleckenstein B, Lopez C, Minson AC, Studdert MJ (1992) The family Herpesviridae: an update. The Herpesvirus Study Group of the International Committee on Taxonomy of Viruses. Arch Virol 123(3–4):425–449

Roush KS, Domiati-Saad RK, Margraf LR, Krisher K, Scheuermann RH, Rogers BB, Dawson DB (2001) Prevalence and cellular reservoir of latent human herpesvirus 6 in tonsillar lymphoid tissue. Am J Clin Pathol 116(5):648–654. https://doi.org/10.1309/Y2HH-B1CK-0F5L-U7B8

Ryckman BJ, Jarvis MA, Drummond DD, Nelson JA, Johnson DC (2006) Human cytomegalovirus entry into epithelial and endothelial cells depends on genes UL128 to UL150 and occurs by endocytosis and low-pH fusion. J Virol 80(2):710–722. https://doi.org/10.1128/JVI.80.2.710-722.2006

Sadaoka T, Yamanishi K, Mori Y (2006) Human herpesvirus 7 U47 gene products are glycoproteins expressed in virions and associate with glycoprotein H. J Gen Virol 87(Pt 3):501–508. https://doi.org/10.1099/vir.0.81374-0

Saeed MF, Kolokoltsov AA, Albrecht T, Davey RA (2010) Cellular entry of ebola virus involves uptake by a macropinocytosis-like mechanism and subsequent trafficking through early and late endosomes. PLoS Pathog 6(9):e1001110. https://doi.org/10.1371/journal.ppat.1001110

Salahuddin SZ, Ablashi DV, Markham PD, Josephs SF, Sturzenegger S, Kaplan M, Halligan G, Biberfeld P, Wong-Staal F, Kramarsky B et al (1986) Isolation of a new virus, HBLV, in patients with lymphoproliferative disorders. Science 234(4776):596–601

Santoro F, Kennedy PE, Locatelli G, Malnati MS, Berger EA, Lusso P (1999) CD46 is a cellular receptor for human herpesvirus 6. Cell 99(7):817–827

Santoro F, Greenstone HL, Insinga A, Liszewski MK, Atkinson JP, Lusso P, Berger EA (2003) Interaction of glycoprotein H of human herpesvirus 6 with the cellular receptor CD46. J Biol Chem 278(28):25964–25969. https://doi.org/10.1074/jbc.M302373200

Saxinger C, Polesky H, Eby N, Grufferman S, Murphy R, Tegtmeir G, Parekh V, Memon S, Hung C (1988) Antibody reactivity with HBLV (HHV-6) in U.S. populations. J Virol Methods 21(1–4):199–208

Schiewe U, Neipel F, Schreiner D, Fleckenstein B (1994) Structure and transcription of an immediate-early region in the human herpesvirus 6 genome. J Virol 68(5):2978–2985

Schmiedel D, Tai J, Levi-Schaffer F, Dovrat S, Mandelboim O (2016) Human Herpesvirus 6 down-regulates the expression of activating ligands during lytic infection to escape elimination by natural killer cells. J Virol 90:9608–9617. https://doi.org/10.1128/JVI.01164-16

Scrivano L, Esterlechner J, Muhlbach H, Ettischer N, Hagen C, Grunewald K, Mohr CA, Ruzsics Z, Koszinowski U, Adler B (2010) The m74 gene product of murine cytomegalovirus (MCMV) is a functional homolog of human CMV gO and determines the entry pathway of MCMV. J Virol 84(9):4469–4480. https://doi.org/10.1128/JVI.02441-09

Secchiero P, Sun D, De Vico AL, Crowley RW, Reitz MS Jr, Zauli G, Lusso P, Gallo RC (1997) Role of the extracellular domain of human herpesvirus 7 glycoprotein B in virus binding to cell surface heparan sulfate proteoglycans. J Virol 71(6):4571–4580

Seya T, Hara T, Matsumoto M, Sugita Y, Akedo H (1990) Complement-mediated tumor cell damage induced by antibodies against membrane cofactor protein (MCP, CD46). J Exp Med 172(6):1673–1680

Shaw ML, Stone KL, Colangelo CM, Gulcicek EE, Palese P (2008) Cellular proteins in influenza virus particles. PLoS Pathog 4(6):e1000085. https://doi.org/10.1371/journal.ppat.1000085

Skehel JJ, Wiley DC (2000) Receptor binding and membrane fusion in virus entry: the influenza hemagglutinin. Annu Rev Biochem 69:531–569. https://doi.org/10.1146/annurev.biochem.69.1.531

Skrincosky D, Hocknell P, Whetter L, Secchiero P, Chandran B, Dewhurst S (2000) Identification and analysis of a novel heparin-binding glycoprotein encoded by human herpesvirus 7. J Virol 74(10):4530–4540

Soroceanu L, Akhavan A, Cobbs CS (2008) Platelet-derived growth factor-alpha receptor activation is required for human cytomegalovirus infection. Nature 455(7211):391–395. https://doi.org/10.1038/nature07209

Spear PG (2004) Herpes simplex virus: receptors and ligands for cell entry. Cell Microbiol 6(5):401–410. https://doi.org/10.1111/j.1462-5822.2004.00389.x

Stamey FR, Dominguez G, Black JB, Dambaugh TR, Pellett PE (1995) Intragenomic linear amplification of human herpesvirus 6B oriLyt suggests acquisition of oriLyt by transposition. J Virol 69(1):589–596

Stegen C, Yakova Y, Henaff D, Nadjar J, Duron J, Lippe R (2013) Analysis of virion-incorporated host proteins required for herpes simplex virus type 1 infection through a RNA interference screen. PLoS One 8(1):e53276. https://doi.org/10.1371/journal.pone.0053276

Steinhauer DA (1999) Role of hemagglutinin cleavage for the pathogenicity of influenza virus. Virology 258(1):1–20. https://doi.org/10.1006/viro.1999.9716

Stiasny K, Fritz R, Pangerl K, Heinz FX (2011) Molecular mechanisms of flavivirus membrane fusion. Amino Acids 41(5):1159–1163. https://doi.org/10.1007/s00726-009-0370-4

Tal-Singer R, Peng C, Ponce De Leon M, Abrams WR, Banfield BW, Tufaro F, Cohen GH, Eisenberg RJ (1995) Interaction of herpes simplex virus glycoprotein gC with mammalian cell surface molecules. J Virol 69(7):4471–4483

Tanaka Y, Suenaga T, Matsumoto M, Seya T, Arase H (2013) Herpesvirus 6 glycoproteins B (gB), gH, gL, and gQ are necessary and sufficient for cell-to-cell fusion. J Virol 87(19):10900–10903. https://doi.org/10.1128/JVI.01427-13

Tang H, Mori Y (2015) Determinants of Human CD134 Essential for Entry of Human Herpesvirus 6B. J Virol 89(19):10125–10129. https://doi.org/10.1128/JVI.01606-15

Tang H, Hayashi M, Maeki T, Yamanishi K, Mori Y (2011) Human herpesvirus 6 glycoprotein complex formation is required for folding and trafficking of the gH/gL/gQ1/gQ2 complex and its cellular receptor binding. J Virol 85(21):11121–11130. https://doi.org/10.1128/JVI.05251-11

Tang H, Serada S, Kawabata A, Ota M, Hayashi E, Naka T, Yamanishi K, Mori Y (2013) CD134 is a cellular receptor specific for human herpesvirus-6B entry. Proc Natl Acad Sci U S A 110(22):9096–9099. doi:10.1073/pnas.1305187110 1305187110 [pii]

Tang H, Wang J, Mahmoud NF, Mori Y (2014) Detailed study of the interaction between human herpesvirus 6B glycoprotein complex and its cellular receptor, human CD134. J Virol 88(18):10875–10882. https://doi.org/10.1128/JVI.01447-14

Tang H, Mahmoud NF, Mori Y (2015) Maturation of human herpesvirus 6A glycoprotein O requires coexpression of glycoprotein H and glycoprotein L. J Virol 89(9):5159–5163. https://doi.org/10.1128/JVI.00140-15

Tanner A, Carlson SA, Nukui M, Murphy EA, Berges BK (2013) Human herpesvirus 6A infection and immunopathogenesis in humanized Rag2(-)/(-) gammac(-)/(-) mice. J Virol 87(22):12020–12028. https://doi.org/10.1128/JVI.01556-13

Torrisi MR, Gentile M, Cardinali G, Cirone M, Zompetta C, Lotti LV, Frati L, Faggioni A (1999) Intracellular transport and maturation pathway of human herpesvirus 6. Virology 257(2):460–471. https://doi.org/10.1006/viro.1999.9699

Trybala E, Liljeqvist JA, Svennerholm B, Bergstrom T (2000) Herpes simplex virus types 1 and 2 differ in their interaction with heparan sulfate. J Virol 74(19):9106–9114

Vanarsdall AL, Johnson DC (2012) Human cytomegalovirus entry into cells. Curr Opin Virol 2(1):37–42. https://doi.org/10.1016/j.coviro.2012.01.001

Vanarsdall AL, Chase MC, Johnson DC (2011) Human cytomegalovirus glycoprotein gO complexes with gH/gL, promoting interference with viral entry into human fibroblasts but not entry into epithelial cells. J Virol 85(22):11638–11645. https://doi.org/10.1128/JVI.05659-11

Voorhees P, Deignan E, van Donselaar E, Humphrey J, Marks MS, Peters PJ, Bonifacino JS (1995) An acidic sequence within the cytoplasmic domain of furin functions as a determinant of trans-Golgi network localization and internalization from the cell surface. EMBO J 14(20):4961–4975

Weinberg AD, Bourdette DN, Sullivan TJ, Lemon M, Wallin JJ, Maziarz R, Davey M, Palida F, Godfrey W, Engleman E, Fulton RJ, Offner H, Vandenbark AA (1996) Selective depletion of myelin-reactive T cells with the anti-OX-40 antibody ameliorates autoimmune encephalomyelitis. Nat Med 2(2):183–189

Wille PT, Knoche AJ, Nelson JA, Jarvis MA, Johnson DC (2010) A human cytomegalovirus gO-null mutant fails to incorporate gH/gL into the virion envelope and is unable to enter fibroblasts and epithelial and endothelial cells. J Virol 84(5):2585–2596. https://doi.org/10.1128/JVI.02249-09

Williams RK, Straus SE (1997) Specificity and affinity of binding of herpes simplex virus type 2 glycoprotein B to glycosaminoglycans. J Virol 71(2):1375–1380

Wyatt LS, Balachandran N, Frenkel N (1990) Variations in the replication and antigenic properties of human herpesvirus 6 strains. J Infect Dis 162(4):852–857

Yamanishi K, Okuno T, Shiraki K, Takahashi M, Kondo T, Asano Y, Kurata T (1988) Identification of human herpesvirus-6 as a causal agent for exanthem subitum. Lancet 1(8594):1065–1067

Zhang N, Yan J, Lu G, Guo Z, Fan Z, Wang J, Shi Y, Qi J, Gao GF (2011) Binding of herpes simplex virus glycoprotein D to nectin-1 exploits host cell adhesion. Nat Commun 2:577. https://doi.org/10.1038/ncomms1571

Zhou M, Lanchy JM, Ryckman BJ (2015) Human Cytomegalovirus gH/gL/gO promotes the fusion step of entry into all cell types, whereas gH/gL/UL128-131 Broadens Virus tropism through a distinct mechanism. J Virol 89(17):8999–9009. https://doi.org/10.1128/JVI.01325-15

Chapter 9
Betaherpesvirus Virion Assembly and Egress

William L. Close, Ashley N. Anderson, and Philip E. Pellett

Abstract Virions are the vehicle for cell-to-cell and host-to-host transmission of viruses. Virions need to be assembled reliably and efficiently, be released from infected cells, survive in the extracellular environment during transmission, recognize and then trigger entry of appropriate target cells, and disassemble in an orderly manner during initiation of a new infection. The betaherpesvirus subfamily includes four human herpesviruses (human cytomegalovirus and human herpesviruses 6A, 6B, and 7), as well as viruses that are the basis of important animal models of infection and immunity. Similar to other herpesviruses, betaherpesvirus virions consist of four main parts (in order from the inside): the genome, capsid, tegument, and envelope. Betaherpesvirus *genomes* are dsDNA and range in length from ~145 to 240 kb. Virion *capsids* (or nucleocapsids) are geometrically well-defined vessels that contain one copy of the dsDNA viral genome. The *tegument* is a collection of several thousand protein and RNA molecules packed into the space between the envelope and the capsid for delivery and immediate activity upon cellular entry at the initiation of an infection. Betaherpesvirus *envelopes* consist of lipid bilayers studded with virus-encoded glycoproteins; they protect the virion during transmission and mediate virion entry during initiation of new infections. Here, we summarize the mechanisms of betaherpesvirus virion assembly, including how infection modifies, reprograms, hijacks, and otherwise manipulates cellular processes and pathways to produce virion components, assemble the parts into infectious virions, and then transport the nascent virions to the extracellular environment for transmission.

W. L. Close
Department of Microbiology & Immunology, University of Michigan School of Medicine, Ann Arbor, MI, USA

Department of Biochemistry, Microbiology, & Immunology, Wayne State University School of Medicine, Detroit, MI, USA

A. N. Anderson · P. E. Pellett (✉)
Department of Biochemistry, Microbiology, & Immunology, Wayne State University School of Medicine, Detroit, MI, USA
e-mail: ppellett@med.wayne.edu

The *envelope* is a lipid bilayer that is studded with virus-encoded glycoproteins and some cellular proteins of uncertain significance. Virion envelopment occurs in the cytoplasm at specialized membranes of organelles associated with exocytic transport. The envelope is essential because it carries the glycoproteins and other signaling molecules needed to protect virions during extracellular phases of transmission and to initiate new infections (Ryckman et al. 2006; Spear and Longnecker 2003). Each of the several glycoprotein species is present at tens to hundreds of copies per virion, for a total of more than 2000 glycoprotein molecules per virion.

In sum, individual, mature, infectious herpesvirus virions are complex, but ordered structures that include well over 5000 protein molecules, one genome, bioactive RNAs, lipids, and various other small molecules.

In addition to infectious virions, other particles are produced in infected cells. Capsids lacking a viral genome can undergo tegumentation and envelopment during nucleocapsid maturation, resulting in the production of noninfectious, enveloped particles (NIEPs). Additionally, capsid-free enveloped structures filled with tegument proteins are called dense bodies; their relative abundance varies depending on the virus strain and multiplicity of infection (Mocarski et al. 2013; Benyesh-Melnick et al. 1966).

Efficient production of structures with the complexity of virions has many parallels in the manufacture of goods such as automobiles. Detailed plans are required. The factory needs to be constructed before the first car can be assembled. Diverse raw materials are needed, some of which may need to be imported; others may already be locally available. Special formulations might need to be concocted from the raw materials. Subassemblies can be built at satellite facilities and then transported to site of final assembly. Regulatory processes are required to control the flow of materials, subassemblies, final assembly, and shipment of finished goods. Quality control must be exercised at every step. Security systems are needed to protect the facilities and processes from external and internal threats. Construction and operation of the assembly plant has dramatic effects on the local environment, effects that can linger long after the plant has ceased production. A plethora of governmental and other oversight agencies have responsibilities for regulating the relationship between the factory and the surrounding environment and society, sometimes with the authority to shut down and even dismantle the plant. With such an intricate system, disruptions at any step have downstream consequences that can result in decreased quality, reduced output, or complete elimination of production.

HCMV replication and virion assembly occur in infected cells in which the virus has modified, reprogrammed, hijacked, or otherwise manipulated many cellular processes and pathways. By organizing materials along a complex and customized pathway for virion assembly that is not present prior to infection, virion components are able to be manufactured and assembled. Some of the modified pathways are directly involved in virion assembly, such as ribosome biogenesis, transcriptional regulation, phospholipid and sterol biosynthesis, and intracellular vesicular transport (Tirosh et al. 2015). Modifications of host intrinsic, innate, and adaptive immune responses (Mocarski et al. 2013) contribute to virion assembly by prolonging the life of infected cells.

A societally important objective of studies regarding virion assembly is the development of novel antiviral compounds that disrupt individual steps in the virion manufacturing process. This is important because all currently licensed antivirals against betaherpesviruses target virus genome replication. For human immunodeficiency virus (HIV), antivirals have been developed that target virtually every identified step in virion production. The value of targeting multiple steps in virus replication pathways has been demonstrated by the enhanced clinical efficacy from treatment of HIV infections with cocktails of drugs that target separate critical control points during HIV replication. We anticipate that methods will continue to improve for synthesizing candidate antivirals, predicting structures of potential inhibitors, and preclinical and clinical evaluation of candidates. Thus, it is important to continue to identify steps in the process of viral replication that might serve as targets of antiviral intervention.

9.3 Nuclear Events: Capsid Assembly and Genome Encapsidation

While many of the viral proteins required for betaherpesvirus virion assembly are conserved across the herpesviruses, some are unique to betaherpesviruses. In the descriptions of betaherpesvirus assembly that follow, HCMV protein nomenclature will be used because HCMV is the most studied of the betaherpesviruses. Specifics pertaining to roseoloviruses will also be provided.

9.3.1 Capsid Types

Three different capsid types are produced that correspond to different stages of capsid assembly and maturation: A, B, and C capsids. In electron micrographs, A capsids appear to be empty. B capsids contain scaffold proteins but not the virus genome. C capsids contain DNA and are thought to be the mature form. All three capsid forms are present in the nuclei of infected cells, but C capsids typically outnumber A and B capsids in the cytoplasm. A and B capsids are thought to be either intermediates formed during nucleocapsid assembly or abortive forms that fail to undergo DNA encapsidation (DeRussy and Tandon 2015; Tandon et al. 2015). B capsids that are successfully filled with viral DNA transition into C capsids and may further mature into virions (Mocarski et al. 2013; Benyesh-Melnick et al. 1966). Across the betaherpesviruses, there is variation in capsid and virion diameters (Table 9.1) (Britt 2007; Klussmann et al. 1997; Maeki and Mori 2012; Krug and Pellett 2014). This may be due to differences in virus genome lengths and/or composition of their teguments (Klussmann et al. 1997; Britt 2007; Maeki and Mori 2012; Mocarski et al. 2013; Guo et al. 2010).

Table 9.1 Comparison of betaherpesvirus capsid and virion diameters and genome lengths

Virus	Capsid diameter (Å)	Virion diameter (Å)	Genome length (kbp)[a]	
			Wild virus	Passaged virus
HCMV	~1300	~2000	~236	~230
HHV-6A	~800	~2000	Unknown	~159
HHV-6B	~800	~2000	~170	~159–162
HHV-7	~900–950	~1700	Unknown	~145

[a]Genome lengths vary for some viruses, and can change during passage in cultured cells. The table was adapted from Krug and Pellett (2014) using information from Britt (2007), Klussmann et al. (1997), and Maeki and Mori (2012)

9.3.2 Capsid Components and Assembly

After viral DNA replication, late genes necessary for capsid assembly are transcribed, translated, and then shuttled to the nucleus (Fig. 9.3) (Mocarski et al. 2013). During capsid assembly, a protein scaffold, consisting of the HCMV assembly protease (PR-pUL80a, Table 9.2) (Pellett and Roizman 2013) and assembly protein precursor, (pAP-pUL80.5, Table 9.2) forms. The protease and assembly protein precursor are encoded by the same gene and share the same carboxyl terminus, which contains domains necessary for interaction with the major capsid protein (MCP-pUL86, Table 9.2). These proteins also have conserved amino terminal domains important for self-assembly into a scaffold, around which the capsid proteins assemble (Britt 2007; Mocarski et al. 2013).

Capsid proteins are the building blocks of capsids with the major capsid protein (MCP, Table 9.2) as the major component. During assembly, MCP forms both pentameric and hexameric capsomeres, the major subunits of capsids (Fig. 9.4) (Zhou et al. 1994). Capsomeres are cylindrical and have a pore that runs along their length, which is ~160 Å. Capsomeres self-assemble onto the scaffold to form a capsid with T = 16 icosahedral symmetry. In this arrangement, 150 hexons form the triangular faces of the capsid and 11 pentons make up the vertices of those triangles (Fig. 9.2) (Dai et al. 2013; Yu et al. 2011). Given that there are 6 molecules of MCP in each of the 150 hexons and 5 molecules of MCP in each of the 11 pentons, there are a total of 950 molecules of MCP per capsid. The twelfth pentonal position is occupied by a complex formed by 12 copies of the portal protein (pUL104 or PORT, Table 9.2), which provides the channel necessary for packaging the viral genome into the capsid during assembly and for its release into the nucleus at the initiation of infection. During DNA encapsidation, the portal complex interacts with the terminase complex, which consists of ATPase (pUL89 or TER 1, Table 9.2) and DNA recognition (pUL56 or TER 2, Table 9.2) subunits (Britt 2007; Chen et al. 1999; Dai et al. 2013; Mocarski et al. 2013).

The small capsid protein (SCP-pUL48.5, Table 9.2) is the second most abundant protein in the HCMV capsid with 900 molecules per capsid. SCP is only found associated with the hexons, decorating their tips. In the absence of SCP, nascent

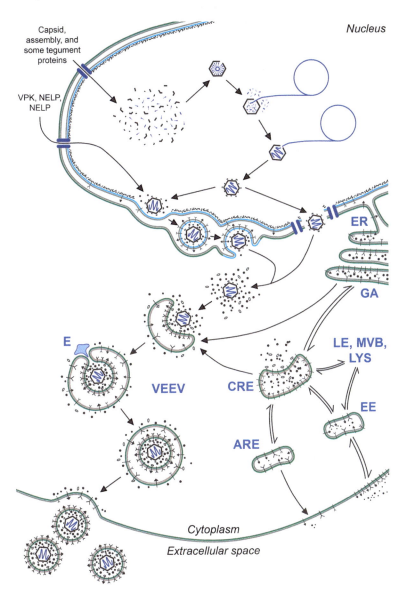

Fig. 9.3 Pathway of betaherpesvirus virion assembly. The diagram illustrates capsid formation and genome packaging in the nucleus, capsid egress from the nucleus to the cytoplasm via primary envelopment and de-envelopment (major path; left branch), and via a disrupted nuclear membrane (less common path; right branch). In the cytoplasm, envelope glycoproteins are translated on ER-associated ribosomes, with mature glycoproteins being delivered to VEEVs directly, or via CREs, either directly or after retrieval from the cell surface. Tegument proteins associate with capsids, each other, and with cytoplasmic domains of virion glycoproteins before envelopment at VEEVs. Infectious mature virions are transported inside VEEVs to the plasma membrane for release into the extracellular space. Abbreviations: *AE* apical recycling endosome, *CRE* common recycling endosome, *E* ESCRT machinery or a functional analog, *EE* early endosome, *ER* endoplasmic reticulum, *GA* Golgi apparatus, *LE* late endosome, *LYS* lysosome, *MVB* multivesicular body, *VEEV* virion envelopment and egress vesicle. Adapted from Pellett and Roizman (2013)

Table 9.2 Betaherpesvirus genes conserved across the *Herpesviridae*

Function (gene name)	HCMV	HHV-6A, HHV-6B, and HHV-7	HSV
Gene regulation			
Multifunctional regulator of expression (MRE)	UL69	U42	UL54
Nucleotide metabolism			
Ribonucleotide reductase, large subunit (RR1)	UL45	U28	UL39
Uracil-DNA glycosylase (UNG)	UL114	U81	UL2
Deoxyuridine triphosphatase (dUTPASE)	UL72	U45	UL50
DNA replication			
Helicase/primase complex			
ATPase subunit (HP1)	UL105	U77	UL5
RNA pol subunit (HP2)	UL70	U43	UL52
subunit C (HP3)	UL102	U74	UL8
DNA polymerase (POL)	UL54	U38	UL30
ssDNA-binding protein (SSB)	UL57	U41	UL29
DNA polymerase processivity subunit (PPS)	UL44	U27	UL42
Nonstructural; roles in virion maturation			
Alkaline exonuclease (NUC)	UL98	U70	UL12
Capsid transport nuclear protein (CTNP)	UL52	U36	UL32
Terminase-binding protein (TERbp)	UL51	U35	UL33
Terminase (TER)			
TER ATPase subunit (TER1)	UL89	U66	UL15
TER DNA recognition subunit (TER2)	UL56	U40	UL28
Assembly protease (PR)	UL80a	U53	UL26
Assembly protein precursor (pAP)	UL80.5	U53a	UL26.5
Capsid nuclear egress complex			
Nuclear egress membrane protein (NEMP)	UL50	U34	UL34
Nuclear egress lamina protein (NELP)	UL53	U37	UL31
Capsid			
Major capsid protein (pentons and hexons; MCP)	UL86	U57	UL19
Portal protein (PORT)	UL104	U76	UL6
Portal capping protein (PCP)	UL77	U50	UL25
Capsid triplex			
Monomer (TRI1)	UL46	U29	UL38
Dimer (TRI2)	UL85	U56	UL18
Small capsid protein (SCP) at hexon tips	UL48.5	U32	UL35
Tegument			
Encapsidation and egress protein (EEP)	UL103	U75	UL7
Myristoylated/palmitoylated cytoplasmic egress tegument protein (CETP)	UL99	U71	UL11
Virion protein kinase (VPK)	UL97	U69	UL13
Encapsidation chaperone protein (ECP)	UL95	U67	UL14
CETP-binding protein (CETPbp)	UL94	U65	UL16

(continued)

Table 9.2 (continued)

Function (gene name)	HCMV	HHV-6A, HHV-6B, and HHV-7	HSV
Capsid transport tegument protein (CTTP)	UL93	U64	UL17
Cytoplasmic egress facilitator 2 (CEF2)	UL87	U58	UL21
Cell-to-cell fusion inhibitor	UL76	U49	UL24
Large tegument protein (LTP)	UL48	U31	UL36
LTP-binding protein (LTPbp)	UL47	U30	UL37
Cytoplasmic egress facilitator 1 (CEF1)	UL71	U44	UL51
Envelope			
Glycoprotein B (gB)	UL55	U39	UL27
Glycoprotein H (gH)	UL75	U48	UL22
Glycoprotein L (gL)	UL115	U82	UL1
Glycoprotein M (gM)	UL100	U72	UL10
Glycoprotein N (gN)	UL73	U46	UL49.5

Table adapted from information in Pellett and Roizman (2013) and Nicholas (1996). Proteins named according to Mocarski (2007)

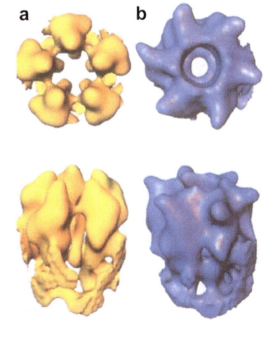

Fig. 9.4 Three-dimensional reconstructions from cryoelectron microscopy images of herpes simplex virus type 1 pentons and hexons. (**a**) Top and side views of penton. (**b**) Top and side views of hexon. Images are representative of the analogous structures of HCMV. (From Zhou et al. (1994) with permission)

capsids are devoid of viral DNA and the capsid-associated tegument protein pUL32 (pp150; "pp" denotes phosphoprotein). In contrast, SCP is dispensable for HSV-1 viral growth, demonstrating virus-specific structural and functional roles for this protein (Britt 2007; Dai et al. 2013; Mocarski et al. 2013). Between the hexons and pentons reside small triplexes consisting of a dimer of the minor capsid protein (MnCP-pUL85, Table 9.2) and a monomer of the minor capsid-binding protein (MnCP-bp-pUL46, Table 9.2). These triplexes are important for stabilizing the nucleocapsid. The portal capping protein (PCP-UL77, Table 9.2) and the capsid transport tegument protein (CTTP-pUL93) form the capsid vertex-capping (CVC) complex, also known as the capsid vertex-specific component (CVSC), which decorates pentons. The terminase-binding protein (TERbp-pUL51, Table 9.2) and the capsid transport nuclear protein (CTNP-pUL52, Table 9.2) associate with this complex as well. The role(s) of this complex and the associated proteins have not been determined, although they are thought to be involved with nucleocapsid stability by helping the capsid to withstand internal pressure from the DNA during encapsidation. They are also thought to be involved in the release of viral DNA into the nucleus during initiation of infection (Britt 2007; Mocarski et al. 2013; Tandon et al. 2015; DeRussy and Tandon 2015; Bigalke and Heldwein 2016). In addition, the CVC and its associated proteins have been shown to be important for cleavage of concatemeric viral DNA into unit length genomes (Borst et al. 2008, 2013, 2016).

9.3.3 Encapsidation of Virus Genomes

Single copies of the virus genome are packed into newly formed capsids. The major events during genome encapsidation are assembly and then degradation of the scaffold, packaging of the virus genome into the rigidly constrained space inside the capsid, cleavage of the genome precisely at its termini, and closure (or corking) at the portal of the highly pressurized capsid (Fig. 9.3). The virus accomplishes the process of DNA encapsidation rapidly, within fractions of a second. As the protease disassembles the scaffold, a virus genome is threaded into the capsid, short terminus first, through the channel formed by the portal complex. The ATPase subunit of the terminase is the motor for translocating viral DNA into capsids (Tandon et al. 2015; DeRussy and Tandon 2015). For HCMV, the DNA recognition subunit of the terminase recognizes the intact packaging and cleavage sequence that is formed when *pac-1* and *pac-2* sequences at the genomic termini are juxtaposed in covalently circularized or concatemeric genomes. Precise cleavage by the terminase results in packaging one complete virus genome per capsid. Other viral proteins implicated in this process are pUL56 and pUL89. pUL56 binds to AT-rich sequences within the *pac* sequences and has nuclease activity. pUL89 also has DNA cleavage activity; in addition to the terminase, it may be responsible for cleavage of the virus genome during encapsidation (Britt 2007; Mocarski et al. 2013).

Once the capsid is assembled, it becomes partially tegumented in the nucleus (Fig. 9.3). Tegument protein pp150 (pUL32) interacts with the outside of the capsid in a manner that adds stability to the capsid, helping it to withstand the high internal pressure associated with the tightly packed genome (Mocarski et al. 2013).

9.3.4 Nuclear Egress

Before capsids can enter the cytoplasm, they must pass through the nuclear envelope, which consists of the inner nuclear membrane (INM), the perinuclear space that is contiguous with the endoplasmic reticulum (ER), and the outer nuclear membrane (ONM) (Fig. 9.3). In uninfected cells, the distance across the nuclear envelope, from inner membrane to outer membrane, is typically ~50 nm. Nuclear envelope integrity is maintained by the linker of nucleoskeleton and cytoskeleton (LINC) complex, a multiprotein complex that directly connects the nuclear skeleton to the cytoskeleton. The complex includes Sad1p, UNC-84 (SUN) and Klarsicht, ANC-1, Syne homology (KASH) domain proteins. Interprotein interactions via SUN and KASH domains are important for nuclear membrane stability and for maintaining proper spacing between the inner and outer nuclear membranes (Mocarski et al. 2013; Bigalke and Heldwein 2016; Alwine 2012).

The nuclear envelope is further stabilized by the nuclear lamina, a proteinaceous network that lines the inner nuclear membrane. Lamin proteins A/C and B associate with each other to form intermediate filaments that provide structural support to the cell. Lamin B also interacts with integral membrane proteins such as the lamin B receptor, which helps to tether the nuclear lamina to the inner nuclear membrane (Mocarski et al. 2013; Bigalke and Heldwein 2016; Alwine 2012).

To get past the nuclear lamina, herpesviruses employ conserved proteins that make up the nuclear egress complex (NEC, pUL50-nuclear egress membrane protein (NEMP), pUL53- nuclear egress lamina protein (NELP), Table 9.2). In HCMV-infected cells, these proteins localize on the interior of the nucleus on the side that faces the cytoplasmic virion assembly complex (cVAC, described in the next section). The NEC competes with lamin proteins A/C and B for binding to each other, disrupting the lamina's fibrillary network. In addition, the NEC disrupts the binding of lamins to the nuclear envelope protein emerin, which links the nuclear lamina to the inner nuclear membrane (Milbradt et al. 2010; Camozzi et al. 2008; Bigalke and Heldwein 2016). pUL50 also interacts with the HCMV-encoded nuclear rim-associated cytomegaloviral protein (RASCAL) that is thought to be important for NEC-dependent degradation of the nuclear lamina. Additionally, the NEC recruits other viral and cellular proteins such as the viral protein kinase (pUL97) and cellular protein kinase C (PKC), both of which phosphorylate nuclear lamins, leading to destabilization and reorganization of the nuclear lamina. pUL97 has also been implicated in the phosphorylation of p32, a nuclear lamina component that interacts with the lamin B receptor, further contributing to nuclear lamina destabilization. These changes in the nuclear lamina facilitate egress of the capsids by allowing capsids to bud from the nucleus into the perinuclear space during primary envelopment (Fig. 9.3) (Milbradt et al. 2010; Alwine 2012; Walzer et al. 2015; Camozzi et al. 2008). In HSV-1, the NEC forms a hexagonal lattice along the INM leading to formation of invaginations in the membrane and budding of capsids into the perinuclear space (Bigalke and Heldwein 2015). X-ray crystallography showed that the HCMV NEC also forms hexameric rings, suggesting conservation of this invagination and budding activity across the herpesviruses. After enveloped capsids enter the perinuclear space, their membranes fuse with the ONM, and the capsids are released into the cytoplasm (Fig. 9.3) (Bigalke and Heldwein 2016; Mocarski et al. 2013).

9.4 Post-nuclear Assembly: Tegumentation, Envelopment, and Egress

9.4.1 Overview

As described above, capsids are formed and filled in the nucleus, acquire a primary envelope at the inner nuclear membrane, bud into the lumen of the nuclear membrane, and then are de-enveloped as they enter the cytoplasm. In this section, we explore the post-nuclear aspects of virion assembly and egress, which include tegumentation, secondary envelopment, and transport of mature virions to the plasma membrane for release. For at least HCMV and MCMV, these steps of virion assembly take place in the cVAC (Fig. 9.5), a specialized cellular compartment whose biogenesis is triggered by viral proteins and miRNAs. Although it is not strictly necessary, formation of the cVAC appears to facilitate efficient production of nascent particles (Das et al. 2014; Alwine 2012).

Through consideration of interactions between viral and host factors, and the resulting modification of host cell functions, we outline a map of the complex and often interwoven processes critical for virion maturation. Progressively finer mapping of these processes during HCMV replication will enable identification of critical control points which can then be exploited for the development of novel antiviral treatments.

9.4.2 HCMV cVAC

The cVAC is a complex, juxtanuclear, cytoplasmic structure about the size of a nucleus but without a clearly defined border (Fig. 9.5) (Das and Pellett 2011; Das et al. 2007; Tandon et al. 2015; Alwine 2012). Infected cells harbor a single cVAC, even cells with multiple nuclei (syncytia).

Within the cVAC, developing virions acquire most of their tegument, become enveloped, and are then transported to the cell surface for release (Alwine 2012; Tandon and Mocarski 2012). It is not enough that the virion structural components are made in sufficient quantities; they must also be distributed to the correct loca-

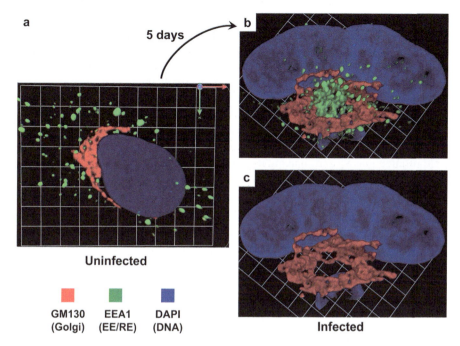

Fig. 9.5 The HCMV cVAC. Three-dimensional, confocal microscopic reconstructions of single, human, lung fibroblasts stained with antibodies against markers of the Golgi apparatus (GM130), early and recycling endosomes (EE/RE; EEA1), and a marker for DNA (DAPI). (**a**) An uninfected cell displaying normal morphology. The tubular Golgi apparatus is located adjacent to the nucleus and the early/recycling endosomes are present in a widely distributed "starry night" pattern. (**b** and **c**) Infected cells displaying typical cVAC structures. Five days after infection with HCMV strain AD169, the Golgi apparatus is remodeled into a cylindrical, manifold, tubular structure. The cell shown has two nuclei, the likely result of post-mitotic arrest prior to segregation of the daughter cells or syncytium formation. Grid spacing in **b** and **c** is 3.31 μm. (Adapted from Das and Pellett 2011)

tions in the appropriate temporal sequence. Multiple levels and forms of cellular rearrangement are needed to efficiently produce virions. The complex orchestration and interactions between host and viral factors underscore the complexity of activities in the cVAC.

The cVAC is constructed by remodeling the nucleus and major cytoplasmic components of the host secretory apparatus during the first 2–4 days after infection (Das et al. 2007; Das and Pellett 2011; Buchkovich et al. 2010). Structurally, the cVAC is arranged as a set of nested cylinders or rings, centered on a microtubule-organizing center (Fig. 9.5). The outer cylinder consists of networks of tubular vesicles derived from the Golgi apparatus and the trans-Golgi network (TGN); the inner cylinder harbors vesicles derived from early/recycling endosomal vesicles. Components of the late endocytic compartment are also present, often with their highest abundance just outside the Golgi ring (Das and Pellett 2011). The nucleus (or nuclei in some cells) is bent into a reniform configuration around one side of the cVAC, the nuclear membrane adjacent to the cVAC becomes more porous, and the distance between the inner and outer nuclear membranes increases, at least in part due to a reduction in the abundance of SUN-domain proteins, which are involved in tethering the inner and outer nuclear membranes (Buchkovich et al. 2010).

cVAC formation is dependent on viral genome replication, indicating that viral late genes are involved, in addition to expression of several HCMV miRNAs (Hook et al. 2014). Consistent with this, inhibition of the expression or stability of HCMV tegument proteins pUL48, pUL94, or pUL103 (all expressed from late genes) adversely affects cVAC biogenesis. Disruption of the cVAC greatly reduces the yield of infectious virions, making its biogenesis and operation rational targets for development of novel antivirals (Das et al. 2014).

9.4.3 Tegument Proteins and Tegumentation

Tegument proteins are important for many processes throughout infection, including disassembly of virions, transcriptional regulation, modulation of cellular responses, and virion maturation (Mocarski et al. 2013; Smith et al. 2014). Many, if not most tegument proteins, perform multiple, distinct functions. At least 38 different proteins (including some of cellular origin) are present in various quantities in HCMV teguments (Guo et al. 2010; Varnum et al. 2004; Smith et al. 2014). Although it lacks a well-defined structure, there are clear elements of structural order throughout the tegument (Smith et al. 2014).

Tegument protein pUL83 (pp65) is the most abundant protein in HCMV virions and dense bodies. While not required for replication, pp65 is important for the initiation of infection by enhancing activation of the major immediate-early promoter and is important for recruiting proteins into the virion during assembly. pUL48 (LTP, Table 9.2) is the largest and second most abundant HCMV tegument protein. It is essential for replication and is involved in the release of viral DNA from the capsid. pUL32 (pp150) is the third most abundant tegument protein. It is essential for viral replication and plays a role in stabilizing DNA-containing capsids during virion maturation (Yu et al. 2017; Mocarski et al. 2013; Smith et al. 2014). pUL103 is a low-abundance tegument protein that plays important roles late in HCMV infection, including cVAC biogenesis, cell-to-cell spread, and virion maturation (Das et al. 2014; Varnum et al. 2004). pUL103 may have roles early in the HCMV replication cycle due to its presence in nuclei and interactions with several proteins of the innate immune system (Ortiz et al. 2016). HHV-6A U14 (homolog of HCMV pUL25) is important for virion maturation, interacts with the tumor suppressor protein p53, and arrests the cell cycle during infection (Mori et al. 2015a, b; Takemoto et al. 2005).

The precise order of addition of tegument proteins to maturing virions is unknown, but some information is available (Mettenleiter et al. 2013). Tegumentation begins in the nucleus with the inner tegument, an organized netlike layer that encloses the capsid shell, consisting mostly of pUL32 and pUL48. Inner tegument proteins tightly associate with the capsid in a manner that is resistant to treatment with various detergents. The majority of the tegument, the outer layer, is added in the cVAC (Guo et al. 2010).

Acquisition of the outer tegument occurs through several mechanisms (Figs. 9.3 and 9.6). Some tegument proteins aggregate in the cytoplasmic milieu and form complexes prior to incorporation in nascent virions. Others accumulate on mem-

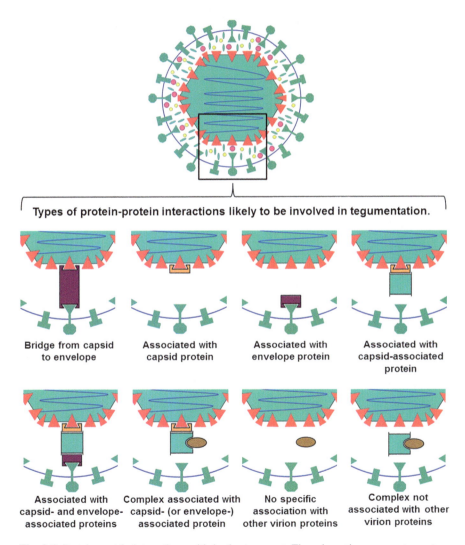

Fig. 9.6 Protein-protein interactions with in the tegument. The schematics represent a mature enveloped virion and the types of protein-protein interactions proven or predicted to occur during tegumentation

branes destined for secondary envelopment by binding the cytosolic domains of embedded virion glycoproteins (Guo et al. 2010; Smith et al. 2014; Mettenleiter 2006; Mori et al. 2008). The marriage of partially tegumented capsids and membrane-associated tegument proteins is consummated during secondary envelopment, which occurs throughout the cVAC (Schauflinger et al. 2013).

9.4.4 Secondary Envelopment in Host-Derived Compartments

Due to the dramatic restructuring of infected cells and the resultant shifts in organelle identity (Das and Pellett 2011), the precise identity of the organelle where secondary envelopment occurs remains uncertain (Jean Beltran et al. 2016). Following the cVAC model, nascent particles are enveloped shortly after microtubule-mediated translocation through the Golgi-derived ring of the cVAC and upon entrance into the region of the cVAC that is predominantly occupied by vesicles that bear markers of early and recycling endosomes (Das et al. 2007; Das and Pellett 2011; Alwine 2012). Although the arrangements of the organelle-derived structures differ, the nature of the vesicles is similar to what has been seen for other herpesviruses (Schauflinger et al. 2013; Das and Pellett 2011; Turcotte et al. 2005; Lee et al. 2006). The cellular pathways usurped by maturing particles for envelopment and egress appear to be those involved in the vesicle-mediated recycling of materials important for normal cell homeostasis.

Immunofluorescence imaging and immunoelectron microscopy revealed that membranes targeted for HCMV envelopment based on glycoprotein accumulation colocalize with typical Golgi-derived markers (TGN46, mannosidase II, Rab3, syntaxin 5) and endosomal markers (CD63, EEA1, Rab11) but not lysosomal markers (LAMP1) in human foreskin fibroblasts (HFFs) (Homman-Loudiyi et al. 2003; Cepeda et al. 2010; Das et al. 2007; Das and Pellett 2011; Sanchez et al. 2000a, b; Fish et al. 1996; Cruz et al. 2017). Supporting the role of recycling endosomes in herpesvirus replication, HSV-1- and HCMV-infected cells labeled with horseradish peroxidase, a fluid phase marker of uptake and release through the endocytic recycling compartment, accumulated horseradish peroxidase in the interstitial space between the vesicle membrane and the membrane of enveloped virions (Tooze et al. 1993; Hollinshead et al. 2012).

The transferrin receptor normally recycles back to the plasma membrane following signal-induced internalization. In HCMV-infected cells, it is sequestered in a perinuclear zone that coincides with markers of the cVAC (Cepeda et al. 2010). Formation of this secretory trap can be partially recapitulated by HCMV microRNAs that downregulate recycling activity by targeting host genes involved in trafficking along vesicle-mediated recycling pathways (Hook et al. 2014), a pattern also seen for Epstein-Barr virus (EBV, *Human gammaherpesvirus 4*) and Kaposi's sarcoma-associated herpesvirus (KSHV, *Human gammaherpesvirus 8*) (Gottwein et al. 2011; Skalsky et al. 2012). Characterization of the HCMV transcriptome late in infection revealed effects that extend well beyond that predicted from HCMV miRNAs alone, with >100 host vesicular trafficking genes being differentially modulated (Hertel and Mocarski 2004; Grey and Nelson 2008; Hook et al. 2014). This activity has multiple effects, including interfering with innate immune signaling by sequestering cytokines such as IL-6 or TNF-α in the secretory trap and enabling perinuclear accumulation of virion components, thus contributing to cVAC formation (Hertel and Mocarski 2004; Hook et al. 2014; Lucin et al. 2015).

The immediate-early HCMV protein, pUL37x1 (vMIA), contributes to altered host morphology and cVAC development by potentiating actin remodeling. pUL37x1 is a multifunctional protein responsible for releasing Ca^{2+} stores from the endoplasmic reticulum before traveling to mitochondria where it inhibits apoptosis (Sharon-Friling et al. 2006; Sharon-Friling and Shenk 2014; Williamson et al. 2011). The Ca^{2+} efflux activates PKCα which remodels actin along with RhoB (Goulidaki et al. 2015), leading to altered cytoskeletal morphology and membrane arrangement. In addition, the calcium efflux triggers accumulation of large cytoplasmic vesicles approximately 0.5–5 μm in diameter through a process that requires synthesis and elongation of fatty acids (Poncet et al. 2006; Sharon-Friling et al. 2006; Sharon-Friling and Shenk 2014). When pUL37x1 is not expressed, the cVAC is disrupted, and there is a buildup of nonenveloped particles in the perinuclear region (Sharon-Friling and Shenk 2014).

9.4.5 The Intersection of Tegumentation and Envelopment

The capsid-associated tegument layer provides a scaffold-like interface for membrane-associated tegument proteins and glycoproteins to adhere to during envelopment. Alterations in tegument composition can lead to defective envelopment.

As an example, products of the HCMV *UL35* open reading frame (ORF) have been implicated as having a role in tegument recruitment (Schierling et al. 2005). At early time points, both ppUL35^A and ppUL35 localize to the nucleus where they interact with ppUL82 and activate the major IE promoter. At late time points, however, the longer form, ppUL35, helps shuttle ppUL82 and pUL83 (pp65) out of the nucleus as it translocates to the cytoplasm for incorporation into the tegument (Liu and Biegalke 2002; Schierling et al. 2004; Varnum et al. 2004; Schierling et al. 2005). If the UL35 ORF is deleted, nonenveloped capsids accumulate in the cytoplasmic space and infectious output is reduced tenfold, most likely due to improper tegument structure and inability to bind lipid-recruiting molecules (Schierling et al. 2005).

HCMV pUL103 is required for cVAC biogenesis and efficient release of nascent virions (Ahlqvist and Mocarski 2011; Das et al. 2014; Yu et al. 2003). Using the pUL103-Stop-F/S deletion mutant or pUL103-FKBP destabilization mutant, decreased pUL103 expression correlated with altered trafficking of the viral protein pUL99 (pp28) in addition to cellular golgin-97, GM130, and CD63. These changes were also accompanied by decreased plaque size (Das et al. 2014; Ahlqvist and

Mocarski 2011). In addition, electron micrographs of cells infected under conditions of pUL103 destabilization reveal increased numbers of virions stalled during envelopment or with abnormal structures accumulating in the perinuclear region (Fig. 9.7) (Das et al. 2014). Because pUL103 has several interacting partners during infection (Ortiz et al. 2016), it is uncertain which process leads to the observed phenotypes, but it appears to be linked to the two C-terminal ALIX-binding motifs. ALIX-binding motifs are also important in the maturation of other enveloped viruses including during primary envelopment of EBV (Bardens et al. 2011; Zhai et al. 2011; Lee et al. 2012). Alternatively, the defective phenotypes may be caused by the inability of pUL103 to interact with pUL71 (Fischer 2012; Ortiz et al. 2016).

HCMV-infected patients mount a B-cell response against tegument protein pUL71, suggesting it may be exposed on virions or it is released from infected cells (Beghetto et al. 2008; Varnum et al. 2004). In cells infected with pUL71-deficient virus, viral proteins are aberrantly localized in the cVAC, and large LAMP1/CD63-positive multivesicular bodies form near the cVAC in infected cells (Schauflinger et al. 2011; Womack and Shenk 2010). Ultrastructural analysis of the pUL71 null

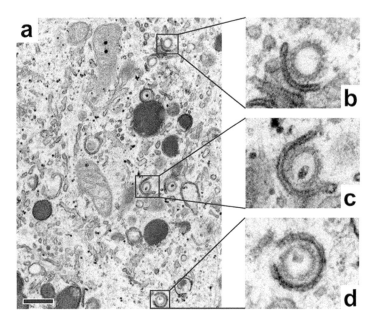

Fig. 9.7 Ultrastructure of capsids undergoing secondary envelopment in the cytoplasm. HFFs were infected at a multiplicity of infection of 0.3 with HCMV UL103-FKBP, a recombinant version of HCMV strain AD169 in which a protein destabilization domain is fused to the C-terminus of UL103 (Das et al. 2014). Under the conditions of infection, pUL103 was destabilized, resulting in a defect in completion of envelopment, several examples of which are visible in panel **a**, with close-ups of three virions at different stages during envelopment in panels **b**, **c**, and **d**, respectively. The scale bar in panel **a** represents 0.5 μm. We thank Dr. Hong Yi of the Robert P. Apkarian Integrated Microscopy Core of Emory University for help with the electron microscopy

mutants, TBstop71 and BAD*in*UL71STOP, showed the accumulation of HCMV particles unable to complete envelopment on the cytoplasmic side of multivesicular bodies (Womack and Shenk 2010; Schauflinger et al. 2011). During TBstop71 infection, 27% of particles were enveloped and 70% were budding compared to 87% and 13%, respectively, during wild-type infection (Meissner et al. 2012). This behavior was recapitulated by expressing pUL71 with a mutated basic leucine zipper (bZIP)-like domain, suggesting that oligomerization is necessary for pUL71 to function properly (Meissner et al. 2012; To et al. 2011). Positional homologs of pUL71 are involved in envelopment and are conserved among other herpesviruses including pUL51 in HSV-1 (Nozawa et al. 2005), pUL51 in pseudorabies virus (PRV, *Suid alphaherpesvirus 1*) (Klupp et al. 2005), and GP71 in guinea pig CMV (GPCMV) (Schleiss et al. 2008). The observed phenotypes of pUL71 suggest that it is involved in membrane scission events during envelopment.

Despite being a low-abundance virion protein (Varnum et al. 2004), the outer tegument protein, pUL99, is important for envelopment as well (Silva et al. 2003). After myristoylation of the amino terminus of pUL99, it attaches to target membranes before localizing to the cVAC and forming multimers late in infection (Seo and Britt 2006; Sanchez et al. 2000a, b; Seo and Britt 2008). When the first 50 residues at the amino terminus of pUL99 are absent, it is aberrantly trafficked, nonenveloped particles accumulate in the cytoplasm, and infectious yield is hindered (Seo and Britt 2007; Jones and Lee 2004). Irregular trafficking of mutated pUL99 does not affect levels of other tegument proteins incorporated into mature virions, suggesting that it is part of the outermost layer of tegument proteins (Seo and Britt 2007). The amino terminal domain of pUL99 is necessary and sufficient for reconstituting infectious output as seen by accumulation of pUL99 in mature virions and proper envelopment (Jones and Lee 2004). The second amino residue, a glycine, is the site of myristoylation (Sanchez et al. 2000b), amino acids 26–43 are responsible for multimerization in the cVAC (Seo and Britt 2008), and amino acids 37–39 enable interaction with the cysteine residue at position 250 of the viral protein pUL94 (Phillips et al. 2012; Phillips and Bresnahan 2011). The ability of pUL99 to be incorporated into maturing virions is dependent on its interaction with pUL94 which serves as a scaffold on the outside of the tegument (Phillips and Bresnahan 2011; Phillips et al. 2012). Without pUL94, secondary envelopment and cVAC formation is hindered (Phillips and Bresnahan 2012). In HSV-1, KSHV, MCMV, and mouse herpesvirus 68 (MHV-68, *Murid gammaherpesvirus 4*), homologs of pUL99 and pUL94 play analogous roles, but in contrast to their counterparts in HCMV, they are not essential for virus replication in cultured cells (Chadha et al. 2012; Baines and Roizman 1992; Maninger et al. 2011; Wu et al. 2015; Guo et al. 2009).

Genes within the HCMV U_L/b' region influence maturation through cell type-specific mechanisms. Within this region, the *UL133–UL138* (*UL133/8*) locus is nonessential for growth in fibroblasts but is required for replication in other cell types (Murphy et al. 2003). Proteins produced by HCMV *UL135* and *UL136* are transcribed as part of *UL133/8* polycistronic mRNAs (Grainger et al. 2010), localize to Golgi membrane structures (Liao et al. 2014; Umashankar et al. 2011), and are required for latency and virion maturation (Bughio et al. 2013; Umashankar et al.

2011; Grainger et al. 2010; Bughio et al. 2015; Umashankar et al. 2014; Caviness et al. 2014, 2016). Mutation of ORFs *UL135* and *UL136* resulted in dispersal of cVAC markers and abnormal particle formation when human lung microvascular endothelial cells (HMVECs), but not fibroblasts, were infected with a TB40/E-*UL133–UL138*$_{NULL}$ virus (Umashankar et al. 2011; Bughio et al. 2013, 2015; Umashankar et al. 2014). When HMVECs were infected with a TB40/E-*UL135*$_{STOP}$ mutant defective in *UL135* expression, only 27% of virions had normal morphology, with the remainder being noninfectious enveloped particles or aberrantly enveloped particles (Bughio et al. 2015). *UL135* mutation also resulted in smaller dense bodies, which was linked to a two- to threefold decrease in pUL83 and pUL32 expression. Furthermore, dense bodies were excluded from multivesicular bodies where they normally aggregate with progeny virus in endothelial cells (Bughio et al. 2013, 2015). In fibroblasts, the only phenotype of infection with TB40/E-*UL135*$_{STOP}$ was a slight increase in NIEPs relative to wild type (Umashankar et al. 2014). Using similar methodology as the *UL135* studies, a TB40/E-*UL136*$_{GalK}$ mutant with a disrupted *UL136* ORF produced aberrantly enveloped virions 65% of the time and dense bodies that were 2.5 times larger on average despite having comparable levels of tegument proteins compared to wild type (Bughio et al. 2015). Of the several different-sized proteins encoded by *UL136* splice variants, the 26 kDa and 33 kDa products were the most important for facilitating normal cVAC biogenesis and particle formation (Caviness et al. 2014, 2016). Although specific mechanisms have yet to be determined, HCMV pUL135 appears to direct maturation and envelopment through interactions with other tegument proteins, while *UL136* ORF isoforms are needed for interactions with target membranes. The endothelial-dependent phenotypes exhibited by *UL133/8* locus mutations are an important example of how HCMV manipulates activities in divergent cell populations to its advantage.

Apart from being structural proteins that interact with capsids, each other, and envelope-associated glycoproteins, it is evident that tegument proteins play numerous other important roles during tegumentation and secondary envelopment.

9.4.6 *Secondary Envelopment and Modulation of Host Cell Specificity by Virion Glycoprotein Composition*

The variability in cellular tropism of nascent virions produced in cells of different types suggests cell type differences in events leading up to envelopment. As for other herpesviruses, including GPCMV and rhesus CMV (RhCMV, *Macacine betaherpesvirus 3*), HCMV glycoprotein composition plays a role in establishing sites of envelopment (Coleman et al. 2015; Britt and Mach 1996; Ryckman et al. 2008; Hutt-Fletcher 2015; Albecka et al. 2016; Maresova et al. 2005; Bowman et al. 2011). The most abundant HCMV glycoproteins are gM (pUL100), gN (pUL73), gB (pUL55), gH (pUL75), gL (pUL115), gO (pUL74), and pUL128–131 (Varnum et al. 2004). Most of the over 60 other predicted HCMV glycoproteins have not been detected in virions, suggesting nonstructural roles (Chee et al. 1990; Cha et al. 1996; Britt and Mach 1996).

The gM/gN complex is the most abundant glycoprotein complex in mature HCMV virion envelopes; if either is deleted, the virus is unable to replicate (Hobom et al. 2000). Similar to its EBV homologs (Lake and Hutt-Fletcher 2000; Lake et al. 1998), HCMV gM and gN must form a complex (gM/gN) when present in the ER before they can be trafficked to cytoplasmic vesicles and colocalize with other markers of the cVAC (Mach et al. 2000, 2005). Translocation of HCMV gM to the cVAC occurs when the cytoplasmic region of gM interacts with cellular FIP4, FIP4 binds Rab11, and then Rab11 recruits further effector proteins until gM is transported in complex with gN (Krzyzaniak et al. 2009). gM and gN also contain other C-terminal endocytic trafficking motifs, including an acidic cluster used for binding cellular transport proteins, such as PACS-1, and a YXXΦ tyrosine motif (Crump et al. 2003; Krzyzaniak et al. 2007; Mach et al. 2007). The highly conserved nature of the C-terminal acidic clusters in herpesvirus glycoproteins suggests a common mechanism for direct transport to the site of virion envelopment at TGN-derived membranes (Chiu et al. 2012; May et al. 2008; Heineman and Hall 2002; Olson and Grose 1997; Alconada et al. 1996). YXXΦ motifs are also conserved across all subfamilies of the *Herpesviridae* and allow various envelope proteins to be retrieved from the plasma membrane through interactions with the AP-2 complex, leading to clathrin-mediated, dynamin-dependent, endocytosis and accumulation in endosomes or the TGN (Ohno et al. 1996; Songyang et al. 1993; Radsak et al. 1996; Albecka et al. 2016; Archer et al. 2017).

During HSV-1 infection, cell-to-cell transmission is dependent on interaction between HSV-1 pUL51 and gE (HCMV pUL71 and US8) (Albecka et al. 2017; Roller et al. 2014) before both are transported to the site of envelopment through use of terminal YXXΦ tyrosine motifs in pUL51 (Alconada et al. 1999; Roller et al. 2014; Nozawa et al. 2005; Nozawa et al. 2003; Tirabassi and Enquist 1998, 1999). When the motif is mutated in pUL51, neither pUL51 nor gE is incorporated into nascent virions, and spread is hindered in Hep-2 human epithelial cells, but not Vero monkey epithelial cells, suggesting cell type-dependent mechanisms for spread (Roller et al. 2014). Additionally, HSV-1 pUL20 helps to chaperone gK and gE from the ER to Golgi; in its absence, neither glycoprotein is incorporated into virions, resulting in accumulation of nonenveloped particles and inability to form syncytia (Foster et al. 2004; Chouljenko et al. 2016).

Inclusion of both the acidic cluster and YXXΦ motifs in most envelope proteins ensures proper localization to the assembly compartment. Neither trafficking pattern is essential for virus production, but both augment infectious output as seen for HCMV gB, gM/gN, and gpUL132 (Kropff et al. 2010; Spaderna et al. 2005; Crump et al. 2003; Radsak et al. 1996; Jarvis et al. 2002, 2003); varicella-zoster virus (VZV, *Human herpesvirus 3*) gE, gH, and gB (Maresova et al. 2005); HSV-1 gB, pUL51, and gE (Beitia Ortiz de Zarate et al. 2007; Albecka et al. 2016; Roller et al. 2014); or PRV gB (Favoreel et al. 2002).

Incorporation of gH/gL complexes into the envelope of nascent virions is another example of a maturation event that defines HCMV cell tropism (Schultz et al. 2015;

Vanarsdall et al. 2011; Wang and Shenk 2005; Kinzler et al. 2002). HCMV and GPCMV virions require the gH/gL/gO complex for entering fibroblasts by fusion at the plasma membrane and need the gH/gL/UL128–131 pentameric complex for entry into epithelial and endothelial cells by pH-dependent endocytosis (Coleman et al. 2015, 2016; Ryckman et al. 2006, 2008). This dichotomy is exemplified by several laboratory strains which are fibroblast-restricted because serial passaging led to loss of functional UL128–UL131 and a compensatory increase in gH/gL/gO abundance (Dunn et al. 2003; Wille et al. 2010; Dolan et al. 2004). The ratio between gH/gL/gO and gH/gL/pUL128–131 complexes is determined within the ER prior to transport to Golgi or post-Golgi membranes for use in envelopment (Ryckman et al. 2008; Theiler and Compton 2002; Kinzler et al. 2002; Zhou et al. 2013). After gH and gL interact and stabilize each other in the ER (Molesworth et al. 2000), a single gH/gL complex can either form a disulfide bond with gO or a noncovalent bond with pUL128–131, but not both (Adler et al. 2006; Ryckman et al. 2008). During formation of the pentameric complex, pUL128, pUL130, and pUL131 are each capable of binding to gH/gL and help recruit the remaining components of the pentameric complex (Ryckman et al. 2008). To a lesser degree, the glycoprotein pUL116 also appears to compete for gH binding in the ER, but its role is still unknown (Calo et al. 2016). Only after the gH/gL/gO or gH/gL/pUL128–131 complexes are formed do they migrate to the Golgi where their glycosylation matures and they become ready for incorporation into virions (Ryckman et al. 2008; Theiler and Compton 2002; Jean Beltran and Cristea 2014).

Glycoprotein complex formation and incorporation in HCMV-infected cells is driven by several viral proteins including pUS16, pUS17, pUL148, and gO (Li et al. 2015; Jiang et al. 2008; Gurczynski et al. 2014; Luganini et al. 2017). As an example, when the immune modulatory transmembrane protein US17 was deleted, gH was mislocalized, and there was a threefold decrease in the level of gH found in virions (Gurczynski et al. 2014).

The glycoprotein pUL148 contains an RXR motif that retains it in the ER where it appears to bind and sequester gH/gL/pUL130 or gH/gL/pUL131, thus reducing the formation and trafficking of completed gH/gL/pUL128–131 complexes to the cVAC. This leads to an enrichment of gH/gL/gO in virions (Li et al. 2015). Using the TB40/E deletion strain TB_Δ148, high-multiplicity infections produced similar levels of virus in fibroblasts but yielded 100x more infectious output in human retinal pigment epithelial cells (Li et al. 2015). The deletion also caused substantially fewer gH/gL/gO complexes to form (Li et al. 2015). Insertion of pUL148 into the laboratory strain ADr131 which previously lacked it decreased the epithelial cell tropism fourfold (Li et al. 2015). A related tropism effect was seen when comparing B cell derived to epithelial cell-derived EBV, suggesting that mechanisms for selecting envelope glycoprotein complexes may be a conserved feature of herpesviruses (Molesworth et al. 2000).

Consistent with this, positional homologs of HCMV gO have approximately 40% amino acid similarity on average and are maintained in HHV-6A (U47), HHV-6B (KA8L), HHV-7 (U47), and MCMV (M74) (Huber and Compton 1998; Mori et al. 2003). Binding properties of HCMV gO vary in efficiency and are strain dependent (Zhou et al. 2013). Aligned amino acid sequences of HCMV gO from 40 clinical and 6 laboratory strains grouped into 8 diverse clades (Rasmussen et al. 2002; Zhou et al. 2013). gO isoforms from HCMV strains Towne, TR, Merlin, TB40/E, and AD169 were all able to form disulfide bonds with gH/gL of strain TR in fibroblasts. Virions produced in the presence of Merlin gO incorporated significantly more gH/gL/pUL128–131 than gH/gL/gO, in contrast to the other strains, for which the ratios were reversed (Zhou et al. 2013). Infecting fibroblasts with strains that do not express gO led to accumulation of nonenveloped cytoplasmic particles and mature virions with increased levels of gH/gL/pUL128–131, but 50% less gH/gL (Wille et al. 2010; Jiang et al. 2008). Interestingly, when human umbilical vein endothelial cells (HUVECs) were infected with a pUL131 deletion mutant, a related phenotype was observed, but virions had higher levels of gH/gL/gO. This emphasizes the cell-specific pathways and competitive nature of glycoprotein complex selection in virion maturation.

As illustrated here, herpesvirus glycoproteins serve crucial roles as bridges between the lipid membrane and the tegument. Importantly, by modulating glycoprotein expression and complex formation during infection, HCMV is able to produce virions optimized for growth in diverse cell types, thereby broadening its pathogenic range while enhancing its evolutionary survival.

9.4.7 Fatty Acid Metabolism as a Driver of Envelopment

As part of the process that leads to cVAC formation, envelopment, and egress, HCMV induces significant alterations in the metabolic profile of host cells (Vastag et al. 2011). As opposed to HSV-1 infection which upregulates pyrimidine nucleotide synthesis, HCMV induces a metabolic shift that favors synthesis of saturated long-chain fatty acids and increases membrane curvature at sites of envelopment (Munger et al. 2008; Spencer et al. 2011; Koyuncu et al. 2013; Vastag et al. 2011). As seen with other enveloped viruses, increased curvature promotes envelopment through concentration of membrane-bound viral proteins and decreased net energy cost during membrane budding (Chlanda et al. 2015; Roller et al. 2010; Schnee et al. 2006; McMahon and Gallop 2005).

The first committed step of fatty acid synthesis is catalyzed by acetyl-CoA carboxylase, which generates malonyl-CoA from by-products of glycolysis and the tricarboxylic acid (TCA) cycle (Wakil et al. 1983; Tong 2005). Malonyl-CoA is then used as a substrate by cellular acyl-CoA synthetases and elongases to create long-chain fatty acids (14–21 C chains) and very long-chain fatty acids (>21 C chains) (Wakil et al. 1983; Tong 2005).

Following HCMV infection, uptake of cellular glucose required for glycolysis increases as a downstream result of activation of the antiviral protein viperin (Chin and Cresswell 2001; Seo and Cresswell 2013; Landini 1984). Likewise, expression of cellular acetyl-CoA carboxylase also increases as a result of infection. By raising intracellular glucose and acetyl-CoA carboxylase levels, host malonyl-CoA production capacity increases and drives downstream fatty acid synthesis. If this process is blocked, such as by inhibition of acetyl-CoA carboxylase using siRNA or the inhibitor TOFA, virion output is decreased by 10- to 100-fold (Spencer et al. 2011; Munger et al. 2008).

An extensive siRNA screen identified 172 cellular enzymes associated with fatty acid metabolism and adipogenesis as having a role in HCMV replication (Koyuncu et al. 2013). From the screen, several acyl-CoA synthetases (ACSM2A, 3–5; ACSBG1–2, ACSL1, 3–6; and SLC27A1–6) plus the ELOVL1-7 family of elongases were found to be important for HCMV biogenesis (Koyuncu et al. 2013). Pharmacological inhibition of either set of enzymes reduced infectious output, with elongase inhibitors delaying expression of viral genes and causing a reduction in overall abundance of the tegument protein pUL99 (Koyuncu et al. 2013). In addition to acyl-CoA synthetases and elongases, class III phosphatidylinositol 3-kinase (Vps34) was also identified in the screen as being required for growth; without it, nonenveloped virions accumulate in the cytoplasm (Sharon-Friling and Shenk 2014; Koyuncu et al. 2013).

During HCMV infection, Vps34, low-density lipoprotein-related receptor 1 (LRP1), and acetyl-CoA carboxylase cooperatively form large cytoplasmic vesicles that are presumed to be sites of virion envelopment and act downstream of the viral protein pUL37x1 in the process of envelopment (Sharon-Friling and Shenk 2014). Through carbon labeling and mass spectrometry, increased acyl-CoA synthetase, elongase, and Vps34 expression were associated with upregulation of the abundance of saturated very long-chain fatty acids in the viral envelope; this was due to C18 fatty acid elongation, not *de novo* synthesis (Koyuncu et al. 2013). The lack of *de novo* synthesis suggested that HCMV uses preexisting stores of fatty acids, later found to be lipid droplets, to generate very long-chain fatty acids for virion envelopes (Koyuncu et al. 2013).

In opposition to the adipogenic metabolic profile of infected cells, LRP1 levels are also upregulated at the plasma membrane as part of an antiviral response (Gudleski-O'Regan et al. 2012). The receptor causes depletion of cellular- and virion-associated cholesterol because of the amplified fatty acid synthesis during infection. Inhibition of LRP1 by siRNA or antibody binding was sufficient to increase cholesterol concentration leading to a corresponding increase in the infectivity of nascent virions (Gudleski-O'Regan et al. 2012).

The ability to maintain elevated lipogenesis during HCMV infection is dependent on cleavage of the cellular sterol regulatory element-binding protein 1 (SREBP1) by its activation protein (SCAP) (Yu et al. 2012). The interaction of SREBP1 and SCAP is normally inhibited by increased sterol formation, but HCMV overrides this failsafe through expression of pUL38 (Yu et al. 2012; Purdy et al. 2015). pUL38 removes a repressor of mTOR activity which is sufficient for maintaining cleavage and activation of SREBP1, thereby inducing the elongase ELOVL7 needed for synthesis of the very long-chain fatty acids required for HCMV virion envelopment (Purdy et al. 2015; Lewis et al. 2011; Laplante and Sabatini 2009).

HCMV thus employs multiple virus-driven mechanisms to ensure an adequate supply of lipids with structures that are optimal for efficient production of infections virions.

9.4.8 Completion of Envelopment by Membrane Scission

Topologically, membrane budding events can be classified into two categories based on their direction:

Inward-bound buds extend from the extracellular environment through the plasma membrane to ultimately deliver membrane-bound vesicles to the interior of the cell (endocytosis). Cytoplasmic dynamin-dependent machinery employs a cinch-like mechanism to scission such vesicles in a manner that preserves the integrity of the plasma membrane and of the budded vesicle.

Outward-bound buds extend from the cytoplasmic space through the plasma membrane toward the outside of the cell, which is topologically similar to budding from the cytoplasm into the interior of a multivesicular body vesicle. Scission of such outward-bound buds is mediated by endosomal sorting complex required for transport (ESCRT)-dependent mechanisms (Raiborg and Stenmark 2009; Hurley and Hanson 2010), which operate from within the cytoplasmic space using a ratchet-like mechanism.

ESCRT machinery is comprised of five main cytoplasmic complexes, ESCRT-0, ESCRT-I, ESCRT-II,ESCRT- III, and Vps4-Vta1. It assists in both budding and scission of cellular vesicles through recognition of ubiquitin signals (Shields et al. 2009; Hurley 2008; Hurley and Ren 2009; Raiborg and Stenmark 2009). There is some conflicting evidence, but ESCRT-0, ESCRT-I, and ESCRT-II appear to act in parallel, not sequentially, to facilitate budding (Hurley and Ren 2009; Hurley 2008; Shields et al. 2009; Raiborg and Stenmark 2009). ESCRT-III and Vps4-Vta1 then act downstream of the other complexes to control scission and release events, respectively (Raiborg and Stenmark 2009; Hurley and Hanson 2010).

Following cVAC formation during HCMV infection, components of the ESCRT machinery are intermingled with Golgi and endosomal markers near sites of envelopment (Das and Pellett 2011; Tandon et al. 2009). During infection of retinal pigment epithelial cells with a GFP-labeled variant of AD169, siRNA silencing of Tsg101, a component of ESCRT-I, and ALIX, which helps recruit ESCRT-III, did not reduce virion output (Fraile-Ramos et al. 2007). In contrast, siRNA silencing of Vps4A/B resulted in increased infectious output suggesting that ESCRT recruitment was nonessential and potentially inhibitory to virion maturation (Fraile-Ramos et al. 2007). In a separate study, HFFs were infected with the HCMV strain Towne followed by transfection with dominant negative (DN) forms of Vps4, Tsg101, and CHMP1 (a component of ESCRT-III) (Tandon et al. 2009). CHMP1$_{DN}$ and Vps4$_{DN}$ reduced infectious output in contrast to the previous study, highlighting the need for further study (Tandon et al. 2009).

Combined with other observations, these studies suggest that HCMV utilizes ESCRT machinery for membrane budding and scission, but functional redundancies may allow it to bypass some requirements for recruitment of upstream ESCRT complexes (Bughio et al. 2015; Fraile-Ramos et al. 2007; Schauflinger et al. 2011; Tandon et al. 2009). For example, HCMV pUL103 contains ALIX-binding domains, so through interactions with other tegument proteins, including membrane-associated pUL71 and ALIX, pUL103 may be able to recruit ESCRT-III to membranes for envelopment without involving ESCRT-0, ESCRT-I, or ESCRT-II (Das et al. 2014; Schauflinger et al. 2011; Fischer 2012; Ortiz et al. 2016; Womack and Shenk 2010). Another possible explanation for the discordant results is that ESCRT complexes are multiprotein formations so experiments that target single proteins may be confounded when complementary paths are available.

As further evidenced by electron microscopy, HCMV virions accumulate in intracellular vesicles we define as virion envelopment and egress vesicles (VEEVs). VEEVs resemble multivesicular bodies, suggesting use of either ESCRT-associated pathways or an analogous pathway mediated by viral gene products to complete vesicle formation (Bughio et al. 2015; Schauflinger et al. 2011; Hurley and Hanson 2010; Raiborg and Stenmark 2009).

HSV-1 and HHV-6 virion envelopment are both dependent on CD63-positive MVB formation (Mori et al. 2008). HSV-1 utilizes a Vps4-dependent mechanism (Kharkwal et al. 2016; Calistri et al. 2007; Crump et al. 2007; Mori et al. 2008), but like HCMV, it is not dependent on either Tsg101 or ALIX (Pawliczek and Crump 2009). It is also possible that the observed defects following targeting of ESCRT machinery may be related to events required for entry into new cells, as seen during KSHV infection (Kumar and Chandran 2016; Kumar et al. 2016; Veettil et al. 2016), or membrane remodeling as seen with EBV (Lee et al. 2012).

9.4.9 Virion Egress Completes HCMV Replication

Following successful envelopment, fully matured virions must be exported out of host cells. For HCMV and other herpesviruses, the mechanism of viral egress is poorly understood.

The various pathways involved in vesicle-mediated transport depend on unique lipid and protein signatures. The concentration of particular fatty acids in a given region of a membrane defines its physiology and restricts the array of interacting proteins (van Spriel et al. 2015; Brown 2000; Resh 2004a, b).

To identify potential pathways of HCMV envelopment, liquid chromatography-mass spectrometry was used to analyze the lipidome of HCMV-infected fibroblasts by measuring the relative abundance of 146 unique glycerophospholipid species with chain lengths of 30–42 carbons (Liu et al. 2011). Except for a fourfold enrichment of phosphatidic acid during infection, the cellular glycerophospholipid profile did not deviate greatly from mock-infected cells (Liu et al. 2011). However, the glycerophospholipid composition of virions was markedly different compared to

HCMV- or mock-infected cells and was dominated by phosphatidylcholine and phosphatidylethanolamine species (Liu et al. 2011). When compared to known subcellular compartments, the lipid composition of virion envelopes most closely matched profiles seen in neuronal synaptic vesicle membranes, suggesting HCMV particles follow a related secretory vesicle pathway that operates in non-neuronal cells (Liu et al. 2011).

Secretion via secretory vesicles is dependent on a highly conserved trafficking pipeline used in cells from diverse lineages, including mast cells in the immune system and β cells in the pancreas (Mizuno et al. 2007; Higashio et al. 2016; Kimura and Niki 2011; Yi et al. 2002; Fukuda 2013). Usage of such a widely available pathway would likely contribute to the broad host cell tropism of HCMV. Starting at the TGN, the secretory vesicle pathway is regulated by several factors that act in sequence. These include Rab GTPases, cytoskeletal motors, and SNAP/SNARE complexes that associate with cargo-bearing vesicles and relay them toward the plasma membrane where fusion occurs by a Ca^{2+}-dependent mechanism (Sudhof 2013; Sheng et al. 1996; Schiavo et al. 1997; Rizo and Rosenmund 2008; Sudhof 1995; McMahon and Sudhof 1995).

Several proteins involved in secretory vesicle transport have been implicated in exocytosis of HCMV and other herpesviruses. Various Rab GTPases control specific intracellular vesicle transport pathways (Stenmark 2009; Grosshans et al. 2006). For secretory vesicle exocytosis, Rab3 and Rab27 work cooperatively to regulate transport and docking of vesicles at the plasma membrane prior to fusion (Tsuboi and Fukuda 2006; Handley et al. 2007; Fukuda 2013). As shown by immunoelectron microscopy, maturing HCMV and HSV-1 virions associate with Rab3-containing membranes (Homman-Loudiyi et al. 2003; Miranda-Saksena et al. 2009). In addition, infectious output was dramatically reduced in Rab27A-deficient cells when infected with HCMV and HSV-1 (Bello-Morales et al. 2012; Fraile-Ramos et al. 2010).

Once docked at the plasma membrane, secretory vesicles require SNAP/SNARE complexes to mediate membrane fusion. Syntaxin 3 (STX3) is one of several SNARE proteins capable of initiating secretory vesicle-plasma membrane fusion events (Mazelova et al. 2009). During HCMV infection, STX3 expression is highly upregulated and localizes to the cVAC. When knocked down using shRNA, production of infectious virions was reduced fourfold (Cepeda and Fraile-Ramos 2011). In neurons, SNAP25 is the major SNAP protein involved in SNAP/SNARE-mediated secretory vesicle exocytosis (McMahon and Sudhof 1995; Schiavo et al. 1997), but in other cell types, including fibroblasts, SNAP23, a homolog of SNAP25, is more widely used for exocytic events. Although SNAP23 abundance was unaffected by HCMV infection, shRNA knockdown of SNAP23 decreased infectious output 1000-fold (Liu et al. 2011).

While several lines of evidence are consistent with HCMV using a secretory vesicle-like pathway for final exocytosis of progeny virions, inhibition of key secretory pathway regulators does not completely eliminate HCMV egress. This suggests that either the virus can exploit alternative pathways for virion egress, or the secretory vesicle pathway may only be important for some aspect of virion replication distinct from egress. For example, although data supports the importance of secretory vesicle-like pathways during HSV-1 and PRV infection of neurons, infectious particles associate with and are released via a Rab6A-/Rab8A-/Rab11A-dependent recycling pathway as opposed to the Rab3A-/Rab27A-staining secretory vesicle pathway (Hogue et al. 2014, 2016; Miranda-Saksena et al. 2009; Johns et al. 2014).

The complexity of studying cellular trafficking events has complicated experiments analyzing herpesvirus egress. While a subject of much debate, it is clear that several herpesviruses employ similarly themed secretory pathways for escaping host cells, even across subfamilies.

9.5 Final Words

Herpesvirus replication relies heavily on remodeling host environments to enable efficient envelopment and egress of nascent particles. While similar high-order requirements are shared among the different subfamilies of the *Herpesviridae*, details of virion replication differ based on the repertoire of cells infected by each virus. Differences include development of a well-defined assembly compartment for HCMV, novel functions of some virion structural proteins, distinct composition of virion envelopes, and the mechanisms used for membrane scission and virion egress.

To enable virion assembly, herpesviruses hijack and modify host metabolic and trafficking pathways. For HCMV, this leads to formation of a secretory trap at the center of the cVAC and an eventual accumulation of the building blocks required to construct a nascent virion. Following construction of the capsid, packaging of the genome, export of the capsid to the cytoplasm, addition of the tegument, envelopment in host-derived membranes, and export along host secretory pathways, fully matured viral particles are finally released.

Modification of cellular trafficking pathways and mechanisms is a product of transcriptional regulation, a complex network of viral protein interactions, and modifications to the metabolic profile of host cells. Key proteins have been identified as having a role in virion maturation, but definitive mechanistic conclusions remain elusive due to functional redundancies and downstream cascades of interactions. While many questions remain unanswered, with persistent effort and the advent of new and improved experimental approaches, we will be able to probe deeper into the mechanistic depths of virion replication, expanding our knowledge one particle at a time.

References

Adler B, Scrivano L, Ruzcics Z, Rupp B, Sinzger C, Koszinowski U (2006) Role of human cytomegalovirus UL131A in cell type-specific virus entry and release. J Gen Virol 87(Pt 9):2451–2460. https://doi.org/10.1099/vir.0.81921-0

Ahlqvist J, Mocarski E (2011) Cytomegalovirus UL103 controls virion and dense body egress. J Virol 85(10):5125–5135. https://doi.org/10.1128/jvi.01682-10

Albecka A, Laine RF, Janssen AF, Kaminski CF, Crump CM (2016) HSV-1 glycoproteins are delivered to virus assembly sites through dynamin-dependent endocytosis. Traffic 17(1):21–39. https://doi.org/10.1111/tra.12340

Albecka A, Owen DJ, Ivanova L, Brun J, Liman R, Davies L, Ahmed MF, Colaco S, Hollinshead M, Graham SC, Crump CM (2017) Dual function of the pUL7-pUL51 tegument protein complex in herpes simplex virus 1 infection. J Virol 91(2). https://doi.org/10.1128/jvi.02196-16

Alconada A, Bauer U, Hoflack B (1996) A tyrosine-based motif and a casein kinase II phosphorylation site regulate the intracellular trafficking of the varicella-zoster virus glycoprotein I, a protein localized in the trans-Golgi network. EMBO J 15(22):6096–6110

Alconada A, Bauer U, Sodeik B, Hoflack B (1999) Intracellular traffic of herpes simplex virus glycoprotein gE: characterization of the sorting signals required for its trans-Golgi network localization. J Virol 73(1):377–387

Alwine JC (2012) The human cytomegalovirus assembly compartment: a masterpiece of viral manipulation of cellular processes that facilitates assembly and egress. PLoS Pathog 8(9):e1002878. https://doi.org/10.1371/journal.ppat.1002878

Archer MA, Brechtel TM, Davis LE, Parmar RC, Hasan MH, Tandon R (2017) Inhibition of endocytic pathways impacts cytomegalovirus maturation. Sci Rep 7:46069. https://doi.org/10.1038/srep46069

Baines JD, Roizman B (1992) The UL11 gene of herpes simplex virus 1 encodes a function that facilitates nucleocapsid envelopment and egress from cells. J Virol 66(8):5168–5174

Bardens A, Doring T, Stieler J, Prange R (2011) Alix regulates egress of hepatitis B virus naked capsid particles in an ESCRT-independent manner. Cell Microbiol 13(4):602–619. https://doi.org/10.1111/j.1462-5822.2010.01557.x

Beghetto E, Paolis FD, Spadoni A, Del Porto P, Buffolano W, Gargano N (2008) Molecular dissection of the human B cell response against cytomegalovirus infection by lambda display. J Virol Methods 151(1):7–14. https://doi.org/10.1016/j.jviromet.2008.04.005

Beitia Ortiz de Zarate I, Cantero-Aguilar L, Longo M, Berlioz-Torrent C, Rozenberg F (2007) Contribution of endocytic motifs in the cytoplasmic tail of herpes simplex virus type 1 glycoprotein B to virus replication and cell-cell fusion. J Virol 81(24):13889–13903. https://doi.org/10.1128/JVI.01231-07

Bello-Morales R, Crespillo AJ, Fraile-Ramos A, Tabares E, Alcina A, Lopez-Guerrero JA (2012) Role of the small GTPase Rab27a during herpes simplex virus infection of oligodendrocytic cells. BMC Microbiol 12:265. https://doi.org/10.1186/1471-2180-12-265

Benyesh-Melnick M, Probstmeyer F, McCombs R, Brunschwig JP, Vonka V (1966) Correlation between infectivity and physical virus particles in human cytomegalovirus. J Bacteriol 92(5):1555–1561

Bigalke JM, Heldwein EE (2015) Structural basis of membrane budding by the nuclear egress complex of herpesviruses. EMBO J 34(23):2921–2936. https://doi.org/10.15252/embj.201592359

Bigalke JM, Heldwein EE (2016) Nuclear exodus: herpesviruses lead the way. Annu Rev Virol 3(1):387–409. https://doi.org/10.1146/annurev-virology-110615-042215

Borst EM, Wagner K, Binz A, Sodeik B, Messerle M (2008) The essential human cytomegalovirus gene UL52 is required for cleavage-packaging of the viral genome. J Virol 82(5):2065–2078. https://doi.org/10.1128/jvi.01967-07

Borst EM, Kleine-Albers J, Gabaev I, Babic M, Wagner K, Binz A, Degenhardt I, Kalesse M, Jonjic S, Bauerfeind R, Messerle M (2013) The human cytomegalovirus UL51 protein is essential for viral genome cleavage-packaging and interacts with the terminase subunits pUL56 and pUL89. J Virol 87(3):1720–1732. https://doi.org/10.1128/jvi.01955-12

Borst EM, Bauerfeind R, Binz A, Stephan TM, Neuber S, Wagner K, Steinbruck L, Sodeik B, Lenac Rovis T, Jonjic S, Messerle M (2016) The essential human cytomegalovirus proteins pUL77 and pUL93 are structural components necessary for viral genome encapsidation. J Virol 90(13):5860–5875. https://doi.org/10.1128/jvi.00384-16

Bowman JJ, Lacayo JC, Burbelo P, Fischer ER, Cohen JI (2011) Rhesus and human cytomegalovirus glycoprotein L are required for infection and cell-to-cell spread of virus but cannot complement each other. J Virol 85(5):2089–2099. https://doi.org/10.1128/JVI.01970-10

Britt B (2007) Maturation and egress. In: Arvin A, Campadelli-Fiume G, Mocarski E et al (eds) Human herpesviruses: biology, Therapy, and Immunoprophylaxis. Cambridge University Press c.2007, Cambridge

Britt WJ, Mach M (1996) Human cytomegalovirus glycoproteins. Intervirology 39(5-6):401–412

Brown D (2000) Targeting of membrane transporters in renal epithelia: when cell biology meets physiology. Am J Physiol Renal Physiol 278(2):F192–F201

Buchkovich NJ, Maguire TG, Alwine JC (2010) Role of the endoplasmic reticulum chaperone BiP, SUN domain proteins, and dynein in altering nuclear morphology during human cytomegalovirus infection. J Virol 84(14):7005–7017. https://doi.org/10.1128/JVI.00719-10

Bughio F, Elliott DA, Goodrum F (2013) An endothelial cell-specific requirement for the UL133-UL138 locus of human cytomegalovirus for efficient virus maturation. J Virol 87(6):3062–3075. https://doi.org/10.1128/JVI.02510-12

Bughio F, Umashankar M, Wilson J, Goodrum F (2015) Human cytomegalovirus UL135 and UL136 genes are required for postentry tropism in endothelial cells. J Virol 89(13):6536–6550. https://doi.org/10.1128/JVI.00284-15

Calistri A, Sette P, Salata C, Cancellotti E, Forghieri C, Comin A, Gottlinger H, Campadelli-Fiume G, Palu G, Parolin C (2007) Intracellular trafficking and maturation of herpes simplex virus type 1 gB and virus egress require functional biogenesis of multivesicular bodies. J Virol 81(20):11468–11478. https://doi.org/10.1128/JVI.01364-07

Calo S, Cortese M, Ciferri C, Bruno L, Gerrein R, Benucci B, Monda G, Gentile M, Kessler T, Uematsu Y, Maione D, Lilja AE, Carfi A, Merola M (2016) The human cytomegalovirus UL116 gene encodes an envelope glycoprotein forming a complex with gH independently from gL. J Virol 90(10):4926–4938. https://doi.org/10.1128/JVI.02517-15

Camozzi D, Pignatelli S, Valvo C, Lattanzi G, Capanni C, Dal Monte P, Landini MP (2008) Remodelling of the nuclear lamina during human cytomegalovirus infection: role of the viral proteins pUL50 and pUL53. J Gen Virol 89(Pt 3):731–740. https://doi.org/10.1099/vir.0.83377-0

Caviness K, Cicchini L, Rak M, Umashankar M, Goodrum F (2014) Complex expression of the UL136 gene of human cytomegalovirus results in multiple protein isoforms with unique roles in replication. J Virol 88(24):14412–14425. https://doi.org/10.1128/JVI.02711-14

Caviness K, Bughio F, Crawford LB, Streblow DN, Nelson JA, Caposio P, Goodrum F (2016) Complex interplay of the UL136 isoforms balances cytomegalovirus replication and latency. MBio 7(2):e01986. https://doi.org/10.1128/mBio.01986-15

Cepeda V, Fraile-Ramos A (2011) A role for the SNARE protein syntaxin 3 in human cytomegalovirus morphogenesis. Cell Microbiol 13(6):846–858. https://doi.org/10.1111/j.1462-5822.2011.01583.x

Cepeda V, Esteban M, Fraile-Ramos A (2010) Human cytomegalovirus final envelopment on membranes containing both trans-Golgi network and endosomal markers. Cell Microbiol 12(3):386–404. https://doi.org/10.1111/j.1462-5822.2009.01405.x

Cha TA, Tom E, Kemble GW, Duke GM, Mocarski ES, Spaete RR (1996) Human cytomegalovirus clinical isolates carry at least 19 genes not found in laboratory strains. J Virol 70(1):78–83

Chadha P, Han J, Starkey JL, Wills JW (2012) Regulated interaction of tegument proteins UL16 and UL11 from herpes simplex virus. J Virol 86(21):11886–11898. https://doi.org/10.1128/JVI.01879-12

Chee MS, Bankier AT, Beck S, Bohni R, Brown CM, Cerny R, Horsnell T, Hutchison CA 3rd, Kouzarides T, Martignetti JA et al (1990) Analysis of the protein-coding content of the sequence of human cytomegalovirus strain AD169. Curr Top Microbiol Immunol 154:125–169

Chen DH, Jiang H, Lee M, Liu F, Zhou ZH (1999) Three-dimensional visualization of tegument/capsid interactions in the intact human cytomegalovirus. Virology 260(1):10–16. https://doi.org/10.1006/viro.1999.9791

Chin KC, Cresswell P (2001) Viperin (cig5), an IFN-inducible antiviral protein directly induced by human cytomegalovirus. Proc Natl Acad Sci U S A 98(26):15125–15130. https://doi.org/10.1073/pnas.011593298

Chiu YF, Sugden B, Chang PJ, Chen LW, Lin YJ, Lan YC, Lai CH, Liou JY, Liu ST, Hung CH (2012) Characterization and intracellular trafficking of Epstein-Barr virus BBLF1, a protein involved in virion maturation. J Virol 86(18):9647–9655. https://doi.org/10.1128/JVI.01126-12

Chlanda P, Schraidt O, Kummer S, Riches J, Oberwinkler H, Prinz S, Krausslich HG, Briggs JA (2015) Structural analysis of the roles of influenza A virus membrane-associated proteins in assembly and morphology. J Virol 89(17):8957–8966. https://doi.org/10.1128/JVI.00592-15

Chouljenko DV, Jambunathan N, Chouljenko VN, Naderi M, Brylinski M, Caskey JR, Kousoulas KG (2016) Herpes simplex virus 1 UL37 protein tyrosine residues conserved among all alphaherpesviruses are required for interactions with glycoprotein K, cytoplasmic virion envelopment, and infectious virus production. J Virol 90(22):10351–10361. https://doi.org/10.1128/JVI.01202-16

Coleman S, Hornig J, Maddux S, Choi KY, McGregor A (2015) Viral glycoprotein complex formation, essential function and immunogenicity in the guinea pig model for cytomegalovirus. PLoS One 10(8):e0135567. https://doi.org/10.1371/journal.pone.0135567

Coleman S, Choi KY, Root M, McGregor A (2016) A homolog pentameric complex dictates viral epithelial tropism, pathogenicity and congenital infection rate in guinea pig cytomegalovirus. PLoS Pathog 12(7):e1005755. https://doi.org/10.1371/journal.ppat.1005755

Crump CM, Hung CH, Thomas L, Wan L, Thomas G (2003) Role of PACS-1 in trafficking of human cytomegalovirus glycoprotein B and virus production. J Virol 77(20):11105–11113. https://doi.org/10.1128/jvi.77.20.11105-11113.2003

Crump CM, Yates C, Minson T (2007) Herpes simplex virus type 1 cytoplasmic envelopment requires functional Vps4. J Virol 81(14):7380–7387. https://doi.org/10.1128/JVI.00222-07

Cruz L, Streck NT, Ferguson K, Desai T, Desai DH, Amin SG, Buchkovich NJ (2017) Potent inhibition of human cytomegalovirus by modulation of cellular SNARE syntaxin 5. J Virol 91(1). https://doi.org/10.1128/JVI.01637-16

Dai X, Yu X, Gong H, Jiang X, Abenes G, Liu H, Shivakoti S, Britt WJ, Zhu H, Liu F, Zhou ZH (2013) The smallest capsid protein mediates binding of the essential tegument protein pp 150 to stabilize DNA-containing capsids in human cytomegalovirus. PLoS Pathog 9(8):e1003525. https://doi.org/10.1371/journal.ppat.1003525

Das S, Pellett PE (2011) Spatial relationships between markers for secretory and endosomal machinery in human cytomegalovirus-infected cells versus those in uninfected cells. J Virol 85(12):5864–5879. https://doi.org/10.1128/jvi.00155-11

Das S, Vasanji A, Pellett PE (2007) Three-dimensional structure of the human cytomegalovirus cytoplasmic virion assembly complex includes a reoriented secretory apparatus. J Virol 81(21):11861–11869. https://doi.org/10.1128/jvi.01077-07

Das S, Ortiz DA, Gurczynski SJ, Khan F, Pellett PE (2014) Identification of human cytomegalovirus genes important for biogenesis of the cytoplasmic virion assembly complex. J Virol 88(16):9086–9099. https://doi.org/10.1128/jvi.01141-14

DeRussy BM, Tandon R (2015) Human cytomegalovirus pUL93 Is required for viral genome cleavage and packaging. J Virol 89(23):12221–12225. https://doi.org/10.1128/jvi.02382-15

Dolan A, Cunningham C, Hector RD, Hassan-Walker AF, Lee L, Addison C, Dargan DJ, McGeoch DJ, Gatherer D, Emery VC, Griffiths PD, Sinzger C, McSharry BP, Wilkinson GW, Davison AJ (2004) Genetic content of wild-type human cytomegalovirus. J Gen Virol 85(Pt 5):1301–1312. https://doi.org/10.1099/vir.0.79888-0

Dunn W, Chou C, Li H, Hai R, Patterson D, Stolc V, Zhu H, Liu F (2003) Functional profiling of a human cytomegalovirus genome. Proc Natl Acad Sci U S A 100(24):14223–14228. https://doi.org/10.1073/pnas.2334032100

Favoreel HW, Van Minnebruggen G, Nauwynck HJ, Enquist LW, Pensaert MB (2002) A tyrosine-based motif in the cytoplasmic tail of pseudorabies virus glycoprotein B is important for both antibody-induced internalization of viral glycoproteins and efficient cell-to-cell spread. J Virol 76(13):6845–6851

Fischer D (2012) Dissecting functional motifs of the human cytomegalovirus tegument protein pUL71 (Ph.D. thesis). University of Ulm, Ulm

Fish KN, Britt W, Nelson JA (1996) A novel mechanism for persistence of human cytomegalovirus in macrophages. J Virol 70(3):1855–1862

Foster TP, Melancon JM, Olivier TL, Kousoulas KG (2004) Herpes simplex virus type 1 glycoprotein K and the UL20 protein are interdependent for intracellular trafficking and trans-Golgi network localization. J Virol 78(23):13262–13277. https://doi.org/10.1128/JVI.78.23.13262-13277.2004

Fraile-Ramos A, Pelchen-Matthews A, Risco C, Rejas MT, Emery VC, Hassan-Walker AF, Esteban M, Marsh M (2007) The ESCRT machinery is not required for human cytomegalovirus envelopment. Cell Microbiol 9(12):2955–2967. https://doi.org/10.1111/j.1462-5822.2007.01024.x

Fraile-Ramos A, Cepeda V, Elstak E, van der Sluijs P (2010) Rab27a is required for human cytomegalovirus assembly. PLoS One 5(12):e15318. https://doi.org/10.1371/journal.pone.0015318

Fukuda M (2013) Rab27 effectors, pleiotropic regulators in secretory pathways. Traffic 14(9):949–963. https://doi.org/10.1111/tra.12083

Gottwein E, Corcoran DL, Mukherjee N, Skalsky RL, Hafner M, Nusbaum JD, Shamulailatpam P, Love CL, Dave SS, Tuschl T, Ohler U, Cullen BR (2011) Viral microRNA targetome of KSHV-infected primary effusion lymphoma cell lines. Cell Host Microbe 10(5):515–526. https://doi.org/10.1016/j.chom.2011.09.012

Goulidaki N, Alarifi S, Alkahtani SH, Al-Qahtani A, Spandidos DA, Stournaras C, Sourvinos G (2015) RhoB is a component of the human cytomegalovirus assembly complex and is required for efficient viral production. Cell Cycle 14(17):2748–2763. https://doi.org/10.1080/15384101.2015.1066535

Grainger L, Cicchini L, Rak M, Petrucelli A, Fitzgerald KD, Semler BL, Goodrum F (2010) Stress-inducible alternative translation initiation of human cytomegalovirus latency protein pUL138. J Virol 84(18):9472–9486. https://doi.org/10.1128/JVI.00855-10

Grey F, Nelson J (2008) Identification and function of human cytomegalovirus microRNAs. J Clin Virol 41(3):186–191. https://doi.org/10.1016/j.jcv.2007.11.024

Grosshans BL, Ortiz D, Novick P (2006) Rabs and their effectors: achieving specificity in membrane traffic. Proc Natl Acad Sci U S A 103(32):11821–11827. https://doi.org/10.1073/pnas.0601617103

Gudleski-O'Regan N, Greco TM, Cristea IM, Shenk T (2012) Increased expression of LDL receptor-related protein 1 during human cytomegalovirus infection reduces virion cholesterol and infectivity. Cell Host Microbe 12(1):86–96. https://doi.org/10.1016/j.chom.2012.05.012

Guo H, Wang L, Peng L, Zhou ZH, Deng H (2009) Open reading frame 33 of a gammaherpesvirus encodes a tegument protein essential for virion morphogenesis and egress. J Virol 83(20):10582–10595. https://doi.org/10.1128/JVI.00497-09

Guo H, Shen S, Wang L, Deng H (2010) Role of tegument proteins in herpesvirus assembly and egress. Protein Cell 1(11):987–998. https://doi.org/10.1007/s13238-010-0120-0

Gurczynski SJ, Das S, Pellett PE (2014) Deletion of the human cytomegalovirus US17 gene increases the ratio of genomes per infectious unit and alters regulation of immune and endoplasmic reticulum stress response genes at early and late times after infection. J Virol 88(4):2168–2182. https://doi.org/10.1128/JVI.02704-13

Handley MT, Haynes LP, Burgoyne RD (2007) Differential dynamics of Rab3A and Rab27A on secretory granules. J Cell Sci 120(Pt 6):973–984. https://doi.org/10.1242/jcs.03406

Heineman TC, Hall SL (2002) Role of the varicella-zoster virus gB cytoplasmic domain in gB transport and viral egress. J Virol 76(2):591–599

Hertel L, Mocarski ES (2004) Global analysis of host cell gene expression late during cytomegalovirus infection reveals extensive dysregulation of cell cycle gene expression and induction of Pseudomitosis independent of US28 function. J Virol 78(21):11988–12011. https://doi.org/10.1128/JVI.78.21.11988-12011.2004

Higashio H, Satoh Y, Saino T (2016) Mast cell degranulation is negatively regulated by the Munc13-4-binding small-guanosine triphosphatase Rab37. Sci Rep 6:22539. https://doi.org/10.1038/srep22539

Hobom U, Brune W, Messerle M, Hahn G, Koszinowski UH (2000) Fast screening procedures for random transposon libraries of cloned herpesvirus genomes: mutational analysis of human cytomegalovirus envelope glycoprotein genes. J Virol 74(17):7720–7729

Hogue IB, Bosse JB, Hu JR, Thiberge SY, Enquist LW (2014) Cellular mechanisms of alpha herpesvirus egress: live cell fluorescence microscopy of pseudorabies virus exocytosis. PLoS Pathog 10(12):e1004535. https://doi.org/10.1371/journal.ppat.1004535

Hogue IB, Scherer J, Enquist LW (2016) Exocytosis of alphaherpesvirus virions, light particles, and glycoproteins uses constitutive secretory mechanisms. MBio 7(3). https://doi.org/10.1128/mBio.00820-16

Hollinshead M, Johns HL, Sayers CL, Gonzalez-Lopez C, Smith GL, Elliott G (2012) Endocytic tubules regulated by Rab GTPases 5 and 11 are used for envelopment of herpes simplex virus. EMBO J 31(21):4204–4220. https://doi.org/10.1038/emboj.2012.262

Homman-Loudiyi M, Hultenby K, Britt W, Soderberg-Naucler C (2003) Envelopment of human cytomegalovirus occurs by budding into Golgi-derived vacuole compartments positive for gB, Rab 3, Trans-Golgi Network 46, and Mannosidase II. J Virol 77(5):3191–3203. https://doi.org/10.1128/jvi.77.5.3191-3203.2003

Hook LM, Grey F, Grabski R, Tirabassi R, Doyle T, Hancock M, Landais I, Jeng S, McWeeney S, Britt W, Nelson JA (2014) Cytomegalovirus miRNAs target secretory pathway genes to facilitate formation of the virion assembly compartment and reduce cytokine secretion. Cell Host Microbe 15(3):363–373. https://doi.org/10.1016/j.chom.2014.02.004

Huber MT, Compton T (1998) The human cytomegalovirus UL74 gene encodes the third component of the glycoprotein H-glycoprotein L-containing envelope complex. J Virol 72(10):8191–8197

Hurley JH (2008) ESCRT complexes and the biogenesis of multivesicular bodies. Curr Opin Cell Biol 20(1):4–11. https://doi.org/10.1016/j.ceb.2007.12.002

Hurley JH, Hanson PI (2010) Membrane budding and scission by the ESCRT machinery: it's all in the neck. Nat Rev Mol Cell Biol 11(8):556–566. https://doi.org/10.1038/nrm2937

Hurley JH, Ren X (2009) The circuitry of cargo flux in the ESCRT pathway. J Cell Biol 185(2):185–187. https://doi.org/10.1083/jcb.200903013

Hutt-Fletcher LM (2015) EBV glycoproteins: where are we now? Future Virol 10(10):1155–1162. https://doi.org/10.2217/fvl.15.80

Jarvis MA, Fish KN, Soderberg-Naucler C, Streblow DN, Meyers HL, Thomas G, Nelson JA (2002) Retrieval of human cytomegalovirus glycoprotein B from cell surface Is not required for virus envelopment in astrocytoma cells. J Virol 76(10):5147–5155. https://doi.org/10.1128/jvi.76.10.5147-5155.2002

Jarvis MA, Jones TR, Drummond DD, Smith PP, Britt WJ, Nelson JA, Baldick CJ (2003) Phosphorylation of human cytomegalovirus glycoprotein B (gB) at the acidic cluster casein kinase 2 site (Ser900) is required for localization of gB to the trans-Golgi network and efficient virus replication. J Virol 78(1):285–293. https://doi.org/10.1128/jvi.78.1.285-293.2004

Jean Beltran PM, Cristea IM (2014) The life cycle and pathogenesis of human cytomegalovirus infection: lessons from proteomics. Expert Rev Proteomics 11(6):697–711. https://doi.org/10.1586/14789450.2014.971116

Jean Beltran PM, Mathias RA, Cristea IM (2016) A portrait of the human organelle proteome in space and time during cytomegalovirus infection. Cell Syst 3(4):361–373. e366. https://doi.org/10.1016/j.cels.2016.08.012

McMahon HT, Gallop JL (2005) Membrane curvature and mechanisms of dynamic cell membrane remodelling. Nature 438(7068):590–596. https://doi.org/10.1038/nature04396

McMahon HT, Sudhof TC (1995) Synaptic core complex of synaptobrevin, syntaxin, and SNAP25 forms high affinity alpha-SNAP binding site. J Biol Chem 270(5):2213–2217

Meissner CS, Suffner S, Schauflinger M, von Einem J, Bogner E (2012) A leucine zipper motif of a tegument protein triggers final envelopment of human cytomegalovirus. J Virol 86(6):3370–3382. https://doi.org/10.1128/JVI.06556-11

Mettenleiter TC (2006) Intriguing interplay between viral proteins during herpesvirus assembly or: the herpesvirus assembly puzzle. Vet Microbiol 113(3-4):163–169. https://doi.org/10.1016/j.vetmic.2005.11.040

Mettenleiter TC, Muller F, Granzow H, Klupp BG (2013) The way out: what we know and do not know about herpesvirus nuclear egress. Cell Microbiol 15(2):170–178. https://doi.org/10.1111/cmi.12044

Milbradt J, Webel R, Auerochs S, Sticht H, Marschall M (2010) Novel mode of phosphorylation-triggered reorganization of the nuclear lamina during nuclear egress of human cytomegalovirus. J Biol Chem 285(18):13979–13989. https://doi.org/10.1074/jbc.M109.063628

Miranda-Saksena M, Boadle RA, Aggarwal A, Tijono B, Rixon FJ, Diefenbach RJ, Cunningham AL (2009) Herpes simplex virus utilizes the large secretory vesicle pathway for anterograde transport of tegument and envelope proteins and for viral exocytosis from growth cones of human fetal axons. J Virol 83(7):3187–3199. https://doi.org/10.1128/JVI.01579-08

Mizuno K, Tolmachova T, Ushakov DS, Romao M, Abrink M, Ferenczi MA, Raposo G, Seabra MC (2007) Rab27b regulates mast cell granule dynamics and secretion. Traffic 8(7):883–892. https://doi.org/10.1111/j.1600-0854.2007.00571.x

Mocarski ES (2007) Comparative analysis of herpesvirus-common proteins. In: Arvin A, Campadelli-Fiume G, Mocarski E et al (eds) Human herpesviruses: biology, therapy, and immunoprophylaxis. Cambridge University Press, Copyright (c) Cambridge University Press 2007, Cambridge

Mocarski ES, Shenk T, Griffiths PD, Pass RF (2013) In: Fields BN, Knipe DM, Howley PM (eds) Cytomegaloviruses, Fields virology, vol 2, 6th edn. Wolters Kluwer Health/Lippincott Williams & Wilkins, c2013, Philadelphia

Molesworth SJ, Lake CM, Borza CM, Turk SM, Hutt-Fletcher LM (2000) Epstein-Barr virus gH is essential for penetration of B cells but also plays a role in attachment of virus to epithelial cells. J Virol 74(14):6324–6332

Mori Y, Akkapaiboon P, Yang X, Yamanishi K (2003) The human herpesvirus 6 U100 gene product is the third component of the gH-gL glycoprotein complex on the viral envelope. J Virol 77(4):2452–2458

Mori Y, Koike M, Moriishi E, Kawabata A, Tang H, Oyaizu H, Uchiyama Y, Yamanishi K (2008) Human herpesvirus-6 induces MVB formation, and virus egress occurs by an exosomal release pathway. Traffic 9(10):1728–1742. https://doi.org/10.1111/j.1600-0854.2008.00796.x

Mori J, Kawabata A, Tang H, Tadagaki K, Mizuguchi H, Kuroda K, Mori Y (2015a) Human herpesvirus-6 U14 induces cell-cycle arrest in G2/M phase by associating with a cellular protein, EDD. PLoS One 10(9):e0137420. https://doi.org/10.1371/journal.pone.0137420

Mori J, Tang H, Kawabata A, Koike M, Mori Y (2015b) Human herpesvirus 6A U14 is important for virus maturation. J Virol 90(3):1677–1681. https://doi.org/10.1128/jvi.02492-15

Munger J, Bennett BD, Parikh A, Feng XJ, McArdle J, Rabitz HA, Shenk T, Rabinowitz JD (2008) Systems-level metabolic flux profiling identifies fatty acid synthesis as a target for antiviral therapy. Nat Biotechnol 26(10):1179–1186. https://doi.org/10.1038/nbt.1500

Murphy E, Yu D, Grimwood J, Schmutz J, Dickson M, Jarvis MA, Hahn G, Nelson JA, Myers RM, Shenk TE (2003) Coding potential of laboratory and clinical strains of human cytomegalovirus. Proc Natl Acad Sci U S A 100(25):14976–14981. https://doi.org/10.1073/pnas.2136652100

Nicholas J (1996) Determination and analysis of the complete nucleotide sequence of human herpesvirus. J Virol 70(9):5975–5989

Nozawa N, Daikoku T, Koshizuka T, Yamauchi Y, Yoshikawa T, Nishiyama Y (2003) Subcellular localization of herpes simplex virus type 1 UL51 protein and role of palmitoylation in Golgi apparatus targeting. J Virol 77(5):3204–3216

Nozawa N, Kawaguchi Y, Tanaka M, Kato A, Kato A, Kimura H, Nishiyama Y (2005) Herpes simplex virus type 1 UL51 protein is involved in maturation and egress of virus particles. J Virol 79(11):6947–6956. https://doi.org/10.1128/JVI.79.11.6947-6956.2005

Ohno H, Fournier MC, Poy G, Bonifacino JS (1996) Structural determinants of interaction of tyrosine-based sorting signals with the adaptor medium chains. J Biol Chem 271(46):29009–29015

Olson JK, Grose C (1997) Endocytosis and recycling of varicella-zoster virus Fc receptor glycoprotein gE: internalization mediated by a YXXL motif in the cytoplasmic tail. J Virol 71(5):4042–4054

Ortiz DA, Glassbrook JE, Pellett PE (2016) Protein-protein interactions suggest novel activities of human cytomegalovirus tegument protein pUL103. J Virol 90(17):7798–7810. https://doi.org/10.1128/jvi.00097-16

Pawliczek T, Crump CM (2009) Herpes simplex virus type 1 production requires a functional ESCRT-III complex but is independent of TSG101 and ALIX expression. J Virol 83(21):11254–11264. https://doi.org/10.1128/jvi.00574-09

Pellett PE, Roizman B (2013) In: Knipe DM, Howley PM, Cohen JI et al (eds) The family Herpesviridae: a brief introduction, Fields virology, vol 2, 6th edn, pp 1802–1822

Phillips SL, Bresnahan WA (2011) Identification of binary interactions between human cytomegalovirus virion proteins. J Virol 85(1):440–447. https://doi.org/10.1128/jvi.01551-10

Phillips SL, Bresnahan WA (2012) The human cytomegalovirus (HCMV) tegument protein UL94 is essential for secondary envelopment of HCMV virions. J Virol 86(5):2523–2532. https://doi.org/10.1128/JVI.06548-11

Phillips SL, Cygnar D, Thomas A, Bresnahan WA (2012) Interaction between the human cytomegalovirus tegument proteins UL94 and UL99 is essential for virus replication. J Virol 86(18):9995–10005. https://doi.org/10.1128/JVI.01078-12

Poncet D, Pauleau AL, Szabadkai G, Vozza A, Scholz SR, Le Bras M, Briere JJ, Jalil A, Le Moigne R, Brenner C, Hahn G, Wittig I, Schagger H, Lemaire C, Bianchi K, Souquere S, Pierron G, Rustin P, Goldmacher VS, Rizzuto R, Palmieri F, Kroemer G (2006) Cytopathic effects of the cytomegalovirus-encoded apoptosis inhibitory protein vMIA. J Cell Biol 174(7):985–996. https://doi.org/10.1083/jcb.200604069

Purdy JG, Shenk T, Rabinowitz JD (2015) Fatty acid elongase 7 catalyzes lipidome remodeling essential for human cytomegalovirus replication. Cell Rep 10(8):1375–1385. https://doi.org/10.1016/j.celrep.2015.02.003

Radsak K, Eickmann M, Mockenhaupt T, Bogner E, Kern H, Eis-Hubinger A, Reschke M (1996) Retrieval of human cytomegalovirus glycoprotein B from the infected cell surface for virus envelopment. Arch Virol 141(3-4):557–572

Raiborg C, Stenmark H (2009) The ESCRT machinery in endosomal sorting of ubiquitylated membrane proteins. Nature 458(7237):445–452. https://doi.org/10.1038/nature07961

Rasmussen L, Geissler A, Cowan C, Chase A, Winters M (2002) The genes encoding the gCIII complex of human cytomegalovirus exist in highly diverse combinations in clinical isolates. J Virol 76(21):10841–10848

Resh MD (2004a) Membrane targeting of lipid modified signal transduction proteins. Subcell Biochem 37:217–232

Resh MD (2004b) A myristoyl switch regulates membrane binding of HIV-1 Gag. Proc Natl Acad Sci U S A 101(2):417–418. https://doi.org/10.1073/pnas.0308043101

Rizo J, Rosenmund C (2008) Synaptic vesicle fusion. Nat Struct Mol Biol 15(7):665–674

Roller RJ, Bjerke SL, Haugo AC, Hanson S (2010) Analysis of a charge cluster mutation of herpes simplex virus type 1 UL34 and its extragenic suppressor suggests a novel interaction between pUL34 and pUL31 that is necessary for membrane curvature around capsids. J Virol 84(8):3921–3934. https://doi.org/10.1128/JVI.01638-09

Roller RJ, Haugo AC, Yang K, Baines JD (2014) The herpes simplex virus 1 UL51 gene product has cell type-specific functions in cell-to-cell spread. J Virol 88(8):4058–4068. https://doi.org/10.1128/JVI.03707-13

Ryckman BJ, Jarvis MA, Drummond DD, Nelson JA, Johnson DC (2006) Human cytomegalovirus entry into epithelial and endothelial cells depends on genes UL128 to UL150 and occurs by endocytosis and low-pH fusion. J Virol 80(2):710–722. https://doi.org/10.1128/JVI.80.2.710-722.2006

Ryckman BJ, Rainish BL, Chase MC, Borton JA, Nelson JA, Jarvis MA, Johnson DC (2008) Characterization of the human cytomegalovirus gH/gL/UL128-131 complex that mediates entry into epithelial and endothelial cells. J Virol 82(1):60–70. https://doi.org/10.1128/JVI.01910-07

Sanchez V, Greis KD, Sztul E, Britt WJ (2000a) Accumulation of virion tegument and envelope proteins in a stable cytoplasmic compartment during human cytomegalovirus replication: characterization of a potential site of virus assembly. J Virol 74(2):975–986

Sanchez V, Sztul E, Britt WJ (2000b) Human cytomegalovirus pp28 (UL99) localizes to a cytoplasmic compartment which overlaps the endoplasmic reticulum-golgi-intermediate compartment. J Virol 74(8):3842–3851

Schauflinger M, Fischer D, Schreiber A, Chevillotte M, Walther P, Mertens T, von Einem J (2011) The tegument protein UL71 of human cytomegalovirus is involved in late envelopment and affects multivesicular bodies. J Virol 85(8):3821–3832. https://doi.org/10.1128/JVI.01540-10

Schauflinger M, Villinger C, Mertens T, Walther P, von Einem J (2013) Analysis of human cytomegalovirus secondary envelopment by advanced electron microscopy. Cell Microbiol 15(2):305–314. https://doi.org/10.1111/cmi.12077

Schiavo G, Stenbeck G, Rothman JE, Sollner TH (1997) Binding of the synaptic vesicle v-SNARE, synaptotagmin, to the plasma membrane t-SNARE, SNAP-25, can explain docked vesicles at neurotoxin-treated synapses. Proc Natl Acad Sci U S A 94(3):997–1001

Schierling K, Stamminger T, Mertens T, Winkler M (2004) Human cytomegalovirus tegument proteins ppUL82 (pp71) and ppUL35 interact and cooperatively activate the major immediate-early enhancer. J Virol 78(17):9512–9523. https://doi.org/10.1128/JVI.78.17.9512-9523.2004

Schierling K, Buser C, Mertens T, Winkler M (2005) Human cytomegalovirus tegument protein ppUL35 is important for viral replication and particle formation. J Virol 79(5):3084–3096. https://doi.org/10.1128/JVI.79.5.3084-3096.2005

Schleiss MR, McGregor A, Choi KY, Date SV, Cui X, McVoy MA (2008) Analysis of the nucleotide sequence of the guinea pig cytomegalovirus (GPCMV) genome. Virol J 5:139. https://doi.org/10.1186/1743-422X-5-139

Schnee M, Ruzsics Z, Bubeck A, Koszinowski UH (2006) Common and specific properties of herpesvirus UL34/UL31 protein family members revealed by protein complementation assay. J Virol 80(23):11658–11666. https://doi.org/10.1128/JVI.01662-06

Schultz EP, Lanchy JM, Ellerbeck EE, Ryckman BJ (2015) Scanning mutagenesis of human cytomegalovirus glycoprotein gH/gL. J Virol 90(5):2294–2305. https://doi.org/10.1128/JVI.01875-15

Seo JY, Britt WJ (2006) Sequence requirements for localization of human cytomegalovirus tegument protein pp28 to the virus assembly compartment and for assembly of infectious virus. J Virol 80(11):5611–5626. https://doi.org/10.1128/JVI.02630-05

Seo JY, Britt WJ (2007) Cytoplasmic envelopment of human cytomegalovirus requires the post-localization function of tegument protein pp28 within the assembly compartment. J Virol 81(12):6536–6547. https://doi.org/10.1128/jvi.02852-06

Seo JY, Britt WJ (2008) Multimerization of tegument protein pp28 within the assembly compartment is required for cytoplasmic envelopment of human cytomegalovirus. J Virol 82(13):6272–6287. https://doi.org/10.1128/JVI.02345-07

Seo JY, Cresswell P (2013) Viperin regulates cellular lipid metabolism during human cytomegalovirus infection. PLoS Pathog 9(8):e1003497. https://doi.org/10.1371/journal.ppat.1003497

Sharon-Friling R, Shenk T (2014) Human cytomegalovirus pUL37x1-induced calcium flux activates PKCalpha, inducing altered cell shape and accumulation of cytoplasmic vesicles. Proc Natl Acad Sci U S A 111(12):E1140–E1148. https://doi.org/10.1073/pnas.1402515111

Sharon-Friling R, Goodhouse J, Colberg-Poley AM, Shenk T (2006) Human cytomegalovirus pUL37x1 induces the release of endoplasmic reticulum calcium stores. Proc Natl Acad Sci U S A 103(50):19117–19122. https://doi.org/10.1073/pnas.0609353103

Sheng ZH, Rettig J, Cook T, Catterall WA (1996) Calcium-dependent interaction of N-type calcium channels with the synaptic core complex. Nature 379(6564):451–454. https://doi.org/10.1038/379451a0

Shields SB, Oestreich AJ, Winistorfer S, Nguyen D, Payne JA, Katzmann DJ, Piper R (2009) ESCRT ubiquitin-binding domains function cooperatively during MVB cargo sorting. J Cell Biol 185(2):213–224. https://doi.org/10.1083/jcb.200811130

Silva MC, Yu QC, Enquist L, Shenk T (2003) Human cytomegalovirus UL99-encoded pp28 Is required for the cytoplasmic envelopment of tegument-associated capsids. J Virol 77(19):10594–10605. https://doi.org/10.1128/jvi.77.19.10594-10605.2003

Skalsky RL, Corcoran DL, Gottwein E, Frank CL, Kang D, Hafner M, Nusbaum JD, Feederle R, Delecluse HJ, Luftig MA, Tuschl T, Ohler U, Cullen BR (2012) The viral and cellular microRNA targetome in lymphoblastoid cell lines. PLoS Pathog 8(1):e1002484. https://doi.org/10.1371/journal.ppat.1002484

Smith RM, Kosuri S, Kerry JA (2014) Role of human cytomegalovirus tegument proteins in virion assembly. Viruses 6(2):582–605. https://doi.org/10.3390/v6020582

Songyang Z, Shoelson SE, Chaudhuri M, Gish G, Pawson T, Haser WG, King F, Roberts T, Ratnofsky S, Lechleider RJ et al (1993) SH2 domains recognize specific phosphopeptide sequences. Cell 72(5):767–778

Spaderna S, Kropff B, Kodel Y, Shen S, Coley S, Lu S, Britt W, Mach M (2005) Deletion of gpUL132, a structural component of human cytomegalovirus, results in impaired virus replication in fibroblasts. J Virol 79(18):11837–11847. https://doi.org/10.1128/JVI.79.18.11837-11847.2005

Spear PG, Longnecker R (2003) Herpesvirus entry: an update. J Virol 77(19):10179–10185

Spencer CM, Schafer XL, Moorman NJ, Munger J (2011) Human cytomegalovirus induces the activity and expression of acetyl-coenzyme A carboxylase, a fatty acid biosynthetic enzyme whose inhibition attenuates viral replication. J Virol 85(12):5814–5824. https://doi.org/10.1128/JVI.02630-10

Stenmark H (2009) Rab GTPases as coordinators of vesicle traffic. Nat Rev Mol Cell Biol 10(8):513–525. https://doi.org/10.1038/nrm2728

Sudhof TC (1995) The synaptic vesicle cycle: a cascade of protein-protein interactions. Nature 375(6533):645–653. https://doi.org/10.1038/375645a0

Sudhof TC (2013) A molecular machine for neurotransmitter release: synaptotagmin and beyond. Nat Med 19(10):1227–1231. https://doi.org/10.1038/nm.3338

Takemoto M, Koike M, Mori Y, Yonemoto S, Sasamoto Y, Kondo K, Uchiyama Y, Yamanishi K (2005) Human herpesvirus 6 open reading frame U14 protein and cellular p53 interact with each other and are contained in the virion. J Virol 79(20):13037–13046. https://doi.org/10.1128/jvi.79.20.13037-13046.2005

Tandon R, Mocarski ES (2012) Viral and host control of cytomegalovirus maturation. Trends Microbiol 20(8):392–401. https://doi.org/10.1016/j.tim.2012.04.008

Tandon R, AuCoin DP, Mocarski ES (2009) Human cytomegalovirus exploits ESCRT machinery in the process of virion maturation. J Virol 83(20):10797–10807. https://doi.org/10.1128/jvi.01093-09

Tandon R, Mocarski ES, Conway JF (2015) The A, B, Cs of herpesvirus capsids. Viruses 7(3):899–914. https://doi.org/10.3390/v7030899

Theiler RN, Compton T (2002) Distinct glycoprotein O complexes arise in a post-Golgi compartment of cytomegalovirus-infected cells. J Virol 76(6):2890–2898

Tirabassi RS, Enquist LW (1998) Role of envelope protein gE endocytosis in the pseudorabies virus life cycle. J Virol 72(6):4571–4579

Tirabassi RS, Enquist LW (1999) Mutation of the YXXL endocytosis motif in the cytoplasmic tail of pseudorabies virus gE. J Virol 73(4):2717–2728

Tirosh O, Cohen Y, Shitrit A, Shani O, Le-Trilling VT, Trilling M, Friedlander G, Tanenbaum M, Stern-Ginossar N (2015) The transcription and translation landscapes during human cytomegalovirus infection reveal novel host-pathogen interactions. PLoS Pathog 11(11):e1005288. https://doi.org/10.1371/journal.ppat.1005288

To A, Bai Y, Shen A, Gong H, Umamoto S, Lu S, Liu F (2011) Yeast two hybrid analyses reveal novel binary interactions between human cytomegalovirus-encoded virion proteins. PLoS One 6(4):e17796. https://doi.org/10.1371/journal.pone.0017796

Tong L (2005) Acetyl-coenzyme A carboxylase: crucial metabolic enzyme and attractive target for drug discovery. Cell Mol Life Sci 62(16):1784–1803. https://doi.org/10.1007/s00018-005-5121-4

Tooze J, Hollinshead M, Reis B, Radsak K, Kern H (1993) Progeny vaccinia and human cytomegalovirus particles utilize early endosomal cisternae for their envelopes. Eur J Cell Biol 60(1):163–178

Tsuboi T, Fukuda M (2006) Rab3A and Rab27A cooperatively regulate the docking step of dense-core vesicle exocytosis in PC12 cells. J Cell Sci 119(Pt 11):2196–2203. https://doi.org/10.1242/jcs.02962

Turcotte S, Letellier J, Lippe R (2005) Herpes simplex virus type 1 capsids transit by the trans-Golgi network, where viral glycoproteins accumulate independently of capsid egress. J Virol 79(14):8847–8860. https://doi.org/10.1128/JVI.79.14.8847-8860.2005

Umashankar M, Petrucelli A, Cicchini L, Caposio P, Kreklywich CN, Rak M, Bughio F, Goldman DC, Hamlin KL, Nelson JA, Fleming WH, Streblow DN, Goodrum F (2011) A novel human cytomegalovirus locus modulates cell type-specific outcomes of infection. PLoS Pathog 7(12):e1002444. https://doi.org/10.1371/journal.ppat.1002444

Umashankar M, Rak M, Bughio F, Zagallo P, Caviness K, Goodrum FD (2014) Antagonistic determinants controlling replicative and latent states of human cytomegalovirus infection. J Virol 88(11):5987–6002. https://doi.org/10.1128/JVI.03506-13

van Spriel AB, van den Bogaart G, Cambi A (2015) Editorial: membrane domains as new drug targets. Front Physiol 6:172. https://doi.org/10.3389/fphys.2015.00172

Vanarsdall AL, Chase MC, Johnson DC (2011) Human cytomegalovirus glycoprotein gO complexes with gH/gL, promoting interference with viral entry into human fibroblasts but not entry into epithelial cells. J Virol 85(22):11638–11645. https://doi.org/10.1128/JVI.05659-11

Varnum SM, Streblow DN, Monroe ME, Smith P, Auberry KJ, Pasa-Tolic L, Wang D, Camp DG 2nd, Rodland K, Wiley S, Britt W, Shenk T, Smith RD, Nelson JA (2004) Identification of proteins in human cytomegalovirus (HCMV) particles: the HCMV proteome. J Virol 78(20):10960–10966. https://doi.org/10.1128/jvi.78.20.10960-10966.2004

Vastag L, Koyuncu E, Grady SL, Shenk TE, Rabinowitz JD (2011) Divergent effects of human cytomegalovirus and herpes simplex virus-1 on cellular metabolism. PLoS Pathog 7(7):e1002124. https://doi.org/10.1371/journal.ppat.1002124

Veettil MV, Kumar B, Ansari MA, Dutta D, Iqbal J, Gjyshi O, Bottero V, Chandran B (2016) ESCRT-0 component Hrs promotes macropinocytosis of Kaposi's sarcoma-associated herpesvirus in human dermal microvascular endothelial cells. J Virol 90(8):3860–3872. https://doi.org/10.1128/JVI.02704-15

Wakil SJ, Stoops JK, Joshi VC (1983) Fatty acid synthesis and its regulation. Annu Rev Biochem 52:537–579. https://doi.org/10.1146/annurev.bi.52.070183.002541

Walzer SA, Egerer-Sieber C, Sticht H, Sevvana M, Hohl K, Milbradt J, Muller YA, Marschall M (2015) Crystal structure of the human cytomegalovirus pUL50-pUL53 core nuclear egress complex provides insight into a unique assembly scaffold for virus-host protein interactions. J Biol Chem 290(46):27452–27458. https://doi.org/10.1074/jbc.C115.686527

Wang D, Shenk T (2005) Human cytomegalovirus virion protein complex required for epithelial and endothelial cell tropism. Proc Natl Acad Sci U S A 102(50):18153–18158. https://doi.org/10.1073/pnas.0509201102

Wille PT, Knoche AJ, Nelson JA, Jarvis MA, Johnson DC (2010) A human cytomegalovirus gO-null mutant fails to incorporate gH/gL into the virion envelope and is unable to enter fibroblasts and epithelial and endothelial cells. J Virol 84(5):2585–2596. https://doi.org/10.1128/JVI.02249-09

Williamson CD, Zhang A, Colberg-Poley AM (2011) The human cytomegalovirus protein UL37 exon 1 associates with internal lipid rafts. J Virol 85(5):2100–2111. https://doi.org/10.1128/JVI.01830-10

Womack A, Shenk T (2010) Human cytomegalovirus tegument protein pUL71 is required for efficient virion egress. MBio 1(5). https://doi.org/10.1128/mBio.00282-10

Wu JJ, Avey D, Li W, Gillen J, Fu B, Miley W, Whitby D, Zhu F (2015) ORF33 and ORF38 of Kaposi's sarcoma-associated herpesvirus interact and are required for optimal production of infectious progeny viruses. J Virol 90(4):1741–1756. https://doi.org/10.1128/JVI.02738-15

Yi Z, Yokota H, Torii S, Aoki T, Hosaka M, Zhao S, Takata K, Takeuchi T, Izumi T (2002) The Rab27a/granuphilin complex regulates the exocytosis of insulin-containing dense-core granules. Mol Cell Biol 22(6):1858–1867

Yu D, Silva MC, Shenk T (2003) Functional map of human cytomegalovirus AD169 defined by global mutational analysis. Proc Natl Acad Sci U S A 100(21):12396–12401. https://doi.org/10.1073/pnas.1635160100

Yu X, Shah S, Lee M, Dai W, Lo P, Britt W, Zhu H, Liu F, Zhou ZH (2011) Biochemical and structural characterization of the capsid-bound tegument proteins of human cytomegalovirus. J Struct Biol 174(3):451–460. https://doi.org/10.1016/j.jsb.2011.03.006

Yu Y, Maguire TG, Alwine JC (2012) Human cytomegalovirus infection induces adipocyte-like lipogenesis through activation of sterol regulatory element binding protein 1. J Virol 86(6):2942–2949. https://doi.org/10.1128/JVI.06467-11

Yu X, Jih J, Jiang J, Zhou ZH (2017) Atomic structure of the human cytomegalovirus capsid with its securing tegument layer of pp150. Science 356(6345). https://doi.org/10.1126/science.aam6892

Zhai Q, Landesman MB, Robinson H, Sundquist WI, Hill CP (2011) Identification and structural characterization of the ALIX-binding late domains of simian immunodeficiency virus SIVmac239 and SIVagmTan-1. J Virol 85(1):632–637. https://doi.org/10.1128/JVI.01683-10

Zhou ZH, Prasad BV, Jakana J, Rixon FJ, Chiu W (1994) Protein subunit structures in the herpes simplex virus A-capsid determined from 400 kV spot-scan electron cryomicroscopy. J Mol Biol 242(4):456–469. https://doi.org/10.1006/jmbi.1994.1594

Zhou M, Yu Q, Wechsler A, Ryckman BJ (2013) Comparative analysis of gO isoforms reveals that strains of human cytomegalovirus differ in the ratio of gH/gL/gO and gH/gL/UL128-131 in the virion envelope. J Virol 87(17):9680–9690. https://doi.org/10.1128/JVI.01167-13

Chapter 10
Chromosomal Integration by Human Herpesviruses 6A and 6B

Louis Flamand

Abstract Upon infection and depending on the infected cell type, human herpesvirus 6A (HHV-6A) and 6B (HHV-6B) can replicate or enter a state of latency. HHV-6A and HHV-6B can integrate their genomes into host chromosomes as one way to establish latency. Viral integration takes place near the subtelomeric/telomeric junction of chromosomes. When HHV-6 infection and integration occur in gametes, the virus can be genetically transmitted. Inherited chromosomally integrated HHV-6 (iciHHV-6)-positive individuals carry one integrated HHV-6 copy per somatic cell. The prevalence of iciHHV-6$^+$ individuals varies between 0.6% and 2%, depending on the geographical region sampled. In this chapter, the mechanisms leading to viral integration and reactivation from latency, as well as some of the biological and medical consequences associated with iciHHV-6, were discussed.

Keywords Chromosomal integration · Telomeres · HHV-6 · iciHHV-6 · Chromosomes · Telomeric motifs

10.1 HHV-6 Life Cycle

Like all herpesviruses, the HHV-6A and HHV-6B life cycle has two phases. The first phase is a lytic phase where HHV-6 infects a permissive cell with coordinated expression of most genes (>80), DNA replication, viral assembly, and release of progeny virions. The second phase is characterized by the infection of a cell but instead of initiating the lytic cycle, HHV-6 enters a state of latency. What drives HHV-6 into latency is not really known. Latency is a reversible process that

---

L. Flamand (✉)
Division of Infectious and Immune Diseases, CHU de Québec Research Center, QC, Quebec, Canada

Department of Microbiology, Infectious Disease and Immunology, Faculty of Medicine, Université Laval, QC, Quebec, Canada
e-mail: Louis.Flamand@crchul.ulaval.ca

comprises the maintenance of viral DNA (without replication) in the nucleus, with limited gene expression. During latency of α and γ herpesviruses such as herpes simplex viruses, Epstein-Barr virus, HHV-8, and possibly varicella-zoster virus, viral DNA is maintained as circular episomes (up to hundred copies/cell), tethered or not, to host chromosomes. Experiments conducted more than 25 years ago suggest that HHV-6 can establish latency in monocytes/macrophages (Kondo et al. 1991). During latent infection of macrophages, HHV-6 expresses latency-associated transcripts that contain open reading frames (ORFs) encoding immediate early proteins IE1 and IE2 and both transcripts are expressed in a small proportion of latently infected cells (Kondo et al. 2002). The state of the viral genome during latency (linear, episomal, or integrated) was not investigated. In fact, the presence of HHV-6 episomes during latency (in vivo or in vitro) was never documented.

10.2 HHV-6 Integration as a Mean to Achieve Latency

Shortly after HHV-6A isolation by Salahuddin in 1986 (Salahuddin et al. 1986), Jarrett et al. reported the detection of high levels of HHV-6A DNA in 2 out of 117 non-Hodgkin lymphoma patients (Jarrett et al. 1988). Subsequently, Torelli et al. identified the first two patients with Hodgkin disease (HD) harboring an unexpected high number of HHV-6 DNA sequences, both in pathologic lymph nodes and in normal peripheral blood mononuclear cells (PBMCs) (Torelli et al. 1991). Two years later, Luppi et al. reported the presence of chromosomally integrated HHV-6 (ciHHV-6), in PBMC of two patients with lymphoproliferative disorders and one with multiple sclerosis (Luppi et al. 1993). In 1998, Daibata et al. were the first to demonstrate vertical transmission of HHV-6 DNA over three generations, by showing identical HHV-6 integration sites in a patient with acute lymphoblastic leukemia, his son and his granddaughter, who were otherwise healthy (Daibata et al. 1998a). One year later, the same group described a woman with Burkitt's lymphoma and HHV-6 integration of 22q13 (Daibata et al. 1998b), whose asymptomatic husband had HHV-6 integration of 1q44. They demonstrated that the daughter had HHV-6 integration in both 1q44 and at 22q13 loci. Thus, the authors concluded that the viral genomes were inherited from both parents, and they proposed viral integration as a mode for HHV-6 to achieve latency (Daibata et al. 1999). Hereditary transmission of integrated HHV-6 is referred to as inherited chromosomally integrated HHV-6 (iciHHV-6). Unlike community-acquired HHV-6, iciHHV-6[+] individuals carry one integrated copy of HHV-6/somatic cell (Fig. 10.1). Furthermore, iciHHV-6[+] individuals will transmit ciHHV-6 to 50% of their descendants. A study from the UK reported the identification of five additional HHV-6 integration sites in nine individuals (Nacheva et al. 2008). As reviewed by Morissette and Flamand, several integration sites have been identified, both in patients and in healthy subjects, with a higher prevalence of iciHHV-6B (60–80%) compared to iciHHV-6A (20–40%) (Morissette and Flamand 2010). Although integration can occur in several distinct chromosomes, the virus targets the ends of chromosomes near the

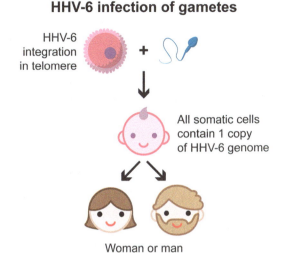

Fig. 10.1 Hereditary transmission of chromosomally integrated HHV-6. HHV-6 infection of gametes (ovum or sperm cell) can lead to the integration of HHV-6 DNA into a cellular chromosome. Upon conception with such gametes, an individual carrying one integrated copy of HHV-6 DNA/somatic cell is born. Such individuals are referred to as *inherited chromosomally integrated HHV-6*+ (iciHHV-6+) subjects. Such individual (man or woman) will transmit its condition (the presence of integrated HHV-6) to 50% of its descendants

subtelomeric/telomeric junction. Interestingly, a recent report by Engdahl et al. provided evidence that HHV-6B induces hypomethylation in chromosome 17p13.3 during acute infection (Engdahl et al. 2017). The telomere region at 17p13.3 has repeatedly been described as a chromosomal integration site for both HHV-6A and HHV-6B (Arbuckle et al. 2010; Morris et al. 1999; Nacheva et al. 2008; Torelli et al. 1995) suggesting that hypomethylation in this region may be predispose to viral integration. Up to 25% of iciHHV-6A/B subjects analyzed contain integration in chromosome 17p (Tweedy et al. 2016, Flamand unpublished observation). The overall conserved in vivo germline integration site structure at chromosome 17p supports a common ancestral integration event (Tweedy et al. 2016).

HHV-6 integration is likely one way through which HHV-6 achieves latency. Most likely, this is not the only way. To be efficient, a latent virus must be able to reactivate and generate progeny virions when conditions are adequate. Several studies indicate that integrated HHV-6 can either be transcriptionally silent, can spontaneously express viral transcripts/proteins, or can be induced to express viral genes (Clark et al. 2006; Strenger et al. 2014; Daibata et al. 1998b). Using recently developed tissue culture systems to study HHV-6 integration in vitro, Gravel et al. reported that viral gene expression is quite variable between individual clones containing ciHHV-6 (Gravel et al. 2017b). The differences in viral gene expression are probably affected by the chromosome targeted for integration, epigenetics, and the degree of chromatin condensation. In vitro reactivation and production from ciHHV-6 has been a challenging task. The conditions used are therefore not optimal to promote viral excision and initiation of the lytic cycle. The cellular environment is a key factor in determining whether reactivation will occur. Integration of HHV-6 in semi- or nonpermissive cells occurs relatively frequent (Gravel et al. 2017b). Interestingly, in highly permissive cells such as T-cells, where lytic infection is predominant, viral integration also occurs (Arbuckle et al. 2010; Engdahl et al. 2017), with a small proportion of surviving cells. Surviving T-cells with ciHHV-6 is a more relevant and suitable model to study reactivation from ciHHV-6. Perhaps the most compelling evidence suggesting viral reactivation from ciHHV-6 is from in vivo data. Evidence of in vivo reactivation of ciHHV-6 first came from Gravel et al., who provided data consistent with transplacentally acquired HHV-6, originating from the transmission of reactivated iciHHV-6 from the mother (Gravel et al. 2013a). A second report was by Endo et al.; this goes a step further to convincingly demonstrate iciHHV-6A reactivation and isolation of infectious HHV-6A from a young Japanese boy afflicted with X-SCID, who inherited ciHHV-6A from his father (Endo et al. 2014). HHV-6A infection is rare in Japan, reducing the possibility of community-acquired infection. HHV-6A infection was confirmed by RT-PCR, DNA viral load and HHV-6 antigen detection, using immunohistochemistry performed on bone marrow biopsies. Antiviral treatments proved effective in reducing HHV-6A viral burden for ongoing active infection. Researchers successfully isolated infectious HHV-6A at different time points with viral sequence analyses confirming that the isolated HHV-6A was identical to the DNA sequence of iciHHV-6A, from the child and his father, but different from other HHV-6A and HHV-6B isolates. These results clearly suggest that integration is one way through which HHV-6 can achieve latency.

10.2.1 Mechanisms of Chromosomal Integration

The genomes of HHV-6A and HHV-6B consist of a single unique segment (U) (~145 kbp) flanked by identical direct repeats (DR) (~9 kbp) (Martin et al. 1991; Dominguez et al. 1999; Gompels and Macaulay 1995; Isegawa et al. 1999) (Fig. 10.2a). The directly repeated regions (DRs) are flanked by *pac1* and *pac2* sequences that play a role in the cleavage and the packaging of the viral genome (Deng and Dewhurst 1998; Thomson et al. 1994). Adjacent to the *pac2* sequence are telomeric repeats (TMR) that are identical to the human telomere sequences (TTAGGG). The number of TMR repeats ranges from 15 to 180 copies in clinical isolates (Achour et al. 2009; Gompels and Macaulay 1995; Kishi et al. 1988; Thomson et al. 1994). In proximity to *pac1* is a second telomere array, consisting of imperfect TMR (impTMR) (Thomson et al. 1994; Gompels and Macaulay 1995).

Cellular chromosome ends have a 3′ single-stranded G-rich (TTAGGG) overhang, which is 30–500 nucleotides in length (Chai et al. 2005; Makarov et al. 1997;

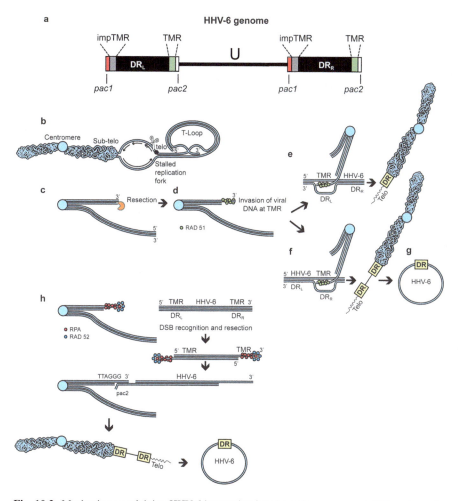

Fig. 10.2 Mechanisms explaining HHV-6 integration into host chromosomes. (**a**) The schematic representation of the HHV-6 genome (not to scale). DR_L direct repeat left, DR_R direct repeat right, *TMR* telomeric motifs, *impTMR* imperfect telomeric motifs, *pac1 and pac2* packaging sequence, *U* unique region. (**b**) Stalled replication fork at subtemolere/telomere junction with subsequent DNA break. (**c–f**) Chromosomal integration mediated through the break-induced replication repair. With the help of Rad51, the resected cellular DNA invades HHV-6 DNA at TMR. (**g**) Recombination event between the viral DRs generating a complete viral episome with a single DR. (**h**) Chromosomal integration mediated by single-strand annealing processes. The linear viral genome is recognized as broken DNA and is resected at its 5′ ends, generating a TMR from DR_R that is complimentary to the resected cellular DNA. With the participation of RPA and Rad52, both strands are annealed with the *pac2* sequence getting clipped off during the process

McElligott and Wellinger 1997). To avoid being recognized as damaged DNA, the 3′ protruding end folds back and invades the duplex telomeric DNA to generate a T-loop structure (Griffith et al. 1999; Nikitina and Woodcock 2004; Raices et al. 2008). A total of six proteins (TRF1, TRF2, TPP1, RAP1, POT1, TIN2), referred to as the shelterin complex, bind and assist with the T-loop formation, stabilize chromosomal ends, and prevent DNA-damage responses (de Lange 2009). Due to the combination of their repetitive G-rich sequence and extensive heterochromatinization, telomeres represent a challenge for the replication machinery. The structural elements and secondary structures of telomeres, such as G-quadruplexes, R-loops, and T-loops, are potential obstacles to the replication fork passage (Lipps and Rhodes 2009; Paeschke et al. 2005). In effect, studies do suggest that telomeric DNA has an inherent ability to pause or stall replication forks (Anand et al. 2012; Verdun and Karlseder 2006).

Two models explaining HHV-6 integration are presented. The first one is based on the DNA repair mechanism, referred to as break-induced replication (BIR); and the second one is known as single-strand annealing (SSA). BIR is a homologous recombination (HR) pathway, which facilitates the repair of DNA breaks that have only one end. It contributes to the repair of broken replication forks and allows the expansion of telomere, in a situation where telomerase is absent. BIR has been seen in various organisms including viruses, bacteria, and eukaryotes [reviewed in (Malkova and Ira 2013)]. Stalled replication forks at telomeres can be resolved in several ways; one of such is through the cleavage of DNA by endonucleases (Fig. 10.2b). Double-stranded breaks that occur or are detected during the S and G2 phases preferentially activate ATM using MRE11-Rad50-NBS1 (MRN) complex. The 5′ extremity will be resected by MRN complex, C-terminal-binding protein interacting protein (CtIP), exonuclease 1 (EXO1), Bloom syndrome protein (BLM), DNA2 nuclease/helicase, and several chromatin remodeling factors (Mimitou and Symington 2009). Such activity will generate a 3′ protruding strand containing TTAGGG repeats that will get wrapped by Rad51 (Fig. 10.2c–d), after which the complex will search for homology pairing and promote strand invasion. When a HHV6 genome is in proximity, strand invasion can occur in the TMR regions of both DRs (Fig. 10.2e–f). In line with this, recent data from Wallaschek et al. provided conclusive evidence that the TMR regions of HHV-6 were required for efficient viral integration (Wallaschek et al. 2016b). When invasion occurs at TMR of DR_R, the entire genome will end up being copied (except for the *pac*2 sequence of DR_R) and fused to the telomere. On the other hand, when invasion occurs in the TMR of DR_L, only the DR_L (minus the *pac*2 sequence of DR_L) will be copied. At present, screening for iciHHV-6 is mostly done using primers/probe located within the unique region of HHV-6. With such assays, the truncated form consisting of only DR_L cannot be detected. Nonetheless, individuals carrying a single DR in the absence of the remaining viral genome do exist (Bell et al. 2015a; Gulve et al. 2017). These data provide evidence that invasion of DR_L TMR, by the 3′ protruding telomeric ssDNA, might occur. Alternatively, deletion of the entire HHV-6 genome via the formation of T-loops might also explain the presence of a single DR in the absence of the rest of the viral genome (Huang et al. 2014; Prusty et al. 2013). At present, inadequate information is available to determine whether strand invasion at both TMR occurs with equal frequency. Individuals carrying more than one copy of the HHV-6 genome also exist. Aside those who inherit a viral copy from both parents (Daibata et al. 1999), individuals with HHV-6A and HHV-6B concatemers (with three or four juxtaposed viral genomes) have been identified [(Bell et al. 2014, 2015b; Gulve et al. 2017; Ohye et al. 2014); Flamand, unpublished observation].

The SSA model also necessitates a break in the cellular DNA, in the subtelomere/telomere region. Additionally, unless HHV-6 can counteract the DNA-damage response, the presence of unprotected linear dsDNA viral genomes is likely to be recognized as broken DNA needing repair. The viral DNA will be resected by the MRN complex as described above and generate single-stranded (ss) viral extremities (Fig. 10.2 h). On the DR_R side, adjacent to the terminal *pac2* sequence, the viral extremity consist of (CCCTAA) replications that are complementary to the 3′ cellular ssDNA $(TTAGGG)_n$ generated following the cleavage of the stalled replication fork. With the participation of RPA and Rad52, the two pieces of DNA (viral genome and chromosome) are united, and in the process, the *pac2* sequence gets clipped off. In contrast to the model involving invasion of the viral DNA at TMR by the 3′ protruding cellular DNA, the SSA model favors integration of HHV-6 using DR_R.

To be viable, the chromosome containing the integrated HHV-6 must restore its telomeric cap. Upon integration and cell divisions, the viral *pac1* sequence is lost due to incomplete replication of DNA ends. Once the viral impTMR constitutes the chromosome end, it can serve as a template for telomere lengthening either by the telomerase or alternative processes (Ohye et al. 2014). Of potential importance, the data reported by Huang et al. also indicate that the telomere on the distal end of the integrated virus is frequently the shortest (Huang et al. 2014).

10.2.2 Reactivation from a Chromosomally Integrated State

Ultimately, the virus needs to reactivate and propagate itself. All herpesviruses use the viral episome as a template to replicate their DNA. The generation of a functional episome can occur via a second homologous recombination event between the telomeric sequences at the chromosome termini and the viral TMR present near the subtelomeric region (Fig. 10.2 g–h). In this light, Huang et al. reported the presence of extrachromosomal circular HHV-6 molecules, some made up of the entire HHV-6 genome with a single DR region flanked with *pac1* and *pac2* sequences; this indicates that once integrated, the virus can excise itself and reconstitute a full length viral episome (Huang et al. 2014). Reconstitution of the viral episome is likely not enough to initiate a productive infection and likely requires the expression of a trans-activating viral proteins, such as the immediate early 2 protein that will favor the expression of many viral genes (Gravel et al. 2003).

It is unlikely that homologous recombination occurs without the participation of cellular or viral factors. This poses the question, which cellular components could possibly contribute to this integration? Proteins involved in HR such as BLM, BRCA2, and Rad51 are most likely involved. Shelterin complex proteins are also possibly involved. Arbuckle et al. hypothesized that telomeric repeat binding factor 2 (TRF2), known to bind the EBV sequences TTAGGGTTA (Deng et al. 2002), might also do the same with the TMR of HHV-6 and facilitate viral integration (Arbuckle and Medveczky 2011). A recent work from Gravel et al. indicates that HHV-6 integration occurs in cells independently of telomerase activity and p53 expression (Gravel et al. 2017b). Biochemical data suggests that the U94 protein could facilitate integration (Morissette and Flamand 2010; Kaufer and Flamand 2014; Trempe et al. 2015). However, despite some growth defects, a HHV-6A U94 deletion mutant proved as efficient as WT HHV-6A at integrating host chromosomes, indicating that U94 is dispensable for HHV-6A integration (Wallaschek et al. 2016a). The quest for cellular and viral candidate proteins involved in integration is ongoing. Also, the cell type (such as fibroblast, endothelial cell, lymphocyte, neuron, etc.) is likely to influence the frequency of viral integration and reactivation. Infection in non-dividing cells such as neurons or macrophages is not expected to result in viral integration considering that events needed for integration, HR, occur during the S phase of the cycle. Thus, actively replicating cells are most likely to integrate HHV-6. Similarly, viral excision in quiescent cells is less likely to occur. Considering the need for large pools of nucleotides to ensure viral DNA replication and amino acids for protein synthesis, efficient reactivation likely occurs in proliferating cells that are naturally permissive to HHV-6 growth, such as T lymphocytes.

10.3 Detection of HHV-6 Active Infection in the Context of iciHHV-6 Infection

Based on several studies, including 2 with ≥20,000 subjects, the estimated prevalence of iciHHV-6 is determined to vary between 0.6% and 2%, depending of the geographical region sampled (Gravel et al. 2013b, 2015, 2017a; Hubacek et al. 2009; Jarrett 2015; Bell et al. 2014). The most frequent encountered issue associated with iciHHV-6$^+$ individuals is the wrongful diagnosis of active HHV-6 infection. Depending on the type of tissue examined, the HHV-6 DNA copy number can vary greatly. For example, iciHHV-6$^+$ individuals can have high levels of HHV-6 DNA in plasma (>3.5 log10 copies/ml) or in cerebrospinal fluid (4.0 log10 copies/ml), which may be misinterpreted as subjects with active HHV-6 infection and prescribed antivirals (Pellett et al. 2012; Ward et al. 2006). A study demonstrated that qualitative or quantitative HHV-6 PCR of plasma is not sufficient to distinguish active viral replication from the chromosomally integrated form of HHV-6 (Caserta et al. 2010). When plasma, serum, or cerebrospinal fluid contains unexpectedly high HHV-6 levels, the most practical way to rule out that a patient is iciHHV-6$^+$ is by quantitative PCR or droplet digital PCR (ddPCR) using whole blood or isolated PBMCs (Pellett et al. 2012; Sedlak et al. 2014). Individuals with iciHHV-6 have significantly higher viral DNA loads in PBMCs and whole blood (>5.5 log10 copies/ml) than non-iciHHV-6 individuals, even those immunocompetent individuals with primary HHV-6 infection or those immunosuppressed subjects, with HHV-6 reactivations (Pellett et al. 2012). It has been recommended by Ljungman et al. for the European Conference on Infections in Leukemia (ECIL) that iciHHV-6 should be excluded before the diagnosis of HHV-6 active disease can be made, irrespective of the immune status of the patients (Ljungman et al. 2008). Furthermore, it has been stated that iciHHV-6 may only mislead the laboratory diagnosis of HHV-6 active disease, inducing the administration of an undesirable and inappropriate antiviral treatment, especially in transplant patients, where iciHHV-6 may solely reflect either the rate of donor/recipient engraftment or the donor/recipient origin of the phenomenon (Ljungman et al. 2008).

Until recently, the molecular detection of iciHHV-6 was mostly done by qPCR (Ward et al. 2006). A new procedure providing absolute, more precise, and generally unequivocal results are obtained using droplet digital PCR (ddPCR). ddPCR partitions the reaction sample into thousands of droplets with each droplet behaving as a mini reaction vessel in which the PCR reaction occurs. After PCR, each droplet is read as positive or negative for the DNA template, allowing absolute quantification of DNA copies without the use of a standard curve (Hindson et al. 2011, 2013). Using primer pairs that detect HHV-6 and a reference cellular gene (such as RPP30), the ratio of HHV-6/cell can be easily calculated. Sedlak et al. developed a ddPCR for the detection of iciHHV-6 and further adapted this technology to study HHV-6B reactivation in the content of iciHHV-6A$^+$ transplant patients (Sedlak et al. 2014, 2016). An example of ddPCR results from iciHHV-6$^-$ and iciHHV-6$^+$ subjects is presented in Fig. 10.3. As presented, most asymptomatic iciHHV-6$^-$ subjects will have very little or no HHV-6 DNA in their PBMCs (Fig. 10.3a). Thus, only the reference cellular gene is detected. In contrast, ddPCR performed on DNA from iciHHV-6+ subject yields amplicons for both HHV-6 and RPP30, with a ratio of HHV-6 per cell (RPP30/2) of approximately one or more (Fig. 10.3b–c). A word of caution is, however, required when ddPCR is performed on limited amounts of starting material. A study from Sedlak et al. implies that, specimens with low genomic DNA content (such as biopsy) may yield inaccurate ratios that will make distinguishing iciHHV-6 from active infection difficult. In cases such as these, it is advisable to obtain additional tissue or cellular samples, to determine the likelihood of iciHHV-6 (Sedlak et al. 2014).

Using plasma/serum as starting material can make it is very difficult to identify bona fide active HHV-6 reactivation/infection, considering the high HHV-6 DNA content in iciHHV-6$^+$ individuals. This is even more problematic in iciHHV-6B$^+$ individuals, knowing that most reactivation/infection events involve HHV-6B. One option would be to include a DNAse step prior to the DNA isolation. Such approach would eliminate all non-encapsidated DNA and eliminate naked DNA originating from spontaneous or induced (due to sheer stress) cellular lysis (Djikeng et al. 2008). However, from a conservative clinical point of view, it is worth considering antiviral therapy even in patients with iciHHV-6 when they show signs and/or symptoms consistent with viral infection and other known pathogens that have been extensively searched and ruled out, although iciHHV-6 may confound the diagnosis of active HHV-6 infection (Potenza et al. 2011). A decision algorithm was recently

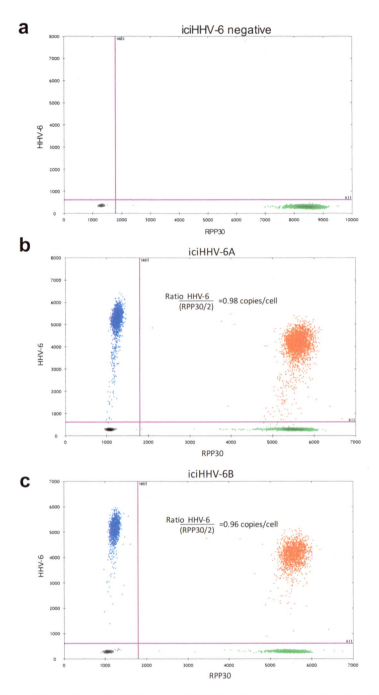

Fig. 10.3 ddPCR analysis of iciHHV-6⁻ and iciHHV-6⁺ individuals. DNA isolated from PBMCs of iciHHV-6⁻ (**a**), iciHHV-6A⁺ (**b**), and iciHHV-6B⁺ (**c**) individuals was analyzed by ddPCR using primers and probes specific for HHV-6 and RPP30 (single copy cellular gene). Bottom left = droplets negative for HHV-6 and RPP30 DNA. Top left = droplets positive for HHV-6 DNA and negative for cellular DNA. Bottom right = droplets negative for HHV-6 DNA and positive for cellular DNA. Top right = droplets positive for HHV-6 and RPP30 DNA

published to help with the decision-making process (Agut et al. 2015). Detection of multiple HHV-6 RNA (including RNA coding for structural proteins) by RT-qPCR, HHV-6 proteins by immunohistochemistry, or viral isolation likely represent the best options to document active HHV-6B infection.

10.4 Impacts of ciHHV-6 on Cell Biology

The absence of knowledge on the biological consequences of having or transmitting iciHHV-6 is of medical concern [for reviews consult (Pellett et al. 2012; Morissette and Flamand 2010; Kaufer and Flamand 2014)]. Does integration lead to proto-oncogene activation or gene alteration that can lead to diseases such as cancer? Does HHV-6 integration within the telomeric region affects chromosome stability and predispose to certain pathologies through accelerated telomere shortening (such as by decreasing the cellular renewal potential and increasing the susceptibility to normally aging diseases)? Does expression of viral proteins trigger immune responses that cause inflammation and tissue destruction? These are some of the fundamental questions awaiting answers.

Compilation of several small independent studies was made. Pellet et al. reported that iciHHV-6 is 2.3X more frequent ($p < 0.001$) in diseased (various diseases) individuals compared to healthy ones, suggesting that iciHHV-6 may represent a risk factor in disease development (Pellett et al. 2012). Much of what is known about the causes of chronic disorders comes from large epidemiological studies (Doll 2001; Willett and Colditz 1998). At present, two large-scale studies ($\geq 20,000$ subjects) were conducted to determine the prevalence of diseases in iciHHV-6$^+$ individuals. Using DNA samples from the CARTaGENE study (Awadalla et al. 2013) and the Generation Scotland: Scottish Family Health Study (Smith et al. 2013), researchers were able to find a relationship between iciHHV-6 and the development of angina pectoris (Gravel et al. 2015; Jarrett 2015).

Angina is a cardiovascular disease (CVD) caused by the narrowing of the lumen of arteries, resulting in reduced oxygen delivery to the heart [reviewed in (Libby 2013)]. Atherosclerosis is the most common cause of stenosis of the heart's arteries and, therefore, angina. The starting point for plaque formation is endothelial dysfunction or activation. At present, the most important contributors of endothelial dysfunction are hemodynamic disturbances, hypercholesterolemia, and inflammation. However, infections, whether they are of bacterial or viral origin (including CMV), also play important roles in atherosclerosis development (Chatzidimitriou et al. 2012). One hypothesis is that endothelial cells (ECs) from subjects express, at some point in time, viral proteins and/or produce virions that can lead to immune activation (of monocytes, lymphocytes, and/or platelets) and cell damage, initiating and fueling the development of atherosclerotic lesions and eventually leading to angina [reviewed in (Libby et al. 2013)]. In support, a recent study indicates reactivation of iciHHV-6 in subjects with myocardial dysfunctions. Virus particles were identified in degenerating myocytes and interstitial cells and antiviral treatment abolished viral mRNA and ameliorated cardiac symptoms (Kuhl et al. 2015). The authors hypothesize that the damage of ECs caused by HHV-6 reactivation in iciHHV-6 patients might give explanation to the complaints of angina as a sequela of myocyte and vascular endothelial dysfunction, as previously reported for Parvovirus B19 (Schmidt-Lucke et al. 2010).

In case control studies, a potential role for iciHHV-6 in the development of breast cancer, Hodgkin lymphomas, and acute lymphoblastic leukemia was ruled out (Bell et al. 2014; Gravel et al. 2013b, 2017a).

iciHHV-6 may play a pathophysiological role in transplant patients. Several case reports have described transplant recipients receiving hematopoietic stem cells, including cord blood from iciHHV-6$^+$ donors (Clark et al. 2006; Kamble et al. 2007; Yamada et al. 2017). The long-term follow-up on these patients is limited to a few months or years, but no apparent problems were noted. Considering that the telomere of the chromosome carrying the integrated HHV-6 is often the shortest (Huang et al. 2014) and that telomere length is proportional to the lifespan of a cell, it remains undetermined whether cells derived from iciHHV-6$^+$ individuals have the same long-term proliferative capacity of cells from iciHHV-6$^-$ subjects. In the recipients, the HHV-6 DNA levels increases concomitantly with engraftment and should not be mistaken with bona fide HHV-6B reactivation, which typically occurs between days 20 and 40 of posttransplant in 30–50% of allogeneic peripheral blood mobilized stem cells and up to 90% of cord blood transplant recipients (Dulery et al. 2012; Imbert-Marcille et al. 2000; Zerr 2006; Zerr et al. 2011, 2012; Aoki et al. 2015; de Pagter et al. 2008, 2013; Quintela et al. 2016).

Experience on the transplantation of solid organs from iciHHV-6$^+$ subjects are more limited. Considering that all transplant recipients over the age of three are immune to HHV-6 and knowing that once integrated, HHV-6 can express some of its genes, the possibility that a transplanted organ from an iciHHV-6$^+$ donor triggers an immunological attack from the recipient's immune system, even in the absence of detectable viral replication, is present. In the absence of HHV-6 replication, the diagnostic is likely to be idiopathic organ rejection. Considering this, the iciHHV-6 status of the donors should be determined prior to organ transplant, and recipients should be carefully monitored for signs of active HHV-6 infection (viral loads) and/or signs of immune (induced by HHV-6 antigens) organ rejection (Das and Munoz 2017; Flamand 2014).

10.5 Conclusion

Much has been learned since the initial report of ciHHV-6 25 years ago. These knowledge include the structures of the integrated genome, the prevalence of iciHHV-6 in different regions of the world, the ability to discriminate between bona fide active infection "false" DNA loads in iciHHV-6$^+$ subjects, the development of in vitro culture models to study integration, as well as documented cases of pathogenic events originating from the reactivation of integrated HHV-6, to name a few.

Considering an iciHHV-6 prevalence of 0.6–2%, a large number of samples are needed to link iciHHV-6 to certain disease. The fact that integration can occur in distinct chromosomes further complicates the analyses. Since the chromosome targeted for integration may influence the outcome, future studies should consider which chromosomes carry the integrated virus. A lot remains unlearned, on the biological consequences linked with chromosomal integration. Also, the fact that integration takes place within the telomeric regions, several questions pertaining to the long-lasting proliferative potential of cells carrying ciHHV-6 are raised.

References

Achour A, Malet I, Deback C, Bonnafous P, Boutolleau D, Gautheret-Dejean A, Agut H (2009) Length variability of telomeric repeat sequences of human herpesvirus 6 DNA. J Virol Methods 159(1):127–130. S0166-0934(09)00104-9 [pii] https://doi.org/10.1016/j.jviromet.2009.03.002

Agut H, Bonnafous P, Gautheret-Dejean A (2015) Laboratory and clinical aspects of human herpesvirus 6 infections. Clin Microbiol Rev 28(2):313–335. https://doi.org/10.1128/CMR.00122-14

Anand RP, Shah KA, Niu H, Sung P, Mirkin SM, Freudenreich CH (2012) Overcoming natural replication barriers: differential helicase requirements. Nucleic Acids Res 40(3):1091–1105. https://doi.org/10.1093/nar/gkr836

Aoki J, Numata A, Yamamoto E, Fujii E, Tanaka M, Kanamori H (2015) Impact of human Herpesvirus-6 reactivation on outcomes of allogeneic hematopoietic stem cell transplantation. Biol Blood Marrow Transplant 21(11):2017–2022. https://doi.org/10.1016/j.bbmt.2015.07.022

Arbuckle JH, Medveczky PG (2011) The molecular biology of human herpesvirus-6 latency and telomere integration. Microbes Infect 13(8–9):731–741. https://doi.org/10.1016/j.micinf.2011.03.006

Arbuckle JH, Medveczky MM, Luka J, Hadley SH, Luegmayr A, Ablashi D, Lund TC, Tolar J, De Meirleir K, Montoya JG, Komaroff AL, Ambros PF, Medveczky PG (2010) The latent human herpesvirus-6A genome specifically integrates in telomeres of human chromosomes in vivo and in vitro. Proc Natl Acad Sci USA 107(12):5563–5568. https://doi.org/10.1073/pnas.0913586107

Awadalla P, Boileau C, Payette Y, Idaghdour Y, Goulet JP, Knoppers B, Hamet P, Laberge C, Project CA (2013) Cohort profile of the CARTaGENE study: Quebec's population-based biobank for public health and personalized genomics. Int J Epidemiol 42(5):1285–1299. https://doi.org/10.1093/ije/dys160

Bell AJ, Gallagher A, Mottram T, Lake A, Kane EV, Lightfoot T, Roman E, Jarrett RF (2014) Germ-line transmitted, chromosomally integrated HHV-6 and classical Hodgkin lymphoma. PLoS One 9(11):e112642. https://doi.org/10.1371/journal.pone.0112642

Bell AJ, Brownlie CA, Gallacher A, Campbell A, Porteous DJ, Smith BH, Hocking L, Padmanabhan S, Jarrett RF (2015a) Prevalence of inherited chromosomally integrated HHV-6 varies by geographical location/nationality within the UK. In: 9th International conference on HHV-6 and HHV-7, Boston, November 9–11, abstract 8–16

Bell AJ, Johnson PDC, Jarrett RF (2015b) Integration and inheritance of HHV-6 genome concatemers. In: 9th International conference on HHV-6 and HHV-7, Boston, November 9–11

Caserta MT, Hall CB, Schnabel K, Lofthus G, Marino A, Shelley L, Yoo C, Carnahan J, Anderson L, Wang H (2010) Diagnostic assays for active infection with human herpesvirus 6 (HHV-6). J Clin Virol 48(1):55–57. https://doi.org/10.1016/j.jcv.2010.02.007

Chai W, Shay JW, Wright WE (2005) Human telomeres maintain their overhang length at senescence. Mol Cell Biol 25(6):2158–2168. https://doi.org/10.1128/MCB.25.6.2158-2168.2005

Chatzidimitriou D, Kirmizis D, Gavriilaki E, Chatzidimitriou M, Malisiovas N (2012) Atherosclerosis and infection: is the jury still not in? Future Microbiol 7(10):1217–1230. https://doi.org/10.2217/fmb.12.87

Clark DA, Nacheva EP, Leong HN, Brazma D, Li YT, Tsao EH, Buyck HC, Atkinson CE, Lawson HM, Potter MN, Griffiths PD (2006) Transmission of integrated human herpesvirus 6 through stem cell transplantation: implications for laboratory diagnosis. J Infect Dis 193(7):912–916

Daibata M, Taguchi T, Sawada T, Taguchi H, Miyoshi I (1998a) Chromosomal transmission of human herpesvirus 6 DNA in acute lymphoblastic leukaemia. Lancet 352(9127):543–544. S0140-6736(05)79251-5 [pii] https://doi.org/10.1016/S0140-6736(05)79251-5

Daibata M, Taguchi T, Taguchi H, Miyoshi I (1998b) Integration of human herpesvirus 6 in a Burkitt's lymphoma cell line. Br J Haematol 102(5):1307–1313

Daibata M, Taguchi T, Nemoto Y, Taguchi H, Miyoshi I (1999) Inheritance of chromosomally integrated human herpesvirus 6 DNA. Blood 94(5):1545–1549

Das BB, Munoz FM (2017) Screening for chromosomally integrated human herpesvirus 6 status in solid-organ donors and recipients. J Heart Lung Transplant 36(4):481. https://doi.org/10.1016/j.healun.2017.01.004

de Lange T (2009) How telomeres solve the end-protection problem. Science 326(5955):948–952. https://doi.org/10.1126/science.1170633

de Pagter PJ, Schuurman R, Meijer E, van Baarle D, Sanders EA, Boelens JJ (2008) Human herpesvirus type 6 reactivation after haematopoietic stem cell transplantation. J Clin Virol 43(4):361–366. https://doi.org/10.1016/j.jcv.2008.08.008

de Pagter PJ, Schuurman R, Keukens L, Schutten M, Cornelissen JJ, van Baarle D, Fries E, Sanders EA, Minnema MC, van der Holt BR, Meijer E, Boelens JJ (2013) Human herpes virus 6 reactivation: important predictor for poor outcome after myeloablative, but not non-myeloablative allo-SCT. Bone Marrow Transplant 48(11):1460–1464. https://doi.org/10.1038/bmt.2013.78

Deng H, Dewhurst S (1998) Functional identification and analysis of cis-acting sequences which mediate genome cleavage and packaging in human herpesvirus 6. J Virol 72(1):320–329

Deng Z, Lezina L, Chen CJ, Shtivelband S, So W, Lieberman PM (2002) Telomeric proteins regulate episomal maintenance of Epstein-Barr virus origin of plasmid replication. Mol Cell 9(3):493–503

Djikeng A, Halpin R, Kuzmickas R, Depasse J, Feldblyum J, Sengamalay N, Afonso C, Zhang X, Anderson NG, Ghedin E, Spiro DJ (2008) Viral genome sequencing by random priming methods. BMC Genomics 9:5. https://doi.org/10.1186/1471-2164-9-5

Doll R (2001) Cohort studies: history of the method. I. Prospective cohort studies. Sozial- und Praventivmedizin 46(2):75–86

Dominguez G, Dambaugh TR, Stamey FR, Dewhurst S, Inoue N, Pellett PE (1999) Human herpesvirus 6B genome sequence: coding content and comparison with human herpesvirus 6A. J Virol 73(10):8040–8052

Dulery R, Salleron J, Dewilde A, Rossignol J, Boyle EM, Gay J, de Berranger E, Coiteux V, Jouet JP, Duhamel A, Yakoub-Agha I (2012) Early human herpesvirus type 6 reactivation after allogeneic stem cell transplantation: a large-scale clinical study. Biol Blood Marrow Transplant 18(7):1080–1089. https://doi.org/10.1016/j.bbmt.2011.12.579

Endo A, Watanabe K, Ohye T, Suzuki K, Matsubara T, Shimizu N, Kurahashi H, Yoshikawa T, Katano H, Inoue N, Imai K, Takagi M, Morio T, Mizutani S (2014) Molecular and Virological evidence of viral activation from chromosomally integrated human herpesvirus 6A in a patient with X-linked severe combined immunodeficiency. Clin Infect Dis 59(4):545–548. https://doi.org/10.1093/cid/ciu323

Engdahl E, Dunn N, Niehusmann P, Wideman S, Wipfler P, Becker AJ, Ekstrom TJ, Almgren M, Fogdell-Hahn A (2017) Human herpesvirus 6B induces hypomethylation on chromosome 17p13.3 correlating with increased gene expression and virus integration. J Virol. https://doi.org/10.1128/JVI.02105-16

Flamand L (2014) Pathogenesis from the reactivation of chromosomally integrated human herpesvirus type 6: facts rather than fiction. Clin Infect Dis 59(4):549–551. https://doi.org/10.1093/cid/ciu326

Gompels UA, Macaulay HA (1995) Characterization of human telomeric repeat sequences from human herpesvirus 6 and relationship to replication. J Gen Virol 76(Pt 2):451–458

Gravel A, Tomoiu A, Cloutier N, Gosselin J, Flamand L (2003) Characterization of the immediate-early 2 protein of human herpesvirus 6, a promiscuous transcriptional activator. Virology 308(2):340–353

Gravel A, Hall CB, Flamand L (2013a) Sequence analysis of Transplacentally acquired human herpesvirus 6 DNA is consistent with transmission of a chromosomally integrated reactivated virus. J Infect Dis 207(10):1585–1589. https://doi.org/10.1093/infdis/jit060

Gravel A, Sinnett D, Flamand L (2013b) Frequency of chromosomally-integrated human herpesvirus 6 in children with acute lymphoblastic leukemia. PLoS One 8(12):e84322. https://doi.org/10.1371/journal.pone.0084322

Gravel A, Dubuc I, Morissette G, Sedlak RH, Jerome KR, Flamand L (2015) Inherited chromosomally integrated human herpesvirus 6 as a predisposing risk factor for the development of angina pectoris. Proc Natl Acad Sci USA 112(26):8058–8063. https://doi.org/10.1073/pnas.1502741112

Gravel A, Dubuc I, Brooks-Wilson A, Aronson KJ, Simard J, Velasquez-Garcia HA, Spinelli JJ, Flamand L (2017a) Inherited chromosomally integrated human herpesvirus 6 and breast cancer. Cancer Epidemiol Biomark Prev 26(3):425–427. https://doi.org/10.1158/1055-9965.EPI-16-0735

Gravel A, Dubuc I, Wallaschek N, Gilbert-Girard S, Collin V, Hall-Sedlak R, Jerome KR, Mori Y, Carbonneau J, Boivin G, Kaufer BB, Flamand L (2017b) Cell culture systems to study human Herpesvirus-6 chromosomal integration. J Virol 91:pii: e00437-17

Griffith JD, Comeau L, Rosenfield S, Stansel RM, Bianchi A, Moss H, de Lange T (1999) Mammalian telomeres end in a large duplex loop. Cell 97(4):503–514

Gulve N, Frank C, Klepsch M, Prusty BK (2017) Chromosomal integration of HHV-6A during non-productive viral infection. Sci Rep 7(1):512. https://doi.org/10.1038/s41598-017-00658-y

Hindson BJ, Ness KD, Masquelier DA, Belgrader P, Heredia NJ, Makarewicz AJ, Bright IJ, Lucero MY, Hiddessen AL, Legler TC, Kitano TK, Hodel MR, Petersen JF, Wyatt PW, Steenblock ER, Shah PH, Bousse LJ, Troup CB, Mellen JC, Wittmann DK, Erndt NG, Cauley TH, Koehler RT, So AP, Dube S, Rose KA, Montesclaros L, Wang S, Stumbo DP, Hodges SP, Romine S, Milanovich FP, White HE, Regan JF, Karlin-Neumann GA, Hindson CM, Saxonov S, Colston BW (2011) High-throughput droplet digital PCR system for absolute quantitation of DNA copy number. Anal Chem 83(22):8604–8610. https://doi.org/10.1021/ac202028g

Hindson CM, Chevillet JR, Briggs HA, Gallichotte EN, Ruf IK, Hindson BJ, Vessella RL, Tewari M (2013) Absolute quantification by droplet digital PCR versus analog real-time PCR. Nat Methods 10(10):1003–1005. https://doi.org/10.1038/nmeth.2633

Huang Y, Hidalgo-Bravo A, Zhang E, Cotton VE, Mendez-Bermudez A, Wig G, Medina-Calzada Z, Neumann R, Jeffreys AJ, Winney B, Wilson JF, Clark DA, Dyer MJ, Royle NJ (2014) Human telomeres that carry an integrated copy of human herpesvirus 6 are often short and unstable, facilitating release of the viral genome from the chromosome. Nucleic Acids Res 42(1):315–327. https://doi.org/10.1093/nar/gkt840

Hubacek P, Muzikova K, Hrdlickova A, Cinek O, Hyncicova K, Hrstkova H, Sedlacek P, Stary J (2009) Prevalence of HHV-6 integrated chromosomally among children treated for acute lymphoblastic or myeloid leukemia in the Czech Republic. J Med Virol 81(2):258–263. https://doi.org/10.1002/jmv.21371

Imbert-Marcille BM, Tang XW, Lepelletier D, Besse B, Moreau P, Billaudel S, Milpied N (2000) Human herpesvirus 6 infection after autologous or allogeneic stem cell transplantation: a single-center prospective longitudinal study of 92 patients. Clin Infect Dis 31(4):881–886. https://doi.org/10.1086/318142

Isegawa Y, Mukai T, Nakano K, Kagawa M, Chen J, Mori Y, Sunagawa T, Kawanishi K, Sashihara J, Hata A, Zou P, Kosuge H, Yamanishi K (1999) Comparison of the complete DNA sequences of human herpesvirus 6 variants A and B. J Virol 73(10):8053–8063

Jarrett R (2015) iciHHV-6 prevalence and disease associations in the generation Scotland study. In: 9th International conference on HHV-6 and HHV-7, abstract 8-3, Boston, November 9–11

Jarrett RF, Gledhill S, Qureshi F, Crae SH, Madhok R, Brown I, Evans I, Krajewski A, O'Brien CJ, Cartwright RA et al (1988) Identification of human herpesvirus 6-specific DNA sequences in two patients with non-Hodgkin's lymphoma. Leukemia 2(8):496–502

Kamble RT, Clark DA, Leong HN, Heslop HE, Brenner MK, Carrum G (2007) Transmission of integrated human herpesvirus-6 in allogeneic hematopoietic stem cell transplantation. Bone Marrow Transplant 40(6):563–566. 1705780 [pii] https://doi.org/10.1038/sj.bmt.1705780

Kaufer BB, Flamand L (2014) Chromosomally integrated HHV-6: impact on virus, cell and organismal biology. Curr Opin Virol 9C:111–118. https://doi.org/10.1016/j.coviro.2014.09.010

Kishi M, Harada H, Takahashi M, Tanaka A, Hayashi M, Nonoyama M, Josephs SF, Buchbinder A, Schachter F, Ablashi DV et al (1988) A repeat sequence, GGGTTA, is shared by DNA of human herpesvirus 6 and Marek's disease virus. J Virol 62(12):4824–4827

Kondo K, Kondo T, Okuno T, Takahashi M, Yamanishi K (1991) Latent human herpesvirus 6 infection of human monocytes/macrophages. J Gen Virol 72(Pt 6):1401–1408

Kondo K, Shimada K, Sashihara J, Tanaka-Taya K, Yamanishi K (2002) Identification of human herpesvirus 6 latency-associated transcripts. J Virol 76(8):4145–4151

Kuhl U, Lassner D, Wallaschek N, Gross UM, Krueger GR, Seeberg B, Kaufer BB, Escher F, Poller W, Schultheiss HP (2015) Chromosomally integrated human herpesvirus 6 in heart failure: prevalence and treatment. Eur J Heart Fail 17(1):9–19. https://doi.org/10.1002/ejhf.194

Libby P (2013) Mechanisms of acute coronary syndromes and their implications for therapy. N Engl J Med 368(21):2004–2013. https://doi.org/10.1056/NEJMra1216063

Libby P, Lichtman AH, Hansson GK (2013) Immune effector mechanisms implicated in atherosclerosis: from mice to humans. Immunity 38(6):1092–1104. https://doi.org/10.1016/j.immuni.2013.06.009

Lipps HJ, Rhodes D (2009) G-quadruplex structures: in vivo evidence and function. Trends Cell Biol 19(8):414–422. https://doi.org/10.1016/j.tcb.2009.05.002

Ljungman P, de la Camara R, Cordonnier C, Einsele H, Engelhard D, Reusser P, Styczynski J, Ward K (2008) Management of CMV, HHV-6, HHV-7 and Kaposi-sarcoma herpesvirus (HHV-8) infections in patients with hematological malignancies and after SCT. Bone Marrow Transplant 42(4):227–240. https://doi.org/10.1038/bmt.2008.162

Luppi M, Marasca R, Barozzi P, Ferrari S, Ceccherini-Nelli L, Batoni G, Merelli E, Torelli G (1993) Three cases of human herpesvirus-6 latent infection: integration of viral genome in peripheral blood mononuclear cell DNA. J Med Virol 40(1):44–52

Makarov VL, Hirose Y, Langmore JP (1997) Long G tails at both ends of human chromosomes suggest a C strand degradation mechanism for telomere shortening. Cell 88(5):657–666

Malkova A, Ira G (2013) Break-induced replication: functions and molecular mechanism. Curr Opin Genet Dev 23(3):271–279. https://doi.org/10.1016/j.gde.2013.05.007

Martin ME, Thomson BJ, Honess RW, Craxton MA, Gompels UA, Liu MY, Littler E, Arrand JR, Teo I, Jones MD (1991) The genome of human herpesvirus 6: maps of unit-length and concatemeric genomes for nine restriction endonucleases. J Gen Virol 72(Pt 1):157–168

McElligott R, Wellinger RJ (1997) The terminal DNA structure of mammalian chromosomes. EMBO J 16(12):3705–3714. https://doi.org/10.1093/emboj/16.12.3705

Mimitou EP, Symington LS (2009) DNA end resection: many nucleases make light work. DNA Repair (Amst) 8(9):983–995. https://doi.org/10.1016/j.dnarep.2009.04.017

Morissette G, Flamand L (2010) Herpesviruses and chromosomal integration. J Virol 84(23):12100–12109. https://doi.org/10.1128/JVI.01169-10

Morris C, Luppi M, McDonald M, Barozzi P, Torelli G (1999) Fine mapping of an apparently targeted latent human herpesvirus type 6 integration site in chromosome band 17p13.3. J Med

Virol 58(1):69–75. doi:10.1002/(SICI)1096-9071(199905)58:1<69::AID-JMV11>3.0.CO;2-3 [pii]

Nacheva EP, Ward KN, Brazma D, Virgili A, Howard J, Leong HN, Clark DA (2008) Human herpesvirus 6 integrates within telomeric regions as evidenced by five different chromosomal sites. J Med Virol 80(11):1952–1958. https://doi.org/10.1002/jmv.21299

Nikitina T, Woodcock CL (2004) Closed chromatin loops at the ends of chromosomes. J Cell Biol 166(2):161–165. https://doi.org/10.1083/jcb.200403118

Ohye T, Inagaki H, Ihira M, Higashimoto Y, Kato K, Oikawa J, Yagasaki H, Niizuma T, Takahashi Y, Kojima S, Yoshikawa T, Kurahashi H (2014) Dual roles for the telomeric repeats in chromosomally integrated human herpesvirus-6. Sci Rep 4:4559. https://doi.org/10.1038/srep04559

Paeschke K, Simonsson T, Postberg J, Rhodes D, Lipps HJ (2005) Telomere end-binding proteins control the formation of G-quadruplex DNA structures in vivo. Nat Struct Mol Biol 12(10):847–854. https://doi.org/10.1038/nsmb982

Pellett PE, Ablashi DV, Ambros PF, Agut H, Caserta MT, Descamps V, Flamand L, Gautheret-Dejean A, Hall CB, Kamble RT, Kuehl U, Lassner D, Lautenschlager I, Loomis KS, Luppi M, Lusso P, Medveczky PG, Montoya JG, Mori Y, Ogata M, Pritchett JC, Rogez S, Seto E, Ward KN, Yoshikawa T, Razonable RR (2012) Chromosomally integrated human herpesvirus 6: questions and answers. Rev Med Virol 22(3):144–155. https://doi.org/10.1002/rmv.715

Potenza L, Barozzi P, Rossi G, Riva G, Vallerini D, Zanetti E, Quadrelli C, Morselli M, Forghieri F, Maccaferri M, Paolini A, Marasca R, Narni F, Luppi M (2011) May the indirect effects of CIHHV-6 in transplant patients be exerted through the reactivation of the viral replicative machinery? Transplantation 92(9):e49–e51. author reply e51-42. https://doi.org/10.1097/TP.0b013e3182339d1a

Prusty BK, Krohne G, Rudel T (2013) Reactivation of chromosomally integrated human herpesvirus-6 by telomeric circle formation. PLoS Genet 9(12):e1004033. https://doi.org/10.1371/journal.pgen.1004033

Quintela A, Escuret V, Roux S, Bonnafous P, Gilis L, Barraco F, Labussiere-Wallet H, Duscastelle-Lepretre S, Nicolini FE, Thomas X, Chidiac C, Ferry T, Frobert E, Morisset S, Poitevin-Later F, Monneret G, Michallet M, Ader F, Lyon HSG (2016) HHV-6 infection after allogeneic hematopoietic stem cell transplantation: from chromosomal integration to viral co-infections and T-cell reconstitution patterns. J Infect 72:214–222. https://doi.org/10.1016/j.jinf.2015.09.039

Raices M, Verdun RE, Compton SA, Haggblom CI, Griffith JD, Dillin A, Karlseder J (2008) C. Elegans telomeres contain G-strand and C-strand overhangs that are bound by distinct proteins. Cell 132(5):745–757. https://doi.org/10.1016/j.cell.2007.12.039

Salahuddin SZ, Ablashi DV, Markham PD, Josephs SF, Sturzenegger S, Kaplan M, Halligan G, Biberfeld P, Wong-Staal F, Kramarsky B et al (1986) Isolation of a new virus, HBLV, in patients with lymphoproliferative disorders. Science 234(4776):596–601

Schmidt-Lucke C, Spillmann F, Bock T, Kuhl U, Van Linthout S, Schultheiss HP, Tschope C (2010) Interferon beta modulates endothelial damage in patients with cardiac persistence of human parvovirus b19 infection. J Infect Dis 201(6):936–945. https://doi.org/10.1086/650700

Sedlak RH, Cook L, Huang ML, Magaret A, Zerr DM, Boeckh M, Jerome KR (2014) Identification of chromosomally integrated human herpesvirus 6 by droplet digital PCR. Clin Chem 60(5):765–772. https://doi.org/10.1373/clinchem.2013.217240

Sedlak RH, Hill JA, Nguyen T, Cho M, Levin G, Cook L, Huang ML, Flamand L, Zerr DM, Boeckh M, Jerome KR (2016) Detection of Human Herpesvirus 6B (HHV-6B) reactivation in hematopoietic cell transplant recipients with inherited chromosomally integrated HHV-6A by droplet digital PCR. J Clin Microbiol 54(5):1223–1227. https://doi.org/10.1128/JCM.03275-15

Smith BH, Campbell A, Linksted P, Fitzpatrick B, Jackson C, Kerr SM, Deary IJ, Macintyre DJ, Campbell H, McGilchrist M, Hocking LJ, Wisely L, Ford I, Lindsay RS, Morton R, Palmer CN, Dominiczak AF, Porteous DJ, Morris AD (2013) Cohort profile: generation Scotland: Scottish Family Health Study (GS:SFHS). The study, its participants and their potential for genetic research on health and illness. Int J Epidemiol 42(3):689–700. https://doi.org/10.1093/ije/dys084

Strenger V, Caselli E, Lautenschlager I, Schwinger W, Aberle SW, Loginov R, Gentili V, Nacheva E, DiLuca D, Urban C (2014) Detection of HHV-6-specific mRNA and antigens in PBMCs of individuals with chromosomally integrated HHV-6 (ciHHV-6). Clin Microbiol Infect 20(10):1027–1032. https://doi.org/10.1111/1469-0691.12639

Thomson BJ, Dewhurst S, Gray D (1994) Structure and heterogeneity of the a sequences of human herpesvirus 6 strain variants U1102 and Z29 and identification of human telomeric repeat sequences at the genomic termini. J Virol 68(5):3007–3014

Torelli G, Marasca R, Luppi M, Selleri L, Ferrari S, Narni F, Mariano MT, Federico M, Ceccherini-Nelli L, Bendinelli M et al (1991) Human herpesvirus-6 in human lymphomas: identification of specific sequences in Hodgkin's lymphomas by polymerase chain reaction. Blood 77(10):2251–2258

Torelli G, Barozzi P, Marasca R, Cocconcelli P, Merelli E, Ceccherini-Nelli L, Ferrari S, Luppi M (1995) Targeted integration of human herpesvirus 6 in the p arm of chromosome 17 of human peripheral blood mononuclear cells in vivo. J Med Virol 46(3):178–188

Trempe F, Gravel A, Dubuc I, Wallaschek N, Collin V, Gilbert-Girard S, Morissette G, Kaufer BB, Flamand L (2015) Characterization of human herpesvirus 6A/B U94 as ATPase, helicase, exonuclease and DNA-binding proteins. Nucleic Acids Res 43(12):6084–6098. https://doi.org/10.1093/nar/gkv503

Tweedy J, Spyrou MA, Pearson M, Lassner D, Kuhl U, Gompels UA (2016) Complete genome sequence of germline chromosomally integrated human herpesvirus 6A and analyses integration sites define a new human endogenous virus with potential to reactivate as an emerging infection. Viruses 8(1):19. https://doi.org/10.3390/v8010019

Verdun RE, Karlseder J (2006) The DNA damage machinery and homologous recombination pathway act consecutively to protect human telomeres. Cell 127(4):709–720. https://doi.org/10.1016/j.cell.2006.09.034

Wallaschek N, Gravel A, Flamand L, Kaufer BB (2016a) The putative U94 integrase is dispensable for human herpesvirus 6 (HHV-6) chromosomal integration. J Gen Virol. https://doi.org/10.1099/jgv.0.000502

Wallaschek N, Sanyal A, Pirzer F, Gravel A, Mori Y, Flamand L, Kaufer BB (2016b) The Telomeric repeats of Human Herpesvirus 6A (HHV-6A) are required for efficient virus integration. PLoS Pathog 12(5):e1005666. https://doi.org/10.1371/journal.ppat.1005666

Ward KN, Leong HN, Nacheva EP, Howard J, Atkinson CE, Davies NW, Griffiths PD, Clark DA (2006) Human herpesvirus 6 chromosomal integration in immunocompetent patients results in high levels of viral DNA in blood, sera, and hair follicles. J Clin Microbiol 44(4):1571–1574. 44/4/1571 [pii] https://doi.org/10.1128/JCM.44.4.1571-1574.2006

Willett WC, Colditz GA (1998) Approaches for conducting large cohort studies. Epidemiol Rev 20(1):91–99

Yamada Y, Osumi T, Imadome KI, Takahashi E, Ohye T, Yoshikawa T, Tomizawa D, Kato M, Matsumoto K (2017) Transmission of chromosomally integrated human herpesvirus 6 via cord blood transplantation. Transpl Infect Dis 19(1). https://doi.org/10.1111/tid.12636

Zerr DM (2006) Human herpesvirus 6 and central nervous system disease in hematopoietic cell transplantation. J Clin Virol 37(Suppl 1):S52–S56

Zerr DM, Fann JR, Breiger D, Boeckh M, Adler AL, Xie H, Delaney C, Huang ML, Corey L, Leisenring WM (2011) HHV-6 reactivation and its effect on delirium and cognitive functioning in hematopoietic cell transplantation recipients. Blood 117(19):5243–5249. https://doi.org/10.1182/blood-2010-10-316083

Zerr DM, Boeckh M, Delaney C, Martin PJ, Xie H, Adler AL, Huang ML, Corey L, Leisenring WM (2012) HHV-6 reactivation and associated sequelae after hematopoietic cell transplantation. Biol Blood Marrow Transplant 18(11):1700–1708. https://doi.org/10.1016/j.bbmt.2012.05.012

Chapter 11
Structural Aspects of Betaherpesvirus-Encoded Proteins

Mitsuhiro Nishimura and Yasuko Mori

Abstract Betaherpesvirus possesses a large genome DNA with a lot of open reading frames, indicating abundance in the variety of viral protein factors. Because the complicated pathogenicity of herpesvirus reflects the combined functions of these factors, analyses of individual proteins are the fundamental steps to comprehensively understand about the viral life cycle and the pathogenicity. In this chapter, structural aspects of the betaherpesvirus-encoded proteins are introduced. Betaherpesvirus-encoded proteins of which structural information is available were summarized and subcategorized into capsid proteins, tegument proteins, nuclear egress complex proteins, envelope glycoproteins, enzymes, and immune-modulating factors. Structure of capsid proteins are analyzed in capsid by electron cryomicroscopy at quasi-atomic resolution. Structural information of teguments is limited, but a recent crystallographic analysis of an essential tegument protein of human herpesvirus 6B is introduced. As for the envelope glycoproteins, crystallographic analysis of glycoprotein gB has been done, revealing the fine-tuned structure and the distribution of its antigenic domains. gH/gL structure of betaherpesvirus is not available yet, but the overall shape and the spatial arrangement of the accessory proteins are analyzed by electron microscopy. Nuclear egress complex was analyzed from the structural perspective in 2015, with the structural analysis of cytomegalovirus UL50/UL53. The category "enzymes" includes the viral protease, DNA polymerase and terminase for which crystallographic analyses have been done. The immune-modulating factors are viral ligands or receptors for immune regulating factors of host immune cells, and their communications with host immune molecules are demonstrated in the aspect of molecular structure.

Keywords Betaherpesvirus · Protein structure · X-ray crystallography · Electron cryomicroscopy · Capsid · Tegument · Glycoprotein · Immune modulation

---

M. Nishimura (✉)
Division of Clinical Virology, Kobe University Graduate School of Medicine, Kobe, Hyogo, Japan
e-mail: mnishimu@med.kobe-u.ac.jp

Y. Mori
Division of Clinical Virology, Center for Infectious Diseases, Kobe University Graduate School of Medicine, Kobe, Hyogo, Japan

© Springer Nature Singapore Pte Ltd. 2018
Y. Kawaguchi et al. (eds.), *Human Herpesviruses*, Advances in Experimental Medicine and Biology 1045, https://doi.org/10.1007/978-981-10-7230-7_11

11.1 Introduction

Betaherpesvirus subfamily includes human cytomegalovirus (HCMV) of cytomegalovirus genus and human herpesvirus 6-A and human herpesvirus 6-B (HHV-6A and HHV-6B) and human herpesvirus 7 (HHV-7) of roseolovirus genus. They have a large genome DNA with a lot of open reading frames (ORF) which encode betaherpesvirus-specific, genus-specific, and individually specific viral proteins in addition to herpesvirus-common proteins. Since the life cycle and pathogenicity of these viruses are established as a result of the concerted functions and behaviors of their viral factors, understanding of each viral protein is a fundamental procedure to research these viruses. Structural analysis of viral protein is an effective approach, because structure and function of a protein is highly correlated. At the present time, there is huge and growing data about protein structures, however, low sequence homology of the viral proteins with these known proteins limits the availability of the information.

In this chapter, the structural researches of betaherpesvirus-encoded proteins are introduced. Although structural analyses of viral proteins of alphaherpesvirus and gammaherpesviruses precede those of betaherpesvirus in many cases, there are a growing number of publications about the structural analysis of betaherpesvirus-encoded proteins during recent years. For example, the crystal structure of herpesvirus-common fusogen, namely, glycoprotein B (gB), has been determined for an alphaherpesvirus human simplex virus 1 (HSV-1) in 2006 (Heldwein et al. 2006) for the first time, followed by for a gammaherpesvirus Epstein-Barr virus (EBV) in 2009 (Backovic et al. 2009), while for betaherpesvirus HCMV gB, the structure has been revealed partially in 2014 (Spindler et al. 2014) and finally determined in 2015 (Burke and Heldwein 2015; Chandramouli et al. 2015). Almost all structural analyses are focusing on the HCMV or animal cytomegalovirus proteins, except for a recent structural study about the roseolovirus tegument protein HHV-6B U14 (Wang et al. 2016), although the structural information is useful to understand the corresponding factors of other betaherpesvirus members. Structural analysis for the herpesvirus-common factors reveals how betaherpesvirus fine-tuned the common machinery for their specific functions.

Table 11.1 summarized the availability of structural information of betaherpesvirus-encoded proteins introduced in this chapter. The viral proteins are arbitrarily categorized into capsid proteins, tegument proteins, nuclear egress complex (NEC) proteins, envelope glycoproteins, enzymes, and immune-modulating factors. Overview of each structure, key functional features, and the relationship with its homologs of other herpesvirus are described in the following sections.

11.2 Capsid Proteins

Capsid is a highly ordered icosahedral protein shell that tightly packs the viral genome DNA. Three-dimensional structure of the HCMV capsid has been analyzed at 35 Å resolution by electron cryomicroscopy in 1998 (Butcher et al. 1998),

11 Structural Aspects of Betaherpesvirus-Encoded Proteins

Table 11.1 Summarized structural information for betaherpesvirus-encoded proteins

Subcategory/ common name	Cytomegalovirus	Roseolovirus	Notes	Accession IDs of representative data
Capsid proteins				
Major capsid protein (MCP)	UL86	U57	Hexon, Penton	EMDB-5695 (capsid)
Minor capsid protein (mCP)	UL85	U56	Triplex	EMDB-5695 (capsid)
Minor capsid-binding protein (mCBP)	UL46	U29	Triplex	EMDB-5695 (capsid)
Smallest capsid protein (SCP)	UL48/49	U32		EMDB-5696 (virion)
Tegument proteins				
pp150 (HCMV), p100 (HHV-6A)	UL32	U11	N-terminal region	EMDB-5696 (virion)
	UL25, UL35	*U14*	N-terminal domain	5B1Q
NEC proteins				
Nuclear egress membrane protein	UL50, M50	U34		5D5N, 5DOB, 5A3G (M50)
Nuclear egress lamina protein	UL53	U37		5D5N, 5DOB
Envelope glycoproteins				
gB	UL55	U39	Ectodomain	5C6T, 5CXF, 4OSU (Dom-II) 5FZ2 (HSV-1, pre-fusion)
gH/gL	gH: UL75, gL: UL115	gH: U48, gL: U82	Ectodomain	3M1C (HSV-1), 3PHF (EBV)
gH/gL/gO	UL74	U47		5T1D (EBV gH/gL/gp42)
gH/gL/UL128/ UL130/UL131A	UL128, UL130, UL131A	–		–
gH/gL/gQ1/gQ2	–	U100		–
Enzymes				
Protease	UL80	U53	N-terminal region	1NJU, 1BIL, 1BIM, 1JQ6, 2WPO, 1WPO, 1CMV, 1LAY
Polymerase, processivity factor	UL44	U27	N-terminal region	1T6L, 1YYP
Terminase, nuclease domain	UL89	U66	C-terminal domain	3N4P
Immune-modulating factors				
Chemokine receptor homolog	US28	–	Complex with CX3CL1	4XT1
IE1	UL122	–	Core domain	4WID

(continued)

Table 11.1 (continued)

Subcategory/common name	Cytomegalovirus	Roseolovirus	Notes	Accession IDs of representative data
IL-10 mimic	UL111A	–	Complex with IL-10R1	1LQS
	UL141	–	Complex with TRAIL-R2	4I9X, 4JM0
Immunoevasin	US2, m04	–	Complex with HLA-A2	1IM3, 4PN6 (m04), 2MIZ (m04)
	UL16	–	Complex with MICBpf	2WY3
MHC class I homolog	*UL18* *m144, m152,* *m153, m157*	–	Complex with LIR-1	3D2U, 1PQZ (m144), 4G59 (m152), 2O5N (m153), 2NYK (m157),

followed by analyses with improved resolution of HCMV capsid (Chen et al. 1999) and simian cytomegalovirus capsid (Trus et al. 1999). Owing to the technical advance in electron cryomicroscopy, structures of the HCMV capsid and murine cytomegalovirus capsid has been determined at the resolution of 6 Å and 9.1 Å, respectively (Dai et al. 2013; Hui et al. 2013), enabling the visualization of the mainchain traces of capsid proteins. For comparison among herpesvirus subfamilies, quasi-atomic structures of Kaposi's sarcoma-associated herpesvirus (KSHV) capsid (Dai et al. 2015) and HSV-1 capsid (Huet et al. 2016) are also available.

In HCMV, the capsid consists of major capsid protein (MCP) UL86, minor capsid protein (mCP) UL85, minor capsid protein-binding protein (mCBP) UL46, and smallest capsid protein (SCP) UL48/49 (Table 11.1). The capsid contains three structural units, namely, hexon, penton, and triplex. One capsid contains 150 hexons, 12 pentons, and 320 triplexes, and they form the icosahedral architecture (Fig. 11.1a). Hexon and penton are self-assembled forms of hexameric and pentameric MCPs (Fig. 11.1a and b). Triplex is a heterotrimeric complex which is comprised of two mCPs and one mCBP (Fig. 11.1b). SCPs are additionally attached at the outer tips of the hexons and show the horn-like appearance (Fig. 11.1c). Overall appearance of the capsid and each structural unit are basically similar to their counterparts of HSV-1 capsid and KSHV capsid. The MCP has three domains, that is, upper domain, middle domain, and lower/floor domains (Fig. 11.1d). As for the upper domain, a crystal structure of HSV-1 MCP upper domain (MCPud) has been determined (Bowman et al. 2003). The HSV-1 MCPud could be fitted in the observed densities of HCMV MCP and murine cytomegalovirus (MCMV) MCP (Dai et al. 2013; Hui et al. 2013). The lower/floor domains of MCP have a structural similarity with bacteriophage HK97 gp5 (Gan et al. 2006). The HK97 gp5 structure was well fitted to the MCMV MCP density, demonstrating the structural similarity (Hui et al. 2013). The MCP and mCP are well conserved between HCMV and HSV-1, while mCBP and SCP are different. The mCBP of HSV-1 is VP19c (UL38), and UL46 is two thirds in size (VP19c, 50 kDa; UL46, 33 kDa), and their sequences are not similar.

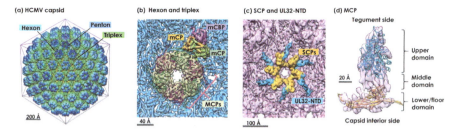

Fig. 11.1 Capsid proteins and tegument protein UL32. (**a**) HCMV capsid structure colored by radial distance from the center. Representatives of hexons, pentons, and triplex are indicated. (**b**) A close-up view of a hexon (MCPs) and a triplex (two mCPs and one mCBP). (**c**) The MCP structure in the hexon. Crystal structures of HSV-1 MCPud (PDB ID: 1NO7) and HK97 gp5 (PDB ID: 2FT1) were fitted around the upper and lower/floor domain in the map, respectively. (**d**) SCP and UL32-N-terminal domain (UL32-NTD). The density map from HCMV virions contains additional densities around the hexons, corresponding to SCP and UL32-NTD. The original data used for these illustrations are as follows: (**a**), (**b**), (**d**) EMDB-5695; (**c**) EMDB-5696. All illustrations in this chapter were prepared by using UCSF Chimera (Pettersen et al. 2004)

Nevertheless, the triplex has similar appearance in each capsid. SCPs of HSV-1 and HCMV have limited sequence homology, although they share molecular characteristics, that is, the small molecule size (approx. 8.6 kDa) and the enriched positively charged residues. Although both SCPs are located on the hexons with similar appearances in HCMV capsid and HSV-1 capsid, their spatial arrangement are slightly different. The inner volume of capsid, in which the genome DNA is packed, is enlarged in HCMV capsid at a ratio of 1.17 to that of HSV-1 capsid (Butcher et al. 1998). This enlarged space is suitable but not enough to pack the larger genome size of cytomegalovirus, implying differences in the DNA packing mechanisms.

There are limited information about the capsid proteins of roseoloviruses. Electron cryomicroscopy of HHV-6A capsid revealed that its appearance is similar to those of HCMV with hexons, pentons, and triplexes [introduced in (Krueger and Ablashi 2006)], although the details have not been discussed. The MCP, mCP, mCBP, and SCP of HHV-6A are U57, U56, U29, and U32, respectively (Table 11.1). The amino acid sequence identity of MCP, mCP, mCBP, and SCP between HHV-6A and HCMV are 44.6, 42.9, 31.3, and 34.3%, respectively. The relatively high sequence homology is consistent with the observed structural similarity, although the conclusion has to wait for a detailed analysis of the roseolovirus capsid.

11.3 Tegument Proteins

Tegument proteins are the constituents of tegument which fills the space between the capsid and envelope of the herpesvirus virion. Tegument proteins are required during the virion maturation steps to promote the transport and envelopment of the capsid. In HCMV and HHV-6A, approximately 20 proteins are annotated as functional sites on the protein surface. It is noteworthy that the residues involved in interaction with U11 on HHV-6B U14-NTD are not conserved in HCMV UL25 and UL35.

11.4 Nuclear Egress Complex (NEC) Proteins

When a newly assembled capsid passes through the nuclear membrane, two conserved nuclear egress complex (NEC) proteins play a major role in the transport. In HCMV, an inner nuclear membrane (INM) protein UL50 and an associating protein UL53 form a heterodimeric NEC on the INM. Numerous copies of UL50/UL53 complex are assembled into a lattice-like network which packs the capsid. The N-terminal part of UL50 homolog of murine CMV, that is, M50, has been analyzed by NMR (Leigh et al. 2015). Following that, the crystal structures of HCMV UL50/UL53 were reported by two research groups at the almost same time (Walzer et al. 2015; Lye et al. 2015), as well as its homolog of HSV-1 UL31/UL34 complex (Bigalke and Heldwein 2015; Zeev-Ben-Mordehai et al. 2015). Both UL50 and UL53 include an α/β two-layered Bergerat fold, combined with different structural elements (Fig. 11.2b). UL53 has a zinc binding site (Cys_3His) and an N-terminal hook-like extension which is encompassed by α-helices of UL50. The Cys and His residues are critical for the function and conserved across herpesviruses. In the crystal lattice of P6 space group, the UL50/UL53 complexes form a ring-like hexamer (Walzer et al. 2015) (Fig. 11.2c). A similar hexameric ring structure was also observed in the HSV-1 UL31/UL34 crystallographic analysis (Bigalke and Heldwein 2015), and it is consistent with the small angle X-ray scattering (SAXS)-derived ab initio model of UL31/UL34 and electron cryotomography/micrography analysis of HSV-1 NEC (Hagen et al. 2015; Bigalke et al. 2014; Zeev-Ben-Mordehai et al. 2015). In roseolovirus, UL50 and UL53 of HCMV correspond to U34 and U37 of HHV-6A, respectively. Although the HCMV UL50 and UL53 have significantly longer C-terminal regions compared to HHV-6A U34 and U37, the amino acid identities at the N-terminal regions are 34% (UL50 and U34) and 43% (UL53 and U37), respectively.

11.5 Envelope Glycoproteins

A herpesvirus virion has a host-derived envelope studded with a lot of viral glycoproteins. Most important and structurally studied glycoproteins are gB and gH/gL, which are conserved among all herpesviruses. gB and gH/gL play critical roles in a series of events upon the viral entry into the target cell cytoplasm. Because of their outermost locations on the virion and the critical functions in infection, the gB and gH/gL are also important as the major target of host immunity.

11.5.1 Fusion Protein gB

gB is a class III fusion protein common for herpesvirus to catalyze the fusion event between target cell membrane and the viral membrane. The first gB structure has been determined for an alphaherpesvirus HSV-1, revealing the homotrimeric structure (Heldwein et al. 2006). Following that, the gB structure of a gammaherpesvirus EBV has been analyzed, and the structural similarity of gB between the two viral subfamilies was demonstrated (Backovic et al. 2009). In 2015, the crystal structures of HCMV gB have been published from two research groups at the almost same time (Burke and Heldwein 2015; Chandramouli et al. 2015). HCMV gB consists of 907 amino acids in full length, and the structural analyses were done for the ectodomain of approximately 700 amino acids, removing the membrane proximal region, the transmembrane domain, and the cytoplasmic domain at the C-terminus. The HCMV gB ectodomain has an trimeric fold with five domains same as those of HSV-1 gB and EBV gB (Fig. 11.3a and b). The respective domains are similar in fold, with a few local variations. Compared to gB structures of HSV-1 and EBV, the HCMV gB has difference in the arrangements of D-IV and D-II relative to D-III/D-V and to D-I/D-III, respectively. The charge distribution at the membrane distal end called "crown" on D-IV is covered by negatively charged electrostatic potential, in contrast to the positively charged electrostatic potential observed in HSV-1 gB and EBV gB. Another feature of HCMV gB is its numerous N-linked glycosylation sites, and the structural mapping of the glycosylation sites indicated that the gB surface is extensively shielded by the glycans, especially around D-II. The glycan layer is considered to protect gB from the host immune surveillance. Mapping analysis of antigenic domains (AD)-1 to AD-5, which are defined as the epitope clusters of antibodies, clearly showed that the AD-1 on DVI is poorly protected by glycans, and it is consistent with the high immunogenicity of AD-1. Among the numerous neutralizing antibodies against HCMV gB, two neutralizing antibodies were analyzed their binding mode in crystal structures. gB complexed with 1G2 Fab domain revealed its interaction mode at the AD-5 on D-I (Chandramouli et al. 2015). The other neutralizing antibody SM5-1 recognizes the AD-4 on D-II, and the complex structure of the SM5-1 Fab domain and a recombinant protein of gB domain II (gB Dom-II) has been determined (Spindler et al. 2014). There are no structural analysis of the roseolovirus gB, although the sequence similarity at the ectodomain (amino acid identity, 43%, HHV-6A U1102 gB) indicates that roseolovirus gB also has similar structural features same as HCMV gB.

All of the gB crystal structures of HSV-1, EBV, and HCMV represent the post-fusion form of gB; thus there is limited information for the pre-fusion form of gB. Pre-fusion forms of gB is simulated by using the pre-fusion structure of vesicular stomatitis virus G protein (VSV-G) (Roche et al. 2007), which is a class III fusion protein same as gB. Although VSV-G and gB have low sequence homology, their shared domain constructs and their arrangements rationalizes such model-based analyses. However, a recent electron cryotomographic analysis of HSV-1 gB enriched on cell-derived vehicles has visualized a putative pre-fusion (or intermediate)

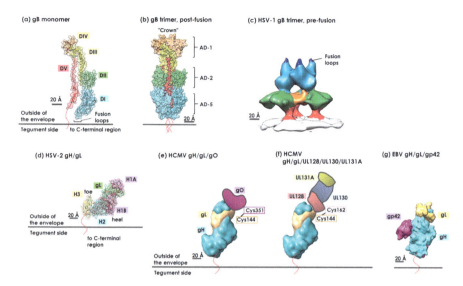

Fig. 11.3 Envelope glycoproteins. (**a**) and (**b**) The crystal structure of HCMV gB in post-fusion conformation was shown as a monomer (**a**) or trimer (**b**). The five domains were colored separately. Approximate locations of three antigenic domains (AD-1, AD-3, and AD-5) are indicated in (**b**). (**c**) HSV-1 gB pre-fusion structure visualized by electron cryotomography. The color codes are the same as (**a**) and (**b**). (**d**) Crystal structure of HSV-2 gH/gL. The four domains were colored separately and the locations of "heel" and "toe" were shown. (**e**) and (**f**) The relative locations of gO (**e**) and UL128/UL130/UL131A (**d**) against gH/gL were illustrated. (**g**) EBV gH/gL/gp42 complex determined by X-ray crystallography. The original data used for these illustrations are as follows: (**a**), (**b**) 5XCF (Burke and Heldwein 2015); (**c**) 5FZ2 (Zeev-Ben-Mordehai et al. 2016); (**d**), (**e**), (**f**) 3M1C (Chowdary et al. 2010); (**g**) 5T1D (Sathiyamoorthy et al. 2016). The image in (**e**) was illustrated by reference to the original literature (Ciferri et al. 2015a)

form of gB in addition of the post-fusion form (Zeev-Ben-Mordehai et al. 2016). The pre-fusion form of gB is largely different from that of VSV-G, in which the fusion peptides in the D-I is sequestered near the membrane proximal side. The HSV-1 gB model has fusion peptides protruding toward the membrane distal side (Fig. 11.3c), suggesting sharp contrast between herpesvirus gB and VSV-G.

11.5.2 HCMV gH/gL/gO and gH/gL/UL128/UL130/UL131A

The heterodimeric complex of gH/gL is also conserved among herpesvirus and serves as a regulator of gB. Normally, gB does not spontaneously exert its fusion activity, but the activation by gH/gL is required [reviewed in (Heldwein 2016)]. The gH/gL activation is coupled with the receptor binding via other viral factors. In HSV-1 and HSV-2, another glycoprotein gD recognizes the host receptors, namely, herpesvirus entry mediator (HVEM) and nectin, and then activates gH/gL. In EBV gH/gL can directly interact with integrin on epithelial cells, but an additional accessory protein of gH/gL, namely, gp42, is required to infect the B cells via human leukocyte antigen (HLA). Similarly to the EBV gH/gL/gp42 complex, gH/gL of cytomegalovirus and roseolovirus are combined with other glycoproteins. In HCMV, UL128, UL130, and UL131A or gO are combined with gH/gL and then forms a pentameric gH/gL/UL128/UL130/UL131A complex or a trimeric gH/gL/gO complex. On the other hand, HHV-6A and HHV-6B have gQ1/gQ2 and gO to form a tetrameric gH/gL/gQ1/gQ2 complex and a trimeric gH/gL/gO complex, respectively (Mori et al. 2004; Akkapaiboon et al. 2004; see also the chapter written by Tang and Mori in this book). These additional factors are directly related to the cell tropisms of these viruses. HCMV gH/gL/gO is used for infection to fibroblast, whereas the gH/gL/UL128/UL130/UL131A drastically promotes infection to epithelial/endothelial cells. Although the role of gH/gL/gO in HHV-6A and HHV-6B is unclear, the tetrameric complex gH/gL/gQ1/gQ2 has been revealed to play a critical role for infection to their target cells. Interestingly, HHV-6A gH/gL/gQ1/gQ2 recognizes the ubiquitously expressed protein CD46 (Santoro et al. 1999), while HHV-6B gH/gL/gQ1/gQ2 recognizes CD134 (also known as OX40) which is specifically expressed on the activated T-cells (Tang et al. 2013).

The crystal structures of gH/gL have been determined for HSV-2 (Chowdary et al. 2010), and EBV (Matsuura et al. 2010), but not yet for betaherpesvirus. HSV-2 gH consists of 838 amino acids and a type I membrane protein, and the ectodomain 48-803 was used for the analysis, removing N-terminal signal sequence and the C-terminal transmembrane domain and cytosolic domain. The gH contains four domains, namely, H1A, H1B, H2, and H3 (Fig. 11.3d). gL is tightly integrated in-between the H1A and H1B domains near the N-terminal region of gH. The overall shape of HSV-2 gH/gL resembles a "boot" with a gentle bend between H1B and H2 domains ("heel"). The crystal structure of EBV gH has a comparable structure with the HSV-2 gH, including four domains. D-I, D-II, D-III, and D-IV of EBV gH corresponds to H1A, H1B, H2, and H3 of HSV-2 gH, respectively. Similarly to the HSV-2 gL, EBV gL resides between D-I and D-II. One critical difference is the overall appearance; EBV gH/gL does not have the boot-like shape because the angle between D-II and D-III is flat.

Although the structural information of gH/gL of betaherpesvirus is limited, negative stain electron microscopy analyses revealed the overall shape of HCMV gH/gL, revealing a boot-like shape with a central bend like the HSV-2 gH/gL (Ciferri et al. 2015a). The overall shapes of an artificial dimer of HCMV gH/gL with the Fab domain of a neutralizing antibody MSL-109, the gH/gL/gO complex, and the gH/gL/UL128/UL130/UL131A complex were also visualized. The averaged electron microscopic images of gH/gL/gO and gH/gL/UL128/UL130/UL131A show clear density for gO or ULs at the membrane-distal end of the gH/gL complex, extending the longitudinal axis of gH/gL (Fig. 11.3e and f). The artificial gH/gL dimer was formed at the membrane-distal end of the gH/gL via Cys144 of gL, and the same Cys144 is the key residue which forms disulfide bond with gO-Cys351 or UL128-Cys162. The shared binding site on gL explained for the mutual exclusion between gO and UL128/UL130/UL131A. This extending mode of gH/gL/gO and gH/gL/UL128/UL130/UL131A is sharp contrast to the EBV gH/gL/gp42 of which gp42 confers affinity to the HLA. The complex structure of gH/gL/gp42 was analyzed by electron microscopy and crystallography (Sathiyamoorthy et al. 2014, 2016). In contrast to the HCMV gH/gL/gO and gH/gL/ UL128/UL130/UL131A, EBV gp42 binds to gH/gL at the side position against the longitudinal axis, and the interaction is exclusively with gH, not with gL (Fig. 11.3g). Thus, gH/gL of betaherpesvirus and that of gammaherpesvirus extend their machinery by different ways. Antibodies against HCMV UL128/UL130/UL131A strongly inhibit the infection to epithelial/endothelial cells, and their binding sites were mapped on the three-dimensional density obtained by negative stain electron microscopy (Ciferri et al. 2015b).

11.6 Enzymes

11.6.1 Protease

All members of herpesvirus possess a serine protease to process the assembly protein precursor required for the capsid assembly. In HCMV, UL80 encodes 708-amino-acid polypeptide, and the N-terminal region 1–256 corresponds to the protease, which is self-cleaved at the position 256–257. The crystal structure of HCMV protease has been determined in 1996 and published by four research groups at the same time (Chen et al. 1996; Qiu et al. 1996; Shieh et al. 1996; Tong et al. 1996). The protease has orthogonally stacked three- and four-stranded β-sheets surrounded by seven (or eight) α-helices (Fig. 11.4a), different from other serine proteases such as chymotrypsin and subtilisin. The catalytic site of HCMV protease does not have the classical Ser-His-Asp catalytic triad, but a unique Ser-His-His catalytic triad. The key residues Ser132, His63, and His157 are conserved across all of the herpesvirus members. The protease forms a dimer and the interaction between monomers have indirect allosteric effect to the conformation of the catalytic sites, thus critical for the activity (Batra et al. 2001). Complex structures of HCMV protease and peptide mimic inhibitors have been also determined, giving a template for structure-based drug design (Khayat et al. 2003; Tong et al. 1998). HHV-6A U53 corresponds to the HCMV UL80, and the protease domain has 39% sequence identity, with the conserved catalytic triad, Ser116, His46, and His135.

11.6.2 DNA Polymerase HCMV UL44

The DNA polymerase of HCMV consists of two factors which are encoded by UL54 (catalytic subunit) and UL44 (processivity factor). The processivity factor is loosely associated with the template DNA and works as a "sliding clamp," enabling the catalytic subunit to slide on the template. The known sliding clamp includes the proliferating cell nuclear antigen (PCNA) of eukaryotic DNA polymerase δ and ε and HSV-1 UL42. The crystal structure of UL44 N-terminal region (1–290),

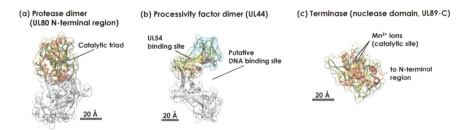

Fig. 11.4 Enzymes. (**a**) The crystal structure of HCMV protease (N-terminal region of UL80). The protease forms a dimer. The unique Ser-His-His catalytic triad composed of Ser132, His63, and His157 was shown. (**b**) The crystal structure of HCMV polymerase processivity factor (UL44). The UL44 forms a C-shaped dimer. The interior of the "C" letter is the DNA binding site and the back side contain the catalytic subunit (UL54) binding site. (**c**) The crystal structure of HCMV terminase nuclease domain, corresponding to the C-terminal domain of UL89 (UL89-C). The essential Mn^{2+} ions at the catalytic site are shown. The original data used for these illustrations are as follows: (**a**) 1NJU (Khayat et al. 2003); (**b**) 1T6L (Appleton et al. 2004); (**c**) 3N4Q (Nadal et al. 2010)

removing the unstructured C-terminal tail, has been determined (Appleton et al. 2004). The UL44 fold is comprised of two similar subdomains from 10–128 to 143–270 with a twofold pseudosymmetry in a monomer (Fig. 11.4b). The overall fold is similar to those of HSV-1 UL42 and PCNA, whereas their oligomeric states are different. UL44 forms a C-clamp-shaped homodimer, while UL42 and PCNA exist as a monomer and trimer, respectively. It has been demonstrated that the dimeric form of UL44 is important for the DNA-binding activity. A crystal structure of UL44 complexed with the C-terminal peptide of the catalytic subunit UL54 has been also reported (Appleton et al. 2006), showing the linkage between the processivity factor and the catalytic subunit. The peptide binding site is similar to the HSV-1 U30 binding site of HSV-1 UL42, and the complex formation induces a wider opening of the C-clamp. The space corresponding to the interior of the "C" letter is expected to be the binding site of DNA (Komazin-Meredith et al. 2008). HCMV UL44 corresponds to HHV-6A U27, and the amino acid identity between UL44 and U27 at the 1–290 region is 41.3%. Residues involved in the dimer interface are conserved in betaherpesviruses.

11.6.3 Terminase

Herpesvirus replicates the genome DNA by the means of the "rolling circle replication." The concatemeric DNA is cleaved to the unit length by viral terminase and then translocated into viral procapsids. The terminase of HCMV is composed of UL89 (HHV-6A U66) and UL56 (HHV-6A U40). UL56 recognizes linearized DNA, while UL89 cleaves the DNA. UL89 consists of 674 amino acids and has two domains, that is, the N-terminal ATPase domain and the C-terminal nuclease domain.

The crystal structure of HCMV UL89 C-terminal nuclease domain (419–674, UL89-C) has been determined (Nadal et al. 2010). UL89-C has an eight stranded parallel-antiparallel mixed β-sheet at the center, sandwiched by α helices (Fig. 11.4c). The protein fold belongs to the RNase H-like superfamily of nuclease and polynucleotidyl transferase, which includes the human immunodeficiency virus (HIV) integrase. The active site has a negative charge electrostatic potential and contains two metal binding sites. The two Mn^{2+} ions bound at the catalytic site are critical for the catalytic activity. As expected from the fact that the UL89-C has a substantial similarity to the HIV integrase, a HIV integrase inhibitor raltegravir can inhibit the UL89-C nuclease activity, while another integrase inhibitor elvitegravir has no effect.

11.7 Immune-Modulating Factors

11.7.1 Chemokine Receptor US28

Chemokines are the ligands of G-protein coupled receptors (GPCRs) and trigger a variety of cell responses, leading to immune cell migration. HCMV encodes US28, a class A GPCR homolog. US28 receives a variety of human chemokines and internalized them, thereby serve as a scavenger of inflammatory chemokines. Recently, crystal structures of US28 liganded with the chemokine domain of a chemokine CX3CL1 (fractalkine) have been reported (Burg et al. 2015). US28 has a helix bundle composed of seven transmembrane helices, with an extra C-terminal α helix at the intracellular side (Fig. 11.5a). The CX3CL1 binds to the cleft formed at the extracellular regions of US28, deeply inserting the N-terminal region. Structural analysis suggested that US28 takes an active conformation in complex with the CX3CL1. However, US28 lacks a key residue to stabilize the inactive conformation and has intrinsic structural features to lock the active conformation; therefore US28 is considered to be constitutively active independent of the ligand.

11.7.2 Immediate Early Protein IE1

Cytomegaloviruses antagonize the function of promyelocytic leukemia protein (PML) and nuclear bodies (NB) by an immediate early protein IE1, and interferes their promotion of intrinsic immunity. HCMV IE1 is encoded at UL122, and consists of 491 amino acids. Structural analysis of IE1 has been done for the core region at the region 20–382 (IE1core) because the N-terminal and C-terminal regions are predicted to be intrinsically disordered (Scherer et al. 2014; Klingl et al. 2015). IE1core protein derived from rhesus cytomegalovirus was crystalized and the structure was determined. The IE1core consists of 11 α helices and adopts an elongated fold with three subdomains, likened to a femur (Fig. 11.5b). Although the overall structure is unique, structural analysis revealed that the coiled-coil part of the

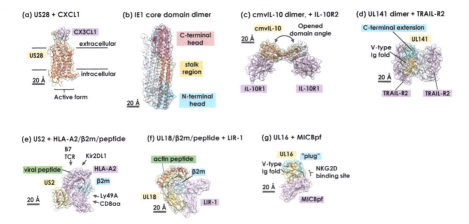

Fig. 11.5 Immune-modulating factors. (**a**) The crystal structure of HCMV US28/CX3CL1 complex. The nanobody included in the crystal structure is not shown for clarity. The intracellular domain has an active-form conformation. (**b**) The crystal structure of rhesus cytomegalovirus IE1 core domain (IE1core). The IE1core has elongated helix coils and forms a dimer. The three subdomains are colored separately. (**c**) The crystal structure of HCMV IL-10 homolog (cmvIL-10) with the receptor IL-10R1. As same as the human IL-10, cmvIL-10 form an intertwisted dimer, although the domain angle is opened. (**d**) The crystal structure of HCMV UL141/TRAIL-R2 complex. The UL141 has a V-type Ig fold with a C-terminal extension, forming a dimer. Two TRAIL-R2 molecules are bound on the dimer. (**e**) The crystal structure of HCMV US2 as a complex with the target molecule HLA-A2/β2m/viral peptide. The binding site of US2 is different from the other HLA-A2 interacting molecules indicated by arrows. (**f**) The crystal structure of HCMV UL18. UL18 forms a complex with β2m and a peptide just like as MHC class I molecules. The structure was determined as a complex with the ligand LIR-1. (**g**) The crystal structure of HCMV UL16 as a complex with the plat form domain of MICB (MICBpf). UL16 has a V-type Ig fold with an N-terminal extension named "plug." The UL16 binding site on the MICB is overlapping with the binding site of the host receptor NKG2D. The original data used for these illustrations are as follows: (**a**) 4XT1 (Burg et al. 2015); (**b**) 4WID (Scherer et al. 2014); (**c**) 1LQS (Jones et al. 2002); (**d**) 4I9X (Nemcovicova et al. 2013); (**e**) 1IM3 (Gewurz et al. 2001); (**f**) 3D2U (Yang and Bjorkman 2008); (**g**) 2WY3 (Muller et al. 2010)

IE1core resembles a TRIM (tripartite motif) family members, TRIM25, and that is also predicted in the PML (PML is also known as TRIM19). IE1core crystal structure forms a homodimer with extensive interactions, and it was implied that IE1 interact with PML (TRIM19) by forming an intermolecular coiled-coil interaction. HHV-6A has a positional homolog U90; however, there is no notable sequence homology.

11.7.3 Cytokine Mimic cmvIL-10

HCMV encodes a homolog of the pleotropic human cytokine IL-10 (hIL-10) to negatively regulate the host immunity. The hIL-10 act on immune cells to block the production of proinflammatory cytokines. HCMV U111A encodes cytomegalovirus

IL-10 (cmvIL-10) and thereby exploits the immune-suppressive system. The cmvIL-10 has relatively low sequence identity, 27%, with hIL-10, nevertheless cmvIL-10 can bind to the cell receptor IL-10R1 and IL-10R2 with high affinity comparable to hIL-10. A crystal structure of the complex of cmvIL-10 and soluble IL-10R1 (sIL-10R1) has been determined at 2.7 Å resolution (Jones et al. 2002). cmvIL-10 forms an intertwined homodimer (Fig. 11.5c). The dimer has two symmetrical domains which consist of five helices, three come from the N-terminal region of one monomer, and two from the C-terminal region of the other monomer. The dimeric fold is similar to that of hIL-10; however, the inter-domain angle is wider in cmvIL-10 (130°), compared to hIL-10 (90°). The interaction mode to sIL-10R1 is almost same as that of hIL-10, and the interaction hot spot is conserved. Although roseolovirus does not have the corresponding factor to cmvIL-10, the gammaherpesvirus EBV has an IL-10 homolog, ebvIL-10. Unlike cmvIL-10, evbIL-10 has extremely high sequence identity, 83%, with hIL-10; however, the affinity to IL-10 receptors is ~1000-fold less.

11.7.4 UL141

HCMV infection leads to a host response of tumor necrosis factor receptor (TNFR) superfamily to induce apoptosis. HCMV blocks such signaling pathway by an encoded factor UL141 which interacts with TRAIL (TNF-related apoptosis inducing ligand) death receptor of T-cell. UL141 can bind to TRAIL-R1 and R2 and then inhibit their surface expression. A complex structure of UL141 and TRAIL-R2 and an unliganded UL141 structure have been solved by crystallography (Nemcovicova et al. 2013; Nemcovicova and Zajonc 2014). UL141 has an N-terminal V-type Ig-like fold, comparable to the HCMV UL16 described below (Sect. 11.7.7), with additional C-terminal β-sheet domain (Fig. 11.5d). The UL141 exhibited a dimeric form in the crystal, and each monomer binds one TRAIL-R2 molecule. In contrast to the trimeric architecture of TRAIL and TRAIL-R2 complex structure, the UL141/TRAIL-R2 exhibited a dimeric architecture, resulting in a large change in the relative arrangement of TRAIL-R2 molecules.

11.7.5 Immunoevasin US2

HCMV US2 encodes a 199-amino-acid protein, which works as an immunoevasin. US2 captures newly synthesized major histocompatibility complex (MHC) class I molecules, specifically HLA-A and HLA-B locus products, promotes degradation by translocating them to the cellular proteasome, and thereby US2 prevents antigen presentation. Crystal structure of US2 endoplasmic reticulum (ER)-luminal domain

(15–140) has been determined as a complex with a MHC class I molecule HLA-A2 and a viral peptide (Gewurz et al. 2001). US2 has a fold with two-layered β-sheets, which is an Ig-like fold similar to the H subtype of Ig superfamily (Fig.11.5e). The US2 use one side of its β-sheet surface to bind to the HLA-A2. The US2 binding site on the HLA-A2 includes a HLA locus divergent site, rationalizing the target specificity of US2. The location of the US2 binding site is different from those of other MHC class I ligands, such as the B7 T-cell receptor, CD8αα, KIR natural killer cell receptor (Kir2DL1), and lectin-like natural killer receptor Ly49A. From the sequence homology among US2 and the other US6 and US11 gene families, similar Ig-like folds are predicted for US3, US11, US9, US8, US7, US10, and US6 proteins of HCMV.

Murine cytomegalovirus encodes m04 which also interact with the newly synthesized MHC class I molecule; however, it does not bring MHC class I molecule to proteasome or lysosome, but escorts it to the cell surface. The crystal structure and solution structure of murine cytomegalovirus m04 have been determined (Sgourakis et al. 2014; Berry et al. 2014), revealing the Ig-like fold similar to US2.

11.7.6 MHC Class I Homolog UL18

HCMV encodes MHC class I-like molecules UL18 and UL142 to regulate host immunity. Among them, structural analysis of UL18 has been published. UL18 is a viral homolog of MHC class I molecule with ~25% sequence identity. UL18 is expressed as a highly glycosylated transmembrane protein of 348 amino acids and forms a complex with the class I MHC light chain, β2-microglobulin (β2m), like the host MHC class I. UL18 binds to an inhibitory receptor, leukocyte immunoglobulin-like receptor (LIR-1), on NK cells or T-cells with higher affinity than the host MHC class I molecules do. The sequence similarity enabled to build a complex model with LIR-1 based on the LIR-1/HLA-1 complex structure to predict the interaction mode (Wagner et al. 2007), and the actual crystal structure of UL18/LIR-1 with β2m has been determined at 2.2 Å resolution (Yang and Bjorkman 2008). The UL18 structure exhibited the α1α2 platform domain and the α3 Ig-domain, which are remarkably similar to those of MHC class I molecule HLA-A2. The bound LIR-1 structure is almost same, although the domain angle of LIR-1 between domain 1 and domain 2 is slightly opened. Increased interaction was observed at one of the interaction sites, consistent with the higher affinity than MHC class I. The interaction mode also explained for the different affinity with LIR-1 and LIR-2 in structural aspect. As it is a MHC class I homolog, UL18 is targeted by other MHC-I receptors including the HCMV US2 described above. However, the glycosylation sites on UL18 indicated that their binding sites are extensively covered by glycans; thereby UL18 can escape from their recognition.

Murine cytomegalovirus also has MHC class I homologs, and the crystal structures of m144 (Natarajan et al. 2006), m157 (Adams et al. 2007) (complex with the Ly49 natural killer cell receptor), m153 (Mans et al. 2007), and m152 (Wang et al. 2012) (complex with the retinoic acid early inducible-1, RAE1) have been published.

11.7.7 UL16

HCMV downregulates MHC class I expression to prevent antigen presentation to T-cells, but it would facilitate the natural killer (NK) cell surveillance, because NK cells sense the MHC class I molecules on cells. A major activating receptor of NK cells and cytotoxic T-cells against stressed cells is the natural killer group 2, member D (NKG2D). To avoid the activation of NKG2D, HCMV expresses UL16 to inactivate a portion of the NKG2D ligands, namely, MIC (MHC class I chain related molecule) and ULBP (UL16-binding protein) families. UL16 is a 50 kDa type I transmembrane protein, and specifically recognizes MICB, ULBP1, and ULBP2. The molecular mechanism has been revealed by structural analysis. UL16 was co-crystallized with the α1α2 platform domain of MICB (MICBpf) which is responsible to the interaction, and the complex structure has been determined at 1.8 Å resolution (Muller et al. 2010). UL16 has a modified version of variable (V-type) Ig-like fold with an additional N-terminal "plug." UL16 binds to MICBpf via its non-glycosylated hydrophobic surface on its β-sheet, and the interface contains a lot of polar and nonpolar interactions, in agreement with the high affinity. Although the structure of UL16 is not similar to NKG2D, the interaction site is overlapping with the NKG2D binding site on the MICB, predicted from the known MICA/NKG2D complex.

11.8 Summary

Structural analysis of viral proteins unveiled their structural properties such as protein fold, oligomeric states, functional sites, and interaction modes with their target molecules. The structures are useful to predict the molecular functions of the viral proteins as well as to interpret the accumulated knowledge from virological assays. The structural analyses of betaherpesvirus-encoded proteins are greatly advanced in recent years; however, there still remain a lot of viral factors to be analyzed, especially the complexes of gH/gL and its accessory factors which determine the cell tropism of the viruses. In addition to the advance in the current X-ray crystallography, the novel approach using X-ray free-electron laser (XFEL) and the surprising improvement in the electron cryomicroscopy/cryotomography reaching atomic resolution will facilitate the structural analysis of betaherpesvirus-encoded proteins.

References

Adams EJ, Juo ZS, Venook RT, Boulanger MJ, Arase H, Lanier LL, Garcia KC (2007) Structural elucidation of the m157 mouse cytomegalovirus ligand for Ly49 natural killer cell receptors. Proc Natl Acad Sci USA 104(24):10128–10133. 0703735104 [pii] https://doi.org/10.1073/pnas.0703735104

Akkapaiboon P, Mori Y, Sadaoka T, Yonemoto S, Yamanishi K (2004) Intracellular processing of human herpesvirus 6 glycoproteins Q1 and Q2 into tetrameric complexes expressed on the viral envelope. J Virol 78(15):7969–7983. https://doi.org/10.1128/JVI.78.15.7969-7983.2004 78/15/7969 [pii]

Appleton BA, Loregian A, Filman DJ, Coen DM, Hogle JM (2004) The cytomegalovirus DNA polymerase subunit UL44 forms a C clamp-shaped dimer. Mol Cell 15(2):233–244. https://doi.org/10.1016/j.molcel.2004.06.018 S1097-2765(04)00351-X [pii]

Appleton BA, Brooks J, Loregian A, Filman DJ, Coen DM, Hogle JM (2006) Crystal structure of the cytomegalovirus DNA polymerase subunit UL44 in complex with the C terminus from the catalytic subunit. Differences in structure and function relative to unliganded UL44. J Biol Chem 281(8):5224–5232. M506900200 [pii] https://doi.org/10.1074/jbc.M506900200

Backovic M, Longnecker R, Jardetzky TS (2009) Structure of a trimeric variant of the Epstein-Barr virus glycoprotein B. Proc Natl Acad Sci USA 106(8):2880–2885. doi:https://doi.org/10.1073/pnas.0810530106 0810530106 [pii]

Batra R, Khayat R, Tong L (2001) Molecular mechanism for dimerization to regulate the catalytic activity of human cytomegalovirus protease. Nat Struct Biol 8(9):810–817. https://doi.org/10.1038/nsb0901-810 nsb0901-810 [pii]

Berry R, Vivian JP, Deuss FA, Balaji GR, Saunders PM, Lin J, Littler DR, Brooks AG, Rossjohn J (2014) The structure of the cytomegalovirus-encoded m04 glycoprotein, a prototypical member of the m02 family of immunoevasins. J Biol Chem 289(34):23753–23763. doi:https://doi.org/10.1074/jbc.M114.584128 M114.584128 [pii]

Bigalke JM, Heldwein EE (2015) Structural basis of membrane budding by the nuclear egress complex of herpesviruses. EMBO J 34(23):2921–2936. https://doi.org/10.15252/embj.201592359 embj.201592359 [pii]

Bigalke JM, Heuser T, Nicastro D, Heldwein EE (2014) Membrane deformation and scission by the HSV-1 nuclear egress complex. Nat Commun 5:4131. https://doi.org/10.1038/ncomms5131 ncomms5131 [pii]

Bowman BR, Baker ML, Rixon FJ, Chiu W, Quiocho FA (2003) Structure of the herpesvirus major capsid protein. EMBO J 22(4):757–765. https://doi.org/10.1093/emboj/cdg086

Burg JS, Ingram JR, Venkatakrishnan AJ, Jude KM, Dukkipati A, Feinberg EN, Angelini A, Waghray D, Dror RO, Ploegh HL, Garcia KC (2015) Structural biology. Structural basis for chemokine recognition and activation of a viral G protein-coupled receptor. Science 347(6226):1113–1117. https://doi.org/10.1126/science.aaa5026 347/6226/1113 [pii]

Burke HG, Heldwein EE (2015) Crystal structure of the human cytomegalovirus glycoprotein B. PLoS Pathog 11(10):e1005227. https://doi.org/10.1371/journal.ppat.1005227 PPATHOGENS-D-15-01758 [pii]

Butcher SJ, Aitken J, Mitchell J, Gowen B, Dargan DJ (1998) Structure of the human cytomegalovirus B capsid by electron cryomicroscopy and image reconstruction. J Struct Biol 124(1):70–76. S1047-8477(98)94055-2 [pii] https://doi.org/10.1006/jsbi.1998.4055

Chandramouli S, Ciferri C, Nikitin PA, Calo S, Gerrein R, Balabanis K, Monroe J, Hebner C, Lilja AE, Settembre EC, Carfi A (2015) Structure of HCMV glycoprotein B in the postfusion conformation bound to a neutralizing human antibody. Nat Commun 6:8176. https://doi.org/10.1038/ncomms9176 ncomms9176 [pii]

Chen P, Tsuge H, Almassy RJ, Gribskov CL, Katoh S, Vanderpool DL, Margosiak SA, Pinko C, Matthews DA, Kan CC (1996) Structure of the human cytomegalovirus protease catalytic domain reveals a novel serine protease fold and catalytic triad. Cell 86(5):835–843 doi:S0092-8674(00)80157-9 [pii]

Chen DH, Jiang H, Lee M, Liu F, Zhou ZH (1999) Three-dimensional visualization of tegument/capsid interactions in the intact human cytomegalovirus. Virology 260(1):10–16. https://doi.org/10.1006/viro.1999.9791 S0042-6822(99)99791-X [pii]

Chowdary TK, Cairns TM, Atanasiu D, Cohen GH, Eisenberg RJ, Heldwein EE (2010) Crystal structure of the conserved herpesvirus fusion regulator complex gH-gL. Nat Struct Mol Biol 17(7):882–888. https://doi.org/10.1038/nsmb.1837 nsmb.1837 [pii]

Ciferri C, Chandramouli S, Donnarumma D, Nikitin PA, Cianfrocco MA, Gerrein R, Feire AL, Barnett SW, Lilja AE, Rappuoli R, Norais N, Settembre EC, Carfi A (2015a) Structural and biochemical studies of HCMV gH/gL/gO and Pentamer reveal mutually exclusive cell entry complexes. Proc Natl Acad Sci USA 112(6):1767–1772. https://doi.org/10.1073/pnas.1424818112 1424818112 [pii]

Ciferri C, Chandramouli S, Leitner A, Donnarumma D, Cianfrocco MA, Gerrein R, Friedrich K, Aggarwal Y, Palladino G, Aebersold R, Norais N, Settembre EC, Carfi A (2015b) Antigenic characterization of the HCMV gH/gL/gO and Pentamer cell entry complexes reveals binding sites for potently neutralizing human antibodies. PLoS Pathog 11(10):e1005230. https://doi.org/10.1371/journal.ppat.1005230 PPATHOGENS-D-15-01403 [pii]

Dai X, Yu X, Gong H, Jiang X, Abenes G, Liu H, Shivakoti S, Britt WJ, Zhu H, Liu F, Zhou ZH (2013) The smallest capsid protein mediates binding of the essential tegument protein pp150 to stabilize DNA-containing capsids in human cytomegalovirus. PLoS Pathog 9(8):e1003525. https://doi.org/10.1371/journal.ppat.1003525 PPATHOGENS-D-12-01355 [pii]

Dai X, Gong D, Xiao Y, Wu TT, Sun R, Zhou ZH (2015) CryoEM and mutagenesis reveal that the smallest capsid protein cements and stabilizes Kaposi's sarcoma-associated herpesvirus capsid. Proc Natl Acad Sci USA 112(7):E649–E656. https://doi.org/10.1073/pnas.1420317112 1420317112 [pii]

Gan L, Speir JA, Conway JF, Lander G, Cheng N, Firek BA, Hendrix RW, Duda RL, Liljas L, Johnson JE (2006) Capsid conformational sampling in HK97 maturation visualized by X-ray crystallography and cryo-EM. Structure 14(11):1655–1665. S0969-2126(06)00392-3 [pii] https://doi.org/10.1016/j.str.2006.09.006

Gewurz BE, Gaudet R, Tortorella D, Wang EW, Ploegh HL, Wiley DC (2001) Antigen presentation subverted: structure of the human cytomegalovirus protein US2 bound to the class I molecule HLA-A2. Proc Natl Acad Sci USA 98(12):6794–6799. https://doi.org/10.1073/pnas.121172898 98/12/6794 [pii]

Hagen C, Dent KC, Zeev-Ben-Mordehai T, Grange M, Bosse JB, Whittle C, Klupp BG, Siebert CA, Vasishtan D, Bauerlein FJ, Cheleski J, Werner S, Guttmann P, Rehbein S, Henzler K, Demmerle J, Adler B, Koszinowski U, Schermelleh L, Schneider G, Enquist LW, Plitzko JM, Mettenleiter TC, Grunewald K (2015) Structural basis of vesicle formation at the inner nuclear membrane. Cell 163(7):1692–1701. https://doi.org/10.1016/j.cell.2015.11.029 S0092-8674(15)01548-2 [pii]

Heldwein EE (2016) gH/gL supercomplexes at early stages of herpesvirus entry. Curr Opin Virol 18:1–8. https://doi.org/10.1016/j.coviro.2016.01.010 S1879-6257(16)00013-4 [pii]

Heldwein EE, Lou H, Bender FC, Cohen GH, Eisenberg RJ, Harrison SC (2006) Crystal structure of glycoprotein B from herpes simplex virus 1. Science 313(5784):217–220. 313/5784/217 [pii] https://doi.org/10.1126/science.1126548

Huet A, Makhov AM, Huffman JB, Vos M, Homa FL, Conway JF (2016) Extensive subunit contacts underpin herpesvirus capsid stability and interior-to-exterior allostery. Nat Struct Mol Biol 23(6):531–539. https://doi.org/10.1038/nsmb.3212

Hui WH, Tang Q, Liu H, Atanasov I, Liu F, Zhu H, Zhou ZH (2013) Protein interactions in the murine cytomegalovirus capsid revealed by cryoEM. Protein Cell 4(11):833–845. https://doi.org/10.1007/s13238-013-3060-7

Jones BC, Logsdon NJ, Josephson K, Cook J, Barry PA, Walter MR (2002) Crystal structure of human cytomegalovirus IL-10 bound to soluble human IL-10R1. Proc Natl Acad Sci USA 99(14):9404–9409. https://doi.org/10.1073/pnas.152147499 152147499 [pii]

Khayat R, Batra R, Qian C, Halmos T, Bailey M, Tong L (2003) Structural and biochemical studies of inhibitor binding to human cytomegalovirus protease. Biochemistry 42(4):885–891. https://doi.org/10.1021/bi027045s

Klingl S, Scherer M, Stamminger T, Muller YA (2015) Controlled crystal dehydration triggers a space-group switch and shapes the tertiary structure of cytomegalovirus immediate-early 1 (IE1) protein. Acta Crystallogr D Biol Crystallogr 71(Pt 7):1493–1504 https://doi.org/10.1107/S1399004715008792 S1399004715008792 [pii]

Komazin-Meredith G, Petrella RJ, Santos WL, Filman DJ, Hogle JM, Verdine GL, Karplus M, Coen DM (2008) The human cytomegalovirus UL44 C clamp wraps around DNA. Structure 16(8):1214–1225. https://doi.org/10.1016/j.str.2008.05.008 S0969-2126(08)00243-8 [pii]

Krueger GRF, Ablashi DV (2006) Human herpesvirus-6 general virology, epidemiology and clinical pathology, Perspectives in medical virology. Elsevier, Boston

Leigh KE, Sharma M, Mansueto MS, Boeszoermenyi A, Filman DJ, Hogle JM, Wagner G, Coen DM, Arthanari H (2015) Structure of a herpesvirus nuclear egress complex subunit reveals an interaction groove that is essential for viral replication. Proc Natl Acad Sci USA 112(29):9010–9015. https://doi.org/10.1073/pnas.1511140112 1511140112 [pii]

Lye MF, Sharma M, El Omari K, Filman DJ, Schuermann JP, Hogle JM, Coen DM (2015) Unexpected features and mechanism of heterodimer formation of a herpesvirus nuclear egress complex. EMBO J 34(23):2937–2952. https://doi.org/10.15252/embj.201592651 embj.201592651 [pii]

Mahmoud NF, Kawabata A, Tang H, Wakata A, Wang B, Serada S, Naka T, Mori Y (2016) Human herpesvirus 6 U11 protein is critical for virus infection. Virology 489:151–157. https://doi.org/10.1016/j.virol.2015.12.011 S0042-6822(15)00532-2 [pii]

Mans J, Natarajan K, Balbo A, Schuck P, Eikel D, Hess S, Robinson H, Simic H, Jonjic S, Tiemessen CT, Margulies DH (2007) Cellular expression and crystal structure of the murine cytomegalovirus major histocompatibility complex class I-like glycoprotein, m153. J Biol Chem 282(48):35247–35258. M706782200 [pii] https://doi.org/10.1074/jbc.M706782200

Matsuura H, Kirschner AN, Longnecker R, Jardetzky TS (2010) Crystal structure of the Epstein-Barr virus (EBV) glycoprotein H/glycoprotein L (gH/gL) complex. Proc Natl Acad Sci USA 107(52):22641–22646. https://doi.org/10.1073/pnas.1011806108 1011806108 [pii]

Mori Y, Akkapaiboon P, Yonemoto S, Koike M, Takemoto M, Sadaoka T, Sasamoto Y, Konishi S, Uchiyama Y, Yamanishi K (2004) Discovery of a second form of tripartite complex containing gH-gL of human herpesvirus 6 and observations on CD46. J Virol 78(9):4609–4616

Mori J, Kawabata A, Tang H, Tadagaki K, Mizuguchi H, Kuroda K, Mori Y (2015a) Human Herpesvirus-6 U14 induces cell-cycle arrest in G2/M phase by associating with a cellular protein, EDD. PLoS One 10(9):e0137420. https://doi.org/10.1371/journal.pone.0137420 PONE-D-15-14067 [pii]

Mori J, Tang H, Kawabata A, Koike M, Mori Y (2015b) Human herpesvirus 6A U14 is important for virus maturation. J Virol 90(3):1677–1681. https://doi.org/10.1128/JVI.02492-15 JVI.02492-15 [pii]

Muller S, Zocher G, Steinle A, Stehle T (2010) Structure of the HCMV UL16-MICB complex elucidates select binding of a viral immunoevasin to diverse NKG2D ligands. PLoS Pathog 6(1):e1000723. https://doi.org/10.1371/journal.ppat.1000723

Nadal M, Mas PJ, Blanco AG, Arnan C, Sola M, Hart DJ, Coll M (2010) Structure and inhibition of herpesvirus DNA packaging terminase nuclease domain. Proc Natl Acad Sci USA 107(37):16078–16083. https://doi.org/10.1073/pnas.1007144107 1007144107 [pii]

Natarajan K, Hicks A, Mans J, Robinson H, Guan R, Mariuzza RA, Margulies DH (2006) Crystal structure of the murine cytomegalovirus MHC-I homolog m144. J Mol Biol 358(1):157–171. S0022-2836(06)00109-4 [pii] https://doi.org/10.1016/j.jmb.2006.01.068

Nemcovicova I, Zajonc DM (2014) The structure of cytomegalovirus immune modulator UL141 highlights structural Ig-fold versatility for receptor binding. Acta Crystallogr D Biol Crystallogr 70(Pt 3):851–862. https://doi.org/10.1107/S1399004713033750 S1399004713033750 [pii]

Nemcovicova I, Benedict CA, Zajonc DM (2013) Structure of human cytomegalovirus UL141 binding to TRAIL-R2 reveals novel, non-canonical death receptor interactions. PLoS Pathog 9(3):e1003224. https://doi.org/10.1371/journal.ppat.1003224 PPATHOGENS-D-12-02360 [pii]

Pettersen EF, Goddard TD, Huang CC, Couch GS, Greenblatt DM, Meng EC, Ferrin TE (2004) UCSF Chimera–a visualization system for exploratory research and analysis. J Comput Chem 25(13):1605–1612. https://doi.org/10.1002/jcc.20084

Qiu X, Culp JS, DiLella AG, Hellmig B, Hoog SS, Janson CA, Smith WW, Abdel-Meguid SS (1996) Unique fold and active site in cytomegalovirus protease. Nature 383(6597):275–279. https://doi.org/10.1038/383275a0

Roche S, Rey FA, Gaudin Y, Bressanelli S (2007) Structure of the prefusion form of the vesicular stomatitis virus glycoprotein G. Science 315(5813):843–848. 315/5813/843 [pii] https://doi.org/10.1126/science.1135710

Santoro F, Kennedy PE, Locatelli G, Malnati MS, Berger EA, Lusso P (1999) CD46 is a cellular receptor for human herpesvirus 6. Cell 99(7):817–827 doi:S0092-8674(00)81678-5 [pii]

Sathiyamoorthy K, Jiang J, Hu YX, Rowe CL, Mohl BS, Chen J, Jiang W, Mellins ED, Longnecker R, Zhou ZH, Jardetzky TS (2014) Assembly and architecture of the EBV B cell entry triggering complex. PLoS Pathog 10(8):e1004309. https://doi.org/10.1371/journal.ppat.1004309 PPATHOGENS-D-14-00440 [pii]

Sathiyamoorthy K, Hu YX, Mohl BS, Chen J, Longnecker R, Jardetzky TS (2016) Structural basis for Epstein-Barr virus host cell tropism mediated by gp42 and gHgL entry glycoproteins. Nat Commun 7:13557. https://doi.org/10.1038/ncomms13557 ncomms13557 [pii]

Scherer M, Klingl S, Sevvana M, Otto V, Schilling EM, Stump JD, Muller R, Reuter N, Sticht H, Muller YA, Stamminger T (2014) Crystal structure of cytomegalovirus IE1 protein reveals targeting of TRIM family member PML via coiled-coil interactions. PLoS Pathog 10(11):e1004512. https://doi.org/10.1371/journal.ppat.1004512 PPATHOGENS-D-14-01596 [pii]

Sgourakis NG, Natarajan K, Ying J, Vogeli B, Boyd LF, Margulies DH, Bax A (2014) The structure of mouse cytomegalovirus m04 protein obtained from sparse NMR data reveals a conserved fold of the m02-m06 viral immune modulator family. Structure 22(9):1263–1273. https://doi.org/10.1016/j.str.2014.05.018 S0969-2126(14)00207-X [pii]

Shieh HS, Kurumbail RG, Stevens AM, Stegeman RA, Sturman EJ, Pak JY, Wittwer AJ, Palmier MO, Wiegand RC, Holwerda BC, Stallings WC (1996) Three-dimensional structure of human cytomegalovirus protease. Nature 383(6597):279–282. https://doi.org/10.1038/383279a0

Spindler N, Diestel U, Stump JD, Wiegers AK, Winkler TH, Sticht H, Mach M, Muller YA (2014) Structural basis for the recognition of human cytomegalovirus glycoprotein B by a neutralizing human antibody. PLoS Pathog 10(10):e1004377. https://doi.org/10.1371/journal.ppat.1004377 PPATHOGENS-D-14-00683 [pii]

Takemoto M, Koike M, Mori Y, Yonemoto S, Sasamoto Y, Kondo K, Uchiyama Y, Yamanishi K (2005) Human herpesvirus 6 open reading frame U14 protein and cellular p53 interact with each other and are contained in the virion. J Virol 79(20):13037–13046. 79/20/13037 [pii] https://doi.org/10.1128/JVI.79.20.13037-13046.2005

Tang H, Serada S, Kawabata A, Ota M, Hayashi E, Naka T, Yamanishi K, Mori Y (2013) CD134 is a cellular receptor specific for human herpesvirus-6B entry. Proc Natl Acad Sci USA 110(22):9096–9099. https://doi.org/10.1073/pnas.1305187110 1305187110 [pii]

Tong L, Qian C, Massariol MJ, Bonneau PR, Cordingley MG, Lagace L (1996) A new serine-protease fold revealed by the crystal structure of human cytomegalovirus protease. Nature 383(6597):272–275. https://doi.org/10.1038/383272a0

Tong L, Qian C, Massariol MJ, Deziel R, Yoakim C, Lagace L (1998) Conserved mode of peptidomimetic inhibition and substrate recognition of human cytomegalovirus protease. Nat Struct Biol 5(9):819–826. https://doi.org/10.1038/1860

Trus BL, Gibson W, Cheng N, Steven AC (1999) Capsid structure of simian cytomegalovirus from cryoelectron microscopy: evidence for tegument attachment sites. J Virol 73(3):2181–2192

Wagner CS, Rolle A, Cosman D, Ljunggren HG, Berndt KD, Achour A (2007) Structural elements underlying the high binding affinity of human cytomegalovirus UL18 to leukocyte immunoglobulin-like receptor-1. J Mol Biol 373(3):695–705. S0022-2836(07)01092-3 [pii] https://doi.org/10.1016/j.jmb.2007.08.020

Walzer SA, Egerer-Sieber C, Sticht H, Sevvana M, Hohl K, Milbradt J, Muller YA, Marschall M (2015) Crystal structure of the human cytomegalovirus pUL50-pUL53 Core nuclear egress complex provides insight into a unique assembly scaffold for virus-host protein interactions. J Biol Chem 290(46):27452–27458. https://doi.org/10.1074/jbc.C115.686527 C115.686527 [pii]

Wang R, Natarajan K, Revilleza MJ, Boyd LF, Zhi L, Zhao H, Robinson H, Margulies DH (2012) Structural basis of mouse cytomegalovirus m152/gp40 interaction with RAE1gamma reveals a paradigm for MHC/MHC interaction in immune evasion. Proc Natl Acad Sci USA 109(51):E3578–E3587. https://doi.org/10.1073/pnas.1214088109 1214088109 [pii]

Wang B, Nishimura M, Tang H, Kawabata A, Mahmoud NF, Khanlari Z, Hamada D, Tsuruta H, Mori Y (2016) Crystal structure of human herpesvirus 6B tegument protein U14. PLoS Pathog 12(5):e1005594. https://doi.org/10.1371/journal.ppat.1005594 PPATHOGENS-D-15-02987 [pii]

Yang Z, Bjorkman PJ (2008) Structure of UL18, a peptide-binding viral MHC mimic, bound to a host inhibitory receptor. Proc Natl Acad Sci USA 105(29):10095–10100. https://doi.org/10.1073/pnas.0804551105 0804551105 [pii]

Zeev-Ben-Mordehai T, Weberruss M, Lorenz M, Cheleski J, Hellberg T, Whittle C, El Omari K, Vasishtan D, Dent KC, Harlos K, Franzke K, Hagen C, Klupp BG, Antonin W, Mettenleiter TC, Grunewald K (2015) Crystal structure of the herpesvirus nuclear egress complex provides insights into inner nuclear membrane remodeling. Cell Rep 13(12):2645–2652. https://doi.org/10.1016/j.celrep.2015.11.008 S2211-1247(15)01295-4 [pii]

Zeev-Ben-Mordehai T, Vasishtan D, Hernandez Duran A, Vollmer B, White P, Prasad Pandurangan A, Siebert CA, Topf M, Grunewald K (2016) Two distinct trimeric conformations of natively membrane-anchored full-length herpes simplex virus 1 glycoprotein B. Proc Natl Acad Sci USA 113(15):4176–4181. https://doi.org/10.1073/pnas.1523234113 1523234113 [pii]

Chapter 12
Betaherpesvirus Complications and Management During Hematopoietic Stem Cell Transplantation

Tetsushi Yoshikawa

Abstract Two of the four betaherpesviruses, *Cytomegalovirus* (CMV) and human herpesvirus 6B (HHV-6B), play an important role in opportunistic infections in hematopoietic stem cell transplant (HSCT) recipients. These viruses are ubiquitous in humans and can latently infect mononuclear lymphocytes, complicating the diagnosis of the diseases they cause. Although the detection of viral DNA in a patient's peripheral blood by real-time PCR is widely used for monitoring viral infection, it is insufficient for the diagnosis of virus-associated disease. Theoretically, end-organ disease should be confirmed by detecting either viral antigen or significant amounts of viral DNA in a tissue sample obtained from the involved organ; however, this is often difficult to perform in clinical practice. The frequency of CMV-associated diseases has decreased gradually as a result of the introduction of preemptive or prophylactic treatments; however, CMV and HHV-6B infections remain a major problem in HSCT recipients. Measurement of viral DNA load in peripheral blood or plasma using real-time PCR is commonly used for monitoring these infections. Additionally, recent data suggest that an assessment of host immune response, particularly cytotoxic T-cell response, may be a reliable tool for predicting these viral infections. The antiviral drugs ganciclovir and foscarnet are used as first-line treatments; however, it is well known that these drugs have side effects, such as bone marrow suppression and nephrotoxicity. Further research is required to develop less-toxic antiviral drugs.

Keywords *Cytomegalovirus* · HHV-6 · Pneumonia · Encephalitis · Real-time PCR · Antigenemia · Antiviral drug

T. Yoshikawa (✉)
Department of Pediatrics, Fujita Health University School of Medicine, Toyoake, Japan
e-mail: tetsushi@fujita-hu.ac.jp

making it difficult to differentiate between CMV gastrointestinal disease and gastroenteritis caused by other etiologies, including graft versus host disease (GVHD) (Bhutani et al. 2015; He et al. 2008). Detection of viral antigen or DNA in peripheral blood is insufficient for diagnosis of CMV-induced gastroenteritis; it is necessary to confirm active viral infection in a tissue sample from the ulcerative regions using immunohistochemical analysis or in situ hybridization. Alternatively, real-time PCR analysis may be used to compare copy numbers of CMV DNA between tissue obtained from the ulcerative region and control tissue. PCR detection of CMV DNA in stool samples is not recommended due to the low sensitivity of this method (Sun et al. 2015).

12.2.3 Retinitis

CMV retinitis generally occurs later in the post-HSCT period than other CMV-associated diseases, and an incidence of the disease has increased gradually (Eid et al. 2008; Song et al. 2008). Common symptoms of the disease are decreased visual acuity and blurred vision, which are not specific to CMV retinitis. The retinitis occurs bilaterally in approximately 60% of the patients (Crippa et al. 2001). Early diagnosis and treatment are essential for preventing visual loss. Lymphopenia is a risk factor for developing CMV retinitis (Jeon et al. 2012). A recent study analyzing pediatric HSCT recipients demonstrated that CMV retinitis occurred as a late-onset disease (median 199 days after HSCT) in comparison to other types of CMV diseases, suggesting that this complication may be induced by immune reconstitution (Hiwarkar et al. 2014). Quantitation of CMV DNA in aqueous humor by real-time PCR can be used to diagnose CMV retinitis.

12.2.4 Others

CMV infection may be associated with other clinical manifestations such as hepatitis, encephalitis, nephritis, and myocarditis. In order to confirm a correlation between those manifestations and CMV infection, detection of viral antigen and/or DNA in an affected tissue sample by immunohistochemical analysis and in situ hybridization analysis is necessary. This analysis should rule out the possibility of other etiological agents. Recent studies have demonstrated that CMV infection may prevent relapse of myeloid malignancies after HSCT in adult transplant recipients (Takenaka et al. 2015; Green et al. 2013; Elmaagacli et al. 2011). Although a similar protective effect against relapse of underlying diseases by CMV infection was observed in pediatric acute leukemia patients, CMV reactivation was also associated with non-relapse mortality caused by opportunistic infection after GVHD (Inagaki et al. 2016). Therefore, overall patient mortality was not always improved. A better understanding of the mechanisms by which CMV infection may inhibit disease relapse, combined with the development of better intervention protocols to prevent CMV infection, should contribute to a decrease in both relapse and nonrelapse patient mortality.

After the introduction of CMV prophylactic and preemptive treatments, the incidence of CMV diseases has gradually declined. However, CMV disease occurring relatively late post-HSCT has become a major problem (Einsele et al. 2000; Wolf et al. 2003; Boeckh et al. 2003).

12.3 Clinical Manifestations of HHV-6B Infection

As described above, HHV-6 is divided into two different species: HHV-6A and HHV-6B. There is evidence that the neurovirulence of HHV-6A is stronger than HHV-6B (Donati et al. 2005), and a fatal case of HHV-6A encephalitis has been reported in HSCT recipients (de Labarthe et al. 2005). Because seroprevalence of HHV-6A is considered to be low in developed countries, morbidity and mortality associated with HHV-6A infection are likely to be correspondingly low in these countries.

An association between HHV-6 infection and pneumonitis has been examined extensively in HSCT recipients, because this virus has a high degree of similarity to CMV. Subsequently, many clinical manifestations, including interstitial pneumonitis, have been proposed as HHV-6-associated manifestations in HSCT recipients (Carrigan and Knox 1994; Drobyski et al. 1993, Kadakia et al. 1996; Belford et al. 2004; Asano et al. 1991; Yoshikawa et al. 1991, 2002b). One of the strongest correlations between HHV-6B infection and clinical manifestations is posttransplant acute limbic encephalitis (PALE).

12.3.1 Encephalitis

HHV-6B is well known to exhibit neurovirulence. The most significant complications at the time of primary HHV-6B infection (exanthema subitum) are central nervous system manifestations such as febrile seizure (Suga et al. 2000) and encephalitis (Suga et al. 1993; Kawamura et al. 2013). Additionally, it has been demonstrated that HHV-6B can infect glial cells or neuronal cells and cause cell damage either directly or indirectly (Yoshikawa et al. 2002a). HHV-6B encephalitis generally occurs between 2 and 4 weeks after transplant. The most striking characteristics of HHV-6B encephalitis are the clinical features of the disease, which demonstrate acute limbic encephalitis (Visser et al. 2005; Seeley et al. 2007; Zerr et al. 2011). Thus, HHV-6B is considered to be the pathogen most closely associated with PALE (Ogata et al. 2010, 2013b; Seeley et al. 2007). The incidence of posttransplant HHV-6 encephalitis has been estimated at about 1–3% in HSCT recipients. Cord blood transplant recipients are at high risk for this complication (Roback 2002). The majority of HHV-6B infection-associated PALE cases appear in adult HSCT recipients; however, the reason for this is unclear. One child case with posttransplant HHV-6 encephalitis demonstrated posterior reversible encephalopathy syndrome (Kawamura et al. 2012). It has been suggested that HHV-6 can spread to the limbic area via the olfactory glove route (Harberts et al. 2011).

The characteristic symptoms are delirium and memory loss in combination with convulsion, which are different from the symptoms observed in HHV-6 encephalitis patients at the time of primary viral infection. Additionally, the patient's cerebrospinal fluid demonstrates generally mild pleocytosis and elevation of protein and contains measurably higher amounts of HHV-6 DNA than an HHV-6 encephalitis patient at the time of primary viral infection (Kawamura et al. 2011). These findings suggest that pathogenesis of HHV-6B encephalitis may be different in a patient with a primary infection and a patient with viral reactivation after HSCT. As suggested, a direct invasion of HHV-6B into brain tissue may play an important role in the pathogenesis of posttransplant HHV-6B encephalitis, supporting the idea that administration of antiviral drugs is an important component of post-HSCT treatment. Neuroimaging analysis, in particular brain MRI, can be useful in the diagnosis of PALE. HHV-6 encephalitis typically shows an MRI signature of hyperintense lesions on T2-weighted, fluid-attenuated inversion recovery imaging and diffusion-weighted imaging of bilateral medial temporal lobes, primarily affecting the hippocampus and amygdala (Seeley et al. 2007). The high morbidity and mortality of this complication are significant problems from a clinical standpoint.

12.3.2   GVHD and GVHD-Like Skin Rash

As described above, HHV-6 infection frequently occurs 2–4 weeks after transplant, which correspond to the timing of engraftment and acute GVHD. Therefore, several investigators have demonstrated an association between HHV-6 infection and acute GVHD (Wang et al. 2008; Jeulin et al. 2013; Brands-Nijenhuis et al. 2011; Aoki et al. 2015; Wilborn et al. 1994) and acute GVHD-like skin rash (Asano et al. 1991; Yoshikawa et al. 1991, 2001, 2002b). Pathological analysis also supports the correlation between HHV-6 and acute GVHD (Appleton et al. 1995). It has been demonstrated that HHV-6 infection induces proinflammatory cytokines in HSCT recipients, which may play an important role in the pathogenesis of acute GVHD (Fujita et al. 2008).

12.3.3   Others

It has been demonstrated that HHV-6 infection increases the risk of other opportunistic infections, including CMV and fungal infections (Aoki et al. 2015), and non-relapse mortality (Zerr et al. 2012). An association between HHV-6 infection and an increase in patient mortality has also been demonstrated in solid organ transplant recipients. Because HHV-6 is a lymphotropic virus, it has been suggested that viral infection can interfere with host immunity. HHV-6 infection may suppress the host immune response in HSCT recipient, which could result in an increase in the frequency of opportunistic infections.

12.4 Clinical Manifestations of HHV-7 Infection

Because HHV-7 has many similarities to HHV-6 and CMV, the possible role of HHV-7 infection in HSCT recipients has been examined by several groups. However, precise clinical manifestations of HHV-7 infection after transplant remain unclear. HHV-7 has been demonstrated as one of the exacerbating factors of CMV disease (Chapenko et al. 2000), and several cases of HHV-7 encephalitis have been reported in HSCT recipients (Yoshikawa et al. 2003; Ward et al. 2002). Like HHV-6, HHV-7 is a ubiquitous virus in humans that is excreted in the saliva of most seropositive adults. HHV-7 DNA has been detected by PCR in almost half of seropositive healthy adults, which makes it difficult to assess active HHV-7 infection in HSCT recipients based on PCR analysis (Boutolleau et al. 2003).

12.5 Diagnosis

12.5.1 Viral Isolation and Serological Assays

Betaherpesviruses can latently infect various cell types, including mononuclear cells. This makes it necessary to distinguish between active and latent viral infection. Although viral isolation from a clinical specimen and serological assay are reliable indicators of active viral infection for CMV, HHV-6B, and HHV-7, the slow growth of these viruses in vitro is a major obstacle to their rapid diagnosis. Additionally, cord blood mononuclear cells are preferable for HHV-6B isolation, which can be problematic to perform in hospital laboratory. The shell vial assay that detects early antigen fluorescent foci after rapid culture of CMV is not sensitive enough for monitoring peripheral blood; however, this assay is useful for the diagnosis of CMV pneumonitis by detection of CMV from bronchoalveolar lavage fluid (Cathomas et al. 1993; Crawford et al. 1988).

Similar to viral isolation, a serological assay is not appropriate for rapid diagnosis, and cross-reaction between HHV-6B and HHV-7 antibodies interferes with the correct diagnosis. Additionally, an impaired host immune response due to immunosuppressive treatments may prevent a robust antibody response in HSCT recipients. Pre-transplant donor and recipient serostatus of CMV and HHV-6 is important for the evaluation of risk of viral infection after transplant (Broers et al. 2000).

12.5.2 Antigenemia Assay

Antigenemia assays have been used for diagnosis of CMV and HHV-6B infection. To diagnose CMV, a monoclonal antibody against the CMV tegument protein (pp65) is generally used. Although detection of viral antigens in peripheral blood is a good indication of active viral infection for these two viruses, assay sensitivity may be insufficient, and counting for antigen-positive cells is cumbersome in comparison to molecular diagnostic methods. In contrast to the CMV assay, the HHV-6 antigenemia assay has not been widely used but is considered to be reliable mainly in solid organ transplant recipients (Lautenschlager et al. 2002).

12.5.3 Molecular Diagnostic Assays

Real-time polymerase chain reaction (PCR) is the most sensitive method for detecting CMV and HHV-6 and is the most widely used method. As described above, these viruses can latently infect peripheral blood mononuclear cells (Kondo et al. 1991; Suga et al. 1998). Therefore, quantitative analysis of viral DNA in peripheral blood samples is necessary to distinguish between active and latent viral infection. Multiplex real-time PCR targeting CMV, HHV-6, and Epstein-Barr virus – an additional significant pathogen for opportunistic infections – is very useful for monitoring viral infections simultaneously (Wada et al. 2007). In addition to peripheral blood or serum samples, bronchoalveolar lavage and cerebrospinal fluid can be analyzed using this method, making it a reliable method for the diagnosis of pneumonia and encephalitis. Thus, real-time PCR is an important tool for informing preemptive treatment. A major problem of this method has been the determination and standardization of a threshold level for discrimination of active viral infection from latency (Gimenez et al. 2014a, b; Lilleri et al. 2007; Griffiths et al. 2016; Emery et al. 2000; Boeckh et al. 2004). Various types of molecular diagnostic methods, sampling schedules, clinical specimens, and cutoff copy numbers are used in several different transplant units in Spain (Solano et al. 2015). In CMV infection, it is recommended that a calibration of copy numbers determined by in-house real-time PCR should be performed according to the World Health Organization International Standard (Fryer et al. 2016). Recent data demonstrates that a cutoff level of 500 international units of CMV DNA per mL of bronchoalveolar lavage fluid is reliable for differentiation between CMV pneumonia and pulmonary shedding without CMV pneumonia (Boeckh et al. 2017).

Similar to CMV, HHV-6 real-time PCR has been widely used for monitoring active HHV-6 infection in HSCT recipients. Species-specific (HHV-6A and HHV-6B) real-time PCR is recommended. Because HHV-6 can latently infect mononuclear lymphocytes, if peripheral blood is used as the clinical specimen, the kinetics of viral load should be monitored to evaluate active viral infection. It may be difficult to determine a cutoff level for HHV-6 DNA to discriminate active viral infection from latency, because differences in assay systems and tested samples may influence cutoff levels. An international collaborative study is now underway to determine international units for HHV-6 DNA load.

A potential diagnostic pitfall of real-time PCR is chromosomally integrated HHV-6 (ciHHV-6). A ciHHV-6 subject has at least one copy of HHV-6 DNA in every cell. This can result in the persistent detection of unusually high copy numbers of viral DNA in peripheral blood and plasma, potentially leading to the misdiagnosis of active HHV-6 infection and unnecessary antiviral drug administration (Clark and Ward 2008).

Monitoring of viral mRNA is another strategy for diagnosis of active viral infection. Although viral DNAs are persistently detected in latently infected cells such as peripheral blood mononuclear cells, viral mRNAs are generally expressed only during active viral replication. Several real-time RT-PCR assays for the detection of either CMV (Hebart et al. 2011; Gerna et al. 2003) or HHV-6 mRNAs have been developed and demonstrated to be reliable for the diagnosis of active viral infection (Pradeau et al. 2006; Ihira et al. 2012).

12.5.4 The Importance of Pathological Analysis

Both the evidence for active viral infection and confirmation of end-organ disease caused by CMV infection are required for the diagnosis of (what we termed) CMV disease. Identification of viral antigen or inclusion bodies in tissue samples obtained from the involved organ on the basis of pathological analysis is the gold standard for diagnosis of end-organ disease due to CMV infection. However, biopsy is not always a feasible option for HSCT recipients.

12.5.5 Caution for Chromosomally Integrated HHV-6

Primary HHV-6B infection typically occurs in infancy and causes exanthema subitum, a common febrile exanthematous disease. Although horizontal transmission, in particular from parent to child, of HHV-6 is considered to be the main route of viral infection, it has been demonstrated that HHV-6 was genetically transmitted from parent to child as an inherited chromosomally integrated human herpesvirus 6 (iciHHV-6) (Kaufer and Flamand 2014). As seen in individuals latently infected with HHV-6B after primary viral infection, it has been suggested that viral reactivation from integrated HHV-6 genome also occurs in immunocompromised patients (Endo et al. 2014; Sedlak et al. 2016) and pregnant women (Hall et al. 2010). Certain chemical compounds appear to be able to induce viral reactivation from ciHHV-6 cells in vitro.

ciHHV-6 individuals demonstrate extremely high copy numbers of viral DNA, generally over 5.5 $\log_{10}$ copies/mL of whole blood. Transient elevation of HHV-6B DNA load in serum occurs upon reactivation of a primary HHV-6B infection. These high viral loads can lead to the misdiagnosis of an active HHV-6 infection, resulting in the unnecessary administration of antiviral drugs (Clark and Ward 2008; Hubacek et al. 2009). The kinetics of HHV-6 DNA loads after transplant are different between a ciHHV-6 donor and ciHHV-6 recipient (Jeulin et al. 2009; Miura et al. 2015). If the donor is ciHHV-6, asymptomatic elevation of HHV-6 DNA corresponding to the timing of donor engraftment is observed. These elevated viral levels will not respond to antiviral treatments. Therefore, if a physician observes extremely high copy numbers of HHV-6 DNA in clinical specimens from HSCT recipients, measurement of HHV-6 DNA load in hair follicle or buccal swab samples should be carried out to exclude ciHHV-6.

12.6 Treatment

12.6.1 General Considerations for Prevention of Viral Infection

CMV serostatus of both donor and recipient is important for evaluating the risk of CMV diseases after HSCT. Blood products obtained from seronegative individuals should be used for CMV-seronegative recipients for the prevention of primary CMV infections via blood products (Nichols et al. 2003). Where available, a CMV-seronegative donor is advised; however because the selection of donor is affected by various other factors such as HLA-matching, it may be difficult to identify a CMV-seronegative donor. Although the serostatus of HHV-6 and HHV-7 should be tested in both the recipient and donor, most individuals older than 1 year are considered to be seropositive for these two viruses.

Although donor CMV serostatus requirements for CMV-seropositive recipients have been less well characterized, a recent large case analysis using the European Group for Blood and Marrow Transplantation database demonstrated that seropositive recipients who had seropositive, unrelated donors had improved overall survival compared with seropositive recipients who had seronegative donors. This effect was observed in recipients receiving myeloablative conditioning (Ljungman et al. 2014), supporting the idea that a CMV-seropositive donor should be selected for CMV-seropositive recipients who received myeloablative conditioning.

Serological screening and DNA amplification assays for HHV-6 and HHV-7 are not generally carried out in the preparation of blood products.

12.6.2 Prophylaxis and Preemptive Treatment for CMV Infection

There are two different antiviral drug strategies used for the prevention of CMV infection in HSCT patients: prophylactic administration of antiviral drugs to all patients or preemptive administration based on the early detection of infection by pp65 antigenemia assay or real-time PCR.

The six antiviral drugs acyclovir, valacyclovir, ganciclovir, valganciclovir, foscarnet, and cidofovir are currently available for prophylactic treatment of CMV infection. Although acyclovir and valacyclovir are considered to have minimal side effects, the other drugs are associated with toxicities including bone marrow suppression and nephrotoxicity.

The efficacy of preemptive treatment depends on the ability to reliably detect CMV during early infection using a pp65 antigenemia or real-time PCR assay. Assays with high sensitivity may result in the unnecessary administration of antiviral drugs, increasing the cost of treatment and toxicity risks, whereas assays with low sensitivity may result in inappropriate delays in treatment. Therefore, a guideline for the detection of CMV infection has been established. It is recommended that patients be monitored for antigenemia or DNAemia once a week from days 10 to 100 post-HSCT. The most widely used criteria for cutoff is 1000 copies/ml of CMV DNA or a fivefold increase over baseline levels as the threshold level for starting preemptive treatments. Ganciclovir is the most commonly used antiviral for preemptive treatment, followed by foscarnet and cidofovir. Ganciclovir should be administered for 2 weeks or until the viral load declines to below the detection limit. Because ganciclovir can cause neutropenia and thrombocytopenia, foscarnet may be preferable for recipients with severe myelosuppression. Nephrotoxicity may become a limitation for the use of foscarnet. In order to reduce the side effects of ganciclovir, the use of half doses has been examined in HSCT recipients with low levels of CMV infection (Ju et al. 2016). If no significant decrease in CMV DNA load or antigenemia was observed after 2 weeks of antiviral drug administration, the emergence of antiviral drug resistance should be considered. Foscarnet or cidofovir can be used for preemptive treatment instead of ganciclovir; however cidofovir is also associated with bone marrow suppression and nephrotoxicity.

12.6.3 Prophylaxis and Preemptive Treatment of HHV-6 Infection

The complications of posttransplant HHV-6 encephalitis patients are significant, making preemptive treatment protocols for this virus especially desirable. Although peripheral blood and plasma have been used as clinical specimens in real-time PCR assays, no appropriate protocol has been delineated for identifying preemptive treatment for HHV-6 infection in HSCT recipients (Ogata et al. 2008). Prophylactic treatments using ganciclovir and foscarnet have been studied (Takenaka et al. 2015; Ogata et al. 2013a; Tokimasa et al. 2002); however, no significant efficacy of these prophylactic treatments has been demonstrated to date. An appropriate duration and dose of antiviral drugs are now under investigation.

12.6.4 Treatment

The standard treatment of CMV diseases after HSCT consists of ganciclovir at induction doses for 2–3 weeks and maintenance doses until signs and symptoms have disappeared. When cytopenia is present, foscarnet can be used. Additional intravenous immunoglobulin administration has been suggested in patients with CMV pneumonia (Boeckh 2011); however, no treatment effect of additional immunoglobulin administration has been demonstrated in patients with other types of complications caused by CMV infection. As described above, if no significant decrease in CMV DNA load or antigenemia is observed after 2 weeks of antiviral drug administration, the emergence of antiviral drug resistance should be considered. To combat drug resistance, different types of antiviral drugs for CMV infection are needed. Furthermore, drug toxicity is a major limitation of most of the antiviral drugs currently in use for CMV infection. Thus, there is much interest in developing new anti-CMV drugs with low toxicity. Clinical trials of several new antiviral drugs such as maribavir, letermovir, and brincidofovir are now underway (Boeckh et al. 2015).

Ganciclovir, foscarnet, and cidofovir have been shown to have an antiviral effect on HHV-6 based on in vitro analysis. No randomized control study has been conducted to evaluate the clinical effect of these drugs; however, a limited number of case reports support the idea that they have antiviral effects in HSCT patients with HHV-6 encephalitis (Tokimasa et al. 2002; Denes et al. 2004). These patients experienced an improvement of clinical symptoms and a decrease in HHV-6 DNA load in cerebrospinal fluid (Hirabayashi et al. 2013). Meanwhile, an emergence of ganciclovir-resistant HHV-6 has been reported (Imataki and Uemura 2015).

12.6.5 Vaccine

As described above, recipients from donors who are CMV seropositive are at high risk for CMV reactivation. Host immunity induced by CMV vaccine may suppress viral reactivation. CMVPepVax is a chimeric peptide vaccine containing a cytotoxic T-cell epitope from CMV pp65 combined with a tetanus T-helper epitope and a Toll-like receptor 9 agonist as an adjuvant. Phase 1b clinical trials of this vaccine have shown no significant impact on the occurrence of adverse events, including GVHD, between the vaccine group and control (Nakamura et al. 2016)3. No vaccine has been developed for preventing HHV-6 and HHV-7 infections.

12.7 Conclusion

Although advancements in diagnostic methods and antiviral treatments have improved the prognosis of patients with betaherpesviruses infections, these viral infections remain a significant problem in HSCT recipients. In addition to the already established clinical manifestations such as CMV pneumonia and HHV-6 encephalitis, unidentified clinical manifestations caused by betaherpesviruses may be discovered in a future clinical study. Furthermore, development of new diagnostic methods with high sensitivity and specificity and new treatment strategies including new antiviral drugs are necessary for improved outcomes in HSCT patients.

References

Abate D, Cesaro S, Cofano S, Fiscon M, Saldan A, Varotto S, Mengoli C, Pillon M, Calore E, Biasolo MA, Cusinato R, Barzon L, Messina C, Carli M, Palu G (2012) Diagnostic utility of human cytomegalovirus-specific T-cell response monitoring in predicting viremia in pediatric allogeneic stem-cell transplant patients. Transplantation 93:536–542

Aoki J, Numata A, Yamamoto E, Fujii E, Tanaka M, Kanamori H (2015) Impact of human Herpesvirus-6 reactivation on outcomes of allogeneic hematopoietic stem cell transplantation. Biol Blood Marrow Transplant 21:2017–2022

Appleton AL, Sviland L, Peiris JS, Taylor CE, Wilkes J, Green MA, Pearson AD, Kelly PJ, Malcolm AJ, Proctor SJ et al (1995) Human herpes virus-6 infection in marrow graft recipients: role in pathogenesis of graft-versus-host disease. Newcastle upon Tyne bone marrow transport group. Bone Marrow Transplant 16:777–782

Asano Y, Yoshikawa T, Suga S, Yazaki T, Kondo k, Yamanishi K (1990) Fatal fulminant hepatitis in an infant with human herpesvirus-6 infection. Lancet 335:862–863

Asano Y, Yoshikawa T, Suga S, Nakashima T, Yazaki T, Fukuda M, Kojima S, Matsuyama T (1991) Reactivation of herpesvirus type 6 in children receiving bone marrow transplants for leukemia. N Engl J Med 324:634–635

Asano Y, Yoshikawa T, Suga S, Kobayashi I, Nakashima T, Yazaki T, kajita Y, Ozaki T (1994) Clinical features of infants with primary human herpesvirus 6 infection (exanthem subitum, roseola infantum). Pediatrics 93:104–108

Aubert G, Hassan-Walker AF, Madrigal JA, Emery VC, Morte C, Grace S, Koh MB, Potter M, Prentice HG, Dodi IA, Travers PJ (2001) Cytomegalovirus-specific cellular immune responses and viremia in recipients of allogeneic stem cell transplants. J Infect Dis 184:955–963

Avetisyan G, Aschan J, Hagglund H, Ringden O, Ljungman P (2007) Evaluation of intervention strategy based on CMV-specific immune responses after allogeneic SCT. Bone Marrow Transplant 40:865–869

Belford A, Myles O, Magill A, Wang J, Myhand RC, Waselenko JK (2004) Thrombotic microangiopathy (TMA) and stroke due to human herpesvirus-6 (HHV-6) reactivation in an adult receiving high-dose melphalan with autologous peripheral stem cell transplantation. Am J Hematol 76:156–162

Bhutani D, Dyson G, Manasa R, Deol A, Ratanatharathorn V, Ayash L, Abidi M, Lum LG, Al-Kadhimi Z, Uberti JP (2015) Incidence, risk factors, and outcome of cytomegalovirus viremia and gastroenteritis in patients with gastrointestinal graft-versus-host disease. Biol Blood Marrow Transplant 21:159–164

Boeckh M (2011) Complications, diagnosis, management, and prevention of CMV infections: current and future. Hematology Am Soc Hematol Educ Program 2011:305–309

Boeckh M, Leisenring W, Riddell SR, Bowden RA, Huang ML, Myerson D, Stevens-Ayers T, Flowers ME, Cunningham T, Corey L (2003) Late cytomegalovirus disease and mortality in recipients of allogeneic hematopoietic stem cell transplants: importance of viral load and T-cell immunity. Blood 101:407–414

Boeckh M, Huang M, Ferrenberg J, Stevens-Ayers T, Stensland L, Nichols WG, Corey L (2004) Optimization of quantitative detection of cytomegalovirus DNA in plasma by real-time PCR. J Clin Microbiol 42:1142–1148

Boeckh M, Murphy WJ, Peggs KS (2015) Reprint of: recent advances in cytomegalovirus: an update on pharmacologic and cellular therapies. Biol Blood Marrow Transplant 21:S19–S24

Boeckh M, Stevens-Ayers T, Travi G, Huang ML, Cheng GS, Xie H, Leisenring W, Erard V, Seo S, Kimball L, Corey L, Pergam SA, Jerome KR (2017) Bronchoalveolar lavage DNA quantification in cytomegalovirus pneumonia after hematopoietic cell transplantation. J Infect Dis 215:1514–1522

Boutolleau D, Fernandez C, Andre E, Imbert-Marcille BM, Milpied N, Agut H, Gautheret-Dejean A (2003) Human herpesvirus (HHV)-6 and HHV-7: two closely related viruses with different infection profiles in stem cell transplantation recipients. J Infect Dis 187:179–186

Brands-Nijenhuis AV, Van Loo IH, Schouten HC, Van Gelder M (2011) Temporal relationship between HHV 6 and graft vs host disease in a patient after haplo-identical SCT and severe T-cell depletion. Bone Marrow Transplant 46:1151–1152

Broers AE, Van Der Holt R, Van Esser JW, Gratama JW, Henzen-Logmans S, Kuenen-Boumeester V, Lowenberg B, Cornelissen JJ (2000) Increased transplant-related morbidity and mortality in CMV-seropositive patients despite highly effective prevention of CMV disease after allogeneic T-cell-depleted stem cell transplantation. Blood 95:2240–2245

Carrigan DR, Knox KK (1994) Human herpesvirus 6 (HHV-6) isolation from bone marrow: HHV-6-associated bone marrow suppression in bone marrow transplant patients. Blood 84:3307–3310

Cathomas G, Morris P, Pekle K, Cunningham I, Emanuel D (1993) Rapid diagnosis of cytomegalovirus pneumonia in marrow transplant recipients by bronchoalveolar lavage using the polymerase chain reaction, virus culture, and the direct immunostaining of alveolar cells. Blood 81:1909–1914

Chapenko S, Folkmane I, Tomsone V, Amerika D, Rozentals R, Murovska M (2000) Co-infection of two beta-herpesviruses (CMV and HHV-7) as an increased risk factor for 'CMV disease' in patients undergoing renal transplantation. Clin Transpl 14:486–492

Clark DA, Ward KN (2008) Importance of chromosomally integrated HHV-6A and -6B in the diagnosis of active HHV-6 infection. Herpes 15:28–32

Crawford SW, Bowden RA, Hackman RC, Gleaves CA, Meyers JD, Clark JG (1988) Rapid detection of cytomegalovirus pulmonary infection by bronchoalveolar lavage and centrifugation culture. Ann Intern Med 108:180–185

Crippa F, Corey L, Chuang EL, Sale G, Boeckh M (2001) Virological, clinical, and ophthalmologic features of cytomegalovirus retinitis after hematopoietic stem cell transplantation. Clin Infect Dis 32:214–219

De Labarthe A, Gauthert-Dejean A, Bossi P, Vernant JP, Dhedin N (2005) HHV-6 variant A meningoencephalitis after allogeneic hematopoietic stem cell transplantation diagnosed by quantitative real-time polymerase chain reaction. Transplantation 80:539

Denes E, Magy L, Pradeau K, Alain S, Weinbreck P, Ranger-Rogez S (2004) Successful treatment of human herpesvirus 6 encephalomyelitis in immunocompetent patient. Emerg Infect Dis 10:729–731

Donati D, Martinelli E, Cassiani-Ingoni R, Ahlqvist J, Hou J, Major EO, Jacobson S (2005) Variant-specific tropism of human herpesvirus 6 in human astrocytes. J Virol 79:9439–9448

Drobyski WR, Dunne WM, Burd EM, Knox KK, Ash RC, Horowitz MM, Flomenberg N, Carrigan DR (1993) Human herpesvirus-6 (HHV-6) infection in allogeneic bone marrow transplant recipients: evidence of a marrow-suppressive role for HHV-6 in vivo. J Infect Dis 167:735–739

Eid AJ, Bakri SJ, Kijpittayarit S, Razonable RR (2008) Clinical features and outcomes of cytomegalovirus retinitis after transplantation. Transpl Infect Dis 10:13–18

Einsele H, Hebart H, Kauffmann-Schneider C, Sinzger C, Jahn G, Bader P, Klingebiel T, Dietz K, Loffler J, Bokemeyer C, Muller CA, Kanz L (2000) Risk factors for treatment failures in patients receiving PCR-based preemptive therapy for CMV infection. Bone Marrow Transplant 25:757–763

Elmaagacli AH, Steckel NK, Koldehoff M, Hegerfeldt Y, Trenschel R, Ditschkowski M, Christoph S, Gromke T, Kordelas L, Ottinger HD, Ross RS, Horn PA, Schnittger S, Beelen DW (2011) Early human cytomegalovirus replication after transplantation is associated with a decreased relapse risk: evidence for a putative virus-versus-leukemia effect in acute myeloid leukemia patients. Blood 118:1402–1412

Emery VC, Sabin CA, Cope AV, Gor D, Hassan-Walker AF, Griffiths PD (2000) Application of viral-load kinetics to identify patients who develop cytomegalovirus disease after transplantation. Lancet 355:2032–2036

Endo A, Watanabe K, Ohye T, Suzuki K, Matsubara T, Shimizu N, Kurahashi H, Yoshikawa T, Katano H, Inoue N, Imai K, Takagi M, Morio T, Mizutani S (2014) Molecular and virological evidence of viral activation from chromosomally integrated human herpesvirus 6A in a patient with X-linked severe combined immunodeficiency. Clin Infect Dis 59:545–548

Epstein LG, Shinnar S, Hesdorffer DC, Nordli DR, Hamidullah A, Benn EK, Pellock JM, Frank LM, Lewis DV, Moshe SL, Shinnar RC, Sun S, Team FS (2012) Human herpesvirus 6 and 7 in febrile status epilepticus: the FEBSTAT study. Epilepsia 53:1481–1488

Erard V, Guthrie KA, Seo S, Smith J, Huang M, Chien J, Flowers ME, Corey L, Boeckh M (2015) Reduced mortality of cytomegalovirus pneumonia after hematopoietic cell transplantation due to antiviral therapy and changes in transplantation practices. Clin Infect Dis 61:31–39

Fryer JF, Heath AB, Minor PD, Collaborative Study Group (2016) A collaborative study to establish the 1st WHO International standard for human cytomegalovirus for nucleic acid amplification technology. Biologicals 44:242–251

Fujino M, Ohashi M, Tanaka K, Kato T, Asano Y, Yoshikawa T (2012) Rhabdomyolysis in an infant with primary human herpesvirus 6 infection. Pediatr Infect Dis J 31:1202–1203

Fujita A, Ihira M, Suzuki R, Enomoto Y, Sugiyama H, Sugata K, Suga S, Asano Y, Yagasaki H, Kojima S, Matsumoto K, Kato K, Yoshikawa T (2008) Elevated serum cytokine levels are associated with human herpesvirus 6 reactivation in hematopoietic stem cell transplantation recipients. J Infect 57:241–248

Gerna G, Lilleri D, Baldanti F, Torsellini M, Giorgiani G, Zecca M, De Stefano P, Middeldorp J, Locatelli F, Revello MG (2003) Human cytomegalovirus immediate-early mRNAemia versus pp65 antigenemia for guiding pre-emptive therapy in children and young adults undergoing hematopoietic stem cell transplantation: a prospective, randomized, open-label trial. Blood 101:5053–5060

Gimenez E, Munoz-Cobo B, Solano C, Amat P, Navarro D (2014a) Early kinetics of plasma cytomegalovirus DNA load in allogeneic stem cell transplant recipients in the era of highly sensitive real-time PCR assays: does it have any clinical value? J Clin Microbiol 52:654–656

Gimenez E, Solano C, Azanza JR, Amat P, Navarro D (2014b) Monitoring of trough plasma ganciclovir levels and peripheral blood cytomegalovirus (CMV)-specific CD8+ T cells to predict CMV DNAemia clearance in preemptively treated allogeneic stem cell transplant recipients. Antimicrob Agents Chemother 58:5602–5605

Green ML, Leisenring WM, Xie H, Walter RB, Mielcarek M, Sandmaier BM, Riddell SR, Boeckh M (2013) CMV reactivation after allogeneic HCT and relapse risk: evidence for early protection in acute myeloid leukemia. Blood 122:1316–1324

Griffiths PD, Rothwell E, Raza M, Wilmore S, Doyle T, Harber M, O'Beirne J, Mackinnon S, Jones G, Thorburn D, Mattes F, Nebbia G, Atabani S, Smith C, Stanton A, Emery VC (2016) Randomized controlled trials to define viral load thresholds for cytomegalovirus pre-emptive therapy. PLoS One 11:e0163722

Hall CB, Caserta MT, Schnabel KC, Shelley LM, Carnahan JA, Marino AS, Yoo C, Lofthus GK (2010) Transplacental congenital human herpesvirus 6 infection caused by maternal chromosomally integrated virus. J Infect Dis 201:505–507

Harberts E, Yao K, Wohler JE, Maric D, Ohayon J, Henkin R, Jacobson S (2011) Human herpesvirus-6 entry into the central nervous system through the olfactory pathway. Proc Natl Acad Sci USA 108:13734–13739

He JD, Liu YL, Wang ZF, Liu DH, Chen H, Chen YH (2008) Colonoscopy in the diagnosis of intestinal graft versus host disease and cytomegalovirus enteritis following allogeneic haematopoietic stem cell transplantation. Chin Med J 121:1285–1289

Hebart H, Lengerke C, Ljungman P, Paya CV, Klingebiel T, Loeffler J, Pfaffenrath S, Lewensohn-Fuchs I, Barkholt L, Tomiuk J, Meisner C, Lunenberg J, Top B, Razonable RR, Patel R, Litzow MR, Jahn G, Einsele H (2011) Prospective comparison of PCR-based vs late mRNA-based preemptive antiviral therapy for HCMV infection in patients after allo-SCT. Bone Marrow Transplant 46:408–415

Hirabayashi K, Nakazawa Y, Katsuyama Y, Yanagisawa T, Saito S, Yoshikawa K, Shigemura T, Sakashita K, Ichikawa M, Koike K (2013) Successful ganciclovir therapy in a patient with human herpesvirus-6 encephalitis after unrelated cord blood transplantation: usefulness of longitudinal measurements of viral load in cerebrospinal fluid. Infection 41:219–223

Hiwarkar P, Gajdosova E, Qasim W, Worth A, Breuer J, Chiesa R, Ridout D, Edelsten C, Moore A, Amrolia P, Veys P, Rao K (2014) Frequent occurrence of cytomegalovirus retinitis during immune reconstitution warrants regular ophthalmic screening in high-risk pediatric allogeneic hematopoietic stem cell transplant recipients. Clin Infect Dis 58:1700–1706

Hubacek P, Virgili A, Ward KN, Pohlreich D, Keslova P, Goldova B, Markova M, Zajac M, Cinek O, Nacheva EP, Sedlacek P, Cetkovsky P (2009) HHV-6 DNA throughout the tissues of two stem cell transplant patients with chromosomally integrated HHV-6 and fatal CMV pneumonitis. Br J Haematol 145:394–398

Ihira M, Enomoto Y, Kawamura Y, Nakai H, Sugata K, Asano Y, Tsuzuki M, Emi N, Goto T, Miyamura K, Matsumoto K, Kato K, Takahashi Y, Kojima S, Yoshikawa T (2012) Development of quantitative RT-PCR assays for detection of three classes of HHV-6B gene transcripts. J Med Virol 84:1388–1395

Imataki O, Uemura M (2015) Ganciclovir-resistant HHV-6 encephalitis that progressed rapidly after bone marrow transplantation. J Clin Virol 69:176–178

Inagaki J, Noguchi M, Kurauchi K, Tanioka S, Fukano R, Okamura J (2016) Effect of cytomegalovirus reactivation on relapse after allogeneic hematopoietic stem cell transplantation in pediatric acute leukemia. Biol Blood Marrow Transplant 22:300–306

Jeon S, Lee WK, Lee Y, Lee DG, Lee JW (2012) Risk factors for cytomegalovirus retinitis in patients with cytomegalovirus viremia after hematopoietic stem cell transplantation. Ophthalmology 119:1892–1898

Jeulin H, Guery M, Clement L, Salmon A, Beri M, Bordigoni P, Venard V (2009) Chromosomally integrated HHV-6: slow decrease of HHV-6 viral load after hematopoietic stem-cell transplantation. Transplantation 88:1142–1143

Jeulin H, Agrinier N, Guery M, Salmon A, Clement L, Bordigoni P, Venard V (2013) Human herpesvirus 6 infection after allogeneic stem cell transplantation: incidence, outcome, and factors associated with HHV-6 reactivation. Transplantation 95:1292–1298

Ju HY, Kang HJ, Hong CR, Lee JW, Kim H, Park KD, Shin HY, Park JD, Choi EH, Lee HJ, Ahn HS (2016) Half-dose ganciclovir preemptive treatment of cytomegalovirus infection after pediatric allogeneic hematopoietic stem cell transplantation. Transpl Infect Dis 18:396–404

Kadakia MP, Rybka WB, Stewart JA, Patton JL, Stamey FR, Elsawy M, Pellett PE, Armstrong JA (1996) Human herpesvirus 6: infection and disease following autologous and allogeneic bone marrow transplantation. Blood 87:5341–5354

Kaufer BB, Flamand L (2014) Chromosomally integrated HHV-6: impact on virus, cell and organismal biology. Curr Opin Virol 9:111–118

Kawamura Y, Sugata K, Ihira M, Mihara T, Mutoh T, Asano Y, Yoshikawa T (2011) Different characteristics of human herpesvirus 6 encephalitis between primary infection and viral reactivation. J Clin Virol 51:12–19

Kawamura Y, Ohashi M, Asahito H, Takahashi Y, Kojima S, Yoshikawa T (2012) Posterior reversible encephalopathy syndrome in a child with post-transplant HHV-6B encephalitis. Bone Marrow Transplant 47:1381–1382

Kawamura Y, Nakai H, Sugata K, Asano Y, Yoshikawa T (2013) Serum biomarker kinetics with three different courses of HHV-6B encephalitis. Brain Dev 35:590–595

Kondo K, Kondo T, Okuno T, Takahashi M, Yamanishi K (1991) Latent human herpesvirus 6 infection of human monocytes/macrophages. J Gen Virol 72(Pt 6):1401–1408

Konoplev S, Champlin RE, Giralt S, Ueno NT, Khouri I, Raad I, Rolston K, Jacobson K, Tarrand J, Luna M, Nguyen Q, Whimbey E (2001) Cytomegalovirus pneumonia in adult autologous blood and marrow transplant recipients. Bone Marrow Transplant 27:877–881

Lautenschlager I, Lappalainen M, Linnavuori K, Suni J, Hockerstedt K (2002) CMV infection is usually associated with concurrent HHV-6 and HHV-7 antigenemia in liver transplant patients. J Clin Virol 25(Suppl 2):S57–S61

Lee SM, Kim YJ, Yoo KH, Sung KW, Koo HH, Kang ES (2017) Clinical usefulness of monitoring cytomegalovirus-specific immunity by Quantiferon-CMV in pediatric allogeneic hematopoietic stem cell transplantation recipients. Ann Lab Med 37:277–281

Li CR, Greenberg PD, Gilbert MJ, Goodrich JM, Riddell SR (1994) Recovery of HLA-restricted cytomegalovirus (CMV)-specific T-cell responses after allogeneic bone marrow transplant: correlation with CMV disease and effect of ganciclovir prophylaxis. Blood 83:1971–1979

Lilleri D, Gerna G, Furione M, Bernardo ME, Giorgiani G, Telli S, Baldanti F, Locatelli F (2007) Use of a DNAemia cut-off for monitoring human cytomegalovirus infection reduces the number of preemptively treated children and young adults receiving hematopoietic stem-cell transplantation compared with qualitative pp65 antigenemia. Blood 110:2757–2760

Ljungman P, Perez-Bercoff L, Jonsson J, Avetisyan G, Sparrelid E, Aschan J, Barkholt L, Larsson K, Winiarski J, Yun Z, Ringden O (2006) Risk factors for the development of cytomegalovirus disease after allogeneic stem cell transplantation. Haematologica 91:78–83

Ljungman P, De La Camara R, Cordonnier C, Einsele H, Engelhard D, Reusser P, Styczynski J, Ward K, European Conference on Infections in Leukemia (2008) Management of CMV, HHV-6, HHV-7 and Kaposi-sarcoma herpesvirus (HHV-8) infections in patients with hematological malignancies and after SCT. Bone Marrow Transplant 42:227–240

Ljungman P, Hakki M, Boeckh M (2011) Cytomegalovirus in hematopoietic stem cell transplant recipients. Hematol Oncol Clin North Am 25:151–169

Ljungman P, Brand R, Hoek J, De La Camara R, Cordonnier C, Einsele H, Styczynski J, Ward KN, Cesaro S, Infectious Diseases Working Party of The European Group for Blood and Marrow Transplantation (2014) Donor cytomegalovirus status influences the outcome of allogeneic stem cell transplant: a study by the European group for blood and marrow transplantation. Clin Infect Dis 59:473–481

Miura H, Kawamura Y, Kudo K, Ihira M, Ohye T, Kurahashi H, Kawashima N, Miyamura K, Yoshida N, Kato K, Takahashi Y, Kojima S, Yoshikawa T (2015) Virological analysis of inherited chromosomally integrated human herpesvirus-6 in three hematopoietic stem cell transplant patients. Transpl Infect Dis 17:728–731

Nakamura R, La Rosa C, Longmate J, Drake J, Slape C, Zhou Q, Lampa MG, O'donnell M, Cai JL, Farol L, Salhotra A, Snyder DS, Aldoss I, Forman SJ, Miller JS, Zaia JA, Diamond DJ (2016) Viraemia, immunogenicity, and survival outcomes of cytomegalovirus chimeric epitope vaccine supplemented with PF03512676 (CMVPepVax) in allogeneic haemopoietic stem-cell transplantation: randomised phase 1b trial. Lancet Haematol 3:e87–e98

Nesher L, Shah DP, Ariza-Heredia EJ, Azzi JM, Siddiqui HK, Ghantoji SS, Marsh LY, Michailidis L, Makedonas G, Rezvani K, Shpall EJ, Chemaly RF (2016) Utility of the enzyme-linked Immunospot interferon-gamma-release assay to predict the risk of cytomegalovirus infection in hematopoietic cell transplant recipients. J Infect Dis 213:1701–1707

Nichols WG, Corey L, Gooley T, Davis C, Boeckh M (2002) High risk of death due to bacterial and fungal infection among cytomegalovirus (CMV)-seronegative recipients of stem cell transplants from seropositive donors: evidence for indirect effects of primary CMV infection. J Infect Dis 185:273–282

Nichols WG, Price TH, Gooley T, Corey L, Boeckh M (2003) Transfusion-transmitted cytomegalovirus infection after receipt of leukoreduced blood products. Blood 101:4195–4200

Ogata M, Satou T, Kawano R, Goto K, Ikewaki J, Kohno K, Ando T, Miyazaki Y, Ohtsuka E, Saburi Y, Saikawa T, Kadota JI (2008) Plasma HHV-6 viral load-guided preemptive therapy against HHV-6 encephalopathy after allogeneic stem cell transplantation: a prospective evaluation. Bone Marrow Transplant 41:279–285

Ogata M, Satou T, Kawano R, Takakura S, Goto K, Ikewaki J, Kohno K, Ikebe T, Ando T, Miyazaki Y, Ohtsuka E, Saburi Y, Saikawa T, Kadota J (2010) Correlations of HHV-6 viral load and plasma IL-6 concentration with HHV-6 encephalitis in allogeneic stem cell transplant recipients. Bone Marrow Transplant 45:129–136

Ogata M, Satou T, Inoue Y, Takano K, Ikebe T, Ando T, Ikewaki J, Kohno K, Nishida A, Saburi M, Miyazaki Y, Ohtsuka E, Saburi Y, Fukuda T, Kadota J (2013a) Foscarnet against human herpesvirus (HHV)-6 reactivation after allo-SCT: breakthrough HHV-6 encephalitis following antiviral prophylaxis. Bone Marrow Transplant 48:257–264

Ogata M, Satou T, Kadota J, Saito N, Yoshida T, Okumura H, Ueki T, Nagafuji K, Kako S, Uoshima N, Tsudo M, Itamura H, Fukuda T (2013b) Human herpesvirus 6 (HHV-6) reactivation and HHV-6 encephalitis after allogeneic hematopoietic cell transplantation: a multicenter, prospective study. Clin Infect Dis 57:671–681

Portolani M, Cermelli C, Meacci M, Pietrosemoli P, Pecorari M, Sabbatini AM, Leoni S, Guerra A, Bonaccorsi G (1997) Primary infection by HHV-6 variant B associated with a fatal case of hemophagocytic syndrome. New Microbiol 20:7–11

Pradeau K, Bordessoule D, Szelag JC, Rolle F, Ferrat P, Le Meur Y, Turlure P, Denis F, Ranger-Rogez S (2006) A reverse transcription-nested PCR assay for HHV-6 mRNA early transcript detection after transplantation. J Virol Methods 134:41–47

Roback JD (2002) CMV and blood transfusions. Rev Med Virol 12:211–219

Sedlak RH, Hill JA, Nguyen T, Cho M, Levin G, Cook L, Huang ML, Flamand L, Zerr DM, Boeckh M, Jerome KR (2016) Detection of human herpesvirus 6B (HHV-6B) reactivation in hematopoietic cell transplant recipients with inherited chromosomally integrated HHV-6A by droplet digital PCR. J Clin Microbiol 54:1223–1227

Seeley WW, Marty FM, Holmes TM, Upchurch K, Soiffer RJ, Antin JH, Baden LR, Bromfield EB (2007) Post-transplant acute limbic encephalitis: clinical features and relationship to HHV6. Neurology 69:156–165

Solano C, De La Camara R, Vazquez L, Lopez J, Gimenez E, Navarro D (2015) Cytomegalovirus infection management in allogeneic stem cell transplant recipients: a national survey in Spain. J Clin Microbiol 53:2741–2744

Song WK, Min YH, Kim YR, Lee SC (2008) Cytomegalovirus retinitis after hematopoietic stem cell transplantation with alemtuzumab. Ophthalmology 115:1766–1770

Suga S, Yoshikawa T, Asano Y, Kozawa T, Nakashima T, Kobayashi I, Yazaki T, Yamamoto H, Kajita Y, Ozaki T et al (1993) Clinical and virological analyses of 21 infants with exanthem subitum (roseola infantum) and central nervous system complications. Ann Neurol 33:597–603

Suga S, Yoshikawa T, Kajita Y, Ozaki T, Asano Y (1998) Prospective study of persistence and excretion of human herpesvirus-6 in patients with exanthem subitum and their parents. Pediatrics 102:900–904

Suga S, Suzuki K, Ihira M, Yoshikawa T, Kajita Y, Ozaki T, Iida K, Saito Y, Asano Y (2000) Clinical characteristics of febrile convulsions during primary HHV-6 infection. Arch Dis Child 82:62–66

Sun YQ, Xu LP, Han TT, Zhang XH, Wang Y, Han W, Wang FR, Wang JZ, Chen H, Chen YH, Yan CH, Chen Y, Liu KY, Huang XJ (2015) Detection of human cytomegalovirus (CMV) DNA in feces has limited value in predicting CMV enteritis in patients with intestinal graft-versus-host disease after allogeneic stem cell transplantation. Transpl Infect Dis 17:655–661

Takenaka K, Nishida T, Asano-Mori Y, Oshima K, Ohashi K, Mori T, Kanamori H, Miyamura K, Kato C, Kobayashi N, Uchida N, Nakamae H, Ichinohe T, Morishima Y, Suzuki R, Yamaguchi T, Fukuda T (2015) Cytomegalovirus reactivation after allogeneic hematopoietic stem cell transplantation is associated with a reduced risk of relapse in patients with acute myeloid leukemia who survived to day 100 after transplantation: the Japan Society for Hematopoietic Cell Transplantation Transplantation-Related Complication Working Group. Biol Blood Marrow Transplant 21:2008–2016

Tokimasa S, Hara J, Osugi Y, Ohta H, Matsuda Y, Fujisaki H, Sawada A, Kim JY, Sashihara J, Amou K, Miyagawa H, Tanaka-Taya K, Yamanishi K, Okada S (2002) Ganciclovir is effective for prophylaxis and treatment of human herpesvirus-6 in allogeneic stem cell transplantation. Bone Marrow Transplant 29:595–598

Van Burik JA, Lawatsch EJ, Defor TE, Weisdorf DJ (2001) Cytomegalovirus enteritis among hematopoietic stem cell transplant recipients. Biol Blood Marrow Transplant 7:674–679

Visser AM, Van Doornum GJ, Cornelissen JJ, Van Den Bent MJ (2005) Severe amnesia due to HHV-6 encephalitis after allogenic stem cell transplantation. Eur Neurol 54:233–234

Wada K, Kubota N, Ito Y, Yagasaki H, Kato K, Yoshikawa T, Ono Y, Ando H, Fujimoto Y, Kiuchi T, Kojima S, Nishiyama Y, Kimura H (2007) Simultaneous quantification of Epstein-Barr virus, cytomegalovirus, and human herpesvirus 6 DNA in samples from transplant recipients by multiplex real-time PCR assay. J Clin Microbiol 45:1426–1432

Wang LR, Dong LJ, Zhang MJ, Lu DP (2008) Correlations of human herpesvirus 6B and CMV infection with acute GVHD in recipients of allogeneic haematopoietic stem cell transplantation. Bone Marrow Transplant 42:673–677

Ward KN, White RP, Mackinnon S, Hanna M (2002) Human herpesvirus-7 infection of the CNS with acute myelitis in an adult bone marrow recipient. Bone Marrow Transplant 30:983–985

Wilborn F, Brinkmann V, Schmidt CA, Neipel F, Gelderblom H, Siegert W (1994) Herpesvirus type 6 in patients undergoing bone marrow transplantation: serologic features and detection by polymerase chain reaction. Blood 83:3052–3058

Wolf DG, Lurain NS, Zuckerman T, Hoffman R, Satinger J, Honigman A, Saleh N, Robert ES, Rowe JM, Kra-Oz Z (2003) Emergence of late cytomegalovirus central nervous system disease in hematopoietic stem cell transplant recipients. Blood 101:463–465

Yoshikawa T, Suga S, Asano Y, Yazaki T, Kodama H, Ozaki T (1989) Distribution of antibodies to a causative agent of exanthem subitum (human herpesvirus-6) in healthy individuals. Pediatrics 84:675–677

Yoshikawa T, Suga S, Asano Y, Nakashima T, Yazaki T, Sobue R, Hirano M, Fukuda M, Kojima S, Matsuyama T (1991) Human herpesvirus-6 infection in bone marrow transplantation. Blood 78:1381–1384

Yoshikawa T, Asano Y, Kobayashi I, Nakashima T, Yazaki T, Suga S, Ozaki T, Wyatt LS, Frenkel N (1993) Seroepidemiology of human herpesvirus 7 in healthy children and adults in Japan. J Med Virol 41:319–323

Yoshikawa T, Ihira M, Ohashi M, Suga S, Asano Y, Miyazaki H, Hirano M, Suzuki K, Matsunaga K, Horibe K, Kojima S, Kudo K, Kato K, Matsuyama T, Nishiyama Y (2001) Correlation between HHV-6 infection and skin rash after allogeneic bone marrow transplantation. Bone Marrow Transplant 28:77–81

Yoshikawa T, Asano Y, Akimoto S, Ozaki T, Iwasaki T, Kurata T, Goshima F, Nishiyama Y (2002a) Latent infection of human herpesvirus 6 in astrocytoma cell line and alteration of cytokine synthesis. J Med Virol 66:497–505

Yoshikawa T, Asano Y, Ihira M, Suzuki K, Ohashi M, Suga S, Kudo K, Horibe K, Kojima S, Kato K, Matsuyama T, Nishiyama Y (2002b) Human herpesvirus 6 viremia in bone marrow transplant recipients: clinical features and risk factors. J Infect Dis 185:847–853

Yoshikawa T, Yoshida J, Hamaguchi M, Kubota T, Akimoto S, Ihira M, Nishiyama Y, Asano Y (2003) Human herpesvirus 7-associated meningitis and optic neuritis in a patient after allogeneic stem cell transplantation. J Med Virol 70:440–443

Zerr DM, Fann JR, Breiger D, Boeckh M, Adler AL, Xie H, Delaney C, Huang ML, Corey L, Leisenring WM (2011) HHV-6 reactivation and its effect on delirium and cognitive functioning in hematopoietic cell transplantation recipients. Blood 117:5243–5249

Zerr DM, Boeckh M, Delaney C, Martin PJ, Xie H, Adler AL, Huang ML, Corey L, Leisenring WM (2012) HHV-6 reactivation and associated sequelae after hematopoietic cell transplantation. Biol Blood Marrow Transplant 18:1700–1708

Chapter 13
Vaccine Development for Cytomegalovirus

Naoki Inoue, Mao Abe, Ryo Kobayashi, and Souichi Yamada

Abstract The development of a cytomegalovirus (CMV) vaccine has become a top priority due to its potential cost-effectiveness and associated public health benefits. However, there are a number of challenges facing vaccine development including the following: (1) CMV has many mechanisms for evading immune responses, and natural immunity is not perfect, (2) the immune correlates for protection are unclear, (3) a narrow range of CMV hosts limits the value of animal models, and (4) the placenta is a specialized organ formed transiently and its immunological status changes with time. In spite of these limitations, several types of CMV vaccine candidate, including live-attenuated, DISC, subunit, DNA, vectored, and peptide vaccines, have been developed or are currently under development. The recognition of the pentameric complex as the major neutralization target and identification of various strategies to block viral immune response evasion mechanisms have opened new avenues to CMV vaccine development. Here, we discuss the immune correlates for protection, the characteristics of the various vaccine candidates and their clinical trials, and the relevant animal models.

Keywords Immune correlates · Congenital infection · Attenuated vaccines · Vectored vaccines · Epitope-based vaccines · gB · Pentameric complex · Neutralizing antibodies · Animal model

13.1 Background

Human CMV (HCMV) is a ubiquitous virus that does not usually cause any diseases in healthy individuals. Like other herpesviruses, it establishes latency or persistency throughout life after primary infection and is reactivated under immunologically suppressed conditions, such as transplantation and HIV infection,

N. Inoue (✉) · M. Abe · R. Kobayashi
Microbiology and Immunology, Gifu Pharmaceutical University, Gifu, Japan
e-mail: inoue@gifu-pu.ac.jp

S. Yamada
Department of Virology I, National Institute of Infectious Diseases, Tokyo, Japan

and can cause severe, sometimes fatal, diseases in solid organ transplant (SOT) and hematopoietic stem cell transplant (HSCT) recipients, as described in Chap. 12. HCMV is also the major cause of congenital CMV (cCMV) infection that occurs in 0.2–2% of all births and is associated with significant clinical consequences, not only at birth but also later as neurologic sequelae, including sensorineural hearing loss (SNHL) and developmental delays (Kenneson and Cannon 2007). Our retrospective studies demonstrated that 12–15% of cases with severe SNHL and 25% of cases with developmental delay of unknown cause were associated with cCMV infection; half of the sequelae were of late onset (Ogawa et al. 2007; Koyano et al. 2009). Although anti-HCMV drugs can be used for the treatment of HCMV diseases in transplant recipients and are potentially applicable for the treatment of cCMV diseases, their use is limited due to potential side effects and the development of viral resistance. Therefore, the development for a CMV vaccine has become a top priority as it is expected to be highly cost-effective and provide significant public health benefits, as reported by the Institute of Medicine (Stratton et al. 2000). The National Vaccine Advisory Committee of the USA also advocated the need for CMV vaccine development (Arvin et al. 2004).

13.2 Endpoints for Vaccine Development

There are a number of critical challenges to be overcome in the development of CMV vaccines. First, there are frequent reinfection and reactivation, indicating that immunity induced by natural infection is not perfect to protect against subsequent infection. Second, in spite of many studies, the immune correlates for protection remain unclear. Finally, endpoints for vaccine development may differ between the transmission of maternally infected CMV to the fetus and reactivation in the context of immunosuppression. In this chapter, our major focus is on vaccine development for cCMV infection.

13.2.1 Immune Correlates for Protection of cCMV Infection

Several studies on immunocompromised patients have identified CMV infections with multiple strains (Coaquette et al. 2004; Puchhammer-Stockl and Gorzer 2006; Ishibashi et al. 2007). cCMV infection often occurs via maternal CMV reinfection of pregnant women, although mixed infection in cCMV-infected newborns is seldom detected (Ross et al. 2010; Ikuta et al. 2013). CMV transmission to the fetus occurs in 1–2% of seropositive pregnant women, although 40% of seronegative pregnant women who are exposed to CMV transmit the virus to the fetus (Fowler et al. 1992, 2003; Adler et al. 1995). These observations indicate that preexisting immunity provides partial protection against CMV transmission to the placenta and fetus, although preexisting immunity is not perfect even in healthy individuals.

Importantly, the clinical outcomes for neonates born to mothers with preexisting immunity are milder than those born to seronegative mothers (Fowler et al. 1992). However, there are arguments against the inverse correlation between the preexisting maternal immunity and clinical outcomes of cCMV infection (Britt 2017). Immune factors for protection of cCMV infection, including innate immunity, neutralizing antibodies (NAbs), and CMV-specific CD4+ T-cell responses, are discussed below.

13.2.1.1 Innate Immunity

Innate immunity, based on the functions of NK cells, macrophages, pattern recognition receptors, and cytokines, is considered to play a crucial role in protection against cCMV infection and diseases. CMV employs several methods by which to evade these host innate immune responses.

The first steps in cCMV infection and amplification take place in the decidua where both maternal and fetal cells are in close contact. During the first trimester of pregnancy, a subset of NK cells, decidual NK (dNK) cells, form the majority of maternal immune cells in the placenta. Although dNK cells have little cytotoxic ability and have pregnancy-specific functions, including the capacity to produce angiogenic molecules to control trophoblast invasion, once they are exposed to CMV-infected autologous decidual fibroblasts, they undergo major functional and phenotypic changes, becoming cytotoxic effectors and co-localizing with infected cells (Siewiera et al. 2013). Therefore, dNK cells may limit the spread of viral infection to fetal tissues. Understanding the mechanisms underlying the regulation of dNK cytotoxic ability may help clarify the key factors involved in the immunopathology of CMV infection and lead to the design of more robust strategies to block viral immune escape.

Fcγ receptors induce a wide range of immune responses, including antibody-dependent cellular cytotoxicity (ADCC) in virus-infected cells by NK cells and cytokine secretion. HCMV encodes glycoproteins gp34 (RL11) and gp68 (UL119–118), which bind to Fcγ receptors to block the IgG-mediated activation of Fcγ receptors and may limit the efficacy of antibody-mediated protection (Corrales-Aguilar et al. 2014).

The presence of viral antigens in macrophages have been demonstrated in term cCMV-infected placentas (McDonagh et al. 2006) and in decidual organ cultures (Weisblum et al. 2011). As macrophages may transmit CMV to trophoblasts, the protection of macrophages from CMV infection would be required to block maternal-fetal transmission.

A TLR2 polymorphism is associated with increased risk of CMV diseases in liver transplant recipients due to differences in the induction of gB-mediated signaling (Kang et al. 2012). CMV-mediated TLR2 signaling in syncytiotrophoblast cultures stimulates TNFα expression and apoptosis (Chaudhuri et al. 2009). On the other hand, miR-UL112-113p, a HCMV encoding microRNA, downregulates TLR2 during the late stages of infection in cell cultures (Landais et al. 2015), suggesting that miRNA-mediated TLR2 downregulation works to prevent gB-mediated innate signaling. Although some studies have reported an association between genetic polymorphisms in TLRs and CMV infection and related diseases, the mechanisms underlying those findings remain unclear (Arav-Boger et al. 2012; Nahum et al. 2012; Taniguchi et al. 2013). Some studies observed that HCMV induced a distinct pattern of cytokine production, which could affect the outcome of cCMV infection (Scott et al. 2012; Weisblum et al. 2015).

HCMV encodes cmvIL-10, an ortholog of cellular interleukin-10 (cIL-10), which functions similarly to cIL-10 in the downregulation of proinflammatory cytokine production and inhibition of antigen-presenting cell functions (Raftery et al. 2004; Chang et al. 2004). Rhesus macaques infected with a rhesus CMV (RhCMV) variant lacking its cmvIL-10, rhcmvIL-10, exhibited increased inflammatory responses and greater humoral and cellular immune responses than did animals infected with the parental virus (Chang and Barry 2010). Macaques immunized with an inactive form of rhcmvIL-10 developed antibodies that neutralized rhcmvIL-10, but not cIL-10, and exhibited reduced viremia and viral shedding in bodily fluids after RhCMV challenge, indicating that neutralization of viral immunomodulation may afford a new strategy for vaccine development (Eberhardt et al. 2013). Similarly, guinea pig CMV (GPCMV) lacking immune evasion genes showed an attenuated phenotypes in vivo and worked as a live vaccine in guinea pigs (see Sect. 13.4.1). To block the evasion of CMV from the immune response mediated by NKG2D, studies using a murine CMV (MCMV) model demonstrated that a recombinant virus encoding the NKG2D ligand RAE-1γ could provide a powerful approach toward the development of a safe attenuated and immunogenic vaccine (Slavuljica et al. 2010). As an alternative approach, expression of NKG2D ligand ULBP2 by a recombinant HCMV activated both innate and adaptive immunity in a humanized mouse model (Tomić et al. 2016).

13.2.1.2 Neutralizing Antibodies

The importance of humoral immunity has been exemplified by the findings that the adaptive transfer of GPCMV-specific antibodies provided protection in guinea pig models (Chatterjee et al. 2001; Auerbach et al. 2014) and the plasma titers of CMV-specific neutralizing IgG with high avidity are inversely correlated with virus transmission rates and with histopathological findings of CMV products in the placenta (Boppana and Britt 1995; Pereira et al. 2003). However, a randomized Phase 2b trial of hyperimmune globulin (HIG) for the prevention of cCMV infection did not demonstrate the expected statistical difference in results (Revello et al. 2014). Two large-scale Phase 3 trials for the evaluation of HIG use in the prevention of cCMV infection are currently underway. One is a NIH-funded trial (NCT01376778) to be completed in 2018 that plans to enroll 800 pregnant women with primary CMV infection and analyze the primary outcomes, defined as fetal loss, fetal CMV infection, neonatal death, or neonatal cCMV infection. The other is a trial sponsored by Biotest AG aimed as clarifying whether their HIG product, Cytotect, can be used as the standard treatment for cCMV infection (EudraCT 2007–004692-19). Its interim analysis, conducted on 7000 of the expected 30,000 enrollments, indicated the efficacy of Cytotect (News release 2017.1).

Until recently NAbs against HCMV were measured in human fibroblasts and were considered to appear at a low titer late after primary infection. Based on the conserved nature and the kinetics of antibody development, gB and gH had been believed to be the major antigens for NAbs. However, after identification of an approximately 13 kb additional segment in a low-passaged Toledo strain (Cha et al. 1996), the requirement of each of the gene products in the UL128–131A locus for endothelial and epithelial cell tropisms and formation of a pentameric complex (Pentamer) of gH, gL, UL128, UL130, and UL131A were demonstrated (Hahn et al. 2004; Wang and Shenk 2005a, b). Importantly, human sera obtained from seropositive individuals and CMV-HIGs have, on average, 48-fold higher neutralizing activities against epithelial cell entry than against fibroblast entry, while sera from individuals vaccinated with gB/MF59 or with Towne, which lacks Pentamer, had low NAb titers against epithelial cell entry, indicating the presence of epithelial cell entry-specific antibodies (Cui et al. 2008). Serial depletion of HIGs with CMV antigens demonstrated that the major NAb response is directed at Pentamer, with only a minor role played by anti-gB antibodies (Fouts et al. 2012). Human monoclonal antibodies prepared from seropositive donors, which neutralize HCMV infection in epithelial, endothelial, and myeloid cells, target gB, gM/gN, gH, and mainly conformational epitopes of Pentamer (Macagno et al. 2010). Immunization of animals with virions of AD169 strain expressing the restored Pentamer or with purified Pentamer induced high NAb titers for the viral infection of endothelial and epithelial cells but not for fibroblasts (Gerna et al. 2008; Fu et al. 2012; Freed et al. 2013; Kabanova et al. 2014). Recent studies have analyzed the precise binding sites of NAbs against Pentamer to clarify the structural characteristics of Pentamer (Ciferri et al. 2015a, b; Loughney et al. 2015; Chiuppesi et al. 2017; Ha et al. 2017; Chandramouli et al. 2017).

Importantly, monoclonal NAbs against Pentamer, but not against gH, blocked the infection of human placental cytotrophoblasts (Chiuppesi et al. 2015), and development of antibodies against most of neutralizing epitopes of Pentamer was delayed in transmitting mothers, indicating that Pentamer is the major target of antibody-mediated maternal immunity (Lilleri et al. 2012, 2013).

Previously, we established a HCMV reporter cell line derived from U373MG, a glioma cell line, and demonstrated its use for the screening of novel anti-CMV compounds (Fukui et al. 2008; Majima et al. 2017). To conduct a high-throughput assay to measure NAbs against Pentamer, we have recently established a similar reporter cell line derived from the epithelial cell line ARPE-19, which expresses luciferase upon HCMV infection. Indeed, AD169rev, the UL131 gene of which was fixed, but not the parental AD169, induced luciferase production in the cell line. The cell line can be replaced with a conventional immunostaining assay as NAb titers in human sera measured by the reporter assay showed a good correlation with the conventional assay (Fig. 13.1).

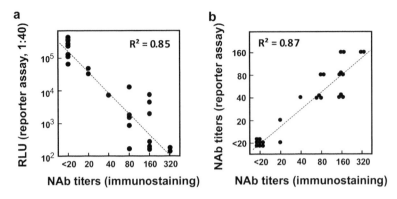

Fig. 13.1 A410F cells, ARPE-19-based CMV reporter cells, were plated in 96-well plates and infected with 80 infectious units (IUs) of AD160rev that had been incubated with diluted sera from 20 seropositive and 6 seronegative volunteers for 30 min. At 48 h after infection, the infected cells were reacted with monoclonal antibodies against IE1/IE2 followed by HRP-conjugated anti-mouse IgG, and the reacted cells were visualized with DAB, and a number of the stained foci were counted (immunostaining). The serum dilution giving 50% of the mean number of foci obtained with seronegative sera was defined as the NAb titer. In parallel, luciferase activities in the infected cells at 48 hrs after infection were measured (reporter assay). To obtain neutralizing antibody (NAb) titers, the relative light units (RLUs) in the cells infected with 20 IUs of AD169rev were used as a cutoff value. Measurements were done in triplicated wells. (**a**) RLUs obtained by the reporter assay of the cells infected after reaction with 40-fold diluted sera vs. NAb titers obtained by immunostaining assay. (**b**) NAb titers obtained by the reporter assay vs. those obtained by immunostaining assay

CMV-specific, high-avidity NAbs from maternal circulation are transcytosed into the fetal bloodstream, contributing to suppression of viral replication in the placenta in seropositive women; however, paradoxically, the non-neutralizing CMV antibodies may enhance CMV transmission to the fetus, as neonatal Fc receptors on syncytiotrophoblasts may facilitate the transcytosis of CMV (Maidji et al. 2006; Nozawa et al. 2009). Antibodies may also have functions other than protection from infection, as HIG was observed to reduce placental thickness in women with a cCMV-infected fetus (La Torre et al. 2006).

13.2.1.3 Cellular Immunity

As most CMV infections occur asymptomatically, it has been difficult to characterize the early phases of cellular immune responses in human individuals. In contrast, several clinical studies have demonstrated that the adaptive transfer of T-cell clones restores viral immunity in HSCT recipients (Riddell et al. 1992; Walter et al. 1995; Peggs et al. 2003).

Although HCMV-specific T-cell immunity to pp65(UL83) and IE-1(UL123) have been well analyzed, Elkington et al. (2003) found CD8+ T-cell responses to a broad range of antigens by the use of bioinformatics and ex vivo functional T-cell assays. Thereafter, Sylwester et al. (2005) conducted a comprehensive study showing that 70% of HCMV ORFs were immunogenic for CD4+ and/or CD8+ T-cells and that the total HCMV-specific T-cell responses in seropositive subjects were enormous, comprising ~10% of both the CD4+ and CD8+ memory compartments in the blood. The total HCMV-specific CD4+ T-cell response could be represented by the six most immunogenic ORFs, including pp65, pp150(UL32), and gB, and the total $CD8^+$ response by the 15 ORFs, including pp65, IE1, and IE2. Still further studies are required to define the roles of CD4+ and CD8+ cells against individual antigens in the protection against CMV infection and reactivation. For example, a high frequency of IE-1- but not pp65-specific CD8 T-cells is correlated with protection from CMV disease in SOT recipients (Bunde et al. 2005). There were also functional differences observed between CTLs recognizing individual epitopes within the IE1 and pp65 antigens in healthy donors and HSCT recipients (Lacey et al. 2006).

Cellular immune responses in the context of cCMV infection are still only poorly understood. HCMV-specific CD4+ T-cell responses, IgM, IgG avidity, and viral DNAemia levels were compared between 74 pregnant and 29 nonpregnant women as well as between 19 transmitter and 21 non-transmitter mothers with primary infection during the first or second trimester (Revello et al. 2006). The study found that (1) pregnancy had little effect on HCMV-specific cell-mediated immune responses; (2) HCMV-specific CD4+ T-cells were detected by cytokine flow cytometry in the absence of lymphoproliferative responses (LPR) to CMV, irrespective of pregnancy; and (3) LPR to HCMV was significantly lowered or delayed in transmitter mothers. Impairment of LPR is consistent with the findings of previous studies regarding the long-lasting suppression of lymphocyte blastogenic responses after primary infection. Another study from the same group analyzed CD4+ and CD8+ T-cell proliferation in different sets of women during the first year after primary infection, confirming a significant delay in the development of the CD4+ T-cell LPR in the transmitter mothers (Lilleri et al. 2007). Non-transmitter mothers showed a significantly higher frequency of HCMV-specific CD4+ T-cells with a IL-7R-positive phenotype than transmitter mothers, both 1 month and 6–12 months after infection, while no difference was observed for HCMV-specific CD8+ T-cells, suggesting that the early appearance of CD4+ T-cells with a long-term memory phenotype is associated with a lower risk of viral transmission to the fetus (Mele et al. 2017). Another recent study demonstrated that memory T-cells with proliferative capacity, mainly CD4+ T-cells, specific for pp65 was significantly lower in transmitter than nontransmitter mothers, while no differences were observed in the response to IE-1 or IE-2 between the two groups (Fornara et al. 2017). These findings suggest the need for strategies to enhance the maternal CD4+ T-cells with a long-term memory phenotype specific for particular viral antigens.

13.2.2 Target Populations and Practical Endpoints

A multidisciplinary meeting held at the FDA in 2012 discussed the pros and cons of potential vaccination strategies targeting adult women, adolescent girls, and/or young children with the aim of reducing cCMV (Krause et al. 2013). The following aspects of the target populations were compared: vaccine coverage (acceptability), purpose of vaccination, duration of protective effect, effects on sexual transmission, and differences in vaccine effect between seronegative and seropositive subjects. Universal vaccination of young children may have an advantage in the rapid reduction of cCMV rates if prevention of infection or shedding can be achieved, while the vaccination of adolescents or adult women could reduce cCMV disease providing immunity prior to pregnancy. The meeting participants agreed that the primary endpoint for clinical trials of CMV vaccines in women should be protection against cCMV infection as it is an essential precursor of cCMV disease and both more practical and acceptable. Prevention of CMV infection in pregnant women was considered a less useful endpoint as such prevention may be less readily achieved than the attenuation of transplacental transmission.

To evaluate cCMV infection as an endpoint in large-scale trials on vaccine efficacy, it is critical to screen cCMV-infected newborns efficiently using cost-effective, time-saving procedures. To this end, assays based on filter papers containing dried body fluids obtained from newborns are useful. As dried blood spots are already collected from all newborns in most countries, it is easy to establish an infrastructure for the collection and analysis of the specimens. However, dried blood spots contain very limited amounts of viral DNA. In contrast, both saliva and urine specimens contain huge amounts of viral DNA. We have routinely used our specifically developed high-throughput screening method based on the collection of urine using a filter paper inserted into the diapers followed by real-time PCR assay using a filter disc cut from the filter paper as a template (Nozawa et al. 2007; Koyano et al. 2011).

13.2.3 Cost-Effectiveness

Using a decision tree, Dempsey et al. (2012) compared the costs, potential clinical impacts, and cost-effectiveness between "no vaccination" and "vaccination" strategies. The model simulated vaccination for a hypothetical 100,000 11-year-old girls who had never been previously pregnant nor vaccinated against CMV. Under base-case conditions, the vaccination strategy would be both less costly ($32.3 million less) and provide greater clinical benefits (resulting in 8 fewer deaths and 5 fewer cases of vision loss, 52 of hearing loss, and 18 of mental retardation) than the no vaccination strategy. The model was most sensitive to variations in vaccine efficacy, with an efficacy of at least 61% required for vaccination to be beneficial. As most vaccines introduced in the US have efficacies of >80%, surpassing the 61% threshold appears to be feasible. One of the limitations of the study was the assumption of

100% vaccine coverage across the population. Although this assumption is overly optimistic, the high level of CMV vaccine acceptance by parents (Petty et al. 2010) is encouraging.

13.3 Vaccines Developed and Under Development

Although many candidate vaccines have been evaluated, several factors, including incomplete understanding of the immune correlates for protection, viral immune evasion and latency, the lack of an animal model that allows HCMV infection and exhibits HCMV diseases, have made the progress of vaccine development slow. Here, we summarize various CMV vaccine candidates that progressed at least to a Phase 1 trial. Table 13.1 lists the clinical trials of CMV vaccines (please note that the Phase 1 trials of some vaccines that progressed to a Phase 2 trial are omitted from the table).

13.3.1 Live-Attenuated Vaccines

Live-attenuated Towne strain, which lacks several genes and does not establish latency, has been administrated safely and found to induce both humoral and cellular responses. However, the Towne vaccine failed to protect seronegative SOT recipients from infection, although it reduced the incidences of severe CMV diseases. In healthy subjects, the Towne vaccine was protective, but protection was dependent on the challenge dose; in other words, it failed to protect against primary infection with low-passage virulent Toledo strain in cases in which immunity was established by natural infection (reviewed in Adler 2008). To clarify whether the replacement of a particular segment of the Towne genome with that of the Toledo genome increases its immunogenicity without reducing its safety, four Towne-Toledo chimeric strains were established. In Phase 1 trials, these chimera vaccines were well tolerated in both seropositive (Heineman et al. 2006) and seronegative men (NCT01195571) (Adler et al. 2016). Among the four chimeras, chimera 2 and 4 induced seroconversion and CD8+ T-cell response more frequently than the other two. The major limitation of Towne and its chimera vaccines is their lack of Pentamer expression, which may fail to induce strong NAbs, although it is not clear whether this is sufficient reason to exclude these well-characterized vaccine strains.

13.3.2 DISC Vaccine

To increase the safety of live-attenuated vaccines, disabled infectious single cycle (DISC), or replication-defective, vaccine strains have been developed for herpes simplex viruses (Dropulic and Cohen 2012). Recently, MSD developed DISC

Table 13.1 Clinical trials for the development of CMV vaccines

Type	Vaccine	Antigen	Adjuvant	Sponsor	Subjects	Serostatus	Phase	Enrolled no.	Randomize	Double-blind	Trial no.	Ref. or status (completion yr)
Live-attenuated	Chimera	Whole virus		CMV Research Foundation	HV-M	Neg	1	36	–	N	NCT01195571	Adler et al. (2016)
DISC	V160	Whole virus	Aluminum phosphate	MSD	HV	Both	1	190	Y	Y	NCT01986010	Adler et al. (2017)
Protein	gB/MF59	gB	MF59	Pass, NIAID, Sanofi	HV-F	Neg	2	464	Y	Y	NCT00125502	Pass et al. (2009)
				NIAID	HV-F	Neg	2	409	Y	Y	NCT00133497/00815165	Bernstein et al. (2016)
				University College London, NIAID	SOT	Both	2	140	Y	Y	NCT00299260/01883206	Griffith et al. (2011)
	GSK1492903A	gB	AS01E (MPL/QS21)	GSK	HV-M	Neg	1	40	N	N	NCT00435396/01357915	HP of GSK clinical trials
eVLP	VBI-1501A	gB	Alum	VBI vaccines	HV	Neg	1	128	Y	Y	NCT02826798	News release (2017)
Peptide	CMVPepVax	pp65-tetanus epitope	PF0351276	City of Hope	HSCT-R	Pos	1	36	Y	N	NCT01588015	Nakamura et al. (2016)
					HSCT-R	Pos	2	96	Y	Y	NCT02396134	Ongoing (2019)

DNA	ASP0133	pp65, gB	CRL1005-BAK	Vical/Astellas	Pos	2	108	Y	Y	NCT00285259	Kharfan-Dabaja et al. (2012)
				Astellas	HSCT-R	NA	2	10	N	N	NCT01903928	Mori et al. (2017)
				Vical/Astellas	HSCT-R	Pos	3	515	Y	Y	NCT01877655	News release (2017)
				Vical/Astellas	SOT	D + R-	2	150	Y	N	NCT01974206	News release (2016)
Prime-boost	VCL_CT02 +Towne	pp65, gB, IE1		Vical, UCSF	HV	Neg	1	17	N	N	NCT00370006	Jacobson et al. (2009)
				Vical, UCSF	HV	Neg	1	12	Y	N	NCT00373412	
Vectored (ALVAC)	ALVAC-CMV	pp65		NIHLBI	HSCT-DR	Both	2	38	N	N	NCT00353977	(2008)
Vectored (MVA)	MVA triplex	pp65, IE1, IE2		City of Hope, NCI	HV	Both	1	24	N	N	NCT01941056	La Rosa et al. (2017)
				City of Hope, NCI, Helocyte	HSCT-R	Pos	2	115	Y	Y	NCT02506933	Ongoing (2018)
Vectored(AV)	AVX601	gB, pp65/IE1		AlphaVax	HV	Neg	1	40	Y	Y	NCT00439803	Bernstein et al. (2009)
Vectored (LCMV)	HB-101	gB, pp65		Hookipa Biotech	HV	Neg	1	54	N	N	NCT02798692	News release (2017)

HV healthy volunteers, *F* female, *M* male, *Pos* positive, *Neg* negative, *D* donors, *R* recipients vaccine strain V160 as a CMV vaccine and demonstrated its immunogenicity in multiple animal models (US Patent 9546355; Wang et al. 2016). First, the frameshift mutation in UL131 of AD169 was fixed to recover epithelial/endothelial cell tropism (Wang et al. 2011). The UL122/123 gene encoding IE1/IE2 and the UL51 gene encoding one of packaging proteins were subsequently modified to express their encoding proteins fused with the destabilizing domain of the FK506-binding protein (FKBP12), respectively, to generate the V160 strain. In the presence of Shield-1, a synthetic compound, the FKBP12-fused proteins are stable, whereas they are proteolytically degraded in its absence. Thus, VP160 viral stock can be prepared in the presence of Shield-1, but once VP160 is administered in vivo, it loses its growth capability. Administration of VP160 induced durable NAbs as well as CD4+ and CD8+ T-cells specific to several CMV antigens in nonhuman primates. A Phase 1 trial (NCT01986010) to evaluate the safety, immunogenicity of various doses, formulations, and routes of administration of V160 in approximately 190 healthy adults demonstrated that V160 had acceptable safety profile and elicited NAbs and cell-mediated immune responses comparable to natural infection (Adler et al. 2017).

13.3.3 gB Protein-Based Vaccines

Since gB is essential for infection and had been believed to be the major target of NAbs, gB protein derived from Towne was expressed in CHO cells and evaluated as a vaccine antigen in several trials in the 1990s. gB with MF59, a squalene in water emulsion, provided a greater enhancement of immunogenicity than did combination with an alum adjuvant, and NAb titers induced by gB/MF59 reached the level observed in seropositive individuals (Pass et al. 1999). A Phase 2 trial (NCT00125502), which involved 464 seronegative young mothers and observed natural infection after vaccination with gB/MF59, demonstrated 50% efficacy of protection in comparison with natural infection (Pass et al. 2009), and rapid waning of antibody titers was observed. The same vaccine was evaluated in a trial (NCT00133497) that enrolled 409 seronegative girls between 12 and 17 years of age and observed seroconversion and viral shedding (Bernstein et al. 2016). The trial demonstrated that the vaccine was safe and immunogenic, although the vaccine efficacy was only 43% and did not reach conventional levels of significance, which is consistent with the results of the trial on adult women.

The gB/MF59 vaccine was also evaluated in a trial (NCT00299260) involving kidney or liver SOT recipients (Griffiths et al. 2011). In the vaccinated seronegative recipients who received organs from seropositive donors, both the duration of viremia and the total number of days of antiviral treatment were significantly reduced. In patients with viremia, the median peak viral load of vaccinated patients was one tenth of that of patients administered the placebo. The antibody titers against gB were inversely correlated with the duration of viremia. There was no difference between the vaccine and placebo recipients in terms of CMV-responsive CD4+

T-cell frequency. These findings suggest that humoral immunity has a significant role in protection, which is in contrast with the present dogma regarding the major role of T-cell responses to CMV diseases in transplant recipients.

GSK1492903A, a recombinant gB protein derived from AD169 strain, adjuvanted with AS01E, a proprietary adjuvant system containing MPL and QS21, was developed by GlaxoSmithKline. A Phase 1 trial involving 40 seronegative adult males (NCT00435396) revealed no serious adverse effects (gsk-clinicalstudyregister.com/study/108890), while a follow-up study (NCT01357915) demonstrated the long-term persistence of vaccine-induced immunity (gsk-clinicalstudyregister.com/study/115429).

13.3.4 VLP-Based Vaccines

VBI Vaccines Inc. developed enveloped virus-like particles (eVLPs) produced in mammalian cells for vaccine production. Mice were immunized with eVLPs containing the full-length gB derived from Towne strain or those containing the extracellular domain of gB fused with the transmembrane and cytoplasmic domains of vesicular stomatitis virus (VSV)-G protein (gB-G), and gB-specific antibodies and T-cell responses were compared (Kirchmeier et al. 2014). gB-G eVLP induced a tenfold higher level of NAbs than did gB eVLP despite comparable levels of antibody binding titers against gB. Importantly, antibodies induced by gB-G show neutralizing activity against epithelial cell infection. As cells transiently transfected with gB-G, but not with gB, formed syncytia, it is plausible that gB-G takes a prefusion conformation, which makes the gB-G more immunogenic. Interim analysis of a Phase 1 trial (NCT0286798), for which 128 seronegative adults were enrolled to compare the safety and immunogenicity of three different doses of the unadjuvanted and alum-adjuvanted eVLP/gB-G (VBI-1501), indicated that the vaccine was well tolerated at all doses, with no safety signals, 100% seroconversion in subjects who received the highest dose, and development of NAbs in some patients after two doses (News release 2017.7).

Redbiotec AG also developed eVLPs based on a baculovirus expression system and presented their results showing the incorporation of various combinations of glycoproteins at a number of international conferences (e.g., the 38th International Herpesvirus Workshop). The technology was purchased by Pfizer in 2015.

13.3.5 Peptide-Based Vaccines

HLA-restricted CTL peptide-based vaccines are feasible for HSCT recipients, as potential responders to the vaccine can be selected in advance based on their HLA type. Peptide vaccines composed of covalently linked CTL and helper T epitopes are expected to show the most efficacy. To this end, the HLA A*0201-restricted pp65 CTL epitopes linked with the synthetic pan-DR epitope peptide (PADRE) or with a tetanus Th epitope were developed as epitope-based peptide vaccines. To evaluate such vaccines, a Phase 1 trial (NCT00722839) was undertaken involving healthy seropositive adults vaccinated with escalating doses of the PADRE- or tetanus-pp65 peptides with and without the CpG DNA adjuvant PF03512676 (La Rosa et al. 2012). The peptide vaccines without the adjuvant were safe and well tolerated. Although addition of the adjuvant exacerbated the mild to moderate cutaneous reaction at the injection site and systemic flu-like symptoms, there were no serious adverse events. Increases in vaccine-elicited T-cells specific to the pp65 peptide was observed in 30% of the subjects treated with the unadjuvanted vaccine and in all subjects treated with the adjuvanted vaccine. Based on the results, a randomized Phase 1b trial (NCT01588015) was undertaken involving 36 eligible seropositive patients with hematologic malignancies who progressed to allogenic HSCT and were randomly assigned to the observation arm and the arm for vaccination with the tetanus-pp65 peptide vaccine adjuvanted with PF03512676, designated as CMVPepVax (Nakamura et al. 2016). Favorable outcomes, including higher relapse-free survival, a twofold increase in pp65-specific CD8 T-cells, and less CMV reactivation and antiviral treatment, were obtained in the vaccinated arm. As the results demonstrated both the safety and immunogenicity of the vaccine, a Phase 2 trial is now ongoing (NCT02396134).

13.3.6 DNA-Based Vaccines

DNA therapeutic vaccine ASP0113 (also known as TransVax and VCL-CB01), which was developed by Vical and is licensed to Astellas, consists of two plasmids encoding pp65 and gB formulated with a poloxamer CRL1005 and benzalkonium chloride delivery system. A Phase 1 trial demonstrated its safety profile and induction of antigen-specific T-cell responses and NAbs in seronegative subjects. The subsequent Phase 2 trial involving enrolled 94 seropositive HSCT recipients and 14 paired donors (NCT00285259) demonstrated that there was no difference in the incidence of common adverse events between the vaccine and placebo groups and that the vaccine significantly reduced the occurrence of viremia and improved the time-to-event for viremia episodes, although there was a lack of any significant reduction in antiviral treatment (Kharfan-Dabaja et al. 2012). A randomized Phase 2 trial to evaluate the safety and efficacy of ASP0113 in 150 seronegative transplant patients receiving kidneys from seropositive donors did not meet its primary endpoint, which was the protection from viremia with >1000 copies/ml, and there were no differences in the secondary endpoints, CMV-associated disease, and requirement of antiviral therapy (News release 2016.9). In addition, in a global Phase 3 trial that enrolled 514 seropositive HSCT recipients (NCT0187765), ASP0113 did not meet its primary composite endpoint of overall mortality and CMV end-organ disease through the first year post-transplantation, nor secondary endpoints of time to first protocol-defined CMV viremia and time to first use of adjudicated CMV-specific antiviral therapy (News release 2018.1).

In a Phase 1 trial, seronegative subjects were vaccinated with VCL-CT02, a trivalent CMV DNA vaccine consisting of pp65-, IE1-, gB-expressing plasmids, as a priming step followed by boosting via the administration of the Towne vaccine (Jacobson et al. 2009). The median time to first CMV-specific memory T-cell responses after Towne vaccine administration was 14 days for the DNA vaccine-primed subjects and 28 days for the subjects without priming. In addition, a greater number of subjects primed with the DNA vaccine developed gB-specific T-cell responses.

Vical is currently developing a prophylactic vaccine, CyMVectin, for prevention of cCMV that consists of gB- and pp65-expressing plasmids with the cationic lipid adjuvant Vaxfectin.

13.3.7 Vectored Vaccines

Several CMV vaccines based on viral expression vectors have been developed. Attenuated canary pox (ALVAC) is one such vaccine candidate vector that can accept multiple large expression cassettes of foreign DNA and elicit protective immune response in non-avian species. ALVAC does not produce progeny in mammalian cells or non-avian species and causes no diseases in healthy individuals or immunosuppressed patients. Although gB-expressing ALVAC (vCP139) did not significantly enhance or induce NAbs among seropositive or seronegative subjects, the administration of Towne strain to the subjects led to the development of significantly high levels of gB-specific antibodies, indicating that ALVAC-gB can prime the immune system to produce a strong neutralizing response (Adler et al. 1999). The administration of four doses of pp65-expressing ALVAC (vCP260) induced pp65-specific CTLs in seronegative adults, and there were no significant safety issues (Berencsi et al. 2001).

Like ALVAC, the modified vaccinia Ankara (MVA) virus grows in cell cultures only abortively, with the exception of chick embryo fibroblasts. Although replication occurs, little or no packaging of the infectious virus takes place in primate and other mammalian cells. MVA was administered as a vaccine against smallpox in >100 thousand individuals, and clinical trials of MVA-based HIV vaccines demonstrated a good safety profile for MVA. The NAb titers developed in mice immunized with gB-expressing MVA were equivalent to those found after natural infection (Wang et al. 2004). Rhesus macaques immunized with MVA expressing RhCMV gB or pp65 demonstrated immunogenicity and protective efficacy (Yue et al. 2007, 2008; Abel et al. 2011). NAbs induced by MVA expressing RhCMV Pentamer inhibited viral entry to both rhesus epithelial/endothelial cells and fibroblasts and reduced plasma viral loads after viral challenge (Wussow et al. 2013). Mice or rhesus macaques immunized with MVA expressing HCMV Pentamer also induced durable and efficacious NAbs to prevent the HCMV infection of macrophages localized in the placenta (Wussow et al. 2014). A Phase 1 trial of MVA Triplex, MVA expressing HCMV pp65, IE1 exon 4, and IE2 exon 5, involving 24 healthy adults, demonstrated that the vaccine was well tolerated, and no serious adverse events were attributed to vaccination (NCT01941056) (La Rosa et al. 2017). The robust, functional, and durable Triplex-driven expansions of CMV-specific T-cells were detected. A multicenter Phase 2 trial of Triplex in seropositive allogeneic HSCT recipients (NCT02506933) is currently ongoing.

A propagation-defective, single-cycle RNA replicon vector system derived from an attenuated strain of Venezuelan equine encephalitis virus, an alphavirus, was used to produce virus-like replicon particles (VRPs) expressing gB or a pp65/IE1 fusion protein. These VRPs induced strong NAb and T-cell responses in mice and rabbits with no adverse effects (Reap et al. 2007). To evaluate AVX601 comprised of VRPs expressing gB and those expressing pp65/IE1, a Phase 1 trial (NCT00439803) involving 40 seronegative adults was undertaken, and three parameters regarding administration, doses, vaccine vs. placebo, and injection procedures were compared (Bernstein et al. 2009). The vaccine was well tolerated, with mild to moderate local and minimal systemic reactogenicities observed. Recipients of the vaccine developed CTL and NAbs to all CMV antigens in the vaccine. After acquisition of this alphavirus-based technology by Novartis in 2008, the company produced VRPs expressing gH and gL (Loomis et al. 2013) and, subsequently, those expressing Pentamer (Wen et al. 2014). These VRPs induced higher NAb titers than did gB-expressing VRPs in mice.

Vaccines based on nonreplicating lymphocytic choriomeningitis virus (rLCMV) vectors have been developed recently. Immunization of mice with gB-expressing rLCMV elicited a comparable gB-binding antibody response and a superior neutralizing response to that elicited by adjuvanted subunit gB. Immunization with pp65-expressing rLCMV elicited robust T-cell responses. A vaccine that combined both vectors, designated as HB-101 Vaxwave, provided comparable immunogenicity to that of the individual monovalent formulations (Schleiss et al. 2017). A Phase 1 trial to obtain data on dose escalation (NCT02798692) is currently ongoing. Interim data indicated that 93% of the subjects in the lowest dose group and 100% of the subjects in the medium and high dose groups developed detectable NAbs specific to CMV after three doses and that all three dose groups of the vaccine induced robust and statistically significant cellular immune responses when compared to placebo (News release 2017.5; presentation at CMV 2017 Conference).

13.3.8 Other Vaccine

Dense bodies are complex, noninfectious particles produced in HCMV-infected cells and contain >20 viral proteins. These dense bodies induce NAbs that prevent the infection of fibroblasts and epithelial cells and cell-mediated immune responses to multiple viral proteins (Pepperl et al. 2000; Cayatte et al. 2013). Combined with the culturing of cells on microcarriers, the addition of the viral terminase inhibitors and purification of dense bodies by tangential flow filtration allowed scalable production without any reduction in purity or safety (Schneider-Ohrum et al. 2016).

13.4 Animal Models for Preclinical Trials

13.4.1 Guinea Pig CMV (GPCMV) Models

GPCMV, but not MCMV, causes infection in utero, which makes GPCMV animal models useful for the demonstration of "proof-of-concept" toward the development of cCMV vaccine strategies. Humans and guinea pigs, but no other rodents, display similar placental structures, i.e., a hemomonochorial maternofetal barrier (Kaufmann and Davidoff 1977). Guinea pigs also show some similarities in placental development, and its relatively long (~10 weeks) gestation period can be classified into trimester-like divisions as in humans. In addition to stillbirths and intrauterine growth restrictions, hearing impairment can also be demonstrated using a guinea pig model of congenital infection. We demonstrated GPCMV transmission from mother to fetus via the placenta and hematogenous viral spread to the perilymph and ganglion in the inner ear of the fetus (Katano et al. 2007). In addition to studies on pathogenesis, the presence of GPCMV homologs of HCMV Pentamer components and their involvement in the infection of macrophages and efficient dissemination in animals make the GPCMV model useful for studies of Pentamer-based vaccine strategies (Nozawa et al. 2008; Yamada et al. 2009, 2014; Auerbach et al. 2014).

GPCMV lacking three potential MHC-I homologs, gp147, gp148, and gp149, was highly attenuated in animals (Crumpler et al. 2009). The virus produced elevated IFN-γ levels and higher antibody titers than did the wild-type, although the deletion of the three genes had no impact on MHC-I downregulation. As a live vaccine, the virus exhibited comparable protection against pup mortality in the congenital infection model. Similarly, GPCMV strains lacking a chemokine MIP homolog and a viral protein kinase R inhibitor, respectively, were also attenuated, and vaccination with these strains resulted in reduced maternal viral loads, pup mortality, and congenital infection rates (Leviton et al. 2013; Schleiss et al. 2015). These studies imply that removal of the immune evasion genes from live vaccine strains may improve the efficacy of vaccines in protecting against cCMV diseases and provide more safety features.

A DNA vaccine expressing GPCMV gB, but not that expressing pp65, protected against virus transmission (Schleiss et al. 2003). A DNA vaccine based on GPCMV BAC, the growth of which was disabled by knocking out its UL48 homolog gene, reduced viral loads and both dam and pup mortality (Schleiss et al. 2006). Various adjuvant systems have also been examined for gB protein-based vaccination using guinea pig models (Schleiss et al. 2004, 2014). To understand the mechanisms for gB vaccine protection, we analyzed the spread of the challenge viruses in the placentas and fetuses of guinea pig dams immunized with recombinant adenoviruses expressing GPCMV gB and β-galactosidase (Hashimoto et al. 2013). Focal localization of viral antigens in the spongiotrophoblast layer suggests cell-to-cell viral spread in the placenta. Even in the presence of high gB antibody titers with high-avidity indices in fetuses, CMV spread to most organs in a small proportion of littermates. Our results suggest that gB antibodies protected against infection mainly at the interface of the placenta rather than from the placenta to the fetus.

The guinea pig model has also been used to examine some vectored vaccines. Alphavirus- and MVA-vectored pp65 induced humoral and cellular immunity and improved pregnancy outcomes (Schleiss et al. 2007; Gillis et al. 2014). Increased pup survival, pup weights, and gestation time were also demonstrated by immunization with rLCMVs expressing either gB or pp65, although the increases were only statistically significant in gB-immunized animals (Cardin et al. 2016). Using the same vector system, combined vaccination with gB and pp65 conferred additive protection (Schleiss et al. 2017), which is inconsistent with the results reported from the same group for the combination of pp65-expressing MVA with gB-expressing MVA, which reduced antibody response against gB (Swanson et al. 2015).

13.4.2 Rhesus CMV Models

Nonhuman primates are considered as excellent models that recapitulate HCMV infection and immunity; however, there are some limitations, including problems with handling, cost, and the unavailability of inbred and knockout (KO) animals. Among nonhuman primates, macaques have been widely used in various studies. Several CMV isolates were obtained from rhesus and cynomolgus macaques, and their whole genome sequences were determined (Hansen et al. 2003; Marsh et al. 2011; Russell et al. 2016). Studies on cross-species infection between rhesus and cynomolgus macaques demonstrated multiple layers of cross-species restriction between the closely related hosts, including cell tropism and the evasion of apoptosis as critical determinants (Burwitz et al. 2016). Genomic and proteomic analyses of RhCMV strains and the genetic restoration or deletion of particular genes sets using BAC clones demonstrated that Pentamer homologs, the genes in UL/b' region, and the genes for immune evasion are required for efficient in vivo CMV replication and dissemination (Lilja and Shenk 2008; Malouli et al. 2012; Assaf et al. 2014).

As described in Sect. 13.3, the vaccination of rhesus macaques with MVA- or DNA-based vaccines expressing RhCMV gB, pp65, IE1, or Pentamer or with inactivated whole virions induced humoral and cellular immunity, and the viral challenge of vaccinated animals showed a reduction in viremia and viral shedding after viral challenge (Yue et al. 2006, 2007, 2008; Abel et al. 2011; Wussow et al. 2013).

Fetal inoculation of rhesus macaques with RhCMV results in neuropathologic outcomes similar to those associated with cCMV diseases in humans (reviewed in Barry et al. 2006). Recently, the first vertical transmission model was established by using RhCMV-free rhesus macaque dams treated by intravenous inoculation of RhCMV during the early second trimester of pregnancy, with or without CD4 + T-cell depletion (Bialas et al. 2015). Vertical transmission of RhCMV was observed in all T-cell-depleted as well as in two of three immunocompetent dams. Higher plasma and amniotic fluid viral loads were observed in the depleted dams. Further, fetal loss or CMV-associated sequelae in infants were observed in all CD4 + T-cell-depleted dams, while the immunocompetent dams carried fetuses to term, indicating that maternal CD4 + T-cell immunity is important for controlling maternal viremia and inducing protective immune response against fetal CMV diseases. This model was also used for the analysis of the characteristics of the early maternal RhCMV-specific humoral immune response (Fan et al. 2017) and for the demonstration of protection of cCMV infection by the adaptive transfer of RhCMV-specific antibodies (Nelson et al. 2017). Although this rhesus macaque cCMV transmission model is a useful precursor to clinical trials, particularly for the confirmation of vaccine efficacy, the high cost, limited availability of the RhCMV-free animals and artificial nature of CD4+ T-cell depletion mean that guinea pig models are still required for the early phase of "proof-of-concept" studies.

13.5 Perspective

Some CMV vaccine candidates demonstrate encouraging results; however, there are still no vaccines for cCMV that have progressed to Phase 3 trials. In spite of the serious need for CMV vaccines, knowledge about CMV infection and diseases is still limited in most societies, and routine surveillance for cCMV has not yet established anywhere, resulting in poor infrastructure and little support from society for large-scale clinical studies. Financial constraints in governmental agencies and private companies have discouraged the development of innovative vaccines and clinical studies of developed vaccines in large-scale trials. We need to break this vicious circle and fill the gap between the scientific efforts and the public conceptions to reach to our final goal, that is, the implementation of practical CMV vaccines.

References

Abel K, Martinez J, Yue Y et al (2011) Vaccine-induced control of viral shedding following rhesus cytomegalovirus challenge in rhesus macaques. J Virol 85:2878–2890
Adler SP (2008) Human CMV vaccine trials: what if CMV caused a rash? J Clin Virol 41:231–236
Adler SP, Starr SE, Plotkin SA et al (1995) Immunity induced by primary human cytomegalovirus-infection protects against secondary infection among women of childbearing age. J Infect Dis 171:26–32
Adler SP, Plotkin SA, Gonczol E et al (1999) A canarypox vector expressing cytomegalovirus (CMV) glycoprotein B primes for antibody responses to a live attenuated CMV vaccine (Towne). J Infect Dis 180:843–846

Adler SP, Manganello AM, Lee R et al (2016) A phase 1 study of 4 live, recombinant human cytomegalovirus Towne/Toledo chimera vaccines in cytomegalovirus-seronegative men. J Infect Dis 214:1341–1348

Adler S, Lewis N, Conlon A et al (2017) Phase 1 clinical trial of a replication-defective human cytomegalovirus (CMV) vaccine. ID Week 2017 presentation

Arav-Boger R, Wojcik GL, Duggal P et al (2012) Polymorphisms in toll-like receptor genes influence antibody responses to cytomegalovirus glycoprotein B vaccine. BMC Res Notes 5:140

Arvin AM, Fast P, Myers M et al (2004) Vaccine development to prevent cytomegalovirus disease: report from the National Vaccine Advisory Committee. Clin Infect Dis 39:233–239

Assaf BT, Mansfield KG, Strelow L et al (2014) Limited dissemination and shedding of the UL128 complex-intact, UL/b'-defective rhesus cytomegalovirus strain 180.92. J Virol 88:9310–9320

Auerbach MR, Yan D, Vij R et al (2014) A neutralizing anti-gH/gL monoclonal antibody is protective in the guinea pig model of congenital CMV infection. PLoS Pathog 10:e1004060

Barry PA, Lockridge KM, Salamat S et al (2006) Nonhuman primate models of intrauterine cytomegalovirus infection. ILAR J 47:49–64

Berencsi K, Gyulai Z, Gönczöl E et al (2001) A canarypox vector-expressing cytomegalovirus (CMV) phosphoprotein 65 induces long-lasting cytotoxic T cell responses in human CMV-seronegative subjects. J Infect Dis 183:1171–1179

Bernstein DI, Reap EA, Katen K et al (2009) Randomized, double-blind, phase 1 trial of an alphavirus replicon vaccine for cytomegalovirus in CMV seronegative adult volunteers. Vaccine 28:484–493

Bernstein DI, Munoz FM, Callahan ST et al (2016) Safety and efficacy of a cytomegalovirus glycoprotein B (gB) vaccine in adolescent girls: a randomized clinical trial. Vaccine 34:313–319

Bialas KM, Tanaka T, Tran D et al (2015) Maternal CD4+ T cells protect against severe congenital cytomegalovirus disease in a novel nonhuman primate model of placental cytomegalovirus transmission. Proc Natl Acad Sci U S A 112:13645–13650

Boppana SB, Britt WJ (1995) Antiviral antibody responses and intrauterine transmission after primary maternal cytomegalovirus infection. J Infect Dis 171:1115–1121

Britt WJ (2017) Congenital human cytomegalovirus infection and the enigma of maternal immunity. J Virol 91:e02392-16

Bunde T, Kirchner A, Hoffmeister B et al (2005) Protection from cytomegalovirus after transplantation is correlated with immediate early 1-specific CD8 T cells. J Exp Med 201:1031–1036

Burwitz BJ, Malouli D, Bimber BN et al (2016) Cross-species rhesus cytomegalovirus infection of cynomolgus macaques. PLoS Pathog 12:e1006014

Cardin RD, Bravo FJ, Pullum DA et al (2016) Replication-defective lymphocytic choriomeningitis virus vectors expressing guinea pig cytomegalovirus gB and pp65 homologs are protective against congenital guinea pig cytomegalovirus infection. Vaccine 34:1993–1999

Cayatte C, Schneider-Ohrum K, Wang Z et al (2013) Cytomegalovirus vaccine strain Towne-derived dense bodies induce broad cellular immune responses and neutralizing antibodies that prevent infection of fibroblasts and epithelial cells. J Virol 87:11107–11120

Cha TA, Tom E, Kemble GW et al (1996) Human cytomegalovirus clinical isolates carry at least 19 genes not found in laboratory strains. J Virol 70:78–83

Chandramouli S, Malito E, Nguyen T et al (2017) Structural basis for potent antibody-mediated neutralization of human cytomegalovirus. Sci Immunol 2:eaan1457

Chang WLW, Barry PA (2010) Attenuation of innate immunity by cytomegalovirus IL-10 establishes a long-term deficit of adaptive antiviral immunity. Proc Natl Acad Sci U S A 107:22647–22652

Chang WL, Baumgarth N, Yu D, Barry PA (2004) Human cytomegalovirus-encoded interleukin-10 homolog inhibits maturation of dendritic cells and alters their functionality. J Virol 78:8720–8731

Chatterjee A, Harrison CJ, Britt WJ, Bewtra C (2001) Modification of maternal and congenital cytomegalovirus infection by anti-glycoprotein b antibody transfer in guinea pigs. J Infect Dis 183:1547–1553

Chaudhuri S, Lowen B, Chan G et al (2009) Human cytomegalovirus interacts with toll-like receptor 2 and CD14 on syncytiotrophoblasts to stimulate expression of TNFα mRNA and apoptosis. Placenta 30:994–1001

Chiuppesi F, Wussow F, Johnson E et al (2015) Vaccine-derived neutralizing antibodies to the human cytomegalovirus gH/gL pentamer potently block primary cytotrophoblast infection. J Virol 89:11884–11898

Chiuppesi F, Kaltcheva T, Meng Z et al (2017) Identification of a continuous neutralizing epitope within UL128 of human cytomegalovirus. J Virol 91:e01857–e01816

Ciferri C, Chandramouli S, Donnarumma D et al (2015a) Structural and biochemical studies of HCMV gH/gL/gO and Pentamer reveal mutually exclusive cell entry complexes. Proc Natl Acad Sci U S A 112:1767–1772

Ciferri C, Chandramouli S, Leitner A et al (2015b) Antigenic characterization of the HCMV gH/gL/gO and Pentamer cell entry complexes reveals binding sites for potently neutralizing human antibodies. PLoS Pathog 11:e1005230

Coaquette A, Bourgeois A, Dirand C et al (2004) Mixed cytomegalovirus glycoprotein B genotypes in immunocompromised patients. Clin Infect Dis 39:155–161

Corrales-Aguilar E, Trilling M, Hunold K et al (2014) Human cytomegalovirus Fcγ binding proteins gp34 and gp68 antagonize Fcγ receptors I, II and III. PLoS Pathog 10:e1004131

Crumpler MM, Choi KY, McVoy MA, Schleiss MR (2009) A live guinea pig cytomegalovirus vaccine deleted of three putative immune evasion genes is highly attenuated but remains immunogenic in a vaccine/challenge model of congenital cytomegalovirus infection. Vaccine 27:4209–4218

Cui X, Meza BP, Adler SP, McVoy MA (2008) Cytomegalovirus vaccines fail to induce epithelial entry neutralizing antibodies comparable to natural infection. Vaccine 26:5760–5766

Dempsey AF, Pangborn HM, Prosser LA (2012) Cost-effectiveness of routine vaccination of adolescent females against cytomegalovirus. Vaccine 30:4060–4066

Dropulic LK, Cohen JI (2012) The challenge of developing a herpes simplex virus 2 vaccine. Expert Rev Vaccines 11:1429–1440

Eberhardt MK, Deshpande A, Chang WLW et al (2013) Vaccination against a virus-encoded cytokine significantly restricts viral challenge. J Virol 87:11323–11331

Elkington R, Walker S, Crough T et al (2003) Ex vivo profiling of CD8+-T-cell responses to human cytomegalovirus reveals broad and multispecific reactivities in healthy virus carriers. J Virol 77:5226–5240

Fan Q, Nelson CS, Bialas KM et al (2017) Plasmablast response to primary rhesus cytomegalovirus infection in a monkey model of congenital CMV transmission. Clin Vaccine Immunol 24:e00510-16

Fornara C, Cassaniti I, Zavattoni M et al (2017) Human cytomegalovirus-specific memory CD4+ T-Cell response and its correlation with virus transmission to the fetus in pregnant women with primary infection. Clin Infect Dis 65:1659–1665

Fouts AE, Chan P, Stephan J-PP et al (2012) Antibodies against the gH/gL/UL128/UL130/UL131 complex comprise the majority of the anti-cytomegalovirus (anti-CMV) neutralizing antibody response in CMV hyperimmune globulin. J Virol 86:7444–7447

Fowler KB, Stagno S, Pass RF et al (1992) The outcome of congenital cytomegalovirus infection in relation to maternal antibody status. N Engl J Med 326:663–667

Fowler KB, Stagno S, Pass RF (2003) Maternal immunity and prevention of congenital cytomegalovirus infection. J Am Med Assoc 289:1008–1011

Freed DC, Tang Q, Tang A et al (2013) Pentameric complex of viral glycoprotein H is the primary target for potent neutralization by a human cytomegalovirus vaccine. Proc Natl Acad Sci U S A 110:E4997–E5005

Fu TM, Wang D, Freed DC et al (2012) Restoration of viral epithelial tropism improves immunogenicity in rabbits and rhesus macaques for a whole virion vaccine of human cytomegalovirus. Vaccine 30:7469–7474

McDonagh S, Maidji E, Chang HT, Pereira L (2006) Patterns of human cytomegalovirus infection in term placentas: a preliminary analysis. J Clin Virol 35:210–215

Mori T, Kanda Y, Takenaka K et al (2017) Safety of ASP0113, a cytomegalovirus DNA vaccine, in recipients undergoing allogeneic hematopoietic cell transplantation: an open-label phase 2 trial. Int J Hematol 105:206–212

Nahum A, Dadi H, Bates A, Roifman CM (2012) The biological significance of TLR3 variant, L412F, in conferring susceptibility to cutaneous candidiasis, CMV and autoimmunity. Autoimmun Rev 11:341–347

Nakamura R, Rosa C, La LJ et al (2016) Viraemia, immunogenicity, and survival outcomes of cytomegalovirus chimeric epitope vaccine supplemented with PF03512676 (CMVPepVax) in allogeneic haemopoietic stem-cell transplantation: randomised phase 1b trial. Lancet Haematol 3:e87–e98

NCT01974206 News release (2016) HP of Vical (www.vical.com/investors/news-releases/News-Release-Details/2016/Vical-and-Astellas-Announce-Topline-Results-from-a-Phase-2-Study-of-Investigational-Cytomegalovirus-CMV-Vaccine-ASP0113-in-Kidney-Transplant-Patients/default.aspx)

NCT01877655 News release (2017) HP of Vical (www.vical.com/investors/news-releases/News-Release-Details/2017/Vical-Announces-Completion-of-the-Phase-3-ASP0113-CMV-Vaccine-Trial/default.aspx)

NCT02798692 News release (2017) HP of Hooki (www.29c5yd3ksizu1pn67922fy74.wpengine.netdna-cdn.com/wp-content/uploads/2017/05/Hookipa-press-release-HB-101-Data-at-CMV-Conference-4-May-2017.pdf)

NCT02826798 News release (2017) HP of VBI vacinnes (www.vbivaccines.com/wire/cmv-phase-i-clinical-study-update-may-2017/)

Nelson CS, Cruz DV, Tran D et al (2017) Preexisting antibodies can protect against congenital cytomegalovirus infection in monkeys. JCI Insight 2:94002

Nozawa N, Koyano S, Yamamoto Y et al (2007) Real-time PCR assay using specimens on filter disks as a template for detection of cytomegalovirus in urine. J Clin Microbiol 45:1305–1307

Nozawa N, Yamamoto Y, Fukui Y et al (2008) Identification of a 1.6 kb genome locus of guinea pig cytomegalovirus required for efficient viral growth in animals but not in cell culture. Virology 379:45–54

Nozawa N, Fang-Hoover J, Tabata T et al (2009) Cytomegalovirus-specific, high-avidity IgG with neutralizing activity in maternal circulation enriched in the fetal bloodstream. J Clin Virol 46:S58–S63

Ogawa H, Suzutani T, Baba Y et al (2007) Etiology of severe sensorineural hearing loss in children: independent impact of congenital cytomegalovirus infection and mutations. J Infect Dis 195:782–788

Pass RF, Duliegè AM, Boppana S et al (1999) A subunit cytomegalovirus vaccine based on recombinant envelope glycoprotein B and a new adjuvant. J Infect Dis 180:970–975

Pass RF, Zhang C, Evans A et al (2009) Vaccine prevention of maternal cytomegalovirus infection. N Engl J Med 360:1191–1199

Peggs KS, Verfuerth S, Pizzey A et al (2003) Adoptive cellular therapy for early cytomegalovirus infection after allogeneic stem-cell transplantation with virus-specific T-cell lines. Lancet 362:1375–1377

Pepperl S, Münster J, Mach M et al (2000) Dense bodies of human cytomegalovirus induce both humoral and cellular immune responses in the absence of viral gene expression. J Virol 74:6132–6146

Pereira L, Maidji E, McDonagh S et al (2003) Human cytomegalovirus transmission from the uterus to the placenta correlates with the presence of pathogenic bacteria and maternal immunity. J Virol 77:13301–13314

Petty TJ, Todd Callahan S, Chen Q et al (2010) Assessment of parental acceptance of a potential cytomegalovirus vaccine for adolescent females. Vaccine 28:5686–5690

Puchhammer-Stockl E, Gorzer I (2006) Cytomegalovirus and Epstein-Barr virus subtypes--the search for clinical significance. J ClinVirol 36:239–248

Raftery MJ, Wieland D, Gronewald S et al (2004) Shaping phenotype, function, and survival of dendritic cells by cytomegalovirus-encoded IL-10. J Immunol 173:3383–3391

Reap EA, Morris J, Dryga SA et al (2007) Development and preclinical evaluation of an alphavirus replicon particle vaccine for cytomegalovirus. Vaccine 25:7441–7449

Revello MG, Lilleri D, Zavattoni M et al (2006) Lymphoproliferative response in primary human cytomegalovirus (HCMV) infection is delayed in HCMV transmitter mothers. J Infect Dis 193:269–276

Revello MG, Lazzarotto T, Guerra B et al (2014) A randomized trial of hyperimmune globulin to prevent congenital cytomegalovirus. N Engl J Med 370:1316–1326

Riddell SR, Watanabe KS, Goodrich JM et al (1992) Restoration of viral immunity in immunodeficient humans by the adoptive transfer of T cell clones. Science 257:238–241

Ross SA, Arora N, Novak Z et al (2010) Cytomegalovirus reinfections in healthy seroimmune women. J Infect Dis 201:386–389

Russell JNH, Marsh AK, Willer DO et al (2016) A novel strain of cynomolgus macaque cytomegalovirus: implications for host-virus co-evolution. BMC Genomics 17:1–17

Schleiss MR, Bourne N, Bernstein DI (2003) Preconception vaccination with a glycoprotein B (gB) DNA vaccine protects against cytomegalovirus (CMV) transmission in the guinea pig model of congenital CMV infection. J Infect Dis 188:1868–1874

Schleiss MR, Bourne N, Stroup G et al (2004) Protection against congenital cytomegalovirus infection and disease in guinea pigs, conferred by a purified recombinant glycoprotein B vaccine. J Infect Dis 189:1374–1381

Schleiss MR, Stroup G, Pogorzelski K, McGregor A (2006) Protection against congenital cytomegalovirus (CMV) disease, conferred by a replication-disabled, bacterial artificial chromosome (BAC)-based DNA vaccine. Vaccine 24:6175–6186

Schleiss MR, Lacayo JC, Belkaid Y et al (2007) Preconceptual administration of an alphavirus replicon UL83 (pp65 homolog) vaccine induces humoral and cellular immunity and improves pregnancy outcome in the guinea pig model of congenital cytomegalovirus infection. J Infect Dis 195:789–798

Schleiss MR, Choi KY, Anderson J et al (2014) Glycoprotein B (gB) vaccines adjuvanted with AS01 or AS02 protect female guinea pigs against cytomegalovirus (CMV) viremia and offspring mortality in a CMV-challenge model. Vaccine 32:2756–2762

Schleiss MR, Bierle CJ, Swanson EC et al (2015) Vaccination with a live attenuated cytomegalovirus devoid of a protein kinase R inhibitory gene results in reduced maternal viremia and improved pregnancy outcome in a guinea pig congenital infection model. J Virol 89:9727–9738

Schleiss MR, Berka U, Watson E et al (2017) Additive protection against congenital cytomegalovirus conferred by combined glycoprotein B/pp65 vaccination using a lymphocytic choriomeningitis virus vector. ClinVaccine Immunol 24:e00300–e00316

Schneider-Ohrum K, Cayatte C, Liu Y et al (2016) Production of cytomegalovirus dense bodies by scalable bioprocess methods maintains immunogenicity and improves neutralizing antibody titers. J Virol 90:10133–10144

Scott GM, Chow SS, Craig ME et al (2012) Cytomegalovirus infection during pregnancy with maternofetal transmission induces a proinflammatory cytokine bias in placenta and amniotic fluid. J Infect Dis 205:1305–1310

Siewiera J, El Costa H, Tabiasco J et al (2013) Human cytomegalovirus infection elicits new decidual natural killer cell effector functions. PLoS Pathog 9:e1003257

Slavuljica I, Busche A, Babić M et al (2010) Recombinant mouse cytomegalovirus expressing a ligand for the NKG2D receptor is attenuated and has improved vaccine properties. J Clin Invest 120:4532–4545

Stratton KR, Durch JS, Lawrence RS (2000) Vaccines for the 21st century. National Academy Press, Washington, DC

Swanson EC, Gillis P, Hernandez-Alvarado N et al (2015) Comparison of monovalent glycoprotein B with bivalent gB/pp65 (GP83) vaccine for congenital cytomegalovirus infection in a guinea pig model: Inclusion of GP83 reduces gB antibody response but both vaccine approaches provide equivalent protection against p. Vaccine 33:4013–4018

Sylwester A, Mitchell B, Edgar J et al (2005) Broadly targeted human cytomegalovirus-specific CD4+ and CD8+ T cells dominate the memory compartments of exposed subjects. J Exp Med 202:673–685

Taniguchi R, Koyano S, Suzutani T et al (2013) Polymorphisms in TLR-2 are associated with congenital cytomegalovirus (CMV) infection but not with congenital CMV disease. Int J Infect Dis 17:e1092–e1097

Tomić A, Varanasi PR, Golemac M et al (2016) Activation of innate and adaptive immunity by a recombinant human cytomegalovirus strain expressing an NKG2D ligand. PLoS Pathog 12:e1006015

Walter EA, Greenberg PD, Gilbert MJ et al (1995) Reconstitution of cellular immunity against cytomegalovirus in recipients of allogeneic bone marrow by transfer of T-cell clones from the donor. N Engl J Med 333:1038–1044

Wang D, Shenk T (2005a) Human cytomegalovirus virion protein complex required for epithelial and endothelial cell tropism. Proc Natl Acad Sci U S A 102:18153–18158

Wang D, Shenk T (2005b) Human cytomegalovirus UL131 open reading frame is required for epithelial cell tropism. J Virol 79:10330–10338

Wang Z, La Rosa C, Maas R et al (2004) Recombinant modified vaccinia virus Ankara expressing a soluble form of glycoprotein B causes durable immunity and neutralizing antibodies against multiple strains of human cytomegalovirus. J Virol 78:3965–3976

Wang D, Li F, Freed DC et al (2011) Quantitative analysis of neutralizing antibody response to human cytomegalovirus in natural infection. Vaccine 29:9075–9080

Wang D, Freed DC, He X et al (2016) A replication-defective human cytomegalovirus vaccine for prevention of congenital infection. Sci Transl Med 8:362ra145

Weisblum Y, Panet A, Zakay-Rones Z et al (2011) Modeling of human cytomegalovirus maternal-fetal transmission in a novel decidual organ culture. J Virol 85:13204–13213

Weisblum Y, Panet A, Zakay-Rones Z et al (2015) Human cytomegalovirus induces a distinct innate immune response in the maternal-fetal interface. Virology 485:289–296

Wen Y, Monroe J, Linton C et al (2014) Human cytomegalovirus gH/gL/UL128/UL130/UL131A complex elicits potently neutralizing antibodies in mice. Vaccine 32:3796–3804

Wussow F, Yue Y, Martinez J et al (2013) A vaccine based on the rhesus cytomegalovirus UL128 complex induces broadly neutralizing antibodies in rhesus macaques. J Virol 87:1322–1332

Wussow F, Chiuppesi F, Martinez J et al (2014) Human cytomegalovirus vaccine based on the envelope gH/gL pentamer complex. PLoS Pathog 10:e1004524

Yamada S, Nozawa N, Katano H et al (2009) Characterization of the guinea pig cytomegalovirus genome locus that encodes homologs of human cytomegalovirus major immediate-early genes, UL128, and UL130. Virology 391:99–106

Yamada S, Fukuchi S, Hashimoto K et al (2014) Guinea pig cytomegalovirus GP129/131/133, homologues of human cytomegalovirus UL128/130/131A, are necessary for infection of monocytes and macrophages. J Gen Virol 95:1376–1382

Yue Y, Kaur A, Zhou SS, Barry PA (2006) Characterization and immunological analysis of the rhesus cytomegalovirus homologue (Rh112) of the human cytomegalovirus UL83 lower matrix phosphoprotein (pp65). J Gen Virol 87:777–787

Yue Y, Kaur A, Eberhardt MK et al (2007) Immunogenicity and protective efficacy of DNA vaccines expressing rhesus cytomegalovirus glycoprotein B, phosphoprotein 65-2, and viral interleukin-10 in rhesus macaques. J Virol 81:1095–1109

Yue Y, Wang Z, Abel K et al (2008) Evaluation of recombinant modified vaccinia Ankara virus-based rhesus cytomegalovirus vaccines in rhesus macaques. Med Microbiol Immunol 197:117–123

Part III
Gammaherpesviruses

Chapter 14
KSHV Genome Replication and Maintenance in Latency

Keiji Ueda

Abstract Kaposi's sarcoma-associated herpesvirus (KSHV), also called human herpesvirus-8 (HHV-8), is the eighth human herpesvirus found by Yuan Chang and Patrick Moore, 1992. It is a *Rhadinovirus* belonging to the gamma herpesvirus subfamily. As known for many gamma herpesviruses, KSHV is also well-correlated to several cancer formations such as Kaposi's sarcoma, primary effusion lymphoma (PEL), and multicentric Castleman's disease. Different from the other herpesvirus subfamily, gamma herpesviruses establish latency as a default infection strategy when they infect to the target cells, as KSHV is present as the latent form in the related cancers. In the latency, the virus expresses a limited number of the genes such as *latency-associated nuclear antigen* (*LANA*), *v-cyclin* (*v-CYC, ORF72*), *v-FLIP* (*K13*), *kaposin* (*K12*), and 25 microRNAs (*K-miRNAs*). The virus replicates according to cellular replication machinery with a viral replication origin (ori-P) and LANA. Then, the replicated genome is segregated equally to daughter cells by appearance to maintain the virus genome copy number per cell. The virus makes the most use of cellular machinery to achieve this end. In this chapter, I would like to review KSHV replication and gene expression in the latency and discuss.

Keywords Kaposi's sarcoma-associated herpesvirus (KSHV) or human herpesvirus-8 (HHV-8) · Latency · ori-P · LANA (ORF73) · v-cyclin (v-CYC, ORF72) · v-FLIP (K13) · Kaposin (K12) · Kaposi's sarcoma (KS) · Primary effusion lymphoma (PEL) · Multicentric Castleman's disease (MCD)

K. Ueda (✉)
Division of Virology, Department of Microbiology and Immunology, Osaka University Graduate School of Medicine, Osaka, Japan
e-mail: kueda@virus.med.osaka-u.ac.jp

© Springer Nature Singapore Pte Ltd. 2018
Y. Kawaguchi et al. (eds.), *Human Herpesviruses*, Advances in Experimental Medicine and Biology 1045, https://doi.org/10.1007/978-981-10-7230-7_14

14.1 Introduction

Kaposi's sarcoma-associated herpesvirus (KSHV) is a human gamma herpesvirus found from Kaposi's sarcoma in 1992 by Moore and Chang (Chang et al. 1994). This is the second γ-herpesvirus to infect human. As known well for many gamma herpesviruses, the virus is well-associated with several cancers such as Kaposi's sarcoma (KS), primary effusion lymphoma (PEL), and multicentric Castleman's disease (MCD) and is the most important etiologic factor for these malignancies (Boshoff and Weiss 1998, 2001; Boshoff and Chang 2001).

KSHV is disseminated worldwide and incidental to high human immunodeficiency virus (HIV)-disseminated area such as south Saharan Africa, where KSHV infection is a kind of endemic disease. In the developed country, KSHV incidence is limited to a few percent, and high incidence is restricted to homosexual men, though classical KS in which KSHV is a real etiologic agent is seen in some district such as Sicily (Boshoff and Weiss 2001).

Infection pathway of KSHV is thought to be by salivary shedding as a major root as seen for the other herpesvirus infection. Some study has shown that the antibody-positive patient against KSHV increased as the age up to adolescent (Hengge et al. 2002; Plancoulaine et al. 2000; Andreoni et al. 2002). In case of homosexual men, KSHV could be sexually transmitted through secretion of sex organs (Martin et al. 1998). Generally, the virus infects epithelial cells and amplifies slightly at the infection site in situ and then goes to infect major target cells. In case of KSHV, the major target cells appear to be B lymphocytes (Ambroziak et al. 1995; Duus et al. 2004) but also endothelial cells, and KSHV establishes latency as a default infection course (Grundhoff and Ganem 2004). It is not known that primary infection of KSHV causes acute viral infection/disease, though there is some report on that (Karass et al. 2017).

KSHV and EBV are well-related to some lymphoma: primary effusion lymphoma for KSHV and Burkitt lymphoma for EBV. These malignancies maintain the viral episome, which can replicate their genome according to cell cycle, and the replicated genomes are apparently segregated and the copy number is maintained at the same number as the cell genome is replicated, segregated, and maintained accurately. That is, there should be a highly sophisticated tactics for the accuracy.

In this chapter, we summarize and discuss especially viral replication and gene expression strategy of KSHV latency.

14.2 γ-herpesvirus Subfamily and Its Related Diseases

KSHV belongs to γ2-herpesvirus subfamily of *Herpesviridae* that is known for establishment of the latency state after infection (Verma and Robertson 2003). α-Herpesvirus such as herpes simplex virus-1 (HSV-1) or human herpesvirus-1 (HHV-1), herpes simplex virus-2 (HSV-2) or human herpesvirus-2 (HHV-2), and varicella-zoster virus (VZV) or human herpesvirus-3 (HHV-3) establish their latency especially in neuronal cells of infected bodies, and β-herpesviruses such as human cytomegalovirus (HCMV) or human herpesvirus-5 (HHV-5), human herpesvirus-6 (HHV-6), and human herpesvirus-7 (HHV-7) seem to establish in peripheral blood mononuclear cells except B lymphocytes (Pellett 2013). These herpesviruses do not appear to establish latency in vitro infection system using cultured cells (Virgin et al. 2009).

On the other hand, γ-herpesvirus such as Epstein-Barr virus (EBV) or human herpesvirus-4 (HHV-4) and Kaposi's sarcoma-associated herpesvirus (KSHV as noted already) or human herpesvirus-8 (HHV-8) usually establish latency as a default process both *in* in vivo and in vitro. Ordinary target cells of EBV and KSHV are B lymphocytes, although EBV infects the other peripheral blood mononuclear cells such as NK cells and gastric epithelial cells (Longnecker et al. 2013) and KSHV infects blood/lymph vessel endothelial cells (Damania and Cesarman 2013), though some reported that KSHV could also infect the mononuclear cells such as monocytes, macrophage, and dendritic cells (see Table 14.1).

14.3 Establishment of KSHV Latency

After primary amplification of the virus at the infection site, the daughter viruses are disseminated around the body and reach the cells to establish latency especially in B lymphocytes and blood/lymph vessel endothelial cells (Damania and Cesarman 2013). Importantly, in KSHV-related diseases such as Kaposi's sarcoma (KS), primary effusion lymphoma (PEL), and multicentric Castleman's disease (MCD), KSHV is in the latency (Cesarman et al. 1995; Soulier et al. 1995).

14.4 KSHV Genome and Genes

The genome of KSHV in the infectious particles is a linearized double-stranded DNA whose size is about 170 kb including terminal repeats at the genome ends (Fig. 14.1). The linearized DNA genome is circularized to be an episome after entry into the cells (Fig. 14.1) (Veettil et al. 2014), the mechanism of which has not been well understood. The episome is maintained in the nucleus, and such form must be important to express viral genes (Knipe et al. 2013).

There are 80 genes or more encoded in the genome, and all the genes are expressed during lytic replication (Dourmishev et al. 2003). In the case of large DNA viruses, genes are determined based on open reading frames, and thus, in a strict sense, it is unknown how many genes were encoded in the genome, and the gene density appears extremely high compared with those of the cellular genome. Therefore, the viral gene expression regulation must be very elaborate. Herpesviruses usually express their genes in a cascade in the lytic replication phase, in which

Table 14.1 Classification of human herpesviruses

Subfamily	Genus	Members	Abbreviation	Primary latency sites	Primary infection	Recurrence or latency associated diseases
Alphaherpesvirinae	Simplexvirus	Herpes simplex 1 virus (human herpesvirus-1)	HSV-1 (HHV-1)	Sensory ganglia	Stomatitis, corneal herpes, herpetic encephalitis	Cold sores
Alphaherpesvirinae	Simplexvirus	Herpes simplex 1 virus (human herpesvirus-2)	HSV-2 (HHV-2)	Sensory ganglia	Genital ulcers	Genital ulcers
Alphaherpesvirinae	Varicellovirus	Varicella-zoster virus (human herpesvirus-3)	VZV (HHV-3)	Sensory ganglia	Chicken pox	Shingles
Betaherpesvirinae	Cytomegalovirus	Cytomegalovirus (human herpesvirus-5)	CMV (HHV-5)	Secretory glands, lymphoreticular cells, kidneys, etc.	Infectious mononucleosis, congenital CMV (CNS[a] involvement, hearing loss, fatal pneumonitis, etc.)	Disseminated diseases in immunocompromised hosts (CMV retinitis, pneumonia, esophagitis, etc.)
Betaherpesvirinae	Roseolovirus	Human herpesvirus-6	HHV-6	Secretory glands, lymphoreticular cells, kidneys, etc.	Exanthum subitum	Multiple sclerosis? Encephalitis?
Betaherpesvirinae	Roseolovirus	Human herpesvirus-7	HHV-7	Secretory glands, lymphoreticular cells, kidneys, etc.	Exanthum subitum	Unknown
Gammaherpesvirinae	Lymphocryptovirus	Epstein-Barr virus (human herpesvirus-4)	EBV (HHV-4)	B lymphocytes	Infectious mononucleosis	Burkitt lymphoma, Hodgkin's lymphoma, gastric carcinoma, chronic lymphoproliferative disease
Gammaherpesvirinae	Rhadinovirus	Kaposi's sarcoma-associated virus (human herpesvirus-8)	KSHV (HHV-8)	B lymphocytes, endothelial cells	KICS[b]?	Kaposi's sarcoma, primary effusion lymphoma, multiple Castleman's disease

[a]*CNS* central nervous system
[b]*KICS* **KSHV** inflammatory cytokine syndrome

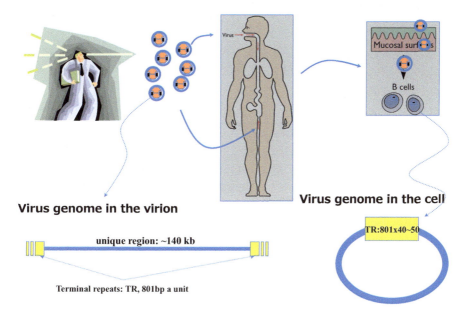

Fig. 14.1 KSHV transmission and genome. KSHV is mainly transmitted through saliva with exception of sexual transmission in homosexual men. The KSHV genome is a double-stranded linear DNA in the particles, and after infected into target cells, the genome is circularized at the end of the genome, which is called episomes and present in the infected nucleus

immediate early (IE) gene expression leads to early (E) gene expression followed by late (L) gene expression (Renne et al. 1996; Sun et al. 1999; Jenner et al. 2001). IE genes encode regulatory genes such as transcription factors.

Viruses usually express their genes just leading to lytic replication to produce daughter viruses after successful infection. In case of KSHV, probably this scenario progresses up to some steps and then quits somehow without reaching daughter virus production. Such abortive infection course seems to be a default process for KSHV (Krishnan et al. 2004), and thus KSHV infection usually goes to establishment of latency.

14.5 KSHV Latency

Latent infection means that virus does not produce any daughter virion and just expresses a limited number of genes required for maintenance of the latency. Otherwise, there is no viral gene expression. In case of KSHV, it seems that the virus establishes latency in B lymphocytes and endothelial cells in the infected body (Grundhoff and Ganem 2004; Krishnan et al. 2004) where latent genes such as latency-associated nuclear antigen (*LANA*), v-cyclin (*v-CYC*, *ORF72*), viral FLICE inhibitory protein *(v-FLIP)*, and *kaposin* (*K12*), 12 *microRNA* (*K-miRNA*), and viral

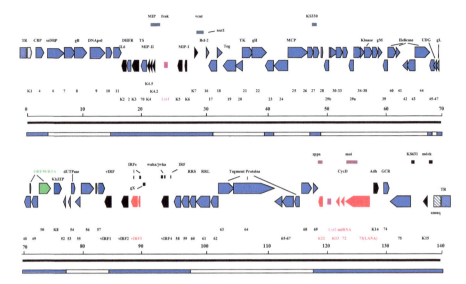

Fig. 14.2 KSHV transcription program in latency. KSHV genes (ORFs) are schematically shown. Red letters and regions show actively expressing genes in latency. There are two loci transcribed actively in latency, a locus around *LANA* and *vIRF3*. The transcription occurs in situ not genome wide. The green and the pink regions show an immediate early gene (*ORF50*, *RTA*) and lytic replication origins, respectively

interferon regulatory factor 3 (*v-IRF3*) are expressed (Burysek and Pitha 2001; Dittmer et al. 1998). Among them, *LANA, v-cyclin, v-FLIP, kaposin,* and 12 *K-miRNA* are clustered in one gene locus, though K-miRNA and kaposin (K12) are transcribed as independent genes (Dittmer et al. 1998; Sarid et al. 1999).

Thus, in the KSHV latency, two limited regions of the KSHV genome are transcriptionally active different from EBV, in which latent genes are transcribed genome-wide (Fig. 14.2) (Longnecker et al. 2013). This difference has not been understood well, but the epigenetic status is completely different, and KSHV gene expression is more tightly regulated (Zhang et al. 2014).

14.6 *LANA*, the Main Regulator of KSHV Latency

LANA is encoded by ORF73 and consists of 1021–1162 amino acids that are divided into the N-terminal region, the central region, and the C-terminal region (Wei et al. 2016). The N-terminal region contains a dominant nuclear localization signal and a histone H2A/H2B-binding region that interacts with host chromosomes. The C-terminal region has a DNA-binding domain interacting with LANA-binding sites (LBS) in the terminal repeat (TR) sequence of the genome and a dimerization domain which could be important for LANA to dimerize and bind with the LBS.

On the other hand, the central region of LANA consisted of very peculiar amino acid sequences that are an aspartate/glutamate-rich region (DDEE) and a glutamine-rich region which could be divided into QQQEP(L), QQQR(Q)EP, QQQDE/QQDE, and LEEQEQ regions. DDEE, QQQEP(L), QQQR(Q)EP, and LEEQEQ are relatively maintained among clones. Some functional motifs such as an Elongin BC-interacting domain (BC-box) and a SUMO-interacting domain (SIM) at the front of the DDEE region are also reported and it is not clear how important the other motifs (P-rich, DE, and Q-rich) are.

Most of LANA-interacting proteins have been reported to bind with N- or C-terminal region which means that the region should be important for function of LANA. Since the peculiar central region is maintained through the long virus evolution, the central region might have some important functions for the virus, though there is some variance of repetitious unit among KSHV clones (Fig. 14.3) (Gao et al. 1999).

LANA is a multifunctional protein and has a pivotal role to maintain latency by controlling transcription, replication, and maintenance of the genome. Especially, LANA binds with terminal repeats (TR) through the C-terminal DNA-binding region and with host chromosomes through the N-terminal histone-binding region. This activity is crucially important for KSHV genome replication followed by segregation and maintenance in latency (Ballestas et al. 1999; Cotter and Robertson 1999). In the KSHV latency, KSHV genome replication is totally dependent on cellular DNA replication cycle and utilizes cellular replication machinery. LANA binds with LBS and recruits cellular replication machinery such as origin recognition complex (ORC) consisting of ORC1, ORC2, ORC3, ORC4, ORC5, and ORC6 on the TR to initiate viral DNA replication (Stedman et al. 2004; Verma et al. 2006). This process is followed by CDC6 and Cdt1 and then MCM2-7 complex recruitment to complete pre-replication complex formation. The recruitment of preRC is carried out by interaction of LANA and ORC proteins through the C-terminal region of LANA. Thus, the multistep cellular DNA replication process and strictly controlled DNA replication licensing system assuring one replication cycle per cell cycle is made the most use by LANA (Sun et al. 2014).

KSHV DNA replication in latency is initiated from a viral replication origin called ori-P present in TR, though some other multiple replication origins were reported recently (ref). The viral ori-P consists of two LBS and 32 bp GC-rich segment (32GC) LBS, where LANA should recruit cellular replication components as mentioned above (Hu and Renne 2005; Ohsaki and Ueda 2012). The detail of the viral DNA replication initiation, however, has not been clarified yet, and thus importance of 32GC is unclear. Thus, there is no report on the real bubbling formation at the viral replication initiation and replication timing during S phase. The latter could be the middle or the late S phase, since TR is probably the state of heterochromatin or heterochromatin-like structure (Sakakibara et al. 2004), which is a rigid DNA conformation and reported to replicate at late S phase (Dimitrova and Gilbert 1999; Klochkov et al. 2009; Schwaiger et al. 2010).

LANA binds with LBS through its C-terminal DNA-binding region as a dimer as mentioned above. LBS is necessary for the viral replication but not sufficient so that

```
KSU75698 LANA.pep          1
MAPPGMRLRSGRSTGAPLTRGSCRKRNRSPERCDLGDDLHLQPRRKHVADSIDGRECGPH    60
U93872 LANA.pep            1 ..................................V........  60
GK18 LANA.pep              1 ..................................V........  60
GQ994935 LANA.pep          1 ..................................V........  60
HQ404500 LANA.pep          1 ......................N...........V........  60
KF588566 BrK#1-4 LANA.pep  1 ..................................V........  60
NC_009333 LANA.pep         1 ..................................V........  60
SPEL LANA.pep              1 ..................................V........  60
ZM004 LANA.pep             1 ..................................V.....P   60
ZM027 LANA.pep             1 ..................................V...D    60
ZM091 LANA.pep             1 ..................................V...D    60
                             *******************************************.***************.***.***.
```

```
KSU75698 LANA.pep          61
TLPIPGSPTVFTSGLPAFVSSPTLPVAPIPSPAPATPLPPPALLPPVTTSSSPIPPSHPV    120
U93872 LANA.pep            61 .................................  120
GK18 LANA.pep              61 .................................  120
GQ994935 LANA.pep          61 .................................  120
HQ404500 LANA.pep          61 ....................S............  120
KF588566 BrK#1-4 LANA.pep  61 .................................  120
NC_009333 LANA.pep         61 .................................  120
SPEL LANA.pep              61 .................................  120
ZM004 LANA.pep             61 ..........L...Q......C...........  120
ZM027 LANA.pep             61 .................................  120
ZM091 LANA.pep             61 .................................  120
                              *********************************.********.****.****.*********
```

```
KSU75698 LANA.pep          121
SPGTTDTHSPSPALPPTQSPESSQRPPLSSPTGRPDSSTPMRPPPSQQTTPPHSPTTPPP    180
U93872 LANA.pep            121 ...............................  180
GK18 LANA.pep              121 ...............................  180
GQ994935 LANA.pep          121 ...............................  180
HQ404500 LANA.pep          121 ...............................  180
KF588566 BrK#1-4 LANA.pep  121 ...............................  180
NC_009333 LANA.pep         121 ...............................  180
SPEL LANA.pep              121 ...............................  180
ZM004 LANA.pep             121 ......E........................  180
ZM027 LANA.pep             121 ......Q...S....................  180
ZM091 LANA.pep             121 ......Q...S....................  180
                               *********.***.*************************************************
```

```
KSU75698 LANA.pep          181
EPPSKSSPDSLAPSTLRSLRKRRLSSPQGPSTLNPICQSPPVSPPRCDFANRSVYPPWAT    240
U93872 LANA.pep            181 ...............................  240
GK18 LANA.pep              181 ...............................  240
GQ994935 LANA.pep          181 ...............................  240
HQ404500 LANA.pep          181 ...............................  240
KF588566 BrK#1-4 LANA.pep  181 ...............................  240
NC_009333 LANA.pep         181 ...............................  240
SPEL LANA.pep              181 ...............................  240
ZM004 LANA.pep             181 ...............................  240
ZM027 LANA.pep             181 ..........C....................  240
ZM091 LANA.pep             181 ..........C....................  240
                               ****************.*******************************************
```

Fig. 14.3 Variation of LANA amino acids sequence. Eleven LANA amino acids sequences are lined up to compare

```
KSU75698 LANA.pep        241
ESPIYVGSSSDGDTPPRQPPTSPISIGSSSPSEGSWGDDTAMLVLLAEIAEEASKNEKEC  300
U93872 LANA.pep          241 .......................................................  300
GK18 LANA.pep            241 .......................................................  300
GQ994935 LANA.pep        241 .......................................................  300
HQ404500 LANA.pep        241 .......................................................  300
KF588566 BrK#1-4 LANA.pep  241 .....................................................  300
NC_009333 LANA.pep       241 .......................................................  300
SPEL LANA.pep            241 .......................................................  300
ZM004 LANA.pep           241 .................T......................A......  300
ZM027 LANA.pep           241 .......................................................  300
ZM091 LANA.pep           241 ..........................................A.....  300
                             *********************.*****************************.******


KSU75698 LANA.pep        301
SENNQAGEDNGDNEISKESQVDKDDNDNKDDEEEQETDEEDEEDDEEDDEEDDEEDDEED  360
U93872 LANA.pep          301 .......................................................  360
GK18 LANA.pep            301 .......................................................  360
GQ994935 LANA.pep        301 .......................................................  360
HQ404500 LANA.pep        301 ..........................D....................  360
KF588566 BrK#1-4 LANA.pep  301 .....................................................  360
NC_009333 LANA.pep       301 .......................................................  360
SPEL LANA.pep            301 .......................................................  360
ZM004 LANA.pep           301 ...................................E.............  360
ZM027 LANA.pep           301 .................A................E..E..EE.DEE..E   360
ZM091 LANA.pep           301 .........................E.....--.D.....  358
                             *****************************.**************.***..**.**......**.


KSU75698 LANA.pep        361 DEEDDEEDDEEDDEEDDEEDDEEDDEEEDEEEDEEEDEEEED----
EEDDDDEDNEDEEDD    417
U93872 LANA.pep          361 ................E...E..........---....---.....  414
GK18 LANA.pep            361 .......E...E...E...----------....---....---......  406
GQ994935 LANA.pep        361 ...........E...E...E.......---.....---....---..  411
HQ404500 LANA.pep        361 ....................................---....--.........  414
KF588566 BrK#1-4 LANA.pep  361 ..........E...E...E........---.....---....---.........  411
NC_009333 LANA.pep       361 .......E...E...E........---------....---....---  406
SPEL LANA.pep            361 ..............E.......ED...DDD..NED....EE..  420
ZM004 LANA.pep           361 .........E...E...E................----....---...........  412
ZM027 LANA.pep           361 ...EED.EE.DEE...E..E...E........ED....DEEED..E.EE..E.....  420
ZM091 LANA.pep           359 ...........E...E..DE..DE..D-----E---...EEKDEE.....  410
                             ***..*..*..**..**..**...........*...  ...  **.....*****


KSU75698 LANA.pep        418 EEEDKKEDEEDGGDGNKTLSIQSSQQQQEPQQ---
QEPQQQEPQQQEPLQEPQQQEPQQQ    474
U93872 LANA.pep          415 ..............................---..........---.......  467
GK18 LANA.pep            407 ..............................---..........---.......  459
GQ994935 LANA.pep        412 ..............................---..........---.......  464
HQ404500 LANA.pep        415 ..............................---..........---L-....  466
KF588566 BrK#1-4 LANA.pep  412 .............................---..........---........  464
NC_009333 LANA.pep       407 ..............................---..........---.......  459
SPEL LANA.pep            421 ..............................---...................  477
ZM004 LANA.pep           413 ............V...........S..EPQ.QEP.........-.QEP.......  471
ZM027 LANA.pep           421 ..................-L.---..Q......---........  472
ZM091 LANA.pep           411 DD..NEDE.D.EEED..E---------DEEDGGDVNKTLSI.SSQ.....-.Q---......  459
                             ..**...*..*...**.............    ...*..*****    ...*****
```

Fig. 14.3 (continued)

```
KSU75698 LANA.pep         475 EPQQQEPLQEPQQQEPQQQEPLQEPQQQEPQQQEPQQQEPQQQEPQQ-
QEPQQQE    533
U93872 LANA.pep           468 ......---....L-..----...........R....R....R...-R....R.    517
GK18 LANA.pep             460 ......---.----....R..----....R....R....R....R...-R....R.    510
GQ994935 LANA.pep         465 ........-.......R..-----...R....R....R....R...-.......    515
HQ404500 LANA.pep         467 ........---.....----..............R......-R.....    517
KF588566 BrK#1-4 LANA.pep 465 ..........---......R..-----..R....R....R....R...-......    515
NC_009333 LANA.pep        460 ......---..----....R..----....R....R....R....R...R.    510
SPEL LANA.pep             478 ........----......-----...............-......    528
ZM004 LANA.pep            472 ....EPQQ..............----....................EPQ..--......    525
ZM027 LANA.pep            473 ........---.........----.............SK.EPQ.EPQQ.EPKQEP...    524
ZM091 LANA.pep            460 ....EPQQ.............----..........-..........-......    512
                              ****..   *****.*.**   **.****.****  *..*..*.  .*.  ..*.**
```

```
KSU75698 LANA.pep         534 PQQQEPQQQEPQQQEPQQREPQQREPQQREPQQREPQQ---REPQQREPQQREP--
-QQR    587
U93872 LANA.pep           518 ...R.........R.......Q....Q....Q......--Q....Q....Q..---..Q    571
GK18 LANA.pep             511 ...R....R....R..........Q....Q....Q......--Q....Q....Q..---..Q    564
GQ994935 LANA.pep         516 .................----Q....Q....Q......--Q....Q....Q..---..Q    564
HQ404500 LANA.pep         518 ...R....R....R.....------..........-Q....Q....Q..---..Q    566
KF588566 BrK#1-4 LANA.pep 516 .................-----Q....Q....Q......--Q....Q....Q..---..Q    564
NC_009333 LANA.pep        511 ...R....R....R..........Q....Q....Q......--Q....Q....Q..---..Q    564
SPEL LANA.pep             529 ......R....R.....................-----........Q....---..Q    582
ZM004 LANA.pep            526 ...-.........EPQ..--EP.QQEP.Q.....Q..K.EPQQ.....Q...Q..KQEP.Q    582
ZM027 LANA.pep            525 ............EPQ.EPQQEPQQEP.Q.....Q....EPQQ.......EPQ.EPQQ---EPQ    581
ZM091 LANA.pep            513 ...-...........K.--EP.Q....Q....Q..K.EPQQ.....Q....Q..-----    563
                              ***  ****.****   ....     .*.****.**.*  ....****...*....  ...
```

```
KSU75698 LANA.pep         588
EPQQREPQQQDEQQQDEQQQDEQQQDEQQQDEQQQDEQQQDEQQQDEQQQDEQQQ   647
U93872 LANA.pep           572 DE..QDE..................E.-------------------QDE..   611
GK18 LANA.pep             565 ....Q.....EP....EP....EP....EP..........-------------------.....   604
GQ994935 LANA.pep         565 ....Q.....EP....EP...............--------------------.....   604
HQ404500 LANA.pep         567 ....Q.....EP....EP...............--------------------......   606
KF588566 BrK#1-4 LANA.pep 565 ....Q.....EP....EP...............--------------------.....   604
NC_009333 LANA.pep        565 ....Q.....EP....EP....EP....EP..........-------------------....   604
SPEL LANA.pep             583 ....Q.....EP...............-------DEQQ..DEQ......   634
ZM004 LANA.pep            583 ....Q.....EP....EP....EP...........E.QDE.EQQDE.QQ.DEQQ..DEQ...   642
ZM027 LANA.pep            582 QEP.Q.....EP....EP.EPQ.EPQQ.EPQQ.EP--Q.EPQQEP.Q.P..EP..EPQ..E   640
ZM091 LANA.pep            564 --..Q.....EP...........----EQE.....E.QDE..QQDE.EQ.DEQQ..DEQ...   617
                              ...*..***..***..**............*..       ...*.
```

```
KSU75698 LANA.pep         648 DEQQQ--
DEQQQDEQQQDEQQQDEQQQDEQQQDEQQQDEQQQDEQQQDEQEQQQDE    705
U93872 LANA.pep           612 QDE..--QQDE.EQ.EEQ...E-..E.Q.E.EQ...EEQE.EL.E.EQ.LEEQ...LEEQ   668
GK18 LANA.pep             605 .....--...............E.QDE...QDE...QQDE.EQ.EEQ..QEEQQ   662
GQ994935 LANA.pep         605 .....--.E.QDE...QDE...--..DEQE..E..E.Q.E.EQQEE.EQ.EEQ...LEEQ   660
HQ404500 LANA.pep         607 .....--......E.QDE..--..DEQ.........QDE.-QQEE.EQ.EEQ...LEEQ   661
KF588566 BrK#1-4 LANA.pep 605 .....--...E.QDE..QDE..--..DEQE..E..E.Q.E.EQQEE.EQ.EEQ...LEEQ   660
NC_009333 LANA.pep        605 .....--...............E.QDE...QDE..QQDE.EQ.EEQ..QEEQQ   662
SPEL LANA.pep             635 QDE.E--QQEE.EQ.---------------------------------   647
ZM004 LANA.pep            643 ....QDEQ.........PQDE.EPQEE.EQELEEQE.EL.E.EQ.LEEQ...LEEQ   702
ZM027 LANA.pep            641 PQ..E--.......E.Q.E..QDE...QDE.EQQDE.QQDE.EQ.EEQQ.DE.EQ   698
ZM091 LANA.pep            618 ....QDEQ.........PQDE.EPQEE.EQELEEQE.EL.E.EQ.LEEQ...LEEQ   677
                              ...*.   ...*............  ..   ...   ... ..................
```

Fig. 14.3 (continued)

```
KSU75698 LANA.pep     706
QEQQDEQEQQDEQQQDEQQQQDEQQQQDEQQQQDEQQQQDEQQQQDEQEQQEEQEQQ---  762
U93872 LANA.pep          669 EQELE....EL.E.-----E.EL.E.E.ELEE.EQ.LEE.EQELEEQ...LE.QEQELEEQ   724
GK18 LANA.pep            663 .DE.QQD...QDE..QDE.E.QDE...........E..E....E............EEQ   722
GQ994935 LANA.pep        661 EQELE....EL.E.-----E.EL.E.E.ELEE.EQ.LEE.EQELEEQ...LE.QEQELEEQ   716
HQ404500 LANA.pep        662 EQELE....EL.E.-----E.EL.E.E.ELEE.EQ.LEE.EQELEEQ...LE.QEQELEEQ   717
KF588566 BrK#1-4 LANA.pep  661 EQELE....EL.E.-----E.EL.E.E.ELEE.EQ.LEE.EQELEEQ...LE.QEQELEEQ   716
NC_009333 LANA.pep       663 .DE.QQD...QDE..QDE.E.QDE...........E..E....E............EEQ   722
SPEL LANA.pep            648 ---------------------------------------------------------   647
ZM004 LANA.pep           703 EQELE....EL----------------------------------------------   713
ZM027 LANA.pep           699 ..E.EQELEEQ..ELE..E.EL.E.E.EL----------------------------   727
ZM091 LANA.pep           678 EQELE....EL----------------------------------------------   688
                         ............    . .....  . . ... ....   . ...
```

```
KSU75698 LANA.pep     763 ----EEQEQELEEQEQELEDQEQELEE-QEQELEE-QEQELEE-QEQELEE-QEQELEE-
813
U93872 LANA.pep          725 EQEL.............E.....-......-.....-......-......-  779
GK18 LANA.pep            723 EQEL.............E.......-......-.....-......-......-  777
GQ994935 LANA.pep        717 EQEL.............E.....-......-.....-......-......-  771
HQ404500 LANA.pep        718 EQEL.............E.....-......-.....-......-......-  772
KF588566 BrK#1-4 LANA.pep  717 EQEL.............E.....-......-.....-......-......-  771
NC_009333 LANA.pep       723 EQEL.............E.....-......-.....-......-......-  777
SPEL LANA.pep            648 ---.........Q....E.......-......-......-......-      698
ZM004 LANA.pep           714 ---.-...........E......VE....VE.....VE....VE....V   769
ZM027 LANA.pep           728 ---.-.............E.......-....VE.....VE.....VE....V   782
ZM091 LANA.pep           689 ---.-.............E......VE.....VE.....VE....V   743
                         .*****.**********.*******.*******.*******.*******.*******
```

```
KSU75698 LANA.pep     814 QEQELEEQEQELEEQEQELEEQE-QELEEQE---VEEQEQEVEEQEQEQEELEEVEEQE   870
U93872 LANA.pep          780 ....-QELE.V......Q....LE.V....QEQ....EQEL.EVE...........-EQ   836
GK18 LANA.pep            778 .....................-.......QEL......L.........LE.V..Q.QEQ   836
GQ994935 LANA.pep        772 .....................-.......QEL......L.........LE.V..Q.QEQ   830
HQ404500 LANA.pep        773 .....................-.......QEL......-------..LE.V..Q.QEQ   824
KF588566 BrK#1-4 LANA.pep  772 .....................-.......QEL......L.........LE.V..Q.QEQ   830
NC_009333 LANA.pep       778 .....................-.......QEL......L.........LE.V..Q.QEQ   836
SPEL LANA.pep            699 .....................-.......QEL......L.........LE.V..Q.QEQ   757
ZM004 LANA.pep           770 E...Q.QE......V.EQEQ...-EQE.QELEE......QGV.QEQE------------  816
ZM027 LANA.pep           783 E...Q.QE......V.EQEQ...--.Q...-EQ....L.EV.EQE...LE.V..Q.Q.G  839
ZM091 LANA.pep           744 E...Q.QEK......V.EQEQ...-EQE.QELEE......QGV..QEQE------------  790
                         .***.....*..**.*...***...*..*......****. ..........
```

```
KSU75698 LANA.pep     871 -QEQE-EQEEQELEEVEEQEE QELEEVEEQEEQELEEVEEQEQQELEEVEEQEQQGVEQQ   928
U93872 LANA.pep          837 -E...L.EV.EQEQQGV..Q..-------------------------------------   858
GK18 LANA.pep            837 -E...L.EV.EQEQ.Q....................E.............   895
GQ994935 LANA.pep        831 -E...L.EV.EQEQ.Q....................E.............   889
HQ404500 LANA.pep        825 -E...L.EV.EQEQ.Q....................E.............   883
KF588566 BrK#1-4 LANA.pep  831 -E...L.EV.EQEQ.Q....................E.............   889
NC_009333 LANA.pep       837 -E...L.EV.EQEQ.Q....................E.............   895
SPEL LANA.pep            758 -E...L.EV.EQEQ.Q....................E.............   816
ZM004 LANA.pep           817 ---------------------------------------------------   816
ZM027 LANA.pep           840 VEQ..Q.--------------------------------------------   846
ZM091 LANA.pep           791 ---------------------------------------------------   790
                         .......................
```

Fig. 14.3 (continued)

```
KSU75698 LANA.pep        929 EQETVEEPIILHGSSSEDEMEVDYPVVSTHEQIASSPPGDNTPDDDPQPGPSREYRYVLR  988
U93872 LANA.pep          859 ---.............................................. 915
GK18 LANA.pep            896 .................................................. 955
GQ994935 LANA.pep        890 .................................................. 949
HQ404500 LANA.pep        884 .................................................. 943
KF588566 BrK#1-4 LANA.pep   890 ............................................... 949
NC_009333 LANA.pep       896 .................................................. 955
SPEL LANA.pep            817 .................................................. 876
ZM004 LANA.pep           817 ---............................................... 873
ZM027 LANA.pep           847 ---............................................... 903
ZM091 LANA.pep           791 ---............................................... 847
                             ...***********************************************

KSU75698 LANA.pep        989 TSPPHRPGVRMRRVPVTHPKKPHPRYQQPPVPYRQIDDCPAKARPQHIFYRRFLGKDGRR 1048
U93872 LANA.pep          916 .................................................... 975
GK18 LANA.pep            956 ....................................................1015
GQ994935 LANA.pep        950 ....................................................1009
HQ404500 LANA.pep        944 ....................................................1003
KF588566 BrK#1-4 LANA.pep   950 .................................................1009
NC_009333 LANA.pep       956 ....................................................1015
SPEL LANA.pep            877 .................................................... 936
ZM004 LANA.pep           874 .................................................... 933
ZM027 LANA.pep           904 ...................S................................ 963
ZM091 LANA.pep           848 .................................................... 907
                             ********************************.******************

KSU75698 LANA.pep        1049 DPKCQWKFAVIFWGNDPYGLKKLSQAFQFGGVKAGPVSCLPHPGPDQSPITYCVYVYCQN 1108
U93872 LANA.pep          976 .................................................... 1035
GK18 LANA.pep            1016 ................................................... 1075
GQ994935 LANA.pep        1010 ................................................... 1069
HQ404500 LANA.pep        1004 ........................L.......................... 1063
KF588566 BrK#1-4 LANA.pep   1010 ................................................ 1069
NC_009333 LANA.pep       1016 ................................................... 1075
SPEL LANA.pep            937 .................................................... 996
ZM004 LANA.pep           934 ....................R.............N................ 993
ZM027 LANA.pep           964 .................................................... 1023
ZM091 LANA.pep           908 ....................R............................... 967
                             *****************************.***************.*******.*********

KSU75698 LANA.pep        1109 KDTSKKVQMARLAWEASHPLAGNLQSSIVKFKKPLPLTQPGENQGPGDSPQEMT 1162
U93872 LANA.pep          1036 ............................................. 1089
GK18 LANA.pep            1076 ............................................. 1129
GQ994935 LANA.pep        1070 ............................................. 1123
HQ404500 LANA.pep        1064 ............................................. 1117
KF588566 BrK#1-4 LANA.pep   1070 .......................................... 1123
NC_009333 LANA.pep       1076 ............................................. 1129
SPEL LANA.pep            997 .............................................. 1050
ZM004 LANA.pep           994 .............................................. 1047
ZM027 LANA.pep           1024 ............................................. 1077
ZM091 LANA.pep           968 .............................................. 1021
                             ******************************************************
```

Fig. 14.3 (continued)

32GC is also required. The full activity of the viral replication needs full length of LANA and at least the N-terminal chromosome-binding region and the C-terminal DNA-binding region including the leucine zipper region. It was reported that LANA without the central region could support the viral replication in the latency (Hu and Renne 2005; Hu et al. 2002). Efficient replication, however, seems to require the full length of LANA. In this term, it is interesting that place of cellular DNA replication takes place at the nuclear matrix and cellular pre-replication complex factors are recruited at the place, since LANA without the central region mainly accumulates in chromatin fraction but not in nuclear matrix with less efficient replication, while full length of LANA is localized mainly at nuclear matrix with full efficiency (Ohsaki et al. 2009; Ohsaki and Ueda 2012). Therefore, the central region of LANA might have a functional role for localizing at nuclear matrix, which has remained to be solved in detail.

LANA should also have an important role for the viral DNA segregation after the viral DNA replication according to cell cycle. LANA binds with LBS of the KSHV genome through its C-terminal DNA-binding region. LANA shows a dot-like structure in the KSHV-infected nucleus, where the KSHV episome is present (Ballestas and Kaye 2001). The KSHV genome copy number appears to be maintained at the same copy number in each KSHV-infected PEL cell line (Ueda et al. 2006; Verma et al. 2007; Ballestas and Kaye 2011). To maintain the same genome copy number as cellular chromosomes are maintained at the same number, there must be a strict mechanism such as a spindle checkpoint. If KSHV genome segregation is governed by cellular spindle checkpoint, there should be viral factors and genome functional units controlled by such. Some report showed that LANA is localized around centromeres of condensed chromosomes with interaction with a cellular spindle checkpoint factor such as Bub1 (Xiao et al. 2010). This fact is very attractive, since replicated KSHV genome should be governed by cellular spindle-assembly checkpoint machinery and correct segregation should be assured to maintain the same viral genome copy number.

From the other observation, however, LANA localization on the condensed chromosomes is very varied and is scattered on the condensed chromosomes and at the chromosome body, around telomere region rather than at the centromere region (Rahayu et al. 2016). Thus, it is unlikely that LANA itself is regulated by the cellular spindle checkpoint. Probably, LANA connects a KSHV genome and a chromosome, and the KSHV genome replicates in situ according to the cellular replication cycle. The replicated KSHV genome positions at the similar region of the chromosome as the chromosome is condensed in the G2-M phase (Fig. 14.4).

LANA is involved in cellular gene expression. LANA upregulates a dozen of cellular genes such as CDKN1A, CDK4, HDAC1, MCL1, BAP1, TNFRSF10B, CDKN2A, JUN, and so on (Wei et al. 2016). In reverse, LANA suppresses several TGFβ-II receptor (TβRII), TNF, etc., modestly (Di Bartolo et al. 2008).

As for viral gene expression, LANA is supposed to be a negative regulator, i.e., LANA inhibits RTA (replication and transcription activator, a key lytic replication inducer) expression by associating recombination signal sequence-binding protein Jκ (RBP-Jκ) (Lan et al. 2004). LANA can interact with heterochromatin protein 1

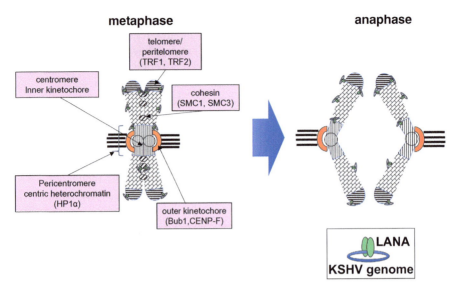

Fig. 14.4 Mechanism of KSHV genome segregation. LANA mediates the segregation of replicated KSHV genomes according to cell cycle. (Retrieved from ref. Rahayu et al. 2016)

(HP1) on LANA-binding sites of the KSHV terminal repeat region to recruit Suv39h1, which is a histone methyltransferase, which recognizes HP1 binding with di-/trimethylated histone H3K9 and could propagate heterochromatin on the viral genome (Sakakibara et al. 2004). As a result, the KSHV genome is generally transcriptionally silent over the genome and a limited region is active for the gene expression. This point is totally different from EBV, and EBV transcription in the latency is genome-wide as mentioned before (Fig. 14.2).

14.7 *v-CYC*, a *Cyclin D* Homolog to Facilitate Cell Cycle

Among the limited genes expressed in KSHV latency, *v-CYC* is transcribed as the same gene as LANA (Fig. 14.5). This is a *cyclin D* (*CYC-D*) homolog (Li et al. 1997) (Fig. 14.6) and could be a viral oncogene. The cellular CYC-D makes a complex with cyclin-dependent kinases such as CDK-4 and CDK-6 and promotes phosphorylation of *Rb* via CDC25A to leave a P1-E2F complex to facilitate expression of S-phase genes such as *cyclin-A* (*CYC-A*), *cyclin-E* (*CYC-E*), and so on (Direkze and Laman 2004).

Sole *v-CYC* overexpression is a kind of toxic to the cells, which means that sole v-CYC does not function as an oncogene and the other gene product should need to weaken the toxic effect (Verschuren et al. 2002; Koopal et al. 2007). In this term, LANA supports function of v-CYC by binding with GSK3ß, which makes an inhibitory phosphorylation on CDK4/6 (Fujimuro et al. 2003).

14 KSHV Genome Replication and Maintenance in Latency 313

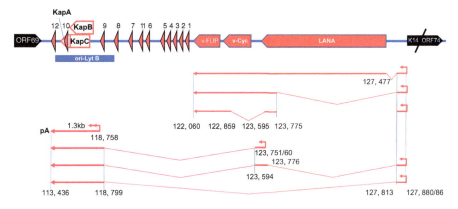

Fig. 14.5 KSHV transcription program in latency. Gene organization of actively transcribed loci is shown. Red color means actively expressed genes of KSHV latency. Triangles are KSHV miRNA

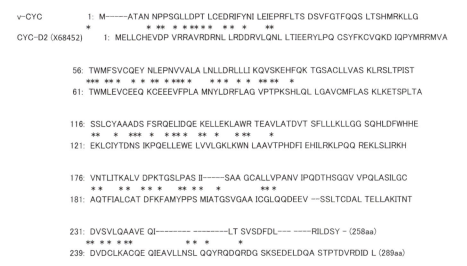

Fig. 14.6 Homology between c-CYC and cellular cyclin D2. Overall, there is about 30% homology

14.8 *v-FLIP*, a Homolog of *FLICE/CASPASE-8 Inhibitory Protein*

v-FLIP is a homolog of *FLIP* (Fig. 14.7) and also a latently expressed gene encoded in the same locus of LANA, and the ORF is K13 (Thome et al. 1997). Classically, this gene was thought to be translated through IRES, but recently it has been shown to be transcribed as a unique alternatively spliced gene (Grundhoff and Ganem 2001; Low et al. 2001). Although there is a specific antibody against v-FLIP and it

```
v-FLIP          1: MATYEVLCEV ARKLGTDDRE VVLFLLNVFI PQPTLAQLIG ALRALKEEGR LTFPLLAECL
                   * **  *  * ** *  ***          * * * *  *** *
c-FLIP (NM_001351592) 1: M-SAEVIHQV EEALDTDEKE MLLFLCRDVA IDVVPPNVRD LLDILRERGK LSVGDLAELL

               61: FRAGRRDLLR DLLHLDPRFL ERHLAGTMSY FSPYQLTVLH VDGELCARDI RSLIFLSKD-
                   * * ***  * * **    **    * *         *  *  ***** **
               60: YRVRRFDLLK RILKMDRKAV ETHLLRNPHL VSDYRVLMAE IGEDLDKSDV SSLIFLMKDY

              120: -TIGSRSTPQ TFLHWVYCME NLDLLGPTDV DALMSMLRSL SRVDL----- ----------
                   * *   ** * * ** *   ** *   * ***
              120: MGRGKISKEK SFLDLVVELE KLNLVAPDQL DLLEKCLKNI HRIDLKTKIQ KYKQSVQGAG

              164: --QRQV-QTL MGLHLSGPS- ---HSQHYRH TP-------- -- (188aa)
                   ***    * **      * * *
              180: TSYRNVLQAA IQKSLKDPSN NFRMITPYAH CPDLKILGNC SM (221aa)
```

Fig. 14.7 Homology between v-FLIP and c-FLIP. There is about 30% homology

has been proved that the open reading frame of *v-FLIP* assuredly encodes v-FLIP by transfection study, there has not been a proof showing the expression in KSHV-infected PEL cells/cell lines and/or Kaposi's sarcoma cells.

Many reports showed that v-FLIP interacted with IKKγ and activated IKKα/IKKβ/IKKγ complex, which results in NFκB activation (Matta et al. 2007a, b) (Fig. 14.8). The activation of NFκB is generally known to be important for cell death control, inflammation, oncogenesis, and so on (Cieniewicz et al. 2016). Thus, v-FLIP itself has an oncogenic activity (Ballon et al. 2011). Furthermore, v-FLIP inhibits oncogenic stress-induced autophagy (Lee et al. 2009) and helps oncogenic activity of *v-CYC* (Leidal et al. 2012; Liang 2012), which makes an oncogenic stress.

NFκB activation by v-FLIP has another aspect for KSHV latency control. KSHV lytic replication is induced by expression of replication and transcription activator (RTA), which is an immediate early gene of the KSHV lytic genes and a homolog of BRLF (Rta) of Epstein-Barr virus (EBV) genes. RTA can activate various kinds of viral and/or cellular genes through the specific DNA-binding and the non-DNA-binding mechanism. In KSHV latency, RTA activation is suppressed basically by NFκB activation (Brown et al. 2003). Thus, v-FLIP-induced NFκB activation controls KSHV reactivation from the latency not to be induced (Fig. 14.8).

14.9 Kaposin

kaposin is encoded next to the *LANA-v-CYC-v-FLIP* locus and translated into not only K12 ORF (kaposin-A) but also into two other frames, kaposin-B (KAP-B) and -C (KAP-C). This locus is reported to generate a transcript called T0.7 which means 0.7 kb mRNA (Sadler et al. 1999), and transcripts covering this region contains GC-rich direct repeat regions (DR1 and DR2), which are components of lytic replication origins. The predominant transcript in PEL cell lines includes the 5′-most

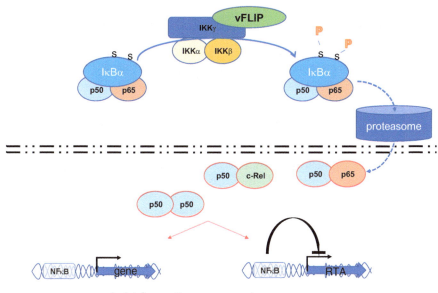

Fig. 14.8 v-FLIP interacts with IKKγ and acts on IKKγ-IKKα-IKKβ function. Canonical pathway of NFkB signal transduction is activated by v-FLIP, and thereafter NFkB responsible genes are activated. As for KSHV latency, the active NFkB suppresses KSHV reactivation

CUG codons and consisted of largely DR1 and DR2 and does not include K12 ORF. The ORF is corresponding to KAP-B (frame 2) and generates 48kd protein in PEL cell lines (Sadler et al. 1999) (Fig. 14.5). The size is much bigger than the predicted size about 18kd, since the coding amino acids are 178aa long and probably it is modified extensively posttranslationally (McCormick and Ganem 2006).

In the functional aspects, the KAP-B activates the p38/MK2 pathway and stabilizes cytokine mRNAs such as GM-CSF and IL-6 (McCormick and Ganem 2005). These features affect maintenance of KSHV latency and oncogenic phenotype of PEL; there has not been a report about knocking down of *kaposin*.

14.10 KSHV MicroRNAs (K-miRNAs)

Twelve pre-miRNAs are encoded in KHSV and processed into 25 mature miRNAs during KSHV latency (Marshall et al. 2007). The genes are located between *kaposin* and *v-FLIP* (Fig. 14.5) (Gottwein 2012). The details of transcription mechanism have not been elucidated yet. Many of them functions as maintenance of KSHV latency and are supposed to contribute KSHV-related pathogenesis (Boss et al. 2009; Lei et al. 2010a, b; Feldman et al. 2014).

14.11 Conclusions

There is no good system to observe how KSHV is involved in the related pathogenesis such as Kaposi's sarcoma (KS), primary effusion lymphoma (PEL), and multicentric Castleman's disease (MCD). Even though it has been reported that KSHV can infect primary endothelial cells, which is an origin of KS, and the other cell lines such as HEK293, Vero, and NIH3T3 cells and even more primary human B lymphocytes, we never observe KSHV transforming activity.

Although oncogenic activity of v-CYC and v-FLIP has been reported, the experimental system was usually a single gene function analysis. As for oncogenesis by KSHV, lytic genes such as *v-GPCR*, *K1*, *K2* (*v-IL6*), *K15*, and *ORF16* (*v-BCL2*) should be functional, and we do not know why KSHV does not represent its oncogenic activity even in vitro. Thus, it is important to understand how KSHV is involved in the related oncogenesis. For this purpose, it is a critical issue to find an experimental system with which we can study KSHV causing malignancies.

References

Ambroziak JA, Blackbourn DJ, Herndier BG, Glogau RG, Gullett JH, McDonald AR, Lennette ET, Levy JA (1995) Herpes-like sequences in HIV-infected and uninfected Kaposi's sarcoma patients. Science 268(5210):582–583

Andreoni M, Sarmati L, Nicastri E, El Sawaf G, El Zalabani M, Uccella I, Bugarini R, Parisi SG, Rezza G (2002) Primary human herpesvirus 8 infection in immunocompetent children. JAMA 287(10):1295–1300

Ballestas ME, Kaye KM (2001) Kaposi's sarcoma-associated herpesvirus latency-associated nuclear antigen 1 mediates episome persistence through cis-acting terminal repeat (TR) sequence and specifically binds TR DNA. J Virol 75(7):3250–3258

Ballestas ME, Kaye KM (2011) The latency-associated nuclear antigen, a multifunctional protein central to Kaposi's sarcoma-associated herpesvirus latency. Future Microbiol 6(12):1399–1413. https://doi.org/10.2217/fmb.11.137

Ballestas ME, Chatis PA, Kaye KM (1999) Efficient persistence of extrachromosomal KSHV DNA mediated by latency-associated nuclear antigen. Science 284(5414):641–644

Ballon G, Chen K, Perez R, Tam W, Cesarman E (2011) Kaposi sarcoma herpesvirus (KSHV) vFLIP oncoprotein induces B cell transdifferentiation and tumorigenesis in mice. J Clin Invest 121(3):1141–1153. https://doi.org/10.1172/JCI44417

Boshoff C, Chang Y (2001) Kaposi's sarcoma-associated herpesvirus: a new DNA tumor virus. Annu Rev Med 52:453–470

Boshoff C, Weiss RA (1998) Kaposi's sarcoma-associated herpesvirus. Adv Cancer Res 75:57–86

Boshoff C, Weiss RA (2001) Epidemiology and pathogenesis of Kaposi's sarcoma-associated herpesvirus. Philos Trans R Soc Lond Ser B Biol Sci 356(1408):517–534

Boss IW, Plaisance KB, Renne R (2009) Role of virus-encoded microRNAs in herpesvirus biology. Trends Microbiol 17(12):544–553. https://doi.org/10.1016/j.tim.2009.09.002

Brown HJ, Song MJ, Deng H, Wu TT, Cheng G, Sun R (2003) NF-kappaB inhibits gammaherpesvirus lytic replication. J Virol 77(15):8532–8540

Burysek L, Pitha PM (2001) Latently expressed human herpesvirus 8-encoded interferon regulatory factor 2 inhibits double-stranded RNA-activated protein kinase. J Virol 75(5):2345–2352

Cesarman E, Chang Y, Moore PS, Said JW, Knowles DM (1995) Kaposi's sarcoma-associated herpesvirus-like DNA sequences in AIDS-related body-cavity-based lymphomas. N Engl J Med 332(18):1186–1191

Chang Y, Cesarman E, Pessin MS, Lee F, Culpepper J, Knowles DM, Moore PS (1994) Identification of herpesvirus-like DNA sequences in AIDS-associated Kaposi's sarcoma. Science 266(5192):1865–1869

Cieniewicz B, Santana AL, Minkah N, Krug LT (2016) Interplay of murine Gammaherpesvirus 68 with NF-kappaB signaling of the host. Front Microbiol 7:1202. https://doi.org/10.3389/fmicb.2016.01202

Cotter MA 2nd, Robertson ES (1999) The latency-associated nuclear antigen tethers the Kaposi sarcoma-associated herpesvirus genome to host chromosomes in body cavity-based lymphoma cells. Virology 264(2):254–264

Damania BA, Cesarman E (2013) In: Knipe DMHPM (ed) Kaposi's sarcoma-associated herpesvirus, Fields virology, vol 2, 6th edn. Lippincott Willamas and Wilkins, Philadelphia, pp 2080–2128

Di Bartolo DL, Cannon M, Liu YF, Renne R, Chadburn A, Boshoff C, Cesarman E (2008) KSHV LANA inhibits TGF-beta signaling through epigenetic silencing of the TGF-beta type II receptor. Blood 111(9):4731–4740

Dimitrova DS, Gilbert DM (1999) The spatial position and replication timing of chromosomal domains are both established in early G1 phase. Mol Cell 4(6):983–993

Direkze S, Laman H (2004) Regulation of growth signalling and cell cycle by Kaposi's sarcoma-associated herpesvirus genes. Int J Exp Pathol 85(6):305–319

Dittmer D, Lagunoff M, Renne R, Staskus K, Haase A, Ganem D (1998) A cluster of latently expressed genes in Kaposi's sarcoma-associated herpesvirus. J Virol 72(10):8309–8315

Dourmishev LA, Dourmishev AL, Palmeri D, Schwartz RA, Lukac DM (2003) Molecular genetics of Kaposi's sarcoma-associated herpesvirus (human herpesvirus-8) epidemiology and pathogenesis. Microbiol Mol Biol Rev 67(2):175–212. table of contents

Duus KM, Lentchitsky V, Wagenaar T, Grose C, Webster-Cyriaque J (2004) Wild-type Kaposi's sarcoma-associated herpesvirus isolated from the oropharynx of immune-competent individuals has tropism for cultured oral epithelial cells. J Virol 78(8):4074–4084

Feldman ER, Kara M, Coleman CB, Grau KR, Oko LM, Krueger BJ, Renne R, van Dyk LF, Tibbetts SA (2014) Virus-encoded microRNAs facilitate gammaherpesvirus latency and pathogenesis in vivo. MBio 5(3):e00981–e00914. https://doi.org/10.1128/mBio.00981-14

Fujimuro M, Wu FY, ApRhys C, Kajumbula H, Young DB, Hayward GS, Hayward SD (2003) A novel viral mechanism for dysregulation of beta-catenin in Kaposi's sarcoma-associated herpesvirus latency. Nat Med 9(3):300–306

Gao SJ, Zhang YJ, Deng JH, Rabkin CS, Flore O, Jenson HB (1999) Molecular polymorphism of Kaposi's sarcoma-associated herpesvirus (Human herpesvirus 8) latent nuclear antigen: evidence for a large repertoire of viral genotypes and dual infection with different viral genotypes. J Infect Dis 180(5):1466–1476

Gottwein E (2012) Kaposi's sarcoma-associated herpesvirus microRNAs. Front Microbiol 3:165. https://doi.org/10.3389/fmicb.2012.00165

Grundhoff A, Ganem D (2001) Mechanisms governing expression of the v-FLIP gene of Kaposi's sarcoma-associated herpesvirus. J Virol 75(4):1857–1863

Grundhoff A, Ganem D (2004) Inefficient establishment of KSHV latency suggests an additional role for continued lytic replication in Kaposi sarcoma pathogenesis. J Clin Invest 113(1):124–136

Hengge UR, Ruzicka T, Tyring SK, Stuschke M, Roggendorf M, Schwartz RA, Seeber S (2002) Update on Kaposi's sarcoma and other HHV8 associated diseases. Part 1: epidemiology, environmental predispositions, clinical manifestations, and therapy. Lancet Infect Dis 2(5):281–292

Hu J, Renne R (2005) Characterization of the minimal replicator of Kaposi's sarcoma-associated herpesvirus latent origin. J Virol 79(4):2637–2642

Hu J, Garber AC, Renne R (2002) The latency-associated nuclear antigen of Kaposi's sarcoma-associated herpesvirus supports latent DNA replication in dividing cells. J Virol 76(22):11677–11687

Jenner RG, Alba MM, Boshoff C, Kellam P (2001) Kaposi's sarcoma-associated herpesvirus latent and lytic gene expression as revealed by DNA arrays. J Virol 75(2):891–902

Karass M, Grossniklaus E, Seoud T, Jain S, Goldstein DA (2017) Kaposi sarcoma inflammatory cytokine syndrome (KICS): a rare but potentially treatable condition. Oncologist 22(5):623–625. https://doi.org/10.1634/theoncologist.2016-0237

Klochkov DB, Gavrilov AA, Vassetzky YS, Razin SV (2009) Early replication timing of the chicken alpha-globin gene domain correlates with its open chromatin state in cells of different lineages. Genomics 93(5):481–486. https://doi.org/10.1016/j.ygeno.2009.01.001

Knipe DM, Lieberman PM, Jung JU, McBride AA, Morris KV, Ott M, Margolis D, Nieto A, Nevels M, Parks RJ, Kristie TM (2013) Snapshots: chromatin control of viral infection. Virology 435(1):141–156. https://doi.org/10.1016/j.virol.2012.09.023

Koopal S, Furuhjelm JH, Jarviluoma A, Jaamaa S, Pyakurel P, Pussinen C, Wirzenius M, Biberfeld P, Alitalo K, Laiho M, Ojala PM (2007) Viral oncogene-induced DNA damage response is activated in Kaposi sarcoma tumorigenesis. PLoS Pathog 3(9):1348–1360

Krishnan HH, Naranatt PP, Smith MS, Zeng L, Bloomer C, Chandran B (2004) Concurrent expression of latent and a limited number of lytic genes with immune modulation and anti-apoptotic function by Kaposi's sarcoma-associated herpesvirus early during infection of primary endothelial and fibroblast cells and subsequent decline of lytic gene expression. J Virol 78(7):3601–3620

Lan K, Kuppers DA, Verma SC, Robertson ES (2004) Kaposi's sarcoma-associated herpesvirus-encoded latency-associated nuclear antigen inhibits lytic replication by targeting Rta: a potential mechanism for virus-mediated control of latency. J Virol 78(12):6585–6594

Lee JS, Li Q, Lee JY, Lee SH, Jeong JH, Lee HR, Chang H, Zhou FC, Gao SJ, Liang C, Jung JU (2009) FLIP-mediated autophagy regulation in cell death control. Nat Cell Biol 11(11):1355–1362. https://doi.org/10.1038/ncb1980

Lei X, Bai Z, Ye F, Huang Y, Gao SJ (2010a) Regulation of herpesvirus lifecycle by viral microRNAs. Virulence 1(5):433–435. https://doi.org/10.4161/viru.1.5.12966

Lei X, Bai Z, Ye F, Xie J, Kim CG, Huang Y, Gao SJ (2010b) Regulation of NF-kappaB inhibitor IkappaBalpha and viral replication by a KSHV microRNA. Nat Cell Biol 12(2):193–199. https://doi.org/10.1038/ncb2019

Leidal AM, Cyr DP, Hill RJ, Lee PW, McCormick C (2012) Subversion of autophagy by Kaposi's sarcoma-associated herpesvirus impairs oncogene-induced senescence. Cell Host Microbe 11(2):167–180. https://doi.org/10.1016/j.chom.2012.01.005

Li M, Lee H, Yoon DW, Albrecht JC, Fleckenstein B, Neipel F, Jung JU (1997) Kaposi's sarcoma-associated herpesvirus encodes a functional cyclin. J Virol 71(3):1984–1991

Liang C (2012) Viral FLIPping autophagy for longevity. Cell Host Microbe 11(2):101–103. https://doi.org/10.1016/j.chom.2012.01.012

Longnecker RM, Kieff E, Cohen JI (2013) In: Knipe DMHPM (ed) Epstein-Barr virus, Fields Virology, vol 2, 6th edn. Lippincott Willamas and Wilkins, Philadelphia, pp 1898–1959

Low W, Harries M, Ye H, Du MQ, Boshoff C, Collins M (2001) Internal ribosome entry site regulates translation of Kaposi's sarcoma-associated herpesvirus FLICE inhibitory protein. J Virol 75(6):2938–2945

Marshall V, Parks T, Bagni R, Wang CD, Samols MA, Hu J, Wyvil KM, Aleman K, Little RF, Yarchoan R, Renne R, Whitby D (2007) Conservation of virally encoded microRNAs in Kaposi sarcoma--associated herpesvirus in primary effusion lymphoma cell lines and in patients with Kaposi sarcoma or multicentric Castleman disease. J Infect Dis 195(5):645–659

Martin JN, Ganem DE, Osmond DH, Page-Shafer KA, Macrae D, Kedes DH (1998) Sexual transmission and the natural history of human herpesvirus 8 infection. N Engl J Med 338(14):948–954

Matta H, Mazzacurati L, Schamus S, Yang T, Sun Q, Chaudhary PM (2007a) Kaposi's sarcoma-associated herpesvirus (KSHV) oncoprotein K13 bypasses TRAFs and directly interacts with the IkappaB kinase complex to selectively activate NF-kappaB without JNK activation. J Biol Chem 282(34):24858–24865

Matta H, Mazzacurati L, Schamus S, Yang T, Sun Q, Chaudhary PM (2007b) KSHV oncoprotein K13 bypasses TRAFs and directly interacts with the Ikappa B kinase complex to selectively activate NF-kappa B without JNK activation. J Biol Chem 282:24858–24865

McCormick C, Ganem D (2005) The kaposin B protein of KSHV activates the p38/MK2 pathway and stabilizes cytokine mRNAs. Science 307(5710):739–741

McCormick C, Ganem D (2006) Phosphorylation and function of the kaposin B direct repeats of Kaposi's sarcoma-associated herpesvirus. J Virol 80(12):6165–6170

Ohsaki E, Ueda K (2012) Kaposi's sarcoma-associated herpesvirus genome replication, partitioning, and maintenance in latency. Front Microbiol 3(7). https://doi.org/10.3389/fmicb.2012.00007

Ohsaki E, Suzuki T, Karayama M, Ueda K (2009) Accumulation of LANA at nuclear matrix fraction is important for Kaposi's sarcoma-associated herpesvirus replication in latency. Virus Res 139(1):74–84. https://doi.org/10.1016/j.virusres.2008.10.011

Pellett PERB (2013) In: Knipe DMHPM (ed) Herpesviridae, Fields virology, vol 2, 6th edn. Lippincott Williams and Wilkins, Philadelphia, pp 1802–1822

Plancoulaine S, Abel L, van Beveren M, Tregouet DA, Joubert M, Tortevoye P, de The G, Gessain A (2000) Human herpesvirus 8 transmission from mother to child and between siblings in an endemic population. Lancet 356(9235):1062–1065. https://doi.org/10.1016/S0140-6736(00)02729-X

Rahayu R, Ohsaki E, Omori H, Ueda K (2016) Localization of latency-associated nuclear antigen (LANA) on mitotic chromosomes. Virology 496:51–58. https://doi.org/10.1016/j.virol.2016.05.020

Renne R, Lagunoff M, Zhong W, Ganem D (1996) The size and conformation of Kaposi's sarcoma-associated herpesvirus (human herpesvirus 8) DNA in infected cells and virions. J Virol 70(11):8151–8154

Sadler R, Wu L, Forghani B, Renne R, Zhong W, Herndier B, Ganem D (1999) A complex translational program generates multiple novel proteins from the latently expressed kaposin (K12) locus of Kaposi's sarcoma-associated herpesvirus. J Virol 73(7):5722–5730

Sakakibara S, Ueda K, Nishimura K, Do E, Ohsaki E, Okuno T, Yamanishi K (2004) Accumulation of heterochromatin components on the terminal repeat sequence of Kaposi's sarcoma-associated herpesvirus mediated by the latency-associated nuclear antigen. J Virol 78(14):7299–7310

Sarid R, Wiezorek JS, Moore PS, Chang Y (1999) Characterization and cell cycle regulation of the major Kaposi's sarcoma-associated herpesvirus (human herpesvirus 8) latent genes and their promoter. J Virol 73(2):1438–1446

Schwaiger M, Kohler H, Oakeley EJ, Stadler MB, Schubeler D (2010) Heterochromatin protein 1 (HP1) modulates replication timing of the Drosophila genome. Genome Res 20(6):771–780. https://doi.org/10.1101/gr.101790.109

Soulier J, Grollet L, Oksenhendler E, Cacoub P, Cazals-Hatem D, Babinet P, d'Agay MF, Clauvel JP, Raphael M, Degos L et al (1995) Kaposi's sarcoma-associated herpesvirus-like DNA sequences in multicentric Castleman's disease. Blood 86(4):1276–1280

Stedman W, Deng Z, Lu F, Lieberman PM (2004) ORC, MCM, and histone hyperacetylation at the Kaposi's sarcoma-associated herpesvirus latent replication origin. J Virol 78(22):12566–12575

Sun R, Lin SF, Staskus K, Gradoville L, Grogan E, Haase A, Miller G (1999) Kinetics of Kaposi's sarcoma-associated herpesvirus gene expression. J Virol 73(3):2232–2242

Sun Q, Tsurimoto T, Juillard F, Li L, Li S, De Leon Vazquez E, Chen S, Kaye K (2014) Kaposi's sarcoma-associated herpesvirus LANA recruits the DNA polymerase clamp loader to mediate efficient replication and virus persistence. Proc Natl Acad Sci U S A 111(32):11816–11821. https://doi.org/10.1073/pnas.1404219111

Thome M, Schneider P, Hofmann K, Fickenscher H, Meinl E, Neipel F, Mattmann C, Burns K, Bodmer JL, Schroter M, Scaffidi C, Krammer PH, Peter ME, Tschopp J (1997) Viral

FLICE-inhibitory proteins (FLIPs) prevent apoptosis induced by death receptors. Nature 386(6624):517–521

Ueda K, Sakakibara S, Ohsaki E, Yada K (2006) Lack of a mechanism for faithful partition and maintenance of the KSHV genome. Virus Res 122(1–2):85–94

Veettil MV, Bandyopadhyay C, Dutta D, Chandran B (2014) Interaction of KSHV with host cell surface receptors and cell entry. Virus 6(10):4024–4046. https://doi.org/10.3390/v6104024

Verma SC, Robertson ES (2003) Molecular biology and pathogenesis of Kaposi sarcoma-associated herpesvirus. FEMS Microbiol Lett 222(2):155–163

Verma SC, Choudhuri T, Kaul R, Robertson ES (2006) Latency-associated nuclear antigen (LANA) of Kaposi's sarcoma-associated herpesvirus interacts with origin recognition complexes at the LANA binding sequence within the terminal repeats. J Virol 80(5):2243–2256

Verma SC, Lan K, Choudhuri T, Cotter MA, Robertson ES (2007) An autonomous replicating element within the KSHV genome. Cell Host Microbe 2(2):106–118

Verschuren EW, Klefstrom J, Evan GI, Jones N (2002) The oncogenic potential of Kaposi's sarcoma-associated herpesvirus cyclin is exposed by p53 loss in vitro and in vivo. Cancer Cell 2(3):229–241

Virgin HW, Wherry EJ, Ahmed R (2009) Redefining chronic viral infection. Cell 138(1):30–50. https://doi.org/10.1016/j.cell.2009.06.036

Wei F, Gan J, Wang C, Zhu C, Cai Q (2016) Cell cycle regulatory functions of the KSHV Oncoprotein LANA. Front Microbiol 7(334). https://doi.org/10.3389/fmicb.2016.00334

Xiao B, Verma SC, Cai Q, Kaul R, Lu J, Saha A, Robertson ES (2010) Bub1 and CENP-F can contribute to Kaposi's sarcoma-associated herpesvirus genome persistence by targeting LANA to kinetochores. J Virol 84(19):9718–9732. https://doi.org/10.1128/JVI.00713-10

Zhang L, Zhu C, Guo Y, Wei F, Lu J, Qin J, Banerjee S, Wang J, Shang H, Verma SC, Yuan Z, Robertson ES, Cai Q (2014) Inhibition of KAP1 enhances hypoxia-induced Kaposi's sarcoma-associated herpesvirus reactivation through RBP-Jkappa. J Virol 88(12):6873–6884. https://doi.org/10.1128/JVI.00283-14

Chapter 15
Signal Transduction Pathways Associated with KSHV-Related Tumors

Tadashi Watanabe, Atsuko Sugimoto, Kohei Hosokawa, and Masahiro Fujimuro

Abstract Signal transduction pathways play a key role in the regulation of cell growth, cell differentiation, cell survival, apoptosis, and immune responses. Bacterial and viral pathogens utilize the cell signal pathways by encoding their own proteins or noncoding RNAs to serve their survival and replication in infected cells. Kaposi's sarcoma-associated herpesvirus (KSHV), also known as human herpesvirus 8 (HHV-8), is classified as a rhadinovirus in the γ-herpesvirus subfamily and was the eighth human herpesvirus to be discovered from Kaposi's sarcoma specimens. KSHV is closely associated with an endothelial cell malignancy, Kaposi's sarcoma, and B-cell malignancies, primary effusion lymphoma, and multicentric Castleman's disease. Recent studies have revealed that KSHV manipulates the cellular signaling pathways to achieve persistent infection, viral replication, cell proliferation, anti-apoptosis, and evasion of immune surveillance in infected cells. This chapter summarizes recent developments in our understanding of the molecular mechanisms used by KSHV to interact with the cell signaling machinery.

Keywords Akt · Apoptosis · Cell cycle · IFN · IRF · KSHV · MAPK · NF-κB · Notch · p53 · Signal transduction pathway · Viral microRNA · Viral noncoding RNA · Wnt/β-catenin

15.1 Introduction

It is well known that viruses can resort to hijacking the cellular machinery for survival and replication. Recently, it has become clear that Kaposi's sarcoma-associated herpesvirus (KSHV), either directly or indirectly, targets the signal transduction pathways that are fundamental to many cellular processes such as proliferation, apoptosis, development, and the immune response.

T. Watanabe · A. Sugimoto · K. Hosokawa · M. Fujimuro (✉)
Department of Cell Biology, Kyoto Pharmaceutical University, Kyoto, Japan
e-mail: fuji2@mb.kyoto-phu.ac.jp

KSHV was isolated from a Kaposi's sarcoma (KS) lesion of a patient with acquired immunodeficiency syndrome (AIDS) (Chang et al. 1994). KSHV is associated with KS and is also linked to AIDS-related lymphoproliferative disorders, such as primary effusion lymphoma (PEL) and plasmablastic variant multicentric Castleman's disease (Moore et al. 1996b; Russo et al. 1996). The neoplastic potential of KSHV has been well established, especially within the context of immunosuppressed patients who are undergoing organ transplant or those who are coinfected with human immunodeficiency virus. KSHV establishes a lifelong infection in its host and exists in either a latent or a lytic state. During latent infection, the KSHV genome circularizes to form an episome in the nucleus, leading to the expression of several latency-associated genes (including vFLIP/ORF71, vCyclin/ORF72, LANA/ORF73, Kaposin/K12, vIL6/K2, viral microRNAs (miRNAs), vIRFs, and so on) that affect cell proliferation and apoptosis through manipulation of the cellular signaling, and contribute to KSHV-associated malignancies (Damania and Cesarman 2013). The latency-associated nuclear antigen (LANA) is also required for the maintenance of KSHV episomal DNA during latency and is expressed in all KSHV-associated malignancies from alternatively spliced transcripts that also encode vFLIP and vCyclin (Dittmer et al. 1998). LANA tethers the viral episomal genome to host chromosomes (Barbera et al. 2006) to ensure viral DNA replication and effective segregation in dividing cells (Hu et al. 2002; Grundhoff and Ganem 2003). KSHV miRNAs, encoded by the KSHV genome, are highly expressed during latency and in KS tumors and are strongly associated with the viral life cycle and development of tumors by directly (or indirectly) targeting gene expression and signaling pathways (Cai et al. 2005; Pfeffer et al. 2005). Upon reactivation, lytic-related genes are tightly regulated in a temporal and sequential manner, which can be divided into three transcriptional stages: immediate early, early, and late. The alternation of KSHV between lytic replication and latency depends on the immediate early gene RTA/ORF50, which triggers transcriptional activation of early genes including transcriptional activators for late genes and replication factors involved in viral DNA replication (Deng et al. 2007). Transcripts of early genes initiate DNA replication from the OliLyt site of the KSHV genome and transcription of late genes encoding structural and functional proteins for producing viral particles (Damania and Cesarman 2013; Deng et al. 2007).

The KSHV genome contains several pirated genes that are homologous with cellular genes such as viral cyclin D2 (vCyclin), viral FLICE inhibitory protein (vFLIP), viral G protein-coupled receptor (vGPCR), viral interferon regulatory factors (vIRF1/K9, vIRF2/K11, vIRF3/K10.5–10.6, vIRF4/K10), viral Bcl-2 (vBcl-2/ORF16), and viral interleukin-6 (vIL6) (Moore and Chang 1998; Choi et al. 2001), which directly (or indirectly) act on the cellular signaling pathways and contribute to KSHV-induced pathogenesis. These viral molecules mimic the cellular molecules for cell proliferation, anti-apoptosis, and immune surveillance. Furthermore, other KSHV-encoded proteins manipulate signal transduction pathways of the cell in various ways for establishment and maintenance of infection and abolishment of cellular defense mechanisms, including apoptosis and the interferon (IFN) response. In particular, latency-expressing KSHV proteins are able to stimulate cell proliferation and improve host cell survival during viral latency, which correlates with proliferation of the infected B-cell populations and the development of B-cell malignancies. In order to hijack the host cell to function in the best interest of the virus, KSHV is able to manipulate proliferation and anti-apoptotic signaling pathways by targeting its regulatory molecules. In this chapter, we describe the strategies used by KSHV at distinct stages of the viral life cycle to control signal transduction pathways and promote oncogenesis and viral persistence.

15.2 KSHV vIRFs and IFN-Related Signal Pathways

IFN-α and IFN-β, classified as the type-I IFNs, are the key cytokines activating antiviral responses via expression of various host genes, which suppress viral replication and induce apoptosis of the infected cells (Taniguchi and Takaoka 2002). The interferon regulatory factor (IRF) family proteins are the essential transcription factors for the expression of type-I IFNs and interferon-stimulated genes (ISGs) (Fig. 15.1). Two of the major pathways responsible for the activation of IRFs are TLR (Toll-like receptor) and RLR (RIG-I-like receptor) signaling pathways. Some TLRs and RLRs are the innate sensor molecules for pathogens including viral DNA

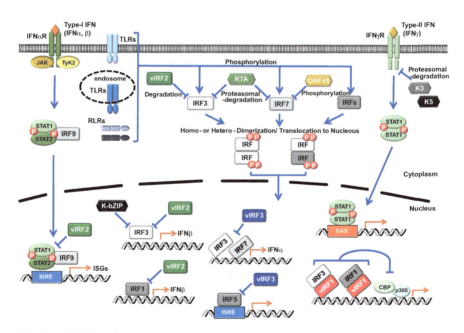

Fig. 15.1 KSHV-mediated dysregulation of IFN signaling
Abbreviations used: *Circled p* phospho, *GAS* IFN-γ-activated site, *IRF* interferon regulatory factor, *ISGs* interferon-stimulated genes, *ISRE* interferon-stimulated response element, *RLRs* RIG-I-like receptors, *TLRs* Toll-like receptors, *vIRF* viral interferon regulatory factor

or RNA. These activation cascades result in phosphorylation of IRFs. Phosphorylated IRF3 and IRF7 form hetero- or homodimers. Another major pathway is the IFN signaling pathway via type-I IFN receptors. Type-I IFN receptor complexes associate with JAK1/Tyk2, recruiting and phosphorylating STAT1 and STAT2. Activated STAT1 and STAT2 form a heterodimer; additionally, the heterodimer forms a complex with IRF9. The complex containing the STATs and IRF9 is termed the ISGF3 complex. Consequently, IRF dimers or the ISGF3 complex translocates to the nucleus, directly binding to ISRE (interferon-stimulated response element), upregulating the transcription of type-I IFN or ISGs harboring an ISRE.

KSHV encodes four homologous genes of cellular IRFs, known as viral IRFs (vIRF1, vIRF2, vIRF3, and vIRF4), at the K9–K11 locus of its genome. While vIRF1, vIRF2, and vIRF4 are classified as lytic genes, vIRF3 (also called LANA-2) is expressed in the nucleus of latent state PEL cells (Jacobs and Damania 2011). vIRFs share comparatively high sequence homology with the DNA-binding domain of cellular IRFs. However, several tryptophan (W) residues critical for DNA binding are not conserved in vIRFs (Tamura et al. 2008). Therefore, it is predicted that some vIRFs are defective or attenuated in their DNA-binding abilities.

vIRF3 has been shown to have a dominant-negative effect on IRF-3- and IRF-7-mediated transcription of the IFN-α promoter (Lubyova and Pitha 2000). Additionally, vIRF3 specifically interacts with IRF7 and inhibits the DNA-binding activity of IRF7. This inhibition results in suppression of IFN-α production and IFN-mediated immunity (Joo et al. 2007). vIRF3 also interacts with IRF5 and inhibits IRF5 binding to ISRE. Thus, vIRF3 counteracts IRF5-mediated signaling pathways, p53-independent apoptosis, and p21-mediated cell cycle arrest (Wies et al. 2009). To account for these results and the DNA-binding disability of vIRFs, vIRF3 plays a role as a decoy molecule to inhibit the formation of functional dimers of cellular IRFs. The IFN-induced protein kinase PKR is a sensor molecule of intracellular double-stranded RNA (dsRNA; e.g., viral infection). After sensing, autophosphorylated PKR enhances eIF-2a phosphorylation causing shutoff of the protein synthesis and activates a caspase cascade leading to apoptosis. vIRF3 antagonizes the PKR-mediated abrogation of protein synthesis and suppresses PKR-induced activation of caspase-3 (Esteban et al. 2003). vIRF3 associates with the c-Myc suppressor MM-1α and enhances c-Myc-dependent cyclin-dependent kinase (Cdk) 4 transcription. This association inhibits interaction between MM-1α and c-Myc. In addition, vIRF3 is recruited to the Cdk4 promoter, and histone acetylation of the promoter is enhanced. These findings suggest that vIRF3 not only cancels out MM-1α-mediated c-Myc suppression but also supports direct c-Myc-mediated transcriptional activation (Lubyova et al. 2007). vIRF3 interacts with and inhibits p53 phosphorylation and counteracts p53 oligomerization and DNA binding. Furthermore, vIRF3 induces p53 polyubiquitination and proteasomal degradation. Thus, vIRF-3 downregulates p53-mediated transcription (Baresova et al. 2014).

vIRF1 inhibits type-I and type-II signal transduction pathways, without disturbance of cellular IRF DNA-binding activities (Zimring et al. 1998; Gao et al. 1997). vIRF1 forms a dimer with IRF1 or IRF3 and inactivates transcriptional activity by inhibiting the histone acetyltransferase (HAT) activity of the CBP/p300 coactivator complex (Burysek et al. 1999; Li et al. 2000; Lin et al. 2001). vIRF1 targets apoptosis-related pathways. vIRF1 associates with pro-apoptotic protein Bim (Choi and Nicholas 2010) and Bid (Choi et al. 2012), which belong to the Bcl2 family of proteins. Bim, Bid, and other BH3 domain-only subfamily proteins activated by various apoptotic stimuli act as initiators of mitochondrial apoptosis. vIRF1 has a specific domain structure resembling the BH3-B domain of Bid, which interacts with the BH3 domain of Bid and contributes to an auto-inhibitory function. Thus, the interaction between vIRF1 and Bim/Bid suppresses mitochondrial apoptosis.

vIRF2 suppresses IRF1- and IRF3-dependent activation of the IFN-β promoter (Fuld et al. 2006). vIRF2 forms a complex with IRF3 and caspase-3 and induces caspase-3-dependent IRF3 degradation. This suggests that the degradation of IRF3 by caspase-3 partially contributes to the attenuation of the IRF3-mediated antiviral response (Areste et al. 2009). Furthermore, vIRF2 attenuates type-I IFNs, inducing signaling. vIRF2 inhibits IFN-α-induced transactivation of ISRE and ISGF3 complex binding to ISRE (Mutocheluh et al. 2011). Similar to vIRF3, vIRF2 interacts with PKR and inhibits autophosphorylation. This inhibition results in blocking the phosphorylation of H2A and eIF-2α, which are the substrates of PKR (Buryýšek and Pitha 2001). vIRF2 has the potency to function as a transcription factor. ChIP-seq revealed vIRF2-binding site mapping in the human genome. vIRF2 interacts with and modulates some specific promoters of *PIK3C3*, *HMGCR*, and *HMGCL*, genes related to autophagosome formation or tumor progression and metastasis (Hu et al. 2015).

Although little is known about the function of vIRF4 in IFN signaling, the effects of vIRF4 on cell proliferation have been revealed. vIRF4 inhibits the association of β-catenin/CBP cofactor at the cyclin D1 promoter, resulting in cell cycle progression and enhancement of viral replication (Lee et al. 2015). For B-cell development and functions, cellular IRF4-mediated c-Myc gene expression is essential. vIRF4 markedly suppresses the expression of cellular IRF4 and c-Myc and competes with cellular IRF4 for the promoter region of the c-Myc gene (Lee et al. 2014). Of the three other vIRFs, only vIRF4 interacts with CBF1/Suppressor of Hairless/Lag-1 and CSL (also known as RBP-Jκ), which functions as a DNA adaptor in Notch signaling pathway. vIRF4 suppresses CSL/CBF1-mediated transcriptional activation, competing with Notch for CSL/CBF1 binding and signaling (Heinzelmann et al. 2010).

RTA, an essential transcriptional activator for lytic infection of KSHV, functions as an E3 ubiquitin ligase. RTA induces ubiquitination and degradation of IRF7 (Yu et al. 2005). Furthermore, RTA assists the proteasomal degradation of IRF3 and IRF7 by another E3 ubiquitin ligase, RAUL (Yu and Hayward 2010). Thus, RTA negatively regulates IFN signaling. K-bZIP/K8, encoded by KSHV early gene and classified in the basic-region leucine zipper (bZIP) family of transcription factors, directly binds to the IFN-β promoter and upregulates basal transcriptional activities at a low level. K-bZIP precludes maximal IFN-β gene transcription by inhibition of IRF3 recruitment to the IFN-β promoter (Lefort et al. 2007). ORF45, an immediate-early gene encoding a viral tegument protein, associates with IRF7 and blocks the phosphorylation and nuclear translocation of IRF7 (Zhu et al. 2002). JAK1 and

JAK2 activated by type-II IFN receptors bind to IFNγ, which phosphorylates STAT1. Activated STAT1 forms a homodimer, translocates to the nucleus, directly binds to GAS (interferon-γ-activated site), and enhances gene transcription. K3 and K5, viral ubiquitin E3 ligases, downregulate the type-II IFN receptor by proteasomal degradation (Li et al. 2007).

15.3 Viral miRNAs and Signaling Pathways

miRNA, a type of small noncoding RNA (approximately 22 nucleotide (nt)), is committed to a broad variety of biological functions via its ability to regulate gene expression (Bartel 2004). Mature miRNA and its protein complex, RNA-induced silencing complex (RISC), binds to the 3′-UTR of target mRNA and downregulates gene expression by inducing the degradation of the mRNA or inhibition of the gene transcription. The precursors of miRNA, pri-miRNA, are the large transcripts (~hundreds to thousands nt) directly transcribed from the coding region of the genome. The pri-miRNA forms a hairpin loop structure via its self-complementary sequence. Pri-miRNA is cleaved by endonuclease Drosha into short stem-loop RNA (approximately 70 nt), named pre-miRNA. Pre-miRNA exported from the nucleus by exportin associates with another endonuclease, Dicer, in the cytoplasm. Dicer converts pre-miRNA into a mature miRNA duplex consisting of dsRNA, by eliminating the loop structure of pre-miRNA. The RISC incorporates one strand of the miRNA duplex into its complex and specifically associates with the 3′-UTR of the mRNA.

It has been shown that the regulatory system of miRNA is not only conserved among eukaryotes but also viruses including KSHV (Strahan et al. 2016). Although KSHV expresses limited latent proteins during its latency, it is likely that KSHV expresses 18 to 25 mature miRNAs (Cai et al. 2005; Samols et al. 2005; Lin et al. 2010). These mature miRNAs are derived from 12 pre-miRNA molecules, which are encoded and clustered on the kaposin K12 locus and its upstream in KSHV genome. These viral miRNAs participate in several cellular events by modulating the expression of cellular and viral factors, in order to survive in an infected cell during de novo, latent, and lytic infection.

KSHV miR-K1 (also known as miR-K12-1) specifically inhibits the expression of tumor suppressor p21, resulting in dysregulation of the cell cycle arrest (Gottwein and Cullen 2010). miR-K1 also downregulates the expression of IκBα and enhances NFκB pathway activities (Lei et al. 2010). Therefore, miR-K1 promotes cell proliferation in the latency state and inhibits the advancement of lytic replication. miR-K3 (miR-K12-3) decreases the expression of nuclear factor I/B, which is a CCAAT box-binding transcription factor. Suppression of nuclear factor I/B inactivates the transcription of RTA, thereby supporting miR-K3 in the maintenance of KSHV latency (Lu et al. 2010a). miR-K4-5p (miR-K12-4-5p) was identified to target Rb-like protein 2 (Rbl2), a transcriptional repressor of DNA methyltransferases (DNMT-3a and DNMT-3b). Rbl2 reduction by miR-K4-5p increases DNMT expression, contributing to maintaining the latency of the KSHV genome epigenetically (Lu et al. 2010b). miR-K3 and miR-K7 (miR-K12-7) associate with the 3'-UTR of C/EBPβ, a bZIP transcription factor (Qin et al. 2010). C/EBPβ is known to regulate the transcriptional activation of IL-6 and IL-10. Consequently, miR-K3 and miR-K7 activate the macrophage cytokine response by enhancing the secretion of these cytokines. miR-K5 (miR-K12-5) and several other viral miRNAs (miR-K9, miR-K10a, and miR-K10b) suppress the expression of Bcl-2-associated transcription factor 1 (BCLAF1), a transcriptional repressor associated with Bcl-2 family proteins (Ziegelbauer et al. 2009). The downregulation of BCLAF1 attenuates the susceptibility to apoptosis. MHC class I polypeptide-related sequence B (MICB), a stress-induced ligand triggering NK cell activation, is not only a direct target of KSHV miR-K7 but also a target of other viral miRNAs encoded in the herpesvirus family (e.g., HCMV and EBV). KSHV miR-K7 contributes to the evasion of host immune systems by the suppression of MICB expression (Nachmani et al. 2009). miR-K10a (miR-K12-10a) robustly downregulates the TNF-like weak inducer of apoptosis (TWEAK) receptor (Abend et al. 2010). The miR-K10a-mediated suppression of the TWEAK receptor evades TWEAK-induced caspase activation and pro-inflammatory cytokine expression. The seed sequences of miR-K11 (miR-K12-11) correspond to the host miRNA miR-155. miR-K11 inhibits the expression of the transcription regulator BACH1, transcription factor FOS, and another known targets of miR-155 (Skalsky et al. 2007; Gottwein et al. 2007).

Profiling of the gene expression in a KSHV miRNA cluster-expressing cell line showed that the expression levels of 81 genes were altered by KSHV miRNAs (Samols et al. 2007). One of the genes, thrombospondin 1 (*THBS1*), is a target of multiple KSHV miRNAs, particularly, miR-K1, miR-K3-3p (miR-K12-3-3p), miR-K6-3p (miR-K12-6-3p), and miR-K11. THBS1 suppresses tumor growth by inhibition of angiogenesis and activation of TGF-β. Decrease in THBS1 by viral miRNA results in reduced TGF-β signaling. On the other hand, proteomics screening has revealed that multiple host genes are suppressed at the protein level, not at the mRNA level (Gallaher et al. 2013). It is confirmed that several host genes identified by the screening (e.g., GRB2, EGF signal mediator; ROCK2, Rho-associated protein kinase; STAT3, signal transducer; and HMGCS1, HMG-CoA synthase) are transcriptionally suppressed.

KSHV miRNAs may contribute to oncogenesis by influencing the differentiation status of infected cells. Tumor cells in KS are poorly differentiated endothelial cells, expressing markers of both lymphatic endothelial cells (LECs) and blood vessel endothelial cells (BECs). KSHV miRNA silences the cellular transcription factor MAF (musculoaponeurotic fibrosarcoma oncogene homolog), which is committed to suppression of BEC-specific genes. Downregulation of MAF by KSHV miRNAs induces the partial differentiation of LECs to BECs (Hansen et al. 2010).

Genome-wide mapping of KSHV transcripts including the 3'-UTR was intensively performed (Arias et al. 2014; McClure et al. 2013; Bai et al. 2014; Zhu et al. 2014; Strahan et al. 2016). These 3'-UTRs of KSHV transcripts affect the KSHV protein expression during the latent and lytic states (McClure et al. 2013). In addition, screening for novel viral targets of KSHV miRNA indicated 28 potential targets of KSHV miRNAs, of which 11 were bicistronic or polycistronic transcripts (Bai et al. 2014). These bicistronic or polycistronic transcripts, identified at several loci of KSHV including ORF30–33 and ORF71–73, have long 3′-UTRs due to 5′-distal ORFs. miR-K3 decreases the expression of ORF31–33 transcripts by binding at two sites in the ORF33 coding region. miR-K10a-3p, miR-K10b-3p, and its variants decrease the rate of ORF71–73 transcripts through distinct binding sites in both 5′-distal ORFs and intergenic regions.

miR-K9-3p (miR-K9*; miR-K12-9-3p) and miR-K7-5p (miR-K12-7-5p) directly target the 3′-UTR of the mRNA encoding RTA, the master transcription factor inducing viral reactivation from latency (Bellare and Ganem 2009; Lin et al. 2011). On the other hand, miR-K5 (miR-K12-5) indirectly targets RTA (Lu et al. 2010b). miR-K5 and miR-K6-3p (miR-K12-6-3p) suppress the expression of ORF56/57 mRNAs (Lin and Ganem 2011). vIL-6 is downregulated by miR-K10a-3p (Haecker et al. 2012).

15.4 Viral Long Noncoding RNAs and Signaling Pathways

KSHV encodes noncoding 1.1-kb RNA, which is a polyadenylated nuclear (PAN) RNA (Conrad 2016; Campbell et al. 2014b). PAN RNA was identified and characterized just after the discovery of KSHV (Sun et al. 1996; Zhong et al. 1996; Zhong and Ganem 1997). PAN RNA is highly expressed during the lytic phase and localized in the nucleoplasm and nuclear speckles. This discovery of PAN RNA was prior to the explosion of recent researches on long noncoding RNA (lncRNA). The major explored functions of lncRNAs involve epigenetic modulation. lncRNAs interact with chromatin-associated proteins to regulate their functions (Wang and Chang 2011). Recently, it has been demonstrated that PAN RNA is essential for KSHV replication, by using knockdown (Borah et al. 2011; Campbell et al. 2014a) or knockout (Rossetto and Pari 2012) approaches. Moreover, the biological functions and the mechanisms of PAN RNA as lncRNA have been unveiled.

Although LANA associates with KSHV episomes in latency, LANA rapidly disassociates from episomes in the lytic state. PAN RNA, abundantly transcribed in reactivation, supports LANA-episome disassociation through an interaction with LANA. Thus, PAN RNA plays a role as a molecular decoy for sequestering LANA from viral episomes (Campbell et al. 2014a). PAN RNA also functions as a decoy for IRF4, an IFN-related transcription factor. The interaction between PAN RNA and IRF4 interferes with the ability of IRF4/PU.1 to activate the IL-4 promoter (Rossetto and Pari 2011). Therefore, PAN RNA modulates immune responses.

One of the major roles of lncRNA in epigenetic gene regulation is the scaffolds. lncRNA has the potential of binding to complementary DNA sequences and proteins. It is known that some of the lncRNAs recruit transcriptional regulators to specific regions of the chromosome to modulate the gene expression. PAN RNA binds to Ezh2 and Suz12, components of the polycomb repression complex 2 (PRC2). PRC2 influences histone modification (methylation) patterns and the subsequent repression of gene expression. Expression of PAN RNA dysregulates the transcription of the genes related to the cell cycle, immune response, and inflammation. In addition, PAN RNA enhances cell proliferation and reduces inflammatory cytokine production (Rossetto et al. 2013). Gene repression, caused by PAN RNA-mediated recruitment of PRC2 to the host genome, may contribute to KSHV oncogenesis. On the other hand, PAN RNA mediates gene activation in the KSHV genome. PAN RNA recruits histone demethylases JMJD3 and UTX to the RTA promoter in the KSHV genome. This association by PAN RNA results in a decrease of the repressive H3K27me3 mark at the RTA promoter and an increase in viral production (Rossetto and Pari 2012). Accordingly, PAN RNA functions as a scaffold molecule for chromatin modifications of the host and viral genome.

15.5 Akt Signaling and KSHV

It is well known that many viruses activate Akt (also known as protein kinase B) signaling, allowing the survival of infected cells and the protection of infected cells from apoptosis. When insulin or growth factors bind their receptor, phosphatidylinositol 3-kinase (PI3K) is activated. PI3K generates phosphatidylinositol 3,4,5-trisphosphate (PIP_3) from phosphatidylinositol 4,5-bisphosphate (PIP_2) by phosphorylation (Fig. 15.2). Membrane phospholipid PIP_3 recruits phosphoinositide-dependent kinase-1 (PDK1) and Akt to the plasma membrane and accelerates Akt phosphorylation at Ser308 by PDK1. In addition to PDK1, rictor-containing mTOR complex 2 (mTORC2) phosphorylates Ser473 of Akt. The phosphorylation of Akt at either Ser308 or Ser473 activates Akt, and Akt promotes the phosphorylation of downstream substrates. Akt inhibits Bad, GSK-3β, NF-κB, caspase-9, FOXO1, and FOXO4 by Akt-mediated phosphorylation, leading to enhanced cell survival and anti-apoptosis, while Akt activates and phosphorylates raptor-containing mTOR complex 1 (mTORC1), leading to enhance mRNA translation and cell proliferation. KSHV infection has been known to activate Akt signaling in infected cells via viral proteins including K1, vGPCR and vIL-6 (Bhatt and Damania 2012).

KSHV gene K1 encodes a transmembrane glycoprotein (Lagunoff and Ganem 1997) and possesses transformation activity (Lee et al. 1998b; Prakash et al. 2002). The cytoplasmic tail of K1 contains an immunoreceptor tyrosine-based activation motif (ITAM), which is essential for the modulation of intracellular signaling pathways. In Akt signaling, K1 expression leads to increased tyrosine phosphorylation of p85, the subunit of PI3K, and results in the activation of signaling downstream of PI3K (Lee et al. 1998a). Furthermore, K1 facilitates Akt signal activation by activation of Akt and inhibition of the phosphatase PTEN, which dephosphorylates PIP_3 required for the association between PDK1 and Akt. Activated Akt by K1 phosphorylates FOXO1 and inhibits FOXO1-mediated transcription. FOXO family proteins play an important role in arresting the cell cycle and inducing apoptosis. To protect the infected cells from apoptosis, K1 promotes cell proliferation and anti-apoptosis via activation of Akt signaling (Tomlinson and Damania 2004).

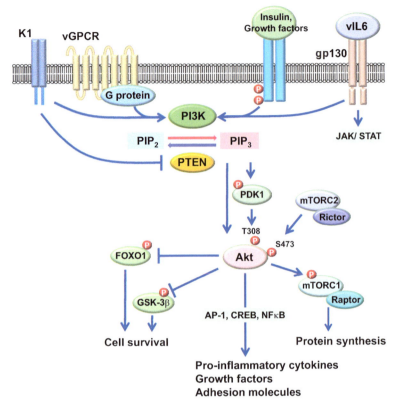

Fig. 15.2 Manipulation of AKT signaling by KSHV for maintenance of latency
Abbreviations used: *Circled p* phospho, *FOXO1* forkhead box protein O1, *GSK-3β* glycogen synthase kinase-3β, *mTORC* mammalian target of rapamycin complex, *PDK1* phosphoinositide-dependent kinase-1, *PI3K* phosphatidylinositide 3-kinase, *PIP2* phosphatidylinositol 4,5-bisphosphate, *PIP3* phosphatidylinositol 3,4,5-trishosohate, *PTEN* phosphatase and tensin homolog, *vGPCR* viral G protein-coupled receptor

vGPCR, a seven-pass transmembrane receptor, has homology with multiple GPCRs, of which the highest homology is with the human chemokine receptor CXCR2 (IL-8 receptor) (Cesarman et al. 1996). The KSHV vGPCR is a constitutively active (agonist-independent) receptor, which stimulates cell proliferation and transformation (Arvanitakis et al. 1997; Bais et al. 1998). vGPCR stimulates the PI3K/Akt pathway through G protein activation in various cell models. However, the target pathways for activated Akt by vGPCR are dependent on the cell type. In PEL cells, vGPCR-induced activation of PI3K/Akt mediates AP-1 and CREB activation, but not NF-κB activation (Cannon and Cesarman 2004). In Human umbilical vein endothelial cells (HUVECs), vGPCR enhances the kinase activity of Akt and promotes translocation of Akt to the plasma membrane. The dominant-negative Akt inhibits vGPCR-mediated NF-κB activation and protection from apoptosis (Montaner et al. 2001). In a KSHV-negative KS cell line, the PI3K/Akt cascade activated by vGPCR phosphorylates GSK-3β and leads to NF-AT activation (Pati et al. 2003). Consequently, these cellular signaling events induced by vGPCR enhance the expression of pro-inflammatory cytokines, growth factors, and adhesion molecules (Couty et al. 2001; Montaner et al. 2001; Schwarz and Murphy 2001; Pati et al. 2003). As secondary effects of vGPCR, vGPCR-induced secreted factors facilitate Akt-related signaling for proliferation and survival in a paracrine manner. Cytokines secreted by vGPCR-positive tumor cells activate multiple pathways including Akt signaling in neighboring cells. Particularly, activations of the PI3K/Akt and NF-κB pathway converge at mTORC1-dependent upregulation of VEGF (Jham et al. 2011).

Viral IL-6 (vIL-6) is a homolog of the human IL-6 (hIL-6) cytokine and shares 24.8% amino acid homology with hIL-6 (Moore et al. 1996a). Though vIL-6 is expressed at lower levels during the latent phase, vIL-6 is highly upregulated by the occurrence of reactivation from the latent phase (Cannon et al. 1999; Staskus et al. 1999; Parravicini et al. 2000; Chandriani et al. 2010). Since vIL-6 mimics the functions of hIL-6, vIL-6 activates cytokine signaling pathways, such as JAK/STAT and MAPK pathways (Moore et al. 1996a; Osborne et al. 1999). vIL-6 binds directly to the IL-6 receptor subunit gp130 and transduces the downstream signal pathways (Chen and Nicholas 2006). The activation of gp130 and its downstream PI3K/Akt signaling by vIL-6 leads to the expression of LECs markers, such as VEGFR, podoplanin, and PROX1. As a result, the activations of gp130 and PI3K/Akt signaling pathways influence the differentiation status of KSHV-infected endothelial cells (Carroll et al. 2004; Morris et al. 2008; Hong et al. 2004; Wang et al. 2004; Morris et al. 2012). Thus, PI3K/Akt activation via gp130 is necessary for the establishment and maintenance of latency.

15.6 MAPK Signaling and KSHV Infection

Mitogen-activated protein kinase (MAPK) signaling pathways are critical for each cell survival stage to control cell proliferation, cell growth, cell migration, cell differentiation, and cell death (Dhillon et al. 2007; Pearson et al. 2001). MAPKs are phosphorylated and activated by MAPK kinases (MKKs), which in turn are activated by MAPK kinase kinases (MKKKs) (Fig. 15.3). MAPK pathways are classified into three major groups: extracellular signal-regulated protein kinases (ERK), p38 mitogen-activated protein kinase (p38MAPK), and c-Jun N-terminal kinases (JNK, also called stress-activated protein kinase/SAPK). In cases of ERK signaling, binding of the tyrosine kinase receptors and their ligands, such as EGF and HGF, induces the phosphorylation and activation of c-Raf through the Grb2-SOS-Ras pathway. Activated c-Raf kinase phosphorylates MEK, and subsequently, MEK meditates ERK phosphorylation, resulting in ERK activation. KSHV is known to hijack MAPK pathways for their life cycle regulation.

During KSHV primary infection, MAPK pathways contribute to establishing KSHV latent infection. To enter target cells, KSHV envelope glycoprotein B (gB)

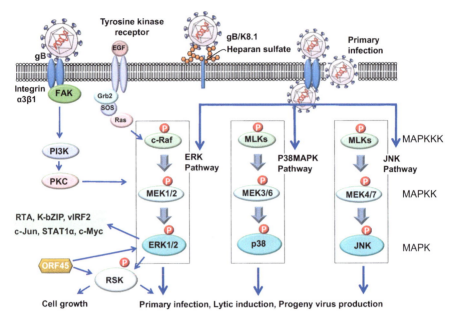

Fig. 15.3 KSHV-induced MAPKs activation during primary infection

Abbreviations used: *Circled p* phospho, *ERK1/2* extracellular signal-regulated kinase 1 and 2, *FAK* focal adhesion kinase, *gB* glycoprotein B, *JNK* c-Jun N-terminal kinase, *MAPK* mitogen-activated protein kinase, *MAPKK* MAPK kinase, *MAPKKK* MAPK kinase kinase, *MEK* MAPK/ERK kinase, *MLK* mixed lineage kinase, *PI3K* phosphatidylinositide 3-kinase, *PKC* protein kinase C, *RSK* ribosomal S6 kinase

(also referred as ORF8) and gpK8.1A, which bind to heparan sulfate molecules that exist on the cell surface (Akula et al. 2001a, b; Wang et al. 2001). In addition, gB also docks with the cellular receptor integrin α3β1. The interaction between gB and integrin α3β1 activates focal adhesion kinase (FAK) (Akula et al. 2002). After infection, KSHV immediately activates the focal adhesion component PI3K. Subsequent to PI3K-mediated PKC activation, the ERK pathway is upregulated. This activation cascade is important for the establishment of viral infection (Naranatt et al. 2003). Furthermore, the protein kinase c-Raf, an upstream transducer of ERK, also enhances KSHV infection in a post-cell attachment step (Akula et al. 2004). KSHV virion-mediated ERK activation enhances the expression of viral genes, such as RTA, K-bZIP/K8, and vIRF2. The activation of ERK enhances the activity of MAPK-regulated transcription factors, such as c-Jun, STAT1α, c-Myc, and c-Fos. gpK8.1A is critical for activation of the MEK/ERK pathway (Sharma-Walia et al. 2005). In addition to the ERK pathway, KSHV activates JNK and p38MAPK pathways during primary infection (Xie et al. 2005). Moreover, JNK, ERK, and p38MAPK pathways are also essential for KSHV infection. Inhibition of ERK, JNK, and p38MAPK reduces KSHV infectivity at the very early stage of infection, partly due to decreasing the expression of RTA, which is essential for the establishment of infection. MAPK signaling pathways modulate the RTA promoter during primary infection. In addition, these MAPK factors affect the production of infectious virions and lytic replication (Pan et al. 2006). Similarly, MEK/ERK, JNK, and p38MAPK pathways were required for 12-O-tetradecanoyl-phorbol-13-acetate (TPA)-induced reactivation from the latent phase, since these pathways activate AP-1 and regulate RTA promoter activity (Cohen et al. 2006; Xie et al. 2008; Ford et al. 2006). Therefore, clinically approved drugs such as celecoxib, which inhibit MAPK signaling pathways, are potentially therapeutic for KSHV-associated cancers to inhibit lytic activation (Chen et al. 2015). During the late stage of lytic replication, KSHV lytic protein ORF45 activates ERK and ribosomal S6 kinase (RSK), which is a major substrate of ERK (Kuang et al. 2008; Kuang et al. 2009). This ORF45-mediated activation is essential for optimal lytic gene expression and virion production (Fu et al. 2015).

15.7 NF-κB Signaling

NF-κB signaling has emerged as a mediator of altered gene programming and as an initiator of the immediate early steps of immune activation and anti-apoptosis. Although certain viruses inhibit NF-κB signaling to disrupt host immunity (Hiscott et al. 2006), KSHV activates NF-κB pathways to promote host cell proliferation, survival, and angiogenesis (Brinkmann and Schulz 2006). Additionally, constitutive activation of NF-κB signaling is thought to be essential for the latent infection of KSHV and survival in PEL cells (Keller et al. 2000; Guasparri et al. 2004). In normal resting cells, the canonical NF-κB signaling pathway is suppressed by IκBα protein (Fig. 15.4 left). NF-κB signaling is further regulated by the IKK complex, consisting of IKKα, β, and a scaffold subunit (IKKγ/NEMO). A molecular chaperone, Hsp90, binds the IKK complex and promotes the kinase activity of the IKK complex. A stimulus, such as TNF-α or IL-1β, induces activation of the IKK complex via TRAF6 (or TRAF2), which acts as an E3 ubiquitin ligase. TRAF6 modifies NEMO protein by K63-linked polyubiquitination. K63-linked polyubiquitinated NEMO can then activate the IKK complex. The activated IKK complex phosphorylates IκBα, and this phosphorylation can be a trigger for K48-linked polyubiquitination, leading to proteasomal degradation. This action releases the active NF-κB heterodimer containing p50 and p65/RelA. There is another upregulatory system for the IKK complex. TGF-β-activated kinase 1 (TAK1) also phosphorylates and activates IKK. TAK1 forms a complex with TAK1-binding protein 1 (TAB1) and TAB2 (or TAB3). TAB2/3 stimulates TAK1 kinase activity and binds to K63-linked polyubiquitinated NEMO (also TRAF6) through the Zn-finger domain of TAB2/3. Moreover, TNF-α stimuli also induce the K63-linked polyubiquitination of receptor-interacting protein (RIP). Polyubiquitinated RIP recruits NEMO to the IKK complex and also TAB2/3 together with TAK1 to the TNF receptor, resulting in NF-κB activation.

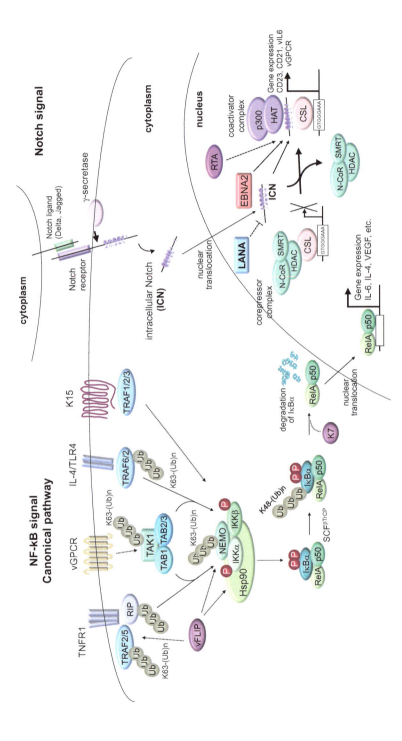

Fig. 15.4 KSHV-induced disruption of NF-κB and Notch signaling

Abbreviations used: *Circled p* phospho, *FLIP* FLICE inhibitory protein, *GPCR* G protein-coupled receptor, *ICN* intracellular Notch, *IFN* interferon, *IKK* IκB kinase, *IRF* interferon regulatory factor, *ITAM* immunoreceptor tyrosine-based activation motif, *LANA* latency-associated nuclear antigen, *RIP* receptor-interacting protein, *TAB1* TAK1-binding protein 1, *TAK1* TGF-β-activated kinase 1, *TLR4* Toll-like receptor 4, *TNFR1* TNF receptor 1, *TRAF* TNF receptor-associated factor, *Ub* ubiquitin

KSHV-encoded vFLIP, K15, vGPCR, K1, K7, and viral miRNA induce NF-κB transcriptional activation by manipulation of its upstream signaling. These viral molecules activate NF-κB signaling through IκBα destabilization. vFLIP is a viral homolog of cellular FLIP and is expressed in latently infected KS spindle cells and PEL cells. The vFLIP protein inhibits apoptosis by activating NF-κB signaling. vFLIP activates the IKK complex by interaction with NEMO of the IKK complex (Chaudhary et al. 1999; Chugh et al. 2005), Hsp90 (Field et al. 2003; Nayar et al. 2013), TRAF2/3 (Guasparri et al. 2006), and RIP (Chaudhary et al. 1999; Liu et al. 2002). These interactions with vFLIP lead to IKK complex activation, causing the activation of NF-κB (p50 and p65/RelA). In addition to viral protein, KSHV miRNA decreases the level of IκBα protein in a ubiquitin-independent manner (Lei et al. 2010; Moody et al. 2013). KSHV miR-K1 reduces IκBα expression by targeting the 3′-UTR of its transcript (Lei et al. 2010). Thus, KSHV miRNA upregulates NF-κB activity and inhibits viral lytic replication in PEL cells. KSHV K15, which shows structural and functional similarities to the EBV LMP1, contains 12 predicted transmembrane domains and a cytoplasmic C-terminal domain, which can activate NF-κB, phospholipase C (PLC), and MAPK (JNK, p38MAPK, and ERK) signaling pathways. The K15 cytoplasmic domain contains a SH3-binding motif, two SH2-binding motifs, and a TRAF-binding site; K15 directly interacts with Src, Lck, Hck, TRAF-1, -2, -3, IKKα/β, and NF-κB-inducing kinase (NIK) which leads to the activation of NF-κB (Brinkmann et al. 2003, 2007 Havemeier et al. 2014; Pietrek et al. 2010). vGPCR, a viral homolog of the hIL-8 receptor, promotes cell proliferation and induces NF-κB activation, resulting in expression of IL-1, IL-6, IL-8, TNF-α, and FGF (Schwarz and Murphy 2001; Pati et al. 2001). vGPCR interacts with TAK1 and recruits TAK1 to the plasma membrane. Furthermore, vGPCR induces TAK1 phosphorylation and K63-linked polyubiquitination, which is related to TAK1 activation and subsequent NF-κB activation (Bottero et al. 2011). K1 has transforming activity. B lymphocytes from K1-expressing transgenic mice were reported to show constitutive NF-κB activation (Prakash et al. 2002), while K1 was reported to suppress vFLIP-mediated NF-κB activation (Konrad et al. 2009). The small membrane (or mitochondrial membrane) protein K7 inhibits caspase-3-induced apoptosis and interacts with ubiquitin/PLIC1. The interaction between K7 and ubiquitin induces activation of NF-κB signaling through IκB degradation (Feng et al. 2004).

15.8 Notch Signaling

The Notch receptor and Notch ligands are single-pass transmembrane proteins that play roles in cell fate decisions, proliferation, and differentiation (Fig. 15.4 right). When a Notch ligand, such as Jagged or Delta, binds the Notch receptor, a γ-secretase complex, such as presenilin, cleaves the intramembrane region of the Notch receptor, Notch-1, Notch-2, Notch-3, and Notch-4, resulting in the release of intracellular Notch (ICN). ICN, which functions as a transcriptional activator, enters the nucleus and binds the DNA-binding protein CSL (also known as CBF1 or RBP-Jκ) (Hayward et al. 2006). In the absence of signaling, DNA-bound CSL is bound to the corepressor complex composed of HDAC, SMRT, and SKIP. Upon Notch signaling, the corepressor complex is displaced by ICN and the coactivator complex composed of HAT and p300, upregulating target genes containing CSL-binding sites. EBNA2 is well known to be CSL-binding protein which is essential for EBV immortalization of B cells. EBNA2 was targeted to the promoters of EBNA2-responsive EBV genes through an interaction with CSL. Both EBNA2 and ICN bound to the transcriptional repression domain of CSL. Binding of the CSL corepressor complex and either the EBNA2 coactivator complex or the ICN coactivator complex is competitive such that there is a conversion from transcriptional repression of the CSL-bound promoter to transcriptional activation in the presence of ICN or EBNA2 (Hayward et al. 2006). In addition to ICN and EBNA2, KSHV RTA interacts with CSL and activates target gene expression in cooperation with coactivators (Liang and Ganem 2003). Moreover, the promoter DNA of vIL-6 and vGPCR contains the CSL-binding sites, indicating that activation of Notch signaling is involved in lytic gene expression including that of vIL-6 and vGPCR (Liang and Ganem 2003, 2004). On the other hand, it was reported that a single CSL-binding site within the LANA promoter plays a role in establishing latency during primary KSHV infection (Lu et al. 2011). Regarding cross talk between NF-κB and Notch, the transcription factor NF-κB forms a complex with CSL and inhibits the binding of CSL and the CSL-target DNA sequence, which indicates that NF-κB contributes to KSHV latency and negatively regulates RTA by antagonizing CSL (Izumiya et al. 2009).

Additionally, KSHV has another way to activate Notch signaling. KSHV-encoded proteins manipulate upstream regulators of cell signaling and exploit the ubiquitin system to alter the stabilization of target proteins. Furthermore, KSHV proteins can directly bind a component of E3 ubiquitin ligase to inhibit the E3 activity for polyubiquitination, leading to stabilization of target substrates. The SCF^{Fbw7} complex, SCF-type E3, is composed of Fbw7 as an F-box protein, which recognizes and binds the intracellular regions of Notch-1, Notch-3, and Notch-4 (Matsumoto et al. 2011; Nakayama and Nakayama 2006). The F-box protein Fbw7, also known as Sel-10 in *Caenorhabditis elegans*, is thus a critical component of SCF^{Fbw7}, while it is a negative regulator of Notch signaling. KSHV LANA binds Fbw7/Sel-10, a component of SCF^{Fbw7}, to inhibit polyubiquitination of intracellular Notch-1 (ICN-1), which leads to the stabilization of ICN-1. The carboxyl-terminus of LANA binds Fbw7/Sel-10, whereas it competes with ICN-1 for interaction with Fbw7/Sel-10, resulting in inhibition of ICN-1 polyubiquitination and degradation (Lan et al. 2007). Thus, stabilization of ICN-1 by LANA is also related to the proliferation of KSHV-infected cells. It is known that LANA represses the RTA promoter to inhibit RTA expression to establish and maintain KSHV latency (Lan et al. 2004). A number of KSHV lytic gene promoters, including the RTA promoter, contain CSL-binding sites. The carboxyl-terminal of LANA also interacts with CSL, and this binding is involved in one of the LANA-mediated RTA repression mechanisms (Jin et al. 2012).

15.9 Wnt/β-Catenin Signaling

Wnt signaling is involved in several critical developmental processes and in tumorigenesis. The Wnt/β-catenin pathway regulates the availability of nuclear β-catenin protein (Fig. 15.5). In the absence of Wnt signaling, β-catenin is held in a complex with Axin, APC, GSK-3β, and CKI. The Axin-APC complex functions as a platform for the association of GSK-3β and β-catenin proteins. CKI provides priming activity for GSK-3β-mediated phosphorylation of β-catenin. Phosphorylated β-catenin is conjugated to K48-linked polyubiquitin chains and then degraded by the 26S proteasome. The Wnt signaling cascade is triggered when the Wnt ligand of a secreted glycoprotein binds to the transmembrane Frizzled receptor and to the Lrp5/6 coreceptor, which leads to recruitment of Axin to LRP5/6 and also recruitment of Dvl and FRAT to the Axin-APC-GSK-3β complex. These events result in dissociation of the Axin-APC-GSK-3β complex and the subsequent stabilization of β-catenin, which translocates to the nucleus and forms a complex with the transcription factor LEF1 (or TCF4/1), stimulating the expression of c-Myc, c-Jun, and cyclin D1. This pathway is deregulated in many human tumors including colorectal cancer, which has mutations in the APC tumor suppressor gene and the β-catenin gene (Su et al. 1993; Morin et al. 1997). Most mutations of APC found in the β-catenin-/Axin-binding domain will result in the expression of a truncated APC protein, which fail to form the complex with β-catenin and Axin. In the case of the β-catenin gene in colorectal cancer cells, mutations usually occur at Ser33, Ser37,

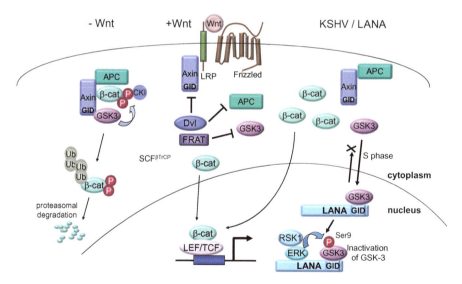

Fig. 15.5 KSHV LANA manipulates the Wnt/β-catenin pathway by stabilizing β-catenin Abbreviations used: *APC* adenomatous polyposis coli, *β-cat* β-catenin, *Circled p* phospho, *CKI* casein kinase I, *Dvl* dishevelled, *FRAT* frequently rearranged in advanced T-cell lymphomas, *GID* GSK-3β interaction domain, *GSK-3β* glycogen synthase kinase-3β, *LANA* latency-associated nuclear antigen, *RSK1* ribosomal S6 kinase 1, *Ub* ubiquitin

Thr41, and Ser45 of the β-catenin protein sequence, which results in the expression of constitutively stable β-catenin.

KSHV dysregulates Wnt signaling by LANA (Fujimuro et al. 2003; Fujimuro and Hayward 2003; Fujimuro et al. 2005; Hayward et al. 2006). The β-catenin protein is very abundant in latently KSHV-infected PEL cells and KS tumor cells. The mechanism of β-catenin dysregulation is linked to the interaction of LANA and GSK-3β. LANA associates with GSK-3β, leading to the nuclear translocation of GSK-3β. Although GSK-3β is localized primarily in the cytoplasm, a small proportion of GSK-3β is known to enter the nucleus during S phase, and LANA increases the number of cells in S phase. Because LANA binds to nuclear GSK-3β, LANA induces depletion of cytoplasmic GSK-3β. In the absence of cytoplasmic GSK-3β, β-catenin accumulates in the cytoplasm and enters the nucleus to stimulate transcriptional activation of downstream target genes. LANA-mediated β-catenin stabilization is dependent on LANA binding to GSK-3β through its C-terminal GSK-3β interaction domain (GID), an analogue of the Axin GID (Webster et al. 2000), and its N-terminal region (Fujimuro et al. 2003; Fujimuro and Hayward 2003). The LANA N-terminal region is phosphorylated by GSK-3β in a manner that is dependent on the activity of priming kinases such as p38MAPK and CKI, and phosphorylation regulates the interaction between GSK-3β and LANA (Fujimuro et al. 2005). Thus, LANA interacts with GSK-3β and relocalizes GSK-3β in a manner that leads to the stabilization of not only β-catenin but also other GSK-3β substrates such as C/EBPβ (Liu et al. 2007a) and c-Myc (Bubman et al. 2007; Liu et al. 2007b). Additionally, Liu et al. have demonstrated the mechanism of LANA-mediated inactivation of nuclear GSK-3β. LANA binds GSK-3β, along with ERK1 and RSK, which participates in the phosphorylation of GSK-3β at Ser9, resulting in inactivation of GSK-3β (Liu et al. 2007a). Destabilization of c-Myc is controlled by the GSK-3β-mediated phosphorylation of c-Myc at Thr58. This phosphorylation is the trigger for the polyubiquitination and degradation of c-Myc. Inactivation of nuclear GSK-3β by LANA can increase the stability and activity of c-Myc and further contribute to LANA-mediated growth dysregulation (Liu et al. 2007b). Furthermore, abnormally stable c-Myc protein has been observed, and LANA is responsible for this stability in PEL cells (Bubman et al. 2007).

15.10 Kaposins and Signaling Pathways

The KSHV K12 locus, the locus encoding kaposins, was initially identified, and the expression of its transcripts was observed in the latent state of PEL cells (Zhong et al. 1996). The mRNA of the K12 locus translated to several viral proteins named kaposins. While translation of Kaposin A initiates at AUG in the ORF K12, translation of other kaposins, typified by Kaposin B and Kaposin C, initiates at CUG codons, located upstream of ORF K12. Therefore, Kaposin B and C are encoded by the amino acid repeats derived from two sets of GC-rich direct repeats (DRs; DR1 and DR2) in the KSHV genome (Sadler et al. 1999). Similar to several other latent genes, the expression of kaposins is upregulated during lytic infection.

Kaposin B induces pro-inflammatory cytokine secretion by stabilization of the mRNAs containing AU-rich elements (AREs) in 3′-UTRs via MK2 (MAPK-associated protein kinase) activation (McCormick and Ganem 2005). The p38MAPK phosphorylates and activates MK2, which in turn inhibits degradation of ARE-containing cytokine mRNAs through MK2-mediated phosphorylation. Kaposin B protein binds and activates MK2 and p38MAPK, and activated MK2 suppresses ARE-dependent mRNA degradation. DR2 and DR1 repeats of Kaposin B are essential for this mRNA stabilization (McCormick and Ganem 2006). In addition, Kaposin B contributes to the dispersion of processing bodies, cytoplasmic foci occurring in ARE-mRNA decay, via activation the Rho-GTPase signaling pathway (Corcoran et al. 2015). Kaposin B also enhances the expression of PROX1, the critical transcription factor for lymphatic endothelial differentiation, by stabilization of PROX1 mRNA (Yoo et al. 2010).

Kaposin A induces tumorigenic transformation (Muralidhar et al. 1998). The direct interaction between Kaposin A and cytohesin-1, a GEF for ARF GTPase and a regulator of integrin-mediated cell adhesion, results in GTP binding of ARF1. The transforming phenotype of Kaposin A is reversed by exchange of the defective form of cytohesin-1. This suggests that Kaposin A-induced transformation is mediated by cytohesin-1 (Kliche et al. 2001).

15.11 Apoptosis Signaling

Apoptosis is a tightly controlled process that plays an important role in homeostasis and antiviral control. A remarkable feature of apoptosis is the activation of caspases, which belong to the cysteine protease family (Hengartner 2000). Caspase activation induces the cleavage of cellular proteins and degradation of the cytoskeleton and nuclear DNA (Lavrik et al. 2005). Apoptosis is induced by the activation of effector caspases, such as caspase-3, caspase-6, and caspase-7, which have been previously activated via either an intrinsic pathway, in which mitochondrial activation (or damage) and caspase-9 are involved, or an extrinsic pathway, in which death signaling and caspase-8 are involved (Fig. 15.6 left). In the extrinsic pathway, when death ligands bind death receptors, death receptors recruit pro-caspase-8, pro-caspase-2, or pro-caspase-10 and adaptor proteins such as TNF receptor-associated death domain (TRADD) and Fas-associated death domain (FADD), resulting in the formation of a ligand-receptor-adaptor protein complex known as the death-inducing signaling complex (DISC). The caspase-8, caspase-2, or caspase-10 activated in DISC cleaves and activates the executioner caspase (caspase-3 and caspase-7), inducing apoptosis. The intrinsic pathway is initiated within the cell by internal stimuli, such as genetic damage, severe oxidative stress, and hypoxia. These stimuli induce cytochrome c release from mitochondria into the cytoplasm (Danial and Korsmeyer 2004). Released cytochrome c binds to Apaf-1, and subsequently the complex of cytochrome c and Apaf-1 are oligomerized, resulting in the formation of the apoptosome, which cleaves and activates pro-caspase-9. In addition, cytochrome

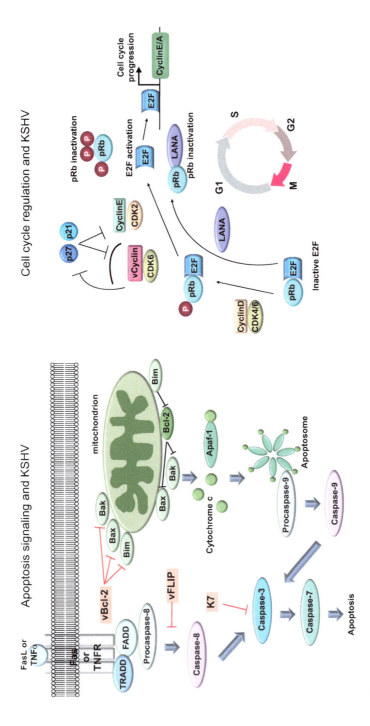

Fig. 15.6 KSHV-mediated modulations of apoptosis and cell cycle
Abbreviations used: *Cdk* cyclin-dependent kinase, *Circled p* phospho, *FADD* Fas-associated death domain, *FLIP* FADD-like interferon-converting enzyme (FLICE) inhibitor protein, *TRADD* TNF receptor-associated death domain

c release is strictly regulated by Bcl-2 family proteins. The Bcl-2 family is comprised of pro-apoptotic Bcl-2 proteins (Bax, Bak, Bad, Bid, Bik, Bim, Noxa, and Puma) and anti-apoptotic Bcl-2 proteins (Bcl-2, Bcl-xL, Bcl-W, Bfl-1, and Mcl-1) (Reed 1997). Anti-apoptotic Bcl-2 proteins suppress apoptosis by inhibiting both the pro-apoptotic Bcl-2 proteins and the release of cytochrome c from mitochondria, whereas pro-apoptotic Bcl-2 proteins induce a change in mitochondrial membrane potential, resulting in the release of cytochrome c (Adams and Cory 1998). In addition to anti-apoptotic Bcl-2 proteins, the IAP (inhibitor of apoptosis protein) family also suppresses apoptosis. IAP contains a BIR domain, which binds to caspase-3, caspase-7, and caspase-9 and inhibits these caspases.

Viral FADD-like interferon-converting enzyme inhibitor protein (vFILP), originally identified as an inhibitor of death receptor-induced apoptosis, is a homolog of cellular FLIP (cFLIP) (Irmler et al. 1997; Thome et al. 1997). vFLIP possesses two death effector domains (DED), which can interact with FADD and TRADD. vFLIP binds to FADD and TRADD through its two tandem DEDs within the DISC, and this interaction inhibits the recruitment and activation of pro-caspase-8, resulting in apoptosis suppression (Krueger et al. 2001). However, recent studies show that vFLIP contributes to activation of NF-κB signaling ((Chaudhary et al. 1999; Chugh et al. 2005) (Field et al. 2003; Nayar et al. 2013), TRAF2/3 (Guasparri et al. 2006),.

KSHV encodes anti-apoptotic protein vBcl-2, the viral homolog of cellular anti-apoptotic protein, Bcl-2. The Bcl-2 family proteins contain multiple Bcl-2 homology (BH) domains, and there are four types of BH domain: BH1, BH2, BH3, and BH4. The anti-apoptotic Bcl-2 proteins tend to contain BH1 and BH2, while the pro-apoptotic Bcl-2 proteins tend to contain BH3. The overall sequence homology between cellular Bcl-2 protein and vBcl-2 protein is low (15–20%); however, the BH1 and BH2 domain of vBcl-2 are highly conserved. The BH1 domain of vBcl-2 contains the NWGR (Asn-Trp-Gly-Arg) sequence, which is believed to be essential for the anti-apoptotic function of Bcl-2 and the formation of heterodimers with other Bcl-2 family members. vBcl-2 conserves the BH3 domain at a lower level than cellular Bcl-2 (Cheng et al. 1997; Huang et al. 2002). The BH3-binding cleft of vBcl-2 binds with high affinity to BH3 domains of the pro-apoptotic Bcl-2 proteins (Loh et al. 2005). vBcl-2 forms a stable complex with Aven, which binds to Apaf-1 and inhibits self-association of Apaf-1. Furthermore, vBcl-2 inhibits the anti-apoptotic function of Aven (Chau et al. 2000). However, unlike cellular Bcl-2, vBcl-2 is not a substrate for vCyclin-CDK6 phosphorylation (Ojala et al. 2000). Expression of vCyclin and CDK6 induces apoptosis by Bcl-2 phosphorylation, leading to the inactivation of Bcl-2. The vCyclin-CDK6 complex phosphorylates the unstructured loop of Bcl-2, and this phosphorylation induces Bcl-2 inactivation. Since vBcl-2 lacks this loop, anti-apoptotic activity of vBcl-2 can be retained even in the presence of vCyclin-CDK6.

K7, a small (mainly mitochondrial) membrane protein, contains a BRI (baculovirus IAP repeat) domain and a BH2 domain and functions as an anti-apoptotic factor (Feng et al. 2002). K7 targets multiple cellular factors, such as calcium-modulating cyclophilin ligand (CAML), Bcl-2, and caspase-3. These target molecules play a role in the regulation of apoptosis. Interaction between K7 and CAML increases the intracellular Ca^{2+} concentration by activating CAML and enhancing ER Ca^{2+} release, resulting in enhanced cytoprotection against mitochondrial damage and apoptosis. K7 also binds to Bcl-2 and caspase-3 via the K7-BH2 domain and K7-BRI domain, respectively (Wang et al. 2002). In particular, the BH2-like domain of K7 inhibits caspase-3 activity through interaction with the caspase-3 BRI domain. K7 bridges Bcl-2 and activated caspase-3, which results in the inhibition of caspase activity. K7 protein is structurally and functionally similar to the human survivin spliced variant. Survivin is known as an IAP family protein and contains a BRI domain as well as K7. Both proteins contain similar functional domains, and these proteins protect against apoptosis (Mahotka et al. 1999; Krieg et al. 2002).

15.12 Cell Cycle Regulation

Cell cycle progression is regulated by Cdk/cyclin complexes. The cyclin D/Cdk4/6 complex phosphorylates pRb at mid-G1 stage, and cyclin E/Cdk2 further phosphorylates pRb at late-G1 (Fig. 15.6 right). Such sequential phosphorylation of pRb leads to release of E2F protein into the cell nucleus to act as a transcriptional factor for S-phase progression. Cyclin/Cdk complexes interact with Cdk inhibitors, namely, those of the INK4 family ($p15^{INK4b}$, $p16^{INK4a}$, $p18^{INK4c}$, and $p19^{INK4d}$) and the Cip/Kip family ($p21^{Cip1}$, $p27^{Kip1}$, and $p57^{Kip2}$). KSHV mimics and uses these regulatory factors. KSHV vCyclin, a human cyclin D2 homolog, interacts with and activates Cdk4 and Cdk6 to achieve phosphorylation of pRb and cellular cyclin D, leading to G1–S phase progression (Chang et al. 1996; Godden-Kent et al. 1997). Unlike the cellular cyclins, the vCyclin/Cdk complex is resistant to the Cdk inhibitors $p16^{INK4a}$, $p21^{Cip1}$, and $p27^{Kip1}$ (Swanton et al. 1997), and therefore this complex is believed to be a constitutively active kinase. KSHV vCyclin D/Cdk6 complex can interact with $p27^{Kip1}$ and phosphorylate $p27^{Kip1}$ at Thr187, which leads to proteasomal degradation (Ellis et al. 1999; Mann et al. 1999; Jarviluoma et al. 2004). This regulation of $p27^{Kip1}$ is normally performed by the cellular cyclin E/Cdk2. The SCF E3 ubiquitin ligase, which contains Skp2 as the F-box protein, can recognize Thr187-phosphorylated $p27^{Kip1}$ and mediates its modification by polyubiquitination, leading to $p27^{Kip1}$ degradation. Interestingly, the vCyclin/Cdk6 complex has a very broad range of substrates; in addition to pRb and $p27^{Kip}$, Orc1, Cdc6, and Bcl-2 are also substrates of vCyclin/Cdk6 (Laman et al. 2001; Ojala et al. 2000). In addition to vCyclin, LANA also leads cell cycle progression from G1 to S phase through interaction with the tumor suppressor pRb (Radkov et al. 2000), which suggests a role for LANA in promoting host cell proliferation via pRb inactivation and E2F release. Regarding interactions between LANA and tumor suppressor gene products, LANA directly binds p53, resulting in abolishment of p53-mediated apoptosis (Friborg et al. 1999). LANA also functions as a potential component within the E3 ubiquitin ligases, which mediates the polyubiquitination of tumor suppressor gene products VHL and p53, resulting in their degradation (Cai et al. 2006). It is known that KSHV encodes vIRFs, which besides LANA, are involved in p53 polyubiquitination for proteasomal degradation. KSHV encodes vIRF4 that binds Mdm2, leading to the proteasomal degradation of p53 in KSHV-infected cells (Lee et al. 2009). vIRF4 interacts with MDM2, leading to the inhibition of MDM2 autoubiquitination, thereby increasing MDM2 stability. In addition, vIRF4 interacts with herpesvirus-associated ubiquitin-specific protease (HAUSP), regulating the stability of p53 and MDM2. The small peptide of vIRF4, responsible for binding to the HAUSP catalytic domain, robustly blocks the de-ubiquitination activity of HAUSP and induces p53-mediated apoptosis and cell cycle arrest (Lee et al. 2011). vIRF1 interacts with p53 and suppresses p53 acetylation and p53-mediated transcriptional activation (Nakamura et al. 2001; Seo et al. 2001). Consequently, inhibition of p53-mediated apoptosis by vIRF1 leads to dysregulation of cell proliferation. Moreover, KSHV-encoded vIRF1 also contributes to destabilization of p53 through manipulating ATM kinase, which is activated by DNA damage (Shin et al. 2006). The ATM phosphorylates Ser15 of p53, and this phosphorylation inhibits the interaction with MDM2, resulting in p53 stabilization. However, vIRF1 suppresses Ser15 phosphorylation of p53 by ATM, resulting in an increase of p53 ubiquitination and degradation by the proteasome. Thus, multiple KSHV proteins are involved in the degradation or inactivation of tumor suppressor gene products; these functions strongly contribute to the oncogenic transformation of host cells.

15.13 Concluding Remarks

Cell signal transduction plays a principal role in controlling gene expression, cell proliferation, apoptosis regulation, and immune responses. As described above, KSHV hijacks the appropriate cell signaling pathways for establishment of infection, persistent infection, prolonging survival, control of cell proliferation, anti-apoptosis, and evasion of immune surveillance in infected cells. KSHV manipulates the appropriate cell signaling pathways in order to downregulate gene expression that exerts an antagonistic effect on virus infection and to upregulate gene expression that creates a favorable environment for KSHV or KSHV-infected cells.

To further elucidate KSHV pathogenesis, it should be determined how KSHV exploits and dysregulates cell signaling pathways. A better understanding of the interaction between KSHV infection and the cell signaling pathways involved could provide new insights into the viral evasion of host immunity and the process of carcinogenesis triggered by KSHV and may provide a theoretical basis for the development of novel therapeutic interventions against KSHV-related cancers. In fact, recent studies have shown that many inhibitors for certain signaling pathways suppress proliferation of PEL cells and induce apoptosis in KSHV-infected tumor cells and PEL cells. Agents that selectively disturb KSHV infection-dependent activated signaling pathways could be promising therapeutic agents for KSHV-associated malignancies in the future. The effectiveness of these agents, in turn, highlights the importance of understanding virus manipulation of cell signal transduction.

References

Abend JR, Uldrick T, Ziegelbauer JM (2010) Regulation of tumor necrosis factor-like weak inducer of apoptosis receptor protein (TWEAKR) expression by Kaposi's sarcoma-associated herpesvirus microRNA prevents TWEAK-induced apoptosis and inflammatory cytokine expression. J Virol 84(23):12139–12151. https://doi.org/10.1128/jvi.00884-10

Adams JM, Cory S (1998) The Bcl-2 protein family: arbiters of cell survival. Sci (NY) 281(5381):1322–1326

Akula SM, Ford PW, Whitman AG, Hamden KE, Shelton JG, McCubrey JA (2004) Raf promotes human herpesvirus-8 (HHV-8/KSHV) infection. Oncogene 23(30):5227–5241. https://doi.org/10.1038/sj.onc.1207643

Akula SM, Pramod NP, Wang FZ, Chandran B (2001a) Human herpesvirus 8 envelope-associated glycoprotein B interacts with heparan sulfate-like moieties. Virology 284(2):235–249. https://doi.org/10.1006/viro.2001.0921

Akula SM, Pramod NP, Wang FZ, Chandran B (2002) Integrin alpha3beta1 (CD 49c/29) is a cellular receptor for Kaposi's sarcoma-associated herpesvirus (KSHV/HHV-8) entry into the target cells. Cell 108(3):407–419

Akula SM, Wang FZ, Vieira J, Chandran B (2001b) Human herpesvirus 8 interaction with target cells involves heparan sulfate. Virology 282(2):245–255. https://doi.org/10.1006/viro.2000.0851

Areste C, Mutocheluh M, Blackbourn DJ (2009) Identification of caspase-mediated decay of interferon regulatory factor-3, exploited by a Kaposi sarcoma-associated herpesvirus immunoregulatory protein. J Biol Chem 284(35):23272–23285. https://doi.org/10.1074/jbc.M109.033290

Arias C, Weisburd B, Stern-Ginossar N, Mercier A, Madrid AS, Bellare P, Holdorf M, Weissman JS, Ganem D (2014) KSHV 2.0: a comprehensive annotation of the Kaposi's sarcoma-associated herpesvirus genome using next-generation sequencing reveals novel genomic and functional features. PLoS Pathog 10(1):e1003847. https://doi.org/10.1371/journal.ppat.1003847

Arvanitakis L, Geras-Raaka E, Varma A, Gershengorn MC, Cesarman E (1997) Human herpesvirus KSHV encodes a constitutively active G-protein-coupled receptor linked to cell proliferation. Nature 385(6614):347–350. https://doi.org/10.1038/385347a0

Bai Z, Huang Y, Li W, Zhu Y, Jung JU, Lu C, Gao SJ (2014) Genomewide mapping and screening of Kaposi's sarcoma-associated herpesvirus (KSHV) 3′ untranslated regions identify bicistronic and polycistronic viral transcripts as frequent targets of KSHV microRNAs. J Virol 88(1):377–392. https://doi.org/10.1128/jvi.02689-13

Bais C, Santomasso B, Coso O, Arvanitakis L, Raaka EG, Gutkind JS, Asch AS, Cesarman E, Gershengorn MC, Mesri EA (1998) G-protein-coupled receptor of Kaposi's sarcoma-associated herpesvirus is a viral oncogene and angiogenesis activator. Nature 391(6662):86–89. https://doi.org/10.1038/34193

Barbera AJ, Chodaparambil JV, Kelley-Clarke B, Joukov V, Walter JC, Luger K, Kaye KM (2006) The nucleosomal surface as a docking station for Kaposi's sarcoma herpesvirus LANA. Sci (NY) 311(5762):856–861. https://doi.org/10.1126/science.1120541

Baresova P, Musilova J, Pitha PM, Lubyova B (2014) p53 tumor suppressor protein stability and transcriptional activity are targeted by Kaposi's sarcoma-associated herpesvirus-encoded viral interferon regulatory factor 3. Mol Cell Biol 34(3):386–399. https://doi.org/10.1128/MCB.01011-13

Bartel DP (2004) MicroRNAs: genomics, biogenesis, mechanism, and function. Cell 116(2):281–297. https://doi.org/10.1016/S0092-8674(04)00045-5

Bellare P, Ganem D (2009) Regulation of KSHV lytic switch protein expression by a virus-encoded microRNA: an evolutionary adaptation that fine-tunes lytic reactivation. Cell Host Microbe 6(6):570–575. https://doi.org/10.1016/j.chom.2009.11.008

Bhatt AP, Damania B (2012) AKTivation of PI3K/AKT/mTOR signaling pathway by KSHV. Front Immunol 3:401. https://doi.org/10.3389/fimmu.2012.00401

Borah S, Darricarrere N, Darnell A, Myoung J, Steitz JA (2011) A viral nuclear noncoding RNA binds re-localized poly(A) binding protein and is required for late KSHV gene expression. PLoS Pathog 7(10):e1002300. https://doi.org/10.1371/journal.ppat.1002300

Bottero V, Kerur N, Sadagopan S, Patel K, Sharma-Walia N, Chandran B (2011) Phosphorylation and polyubiquitination of transforming growth factor beta-activated kinase 1 are necessary for activation of NF-kappaB by the Kaposi's sarcoma-associated herpesvirus G protein-coupled receptor. J Virol 85(5):1980–1993. https://doi.org/10.1128/jvi.01911-10

Brinkmann MM, Glenn M, Rainbow L, Kieser A, Henke-Gendo C, Schulz TF (2003) Activation of mitogen-activated protein kinase and NF-kappaB pathways by a Kaposi's sarcoma-associated herpesvirus K15 membrane protein. J Virol 77(17):9346–9358

Brinkmann MM, Pietrek M, Dittrich-Breiholz O, Kracht M, Schulz TF (2007) Modulation of host gene expression by the K15 protein of Kaposi's sarcoma-associated herpesvirus. J Virol 81(1):42–58. https://doi.org/10.1128/jvi.00648-06

Brinkmann MM, Schulz TF (2006) Regulation of intracellular signalling by the terminal membrane proteins of members of the Gammaherpesvirinae. J Gen Virol 87(Pt 5):1047–1074. https://doi.org/10.1099/vir.0.81598-0

Bubman D, Guasparri I, Cesarman E (2007) Deregulation of c-Myc in primary effusion lymphoma by Kaposi's sarcoma herpesvirus latency-associated nuclear antigen. Oncogene 26(34):4979–4986. https://doi.org/10.1038/sj.onc.1210299

Burysek L, Yeow WS, Lubyova B, Kellum M, Schafer SL, Huang YQ, Pitha PM (1999) Functional analysis of human herpesvirus 8-encoded viral interferon regulatory factor 1 and its association with cellular interferon regulatory factors and p300. J Virol 73(9):7334–7342

Buryýšek L, Pitha PM (2001) Latently expressed human herpesvirus 8-encoded interferon regulatory factor 2 inhibits double-stranded RNA-activated protein kinase. J Virol 75(5):2345–2352. https://doi.org/10.1128/JVI.75.5.2345-2352.2001

Cai Q-L, Knight JS, Verma SC, Zald P, Robertson ES (2006) EC(5)S ubiquitin complex Is recruited by KSHV latent antigen LANA for degradation of the VHL and p53 tumor suppressors. PLoS Pathog 2(10):e116. https://doi.org/10.1371/journal.ppat.0020116

Cai X, Lu S, Zhang Z, Gonzalez CM, Damania B, Cullen BR (2005) Kaposi's sarcoma-associated herpesvirus expresses an array of viral microRNAs in latently infected cells. Proc Natl Acad Sci USA 102(15):5570–5575. https://doi.org/10.1073/pnas.0408192102

Campbell M, Kim KY, Chang PC, Huerta S, Shevchenko B, Wang DH, Izumiya C, Kung HJ, Izumiya Y (2014a) A lytic viral long noncoding RNA modulates the function of a latent protein. J Virol 88(3):1843–1848. https://doi.org/10.1128/jvi.03251-13

Campbell M, Kung HJ, Izumiya Y (2014b) Long non-coding RNA and epigenetic gene regulation of KSHV. Viruses 6(11):4165–4177. https://doi.org/10.3390/v6114165

Cannon JS, Nicholas J, Orenstein JM, Mann RB, Murray PG, Browning PJ, DiGiuseppe JA, Cesarman E, Hayward GS, Ambinder RF (1999) Heterogeneity of viral IL-6 expression in HHV-8-associated diseases. J Infect Dis 180(3):824–828. https://doi.org/10.1086/314956

Cannon ML, Cesarman E (2004) The KSHV G protein-coupled receptor signals via multiple pathways to induce transcription factor activation in primary effusion lymphoma cells. Oncogene 23(2):514–523. https://doi.org/10.1038/sj.onc.1207021

Carroll PA, Brazeau E, Lagunoff M (2004) Kaposi's sarcoma-associated herpesvirus infection of blood endothelial cells induces lymphatic differentiation. Virology 328(1):7–18. https://doi.org/10.1016/j.virol.2004.07.008

Cesarman E, Nador RG, Bai F, Bohenzky RA, Russo JJ, Moore PS, Chang Y, Knowles DM (1996) Kaposi's sarcoma-associated herpesvirus contains G protein-coupled receptor and cyclin D homologs which are expressed in Kaposi's sarcoma and malignant lymphoma. J Virol 70(11):8218–8223

Chandriani S, Xu Y, Ganem D (2010) The lytic transcriptome of Kaposi's sarcoma-associated herpesvirus reveals extensive transcription of noncoding regions, including regions antisense to important genes. J Virol 84(16):7934–7942. https://doi.org/10.1128/JVI.00645-10

Chang Y, Cesarman E, Pessin MS, Lee F, Culpepper J, Knowles DM, Moore PS (1994) Identification of herpesvirus-like DNA sequences in AIDS-associated Kaposi's sarcoma. Science (New York, NY) 266(5192):1865–1869

Chang Y, Moore PS, Talbot SJ, Boshoff CH, Zarkowska T, Godden K, Paterson H, Weiss RA, Mittnacht S (1996) Cyclin encoded by KS herpesvirus. Nature 382(6590):410. https://doi.org/10.1038/382410a0

Chau BN, Cheng EH, Kerr DA, Hardwick JM (2000) Aven, a novel inhibitor of caspase activation, binds Bcl-xL and Apaf-1. Mol Cell 6(1):31–40

Chaudhary PM, Jasmin A, Eby MT, Hood L (1999) Modulation of the NF-kappa B pathway by virally encoded death effector domains-containing proteins. Oncogene 18(42):5738–5746. https://doi.org/10.1038/sj.onc.1202976

Chen D, Nicholas J (2006) Structural requirements for gp80 independence of human herpesvirus 8 interleukin-6 (vIL-6) and evidence for gp80 stabilization of gp130 signaling complexes induced by vIL-6. J Virol 80(19):9811–9821. https://doi.org/10.1128/JVI.00872-06

Chen J, Jiang L, Lan K, Chen X (2015) Celecoxib inhibits the lytic activation of Kaposi's sarcoma-associated herpesvirus through down-regulation of RTA expression by inhibiting the activation of p38 MAPK. Viruses 7(5):2268–2287. https://doi.org/10.3390/v7052268

Cheng EH, Nicholas J, Bellows DS, Hayward GS, Guo HG, Reitz MS, Hardwick JM (1997) A Bcl-2 homolog encoded by Kaposi sarcoma-associated virus, human herpesvirus 8, inhibits apoptosis but does not heterodimerize with Bax or Bak. Proc Natl Acad Sci U S A 94(2):690–694

Choi J, Means RE, Damania B, Jung JU (2001) Molecular piracy of Kaposi's sarcoma associated herpesvirus. Cytokine Growth Factor Rev 12(2-3):245–257

Choi YB, Nicholas J (2010) Bim nuclear translocation and inactivation by viral interferon regulatory factor. PLoS Pathog 6(8):e1001031. https://doi.org/10.1371/journal.ppat.1001031

Choi YB, Sandford G, Nicholas J (2012) Human herpesvirus 8 interferon regulatory factor-mediated BH3-only protein inhibition via Bid BH3-B mimicry. PLoS Pathog 8(6):e1002748. https://doi.org/10.1371/journal.ppat.1002748

Chugh P, Matta H, Schamus S, Zachariah S, Kumar A, Richardson JA, Smith AL, Chaudhary PM (2005) Constitutive NF-kappaB activation, normal Fas-induced apoptosis, and increased incidence of lymphoma in human herpes virus 8 K13 transgenic mice. Proc Natl Acad Sci U S A 102(36):12885–12890. https://doi.org/10.1073/pnas.0408577102

Cohen A, Brodie C, Sarid R (2006) An essential role of ERK signalling in TPA-induced reactivation of Kaposi's sarcoma-associated herpesvirus. The Journal of general virology 87(Pt 4):795–802. https://doi.org/10.1099/vir.0.81619-0

Conrad NK (2016) New insights into the expression and functions of the Kaposi's sarcoma-associated herpesvirus long noncoding PAN RNA. Virus Res 212:53–63. https://doi.org/10.1016/j.virusres.2015.06.012

Corcoran JA, Johnston BP, McCormick C (2015) Viral activation of MK2-hsp27-p115RhoGEF-RhoA signaling axis causes cytoskeletal rearrangements, p-body disruption and ARE-mRNA stabilization. PLoS Pathog 11(1):e1004597. https://doi.org/10.1371/journal.ppat.1004597

Couty JP, Geras-Raaka E, Weksler BB, Gershengorn MC (2001) Kaposi's sarcoma-associated herpesvirus G protein-coupled receptor signals through multiple pathways in endothelial cells. J Biol Chem 276(36):33805–33811. https://doi.org/10.1074/jbc.M104631200

Damania B, Cesarman E (2013) Kaposi's sarcoma–associated herpesvirus. In: Knipe DM, Howley PM (eds) Fields virology, vol 2, 6th edn. Lippincott Williams & Wilkins, Philadelphia, pp 2080–2128

Danial NN, Korsmeyer SJ (2004) Cell death: critical control points. Cell 116(2):205–219

Deng H, Liang Y, Sun R (2007) Regulation of KSHV lytic gene expression. Curr Top Microbiol Immunol 312:157–183

Dhillon AS, Hagan S, Rath O, Kolch W (2007) MAP kinase signalling pathways in cancer. Oncogene 26(22):3279–3290. https://doi.org/10.1038/sj.onc.1210421

Dittmer D, Lagunoff M, Renne R, Staskus K, Haase A, Ganem D (1998) A cluster of latently expressed genes in Kaposi's sarcoma-associated herpesvirus. J Virol 72(10):8309–8315

Ellis M, Chew YP, Fallis L, Freddersdorf S, Boshoff C, Weiss RA, Lu X, Mittnacht S (1999) Degradation of p27(Kip) cdk inhibitor triggered by Kaposi's sarcoma virus cyclin-cdk6 complex. EMBO J 18(3):644–653. https://doi.org/10.1093/emboj/18.3.644

Esteban M, Garcia MA, Domingo-Gil E, Arroyo J, Nombela C, Rivas C (2003) The latency protein LANA2 from Kaposi's sarcoma-associated herpesvirus inhibits apoptosis induced by dsRNA-activated protein kinase but not RNase L activation. The Journal of general virology 84(Pt 6):1463–1470. https://doi.org/10.1099/vir.0.19014-0

Feng P, Park J, Lee BS, Lee SH, Bram RJ, Jung JU (2002) Kaposi's sarcoma-associated herpesvirus mitochondrial K7 protein targets a cellular calcium-modulating cyclophilin ligand to modulate intracellular calcium concentration and inhibit apoptosis. J Virol 76(22):11491–11504

Feng P, Scott CW, Cho NH, Nakamura H, Chung YH, Monteiro MJ, Jung JU (2004) Kaposi's sarcoma-associated herpesvirus K7 protein targets a ubiquitin-like/ubiquitin-associated domain-containing protein to promote protein degradation. Mol Cell Biol 24(9):3938–3948

Field N, Low W, Daniels M, Howell S, Daviet L, Boshoff C, Collins M (2003) KSHV vFLIP binds to IKK-gamma to activate IKK. J Cell Sci 116(Pt 18):3721–3728. https://doi.org/10.1242/jcs.00691

Ford PW, Bryan BA, Dyson OF, Weidner DA, Chintalgattu V, Akula SM (2006) Raf/MEK/ERK signalling triggers reactivation of Kaposi's sarcoma-associated herpesvirus latency. J Gen Virol 87(Pt 5):1139–1144. https://doi.org/10.1099/vir.0.81628-0

Friborg J Jr, Kong W, Hottiger MO, Nabel GJ (1999) p53 inhibition by the LANA protein of KSHV protects against cell death. Nature 402(6764):889–894. https://doi.org/10.1038/47266

Fu B, Kuang E, Li W, Avey D, Li X, Turpin Z, Valdes A, Brulois K, Myoung J, Zhu F (2015) Activation of p90 ribosomal S6 kinases by ORF45 of Kaposi's sarcoma-associated herpesvirus is critical for optimal production of infectious viruses. J Virol 89(1):195–207. https://doi.org/10.1128/jvi.01937-14

Fujimuro M, Hayward SD (2003) The latency-associated nuclear antigen of Kaposi's sarcoma-associated herpesvirus manipulates the activity of glycogen synthase kinase-3beta. J Virol 77(14):8019–8030

Fujimuro M, Liu J, Zhu J, Yokosawa H, Hayward SD (2005) Regulation of the interaction between glycogen synthase kinase 3 and the Kaposi's sarcoma-associated herpesvirus latency-associated nuclear antigen. J Virol 79(16):10429–10441. https://doi.org/10.1128/jvi.79.16.10429-10441.2005

Fujimuro M, Wu FY, ApRhys C, Kajumbula H, Young DB, Hayward GS, Hayward SD (2003) A novel viral mechanism for dysregulation of beta-catenin in Kaposi's sarcoma-associated herpesvirus latency. Nat Med 9(3):300–306. https://doi.org/10.1038/nm829

Fuld S, Cunningham C, Klucher K, Davison AJ, Blackbourn DJ (2006) Inhibition of interferon signaling by the Kaposi's sarcoma-associated herpesvirus full-length viral interferon regulatory factor 2 protein. J Virol 80(6):3092–3097. https://doi.org/10.1128/jvi.80.6.3092-3097.2006

Gallaher AM, Das S, Xiao Z, Andresson T, Kieffer-Kwon P, Happel C, Ziegelbauer J (2013) Proteomic screening of human targets of viral microRNAs reveals functions associated with immune evasion and angiogenesis. PLoS Pathog 9(9):e1003584. https://doi.org/10.1371/journal.ppat.1003584

Gao SJ, Boshoff C, Jayachandra S, Weiss RA, Chang Y, Moore PS (1997) KSHV ORF K9 (vIRF) is an oncogene which inhibits the interferon signaling pathway. Oncogene 15(16):1979–1985. https://doi.org/10.1038/sj.onc.1201571

Godden-Kent D, Talbot SJ, Boshoff C, Chang Y, Moore P, Weiss RA, Mittnacht S (1997) The cyclin encoded by Kaposi's sarcoma-associated herpesvirus stimulates cdk6 to phosphorylate the retinoblastoma protein and histone H1. J Virol 71(6):4193–4198

Gottwein E, Cullen BR (2010) A human herpesvirus microRNA inhibits p21 expression and attenuates p21-mediated cell cycle arrest. J Virol 84(10):5229–5237. https://doi.org/10.1128/jvi.00202-10

Gottwein E, Mukherjee N, Sachse C, Frenzel C, Majoros WH, Chi J-TA, Braich R, Manoharan M, Soutschek J, Ohler U, Cullen BR (2007) A viral microRNA functions as an ortholog of cellular miR-155. Nature 450(7172):1096–1099. https://doi.org/10.1038/nature05992

Grundhoff A, Ganem D (2003) The latency-associated nuclear antigen of Kaposi's sarcoma-associated herpesvirus permits replication of terminal repeat-containing plasmids. J Virol 77(4):2779–2783

Guasparri I, Keller SA, Cesarman E (2004) KSHV vFLIP is essential for the survival of infected lymphoma cells. J Exp Med 199(7):993–1003. https://doi.org/10.1084/jem.20031467

Guasparri I, Wu H, Cesarman E (2006) The KSHV oncoprotein vFLIP contains a TRAF-interacting motif and requires TRAF2 and TRAF3 for signalling. EMBO Rep 7(1):114–119. https://doi.org/10.1038/sj.embor.7400580

Haecker I, Gay LA, Yang Y, Hu J, Morse AM, McIntyre LM, Renne R (2012) Ago HITS-CLIP expands understanding of Kaposi's sarcoma-associated herpesvirus miRNA function in primary effusion lymphomas. PLoS Pathog 8(8):e1002884. https://doi.org/10.1371/journal.ppat.1002884

Hansen A, Henderson S, Lagos D, Nikitenko L, Coulter E, Roberts S, Gratrix F, Plaisance K, Renne R, Bower M, Kellam P, Boshoff C (2010) KSHV-encoded miRNAs target MAF to induce endothelial cell reprogramming. Genes Dev 24(2):195–205. https://doi.org/10.1101/gad.553410

Havemeier A, Gramolelli S, Pietrek M, Jochmann R, Sturzl M, Schulz TF (2014) Activation of NF-kappaB by the Kaposi's sarcoma-associated herpesvirus K15 protein involves recruitment of the NF-kappaB-inducing kinase, IkappaB kinases, and phosphorylation of p65. J Virol 88(22):13161–13172. https://doi.org/10.1128/jvi.01766-14

Hayward SD, Liu J, Fujimuro M (2006) Notch and Wnt signaling: mimicry and manipulation by gamma herpesviruses. Sci STKE Signal Transduction Knowl Environ 2006(335):re4. https://doi.org/10.1126/stke.3352006re4

Heinzelmann K, Scholz BA, Nowak A, Fossum E, Kremmer E, Haas J, Frank R, Kempkes B (2010) Kaposi's sarcoma-associated herpesvirus viral interferon regulatory factor 4 (vIRF4/K10) is a novel interaction partner of CSL/CBF1, the major downstream effector of Notch signaling. J Virol 84(23):12255–12264. https://doi.org/10.1128/jvi.01484-10

Hengartner MO (2000) The biochemistry of apoptosis. Nature 407(6805):770–776. https://doi.org/10.1038/35037710

Hiscott J, Nguyen TL, Arguello M, Nakhaei P, Paz S (2006) Manipulation of the nuclear factor-kappaB pathway and the innate immune response by viruses. Oncogene 25(51):6844–6867. https://doi.org/10.1038/sj.onc.1209941

Hong YK, Foreman K, Shin JW, Hirakawa S, Curry CL, Sage DR, Libermann T, Dezube BJ, Fingeroth JD, Detmar M (2004) Lymphatic reprogramming of blood vascular endothelium by Kaposi sarcoma-associated herpesvirus. Nat Genet 36(7):683–685. https://doi.org/10.1038/ng1383

Hu H, Dong J, Liang D, Gao Z, Bai L, Sun R, Hu H, Zhang H, Dong Y, Lan K (2015) Genome-wide mapping of the binding sites and structural analysis of Kaposi's sarcoma-associated herpesvirus viral interferon regulatory factor 2 reveal that it is a DNA-binding transcription factor. J Virol 90(3):1158–1168. https://doi.org/10.1128/jvi.01392-15

Hu J, Garber AC, Renne R (2002) The latency-associated nuclear antigen of Kaposi's sarcoma-associated herpesvirus supports latent DNA replication in dividing cells. J Virol 76(22):11677–11687

Huang Q, Petros AM, Virgin HW, Fesik SW, Olejniczak ET (2002) Solution structure of a Bcl-2 homolog from Kaposi sarcoma virus. Proc Natl Acad Sci USA 99(6):3428–3433. https://doi.org/10.1073/pnas.062525799

Irmler M, Thome M, Hahne M, Schneider P, Hofmann K, Steiner V, Bodmer JL, Schroter M, Burns K, Mattmann C, Rimoldi D, French LE, Tschopp J (1997) Inhibition of death receptor signals by cellular FLIP. Nature 388(6638):190–195. https://doi.org/10.1038/40657

Izumiya Y, Izumiya C, Hsia D, Ellison TJ, Luciw PA, Kung HJ (2009) NF-kappaB serves as a cellular sensor of Kaposi's sarcoma-associated herpesvirus latency and negatively regulates K-Rta by antagonizing the RBP-Jkappa coactivator. J Virol 83(9):4435–4446. https://doi.org/10.1128/jvi.01999-08

Jacobs SR, Damania B (2011) The viral interferon regulatory factors of KSHV: immunosuppressors or oncogenes? Front Immunol 2:19. https://doi.org/10.3389/fimmu.2011.00019

Jarviluoma A, Koopal S, Rasanen S, Makela TP, Ojala PM (2004) KSHV viral cyclin binds to p27KIP1 in primary effusion lymphomas. Blood 104(10):3349–3354. https://doi.org/10.1182/blood-2004-05-1798

Jham BC, Ma T, Hu J, Chaisuparat R, Friedman ER, Pandolfi PP, Schneider A, Sodhi A, Montaner S (2011) Amplification of the angiogenic signal through the activation of the TSC/mTOR/HIF axis by the KSHV vGPCR in Kaposi's sarcoma. PLoS One 6(4):e19103. https://doi.org/10.1371/journal.pone.0019103

Jin Y, He Z, Liang D, Zhang Q, Zhang H, Deng Q, Robertson ES, Lan K (2012) Carboxyl-terminal amino acids 1052 to 1082 of the latency-associated nuclear antigen (LANA) interact with RBP-Jkappa and are responsible for LANA-mediated RTA repression. J Virol 86(9):4956–4969. https://doi.org/10.1128/jvi.06788-11

Joo CH, Shin YC, Gack M, Wu L, Levy D, Jung JU (2007) Inhibition of interferon regulatory factor 7 (IRF7)-mediated interferon signal transduction by the Kaposi's sarcoma-associated herpesvirus viral IRF homolog vIRF3. J Virol 81(15):8282–8292. https://doi.org/10.1128/jvi.00235-07

Keller SA, Schattner EJ, Cesarman E (2000) Inhibition of NF-kappaB induces apoptosis of KSHV-infected primary effusion lymphoma cells. Blood 96(7):2537–2542

Kliche S, Nagel W, Kremmer E, Atzler C, Ege A, Knorr T, Koszinowski U, Kolanus W, Haas J (2001) Signaling by human herpesvirus 8 kaposin A through direct membrane recruitment of cytohesin-1. Mol Cell 7(4):833–843

Konrad A, Wies E, Thurau M, Marquardt G, Naschberger E, Hentschel S, Jochmann R, Schulz TF, Erfle H, Brors B, Lausen B, Neipel F, Sturzl M (2009) A systems biology approach to identify the combination effects of human herpesvirus 8 genes on NF-kappaB activation. J Virol 83(6):2563–2574. https://doi.org/10.1128/jvi.01512-08

Krieg A, Mahotka C, Krieg T, Grabsch H, Müller W, Takeno S, Suschek CV, Heydthausen M, Gabbert HE, Gerharz CD (2002) Expression of different survivin variants in gastric carcinomas: first clues to a role of survivin-2B in tumour progression. Br J Cancer 86(5):737–743. https://doi.org/10.1038/sj.bjc.6600153

Krueger A, Baumann S, Krammer PH, Kirchhoff S (2001) FLICE-inhibitory proteins: regulators of death receptor-mediated apoptosis. Mol Cell Biol 21(24):8247–8254. https://doi.org/10.1128/mcb.21.24.8247-8254.2001

Kuang E, Tang Q, Maul GG, Zhu F (2008) Activation of p90 ribosomal S6 kinase by ORF45 of Kaposi's sarcoma-associated herpesvirus and its role in viral lytic replication. J Virol 82(4):1838–1850. https://doi.org/10.1128/JVI.02119-07

Kuang E, Wu F, Zhu F (2009) Mechanism of sustained activation of ribosomal S6 kinase (RSK) and ERK by kaposi sarcoma-associated herpesvirus ORF45: multiprotein complexes retain active phosphorylated ERK AND RSK and protect them from dephosphorylation. J Biol Chem 284(20):13958–13968. https://doi.org/10.1074/jbc.M900025200

Lagunoff M, Ganem D (1997) The structure and coding organization of the genomic termini of Kaposi's sarcoma-associated herpesvirus. Virology 236(1):147–154

Laman H, Coverley D, Krude T, Laskey R, Jones N (2001) Viral cyclin-cyclin-dependent kinase 6 complexes initiate nuclear DNA replication. Mol Cell Biol 21(2):624–635. https://doi.org/10.1128/mcb.21.2.624-635.2001

Lan K, Kuppers DA, Verma SC, Robertson ES (2004) Kaposi's sarcoma-associated herpesvirus-encoded latency-associated nuclear antigen inhibits lytic replication by targeting Rta: a potential mechanism for virus-mediated control of latency. J Virol 78(12):6585–6594. https://doi.org/10.1128/jvi.78.12.6585-6594.2004

Lan K, Verma SC, Murakami M, Bajaj B, Kaul R, Robertson ES (2007) Kaposi's sarcoma herpesvirus-encoded latency-associated nuclear antigen stabilizes intracellular activated Notch by targeting the Sel10 protein. Proc Natl Acad Sci U S A 104(41):16287–16292. https://doi.org/10.1073/pnas.0703508104

Lavrik IN, Golks A, Krammer PH (2005) Caspases: pharmacological manipulation of cell death. J Clin Investig 115(10):2665–2672. https://doi.org/10.1172/JCI26252

Lee H, Guo J, Li M, Choi J-K, DeMaria M, Rosenzweig M, Jung JU (1998a) Identification of an immunoreceptor tyrosine-based activation motif of K1 transforming protein of Kaposi's sarcoma-associated herpesvirus. Mol Cell Biol 18(9):5219–5228

Lee H, Veazey R, Williams K, Li M, Guo J, Neipel F, Fleckenstein B, Lackner A, Desrosiers RC, Jung JU (1998b) Deregulation of cell growth by the K1 gene of Kaposi's sarcoma-associated herpesvirus. Nat Med 4(4):435–440

Lee H-R, Choi W-C, Lee S, Hwang J, Hwang E, Guchhait K, Haas J, Toth Z, Jeon YH, Oh T-K, Kim MH, Jung JU (2011) Bilateral inhibition of HAUSP deubiquitinase by a viral interferon regulatory factor protein. Nat Struct Mol Biol 18(12):1336–1344. https://doi.org/10.1038/nsmb.2142

Lee HR, Doganay S, Chung B, Toth Z, Brulois K, Lee S, Kanketayeva Z, Feng P, Ha T, Jung JU (2014) Kaposi's sarcoma-associated herpesvirus viral interferon regulatory factor 4 (vIRF4) targets expression of cellular IRF4 and the Myc gene to facilitate lytic replication. J Virol 88(4):2183–2194. https://doi.org/10.1128/jvi.02106-13

Lee HR, Mitra J, Lee S, Gao SJ, Oh TK, Kim MH, Ha T, Jung JU (2015) Kaposi's sarcoma-associated herpesvirus viral interferon regulatory factor 4 (vIRF4) perturbs the G1-S cell cycle progression via deregulation of the cyclin D1 Gene. J Virol 90(2):1139–1143. https://doi.org/10.1128/jvi.01897-15

Lee HR, Toth Z, Shin YC, Lee JS, Chang H, Gu W, Oh TK, Kim MH, Jung JU (2009) Kaposi's sarcoma-associated herpesvirus viral interferon regulatory factor 4 targets MDM2 to deregulate the p53 tumor suppressor pathway. J Virol 83(13):6739–6747. https://doi.org/10.1128/jvi.02353-08

Lefort S, Soucy-Faulkner A, Grandvaux N, Flamand L (2007) Binding of Kaposi's sarcoma-associated herpesvirus K-bZIP to interferon-responsive factor 3 elements modulates antiviral gene expression. J Virol 81(20):10950–10960. https://doi.org/10.1128/jvi.00183-07

Lei X, Bai Z, Ye F, Xie J, Kim CG, Huang Y, Gao SJ (2010) Regulation of NF-kappaB inhibitor IkappaBalpha and viral replication by a KSHV microRNA. Nat Cell Biol 12(2):193–199. https://doi.org/10.1038/ncb2019

Li M, Damania B, Alvarez X, Ogryzko V, Ozato K, Jung JU (2000) Inhibition of p300 histone acetyltransferase by viral interferon regulatory factor. Mol Cell Biol 20(21):8254–8263

Li Q, Means R, Lang S, Jung JU (2007) Downregulation of gamma interferon receptor 1 by Kaposi's sarcomaassociated herpesvirus K3 and K5. J Virol 81(5):2117–2127. https://doi.org/10.1128/jvi.01961-06

Liang Y, Ganem D (2003) Lytic but not latent infection by Kaposi's sarcoma-associated herpesvirus requires host CSL protein, the mediator of Notch signaling. Proc Natl Acad Sci U S A 100(14):8490–8495. https://doi.org/10.1073/pnas.1432843100

Liang Y, Ganem D (2004) RBP-J (CSL) is essential for activation of the K14/vGPCR promoter of Kaposi's sarcoma-associated herpesvirus by the lytic switch protein RTA. J Virol 78(13):6818–6826. https://doi.org/10.1128/jvi.78.13.6818-6826.2004

Lin H-R, Ganem D (2011) Viral microRNA target allows insight into the role of translation in governing microRNA target accessibility. Proc Natl Acad Sci U S A 108(13):5148–5153. https://doi.org/10.1073/pnas.1102033108

Lin R, Genin P, Mamane Y, Sgarbanti M, Battistini A, Harrington WJ Jr, Barber GN, Hiscott J (2001) HHV-8 encoded vIRF-1 represses the interferon antiviral response by blocking IRF-3 recruitment of the CBP/p300 coactivators. Oncogene 20(7):800–811. https://doi.org/10.1038/sj.onc.1204163

Lin X, Liang D, He Z, Deng Q, Robertson ES, Lan K (2011) miR-K12-7-5p encoded by Kaposi's sarcoma-associated herpesvirus stabilizes the latent state by targeting viral ORF50/RTA. PLoS One 6(1):e16224. https://doi.org/10.1371/journal.pone.0016224

Lin YT, Kincaid RP, Arasappan D, Dowd SE, Hunicke-Smith SP, Sullivan CS (2010) Small RNA profiling reveals antisense transcription throughout the KSHV genome and novel small RNAs. RNA 16(8):1540–1558. https://doi.org/10.1261/rna.1967910

Liu J, Martin H, Shamay M, Woodard C, Tang QQ, Hayward SD (2007a) Kaposi's sarcoma-associated herpesvirus LANA protein downregulates nuclear glycogen synthase kinase 3 activity and consequently blocks differentiation. J Virol 81(9):4722–4731. https://doi.org/10.1128/jvi.02548-06

Liu J, Martin HJ, Liao G, Hayward SD (2007b) The Kaposi's sarcoma-associated herpesvirus LANA protein stabilizes and activates c-Myc. J Virol 81(19):10451–10459. https://doi.org/10.1128/jvi.00804-07

Liu L, Eby MT, Rathore N, Sinha SK, Kumar A, Chaudhary PM (2002) The human herpes virus 8-encoded viral FLICE inhibitory protein physically associates with and persistently activates the Ikappa B kinase complex. J Biol Chem 277(16):13745–13751. https://doi.org/10.1074/jbc.M110480200

Loh J, Huang Q, Petros AM, Nettesheim D, van Dyk LF, Labrada L, Speck SH, Levine B, Olejniczak ET, Virgin HWIV (2005) A surface groove essential for viral Bcl-2 function during chronic infection in vivo. PLoS Pathog 1(1):e10. https://doi.org/10.1371/journal.ppat.0010010

Lu CC, Li Z, Chu CY, Feng J, Feng J, Sun R, Rana TM (2010a) MicroRNAs encoded by Kaposi's sarcoma-associated herpesvirus regulate viral life cycle. EMBO Rep 11(10):784–790. https://doi.org/10.1038/embor.2010.132

Lu F, Stedman W, Yousef M, Renne R, Lieberman PM (2010b) Epigenetic regulation of Kaposi's sarcoma-associated herpesvirus latency by virus-encoded microRNAs that target Rta and the cellular Rbl2-DNMT pathway. J Virol 84(6):2697–2706. https://doi.org/10.1128/jvi.01997-09

Lu J, Verma SC, Cai Q, Robertson ES (2011) The single RBP-Jkappa site within the LANA promoter is crucial for establishing Kaposi's sarcoma-associated herpesvirus latency during primary infection. J Virol 85(13):6148–6161. https://doi.org/10.1128/jvi.02608-10

Lubyova B, Kellum MJ, Frisancho JA, Pitha PM (2007) Stimulation of c-Myc transcriptional activity by vIRF-3 of Kaposi sarcoma-associated herpesvirus. J Biol Chem 282(44):31944–31953. https://doi.org/10.1074/jbc.M706430200

Lubyova B, Pitha PM (2000) Characterization of a novel human herpesvirus 8-encoded protein, vIRF-3, that shows homology to viral and cellular interferon regulatory factors. J Virol 74(17):8194–8201

Mahotka C, Wenzel M, Springer E, Gabbert HE, Gerharz CD (1999) Survivin-deltaEx3 and survivin-2B: two novel splice variants of the apoptosis inhibitor survivin with different antiapoptotic properties. Cancer Res 59(24):6097–6102

Mann DJ, Child ES, Swanton C, Laman H, Jones N (1999) Modulation of p27(Kip1) levels by the cyclin encoded by Kaposi's sarcoma-associated herpesvirus. EMBO J 18(3):654–663. https://doi.org/10.1093/emboj/18.3.654

Matsumoto A, Onoyama I, Sunabori T, Kageyama R, Okano H, Nakayama KI (2011) Fbxw7-dependent degradation of Notch is required for control of "stemness" and neuronal-glial differentiation in neural stem cells. J Biol Chem 286(15):13754–13764. https://doi.org/10.1074/jbc.M110.194936

McClure LV, Kincaid RP, Burke JM, Grundhoff A, Sullivan CS (2013) Comprehensive mapping and analysis of Kaposi's sarcoma-associated herpesvirus 3' UTRs identify differential post-transcriptional control of gene expression in lytic versus latent infection. J Virol 87(23):12838–12849. https://doi.org/10.1128/jvi.02374-13

McCormick C, Ganem D (2005) The kaposin B protein of KSHV activates the p38/MK2 pathway and stabilizes cytokine mRNAs. Science (New York, NY) 307(5710):739–741. https://doi.org/10.1126/science.1105779

McCormick C, Ganem D (2006) Phosphorylation and function of the kaposin B direct repeats of Kaposi's sarcoma-associated herpesvirus. J Virol 80(12):6165–6170. https://doi.org/10.1128/jvi.02331-05

Montaner S, Sodhi A, Pece S, Mesri EA, Gutkind JS (2001) The Kaposi's sarcoma-associated herpesvirus G protein-coupled receptor promotes endothelial cell survival through the activation of Akt/protein kinase B. Cancer Res 61(6):2641–2648

Moody R, Zhu Y, Huang Y, Cui X, Jones T, Bedolla R, Lei X, Bai Z, Gao SJ (2013) KSHV microRNAs mediate cellular transformation and tumorigenesis by redundantly targeting cell growth and survival pathways. PLoS Pathog 9(12):e1003857. https://doi.org/10.1371/journal.ppat.1003857

Moore PS, Boshoff C, Weiss RA, Chang Y (1996a) Molecular mimicry of human cytokine and cytokine response pathway genes by KSHV. Science (New York, NY) 274(5293):1739–1744

Moore PS, Chang Y (1998) Antiviral activity of tumor-suppressor pathways: clues from molecular piracy by KSHV. Trends in genetics : TIG 14(4):144–150

Moore PS, Gao SJ, Dominguez G, Cesarman E, Lungu O, Knowles DM, Garber R, Pellett PE, McGeoch DJ, Chang Y (1996b) Primary characterization of a herpesvirus agent associated with Kaposi's sarcomae. J Virol 70(1):549–558

Morin PJ, Sparks AB, Korinek V, Barker N, Clevers H, Vogelstein B, Kinzler KW (1997) Activation of beta-catenin-Tcf signaling in colon cancer by mutations in beta-catenin or APC. Science (NY) 275(5307):1787–1790

Morris VA, Punjabi AS, Lagunoff M (2008) Activation of Akt through gp130 receptor signaling is required for Kaposi's sarcoma-associated herpesvirus-induced lymphatic reprogramming of endothelial cells. J Virol 82(17):8771–8779. https://doi.org/10.1128/JVI.00766-08

Morris VA, Punjabi AS, Wells RC, Wittkopp CJ, Vart R, Lagunoff M (2012) The KSHV viral IL-6 homolog is sufficient to induce blood to lymphatic endothelial cell differentiation. Virology 428(2):112–120. https://doi.org/10.1016/j.virol.2012.03.013

Muralidhar S, Pumfery AM, Hassani M, Sadaie MR, Kishishita M, Brady JN, Doniger J, Medveczky P, Rosenthal LJ (1998) Identification of kaposin (open reading frame K12) as a human herpesvirus 8 (Kaposi's sarcoma-associated herpesvirus) transforming gene. J Virol 72(6):4980–4988

Mutocheluh M, Hindle L, Areste C, Chanas SA, Butler LM, Lowry K, Shah K, Evans DJ, Blackbourn DJ (2011) Kaposi's sarcoma-associated herpesvirus viral interferon regulatory factor-2 inhibits type 1 interferon signalling by targeting interferon-stimulated gene factor-3. J Gen Virol 92(Pt 10):2394–2398. https://doi.org/10.1099/vir.0.034322-0

Nachmani D, Stern-Ginossar N, Sarid R, Mandelboim O (2009) Diverse herpesvirus microRNAs target the stress-induced immune ligand MICB to escape recognition by natural killer cells. Cell Host Microbe 5(4):376–385. https://doi.org/10.1016/j.chom.2009.03.003

Nakamura H, Li M, Zarycki J, Jung JU (2001) Inhibition of p53 tumor suppressor by viral interferon regulatory factor. J Virol 75(16):7572–7582. https://doi.org/10.1128/jvi.75.16.7572-7582.2001

Nakayama KI, Nakayama K (2006) Ubiquitin ligases: cell-cycle control and cancer. Nat Rev Cancer 6(5):369–381. https://doi.org/10.1038/nrc1881

Naranatt PP, Akula SM, Zien CA, Krishnan HH, Chandran B (2003) Kaposi's sarcoma-associated herpesvirus induces the phosphatidylinositol 3-Kinase-PKC-ζ-MEK-ERK signaling pathway in target cells early during infection: implications for infectivity. J Virol 77(2):1524–1539. https://doi.org/10.1128/JVI.77.2.1524-1539.2003

Nayar U, Lu P, Goldstein RL, Vider J, Ballon G, Rodina A, Taldone T, Erdjument-Bromage H, Chomet M, Blasberg R, Melnick A, Cerchietti L, Chiosis G, Wang YL, Cesarman E (2013) Targeting the Hsp90-associated viral oncoproteome in gammaherpesvirus-associated malignancies. Blood 122(16):2837–2847. https://doi.org/10.1182/blood-2013-01-479972

Ojala PM, Yamamoto K, Castanos-Velez E, Biberfeld P, Korsmeyer SJ, Makela TP (2000) The apoptotic v-cyclin-CDK6 complex phosphorylates and inactivates Bcl-2. Nat Cell Biol 2(11):819–825. https://doi.org/10.1038/35041064

Osborne J, Moore PS, Chang Y (1999) KSHV-encoded viral IL-6 activates multiple human IL-6 signaling pathways. Hum Immunol 60(10):921–927

Pan H, Xie J, Ye F, Gao SJ (2006) Modulation of Kaposi's sarcoma-associated herpesvirus infection and replication by MEK/ERK, JNK, and p38 multiple mitogen-activated protein kinase pathways during primary infection. J Virol 80(11):5371–5382. https://doi.org/10.1128/JVI.02299-05

Parravicini C, Chandran B, Corbellino M, Berti E, Paulli M, Moore PS, Chang Y (2000) Differential viral protein expression in Kaposi's sarcoma-associated herpesvirus-infected diseases: Kaposi's sarcoma, primary effusion lymphoma, and multicentric Castleman's disease. Am J Pathol 156(3):743–749. https://doi.org/10.1016/S0002-9440(10)64940-1

Pati S, Cavrois M, Guo HG, Foulke JS Jr, Kim J, Feldman RA, Reitz M (2001) Activation of NF-kappaB by the human herpesvirus 8 chemokine receptor ORF74: evidence for a paracrine model of Kaposi's sarcoma pathogenesis. J Virol 75(18):8660–8673

Pati S, Foulke JS Jr, Barabitskaya O, Kim J, Nair BC, Hone D, Smart J, Feldman RA, Reitz M (2003) Human herpesvirus 8-encoded vGPCR activates nuclear factor of activated T cells and collaborates with human immunodeficiency virus type 1 Tat. J Virol 77(10):5759–5773

Pearson G, Robinson F, Beers Gibson T, Xu BE, Karandikar M, Berman K, Cobb MH (2001) Mitogen-activated protein (MAP) kinase pathways: regulation and physiological functions. Endocr Rev 22(2):153–183. https://doi.org/10.1210/edrv.22.2.0428

Pfeffer S, Sewer A, Lagos-Quintana M, Sheridan R, Sander C, Grasser FA, van Dyk LF, Ho CK, Shuman S, Chien M, Russo JJ, Ju J, Randall G, Lindenbach BD, Rice CM, Simon V, Ho DD, Zavolan M, Tuschl T (2005) Identification of microRNAs of the herpesvirus family. Nat Methods 2(4):269–276. https://doi.org/10.1038/nmeth746

Pietrek M, Brinkmann MM, Glowacka I, Enlund A, Havemeier A, Dittrich-Breiholz O, Kracht M, Lewitzky M, Saksela K, Feller SM, Schulz TF (2010) Role of the Kaposi's sarcoma-associated herpesvirus K15 SH3 binding site in inflammatory signaling and B-cell activation. J Virol 84(16):8231–8240. https://doi.org/10.1128/jvi.01696-09

Prakash O, Tang ZY, Peng X, Coleman R, Gill J, Farr G, Samaniego F (2002) Tumorigenesis and aberrant signaling in transgenic mice expressing the human herpesvirus-8 K1 gene. J Natl Cancer Inst 94(12):926–935

Qin Z, Kearney P, Plaisance K, Parsons CH (2010) Pivotal advance: Kaposi's sarcoma-associated herpesvirus (KSHV)-encoded microRNA specifically induce IL-6 and IL-10 secretion by macrophages and monocytes. J Leukoc Biol 87(1):25–34

Radkov SA, Kellam P, Boshoff C (2000) The latent nuclear antigen of Kaposi sarcoma-associated herpesvirus targets the retinoblastoma-E2F pathway and with the oncogene Hras transforms primary rat cells. Nat Med 6(10):1121–1127. https://doi.org/10.1038/80459

Reed JC (1997) Bcl-2 family proteins: regulators of apoptosis and chemoresistance in hematologic malignancies. Semin Hematol 34(4 Suppl 5):9–19

Rossetto CC, Pari G (2012) KSHV PAN RNA associates with demethylases UTX and JMJD3 to activate lytic replication through a physical interaction with the virus genome. PLoS Pathog 8(5):e1002680. https://doi.org/10.1371/journal.ppat.1002680

Rossetto CC, Pari GS (2011) Kaposi's sarcoma-associated herpesvirus noncoding polyadenylated nuclear RNA interacts with virus- and host cell-encoded proteins and suppresses expression of genes involved in immune modulation. J Virol 85(24):13290–13297. https://doi.org/10.1128/jvi.05886-11

Rossetto CC, Tarrant-Elorza M, Verma S, Purushothaman P, Pari GS (2013) Regulation of viral and cellular gene expression by Kaposi's sarcoma-associated herpesvirus polyadenylated nuclear RNA. J Virol 87(10):5540–5553. https://doi.org/10.1128/jvi.03111-12

Russo JJ, Bohenzky RA, Chien MC, Chen J, Yan M, Maddalena D, Parry JP, Peruzzi D, Edelman IS, Chang Y, Moore PS (1996) Nucleotide sequence of the Kaposi sarcoma-associated herpesvirus (HHV8). Proc Natl Acad Sci U S A 93(25):14862–14867

Sadler R, Wu L, Forghani B, Renne R, Zhong W, Herndier B, Ganem D (1999) A complex translational program generates multiple novel proteins from the latently expressed kaposin (K12) locus of Kaposi's sarcoma-associated herpesvirus. J Virol 73(7):5722–5730

Samols MA, Hu J, Skalsky RL, Renne R (2005) Cloning and identification of a microRNA cluster within the latency-associated region of Kaposi's sarcoma-associated herpesvirus. J Virol 79(14):9301–9305. https://doi.org/10.1128/jvi.79.14.9301-9305.2005

Samols MA, Skalsky RL, Maldonado AM, Riva A, Lopez MC, Baker HV, Renne R (2007) Identification of cellular genes targeted by KSHV-encoded microRNAs. PLoS Pathog 3(5):e65. https://doi.org/10.1371/journal.ppat.0030065

Schwarz M, Murphy PM (2001) Kaposi's sarcoma-associated herpesvirus G protein-coupled receptor constitutively activates NF-kappa B and induces proinflammatory cytokine and chemokine production via a C-terminal signaling determinant. J Immunol 167(1):505–513

Seo T, Park J, Lee D, Hwang SG, Choe J (2001) Viral interferon regulatory factor 1 of Kaposi's sarcoma-associated herpesvirus binds to p53 and represses p53-dependent transcription and apoptosis. J Virol 75(13):6193–6198. https://doi.org/10.1128/jvi.75.13.6193-6198.2001

Sharma-Walia N, Krishnan HH, Naranatt PP, Zeng L, Smith MS, Chandran B (2005) ERK1/2 and MEK1/2 induced by Kaposi's sarcoma-associated herpesvirus (human herpesvirus 8) early during infection of target cells are essential for expression of viral genes and for establishment of infection. J Virol 79(16):10308–10329. https://doi.org/10.1128/JVI.79.16.10308-10329.2005

Shin YC, Nakamura H, Liang X, Feng P, Chang H, Kowalik TF, Jung JU (2006) Inhibition of the ATM/p53 signal transduction pathway by Kaposi's sarcoma-associated herpesvirus interferon regulatory factor 1. J Virol 80(5):2257–2266. https://doi.org/10.1128/JVI.80.5.2257-2266.2006

Skalsky RL, Samols MA, Plaisance KB, Boss IW, Riva A, Lopez MC, Baker HV, Renne R (2007) Kaposi's sarcoma-associated herpesvirus encodes an ortholog of miR-155. J Virol 81(23):12836–12845. https://doi.org/10.1128/jvi.01804-07

Staskus KA, Sun R, Miller G, Racz P, Jaslowski A, Metroka C, Brett-Smith H, Haase AT (1999) Cellular tropism and viral interleukin-6 expression distinguish human herpesvirus 8 involvement in Kaposi's sarcoma, primary effusion lymphoma, and multicentric Castleman's disease. J Virol 73(5):4181–4187

Strahan R, Uppal T, Verma SC (2016) Next-generation sequencing in the understanding of Kaposi's sarcoma-associated herpesvirus (KSHV) biology. Viruses 8(4):92. https://doi.org/10.3390/v8040092

Su LK, Vogelstein B, Kinzler KW (1993) Association of the APC tumor suppressor protein with catenins. Sci (NY) 262(5140):1734–1737

Sun R, Lin SF, Gradoville L, Miller G (1996) Polyadenylylated nuclear RNA encoded by Kaposi sarcoma-associated herpesvirus. Proc Natl Acad Sci USA 93(21):11883–11888

Swanton C, Mann DJ, Fleckenstein B, Neipel F, Peters G, Jones N (1997) Herpes viral cyclin/Cdk6 complexes evade inhibition by CDK inhibitor proteins. Nature 390(6656):184–187. https://doi.org/10.1038/36606

Tamura T, Yanai H, Savitsky D, Taniguchi T (2008) The IRF family transcription factors in immunity and oncogenesis. Annu Rev Immunol 26:535–584. https://doi.org/10.1146/annurev.immunol.26.021607.090400

Taniguchi T, Takaoka A (2002) The interferon-alpha/beta system in antiviral responses: a multimodal machinery of gene regulation by the IRF family of transcription factors. Curr Opin Immunol 14(1):111–116

Thome M, Schneider P, Hofmann K, Fickenscher H, Meinl E, Neipel F, Mattmann C, Burns K, Bodmer JL, Schroter M, Scaffidi C, Krammer PH, Peter ME, Tschopp J (1997) Viral FLICE-inhibitory proteins (FLIPs) prevent apoptosis induced by death receptors. Nature 386(6624):517–521. https://doi.org/10.1038/386517a0

Tomlinson CC, Damania B (2004) The K1 protein of Kaposi's sarcoma-associated herpesvirus activates the Akt signaling pathway. J Virol 78(4):1918–1927

Wang FZ, Akula SM, Pramod NP, Zeng L, Chandran B (2001) Human herpesvirus 8 envelope glycoprotein K8.1A interaction with the target cells involves heparan sulfate. J Virol 75(16):7517–7527. https://doi.org/10.1128/JVI.75.16.7517-7527.2001

Wang HW, Sharp TV, Koumi A, Koentges G, Boshoff C (2002) Characterization of an anti-apoptotic glycoprotein encoded by Kaposi's sarcoma-associated herpesvirus which resembles a spliced variant of human survivin. EMBO J 21(11):2602–2615. https://doi.org/10.1093/emboj/21.11.2602

Wang HW, Trotter MW, Lagos D, Bourboulia D, Henderson S, Makinen T, Elliman S, Flanagan AM, Alitalo K, Boshoff C (2004) Kaposi sarcoma herpesvirus-induced cellular reprogramming contributes to the lymphatic endothelial gene expression in Kaposi sarcoma. Nat Genet 36(7):687–693. https://doi.org/10.1038/ng1384

Wang KC, Chang HY (2011) Molecular mechanisms of long noncoding RNAs. Mol Cell 43(6):904–914. https://doi.org/10.1016/j.molcel.2011.08.018

Webster MT, Rozycka M, Sara E, Davis E, Smalley M, Young N, Dale TC, Wooster R (2000) Sequence variants of the axin gene in breast, colon, and other cancers: an analysis of mutations that interfere with GSK3 binding. Genes Chromosom Cancer 28(4):443–453

Wies E, Hahn AS, Schmidt K, Viebahn C, Rohland N, Lux A, Schellhorn T, Holzer A, Jung JU, Neipel F (2009) The Kaposi's sarcoma-associated herpesvirus-encoded vIRF-3 inhibits cellular IRF-5. J Biol Chem 284(13):8525–8538. https://doi.org/10.1074/jbc.M809252200

Xie J, Ajibade AO, Ye F, Kuhne K, Gao SJ (2008) Reactivation of Kaposi's sarcoma-associated herpesvirus from latency requires MEK/ERK, JNK and p38 multiple mitogen-activated protein kinase pathways. Virology 371(1):139–154. https://doi.org/10.1016/j.virol.2007.09.040

Xie J, Pan H, Yoo S, Gao SJ (2005) Kaposi's sarcoma-associated herpesvirus induction of AP-1 and interleukin 6 during primary infection mediated by multiple mitogen-activated protein kinase pathways. J Virol 79(24):15027–15037. https://doi.org/10.1128/jvi.79.24.15027-15037.2005

Yoo J, Kang J, Lee HN, Aguilar B, Kafka D, Lee S, Choi I, Lee J, Ramu S, Haas J, Koh CJ, Hong YK (2010) Kaposin-B enhances the PROX1 mRNA stability during lymphatic reprogramming of vascular endothelial cells by Kaposi's sarcoma herpes virus. PLoS Pathog 6(8):e1001046. https://doi.org/10.1371/journal.ppat.1001046

Yu Y, Hayward GS (2010) The ubiquitin E3 ligase RAUL negatively regulates type i interferon through ubiquitination of the transcription factors IRF7 and IRF3. Immunity 33(6):863–877. https://doi.org/10.1016/j.immuni.2010.11.027

Yu Y, Wang SE, Hayward GS (2005) The KSHV immediate-early transcription factor RTA encodes ubiquitin E3 ligase activity that targets IRF7 for proteosome-mediated degradation. Immunity 22(1):59–70. https://doi.org/10.1016/j.immuni.2004.11.011

Zhong W, Ganem D (1997) Characterization of ribonucleoprotein complexes containing an abundant polyadenylated nuclear RNA encoded by Kaposi's sarcoma-associated herpesvirus (human herpesvirus 8). J Virol 71(2):1207–1212

Zhong W, Wang H, Herndier B, Ganem D (1996) Restricted expression of Kaposi sarcoma-associated herpesvirus (human herpesvirus 8) genes in Kaposi sarcoma. Proc Natl Acad Sci U S A 93(13):6641–6646

Zhu FX, King SM, Smith EJ, Levy DE, Yuan Y (2002) A Kaposi's sarcoma-associated herpesviral protein inhibits virus-mediated induction of type I interferon by blocking IRF-7 phosphorylation and nuclear accumulation. Proc Natl Acad Sci U S A 99(8):5573–5578. https://doi.org/10.1073/pnas.082420599

Zhu Y, Huang Y, Jung JU, Lu C, Gao S-J (2014) Viral miRNA targeting of bicistronic and polycistronic transcripts. Current opinion in virology 0:66–72. https://doi.org/10.1016/j.coviro.2014.04.004

Ziegelbauer JM, Sullivan CS, Ganem D (2009) Tandem array-based expression screens identify host mRNA targets of virus-encoded microRNAs. Nat Genet 41(1):130–134. https://doi.org/10.1038/ng.266

Zimring JC, Goodbourn S, Offermann MK (1998) Human herpesvirus 8 encodes an interferon regulatory factor (IRF) homolog that represses IRF-1-mediated transcription. J Virol 72(1):701–707

Chapter 16
Pathological Features of Kaposi's Sarcoma-Associated Herpesvirus Infection

Harutaka Katano

Abstract Kaposi's sarcoma-associated herpesvirus (KSHV, human herpesvirus 8, or HHV-8) was firstly discovered in Kaposi's sarcoma tissue derived from patients with acquired immune deficiency syndrome. KSHV infection is associated with malignancies and certain inflammatory conditions. In addition to Kaposi's sarcoma, KSHV has been detected in primary effusion lymphoma, KSHV-associated lymphoma, and some cases of multicentric Castleman disease (MCD). Recently, KSHV inflammatory cytokine syndrome (KICS) was also defined as a KSHV-associated disease. In KSHV-associated malignancies, such as Kaposi's sarcoma and lymphoma, KSHV latently infects almost all tumor cells, and lytic proteins are rarely expressed. A high titer of KSHV is detected in the sera of patients with MCD and KICS, and the expression of lytic proteins such as ORF50, vIL-6, and vMIP-I and vMIP-II is frequently observed in the lesions of patients with these diseases. Immunohistochemistry of LANA-1 is an important diagnostic tool for KSHV infection. However, much of the pathogenesis of KSHV remains to be elucidated, especially regarding oncogenesis. Some viral proteins have been shown to have transforming activity in mammalian cells; however, these proteins are not expressed in latently KSHV-infected cells. KSHV encodes homologs of cellular proteins in its genome such as cyclin D, G-protein coupled protein, interleukin-6, and macrophage inflammatory protein-1 and -2. Molecular mimicry by these viral proteins may contribute to the establishment of microenvironments suitable for tumor growth. In this review, the virus pathogenesis is discussed based on pathological and experimental findings and clinical aspects.

Keywords Kaposi's sarcoma-associated herpesvirus (KSHV) · HHV-8 · Latency-associated nuclear antigen (LANA-1) · Primary effusion lymphoma (PEL) · Kaposi's sarcoma · Multicentric Castleman disease (MCD) · KSHV inflammatory cytokine syndrome (KICS) · KSHV-associated diffuse large B cell lymphoma, not otherwise specified

---

H. Katano (✉)
Department of Pathology, National Institute of Infectious Diseases, Tokyo, Japan
e-mail: katano@nih.go.jp

16.1 Introduction

More than 20 years have passed from the discovery of Kaposi's sarcoma-associated herpesvirus (KSHV, human herpesvirus 8, or HHV-8) in acquired immune deficiency syndrome (AIDS)-associated Kaposi's sarcoma (KS) (Dittmer and Damania 2016; Goncalves et al. 2017; Gramolelli and Schulz 2015). KSHV was the eighth human herpesvirus to be discovered and is considered a human oncovirus along with Epstein–Barr virus (EBV) (Chang et al. 1994; Ganem 2005). Both EBV and KSHV are gamma herpesvirus with an affinity to human lymphocytes. KSHV infection is associated with relatively rare diseases such as KS, primary effusion lymphoma (PEL) and some cases of multicentric Castleman disease (MCD). Unlike other human herpesviruses, KSHV infection is not common among the general population in most parts of the world, with the exception of African countries and some other regions. Although some important questions remain, recent progress in molecular biology has revealed the unique pathogenesis and virological features of KSHV compared with other herpesviruses. In this article, the unique pathogenesis of KSHV is reviewed from all aspects of pathology.

16.2 Virus Pathogenesis

16.2.1 Electron Microscopic Features of KSHV

Like other human herpesviruses, KSHV infects healthy humans latently. Latency is also predominant in KS and lymphoma. Even in KS and lymphoma lesions, it is difficult to detect viral particles by electron microscopy because KSHV does not usually replicate at these sites. However, virus particles of KSHV can be observed in PEL cell lines stimulated chemically with 12-O-tetradecanoylphorbol-13-acetate. The morphological features and size of KSHV particles are similar to those of other human herpesviruses (Ohtsuki et al. 1999; Orenstein et al. 1997; Renne et al. 1996; Said et al. 1996, 1997). A virus particle consists of a capsid, tegument, and envelope. The size of a whole virus particle including the envelope is 150–200 nm. In the nucleus, virus particles do not possess an envelope, and only capsids of 100 nm in diameter are observed. The capsid contains a viral DNA core and RNA and is covered with tegument and envelope. The envelope is flexible in form and the tegument protein localizes between the capsid and the envelope.

16.2.2 Virus Gene Expression

KSHV encodes around 80 genes in its genome (Russo et al. 1996). Gene expression of KSHV is classified into latent and lytic expression, and lytic transcripts are also categorized into immediate early, early, and late, like other herpesviruses. During latent infection of KSHV, the virus expresses limited viral transcripts encoded within the latency-associated gene cluster. This cluster contains open reading frame (ORF) 73 (latency-associated nuclear antigen 1, LANA-1, LNA, or LNA-1), viral cyclin (ORF72), viral FLICE-inhibitory protein (K13, v-FLIP), Kaposin (K12), and viral-encoded microRNAs. Among them, LANA-1 is the most important protein that is expressed in KSHV-infected cells exclusively. On immunostaining, LANA-1 appears in a dot-like pattern in the nucleus of KSHV-infected cells. Other latent proteins of KSHV are not easily detected in infected cells by immunohistochemistry. Other than the latency-associated gene cluster, some transcripts in the viral interferon regulatory factor (IRF)-encoding region have been suggested to be expressed during the latent phase. The *LANA-2* (*vIRF3* or *K10.5*) gene was identified in PEL cells as another latency-associated nuclear antigen encoded by KSHV (Rivas et al. 2001). Curiously, LANA-2 expression is observed only in PEL cells but not in KS cells. Almost all other genes encoded by KSHV are categorized into lytic transcripts. The most abundant lytic transcript of KSHV may be the *T1.1* gene, an untranslatable transcript encoded by the region between *ORFK7* and *ORF16* (Zhong and Ganem 1997). Its function is unknown, but this gene is expressed highly and detected easily in the early phase of KSHV infection. Replication and transcription activator (RTA or ORF50) is an immediate early protein of KSHV that plays an important role in KSHV replication (Sun et al. 1999). RTA is a transactivator that activates many other viral transcripts. KSHV encodes 12 pre-miRNAs in the K12 region and produces at least 17 mature miRNAs during the latent phase of KSHV infection in KSHV-infected cells and clinical samples (Cai et al. 2005; Samols et al. 2005).

16.2.3 Pathogenesis and Oncogenesis

In healthy persons, KSHV infects circulating B cells (Harrington Jr. et al. 1996). However, the number of KSHV-infected B cells has not been accurately determined. KSHV has not been detected by PCR in KSHV-infected immunocompetent persons, suggesting that the number of KSHV-infected cells in the blood is low. The copy number of KSHV in the blood increases in immunocompromised hosts, such as patients with AIDS. Increased numbers of KSHV-infected B cells interact with endothelial cells, and efficient infection of KSHV into endothelial cells by cell-to-cell transmission has been reported in vitro (Gao et al. 2003; Myoung and Ganem 2011; Sakurada et al. 2001). Although lytic proteins are expressed immediately after transmission of KSHV, their expression reduces after transmission, but LANA-1 expression is stable after transmission (Dupin et al. 1999; Katano et al. 1999b; Kellam et al. 1999). Some KSHV-encoded proteins including K1, vIRF1, and vGPCR have been shown to exert transformation activity on mammalian cells, but these proteins are not usually expressed in KSHV-infected cells (Bais et al. 1998; Gao et al. 1997; Lee et al. 1998; Montaner et al. 2003). LANA-1 is a multi-functional protein, and one of its major roles is to maintain latency in KSHV-infected cells by tethering KSHV DNA to the host chromosomes (Ballestas et al. 1999). LANA-1 binds directly to the KSHV genome and recruits it to the host chromosome. During host cell division, KSHV DNA replicates using the DNA replicative machinery of the host (Sakakibara et al. 2004), and the resulting KSHV DNA is inherited to daughter cells. LANA-1 also binds to p53 and inhibits p53-dependent apoptosis in KSHV-infected cells (Friborg Jr. et al. 1999; Katano et al. 2001). LANA-1 stabilizes beta-catenin by binding to the negative regulator GSK-3beta, promoting cell-cycle induction by nuclear accumulation of GSK-3beta (Fujimuro et al. 2003).

KSHV encodes several homologs of human genes, including those encoding cytokines, cell cycle-associated protein, and apoptosis-associated proteins including IL-6, MIP1, MIP2, BCL2, IRF, cyclin D1, and FLIP (Russo et al. 1996). Some of these mimic the function of human proteins (molecular mimics), and some inhibit the mimic proteins functionally. However, since a large number of these homologs are lytic proteins, KSHV-infected cells do not usually express these proteins, with only small populations of KSHV-infected cells expressing these proteins. However, among them, viral interleukin-6 (vIL-6 or K2) is expressed relatively frequently compared with other lytic proteins and is involved in the pathogenesis of KSHV-associated diseases (Moore et al. 1996). vIL-6 has 62% similarity to human IL-6 in its amino acid sequence. vIL-6 can bind to p80, a subunit of human IL-6 receptor, and induces downstream Stat3 signaling (Molden et al. 1997). vIL-6 also induces expression of vascular endothelial growth factor, resulting in angiogenesis and the growth of KSHV-infected cells (Aoki et al. 1999, 2003). vIL-6 also contributes to evasion from the host immune system by blocking hIL-6R and suppressing p21 expression (Chatterjee et al. 2002).

To date, no transforming genes of note have been reported in KSHV, but recent reports suggest that virus-encoded microRNAs (miRNAs) play important roles in the pathogenesis of KSHV. KSHV encodes at least 12 pre-miRNAs in its genome, generating 24 mature miRNAs in KSHV-infected cells (Cai et al. 2005). These miRNAs have been shown to modulate host gene expression, resulting in dynamic biological changes in KSHV-infected cells. KSHV-encoded miRNA is highly expressed in KSHV-infected cells compared with host miRNA. Surprisingly, in KSHV-infected PEL cell lines, TY-1 and BCBL-1, almost half of the total miRNA originated from KSHV (Hoshina et al. 2016). KSHV-encoded miRNAs are contained in exosomes, small vesicles exported out of cells. miRNAs may be delivered from KSHV-infected cells and lesions via exosomes to exert distal effects. Generally, miRNA binds to host mRNA and modulates expression of target proteins. Each viral miRNA has various functions biologically. BCL2-associated factor BCLAF1 is a target of KSHV-encoded miR-K5, resulting in the inhibition of apoptosis (Lei et al. 2010). miR-K1 targets IkB-alpha, which plays an inhibitory role in NF-kB signaling (Lei et al. 2010). Expression of miR-K1 rescues NF-kappaB activity and inhibit viral lytic replication, maintaining KSHV latent infection in the host cells (Lei et al. 2010; Ziegelbauer et al. 2009). Although there are several reports describing the functions of KSHV-encoded miRNAs, much remains to be determined regarding the functions of miRNAs encoded by KSHV.

16.3 Epidemiology and Transmission

Unlike other human herpesviruses, KSHV infection is not common among the general population in many countries. ELISA using lysates of KSHV viral particles or recombinant viral proteins as antigens has been used as a tool for KSHV serology, as well as immunofluorescence assays using KSHV-infected cells. Serological studies have demonstrated that the seroprevalence of KSHV among the general population differs significantly among regions and countries. KSHV seropositivity in the general population is low (less than 10%) in northern European, American, and Asian countries, 10–30% in the Mediterranean region and Xinjiang region in China, and more than 50% in sub-Saharan African countries (Chatlynne et al. 1998; Davis et al. 1997; He et al. 2007; Katano et al. 2000a; Kedes et al. 1997; Mayama et al. 1998; Rabkin et al. 1998). In western countries and Japan, KSHV infection is more common (8–25%) in men who have sex with men (MSM) than in the general population (Casper et al. 2006; Engels et al. 2007; Grulich et al. 2005; Katano et al. 2013).

Since the saliva contains a high copy number of KSHV in KSHV-infected individuals (Pauk et al. 2000), saliva is thought to be an important route of KSHV transmission. KSHV may be transmitted from mother to child through saliva horizontally in endemic regions such as Africa. Homosexual behavior may also be associated with KSHV transmission between MSM in non-endemic regions. Organ transplantation is another important route of KSHV transmission (Regamey et al. 1998). It is possible that blood transfusion also presents a risk for KSHV transmission (Hladik et al. 2006, 2012).

16.4 KSHV-Related Diseases

KS, PEL, KSHV-associated MCD, KSHV-associated diffuse large B cell lymphoma, and KSHV inflammatory cytokine syndrome (KICS) are now widely accepted as KSHV-related diseases by many researchers (Goncalves et al. 2017). In addition, there are some reports describing the primary infection of KSHV (Andreoni et al. 2002; Wang et al. 2001). Shortly after the discovery of KSHV, it was reported that the KSHV genome could be detected in patients with various diseases such as multiple myeloma, primary pulmonary hypertension, Bowen disease, squamous cell carcinoma, Paget disease, and actinic keratosis. However, the association between KSHV infection and these diseases has become controversial, with many researchers disputing this link.

16.4.1 Primary KSHV Infection

Primary infection with KSHV has been demonstrated by several reports. A febrile maculopapular skin rash was associated with primary KSHV infection, detected through the saliva, in a mass study of immunocompetent children in Egypt (Andreoni et al. 2002; Wang et al. 2001). This infection has been associated with a rash, fever, arthralgia, and lymphadenopathy in immunocompetent persons. In immunocompromised patients, such as peripheral blood stem cell/bone marrow transplantation patients, hepatitis, splenomegaly, and bone marrow failure were reported as illnesses associated with acute infection with KSHV (Luppi et al. 2000).

16.4.2 Kaposi's Sarcoma (KS)

KS is the most frequent KSHV-associated disease. There are four clinical subtypes: classic, AIDS-associated, iatrogenic, and endemic (Antman and Chang 2000). AIDS-associated KS is the most common subtype. Curiously, AIDS-associated KS occurs only in HIV-positive MSM. Although the detailed mechanisms are unknown, higher KSHV seroprevalence among MSM than the general population may be associated with the high incidence of KS in MSM (Casper et al. 2006; Katano et al. 2013).

KS appears more commonly in the skin, oral cavity, and gastrointestinal tract and occasionally in the lymph node and organs such as the lungs and liver (Antman and Chang 2000). The prognosis for pulmonary KS is poor and sometimes fatal. Skin lesions are most common in KS cases. Cases of cutaneous KS are classified into patchy, plaque, and nodular stages clinically. The patchy stage of KS appears as one or a few small red flat lesions in the skin of the four extremities and the trunk. The plaque stage of KS shows as fused patchy lesions and sometimes elevated patchy lesions. The nodular stage of KS is more aggressive than the plaque stage and appears as multiple nodular lesions. Multiple KS lesions in the extremities are often complicated with lymphedema. Histologically, proliferation of spindle cells around the vessel is observed in the patchy stage (Fig. 16.1). Spindle cells form slit-like vascular spaces around the vessel containing extravasated red blood cells. Blood vessels in the patchy stage are dilated and abnormal in shape. In the plaque stage, spindle cell proliferation is prominent and fusion of patchy lesions occurs. Massive nodular proliferation of spindle cells is observed in the nodular stage. Nodules of spindle cells compress the surrounding tissue causing lymphedema. Mitosis is sometimes observed and macrophages with hyaline-globules are found.

Immunohistochemically, LANA-1 is detected exclusively in KS spindle cells regardless of the clinical KS subtype and stage (Dupin et al. 1999; Katano et al. 1999b), indicating that LANA-1 immunohistochemistry is the gold standard for KS diagnosis in histopathology. LANA-1 immunohistochemistry demonstrates a dot-like staining pattern in the nucleus of KS spindle cells. LANA-1 signals are also

16 Pathology of KSHV Infection

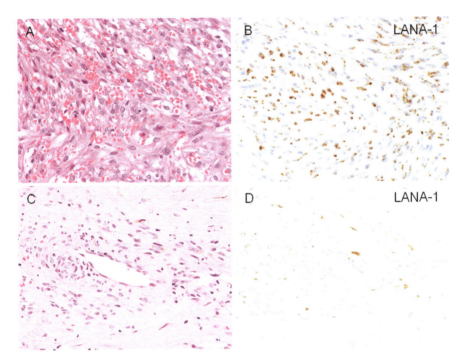

Fig. 16.1 Histology of Kaposi's sarcoma (**a** and **b**). Nodular stage of Kaposi's sarcoma. HE staining demonstrates the proliferation of spindle tumor cells with vascular slits (**a**). LANA-1 immunohistochemistry shows positive signals in the nucleus (**b**). (**c** and **d**) Patchy stage of Kaposi's sarcoma. Spindle cells are found around the vessel (**c**). Some spindle cells around the vessel are positive for LANA-1 by immunohistochemistry (**d**)

found in vascular endothelial cells in KS lesions. Other KSHV-encoded proteins including lytic (RTA, ORF59, K8.1) and latent (vFLIP, v-cyclin D, LANA-2) proteins are difficult to detect or rarely expressed in the pathological samples of KS by immunohistochemistry. Podoplanin (D2–40), a lymphatic marker, is also detected in KS cells; however, this is not specific to KS cells (Weninger et al. 1999).

PCR detection of the KSHV genome is another useful tool for KS diagnosis. KS cells contain 1–2 copies of the KSHV genome per cell (Asahi-Ozaki et al. 2006). Short fragments of the KSHV genome can be detected even in formalin-fixed paraffin-embedded KS tissue by PCR and real-time PCR (Asahi-Ozaki et al. 2006). The KSHV genome is occasionally detected in the sera of KS patients, but not always. The copy number of the KSHV genome in the serum of patients with KS is much lower than that in patients with KSHV-associated MCD. Serum antibody to KSHV is usually positive in patients with KS. However, since all KSHV-infected individuals have serum antibody to KSHV, a positive serum antibody result does not always reflect the presence of KS. LANA-1, ORF59, K8.1, and ORF65 have been reported to be major immunogens of KSHV in patients with KS (Katano et al. 2000b).

For the treatment of AIDS-associated KS, anti-retroviral therapy is effective not only on HIV but also on KS. Regression of KS is sometimes observed by administration of anti-retroviral therapy in the absence of specific treatment for KS. Combination therapy comprising anti-retroviral therapy and chemotherapy of pegylated liposomal doxorubicin is required for aggressive KS cases (Martin-Carbonero et al. 2008). Anti-herpesvirus drugs such as ganciclovir and acyclovir are not effective against KS and there is currently no vaccine available against KSHV. However, animal experiments suggest that vaccination with K8.1 and another KSHV-encoded protein may induce some neutralizing antibodies to KSHV in vivo (Sakamoto et al. 2010).

It is obvious that latent infection of KSHV is predominant in KS lesions. KSHV infects endothelial cells more efficiently through cell-to-cell contact from KSHV-infected B cells to KSHV-uninfected endothelial cells than through cell-free transmission (Myoung and Ganem 2011; Sakurada et al. 2001). KSHV infection induces alterations in gene expression in endothelial cells, resulting in different expression patterns of cellular proteins and spindle forms in endothelial cells (Hong et al. 2004; Wang et al. 2004). LANA-1 plays a central role in the pathogenesis of KS. LANA-1 expression contributes to the transformation and maintenance of KSHV in KS cells. The expression of lytic proteins encoded by KSHV is limited in KS cells. Viral IL-6 is expressed in a small portion of KS cells, and this vIL-6 interacts with hIL-6 receptor. Cellular immunity is important for herpesvirus infection and KSHV displays mechanisms of immune evasion. KSHV-encoded K3 and K5 are homologs of human MARCH8 that downregulates MHC class I expression in KSHV-infected cells (Ishido et al. 2000). Expression of the anti-apoptotic KSHV ORF-K13/v-FLIP also contributes to the tumorigenesis of KS (Thome et al. 1997).

16.4.3 Primary Effusion Lymphoma (PEL)

PEL is defined as KSHV-positive effusion lymphoma developing in body cavity effusion such as the pleural, abdominal, and pericardial effusion (Nador et al. 1996; Said and Cesarman 2008). PEL occurs in immunosuppressed patients, especially HIV-infected MSM (Cesarman et al. 1995; Nador et al. 1996), and accounts for approximately 4% of all cases of AIDS-associated lymphoma (Ota et al. 2014). PEL cells grow in the pleural, abdominal, or pericardial effusion as lymphomatous effusions. Some cases of PEL are complicated by a contiguous tumor mass. Such KSHV-positive solid lymphomas were classified as "extracavitary PEL". Extracavitary PEL cases without PEL lesions are also included into this category.

PEL cells exhibit plasmablastic or large immunoblastic, sometimes anaplastic, large cell morphology (Fig. 16.2). The nuclei of these cells frequently show a spoke-wheel structure with prominent nucleoli. The cytoplasm is abundant and deeply basophilic, and a nuclear halo is sometimes observed in the cytoplasm. These morphological features are similar to plasma cells or plasmablasts. However, PEL cells

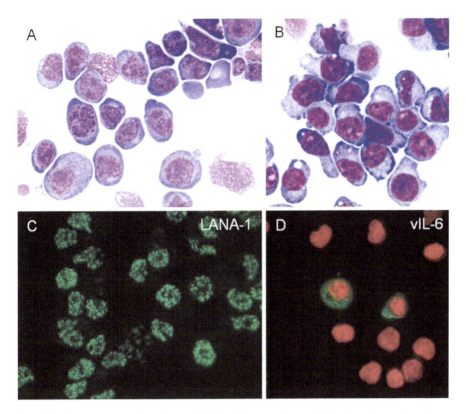

Fig. 16.2 Primary effusion lymphoma (PEL) (**a**) Giemsa staining of PEL. Lymphoblastic cells are observed. (**b**) Giemsa staining of another case of PEL. Lymphoma cells have large nuclei with abundant cytoplasm. (**c**) LANA-1 immunofluorescence staining of PEL cells. Positive signals show a dot-like staining pattern in the nucleus. (**d**) Immunofluorescence assay for vIL-6. vIL-6 is detected in the cytoplasm of some lymphoma cells

are much larger than normal plasma cells, and their nuclei are atypical and more irregular in shape. Binucleated or multinucleated cells, like the Reed–Sternberg cells in Hodgkin lymphoma, are frequently observed. Mitotic cells are also frequently detected. Immunohistochemically, PEL cells express a unique immunophenotype. Almost all cases are positive for CD45, CD38, and CD138. Many cases of PEL are positive for CD30, but not all. PEL cells are usually negative for B cell markers, CD19 and CD20, but immunoglobulin gene rearrangement is detected in all cases, indicating their B cell origin. PEL cells are invariably positive for KSHV-encoded LANA-1; therefore LANA-1 is the most important marker for PEL diagnosis. EBV is detected in >80% of PEL cases. In such cases, EBV and KSHV coinfect PEL cells. PEL cells are positive for EBV-encoded small RNA (EBER), but negative for other EBV-encoded latent proteins such as latent membrane proteins (LMPs) and EBV-encoded nuclear antigens (EBNAs), suggesting the latency I pattern of EBV infection (Horenstein et al. 1997).

Many PEL cell lines have been established including both EBV-positive and EBV-negative cells. To date, 10 KSHV-positive/EBV-negative cell lines have been established from PEL cells (Osawa et al. 2016). Among these, the KSHV-positive/EBV-negative cell line, TY-1, was established from a case of KSHV-positive/EBV-positive PEL (Katano et al. 1999a). This suggests that KSHV is solely responsible for the pathogenesis of PEL. PEL cell lines are the most studied model of KSHV-infected diseases because (1) all PEL cells are infected with KSHV and express KSHV-encoded latent proteins, (2) KSHV is easily activated by chemical induction, and (3) there is no KS cell line that is infected with KSHV continuously. The predominant phase of KSHV infection is latent in PEL cells. PEL cells express not only LANA-1 but also other KSHV-encoded latent proteins such as LANA-2, ORF71, and ORF72. However, it is not easy to detect latent proteins, other than LANA-1, in PEL cells. By contrast, vIL-6, an early protein encoded by KSHV, is relatively easy to detect in PEL cells. Other lytic proteins, such as RTA, ORF59, and K8.1, are rarely expressed in PEL cells.

vIL-6 plays an important role in the pathogenesis of PEL. vIL-6 binds to the IL-6 receptor, gp130, without gp80, another subunit of IL-6 receptor, indicating that vIL-6 induces IL-6 signals in various cells (Chatterjee et al. 2002). The binding to gp130 activates downstream IL-6 signaling and induces secretion of human IL-6, implying an autocrine mechanism between vIL-6 and hIL-6. vIL-6 also induces vascular endothelial growth factor expression in PEL cells, which is associated with indirect proliferation of PEL cells (Aoki et al. 1999). In an animal model of effusion and a solid type of KSHV-associated lymphoma, several adhesion molecules, including LFA1 and 3, were downregulated in PEL but not in KSHV-associated solid lymphoma (Yanagisawa et al. 2006).

16.4.4 *KSHV-Associated Diffuse Large B Cell Lymphoma, Not Otherwise Specified*

"KSHV-associated diffuse large B cell lymphoma, not otherwise specified (KSHV-DLBCL, NOS)" is a new category of lymphoma in the revised 4th edition of the WHO classification (Swerdlow et al. 2016). This category is thought to contain cases of formerly large B cell lymphoma arising in KSHV-associated MCD, and KSHV-associated plasmablastic lymphoma (Isaacson et al. 2008; Said and Cesarman 2008). This type of lymphoma is characterized by monoclonal proliferation of KSHV-infected lymphoid cells resembling plasmablasts (Fig. 16.3). In cases with MCD in the lymph node, lymphoma cells express IgM (Dupin et al. 2000; Oksenhendler et al. 2002). Small confluent sheets of LANA-1$^+$ plasmablasts are seen in the interfollicular zone of KSHV-associated MCD. This type of lymphoma occurs in the lymph node or spleen with generalized lymphadenitis and/or massive splenomegaly. The expression pattern of viral proteins is similar to that of PEL, and the expression of vIL-6 is frequently observed in lymphoma cells. EBV is not detected in this type of lymphoma.

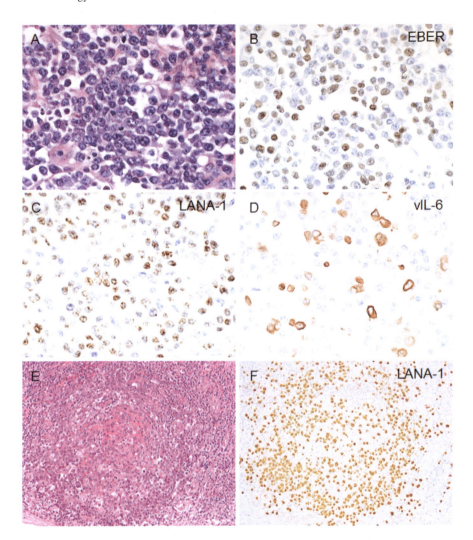

Fig. 16.3 Extracavitary PEL and KSHV-associated diffuse large B cell lymphoma, not otherwise specified. (**a–d**) A case of extracavitary PEL occurring in the colon of a patient with AIDS. Large lymphoblastic cells are detected by HE staining (**a**). These lymphoma cells are positive for EBER (**b**) and LANA-1 (**c**) on immunohistochemistry. vIL-6 is expressed in the cytoplasm of some lymphoma cells (**d**). (**e** and **f**) A case of KSHV-associated diffuse large B cell lymphoma, not otherwise specified. HE staining shows that lymphoma cells infiltrate the germinal center of MCD (**e**). These lymphoma cells are positive for LANA-1 on immunohistochemistry (**f**)

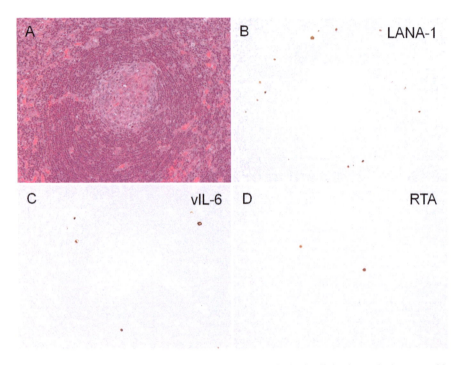

Fig. 16.4 KSHV-associated MCD (**a**) HE staining reveals the hyalinized germinal center with proliferation of plasma cells. (**b**) LANA-1 immunohistochemistry. LANA-1 is detected in some B cells in the mantle zone of the germinal center. (**c** and **d**) Immunohistochemistry for KSHV-encoded lytic proteins. vIL-6 (**c**) and RTA (**d**) are detected in a small population of lymphocytes in the mantle zone of the germinal center

16.4.5 Multicentric Castleman Disease (MCD)

MCD is characterized by generalized lymphadenopathy with polyclonal hypergammaglobulinemia and a high level of IL-6 in the serum (Frizzera et al. 1983). Histologically, follicular hyperplasia and hyaline vascular change in the follicle and interfollicular zone are observed (Fig. 16.4). The proliferation of plasma cells is predominant in the interfollicular area. KSHV is detected in a proportion of MCD cases (Soulier et al. 1995); it is frequently detected in cases of AIDS-associated MCD, but rarely in non-HIV-infected individuals (Suda et al. 2001). KSHV-encoded proteins such as LANA-1 are detected in large blastic B cells around the germinal center (Dupin et al. 1999; Katano et al. 2000b; Parravicini et al. 2000). These KSHV-positive cells are found in the mantle zone of the germinal center and the interfollicular area. Importantly, these KSHV-positive cells express not only LANA-1 but also many KSHV-encoded lytic proteins such as vIL-6, RTA, ORF59, ORF65, and K8.1 (Katano et al. 2000a). A high titer of KSHV is usually detected in the serum of KSHV-associated MCD, and the copy number of serum KSHV

Table 16.1 Case definition of KSHV-inflammatory cytokine syndrome (KICS) (Uldrick et al. 2010)

1. Clinical manifestations
a. Symptoms: fever, fatigue, edema, cachexia, respiratory symptoms, gastrointestinal disturbance, athralgia and myalgia, altered mental state, neuropathy with or without pain b. Laboratory abnormalities: Anemia, thrombocytopenia, hypoalbuminemia, hyponatremia c. Radiographic abnormalities: Lymphadenopathy, splenomegaly, hepatomegaly, body cavity effusions
2. Evidence of systemic inflammation
Elevated C-reactive protein (≥ 3 g/dL)
3. Evidence of KSHV viral activity
Elevated KSHV viral load in plasma (≥ 1000 copies/mL) or peripheral blood mononuclear cells (≥ 100 copies/10^6 cells)
4. No evidence of KSHV-associated multicentric Castleman disease
Exclusion of MCD requires histopathologic assessment of lymphadenopathy if present
The working case definition of KICS requires the presence of at least two clinical manifestations drawn from at least two categories (1a, b, and c), together with each of the criteria in 2, 3, and 4

corresponds to the clinical course in these patients (Oksenhendler et al. 2000). These data suggest that KSHV lytic infection occurs in KSHV-associated MCD. vIL-6 is thought to play an important role in the pathogenesis of KSHV-associated MCD, and a high level of vIL-6 is detected in the serum of MCD patients (Aoki et al. 2001). vIL-6 induces the expression of human IL-6 in vitro, and both vIL-6 and hIL-6 induce the proliferation of plasma cells via an autocrine mechanism.

16.4.6 KSHV Inflammatory Cytokine Syndrome (KICS)

KICS was a new category of disease distinct from cases of KSHV-associated MCD, which was proposed in 2012 (Polizzotto et al. 2012; Uldrick et al. 2010). Almost all KICS cases have been reported in HIV-positive patients with KS and/or PEL. Multiple severe symptoms such as gastrointestinal disturbance, edema, respiratory disturbance, and effusions are observed (Table 16.1). Blood tests reveal high C-reactive protein, high vIL-6 and hIL-6, high IL-10, anemia, hypoalbuminemia, and thrombocytopenia. Importantly, a high viral load of KSHV is detected in the serum of KICS patients. These data suggest that KICS is strongly associated with KSHV replication following a severe inflammatory response. The clinical manifestations of KICS resemble those of KSHV-associated MCD; however, cases complicated by MCD are excluded from the category of KICS. In KICS, lymphadenopathy is not prominent and pathological evidence of MCD is absent. The prognosis for KICS is poor, with an overall survival rate of less than 60% despite intensive treatment of KSHV-associated tumors and KSHV replication.

16.5 Conclusion

During the past decade, KSHV has been intensively investigated using molecular biological techniques, and some of the molecular mechanisms of KSHV infection have been elucidated. Based on such investigations, the classification of KSHV-associated diseases has been reconsidered and revised. Immunohistochemistry to detect the LANA-1 is now a gold standard for the pathological diagnosis of KSHV infection. Many diseases with an unconfirmed association with KSHV infection were excluded by LANA-1 immunohistochemistry. KSHV-associated lymphoma was redefined and KICS was added as a new KSHV-associated disease. These modifications to the classification of KSHV-associated diseases are the result of meticulous investigations by clinicians and researchers. However, many unsolved questions remain in the pathogenesis of KSHV infection, especially regarding oncogenesis. It is clear that LANA-1 plays a unique role in the pathogenesis of KSHV, and the extremely high expression of viral miRNAs in KSHV-infected cells suggests that viral miRNAs might also play a major role in the pathogenesis of this virus, as well as KSHV oncogenesis. Further studies will reveal the role of viral miRNAs in KSHV-associated diseases.

Acknowledgments The authors thank Ms. Yuko Sato for her excellent technical assistance. This work was financially supported by the Research Program on HIV/AIDS (grant numbers JP17fk0410207, JP17fk0410208, and JP17fk0410309) from the Japan Agency for Medical Research and Development and Grants-in-Aid for Scientific Research (C) from the Japan Society for the Promotion of Science (grant number 15K08509).

Conflict of Interest The author declares no conflict of interest.

References

Andreoni M, Sarmati L, Nicastri E, El Sawaf G, El Zalabani M, Uccella I, Bugarini R, Parisi SG, Rezza G (2002) Primary human herpesvirus 8 infection in immunocompetent children. JAMA 287(10):1295–1300. doi:jpc20003 [pii]

Antman K, Chang Y (2000) Kaposi's sarcoma. N Engl J Med 342(14):1027–1038. https://doi.org/10.1056/NEJM200004063421407

Aoki Y, Jaffe ES, Chang Y, Jones K, Teruya-Feldstein J, Moore PS, Tosato G (1999) Angiogenesis and hematopoiesis induced by Kaposi's sarcoma-associated herpesvirus-encoded interleukin-6. Blood 93(12):4034–4043

Aoki Y, Yarchoan R, Wyvill K, Okamoto S, Little RF, Tosato G (2001) Detection of viral interleukin-6 in Kaposi sarcoma-associated herpesvirus-linked disorders. Blood 97(7):2173–2176

Aoki Y, Feldman GM, Tosato G (2003) Inhibition of STAT3 signaling induces apoptosis and decreases survivin expression in primary effusion lymphoma. Blood 101(4):1535–1542. https://doi.org/10.1182/blood-2002-07-2130

Asahi-Ozaki Y, Sato Y, Kanno T, Sata T, Katano H (2006) Quantitative analysis of Kaposi sarcoma-associated herpesvirus (KSHV) in KSHV-associated diseases. J Infect Dis 193(6):773–782. https://doi.org/10.1086/500560

Bais C, Santomasso B, Coso O, Arvanitakis L, Raaka EG, Gutkind JS, Asch AS, Cesarman E, Gershengorn MC, Mesri EA (1998) G-protein-coupled receptor of Kaposi's sarcoma-associated herpesvirus is a viral oncogene and angiogenesis activator. Nature 391(6662):86–89. https://doi.org/10.1038/34193

Ballestas ME, Chatis PA, Kaye KM (1999) Efficient persistence of extrachromosomal KSHV DNA mediated by latency-associated nuclear antigen. Science 284(5414):641–644

Cai X, Lu S, Zhang Z, Gonzalez CM, Damania B, Cullen BR (2005) Kaposi's sarcoma-associated herpesvirus expresses an array of viral microRNAs in latently infected cells. Proc Natl Acad Sci U S A 102(15):5570–5575. https://doi.org/10.1073/pnas.0408192102

Casper C, Carrell D, Miller KG, Judson FD, Meier AS, Pauk JS, Morrow RA, Corey L, Wald A, Celum C (2006) HIV serodiscordant sex partners and the prevalence of human herpesvirus 8 infection among HIV negative men who have sex with men: baseline data from the EXPLORE study. Sex Transm Infect 82(3):229–235. https://doi.org/10.1136/sti.2005.016568

Cesarman E, Chang Y, Moore PS, Said JW, Knowles DM (1995) Kaposi's sarcoma-associated herpesvirus-like DNA sequences in AIDS-related body-cavity-based lymphomas. N Engl J Med 332(18):1186–1191. https://doi.org/10.1056/NEJM199505043321802

Chang Y, Cesarman E, Pessin MS, Lee F, Culpepper J, Knowles DM, Moore PS (1994) Identification of herpesvirus-like DNA sequences in AIDS-associated Kaposi's sarcoma. Science 266(5192):1865–1869

Chatlynne LG, Lapps W, Handy M, Huang YQ, Masood R, Hamilton AS, Said JW, Koeffler HP, Kaplan MH, Friedman-Kien A, Gill PS, Whitman JE, Ablashi DV (1998) Detection and titration of human herpesvirus-8-specific antibodies in sera from blood donors, acquired immunodeficiency syndrome patients, and Kaposi's sarcoma patients using a whole virus enzyme-linked immunosorbent assay. Blood 92(1):53–58

Chatterjee M, Osborne J, Bestetti G, Chang Y, Moore PS (2002) Viral IL-6-induced cell proliferation and immune evasion of interferon activity. Science 298(5597):1432–1435. https://doi.org/10.1126/science.1074883

Davis DA, Humphrey RW, Newcomb FM, O'Brien TR, Goedert JJ, Straus SE, Yarchoan R (1997) Detection of serum antibodies to a Kaposi's sarcoma-associated herpesvirus-specific peptide. J Infect Dis 175(5):1071–1079

Dittmer DP, Damania B (2016) Kaposi sarcoma-associated herpesvirus: immunobiology, oncogenesis, and therapy. J Clin Invest 126(9):3165–3175. https://doi.org/10.1172/JCI84418

Dupin N, Fisher C, Kellam P, Ariad S, Tulliez M, Franck N, van Marck E, Salmon D, Gorin I, Escande JP, Weiss RA, Alitalo K, Boshoff C (1999) Distribution of human herpesvirus-8 latently infected cells in Kaposi's sarcoma, multicentric Castleman's disease, and primary effusion lymphoma. Proc Natl Acad Sci U S A 96(8):4546–4551

Dupin N, Diss TL, Kellam P, Tulliez M, Du MQ, Sicard D, Weiss RA, Isaacson PG, Boshoff C (2000) HHV-8 is associated with a plasmablastic variant of Castleman disease that is linked to HHV-8-positive plasmablastic lymphoma. Blood 95(4):1406–1412

Engels EA, Atkinson JO, Graubard BI, McQuillan GM, Gamache C, Mbisa G, Cohn S, Whitby D, Goedert JJ (2007) Risk factors for human herpesvirus 8 infection among adults in the United States and evidence for sexual transmission. J Infect Dis 196(2):199–207. https://doi.org/10.1086/518791

Friborg J Jr, Kong W, Hottiger MO, Nabel GJ (1999) p53 inhibition by the LANA protein of KSHV protects against cell death. Nature 402(6764):889–894. https://doi.org/10.1038/47266

Frizzera G, Banks PM, Massarelli G, Rosai J (1983) A systemic lymphoproliferative disorder with morphologic features of Castleman's disease. Pathological findings in 15 patients. Am J Surg Pathol 7(3):211–231

Fujimuro M, Wu FY, ApRhys C, Kajumbula H, Young DB, Hayward GS, Hayward SD (2003) A novel viral mechanism for dysregulation of beta-catenin in Kaposi's sarcoma-associated herpesvirus latency. Nat Med 9(3):300–306. https://doi.org/10.1038/nm829

Ganem D (2005) Kaposi's sarcoma-associated herpesvirus, Fields virology, vol 2, 5th edn. Lippincott Williams & Wilkins, Philadelphia

Gao SJ, Boshoff C, Jayachandra S, Weiss RA, Chang Y, Moore PS (1997) KSHV ORF K9 (vIRF) is an oncogene which inhibits the interferon signaling pathway. Oncogene 15(16):1979–1985. https://doi.org/10.1038/sj.onc.1201571

Gao SJ, Deng JH, Zhou FC (2003) Productive lytic replication of a recombinant Kaposi's sarcoma-associated herpesvirus in efficient primary infection of primary human endothelial cells. J Virol 77(18):9738–9749

Goncalves PH, Ziegelbauer J, Uldrick TS, Yarchoan R (2017) Kaposi sarcoma herpesvirus-associated cancers and related diseases. Curr Opin HIV AIDS 12(1):47–56. https://doi.org/10.1097/COH.0000000000000330

Gramolelli S, Schulz TF (2015) The role of Kaposi sarcoma-associated herpesvirus in the pathogenesis of Kaposi sarcoma. J Pathol 235(2):368–380. https://doi.org/10.1002/path.4441

Grulich AE, Cunningham P, Munier ML, Prestage G, Amin J, Ringland C, Whitby D, Kippax S, Kaldor JM, Rawlinson W (2005) Sexual behaviour and human herpesvirus 8 infection in homosexual men in Australia. Sex Health 2(1):13–18

Harrington WJ Jr, Bagasra O, Sosa CE, Bobroski LE, Baum M, Wen XL, Cabral L, Byrne GE, Pomerantz RJ, Wood C (1996) Human herpesvirus type 8 DNA sequences in cell-free plasma and mononuclear cells of Kaposi's sarcoma patients. J Infect Dis 174(5):1101–1105

He F, Wang X, He B, Feng Z, Lu X, Zhang Y, Zhao S, Lin R, Hui Y, Bao Y, Zhang Z, Wen H (2007) Human herpesvirus 8: seroprevalence and correlates in tumor patients from Xinjiang, China. J Med Virol 79(2):161–166. https://doi.org/10.1002/jmv.20730

Hladik W, Dollard SC, Mermin J, Fowlkes AL, Downing R, Amin MM, Banage F, Nzaro E, Kataaha P, Dondero TJ, Pellett PE, Lackritz EM (2006) Transmission of human herpesvirus 8 by blood transfusion. N Engl J Med 355(13):1331–1338. https://doi.org/10.1056/NEJMoa055009

Hladik W, Pellett PE, Hancock J, Downing R, Gao H, Packel L, Mimbe D, Nzaro E, Mermin J (2012) Association between transfusion with human herpesvirus 8 antibody-positive blood and subsequent mortality. J Infect Dis 206(10):1497–1503. https://doi.org/10.1093/infdis/jis543

Hong YK, Foreman K, Shin JW, Hirakawa S, Curry CL, Sage DR, Libermann T, Dezube BJ, Fingeroth JD, Detmar M (2004) Lymphatic reprogramming of blood vascular endothelium by Kaposi sarcoma-associated herpesvirus. Nat Genet 36(7):683–685. https://doi.org/10.1038/ng1383

Horenstein MG, Nador RG, Chadburn A, Hyjek EM, Inghirami G, Knowles DM, Cesarman E (1997) Epstein-Barr virus latent gene expression in primary effusion lymphomas containing Kaposi's sarcoma-associated herpesvirus/human herpesvirus-8. Blood 90(3):1186–1191

Hoshina S, Sekizuka T, Kataoka M, Hasegawa H, Hamada H, Kuroda M, Katano H (2016) Profile of Exosomal and intracellular microRNA in gamma-herpesvirus-infected lymphoma cell lines. PLoS One 11(9):e0162574. https://doi.org/10.1371/journal.pone.0162574

Isaacson PG, Campo E, Harris NL (2008) Large B-cell lymphoma arising in HHV8-associated multicentric Castleman disease. In: Swerdlow SH, Campo E, Harris NL et al (eds) WHO classification of tumours of haematopoetic and lymphoid tissues, 4th edn. International Agency for Research on Cancer (IARC), Lyon, pp 258–259

Ishido S, Wang C, Lee BS, Cohen GB, Jung JU (2000) Downregulation of major histocompatibility complex class I molecules by Kaposi's sarcoma-associated herpesvirus K3 and K5 proteins. J Virol 74(11):5300–5309

Katano H, Hoshino Y, Morishita Y, Nakamura T, Satoh H, Iwamoto A, Herndier B, Mori S (1999a) Establishing and characterizing a CD30-positive cell line harboring HHV-8 from a primary effusion lymphoma. J Med Virol 58(4):394–401. https://doi.org/10.1002/(SICI)1096-9071(199908)58:4<394::AID-JMV12>3.0.CO;2-H[pii]

Katano H, Sato Y, Kurata T, Mori S, Sata T (1999b) High expression of HHV-8-encoded ORF73 protein in spindle-shaped cells of Kaposi's sarcoma. Am J Pathol 155(1):47–52. https://doi.org/10.1016/S0002-9440(10)65097-3

Katano H, Iwasaki T, Baba N, Terai M, Mori S, Iwamoto A, Kurata T, Sata T (2000a) Identification of antigenic proteins encoded by human herpesvirus 8 and seroprevalence in the general population and among patients with and without Kaposi's sarcoma. J Virol 74(8):3478–3485

Katano H, Sato Y, Kurata T, Mori S, Sata T (2000b) Expression and localization of human herpesvirus 8-encoded proteins in primary effusion lymphoma, Kaposi's sarcoma, and multicentric Castleman's disease. Virology 269(2):335–344. https://doi.org/10.1006/viro.2000.0196

Katano H, Sato Y, Sata T (2001) Expression of p53 and human herpesvirus 8 (HHV-8)-encoded latency-associated nuclear antigen (LANA) with inhibition of apoptosis in HHV-8-associated malignancies. Cancer 92:3076–3084

Katano H, Yokomaku Y, Fukumoto H, Kanno T, Nakayama T, Shingae A, Sugiura W, Ichikawa S, Yasuoka A (2013) Seroprevalence of Kaposi's sarcoma-associated herpesvirus among men who have sex with men in Japan. J Med Virol 85(6):1046–1052. https://doi.org/10.1002/jmv.23558

Kedes DH, Ganem D, Ameli N, Bacchetti P, Greenblatt R (1997) The prevalence of serum antibody to human herpesvirus 8 (Kaposi sarcoma-associated herpesvirus) among HIV-seropositive and high-risk HIV-seronegative women. JAMA 277(6):478–481

Kellam P, Bourboulia D, Dupin N, Shotton C, Fisher C, Talbot S, Boshoff C, Weiss RA (1999) Characterization of monoclonal antibodies raised against the latent nuclear antigen of human herpesvirus 8. J Virol 73(6):5149–5155

Lee H, Veazey R, Williams K, Li M, Guo J, Neipel F, Fleckenstein B, Lackner A, Desrosiers RC, Jung JU (1998) Deregulation of cell growth by the K1 gene of Kaposi's sarcoma-associated herpesvirus. Nat Med 4(4):435–440

Lei X, Bai Z, Ye F, Xie J, Kim CG, Huang Y, Gao SJ (2010) Regulation of NF-kappaB inhibitor IkappaBalpha and viral replication by a KSHV microRNA. Nat Cell Biol 12(2):193–199. https://doi.org/10.1038/ncb2019

Luppi M, Barozzi P, Schulz TF, Trovato R, Donelli A, Narni F, Sheldon J, Marasca R, Torelli G (2000) Nonmalignant disease associated with human herpesvirus 8 reactivation in patients who have undergone autologous peripheral blood stem cell transplantation. Blood 96(7):2355–2357

Martin-Carbonero L, Palacios R, Valencia E, Saballs P, Sirera G, Santos I, Baldobi F, Alegre M, Goyenechea A, Pedreira J, Gonzalez del Castillo J, Martinez-Lacasa J, Ocampo A, Alsina M, Santos J, Podzamczer D, Gonzalez-Lahoz J, Caelyx/Kaposi's Sarcoma Spanish G (2008) Long-term prognosis of HIV-infected patients with Kaposi sarcoma treated with pegylated liposomal doxorubicin. Clin Infect Dis 47(3):410–417. https://doi.org/10.1086/589865

Mayama S, Cuevas LE, Sheldon J, Omar OH, Smith DH, Okong P, Silvel B, Hart CA, Schulz TF (1998) Prevalence and transmission of Kaposi's sarcoma-associated herpesvirus (human herpesvirus 8) in Ugandan children and adolescents. Int J Cancer 77(6):817–820. https://doi.org/10.1002/(SICI)1097-0215(19980911)77:6<817::AID-IJC2>3.0.CO;2-X[pii]

Molden J, Chang Y, You Y, Moore PS, Goldsmith MA (1997) A Kaposi's sarcoma-associated herpesvirus-encoded cytokine homolog (vIL-6) activates signaling through the shared gp130 receptor subunit. J Biol Chem 272(31):19625–19631

Montaner S, Sodhi A, Molinolo A, Bugge TH, Sawai ET, He Y, Li Y, Ray PE, Gutkind JS (2003) Endothelial infection with KSHV genes in vivo reveals that vGPCR initiates Kaposi's sarcomagenesis and can promote the tumorigenic potential of viral latent genes. Cancer Cell 3(1):23–36. doi:S1535610802002374 [pii]

Moore PS, Boshoff C, Weiss RA, Chang Y (1996) Molecular mimicry of human cytokine and cytokine response pathway genes by KSHV. Science 274(5293):1739–1744

Myoung J, Ganem D (2011) Infection of lymphoblastoid cell lines by Kaposi's sarcoma-associated herpesvirus: critical role of cell-associated virus. J Virol 85(19):9767–9777. https://doi.org/10.1128/JVI.05136-11

Nador RG, Cesarman E, Chadburn A, Dawson DB, Ansari MQ, Sald J, Knowles DM (1996) Primary effusion lymphoma: a distinct cliniocopathologic entity associated with the Kaposi's sarcoma-associated herpes virus. Blood 88(2):645–656

Ohtsuki Y, Iwata J, Furihata M, Takeuchi T, Sonobe H, Miyoshi I, Ohtsuki Y, Iwata J, Furihata M, Takeuchi T, Sonobe H, Miyoshi I (1999) Ultrastructure of Kaposi's sarcoma-associated herpesvirus (KSHV)/human herpesvirus-8 (HHV-8) in a primary effusion lymphoma cell line treated with tetradecanoyl phorbol acetate (TPA). Med Electron Microsc 32(2):94–99. https://doi.org/10.1007/s007959900029

Oksenhendler E, Carcelain G, Aoki Y, Boulanger E, Maillard A, Clauvel JP, Agbalika F (2000) High levels of human herpesvirus 8 viral load, human interleukin-6, interleukin-10, and C reactive protein correlate with exacerbation of multicentric castleman disease in HIV-infected patients. Blood 96(6):2069–2073

Oksenhendler E, Boulanger E, Galicier L, Du MQ, Dupin N, Diss TC, Hamoudi R, Daniel MT, Agbalika F, Boshoff C, Clauvel JP, Isaacson PG, Meignin V (2002) High incidence of Kaposi sarcoma-associated herpesvirus-related non-Hodgkin lymphoma in patients with HIV infection and multicentric Castleman disease. Blood 99(7):2331–2336

Orenstein JM, Alkan S, Blauvelt A, Jeang KT, Weinstein MD, Ganem D, Herndier B (1997) Visualization of human herpesvirus type 8 in Kaposi's sarcoma by light and transmission electron microscopy. AIDS 11(5):F35–F45

Osawa M, Mine S, Ota S, Kato K, Sekizuka T, Kuroda M, Kataoka M, Fukumoto H, Sato Y, Kanno T, Hasegawa H, Ueda K, Fukayama M, Maeda T, Kanoh S, Kawana A, Fujikura Y, Katano H (2016) Establishing and characterizing a new primary effusion lymphoma cell line harboring Kaposi's sarcoma-associated herpesvirus. Infect Agent Cancer 11:37. https://doi.org/10.1186/s13027-016-0086-5

Ota Y, Hishima T, Mochizuki M, Kodama Y, Moritani S, Oyaizu N, Mine S, Ajisawa A, Tanuma J, Uehira T, Hagiwara S, Yajima K, Koizumi Y, Shirasaka T, Kojima Y, Nagai H, Yokomaku Y, Shiozawa Y, Koibuchi T, Iwamoto A, Oka S, Hasegawa H, Okada S, Katano H (2014) Classification of AIDS-related lymphoma cases between 1987 and 2012 in Japan based on the WHO classification of lymphomas, fourth edition. Cancer Med 3(1):143–153. https://doi.org/10.1002/cam4.178

Parravicini C, Chandran B, Corbellino M, Berti E, Paulli M, Moore PS, Chang Y (2000) Differential viral protein expression in Kaposi's sarcoma-associated herpesvirus-infected diseases: Kaposi's sarcoma, primary effusion lymphoma, and multicentric Castleman's disease. Am J Pathol 156(3):743–749

Pauk J, Huang ML, Brodie SJ, Wald A, Koelle DM, Schacker T, Celum C, Selke S, Corey L (2000) Mucosal shedding of human herpesvirus 8 in men. N Engl J Med 343(19):1369–1377. https://doi.org/10.1056/NEJM200011093431904

Polizzotto MN, Uldrick TS, Hu D, Yarchoan R (2012) Clinical manifestations of Kaposi Sarcoma Herpesvirus Lytic Activation: Multicentric Castleman Disease (KSHV-MCD) and the KSHV inflammatory cytokine syndrome. Front Microbiol 3:73. https://doi.org/10.3389/fmicb.2012.00073

Rabkin CS, Schulz TF, Whitby D, Lennette ET, Magpantay LI, Chatlynne L, Biggar RJ (1998) Interassay correlation of human herpesvirus 8 serologic tests. HHV-8 Interlaboratory Collaborative Group. J Infect Dis 178(2):304–309

Regamey N, Tamm M, Wernli M, Witschi A, Thiel G, Cathomas G, Erb P (1998) Transmission of human herpesvirus 8 infection from renal-transplant donors to recipients. N Engl J Med 339(19):1358–1363. https://doi.org/10.1056/NEJM199811053391903

Renne R, Zhong W, Herndier B, McGrath M, Abbey N, Kedes D, Ganem D (1996) Lytic growth of Kaposi's sarcoma-associated herpesvirus (human herpesvirus 8) in culture. Nat Med 2(3):342–346

Rivas C, Thlick AE, Parravicini C, Moore PS, Chang Y (2001) Kaposi's sarcoma-associated herpesvirus LANA2 is a B-cell-specific latent viral protein that inhibits p53. J Virol 75(1):429–438. https://doi.org/10.1128/JVI.75.1.429-438.2001

Russo JJ, Bohenzky RA, Chien MC, Chen J, Yan M, Maddalena D, Parry JP, Peruzzi D, Edelman IS, Chang Y, Moore PS (1996) Nucleotide sequence of the Kaposi sarcoma-associated herpesvirus (HHV8). Proc Natl Acad Sci U S A 93(25):14862–14867

Said J, Cesarman E (2008) Primary effusion lymphoma. In: Swerdlow SH, Campo E, Harris NL et al (eds) WHO classification of tumours of haematopoetic and lymphoid tissues, 4th edn. International Agency for Research on Cancer (IARC), Lyon, pp 260–261

Said W, Chien K, Takeuchi S, Tasaka T, Asou H, Cho SK, de Vos S, Cesarman E, Knowles DM, Koeffler HP (1996) Kaposi's sarcoma-associated herpesvirus (KSHV or HHV8) in primary effusion lymphoma: ultrastructural demonstration of herpesvirus in lymphoma cells. Blood 87(12):4937–4943

Said JW, Chien K, Tasaka T, Koeffler HP (1997) Ultrastructural characterization of human herpesvirus 8 (Kaposi's sarcoma-associated herpesvirus) in Kaposi's sarcoma lesions: electron microscopy permits distinction from cytomegalovirus (CMV). J Pathol 182(3):273–281. https://doi.org/10.1002/(SICI)1096-9896(199707)182:3<273::AID-PATH835>3.0.CO;2-P

Sakakibara S, Ueda K, Nishimura K, Do E, Ohsaki E, Okuno T, Yamanishi K (2004) Accumulation of heterochromatin components on the terminal repeat sequence of Kaposi's sarcoma-associated herpesvirus mediated by the latency-associated nuclear antigen. J Virol 78(14):7299–7310. https://doi.org/10.1128/JVI.78.14.7299-7310.2004

Sakamoto K, Asanuma H, Nakamura T, Kanno T, Sata T, Katano H (2010) Immune response to intranasal and intraperitoneal immunization with Kaposi's sarcoma-associated herpesvirus in mice. Vaccine 28(19):3325–3332. https://doi.org/10.1016/j.vaccine.2010.02.091

Sakurada S, Katano H, Sata T, Ohkuni H, Watanabe T, Mori S (2001) Effective human herpesvirus 8 infection of human umbilical vein endothelial cells by cell-mediated transmission. J Virol 75(16):7717–7722. https://doi.org/10.1128/JVI.75.16.7717-7722.2001

Samols MA, Hu J, Skalsky RL, Renne R (2005) Cloning and identification of a microRNA cluster within the latency-associated region of Kaposi's sarcoma-associated herpesvirus. J Virol 79(14):9301–9305. https://doi.org/10.1128/JVI.79.14.9301-9305.2005

Soulier J, Grollet L, Oksenhendler E, Cacoub P, Cazals-Hatem D, Babinet P, d'Agay MF, Clauvel JP, Raphael M, Degos L, Sigaux F (1995) Kaposi's sarcoma-associated herpesvirus-like DNA sequences in multicentric Castleman's disease. Blood 86(4):1276–1280

Suda T, Katano H, Delsol G, Kakiuchi C, Nakamura T, Shiota M, Sata T, Higashihara M, Mori S (2001) HHV-8 infection status of AIDS-unrelated and AIDS-associated multicentric Castleman's disease. Pathol Int 51(9):671–679

Sun R, Lin SF, Staskus K, Gradoville L, Grogan E, Haase A, Miller G (1999) Kinetics of Kaposi's sarcoma-associated herpesvirus gene expression. J Virol 73(3):2232–2242

Swerdlow SH, Campo E, Pileri SA, Harris NL, Stein H, Siebert R, Advani R, Ghielmini M, Salles GA, Zelenetz AD, Jaffe ES (2016) The 2016 revision of the World Health Organization classification of lymphoid neoplasms. Blood 127(20):2375–2390. https://doi.org/10.1182/blood-2016-01-643569

Thome M, Schneider P, Hofmann K, Fickenscher H, Meinl E, Neipel F, Mattmann C, Burns K, Bodmer JL, Schroter M, Scaffidi C, Krammer PH, Peter ME, Tschopp J (1997) Viral FLICE-inhibitory proteins (FLIPs) prevent apoptosis induced by death receptors. Nature 386(6624):517–521. https://doi.org/10.1038/386517a0

Uldrick TS, Wang V, O'Mahony D, Aleman K, Wyvill KM, Marshall V, Steinberg SM, Pittaluga S, Maric I, Whitby D, Tosato G, Little RF, Yarchoan R (2010) An interleukin-6-related systemic inflammatory syndrome in patients co-infected with Kaposi sarcoma-associated herpesvirus and HIV but without Multicentric Castleman disease. Clin Infect Dis 51(3):350–358. https://doi.org/10.1086/654798

Wang QJ, Jenkins FJ, Jacobson LP, Kingsley LA, Day RD, Zhang ZW, Meng YX, Pellett PE, Kousoulas KG, Baghian A, Rinaldo CR Jr (2001) Primary human herpesvirus 8 infection generates a broadly specific CD8(+) T-cell response to viral lytic cycle proteins. Blood 97(8):2366–2373

Wang HW, Trotter MW, Lagos D, Bourboulia D, Henderson S, Makinen T, Elliman S, Flanagan AM, Alitalo K, Boshoff C (2004) Kaposi sarcoma herpesvirus-induced cellular reprogramming contributes to the lymphatic endothelial gene expression in Kaposi sarcoma. Nat Genet 36(7):687–693. https://doi.org/10.1038/ng1384

Weninger W, Partanen TA, Breiteneder-Geleff S, Mayer C, Kowalski H, Mildner M, Pammer J, Sturzl M, Kerjaschki D, Alitalo K, Tschachler E (1999) Expression of vascular endothelial growth factor receptor-3 and podoplanin suggests a lymphatic endothelial cell origin of Kaposi's sarcoma tumor cells. Lab Investig 79(2):243–251

Yanagisawa Y, Sato Y, Asahi-Ozaki Y, Ito E, Honma R, Imai J, Kanno T, Kano M, Akiyama H, Sata T, Shinkai-Ouchi F, Yamakawa Y, Watanabe S, Katano H (2006) Effusion and solid lymphomas have distinctive gene and protein expression profiles in an animal model of primary effusion lymphoma. J Pathol 209(4):464–473

Zhong W, Ganem D (1997) Characterization of ribonucleoprotein complexes containing an abundant polyadenylated nuclear RNA encoded by Kaposi's sarcoma-associated herpesvirus (human herpesvirus 8). J Virol 71(2):1207–1212

Ziegelbauer JM, Sullivan CS, Ganem D (2009) Tandem array-based expression screens identify host mRNA targets of virus-encoded microRNAs. Nat Genet 41(1):130–134. https://doi.org/10.1038/ng.266

Chapter 17
EBV-Encoded Latent Genes

Teru Kanda

Abstract Epstein-Barr virus (EBV) is one of the most widespread human pathogens. EBV infection is usually asymptomatic, and it establishes life-long latent infection. EBV latent infection sometimes causes various tumorigenic diseases, such as EBV-related lymphoproliferative diseases, Burkitt lymphomas, Hodgkin lymphomas, NK/T-cell lymphomas, and epithelial carcinomas. EBV-encoded latent genes are set of viral genes that are expressed in latently infected cells. They include virally encoded proteins, noncoding RNAs, and microRNAs. Different latent gene expression patterns are noticed in different types of EBV-infected cells. Viral latent gene products contribute to EBV-mediated B cell transformation and likely contribute to lymphomagenesis and epithelial carcinogenesis as well. Many biological functions of viral latent gene products have been reported, making difficult to understand a whole view of EBV latency. In this review, we will focus on latent gene functions that have been verified by genetic experiments using EBV mutants. We will also summarize how viral latent genes contribute to EBV-mediated B cell transformation, Burkitt lymphomagenesis, and epithelial carcinogenesis.

Keywords Latent genes · EBNA · LMP · microRNA · Burkitt lymphoma · Epithelial carcinogenesis

17.1 Introduction

When cells are latently infected with EBV, a set of viral genes are expressed, which are designated as viral latent genes. There are several different forms of EBV latency, and those are schematically illustrated (Fig. 17.1) (Rickinson and Kieff 2007). Lymphoblastoid cell lines (LCLs), which are immortalized B lymphocytes established by in vitro EBV infection, express all sets of proteins and noncoding RNAs

---

T. Kanda (✉)
Division of Microbiology, Faculty of Medicine, Tohoku medical and Pharmaceutical University, Sendai, Japan
e-mail: tkanda@tohoku-mpu.ac.jp

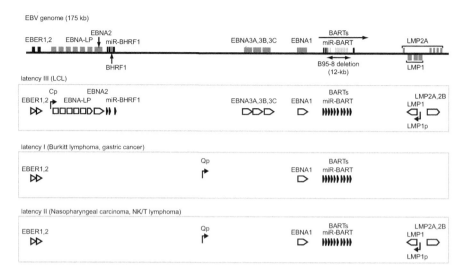

Fig. 17.1 Patterns of EBV latent gene expression in different forms of viral latency. The top panel represents a schematic illustration of the EBV genome, while panels underneath illustrate latent gene expression patterns in different forms of latency. Various viral latent gene products, including proteins, noncoding RNAs, and microRNAs, are illustrated. Viral latent gene promoters are indicated as arrows. Details of mRNA splicing are not faithfully represented

indicated in the figure. Specifically, LCLs express viral latent proteins, consisting of six EBV nuclear antigens (EBNAs 1, 2, 3A, 3B, 3C, and EBNA-LP) and three latent membrane proteins (LMPs 1, 2A, and 2B). They also express viral noncoding EBER RNAs (EBER1 and EBER2) and EBV-encoded microRNAs (miR-BHRF1 and miR-BART). Transcripts from the BamHI A region of the genome (BART transcripts, BARTs) are also expressed. The functions of BART transcripts have been enigmatic for a long time, but it has now become clear that they are precursors of BART miR-NAs. This pattern of viral gene expression is referred to as latency type III (Rickinson and Kieff 2007). On the other hand, typical EBV-positive Burkitt lymphoma cells express only EBNA1, EBERs, and BART transcripts, and such latency type is referred to as latency type I. In latency type II, in addition to the genes expressed in latency I, LMP1 and LMP2 are expressed. Latency type II can be observed in various EBV-associated tumors, including nasopharyngeal carcinoma cells and NK/T lymphoma cells (Rickinson and Kieff 2007). One additional latency type is called Wp-restricted latency observed in Burkitt lymphoma cells, which will be discussed in Sect. 17.3.2.

Molecular functions of viral latent gene products have been extensively studied and characterized, and the results are thoroughly described in textbooks and review articles (Epstein Barr Virus Volume 1 One Herpes Virus: Many Diseases 2015a; Epstein Barr Virus Volume 2 One Herpes Virus: Many Diseases 2015b; Kang and Kieff 2015; Kieff and Rickinson 2007; Young and Rickinson 2004; Young et al. 2016). In these publications, many biological functions have been described for each latent gene product, and it is not our aim to describe all of them. Instead, this review will focus on latent gene functions that have been verified by genetic experiments using EBV mutants. B cell immortalization assay was used by most of the genetic studies for its ease of use. Thus, the results of in vitro transformation assay using naturally occurring as well as artificially created mutants (gene knockout viruses) will be described. This review will also describe how some of the latent gene products contribute to Burkitt lymphomagenesis as well as epithelial carcinogenesis. Finally, since recent studies indicate that clues to understand viral gene functions can be obtained from viral gene heterogeneity, viral latent gene heterogeneity affecting viral phenotypes will be discussed.

17.2 Viral Latent Gene Products Expressed in LCLs

17.2.1 *EBNA2 and EBNA-LP*

At the initial phase of EBV infection to B cells, EBNA proteins are known to be expressed under complicated usage of viral promoters. Transcripts encoding EBNA2 and EBNA-LP are the first to be expressed from viral promoter Wp (BamHI W promoter) (Kieff and Rickinson 2007). EBNA2 and EBNA-LP proteins then regulate transcription starting from both viral and cellular promoters. EBNA2 contributes to the activation of viral latent gene promoter Cp (BamHI C promoter), which drives viral latent genes including EBNA1 and EBNA3s (3A, 3B, and 3C). As a result, promoter usage changes from Wp to Cp. EBNA2 is also necessary for transactivating LMP1 promoter (LMP1p). Thus, EBNA2 turns on the expression of EBNA1, EBNA3s, and LMP1, which are indispensable for B cell transformation. When EBNA2 transactivates Cp and LMP1p, EBNA2 does not bind directly to the promoter region. Rather, EBNA2 is recruited to the promoter regions via a cellular DNA binding protein RBP-Jκ (Kieff and Rickinson 2007). EBNA2 also transactivates cellular promoters, such as those of c-Myc, CD23, and CD21 (Kieff and Rickinson 2007).

EBNA2 is the first viral gene whose function was predicted by the phenotype of a naturally occurring EBV mutant. Burkitt lymphoma cell line P3HR-1 is a well-known virus producing cell line. P3HR-1 strain EBV is a laboratory strain, and it is a derivative of another Burkitt lymphoma-derived EBV strain, Jijoye. Importantly, P3HR-1 virus is incompetent for B cell immortalization, while its parental strain Jijoye can transform B lymphocytes (Rabson et al. 1982). Since P3HR-1 strain EBV has a deletion in EBNA2 region, it was conceivable that EBNA2 was necessary for EBV-mediated B cell immortalization. Later studies demonstrated that EBNA2 is actually indispensable for immortalization (Cohen et al. 1989) (Table 17.1).

EBNA-LP is expressed along with EBNA2 shortly after EBV infection in primary B lymphocytes. EBNA-LP cooperates with EBNA2 in activating viral and cellular gene expression. (Kieff and Rickinson 2007). Making EBNA-LP knockout EBV is technically difficult due to its repetitive nature. Still, a study indicates that EBNA-LP plays critical role in B cell transformation (Mannick et al. 1991).

Table 17.1 EBV-encoded latent genes, genetic studies, and naturally occurring mutants

Classification	Latent genes	Necessity for B cell immortalization revealed by genetic studies	Naturally occurring mutants
Latent proteins	EBNA2	*Essential* Cohen et al. (1989)	Deleted in Burkitt lymphoma-derived strains; P3HR-1, Daudi, Sal, Oku, Ava Jones et al. (1984), Kelly et al. (2002), and Rabson et al. (1982)
	EBNA-LP	*Essential* Mannick et al. (1991)	
	EBNA3A	*Essential* Tomkinson et al. (1993)	
	EBNA3B	Dispensable Tomkinson and Kieff (1992)	Mutations in diffuse large B cell lymphoma and in Hodgkin lymphoma White et al. (2012)
	EBNA3C	*Essential* Tomkinson et al. (1993)	
	EBNA1	*Essential* Humme et al. (2003)	
	LMP1	*Essential* Dirmeier et al. (2003) and Kaye et al. (1993)	
	LMP2	Dispensable Longnecker et al. (1993)	
Noncoding RNAs	EBER1	Dispensable Swaminathan et al. (1991) and Yajima et al. (2005)	
	EBER2		
microRNAs	BHRF1 microRNAs	Dispensable Seto et al. (2010)	
	BART microRNAs	Dispensable Robertson et al. (1994)	Deleted in B95-8 strain Raab-Traub et al. (1980)

17.2.2 EBNA3 Proteins

The EBNA3 genes are tandemly located in the EBV genome, and they are transcribed from far upstream promoters, Cp or Wp (Fig. 17.1). Amino acid sequences of EBNA3 proteins are known to be heterogeneous among various EBV strains (Kanda et al. 2016; Palser et al. 2015). Like EBNA2, EBNA3s also contribute to transcriptional regulation of both viral and cellular promoters. Contribution of EBNA3s on EBV-mediated B cell transformation was examined by series of genetic experiments using P3HR-1 strain EBV. The studies employed rescuing B cell immortalizing activity of P3HR-1 EBV by repairing EBNA2 gene via homologous recombination and simultaneously introducing second-site homologous recombination in other viral genes. P3HR-1 EBV genome with its EBNA2 gene repaired should become transformation-competent and immortalize B lymphocytes. If this transformation-competent virus has acquired a second-site mutation, one can readily conclude that the viral gene mutated by second-site homologous recombination is dispensable for B cell immortalization.

The above experimental strategy revealed that EBNA3B is dispensable for B cell immortalization (Tomkinson and Kieff 1992). By contrast, both EBNA3A and EBNA3C are indispensable (Tomkinson et al. 1993) (Table 17.1). Molecular basis of such difference remains unknown.

EBNA3A and EBNA3C are known to counteract p16 protein (Maruo et al. 2011; Skalska et al. 2010). Downregulating either EBNA3A or EBNA3C expression in the established LCLs results in accumulation of p16 and ceasing of LCL proliferation (Maruo et al. 2011). Thus, EBNA3A and EBNA3C are also necessary to support the continuous growth of the established LCLs. EBNA3A and EBNA3C also epigenetically suppress a proapoptotic tumor-suppressor protein Bim (Paschos et al. 2009). Suppression of Bim appears to be critical during the course of Burkitt lymphomagenesis. This will be discussed further in Sect. 17.3.1.

Although EBNA3A, 3B, and 3C constitute a gene family, EBNA3B has markedly different biological functions compared to EBNA3A and 3C. A recent study indicates that when EBNA3B gene is artificially deleted by gene knockout, the resultant EBNA3B-negative EBV exhibits enhanced B cell transformation ability as well as enhanced lymphomagenesis in humanized mouse model (White et al. 2012). Importantly, the study identified EBNA3B mutations in EBV-positive lymphomas, including diffuse large B cell lymphomas and Hodgkin lymphomas (Table 17.1). Thus, EBNA3B appears to suppress B cell transforming ability. In this viewpoint, while EBNA3A and EBNA3C are viral oncogenes, EBNA3B serves as a tumor suppressor (White et al. 2012).

17.2.3 EBNA1

EBNA1 is an essential molecule to establish and stably maintain EBV latency for multiple reasons. First, EBNA1 is a key molecule to keep EBV genomes as multicopy circular double-stranded DNA molecules (called episomes) in latently infected cells (Leight and Sugden 2000). EBNA1 binds to viral oriP sequence in a sequence-dependent manner. Multiple EBNA1 binding sites (more than 20 copies; copy number varies between EBV strains) are clustered within the oriP sequence. Simultaneously, EBNA1 binds to host cell chromosomes, which can be observed by cytological analyses. This is enabled by a modular domain structure of EBNA1 protein; its C-terminal region contains a sequence-specific oriP DNA binding domain, while its N-terminal region contains chromosome binding domains (Leight and Sugden 2000) (Fig. 17.2). As a result, EBNA1 tethers EBV episomes onto host cell chromosomes and enables their segregation during mitosis.

Molecular mechanisms by which EBNA1 binds to chromosomes have been attracting interests of EBV researchers. An initially proposed hypothesis is that a cellular chromosome binding protein tethers EBNA1 on host chromosome (Shire et al. 1999). Another hypothesis is that EBNA1 protein has amino acid motifs for

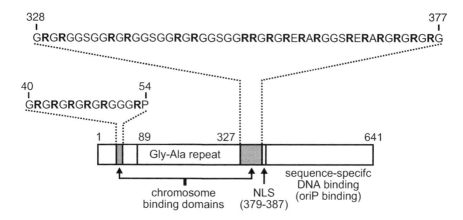

Fig. 17.2 Shown is the amino acid (a.a.) sequence of the chromosome binding domains (a.a. 40–54 and a.a .328–377) of EBNA1 of B95-8 strain EBV. Arginine residues in chromosome binding domains are in bold letters. The positions of nuclear localization signal (NLS), Glycine-Alanine repeat (Gly-Ala), and sequence-specific DNA binding (oriP binding) domain are shown

binding to minor groove of chromosomal DNA (Sears et al. 2004). A recent study demonstrated that interaction between EBNA1 and chromosomes is likely to be electrostatic (Kanda et al. 2013). There are total 24 arginine residues within the chromosome binding domains of EBNA1 (Fig. 17.2), and the regions are positively charged. Series of amino acid substitution mutations were introduced to these arginine residues. Alanine substitution of the arginine residues results in loss of chromosome binding, while substituting arginine with lysine (positively charged, just like arginine) does not affect chromosome binding (Kanda et al. 2013). Thus, positive charges of the chromosome binding domains appear to be primarily important.

Second, EBNA1 works as a transactivator of viral gene expression. EBNA1 binding to oriP sequence contributes to the activation of Cp, which drives the expressions of EBNA2, EBNA3s, and EBNA1 itself in latency type III (Fig. 17.1).

A recombinant virus lacking EBNA1 gene can still transform B lymphocytes, but the transformation efficiency is several thousandfold less compared to the wild-type virus (Humme et al. 2003) (Table 17.1). In the absence of EBNA1, EBV genome cannot be episomally maintained, and, as a result, EBV genomes should integrate into host chromosomes in order to stably express all the viral latent proteins required for B cell transformation. The frequency of getting integrated into host chromosomes is much less than the frequency of forming multi-copy episomes. This explains why the transformation efficiency of EBNA1-negative EBV is so low.

17.2.4 LMP1

LMP1 is a membrane protein with six transmembrane domains and C-terminal intracellular signaling domain, to which cellular signaling molecules bind (Kieff and Rickinson 2007). LMP1 belongs to tumor necrosis factor (TNF) receptor family protein, and it is a constitutively active mimic of cellular CD40. LMP1 activates NF-κB, c-jun N-terminal kinase (JNK), and p38/MAPK pathways. LMP1-mediated activation of NF-κB pathway is necessary to maintain LCL growth, since inhibiting NF-κB pathway in established LCLs results in apoptotic cell death (Cahir-McFarland et al. 2000). LMP1 is also strongly expressed in EBV-positive Hodgkin lymphomas and nasopharyngeal carcinomas.

Genetic studies indicate that LMP1 is either essential (Kaye et al. 1993) or critically important (Dirmeier et al. 2003) for B cell transformation (Table 17.1). LMP1 expression in B cells causes B cell activation phenotypes, including upregulation of cell surface markers (like CD23, CD39, CD40) and cell adhesion molecules (like ICAM-1, LFA-1). LMP1 is a classical viral oncogene, which can transform rodent fibroblasts (Wang et al. 1985). Thus, LMP1 is not only important for B cell transformation but also most likely contributes to various EBV-related tumorigenic diseases as well.

17.2.5 LMP2 Proteins

LMP2A and LMP2B are expressed in latency III infection. LMP2A is an activating molecule with N-terminal intracellular domain, 12 transmembrane domains, and C-terminal intracellular domain. N-terminal intracellular domain contributes to mimic B cell receptor signaling, while the C-terminal region constitutively blocks B cell receptor signaling (Kieff and Rickinson 2007). LMP2A and LMP2B are dispensable for B cell transformation (Longnecker et al. 1993) (Table 17.1). In transgenic mice, LMP2A allows immunoglobulin (Ig)-negative cells to colonize peripheral lymphoid organs (Caldwell et al. 1998), indicating that LMP2A can substitute Ig-mediated signaling. Thus, LMP2 also provides survival and anti-differentiation signals to B cells (Kieff and Rickinson 2007).

A recent study indicates that germinal center-specific expression of LMP2A results in altered humoral immune responses through selecting low-affinity antibody-producing B cells (Minamitani et al. 2015). This observation implies a molecular mechanism of EBV-associated autoimmune diseases, like systemic lupus erythematosus.

17.2.6 EBERs

EBER1 and EBER2 are viral non-polyadenylated RNA molecules that are expressed at very high levels in the majority of EBV-infected cells. EBER in situ hybridization is the most reliable and sensitive method to detect EBV infection, and the method is commonly used to prove EBV infection in tissues. EBERs form secondary structure and are partially double-stranded, and EBERs are known to interact with various cellular proteins to exert various biological activities (Nanbo and Takada 2002).

Using P3HR-1 second-site homologous recombination system, it was demonstrated that an EBV mutant lacking EBERs could still transform B lymphocytes (Swaminathan et al. 1991) (Table 17.1). The result was verified by using a pure recombinant virus lacking EBERs derived from Akata strain EBV (Yajima et al. 2005). Although EBER-deleted EBV could still transform B lymphocytes, they exhibited impaired B cell transformation efficiency when infected B lymphocytes were plated at low cell density (Yajima et al. 2005). EBERs are highly expressed in epithelial cells as well. Thus, contribution of EBERs on epithelial cell tumorigenesis has been attracting interests. However, the roles of EBERs on epithelial tumorigenesis remain unclear.

EBER function in vivo was recently investigated in a humanized mouse model (Gregorovic et al. 2015). The results revealed that wild-type EBV and EBER-deficient EBV exhibited approximately equal abilities to infect immunodeficient mice reconstituted with human hematopoietic system. However, this experimental system does not involve normal immune response against EBV-infected cells. Physiological phenotypes for the EBERs may only be revealed by in vivo infection challenged by normal immune responses (Gregorovic et al. 2015).

17.2.7 EBV-Encoded microRNAs (miRNAs) and BART Transcripts

Most of viral latent proteins are highly immunogenic and stimulate host immune response. By contrast, viral miRNAs can affect gene expression in host cells without stimulating any immune response. Therefore, encoding viral miRNA is advantageous to the virus.

EBV-encoded miRNAs are the very first virally encoded miRNAs ever identified (Pfeffer et al. 2004). The study used B95-8 strain EBV, a deletion derivative strain of EBV (Raab-Traub et al. 1980), and identified only 5 miRNAs: 3 miRNAs in BHRF1 region and 2 miRNAs in the BART region (Pfeffer et al. 2004). Later studies revealed that there are far more miRNAs in the "wild-type" EBV, as the 12-kb region that is deleted in B95-8 strain is rich in pre-miRNA genes. Currently, 44 mature miRNAs are encoded at 2 different loci in the EBV genome: 4 mature miRNA encoded at the BHRF1 locus (BHRF1 miRNAs) and 40 mature miRNAs encoded at the BART locus (BART miRNAs) (www.mirbase.org) (Fig. 17.3).

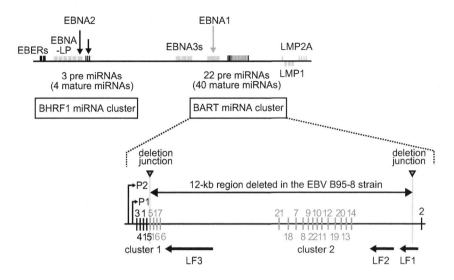

Fig. 17.3 Schematic representation of EBV miRNA genes. The EBV genome is illustrated on the top, and the positions of clustered miRNA genes (BHRF1 and BART clusters) are indicated with numbers of pre-miRNAs and mature miRNAs. BART miRNA cluster is illustrated in detail at the bottom with the positions of miRNA clusters 1 and 2 and B95-8 deletion. Pre-miRNA genes that are deleted in B95-8 strain EBV are colored in gray. Positions of leftward open reading frames are indicated. LF3 may not be a functional open reading frame

Various EBV miRNA are expressed at markedly different levels among different cell lines. BHRF1 miRNAs are highly expressed in cells with latency type III, but they are undetectable in B cells or epithelial cells with latency type I. BHRF1 miRNAs accelerate cell cycle progression and exert antiapoptotic activity of B lymphocyte shortly after infection (Seto et al. 2010).

By contrast, BART miRNAs are expressed not only in B cells with type III latency but also in epithelial cells with type I latency (Klinke et al. 2014). It is now well established that BART transcripts serve as precursors of BART miRNAs. Abundant BART transcripts in latently infected epithelial cells likely cause high BART miRNA expression levels in epithelial cells. In epithelial carcinoma tissues, such as those of gastric cancer cells, especially high levels of BART miRNAs are expressed (Kim et al. 2013).

It is still unclear whether BART transcripts encode functional proteins or not in EBV-infected cells. Although open reading frames (BARF0, RK-BARF0, RPMS1, and A73) have been identified in the BART transcripts (Kieff and Rickinson 2007), their protein products have never been detected even in EBV-infected epithelial cells, in which abundant BART transcripts are expressed. A recombinant EBV with 58-kilobase-pair deletion encompassing the entire BART region could still transform B lymphocytes (Robertson et al. 1994), indicating that BART transcripts and putative protein products, if any, are dispensable for B cell transformation (Robertson et al. 1994). Curiously, all the pre-miRNA genes are encoded in the intronic regions of BART transcripts, and processing of the pre-miRNAs precedes the completion of the splicing reaction (Edwards et al. 2008). Biological meaning of this finding remains unknown.

Viral miRNAs can either target other EBV transcripts or transcripts of host cells. Many viral and cellular targets of EBV miRNAs have been identified (Klinke et al. 2014; Kuzembayeva et al. 2014). However, most of the results have never been verified by EBV mutants. EBV encodes more than 40 mature miRNAs, and it is expected that these miRNAs cooperatively regulate viral and host gene expression. Therefore, knocking out each specific miRNA is expected to confer minimal effect. Instead, either knocking out or reconstitution of entire BART miRNA cluster has been tried (Kanda et al. 2015; Lin et al. 2015; Seto et al. 2010) (Table 17.1). One of such studies implies that, by affecting expression of multiple target genes, BART miRNAs contribute to facilitate long-term persistence of the virus in infected host (Lin et al. 2015).

17.3 EBV Latent Gene Products and Burkitt Lymphomagenesis

17.3.1 EBV Latency Type I to Cope with Myc Translocation

Burkitt lymphomas in endemic area are almost 100% EBV-positive. By contrast, in non-endemic area, only 25% of them are EBV-positive. Similarly, few AIDS-associated Burkitt lymphoma is EBV-positive (Rickinson and Kieff 2007; Thorley-Lawson and Allday 2008). Thus, the presence of EBV infection is not mandatory for Burkitt lymphomagenesis.

Burkitt lymphoma cells are derived from germinal center B cells, where chromosomal translocation involving Myc and Ig occurs (Thorley-Lawson and Allday 2008). Myc-Ig gene translocation is found in all cases of Burkitt lymphoma cells. When Myc-Ig translocation occurs in germinal center B cells, Myc-driven growth program turns on. Cells overexpressing Myc are usually eliminated due to Myc-induced apoptosis. However, if germinal center B cells are infected with EBV prior to the onset of Myc-Ig translocation, cells with Myc translocation escape from apoptotic cell death.

It is known that Myc-induced growth program is not compatible with full EBV latent gene expression (Kelly et al. 2002). After primary infection of B cells with EBV, EBNA2 together with EBNA-LP induce EBNA3A and EBNA3C expression, which in turn suppress proapoptotic protein BIM by epigenetic mechanism (Paschos et al. 2009). Subsequently, EBNA2 is downregulated via epigenetic silencing, and type I latency is established, in which only EBNA1 protein is expressed under the control of viral Qp promoter (Kieff and Rickinson 2007) (Fig. 17.1). EBNA2-negative cells tolerate Myc overexpression, and epigenetic silencing of Bim thereby enables proliferation of cells harboring Myc-Ig translocation (Fig. 17.4). This type

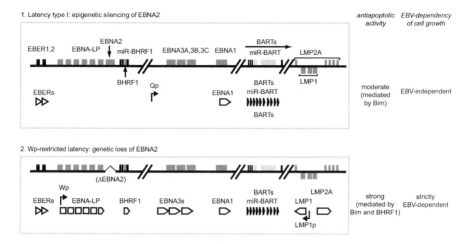

Fig. 17.4 Two distinct mechanisms of EBNA2 downregulation in Burkitt lymphoma cells. EBNA2 expression is not compatible with Myc-induced cell growth, thus EBNA2 somehow should be downregulated. First mechanism involves establishment of latency type I via epigenetic silencing of EBNA2 (top). Second mechanism is designated as Wp-restricted latency, which involves genetic loss of EBNA2 gene from the EBV genome (bottom). EBNA2 deletion, in turn, results in the expression of viral antiapoptotic protein BHRF1, and BHRF1 expression exerts strong antiapoptotic activity. In Wp-restricted latency, cell growth is strictly dependent on the presence of EBV genome

of Burkitt lymphoma cells can proliferate even when EBV genomes are lost from the cells, as their growth is mediated by altered host gene expression including Myc overexpression (Fig. 17.4).

17.3.2 Wp-Restricted Latency and BHRF1 Expression

Rather than epigenetic silencing of EBNA2, there is another mechanism to downregulate EBNA2 expression, which involves genetic loss of EBNA2 gene from EBV genome (Fig. 17.4).

As described above, Burkitt lymphoma-derived cell line P3HR-1 harbors EBNA2-deleted EBV. Similarly, several additional Burkitt lymphoma-derived cell lines (Daudi, Sal, Oku, Ava) harbor EBNA2-deleted EBV genomes (Jones et al. 1984; Kelly et al. 2002). Thus, EBV strains with EBNA2 deletion may somehow be selected in Burkitt lymphoma cells. Supporting the idea, EBNA2-deleted EBV were found to establish a special latency program, designated as Wp-restricted latency (Kelly et al. 2006, 2009; Oudejans et al. 1995) (Figs. 17.1, 17.4). Rather than using Qp promoter, which is a commonly used promoter in Burkitt lymphoma cells, viral latent gene expression is mediated by Wp promoter in Wp-restricted latency. Wp promoter drives expression of viral latent proteins EBNA-LP, EBNA3s (3A, 3B and 3C), and EBNA1. In addition, importantly, BHRF1 gene is expressed (Kelly et al.

2006, 2009; Oudejans et al. 1995) (Fig. 17.4). BHRF1 protein, previously assumed to be a lytic gene product, is expressed as a latent protein in Wp-restricted gene expression pattern. BHRF1 is a homolog of cellular antiapoptotic protein, bcl-2 (Henderson et al. 1993), and BHRF1 expression appears to exert strong antiapoptotic effect in cells harboring Myc-Ig translocation. Supporting the idea, out of EBV latent proteins expressed in P3HR-1 Burkitt lymphoma cells, BHRF1 serves as a survival factor (Watanabe et al. 2010), and the growth of P3HR-1 cells is strictly dependent on the presence of EBV, as BHRF1 expression is necessary for the cell growth (Fig. 17.4).

Therefore, deletion of EBNA2 is not merely a loss-of-function mutation. Rather, it contributes to EBV-mediated Burkitt lymphomagenesis (Bornkamm 2009; Rowe et al. 2009). In this point of view, EBNA2 deletion is also a gain-of-function mutation in a certain fraction of Burkitt lymphoma cells.

17.4 EBV Latent Gene Products and Epithelial Carcinogenesis

EBV is also associated with various epithelial cancers. Nasopharyngeal carcinoma cells are more than 95% EBV-positive, thus causal relationship between EBV infection and carcinogenesis is highly suspected. On the other hand, in case of gastric cancers, only 10% of them are EBV-positive (Rickinson and Kieff 2007). EBV-positive gastric cancer cells are categorized as one of the four types of gastric cancers, due to their prominent CpG island methylation phenotype (Cancer Genome Atlas Research 2014).

It has never been demonstrated that EBV infection causes transformation of primary epithelial cells. EBV infection to primary epithelial cells is usually cytostatic; and infected primary epithelial cells stop to proliferate. A previous study indicates that EBV can establish persistent infection only in precancerous epithelial cells with aberrant CDK4-cyclin D expression (Tsang et al. 2012). This may reflect in vivo situation of epithelial cell infection. Additional host gene mutations are necessary to convert EBV- infected epithelial cells to cancer cells (Tsao et al. 2015).

EBV-positive epithelial cancers are in either latency type II (Nasopharyngeal carcinoma cells) or in latency type I (gastric cancer cells). In latency type I, only a few viral gene products (EBNA1, EBERs, and BART miRNAs) are expressed. Although oncogenic potential of these viral gene products has not been proven, recent studies imply that BART miRNAs significantly contribute to EBV-mediated epithelial carcinogenesis (Tsao et al. 2015).

A study demonstrated that all the BART miRNAs were upregulated when EBV-positive epithelial tumor cells were grafted into mice (Qiu et al. 2015). The study also indicates that BART miRNAs potentiate the growth of epithelial tumor cells in vivo, but not in vitro (Qiu et al. 2015). These results fit well with very high BART miRNA expression levels in human epithelial tumor specimens, implicating that there is a selective pressure to enrich tumor cells with high BART miRNA expression in vivo.

There is also an attempt to clarify the roles of BART miRNAs during epithelial carcinogenesis by means of genetic study. Using B95-8 strain EBV genome cloned in a bacterial artificial chromosome vector, a pair of recombinant EBV, either having the restored BART region or the BART region deleted, was obtained. Comparison of host gene expression in recombinant virus-infected epithelial cells, NDRG1 (N-Myc downstream regulated gene-1) was identified as a gene that was downregulated by BART miRNA cluster 2 (Kanda et al. 2015) (Fig. 17.3). Interestingly, NDRG1 has been characterized as an epithelial cell-specific metastatic suppressor, so it is tempting to speculate that NDRG1 downregulation is somehow related to metastatic potential of nasopharyngeal carcinoma cells.

17.5 Latent Gene Variation among EBV Strains

The major variation in EBV genome was type 1 or type 2 (type A or type B) classification based on differences in EBNA2 and EBNA3s (Rickinson and Kieff 2007). It is known that type 1 EBV can transform B lymphocyte much more efficiently than type 2 EBV. A recent genetic study indicates that the difference is due to single amino acid difference between type 1 EBNA2 and type 2 EBNA2 (Tzellos et al. 2014). By contrast, although amino acid sequences of EBNA3s between type 1 and type 2 EBV are quite divergent, their functions are highly conserved between type 1 and type 2 EBV.

Naturally occurring latent gene variation also gives us clues to understand the functions of the genes (Table 17.1). Regarding genetic variation of EBNA2, as mentioned above, deletion of EBNA2 gene contributes to Burkitt lymphomagenesis. Genetic variation of EBNA3s remains obscure, but mutations of EBNA3B appear to be critical for development of diffuse large B cell lymphomas and Hodgkin lymphomas (White et al. 2012). BART miRNA region is apparently dispensable for B cell transformation, as B95-8 strain EBV, having majority of BART miRNA genes deleted, retains transforming activity.

It is well-known that EBV LMP1 gene is quite heterogeneous among EBV strains. Thus, LMP1 heterogeneity has been analyzed extensively prior to the era of deep sequencing technology. For example, there is apparent functional difference between B cell-derived LMP1 (B95-8 LMP1) and nasopharyngeal carcinoma-derived CAO-LMP1 (Dawson et al. 2000); CAO-LMP1 is less cytostatic compared to that of B95-8 LMP1 (Johnson et al. 1998). Initially, 10 a.a. (30 nucleotides) deletion in the C-terminal signaling domain in CAO-LMP1 was considered to be responsible for the difference. However, it is now considered that amino acid differences in the transmembrane domain appear to be responsible for the difference (Johnson et al. 1998; Nitta et al. 2004).

A recent development of deep sequencing technology enabled determining entire EBV genome sequencing of various EBV strains. Sequences of more than 70 EBV genomes were recently determined (Palser et al. 2015). The result revealed that sequence variation mainly resides in viral latent genes. With deep sequencing technology, one can identify multiple latent gene heterogeneity at once.

A study demonstrated that M81 strain of EBV, derived from nasopharyngeal carcinoma cells, has different properties compared to B cell-derived B95-8 strain EBV (Tsai et al. 2013). M81 strain EBV infects epithelial cells more efficiently than B95-8 strain EBV and has a tendency to spontaneously enter lytic replication cycle in LCLs (Tsai et al. 2013). There are multiple sequence variations between the two strains, and sequence variation responsible for the phenotypic difference has not been identified. Still, the result implicates that there may be a cancer-specific EBV strain. It was previously reported that multiple EBV strains exist even in individuals (Walling et al. 2003), and that EBV strains derived from oropharyngeal fluids were distinct from those derived from peripheral blood B cells of the same individual (Renzette et al. 2014). These results support the idea that multiple strains of EBV are maintained in each individual, and that a specific EBV strain can be selected to generate EBV-associated diseases. Since viral gene heterogeneity mainly resides in latent genes (Palser et al. 2015), latent gene heterogeneity may be responsible for causing difference of pathogenicity.

17.6 Future Perspectives

Viral latent gene functions have been most extensively analyzed by B cell immortalization assay for its ease of use. On the other hand, Burkitt lymphomagenesis and epithelial carcinogenesis are still difficult to reproduce in vitro. We definitely need a better in vitro infection system to investigate how viral latent gene products contribute to epithelial carcinogenesis. We also need an experimental system in which in vitro EBV infection cooperates with host gene mutations to transform primary epithelial cells. When such experimental system becomes available, one can now examine the contribution of viral latent gene products to epithelial tumorigenesis using EBV mutants.

Regarding latent gene heterogeneity in relation to pathogenicity, we should accumulate viral genome sequences of various EBV strains derived from healthy and disease-associated individuals (Farrell 2015; Feederle et al. 2015). Recently, a CRISPR-Cas9 genome editing technology was successfully applied to clone EBV strains from gastric cancer-derived cell lines (Kanda et al. 2016). This technology should accelerate the accumulation of EBV genome sequences derived from various origins. The technology also enables investigating the biological behaviors of the viruses by reconstituting infectious viruses. Latent gene variations that critically affect viral phenotypes may be identified by applying such an experimental approach to various disease-associated EBV strains.

References

Bornkamm GW (2009) Epstein-Barr virus and the pathogenesis of Burkitt's lymphoma: more questions than answers. Int J Cancer 124:1745–1755. https://doi.org/10.1002/ijc.24223

Cahir-McFarland ED, Davidson DM, Schauer SL, Duong J, Kieff E (2000) NF-kappa B inhibition causes spontaneous apoptosis in Epstein-Barr virus-transformed lymphoblastoid cells. Proc Natl Acad Sci U S A 97:6055–6060. https://doi.org/10.1073/pnas.100119497

Caldwell RG, Wilson JB, Anderson SJ, Longnecker R (1998) Epstein-Barr virus LMP2A drives B cell development and survival in the absence of normal B cell receptor signals. Immunity 9:405–411

Cancer Genome Atlas Research N (2014) Comprehensive molecular characterization of gastric adenocarcinoma. Nature 513:202–209. https://doi.org/10.1038/nature13480

Cohen JI, Wang F, Mannick J, Kieff E (1989) Epstein-Barr virus nuclear protein 2 is a key determinant of lymphocyte transformation. Proc Natl Acad Sci U S A 86:9558–9562

Dawson CW, Eliopoulos AG, Blake SM, Barker R, Young LS (2000) Identification of functional differences between prototype Epstein-Barr virus-encoded LMP1 and a nasopharyngeal carcinoma-derived LMP1 in human epithelial cells. Virology 272:204–217. https://doi.org/10.1006/viro.2000.0344

Dirmeier U, Neuhierl B, Kilger E, Reisbach G, Sandberg ML, Hammerschmidt W (2003) Latent membrane protein 1 is critical for efficient growth transformation of human B cells by epstein-barr virus. Cancer Res 63:2982–2989

Edwards RH, Marquitz AR, Raab-Traub N (2008) Epstein-Barr virus BART microRNAs are produced from a large intron prior to splicing. J Virol 82:9094–9106. https://doi.org/10.1128/JVI.00785-08

Farrell PJ (2015) Epstein-Barr virus strain variation. Curr Top Microbiol Immunol 390:45–69. https://doi.org/10.1007/978-3-319-22822-8_4

Feederle R, Klinke O, Kutikhin A, Poirey R, Tsai MH, Delecluse HJ (2015) Epstein-Barr virus: from the detection of sequence polymorphisms to the recognition of viral types. Curr Top Microbiol Immunol 390:119–148. https://doi.org/10.1007/978-3-319-22822-8_7

Gregorovic G et al (2015) Epstein-Barr viruses (EBVs) deficient in EBV-encoded RNAs have higher levels of latent membrane protein 2 RNA expression in Lymphoblastoid cell lines and efficiently establish persistent infections in humanized mice. J Virol 89:11711–11714. https://doi.org/10.1128/JVI.01873-15

Henderson S, Huen D, Rowe M, Dawson C, Johnson G, Rickinson A (1993) Epstein-Barr virus-coded BHRF1 protein, a viral homologue of Bcl-2, protects human B cells from programmed cell death. Proc Natl Acad Sci U S A 90:8479–8483

Humme S, Reisbach G, Feederle R, Delecluse HJ, Bousset K, Hammerschmidt W, Schepers A (2003) The EBV nuclear antigen 1 (EBNA1) enhances B cell immortalization several thousand-fold. Proc Natl Acad Sci U S A 100:10989–10994. https://doi.org/10.1073/pnas.1832776100

Johnson RJ, Stack M, Hazlewood SA, Jones M, Blackmore CG, Hu LF, Rowe M (1998) The 30-base-pair deletion in Chinese variants of the Epstein-Barr virus LMP1 gene is not the major effector of functional differences between variant LMP1 genes in human lymphocytes. J Virol 72:4038–4048

Jones MD, Foster L, Sheedy T, Griffin BE (1984) The EB virus genome in Daudi Burkitt's lymphoma cells has a deletion similar to that observed in a non-transforming strain (P3HR-1) of the virus. EMBO J 3:813–821

Kanda T et al (2013) Interaction between basic residues of Epstein-Barr virus EBNA1 protein and cellular chromatin mediates viral plasmid maintenance. J Biol Chem 288:24189–24199. https://doi.org/10.1074/jbc.M113.491167

Kanda T, Miyata M, Kano M, Kondo S, Yoshizaki T, Iizasa H (2015) Clustered microRNAs of the Epstein-Barr virus cooperatively downregulate an epithelial cell-specific metastasis suppressor. J Virol 89:2684–2697. https://doi.org/10.1128/JVI.03189-14

Kanda T, Furuse Y, Oshitani H, Kiyono T (2016) Highly efficient CRISPR/Cas9-mediated cloning and functional characterization of gastric cancer-derived Epstein-Barr virus strains. J Virol 90:4383–4393. https://doi.org/10.1128/JVI.00060-16

Kang MS, Kieff E (2015) Epstein-Barr virus latent genes. Exp Mol Med 47:e131. https://doi.org/10.1038/emm.2014.84

Kaye KM, Izumi KM, Kieff E (1993) Epstein-Barr virus latent membrane protein 1 is essential for B-lymphocyte growth transformation. Proc Natl Acad Sci U S A 90:9150–9154

Kelly G, Bell A, Rickinson A (2002) Epstein-Barr virus-associated Burkitt lymphomagenesis selects for downregulation of the nuclear antigen EBNA2. Nat Med 8:1098–1104. https://doi.org/10.1038/nm758

Kelly GL, Milner AE, Baldwin GS, Bell AI, Rickinson AB (2006) Three restricted forms of Epstein-Barr virus latency counteracting apoptosis in c-myc-expressing Burkitt lymphoma cells. Proc Natl Acad Sci U S A 103:14935–14940. https://doi.org/10.1073/pnas.0509988103

Kelly GL et al (2009) An Epstein-Barr virus anti-apoptotic protein constitutively expressed in transformed cells and implicated in burkitt lymphomagenesis: the Wp/BHRF1 link. PLoS Pathog 5:e1000341. https://doi.org/10.1371/journal.ppat.1000341

Kieff E, Rickinson AB (2007) In: Knipe DM, Howley PM (eds) Epstein-Barr virus and its replication, Fields virology, vol 2, 5th edn. Lippincott Williams & Wilkins, Philadelphia, pp 2603–2654

Kim DN et al (2013) Characterization of naturally Epstein-Barr virus-infected gastric carcinoma cell line YCCEL1. J Gen Virol 94:497–506. https://doi.org/10.1099/vir.0.045237-0

Klinke O, Feederle R, Delecluse HJ (2014) Genetics of Epstein-Barr virus microRNAs. Semin Cancer Biol 26C:52–59. https://doi.org/10.1016/j.semcancer.2014.02.002

Kuzembayeva M, Hayes M, Sugden B (2014) Multiple functions are mediated by the miRNAs of Epstein-Barr virus. Curr Opin Virol 7:61–65. https://doi.org/10.1016/j.coviro.2014.04.003

Leight ER, Sugden B (2000) EBNA-1: a protein pivotal to latent infection by Epstein-Barr virus. Rev Med Virol 10:83–100

Lin X et al (2015) The Epstein-Barr virus BART miRNA cluster of the M81 strain modulates multiple functions in primary B cells. PLoS Pathog 11:e1005344. https://doi.org/10.1371/journal.ppat.1005344

Longnecker R, Miller CL, Miao XQ, Tomkinson B, Kieff E (1993) The last seven transmembrane and carboxy-terminal cytoplasmic domains of Epstein-Barr virus latent membrane protein 2 (LMP2) are dispensable for lymphocyte infection and growth transformation in vitro. J Virol 67:2006–2013

Mannick JB, Cohen JI, Birkenbach M, Marchini A, Kieff E (1991) The Epstein-Barr virus nuclear protein encoded by the leader of the EBNA RNAs is important in B-lymphocyte transformation. J Virol 65:6826–6837

Maruo S, Zhao B, Johannsen E, Kieff E, Zou J, Takada K (2011) Epstein-Barr virus nuclear antigens 3C and 3A maintain lymphoblastoid cell growth by repressing p16INK4A and p14ARF expression. Proc Natl Acad Sci U S A 108:1919–1924. https://doi.org/10.1073/pnas.1019599108

Minamitani T et al (2015) Evasion of affinity-based selection in germinal centers by Epstein-Barr virus LMP2A. Proc Natl Acad Sci U S A 112:11612–11617. https://doi.org/10.1073/pnas.1514484112

Munz C (ed) (2015a) Epstein Barr virus volume 1 one herpes virus: many diseases, vol 390. Curr Top Microbiol. https://doi.org/10.1007/978-3-319-22822-8

Munz C (ed) (2015b) Epstein Barr virus volume 2 one herpes virus: many diseases, vol 390. Curr Top Microbiol. https://doi.org/10.1007/978-3-319-22834-1

Nanbo A, Takada K (2002) The role of Epstein-Barr virus-encoded small RNAs (EBERs) in oncogenesis. Rev Med Virol 12:321–326. https://doi.org/10.1002/rmv.363

Nitta T, Chiba A, Yamamoto N, Yamaoka S (2004) Lack of cytotoxic property in a variant of Epstein-Barr virus latent membrane protein-1 isolated from nasopharyngeal carcinoma. Cell Signal 16:1071–1081. https://doi.org/10.1016/j.cellsig.2004.03.001

Oudejans JJ et al (1995) BHRF1, the Epstein-Barr virus (EBV) homologue of the BCL-2 proto-oncogene, is transcribed in EBV-associated B-cell lymphomas and in reactive lymphocytes. Blood 86:1893–1902

Palser AL et al (2015) Genome diversity of Epstein-Barr virus from multiple tumor types and normal infection. J Virol 89:5222–5237. https://doi.org/10.1128/JVI.03614-14

Paschos K, Smith P, Anderton E, Middeldorp JM, White RE, Allday MJ (2009) Epstein-barr virus latency in B cells leads to epigenetic repression and CpG methylation of the tumour suppressor gene Bim. PLoS Pathog 5:e1000492. https://doi.org/10.1371/journal.ppat.1000492

Pfeffer S et al (2004) Identification of virus-encoded microRNAs. Science 304:734–736

Qiu J, Smith P, Leahy L, Thorley-Lawson DA (2015) The Epstein-Barr virus encoded BART miR-NAs potentiate tumor growth in vivo. PLoS Pathog 11:e1004561. https://doi.org/10.1371/journal.ppat.1004561

Raab-Traub N, Dambaugh T, Kieff E (1980) DNA of Epstein-Barr virus VIII: B95-8, the previous prototype, is an unusual deletion derivative. Cell 22:257–267

Rabson M, Gradoville L, Heston L, Miller G (1982) Non-immortalizing P3J-HR-1 Epstein-Barr virus: a deletion mutant of its transforming parent, Jijoye. J Virol 44:834–844

Renzette N et al (2014) Epstein-Barr virus latent membrane protein 1 genetic variability in peripheral blood B cells and oropharyngeal fluids. J Virol 88:3744–3755. https://doi.org/10.1128/JVI.03378-13

Rickinson AB, Kieff E (2007) In: Knipe M, Howley PM (eds) Epstein-Barr virus, Fields virology, vol 2, 5th edn. Lippincott Williams & Wilkins, Philadelphia, pp 2655–2700

Robertson ES, Tomkinson B, Kieff E (1994) An Epstein-Barr virus with a 58-kilobase-pair deletion that includes BARF0 transforms B lymphocytes in vitro. J Virol 68:1449–1458

Rowe M, Kelly GL, Bell AI, Rickinson AB (2009) Burkitt's lymphoma: the Rosetta Stone deciphering Epstein-Barr virus biology. Semin Cancer Biol 19:377–388. https://doi.org/10.1016/j.semcancer.2009.07.004

Sears J, Ujihara M, Wong S, Ott C, Middeldorp J, Aiyar A (2004) The amino terminus of Epstein-Barr virus (EBV) nuclear antigen 1 contains AT hooks that facilitate the replication and partitioning of latent EBV genomes by tethering them to cellular chromosomes. J Virol 78:11487–11505. https://doi.org/10.1128/JVI.78.21.11487-11505.2004

Seto E, Moosmann A, Gromminger S, Walz N, Grundhoff A, Hammerschmidt W (2010) Micro RNAs of Epstein-Barr virus promote cell cycle progression and prevent apoptosis of primary human B cells. PLoS Pathog 6:e1001063

Shire K, Ceccarelli DF, Avolio-Hunter TM, Frappier L (1999) EBP2, a human protein that interacts with sequences of the Epstein-Barr virus nuclear antigen 1 important for plasmid maintenance. J Virol 73:2587–2595

Skalska L, White RE, Franz M, Ruhmann M, Allday MJ (2010) Epigenetic repression of p16(INK4A) by latent Epstein-Barr virus requires the interaction of EBNA3A and EBNA3C with CtBP. PLoS Pathog 6:e1000951. https://doi.org/10.1371/journal.ppat.1000951

Swaminathan S, Tomkinson B, Kieff E (1991) Recombinant Epstein-Barr virus with small RNA (EBER) genes deleted transforms lymphocytes and replicates in vitro. Proc Natl Acad Sci U S A 88:1546–1550

Thorley-Lawson DA, Allday MJ (2008) The curious case of the tumour virus: 50 years of Burkitt's lymphoma. Nat Rev Microbiol 6:913–924. https://doi.org/10.1038/nrmicro2015

Tomkinson B, Kieff E (1992) Use of second-site homologous recombination to demonstrate that Epstein-Barr virus nuclear protein 3B is not important for lymphocyte infection or growth transformation in vitro. J Virol 66:2893–2903

Tomkinson B, Robertson E, Kieff E (1993) Epstein-Barr virus nuclear proteins EBNA-3A and EBNA-3C are essential for B-lymphocyte growth transformation. J Virol 67:2014–2025

Tsai MH et al (2013) Spontaneous lytic replication and epitheliotropism define an Epstein-Barr virus strain found in carcinomas. Cell Rep 5:458–470. https://doi.org/10.1016/j.celrep.2013.09.012

Tsang CM et al (2012) Cyclin D1 overexpression supports stable EBV infection in nasopharyngeal epithelial cells. Proc Natl Acad Sci U S A 109:E3473–E3482. https://doi.org/10.1073/pnas.1202637109

Tsao SW, Tsang CM, To KF, Lo KW (2015) The role of Epstein-Barr virus in epithelial malignancies. J Pathol 235:323–333. https://doi.org/10.1002/path.4448

Tzellos S et al (2014) A single amino acid in EBNA-2 determines superior B lymphoblastoid cell line growth maintenance by Epstein-Barr virus type 1 EBNA-2. J Virol 88:8743–8753. https://doi.org/10.1128/JVI.01000-14

Walling DM, Brown AL, Etienne W, Keitel WA, Ling PD (2003) Multiple Epstein-Barr virus infections in healthy individuals. J Virol 77:6546–6550

Wang D, Liebowitz D, Kieff E (1985) An EBV membrane protein expressed in immortalized lymphocytes transforms established rodent cells. Cell 43:831–840

Watanabe A, Maruo S, Ito T, Ito M, Katsumura KR, Takada K (2010) Epstein-Barr virus-encoded Bcl-2 homologue functions as a survival factor in Wp-restricted Burkitt lymphoma cell line P3HR-1. J Virol 84:2893–2901. https://doi.org/10.1128/JVI.01616-09

White RE et al (2012) EBNA3B-deficient EBV promotes B cell lymphomagenesis in humanized mice and is found in human tumors. J Clin Invest 122:1487–1502. https://doi.org/10.1172/JCI58092

Yajima M, Kanda T, Takada K (2005) Critical role of Epstein-Barr virus (EBV)-encoded RNA in efficient EBV-induced B-lymphocyte growth transformation. J Virol 79:4298–4307. https://doi.org/10.1128/JVI.79.7.4298-4307.2005

Young LS, Rickinson AB (2004) Epstein-Barr virus: 40 years on. Nat Rev Cancer 4:757–768. https://doi.org/10.1038/nrc1452

Young LS, Yap LF, Murray PG (2016) Epstein-Barr virus: more than 50 years old and still providing surprises. Nat Rev Cancer 16:789–802. https://doi.org/10.1038/nrc.2016.92

Chapter 18
Encyclopedia of EBV-Encoded Lytic Genes: An Update

Takayuki Murata

Abstract In addition to latent genes, lytic genes of EBV must also be of extreme significance since propagation of the virus can be achieved only through execution of lytic cycle. Research on EBV lytic genes may thus prevent spreading of the virus and alleviate disorders, such as infectious mononucleosis and oral hairy leukoplakia, which are highly associated with EBV lytic infection. Moreover, recent advancements have been demonstrating that at least several lytic genes are expressed to some extent even during latent state. It is also demonstrated now that upon de novo infection, EBV expresses lytic genes in addition to latent genes before establishment of latency (this phase is called "pre-latent abortive lytic state"). In those cases, lytic genes also play important roles in cell proliferation of EBV-positive cells. However, many lytic gene products have not been identified yet nor studied thoroughly, and even worse, some have been misidentified in the literature. Here, I would like to give a detailed up-to-date review on EBV lytic genes.

Keywords EBV; Epstein-Barr virus · Gene expression · Replication · Assembly · Capsid · Tegument · Glycoprotein

18.1 Introduction

EBV lytic genes are categorized into Immediate-Early, Early, and Late genes and are expressed in a coordinated cascade manner. Two Immediate-Early genes induce expression of Early genes, followed by viral DNA synthesis. Then Late genes, encoding viral structural proteins and glycoproteins, are expressed, and viral nucleocapsids are assembled in the nucleus. The nucleocapsids bud into inner nuclear

---

T. Murata (✉)
Department of Virology, Nagoya University Graduate School of Medicine, Nagoya, Japan

Department of Virology and Parasitology, Fujita Health University School of Medicine, Toyoake, Japan
e-mail: tmurata@fujita-hu.ac.jp

membrane and take off the envelope at the outer nuclear membrane. After that, the nucleocapsids put on tegument proteins when they are enveloped again at a certain cytoplasmic membrane structure, presumably TGN or endosome (analogy from α-herpesvirus), which are then released into extracellular space.

Expression kinetics of lytic genes may seem clear, but in some cases, categorization of EBV genes is actually not very simple. Upon de novo infection, α- or β-herpesviruses choose lytic program by default, while EBV generally takes latent infection, and lytic program can be usually induced from latent state, indicating that the nuance of Immediate-Early gene is slightly different. Moreover, Late gene is typically accompanied with noncanonical TATA as mentioned later in this review, but a subset of Late genes bear canonical TATA, and some Early genes carry noncanonical TATA. Those genes may have to be categorized as Leaky-Late or Delayed-Early class, but such classification is not common in EBV. Otherwise, kinetics may have to be determined for each TATA but not for each gene if a gene has both canonical and noncanonical TATA motifs. So, I suggest kinetics of EBV lytic genes be taken only as a kind of rough indication.

EBV encodes >80 genes. Among them, EBNA1, 2, 3A, 3B, 3C, LP, LMP1, 2A, 2B, EBER1, 2, and microRNAs (and presumed ORFs in the *Bam*HI-A region) are accepted as latent genes. I here deal with other EBV genes as listed in the Table 18.1.

18.2 Activators of Viral Gene Expression

It is regarded that EBV encodes at least two Immediate-Early genes, BZLF1 (Zta, Z, ZEBRA, EB1) and BRLF1 (Rta, R). They are both molecular switches or triggers of EBV lytic cycle, because they are necessary and sufficient for lytic induction. We have an impression that the induction capacity of BZLF1 is more potent than that of BRLF1. These two Immediate-Early genes are positioned adjoiningly in the huge EBV genome, indicating that transcription of the Immediate-Early genes is epigenetically regulated together in the same window of relatively active chromatin where CpG methylation level is lower (Murata et al. 2013).

BZLF1 protein is a b-Zip-type transcription factor that can be modified by phosphorylation and SUMO (Sinclair 2003). It predominantly localizes to the nucleus in infected cells. The transcription factor preferentially forms homodimer and activates (or in some cases represses) viral and cellular promoters by recruiting a histone acetyltransferase, CREB-binding protein (CBP), and stabilizing TATA-binding protein (TBP) complex. It is of particular interest that BZLF1 protein preferentially binds a subset of CpG-methylated BZLF1-responsive elements rather than unmethylated ones (Bhende et al. 2004), which is very convincing because BZLF1 needs to activate lytic gene promoters that are deeply silenced by heavy CpG methylation. BZLF1 is essential for lytic replication (Feederle et al. 2000) and influences on transformation of primary B cells at least under a particular condition (Katsumura et al. 2012; Kalla et al. 2010). BZLF1 is reported to disrupt promyelocytic leukemia (PML) bodies like HSV ICP0, although in the ubiquitin-independent mechanism. In

18 Encyclopedia of EBV-Encoded Lytic Genes: An Update

Table 18.1 Lytic genes of EBV

Gene name	Alternative name	Kinetics	HSV	HCMV	KSHV	Properties
BNRF1	MTP, p140	L			ORF75	Involved in ingress and viral gene expression in infected B cells
BCRF1	vIL-10	L				Immune regulator, presumed to increase immortalization in vivo
BHLF1		E				Transcript is very abundant and involved in viral DNA synthesis, ORF is dispensable
BHRF1	vBcl2, EA-R p17	E				Dispensable for lytic replication, but likely protect cells from cell death under some conditions
BFLF2		E	UL31	UL53	ORF69	Interacts with BFRF1 and regulates nuclear egress of nucleocapsid
BFLF1		E	UL32	UL52	ORF68	Packaging of viral DNA
BFRF1A, BFRF0.5			UL33	UL51	ORF67A	Packaging of viral DNA
BFRF1		E	UL34	UL50	ORF67	Interacts with BFLF2 and regulates nuclear egress of nucleocapsid
BFRF2		E		UL49	ORF66	Transactivator for late genes
BFRF3	VCA-p18	L	UL35	UL48A	ORF65	Small capsid protein, attaches to hexon tips
BPLF1	LTP	L	UL36	UL48	ORF64	Large tegument protein, deubiquitinase/deneddylase
BOLF1	LTPBP	L	UL37	UL47	ORF63	Presumed to associate with BPLF1
BORF1	mCPBP	L	UL38	UL46	ORF62	One of the capsid triplex proteins
BORF2	RR large subunit, EA-R p85	E	UL39	UL45	ORF61	Large subunit of ribonucleotide reductase
BaRF1	RR small subunit, EA-R p30	E	UL40		ORF60	Small subunit of ribonucleotide reductase
BMRF1	EA-D, processivity factor	E	UL42	UL44	ORF59	Processibility subunit of DNA polymerase BALF5
BMRF2		E/L			ORF58	Membrane protein, attaches to integrins, interacts with BDLF2
BSLF2/BMLF1	EB2, SM, Mta	E	UL54	*UL69*	ORF57	Post-transcriptional regulator, produced by splicing bringing two possible ORFs together
BSLF1	Primase	E	UL52	UL70	ORF56	Primase for viral DNA synthesis
BSRF1	PalmP	L	UL51	UL71	ORF55	Tegument protein

(continued)

Table 18.1 (continued)

Gene name	Alternative name	Kinetics	HSV	HCMV	KSHV	Properties
BLLF3	dUTPase	E	UL50	UL72	ORF54	Hydrolyzes of dUTP to prevent misincorporation
BLRF1	gN	L	UL49.5	UL73	ORF53	Involved in assembly, egress and infection, interaction with gM
BLRF2		L			ORF52	Tegument protein
BLLF1a/b	gp350/220	L				Attaches to CD21 and CD35, knockout virus exhibited only mild effect, gp220 is spliced form
BZLF2	gp42	L				Forms complex with gH/gL and interact with HLA II, essential for B cell fusion
BZLF1	Zta, ZEBRA, EB1	IE			K8	Transcription factor, transactivation of lytic genes, oriLyt binding protein
BRLF1	Rta	IE			ORF50	Transcriptional activator
BRRF1	Na	E			ORF49	Dispensable for lytic replication
BRRF2		L			ORF48	Tegument, disruption affects mildly on virus production
BKRF2	gL, gp25	L	UL1	UL115	ORF47	Forms complex with gH (and gp42) and interact with integrins (and HLA II), essential for fusion
BKRF3	Uracil-DNA glycosylase	E	UL2	UL114	ORF46	Removes misincorporated uracil from DNA, needed for efficient viral DNA synthesis
BKRF4		L			ORF45	Tegument protein, involved in progeny production
BBLF4	Helicase	E	UL5	UL105	ORF44	Helicase for viral DNA synthesis
BBRF1	Portal	L	UL6	UL104	ORF43	Portal protein through which viral DNA genome is packaged into capsid
BBRF2		L	UL7	UL103	ORF42	Possible tegument protein
BBLF2/3	Primase-binding protein	E	UL8	UL102	ORF41	Binds to BSLF1 primase, produced by splicing joining two possible ORFs
BBRF3	gM	L	UL10	UL100	ORF39	Interacts with gN, needed for authentic processing of gN
BBLF1	MyrP	L	UL11	UL99	ORF38	Myristoylated and palmitoylated tegument protein, involved in virus particle production
BGLF5	DNase	E	UL12	UL98	ORF37	Alkaline exonuclease, host shutoff
BGLF4	PK	E	UL13	U97	ORF36	Phosphorylates viral and cellular proteins, and ganciclovir, involved in progeny production

Gene	Alt name	Class	UL	ORF	Description
BGLF3.5		E	UL14	ORF35	Possible tegument protein, dispensable for lytic replication
BGLF3		E		ORF34	Transactivator for late genes
BGRF1/BDRF1	Terminase small subunit	L	UL15	ORF29	Terminase small subunit, produced by splicing joining two possible ORFs
BGLF2	MyrPBP	L	UL16	ORF33	Tegument protein, activates AP1, involved in progeny production
BGLF1		L	UL17	ORF32	Tegument protein, presumably binds to cork (BVRF1)
BDLF4		E		ORF31	Transactivator for late genes
BDLF3.5		E		UL91	Transactivator for late genes
BDLF3	gp150, gp117	L		ORF28	Dispensable for lytic replication, inhibits antigen presentation by HLA I, II, and CD1d
BDLF2		L		ORF27	Detected in tegument, type II membrane protein, associates BMRF2
BDLF1	mCP	L	UL18	ORF26	Minor capsid protein, one of the capsid triplex proteins
BcLF1	MCP	L	UL19	ORF25	Major capsid protein, forms hexons and pentons
BcRF1	Noncanonical TBP	E		ORF24	Transactivator for late genes, noncanonical TATA (TATT)-binding protein
BTRF1		L		UL88	Dispensable in MHV-68
BXLF2	gH, gp85	L	UL22	ORF22	Forms complex with gL (and gp42) and interacts with integrins (and HLA II), essential for entry
BXLF1	TK	E	UL23	ORF21	Thymidine kinase, dispensable for lytic replication
BXRF1		L	UL24	ORF20	Not reported, dispensable in MHV-68
BVRF1	Portal plug, cork	E/L	UL25	ORF19	Cork protein that confines viral genome DNA in the capsid
BVLF1	BVLF1.5	E		ORF18	Transactivator for late genes
BVRF2	Protease/scaffold	L	UL26	ORF17	Scaffold protein with protease activity
BdRF1	VCA-p40, scaffold	L	UL26.5	ORF17.5	Scaffold protein, expressed from C-terminal part of BVRF2, does not retain protease activity
BILF2	gp55/80, gp78	L			Glycoprotein
LF3		E			Transcript is very abundant and presumably involved in viral DNA synthesis

(continued)

Table 18.1 (continued)

Gene name	Alternative name	Kinetics	HSV	HCMV	KSHV	Properties
LF2						Represses lytic replication, regulate interferon signaling
LF1						Not reported
BILF1	gp64, GPCR	L				Viral G protein-coupled receptor, immune evasion
BALF5	DNA polymerase	E	UL30	UL54	ORF9	Viral DNA polymerase catalytic subunit
BALF4	gB, gP110, gp125	L	UL27	UL55	ORF8	Essential for fusion, also involved in assembly and budding of the virion
BALF3	Terminase large subunit	E	UL28	UL56	ORF7	Terminase large subunit
BALF2	ssDNABP	E	UL29	UL57	ORF6	Single-stranded DNA-binding protein, essential for lytic replication
BALF1	vBcl2	E				Dispensable for lytic replication, but likely protect cells from cell death under some conditions
BARF1		E				Soluble decoy for CSF-1, a role in cell immortalization
BNLF2b		E				Not reported
BNLF2a		E				Downregulates HLA through inhibition of TAP

addition, BZLF1 also functions as an origin-binding protein for lytic DNA replication.

BRLF1 is another transcriptional activator essential for EBV lytic cycle (Feederle et al. 2000), which can be sumoylated, but its mode of activation is more complicated and diverse than that of BZLF1. BRLF1 activates transcription of target genes by binding directly to BRLF1-responsive elements through its N-terminal DNA-binding domain, by indirectly binding to DNA through interaction with other transcription factors, and even by enhancing certain signal transduction pathways, such as MAPK. BRLF1 is also able to associate with CBP and TBP complex.

In addition to the two transcriptional activators, an RNA-binding protein BSLF2/BMLF1 (SM, EB2) is an essential posttranscriptional regulator (Gruffat et al. 2002). SM protein is derived from the spliced mRNA, joining a part of BSLF2 into BMLF1. It reinforces posttranscriptionally a subset of viral gene expression and has positive and negative effects on cellular gene expression by modulating nuclear export, splicing, processing, and stability of mRNAs.

18.3 Nucleotide Metabolism and DNA Replication

Lytic EBV genome multiplication occurs at discrete sites in the nucleus, and such scattered spots later merge and grow into a large size, typically shaping a horseshoe. Such nuclear architecture is called replication compartments, where viral genome DNA and a number of both viral and cellular factors accumulate (Tsurumi et al. 2005). EBV lytic DNA replication starts from two particular origins called oriLyt (Hammerschmidt and Sugden 1988). B95-8 and P3HR-1 viruses lack either of the two lytic origins, but the viruses can still replicate efficiently, indicating redundancy. By using cotransfection assays, at least seven viral gene products, BZLF1, BALF5 (Pol), BMRF1 (EA-D, early antigen, diffuse, processivity factor), BALF2 (ssDNABP, single-stranded DNA-binding protein), BBLF4 (helicase), BSLF1 (primase), and BBLF2/3 (primase-associating factor), were identified to be essential for DNA replication from the oriLyt (Fixman et al. 1992; Schepers et al. 1993).

The lytic switch transcription factor BZLF1 binds to the BZLF1-responsive elements in the oriLyt and recruits those necessary factors through molecular associations (Fixman et al. 1995; Schepers et al. 1993).

BALF5 protein is a catalytic subunit of DNA polymerase, which interacts with the accessory factor BMRF1 and helicase/primase complex. Viral DNA synthesis does not initiate in the BALF5 knockout virus. It has a regulatory domain in the N-terminus and highly conserved DNA polymerase and exonuclease domains in the C-terminal half. This polymerase is used in the synthesis of both leading and lagging strands at the replication fork (Tsurumi et al. 1994), unlike replication machinery of the host cells. When expressed alone, BALF5 protein accumulates in the cytoplasm, but it can be efficiently transported to the nucleus when expressed with BMRF1 (Kawashima et al. 2013).

BMRF1 is an essential, extensively phosphorylated, abundant nuclear protein that associates with and enhances the polymerase processivity and exonuclease activity of BALF5. It is similar to host PCNA, in terms of function, structure, and homo-oligomer formation (Murayama et al. 2009; Nakayama et al. 2010). Some reports indicate that BMRF1 functions as a transcriptional coactivator, too.

Another abundant replication factor BALF2, a ssDNABP, is also essential for viral genome replication. It augments viral DNA replication provably through reducing pause of the polymerase by keeping the single-stranded DNA template into the optimal conformation for DNA elongation (Tsurumi et al. 1998).

BBLF4, BSLF1, and BBLF2/3 form helicase/primase complex and associate with BALF5. The helicase presumably unwinds double-stranded DNA ahead of the replication fork, providing open structure for DNA synthesis. The primase complex catalyzes synthesis of an RNA primer for the leading strand, which is used as a primer for polymerase reaction. For the lagging strand, discontinuous RNA primer fragments are synthesized at intervals, and the gap is filled up by the polymerase. Such RNA primers are destined for degradation by exonuclease activity and are filled up with DNA.

In addition, viral uracil DNA glycosylase encoded by BKRF3 gene is also essential for efficient lytic DNA synthesis (Fixman et al. 1995; Su et al. 2014). When deamination of cytosine to uracil occurs or uracil is somehow mis-incorporated into DNA, BKRF3 catalyzes removal of the uracil from DNA, which can be repaired later. It localizes in the cytoplasm when expressed alone, but coexpression of BMRF1 facilitates nuclear localization.

Besides such essential replication proteins, the transcript of BHLF1 gene (and possibly its paralog LF3) forms stable RNA-DNA hybrid at oriLyt, which appears essential for initiation of lytic DNA replication (Rennekamp and Lieberman 2011). The BHLF1 and LF3 genes are adjacent to each oriLyt, and their highly GC-rich mRNAs are most abundantly expressed. Disruption of BHLF1 ORF or mutation of the initiation codon had no effect on replication, but mutation of the TATA signal significantly reduced viral DNA synthesis. The BHLF1 and LF3 genes potentially encode highly repetitive proteins, and despite abundance of the transcripts, presence of the proteins is still controversial.

EBV thymidine kinase is encoded by BXLF1 gene. The enzyme is the major player of the salvage pathway of pyrimidine metabolism by phosphorylating thymidine into thymidine monophosphate. The gene is not essential for lytic replication at least in rapidly growing cells like HEK293T (Meng et al. 2010), but may play an important role in DNA replication in nondividing cells, such as memory B cells, in which dNTP pool is limited.

Ribonucleotide (diphosphate) reductase (RR, RNR) of EBV is composed of two subunits: BORF2 and BaRF1 proteins. The complex catalyzes de novo synthesis of deoxyribonucleotides from ribonucleotides by reducing NDPs in order to increase the dNTP pool.

The BLLF3 gene encodes deoxyuridine nucleotidohydrolase (dUTPase). This enzyme catalyzes hydrolysis of dUTP into dUMP and thereby prevents dUTP incorporation into DNA. In addition, EBV dUTPase was reported to elicit innate immune response by activating NF-κB (Ariza et al. 2009).

18.4 Regulators of Late Gene Transcription

β- and γ-herpesviruses have developed a common mechanism of viral Late gene expression. The promoter of Late gene commonly has noncanonical TATA box (TATT), and viral Late gene regulator complex, composed of BcRF1, BDLF3.5, BDLF4, BFRF2, BGLF3, and BVLF1, which binds to the noncanonical TATA and perhaps recruits factors that induce Late gene transcription (Gruffat et al. 2012; Watanabe et al. 2015b; Aubry et al. 2014; Djavadian et al. 2016). The mechanism of how the complex activates Late gene transcription remains unclear, but BcRF1 is thought to be the noncanonical TBP, and some of the components or the homologs reportedly interact with RNA polymerase II, thereby enhancing transcription from the noncanonical TATA.

18.5 Other Regulatory Genes

EBV encodes two homologs of Bcl-2, an anti-apoptotic cellular protein: BHRF1 and BALF1. They are expressed in the lytic cycle with Early kinetics and thus speculated to protect lytic cells from apoptosis, but not needed for viral replication in cell culture. Upon primary infection, they are expressed transiently before establishment of latency (termed "pre-latent abortive lytic" state), and double knockout of the two vBcl-2 genes resulted in significant reduction of B cell transformation efficiency although knockout of either of them exhibited no obvious phenotype (Altmann and Hammerschmidt 2005). In "Wp-restricted Burkitt lymphoma" cells, in which EBNA2 is deleted, BHRF1 is expressed from the Wp and plays a pivotal role in survival and proliferation of the host (Kelly et al. 2009).

The BARF1 gene encodes soluble, phosphorylated, and glycosylated protein, expressed with Early kinetics. It can bind with colony-stimulating factor 1 (CSF-1) although structural analysis indicated that the mode of CSF-1 binding to BARF1 has to be principally different from that to CSF-1 receptor. BARF1 serves as a decoy for CSF-1, which is known to elicit macrophage proliferation and IFN-α production. Therefore, BARF1 can inhibit immune response against EBV. This gene is dispensable for lytic replication of the virus and B cell transformation (Cohen and Lekstrom 1999) but may be involved in efficient transformation, possibly in vivo. It was also reported that, when overexpressed, BARF1 induced morphological change, anchorage-independent growth, and tumorigenic transformation (Wei and Ooka 1989; Sheng et al. 2003). In addition, this gene is expressed in some epithelial and B cells latently infected with EBV.

A seven-transmembrane G protein-coupled receptor (GPCR) of EBV is encoded by the BILF1 gene. It has similarity to CXCR4, forms heterodimers with cellular GPCRs, and alters signaling pathways (Nijmeijer et al. 2010; Vischer et al. 2008). It was also reported to transform NIH3T3 cells when exogenously expressed. In addition, BILF1 was found to play a role in immune evasion by targeting HLA class I for enhanced turnover and lysosomal degradation (Zuo et al. 2009).

BCRF1 (vIL-10) protein is highly similar to an anti-inflammatory cytokine, IL-10. On one hand, vIL-10 enhances B cell proliferation. On the other hand, it serves to repress Th1 immunity. It is not essential for virus replication nor B cell transformation at cell culture levels (Swaminathan et al. 1993), but may be involved in immune regulation or B cell growth in vivo. It was also reported that vIL-10 is expressed immediately after infection during pre-latent phase and impairs NK and T-cell activities.

BNLF2a encodes tail-anchored polypeptide of only 60 residues unique to lymphocryptoviruses. BNLF2a also is involved in immune evasion; it downregulates HLA class I-dependent antigen presentation through blocking peptide transport by TAP (Hislop et al. 2007). Knockout of this gene seemingly had no effect on lytic replication and B cell transformation, but viral Immediate-Early and Early antigens are recognized more efficiently by CD8+ T-cells. In addition, BNLF2a could also be expressed during pre-latent phase upon de novo infection and reduces antigen presentation to EBV-specific CD8+ T-cells. Double knockout of vIL-10 and BNLF2a therefore had synergistic effect on protection from innate and acquired immune responses after de novo infection (Jochum et al. 2012). Moreover, BNLF2a is produced even during latent phase at least in gastric carcinoma cells. On the other hand, BNLF2b, a downstream ORF of BNLF2a, has not been reported yet.

EBV expresses a protein kinase, encoded by BGLF4. BGLF4 mRNA translation is intricately regulated presumably by termination/reinitiation process, in which short upstream ORFs of BGLF3.5 play critical roles (Watanabe et al. 2015a). This viral serine/threonine kinase predominantly localizes to the nucleus and can be incorporated into the tegument of virion. Targets of the kinase include a number of viral and cellular proteins, such as BGLF4 itself, BMRF1, BZLF1, EBNA-LP, EBNA2, EBNA1, EF-1δ, IRF3, MCM, p27Kip1, stathmin, lamins, Tip60, and even nucleotide analogs, ganciclovir and maribavir. The PK is also reported to induce chromosome condensation and DNA damage response. Analysis of phosphorylation motifs of BGLF4 target proteins indicates that it has similarity to cellular Cdk1/Cdc2. Knockout virus of the BGLF4 gene exhibited reduced viral DNA replication; production of viral genes, especially Late genes; and maturation of progeny virus (Feederle et al. 2009b; Murata et al. 2009).

Located posterior to the BGLF4 gene is BGLF5, encoding deoxyribonuclease (DNase) that localizes predominantly in the nucleus and exhibits nuclease activities with alkaline pH preference. Disruption of the BGLF5 gene caused moderate inhibition of progeny production (Feederle et al. 2009a). Electrophoresis of the viral genome and results from other herpesvirus counterparts suggested that it is involved in trimming of complicated, entangled structure of viral DNA for better accessibility of capsid/terminase complex. It should also be noted that BGLF5 protein has a host shutoff function by destabilizing mRNAs, including HLA I, HLA II, and TLR9, thus contributing to evasion from immunity (Rowe et al. 2007). In addition, BGLF5 enhances nuclear translocalization of cytoplasmic poly(A)-binding protein, and its counterparts in KSHV and MHV-68 induce hyperpolyadenylation and retention of mRNAs in the nucleus (Park et al. 2014; Kumar and Glaunsinger 2010).

BPLF1 is the largest protein of EBV, incorporated into tegument (thus also named as LTP, large tegument protein). It localizes both cytoplasm and nucleus, and N-terminal part is efficiently transported to the nucleus after cleavage. In HSV, its homolog UL36 interacts with UL37 (homolog of BOLF1) and is involved in the transport and secondary envelopment of the nucleocapsid and correct formation of capsid/tegument interface at the capsid vertex. Although such functions have not been tested yet for EBV BPLF1, knockout of this gene caused slight decrease of viral DNA replication and significant loss of progeny production and B cell transformation (Saito et al. 2013; Whitehurst et al. 2015). Besides its structural function as a tegument protein, N-terminal part of BPLF1 retains deubiquitinase/deneddylase activity (Gastaldello et al. 2010). Targets of the enzyme identified to date include ubiquitinated BORF2, PCNA, TRAF6, Rad6/18, and neddylated cullins.

The BFLF2 and BFRF1, EBV homologs of HSV UL31 and UL34, respectively, alter morphology of inner nuclear meshwork composed of lamins and support primary budding of nucleocapsids into the inner nuclear membrane. BFLF2 and BFRF1 are phosphoproteins and are not incorporated into progeny virus particles. When expressed alone, BFLF2 protein localizes diffusely in the nucleus, and BFRF1 is located in the cytoplasm and perinuclear region, but when expressed together, they form a complex and colocalize at the nuclear rim with lamins. Knockout of either of the genes notably decreased progeny viral yield (Farina et al. 2005; Gonnella et al. 2005).

18.6 Capsid and Related Proteins

At least five viral proteins could be detected in the mature EBV capsid: BcLF1 (MCP, major capsid protein), BDLF1 (mCP, minor capsid protein), BORF1 (mCPBP, mCP-binding protein), BFRF3 (small capsid protein, VCA-p18), and BBRF1 (Heilmann et al. 2012; Johannsen et al. 2004). One capsid is composed of 162 capsomers (12 pentons and 150 hexons). Six molecules of MCP make up a hexon. Five copies of MCP form a penton and pentons are settled at the icosahedral vertices. A homo-oligomer composed of 12 of portal proteins configures a ring at a unique vertex, through which viral genome DNA enters. Two molecules of BDLF1 protein and one BORF1 protein make up a triplex, and 320 triplex complexes lie on the capsid floor, connecting capsomers. Small capsid protein BFRF3 is attached to the distal tips of the hexon MCPs.

For initial self-assembly of EBV capsid, two scaffold proteins BVRF2 (protease/scaffold) and BdRF1 (scaffold, VCA-p40) are needed to solidify the capsid structure (Henson et al. 2009). BdRF1 protein is identical to the C-terminal part of BVRF2 ORF; it is expressed from downstream initiation codon of BVRF2. Self-cleavage of BVRF2 and BdRF1 triggers collapse and ejection of the scaffold proteins, which is coupled to internalization of EBV genome DNA through the portal complex.

The BALF3 and BGRF1/BDRF1 genes encode terminase large and small subunits, respectively, that cleave newly synthesized viral DNA into unit length upon incorporation of the viral genome through the portal (Chiu et al. 2014).

Two EBV proteins BFLF1 and BFRF1A (BFRF0.5) are involved in packaging of viral DNA into capsid (Pavlova et al. 2013).

18.7   Tegument Proteins

At least 13 viral tegument proteins could be detected from purified EBV particles: BVRF1 (cork, portal plug), BGLF1, BNRF1 (MTP, major tegument protein, p140), BPLF1 (LTP), BOLF1 (LTPBP, LTP-binding protein), BGLF4 PK, BBLF1 (MyrP, myristoylated protein), BGLF2 (MyrPBP, MyrP-binding protein), BSRF1 (PalmP, palmitoylated protein), BKRF4, BDLF2, BRRF2, and BLRF2 (Johannsen et al. 2004).

In HSV, UL25 and UL17 (homolog of BVRF1 cork and BGLF1) form a complex, attach to capsid vertices, and are required for stable retention of viral DNA.

BNRF1 appeared dispensable for viral progeny production in HEK293 but is required for efficient transportation of virus particle to the nucleus, and it supports viral gene expression through interaction with Daxx (Feederle et al. 2006; Tsai et al. 2011).

Being a homolog of HSV UL37, BOLF1 is assumed to associate with BPLF1 and incorporated into tegument. Deletion of murid herpesvirus 4 (MuHV-4) ORF63, homolog of BOLF1, hampered nuclear transportation of incoming nucleocapsids (Latif et al. 2015).

BBLF1 protein is myristoylated and palmitoylated, like HSV counterpart UL11, and localizes to TGN, where secondary envelopment takes place. Knockdown of BBLF1 caused reduction in virus particle production (Chiu et al. 2012). BGLF2 is speculated to associate with BBLF1, an analogy of HSV counterparts, UL16 and UL11. Reporter assays and knockdown of BGLF2 revealed that it promotes EBV reactivation by activating p38 MAPK pathway (Liu and Cohen 2016). Knockout of BGLF2 caused reduced progeny production in HEK293.

BSRF1 has not been reported yet, but its HSV homolog UL51 is palmitoylated and involved in maturation and egress of virus particles.

Our experiments indicate that BKRF4 is needed for efficient progeny production, and its interaction with BGLF2 is important for the role (Masud et al. 2017).

Although the BDLF2 gene product has been described as a tegument protein, recent reports indicate that it is a glycosylated, type II membrane protein, whose authentic localization and processing are dependent on the association with another membrane protein BMRF2 (Gore and Hutt-Fletcher 2009).

BRRF2 is a phosphorylated protein and localized in the cytoplasm. Disruption of this gene did not affect viral gene expression and DNA replication, but mildly decreased virus production (Watanabe et al. 2015c).

EBV BLRF2 has not been characterized well. Progeny production of MHV-68 ORF52 (homolog of BLRF2)-null mutant virus was significantly impaired, and the reduced titer of MHV-68 knockout could be rescued by ectopic supply of EBV BLRF2 (Duarte et al. 2013).

In the case of HSV, in addition to the homologs of those proteins listed above, UL7 and UL14 are identified as tegument proteins, although their homologs of EBV, BBRF2 and BGLF3.5, respectively, could not be detected. BBRF2 has not been reported yet, but its counterpart in MHV-68 (ORF42) appeared essential (Song et al. 2005). Disruption of BGLF3.5 exhibited no phenotype, compared to wild-type (Watanabe et al. 2015a), but its counterpart in MHV-68 (ORF35) was reported to be essential.

18.8 Glycoproteins

Mature EBV virion envelope is associated with membrane proteins, such as BLLF1a/b (gp350/220), BALF4 (gB, gp110), BXLF2 (gH, gp85), BKRF2 (gL, gp25), BZLF2 (gp42), BILF2 (gp78), BDLF3 (gp150), BBRF3 (gM), and BLRF1 (gN) (O'Regan et al. 2010; Johannsen et al. 2004).

EBV gp350 (and its spliced variant gp220) bind to CD21/CR2 and CD35/CR1, which initiates attachment of the virus to cell surface. Disruption of gp350 had only mild effect on infection to B cells, indicating presence of gp350-independent infection mechanism (Janz et al. 2000).

It is also reported that the BMRF2, a membrane glycoprotein that interacts with αv, $\alpha 3$, $\alpha 5$, and $\beta 1$ integrins, also reinforces viral attachment to cells.

Among the envelope glycoproteins, gB, gH, and gL are needed for fusion with epithelial cells, whereas gp42 is also required for B cell fusion. Receptor for gB in B cells has not been identified yet, but gB interaction with neuropilin 1 was reported recently to enhance entry of EBV to nasopharyngeal epithelial cells (Wang et al. 2015). It is also speculated that EBV gB plays a role in assembly and budding of the virion. The gH/gL complex is able to bind integrins $\alpha v \beta 5$, $\alpha v \beta 6$, or $\alpha v \beta 8$, and gH/gL/gp42 complex binds to HLA II, both of which trigger membrane fusion.

BILF2 gene product gp78 has not been studied well.

gM and gN form a complex and coexpression of gM is needed for authentic processing of gN. Disruption of gN caused impairment in egress, appropriate envelopment, and nucleocapsid release from the envelope upon de novo infection (Lake and Hutt-Fletcher 2000).

Highly glycosylated protein gp150 is encoded by the BDLF3 gene. Receptor for this glycoprotein is unknown, and EBV lacking the gene did not exhibit any defect (Borza and Hutt-Fletcher 1998). Novel role of gp150 as an immune regulator has been reported; it appears to inhibit antigen presentation by HLA I, HLA II, and CD1d (Gram et al. 2016; Quinn et al. 2016).

18.9 Other Poorly Studied Lytic Genes

An Early gene BRRF1 (Na) is located in the splicing intron of BRLF1. According to an analogy from its homolog in KSHV and MHV-68, BRRF1 has been implicated with transcriptional activation of viral genes; however, our advertent experiments using BRRF1 knockout EBV indicate that this gene is dispensable for viral lytic replication at least in HEK293 (Yoshida et al. 2017).

Three potential ORFs, LF1, 2, and 3, lie in the *Bam*HI-I region, and there is one of the two copies of oriLyt between LF2 and 3. Transcript of LF3 is assumed to play a role in initiation of viral DNA synthesis as mentioned above. LF2 was shown to inhibit lytic replication possibly through interaction with BRLF1 and regulate type I interferon signaling. LF1 has not been reported yet.

The BLLF2, BTFR1, and BXRF1 genes have not been reported to date. Homologs of BTRF1 (ORF23) and BXRF1 (ORF20) in MHV-68 are reportedly nonessential for viral replication.

18.10 Conclusions

Significant number of lytic genes has been left unidentified or uncharacterized. The standard method for such analysis is to prepare knockout viruses, but construction and phenotyping of recombinant EBV take long time and are tricky. Creation of a new efficient strategy after B95-8 BAC system, such as CRISPR/Cas9 system (Masud et al. 2017), is desired. Exhaustive understanding of EBV genes may provide novel targets for development of drug or vaccine.

References

Altmann M, Hammerschmidt W (2005) Epstein-Barr virus provides a new paradigm: a requirement for the immediate inhibition of apoptosis. PLoS Biol 3(12):e404. https://doi.org/10.1371/journal.pbio.0030404

Ariza ME, Glaser R, Kaumaya PT, Jones C, Williams MV (2009) The EBV-encoded dUTPase activates NF-kappa B through the TLR2 and MyD88-dependent signaling pathway. J Immunol 182(2):851–859

Aubry V, Mure F, Mariame B, Deschamps T, Wyrwicz LS, Manet E, Gruffat H (2014) Epstein-Barr virus late gene transcription depends on the assembly of a virus-specific preinitiation complex. J Virol 88(21):12825–12838. https://doi.org/10.1128/JVI.02139-14

Bhende PM, Seaman WT, Delecluse HJ, Kenney SC (2004) The EBV lytic switch protein, Z, preferentially binds to and activates the methylated viral genome. Nat Genet 36(10):1099–1104. https://doi.org/10.1038/ng1424

Borza CM, Hutt-Fletcher LM (1998) Epstein-Barr virus recombinant lacking expression of glycoprotein gp150 infects B cells normally but is enhanced for infection of epithelial cells. J Virol 72(9):7577–7582

Chiu YF, Sugden B, Chang PJ, Chen LW, Lin YJ, Lan YC, Lai CH, Liou JY, Liu ST, Hung CH (2012) Characterization and intracellular trafficking of Epstein-Barr virus BBLF1, a protein involved in virion maturation. J Virol 86(18):9647–9655. https://doi.org/10.1128/JVI.01126-12

Chiu SH, Wu MC, Wu CC, Chen YC, Lin SF, Hsu JT, Yang CS, Tsai CH, Takada K, Chen MR, Chen JY (2014) Epstein-Barr virus BALF3 has nuclease activity and mediates mature virion production during the lytic cycle. J Virol 88(9):4962–4975. https://doi.org/10.1128/JVI.00063-14

Cohen JI, Lekstrom K (1999) Epstein-Barr virus BARF1 protein is dispensable for B-cell transformation and inhibits alpha interferon secretion from mononuclear cells. J Virol 73(9):7627–7632

Djavadian R, Chiu YF, Johannsen E (2016) An Epstein-Barr virus-encoded protein complex requires an origin of lytic replication in cis to mediate late gene transcription. PLoS Pathog 12(6):e1005718. https://doi.org/10.1371/journal.ppat.1005718

Duarte M, Wang L, Calderwood MA, Adelmant G, Ohashi M, Roecklein-Canfield J, Marto JA, Hill DE, Deng H, Johannsen E (2013) An RS motif within the Epstein-Barr virus BLRF2 tegument protein is phosphorylated by SRPK2 and is important for viral replication. PLoS One 8(1):e53512. https://doi.org/10.1371/journal.pone.0053512

Farina A, Feederle R, Raffa S, Gonnella R, Santarelli R, Frati L, Angeloni A, Torrisi MR, Faggioni A, Delecluse HJ (2005) BFRF1 of Epstein-Barr virus is essential for efficient primary viral envelopment and egress. J Virol 79(6):3703–3712. https://doi.org/10.1128/JVI.79.6.3703-3712.2005

Feederle R, Kost M, Baumann M, Janz A, Drouet E, Hammerschmidt W, Delecluse HJ (2000) The Epstein-Barr virus lytic program is controlled by the co-operative functions of two transactivators. EMBO J 19(12):3080–3089. https://doi.org/10.1093/emboj/19.12.3080

Feederle R, Neuhierl B, Baldwin G, Bannert H, Hub B, Mautner J, Behrends U, Delecluse HJ (2006) Epstein-Barr virus BNRF1 protein allows efficient transfer from the endosomal compartment to the nucleus of primary B lymphocytes. J Virol 80(19):9435–9443. https://doi.org/10.1128/JVI.00473-06

Feederle R, Bannert H, Lips H, Muller-Lantzsch N, Delecluse HJ (2009a) The Epstein-Barr virus alkaline exonuclease BGLF5 serves pleiotropic functions in virus replication. J Virol 83(10):4952–4962. https://doi.org/10.1128/JVI.00170-09

Feederle R, Mehl-Lautscham AM, Bannert H, Delecluse HJ (2009b) The Epstein-Barr virus protein kinase BGLF4 and the exonuclease BGLF5 have opposite effects on the regulation of viral protein production. J Virol 83(21):10877–10891. https://doi.org/10.1128/JVI.00525-09

Fixman ED, Hayward GS, Hayward SD (1992) trans-acting requirements for replication of Epstein-Barr virus ori-Lyt. J Virol 66(8):5030–5039

Fixman ED, Hayward GS, Hayward SD (1995) Replication of Epstein-Barr virus oriLyt: lack of a dedicated virally encoded origin-binding protein and dependence on Zta in cotransfection assays. J Virol 69(5):2998–3006

Gastaldello S, Hildebrand S, Faridani O, Callegari S, Palmkvist M, Di Guglielmo C, Masucci MG (2010) A deneddylase encoded by Epstein-Barr virus promotes viral DNA replication by regulating the activity of cullin-RING ligases. Nat Cell Biol 12(4):351–361. https://doi.org/10.1038/ncb2035

Gonnella R, Farina A, Santarelli R, Raffa S, Feederle R, Bei R, Granato M, Modesti A, Frati L, Delecluse HJ, Torrisi MR, Angeloni A, Faggioni A (2005) Characterization and intracellular localization of the Epstein-Barr virus protein BFLF2: interactions with BFRF1 and with the nuclear lamina. J Virol 79(6):3713–3727. https://doi.org/10.1128/JVI.79.6.3713-3727.2005

Gore M, Hutt-Fletcher LM (2009) The BDLF2 protein of Epstein-Barr virus is a type II glycosylated envelope protein whose processing is dependent on coexpression with the BMRF2 protein. Virology 383(1):162–167. https://doi.org/10.1016/j.virol.2008.10.010

Gram AM, Oosenbrug T, Lindenbergh MF, Bull C, Comvalius A, Dickson KJ, Wiegant J, Vrolijk H, Lebbink RJ, Wolterbeek R, Adema GJ, Griffioen M, Heemskerk MH, Tscharke DC, Hutt-Fletcher LM, Wiertz EJ, Hoeben RC, Ressing ME (2016) The Epstein-Barr virus glycoprotein gp150 forms an immune-evasive glycan shield at the surface of infected cells. PLoS Pathog 12(4):e1005550. https://doi.org/10.1371/journal.ppat.1005550

Gruffat H, Batisse J, Pich D, Neuhierl B, Manet E, Hammerschmidt W, Sergeant A (2002) Epstein-Barr virus mRNA export factor EB2 is essential for production of infectious virus. J Virol 76(19):9635–9644

Gruffat H, Kadjouf F, Mariame B, Manet E (2012) The Epstein-Barr virus BcRF1 gene product is a TBP-like protein with an essential role in late gene expression. J Virol 86(11):6023–6032. https://doi.org/10.1128/JVI.00159-12

Hammerschmidt W, Sugden B (1988) Identification and characterization of oriLyt, a lytic origin of DNA replication of Epstein-Barr virus. Cell 55(3):427–433

Heilmann AM, Calderwood MA, Portal D, Lu Y, Johannsen E (2012) Genome-wide analysis of Epstein-Barr virus Rta DNA binding. J Virol 86(9):5151–5164. https://doi.org/10.1128/JVI.06760-11

Henson BW, Perkins EM, Cothran JE, Desai P (2009) Self-assembly of Epstein-Barr virus capsids. J Virol 83(8):3877–3890. https://doi.org/10.1128/JVI.01733-08

Hislop AD, Ressing ME, van Leeuwen D, Pudney VA, Horst D, Koppers-Lalic D, Croft NP, Neefjes JJ, Rickinson AB, Wiertz EJ (2007) A CD8+ T cell immune evasion protein specific to Epstein-Barr virus and its close relatives in Old World primates. J Exp Med 204(8):1863–1873. https://doi.org/10.1084/jem.20070256

Janz A, Oezel M, Kurzeder C, Mautner J, Pich D, Kost M, Hammerschmidt W, Delecluse HJ (2000) Infectious Epstein-Barr virus lacking major glycoprotein BLLF1 (gp350/220) demonstrates the existence of additional viral ligands. J Virol 74(21):10142–10152

Jochum S, Moosmann A, Lang S, Hammerschmidt W, Zeidler R (2012) The EBV immunoevasins vIL-10 and BNLF2a protect newly infected B cells from immune recognition and elimination. PLoS Pathog 8(5):e1002704. https://doi.org/10.1371/journal.ppat.1002704

Johannsen E, Luftig M, Chase MR, Weicksel S, Cahir-McFarland E, Illanes D, Sarracino D, Kieff E (2004) Proteins of purified Epstein-Barr virus. Proc Natl Acad Sci U S A 101(46):16286–16291. https://doi.org/10.1073/pnas.0407320101

Kalla M, Schmeinck A, Bergbauer M, Pich D, Hammerschmidt W (2010) AP-1 homolog BZLF1 of Epstein-Barr virus has two essential functions dependent on the epigenetic state of the viral genome. Proc Natl Acad Sci U S A 107(2):850–855. https://doi.org/10.1073/pnas.0911948107

Katsumura KR, Maruo S, Takada K (2012) EBV lytic infection enhances transformation of B-lymphocytes infected with EBV in the presence of T-lymphocytes. J Med Virol 84(3):504–510. https://doi.org/10.1002/jmv.23208

Kawashima D, Kanda T, Murata T, Saito S, Sugimoto A, Narita Y, Tsurumi T (2013) Nuclear transport of Epstein-Barr virus DNA polymerase is dependent on the BMRF1 polymerase processivity factor and molecular chaperone Hsp90. J Virol 87(11):6482–6491. https://doi.org/10.1128/JVI.03428-12

Kelly GL, Long HM, Stylianou J, Thomas WA, Leese A, Bell AI, Bornkamm GW, Mautner J, Rickinson AB, Rowe M (2009) An Epstein-Barr virus anti-apoptotic protein constitutively expressed in transformed cells and implicated in burkitt lymphomagenesis: the Wp/BHRF1 link. PLoS Pathog 5(3):e1000341. https://doi.org/10.1371/journal.ppat.1000341

Kumar GR, Glaunsinger BA (2010) Nuclear import of cytoplasmic poly(A) binding protein restricts gene expression via hyperadenylation and nuclear retention of mRNA. Mol Cell Biol 30(21):4996–5008. https://doi.org/10.1128/MCB.00600-10

Lake CM, Hutt-Fletcher LM (2000) Epstein-Barr virus that lacks glycoprotein gN is impaired in assembly and infection. J Virol 74(23):11162–11172

Latif MB, Machiels B, Xiao X, Mast J, Vanderplasschen A, Gillet L (2015) Deletion of murid herpesvirus 4 ORF63 affects the trafficking of incoming capsids toward the nucleus. J Virol 90(5):2455–2472. https://doi.org/10.1128/JVI.02942-15

Liu X, Cohen JI (2016) Epstein-Barr virus (EBV) tegument protein BGLF2 promotes EBV reactivation through activation of the p38 mitogen-activated protein kinase. J Virol 90(2):1129–1138. https://doi.org/10.1128/JVI.01410-15

Masud HMA, Watanabe T, Yoshida M, Sato Y, Goshima F, Kimura H, Murata T (2017) Epstein-Barr virus BKRF4 gene product is required for efficient progeny production. J Virol 91(23):e00975-17

Meng Q, Hagemeier SR, Fingeroth JD, Gershburg E, Pagano JS, Kenney SC (2010) The Epstein-Barr virus (EBV)-encoded protein kinase, EBV-PK, but not the thymidine kinase (EBV-TK), is required for ganciclovir and acyclovir inhibition of lytic viral production. J Virol 84(9):4534–4542. https://doi.org/10.1128/JVI.02487-09

Murata T, Isomura H, Yamashita Y, Toyama S, Sato Y, Nakayama S, Kudoh A, Iwahori S, Kanda T, Tsurumi T (2009) Efficient production of infectious viruses requires enzymatic activity of Epstein-Barr virus protein kinase. Virology 389(1–2):75–81. https://doi.org/10.1016/j.virol.2009.04.007

Murata T, Narita Y, Sugimoto A, Kawashima D, Kanda T, Tsurumi T (2013) Contribution of myocyte enhancer factor 2 family transcription factors to BZLF1 expression in Epstein-Barr virus reactivation from latency. J Virol 87(18):10148–10162. https://doi.org/10.1128/JVI.01002-13

Murayama K, Nakayama S, Kato-Murayama M, Akasaka R, Ohbayashi N, Kamewari-Hayami Y, Terada T, Shirouzu M, Tsurumi T, Yokoyama S (2009) Crystal structure of Epstein-Barr virus DNA polymerase processivity factor BMRF1. J Biol Chem 284(51):35896–35905. https://doi.org/10.1074/jbc.M109.051581

Nakayama S, Murata T, Yasui Y, Murayama K, Isomura H, Kanda T, Tsurumi T (2010) Tetrameric ring formation of Epstein-Barr virus polymerase processivity factor is crucial for viral replication. J Virol 84(24):12589–12598. https://doi.org/10.1128/JVI.01394-10

Nijmeijer S, Leurs R, Smit MJ, Vischer HF (2010) The Epstein-Barr virus-encoded G protein-coupled receptor BILF1 hetero-oligomerizes with human CXCR4, scavenges Galphai proteins, and constitutively impairs CXCR4 functioning. J Biol Chem 285(38):29632–29641. https://doi.org/10.1074/jbc.M110.115618

O'Regan KJ, Brignati MJ, Murphy MA, Bucks MA, Courtney RJ (2010) Virion incorporation of the herpes simplex virus type 1 tegument protein VP22 is facilitated by trans-Golgi network localization and is independent of interaction with glycoprotein E. Virology 405(1):176–192. https://doi.org/10.1016/j.virol.2010.06.007

Park R, El-Guindy A, Heston L, Lin SF, Yu KP, Nagy M, Borah S, Delecluse HJ, Steitz J, Miller G (2014) Nuclear translocation and regulation of intranuclear distribution of cytoplasmic poly(A)-binding protein are distinct processes mediated by two Epstein Barr virus proteins. PLoS One 9(4):e92593. https://doi.org/10.1371/journal.pone.0092593

Pavlova S, Feederle R, Gartner K, Fuchs W, Granzow H, Delecluse HJ (2013) An Epstein-Barr virus mutant produces immunogenic defective particles devoid of viral DNA. J Virol 87(4):2011–2022. https://doi.org/10.1128/JVI.02533-12

Quinn LL, Williams LR, White C, Forrest C, Zuo J, Rowe M (2016) The missing link in Epstein-Barr virus immune evasion: the BDLF3 gene induces ubiquitination and downregulation of major histocompatibility complex class I (MHC-I) and MHC-II. J Virol 90(1):356–367. https://doi.org/10.1128/JVI.02183-15

Rennekamp AJ, Lieberman PM (2011) Initiation of Epstein-Barr virus lytic replication requires transcription and the formation of a stable RNA-DNA hybrid molecule at OriLyt. J Virol 85(6):2837–2850. https://doi.org/10.1128/JVI.02175-10

Rowe M, Glaunsinger B, van Leeuwen D, Zuo J, Sweetman D, Ganem D, Middeldorp J, Wiertz EJ, Ressing ME (2007) Host shutoff during productive Epstein-Barr virus infection is mediated by BGLF5 and may contribute to immune evasion. Proc Natl Acad Sci U S A 104(9):3366–3371. https://doi.org/10.1073/pnas.0611128104

Saito S, Murata T, Kanda T, Isomura H, Narita Y, Sugimoto A, Kawashima D, Tsurumi T (2013) Epstein-Barr virus deubiquitinase downregulates TRAF6-mediated NF-kappaB signaling during productive replication. J Virol 87(7):4060–4070. https://doi.org/10.1128/JVI.02020-12

Schepers A, Pich D, Hammerschmidt W (1993) A transcription factor with homology to the AP-1 family links RNA transcription and DNA replication in the lytic cycle of Epstein-Barr virus. EMBO J 12(10):3921–3929

Sheng W, Decaussin G, Ligout A, Takada K, Ooka T (2003) Malignant transformation of Epstein-Barr virus-negative Akata cells by introduction of the BARF1 gene carried by Epstein-Barr virus. J Virol 77(6):3859–3865

Sinclair AJ (2003) bZIP proteins of human gammaherpesviruses. J Gen Virol 84(Pt 8):1941–1949. https://doi.org/10.1099/vir.0.19112-0

Song MJ, Hwang S, Wong WH, Wu TT, Lee S, Liao HI, Sun R (2005) Identification of viral genes essential for replication of murine gamma-herpesvirus 68 using signature-tagged mutagenesis. Proc Natl Acad Sci U S A 102(10):3805–3810. https://doi.org/10.1073/pnas.0404521102

Su MT, Liu IH, Wu CW, Chang SM, Tsai CH, Yang PW, Chuang YC, Lee CP, Chen MR (2014) Uracil DNA glycosylase BKRF3 contributes to Epstein-Barr virus DNA replication through physical interactions with proteins in viral DNA replication complex. J Virol 88(16):8883–8899. https://doi.org/10.1128/JVI.00950-14

Swaminathan S, Hesselton R, Sullivan J, Kieff E (1993) Epstein-Barr virus recombinants with specifically mutated BCRF1 genes. J Virol 67(12):7406–7413

Tsai K, Thikmyanova N, Wojcechowskyj JA, Delecluse HJ, Lieberman PM (2011) EBV tegument protein BNRF1 disrupts DAXX-ATRX to activate viral early gene transcription. PLoS Pathog 7(11):e1002376. https://doi.org/10.1371/journal.ppat.1002376

Tsurumi T, Daikoku T, Nishiyama Y (1994) Further characterization of the interaction between the Epstein-Barr virus DNA polymerase catalytic subunit and its accessory subunit with regard to the 3′-to-5′ exonucleolytic activity and stability of initiation complex at primer terminus. J Virol 68(5):3354–3363

Tsurumi T, Kishore J, Yokoyama N, Fujita M, Daikoku T, Yamada H, Yamashita Y, Nishiyama Y (1998) Overexpression, purification and helix-destabilizing properties of Epstein-Barr virus ssDNA-binding protein. J Gen Virol 79(Pt 5):1257–1264. https://doi.org/10.1099/0022-1317-79-5-1257

Tsurumi T, Fujita M, Kudoh A (2005) Latent and lytic Epstein-Barr virus replication strategies. Rev Med Virol 15(1):3–15. https://doi.org/10.1002/rmv.441

Vischer HF, Nijmeijer S, Smit MJ, Leurs R (2008) Viral hijacking of human receptors through heterodimerization. Biochem Biophys Res Commun 377(1):93–97. https://doi.org/10.1016/j.bbrc.2008.09.082

Wang HB, Zhang H, Zhang JP, Li Y, Zhao B, Feng GK, Du Y, Xiong D, Zhong Q, Liu WL, Du H, Li MZ, Huang WL, Tsao SW, Hutt-Fletcher L, Zeng YX, Kieff E, Zeng MS (2015) Neuropilin 1 is an entry factor that promotes EBV infection of nasopharyngeal epithelial cells. Nat Commun 6:6240. https://doi.org/10.1038/ncomms7240

Watanabe T, Fuse K, Takano T, Narita Y, Goshima F, Kimura H, Murata T (2015a) Roles of Epstein-Barr virus BGLF3.5 gene and two upstream open reading frames in lytic viral replication in HEK293 cells. Virology 483:44–53. https://doi.org/10.1016/j.virol.2015.04.007

Watanabe T, Narita Y, Yoshida M, Sato Y, Goshima F, Kimura H, Murata T (2015b) The Epstein-Barr virus BDLF4 gene is required for efficient expression of viral late lytic genes. J Virol 89(19):10120–10124. https://doi.org/10.1128/JVI.01604-15

Watanabe T, Tsuruoka M, Narita Y, Katsuya R, Goshima F, Kimura H, Murata T (2015c) The Epstein-Barr virus BRRF2 gene product is involved in viral progeny production. Virology 484:33–40. https://doi.org/10.1016/j.virol.2015.05.010

Wei MX, Ooka T (1989) A transforming function of the BARF1 gene encoded by Epstein-Barr virus. EMBO J 8(10):2897–2903

Whitehurst CB, Li G, Montgomery SA, Montgomery ND, Su L, Pagano JS (2015) Knockout of Epstein-Barr virus BPLF1 retards B-cell transformation and lymphoma formation in humanized mice. mBio 6(5):e01574–e01515. https://doi.org/10.1128/mBio.01574-15

Yoshida M, Watanabe T, Narita Y, Sato Y, Goshima F, Kimura H, Murata T (2017) The Epstein-Barr virus BRRF1 gene is dispensable for viral replication in HEK293 cells and transformation. Sci Rep 7(1):6044

Zuo J, Currin A, Griffin BD, Shannon-Lowe C, Thomas WA, Ressing ME, Wiertz EJ, Rowe M (2009) The Epstein-Barr virus G-protein-coupled receptor contributes to immune evasion by targeting MHC class I molecules for degradation. PLoS Pathog 5(1):e1000255. https://doi.org/10.1371/journal.ppat.1000255

Chapter 19
Animal Models of Human Gammaherpesvirus Infections

Shigeyoshi Fujiwara

Abstract Humans are the only natural host of both Epstein-Barr virus (EBV) and Kaposi's sarcoma-associated herpesvirus (KSHV), and this strict host tropism has hampered the development of animal models of these human gammaherpesviruses. To overcome this difficulty and develop useful models for these viruses, three main approaches have been employed: first, experimental infection of laboratory animals [mainly new-world non-human primates (NHPs)] with EBV or KSHV; second, experimental infection of NHPs (mainly old-world NHPs) with EBV- or KSHV-related gammaherpesviruses inherent to respective NHPs; and third, experimental infection of humanized mice, i.e., immunodeficient mice engrafted with functional human cells or tissues (mainly human immune system components) with EBV or KSHV. These models have recapitulated diseases caused by human gammaherpesviruses, their asymptomatic persistent infections, as well as both innate and adaptive immune responses to them, facilitating the development of novel therapeutic and prophylactic measures against these viruses.

Keywords Animal model · Human gammaherpesvirus · Epstein-Barr virus · Kaposi's sarcoma-associated herpesvirus · Rhesus lymphocryptovirus · Rhesus rhadinovirus · Murine gammaherpesvirus 68 · Humanized mouse

---

S. Fujiwara (✉)
Department of Allergy and Clinical Immunology, National Research Institute for Child Health and Development, Tokyo, Japan

Division of Hematology and Rheumatology, Department of Medicine, Nihon University School of Medicine, Tokyo, Japan
e-mail: fujiwara-s@ncchd.go.jp

19.1 Introduction

The two human gammaherpesviruses, Epstein-Barr virus (EBV) and Kaposi's sarcoma-associated herpesvirus (KSHV), belong to the genera *Lymphocryptovirus* (LCV) and *Rhadinovirus* (RV), respectively. Primate rhadinoviruses can be divided into the two groups rhadinovirus 1 (RV1) and RV2, with KSHV belonging to the RV1 group. As gammaherpesviruses, EBV and KSHV share a number of common properties including affinity to B cells and association with malignant diseases, in addition to other properties common to all human herpesviruses such as lifelong persistent infection and involvement in opportunistic infections. LCVs have been so far found only in primates and are characterized by the unique ability to immortalize B cells, whereas RVs are prevalent more widely throughout the mammals and lack the ability to immortalize mature B cells. Humans are the only natural host for both EBV and KSHV, and modeling their life cycle and pathogenesis in laboratory animals has been hampered by this strict host specificity. However, advances in the research on gammaherpesviruses of various non-human host species, as well as the development of severely immunodeficient mouse strains, enabled the generation of unique animal models for human gammaherpesvirus infections (Fig. 19.1, Table 19.1).

EBV, the first human oncogenic virus identified, is essentially a ubiquitous virus prevalent in >90% of adult population in the world [reviewed in Longnecker et al. (2013)]. While most EBV infections are asymptomatic, primary infection in adolescents and young adults may result in infectious mononucleosis (IM), characterized by transient EBV-induced B-cell lymphoproliferation complicated with excessive T-cell responses specific to the virus. As a typical LCV, EBV has the ability to immortalize human B cells and establish lymphoblastoid cell lines (LCLs). In immunocompromised hosts, including HIV-infected individuals and transplant recipients, EBV-immortalized lymphoblastoid cells may proliferate unlimitedly to induce lymphoproliferative disease (LPD), indicating the critical importance of immune responses, especially those by T-cells, in the control of EBV infection [reviewed in Hislop et al. (2007)]. EBV causes a wide variety of lymphoid and epithelial malignancies including Burkitt's lymphoma, Hodgkin's lymphoma, nasopharyngeal carcinoma, and gastric carcinoma. Every year in the world, about 120,000 new cases of malignancies are attributed to EBV (Plummer et al. 2016). Although B cells and epithelial cells are the main targets of EBV infection, it can also infect T-cells and NK cells and in rare occasions induces their systemic proliferation such as chronic active EBV (CAEBV) infection [reviewed in Fujiwara et al. (2014)].

KSHV is the most recently identified human tumor virus and causes various malignancies and LPDs, including Kaposi's sarcoma (KS), primary effusion lymphoma (PEL), and multicentric Castleman's disease (MCD) [reviewed in Damania and Cesarman (2013)]. These KSHV-associated diseases are observed mainly in association with immunodeficiency caused by HIV infection or transplantation therapy. KS is a malignant tumor derived from vascular endothelial cells, whereas

19 Animal Models of Human Gammaherpesvirus Infections

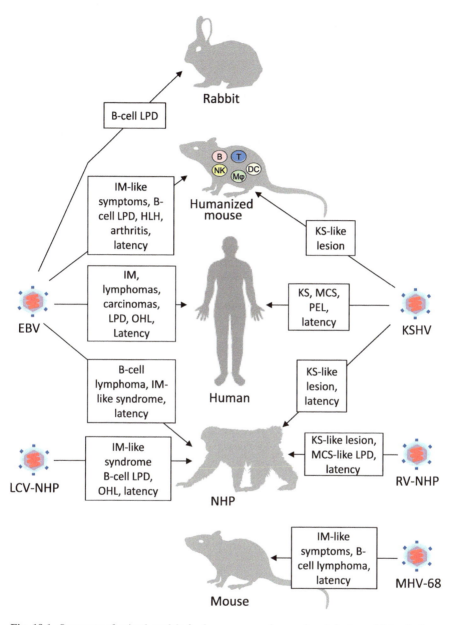

Fig. 19.1 Summary of animal models for human gammaherpesvirus infections. Major findings obtained from each animal model of human gammaherpesviruses, as well as representative features of human infection with EBV and KSHV, are shown. Horizontal arrows represent infections to natural hosts, whereas oblique arrows represent those to nonnatural hosts. See text and Table 19.1 for details. *LCV* lymphocryptovirus, *RV* rhadinovirus, *NHP* non-human primate, *IM* infectious mononucleosis, *LPD* lymphoproliferative disease, *HLH* hemophagocytic lymphohistiocytosis, *OHL* oral hairy leukoplakia, *KS* Kaposi's sarcoma, *MCS* multicentric Castleman's disease, *PE* primary effusion lymphoma, *MHV-68* murine gammaherpesvirus 68

Table 19.1 Representative animal models of human gammaherpesvirus infections

Animal species	Viral species	Features of EBV or KSHV infection reproduced	Reference
Mouse (C.B-17 scid)	Epstein-Barr virus (EBV), *Human gammaherpesvirus 4*	B-cell LPD following ip injection of PBMCs from EBV-infected individuals (*Scid*-hu PBL)	Mosier et al. (1988), Rowe et al. (1991), Johannessen and Crawford (1999)
Mouse (C.B-17 scid)	Kaposi's sarcoma-associated herpesvirus (KSHV), *Human gammaherpesvirus 8*	Latent and lytic infections with viral replication predominantly observed in B cells (*Scid*-hu thy/liv)	Dittmer et al. (1999)
NOG, BRG, NSG mice engrafted with human HSCs	EBV	B-cell LPD, Hodgkin-like lymphoma, EBV-HLH, RA-like arthritis, persistent infection, innate and adaptive immune responses	Traggiai et al. (2004), Melkus et al. (2006), Yajima et al. (2008), Strowig et al. (2009), Lee et al. (2015), Chijioke et al. (2013)
BLT-NSG mice	KSHV	Infection to B cells and macrophages	Wang et al. (2014)
Cotton-top tamarin	EBV	B-cell lymphoma (used in the evaluation of an experimental EBV vaccine)	Shope et al. (1973), Epstein et al. (1985), Johannessen and Crawford (1999)
Common marmoset	EBV	IM-like primary infection and persistent infection	Falk et al. (1976), Johannessen and Crawford (1999)
Common marmoset	KSHV	KS-like lesion	Chang et al. (2009)
Rabbit	EBV	B-cell LPD, peripheral blood EBV DNA, virus-specific antibodies	Takashima et al. (2008), Okuno et al. (2010), Khan et al. (2015)
Rhesus macaque	Rhesus LCV (rhLCV), *Macacine gammaherpesvirus 4*	IM-like syndrome, persistent infection, B-cell LPD, oral hairy leukoplakia (OHL) in immunocompromised hosts	Moghaddam et al. (1997), Rivailler et al. (2004,2002b), Mühe and Wang (2015)
Rhesus macaque	Rhesus rhadinovirus (RRV), *Macacine gammaherpesvirus 5*	Persistent infection, B-cell LPD, KS-like lesion in immunocompromised hosts	Wong et al. (1999), Searles et al. (1999), Estep and Wong (2013)
Common marmoset	Marmoset LCV, *Callitrichine gammaherpesvirus 3*	B-cell lymphoma	Ramer et al. (2000), Rivailler et al. (2002a)
Mouse	Murine herpesvirus 68 (MHV-68), *Murid gammaherpesvirus 4*	B-cell lymphoma, arteritis, fibrosis, persistent infection	Virgin et al. (1997), Barton et al. (2011)

PEL is a B-cell lymphoma of the post-germinal center origin with the expression of a characteristic set of markers. MCD is a B-cell LPD with prominent inflammatory symptoms caused by hypercytokinemia. In contrast to EBV, KSHV is not ubiquitous; its seroprevalence varies from 0–5% in Japan, North and South America, and northern European countries to around 10% in Mediterranean countries and 30–50% in sub-Saharan African countries [reviewed in Rohner et al. (2014)]. This uneven distribution of KSHV infection is consistent with the preferential occurrence of KS and other KSHV-associated diseases in limited areas of the world. Although it appears obvious that cell-mediated immune responses play a central role in the control of KSHV infection, relatively few is known about the detail of anti-KSHV immune responses.

19.2 Experimental Infection of Laboratory Animals with Human Gammaherpesviruses

Only limited species of laboratory animals are susceptible to infection with human gammaherpesviruses. Cotton-top tamarins (*Saguinus oedipus*) can be infected with EBV via parenteral routes and develop B-cell lymphomas (Shope et al. 1973), and this model has been used for the evaluation of an early experimental EBV vaccine (Epstein et al. 1985) [reviewed in Johannessen and Crawford (1999)]. Common marmosets (*Callithrix jacchus*) can be infected with both EBV and KSHV; while EBV induces non-specific IM-like symptoms (Falk et al. 1976), a recent report showed that KSHV can cause persistent infection and induce KS-like lesions with characteristic spindle-shaped cells in a fraction of infected animals (Chang et al. 2009). Although rhesus macaques could be infected experimentally with KSHV, it replicated in very low levels, and no viral gene expression was demonstrated (Renne et al. 2004).

Rabbits inoculated parenterally with EBV exhibited signs of EBV infection, including EBV DNA load in peripheral blood lymphocytes and serum antibodies specific to the virus (Takashima et al. 2008). Subsequent studies revealed that rabbits could be infected also by the oral route (Okuno et al. 2010), the primary natural route of human EBV infection. Although these early studies did not reproduce any specific disease conditions resembling human EBV-associated diseases, a recent report described EBV-induced lymphoproliferation with the latency III type viral gene expression (i.e., expression of EBNAs 1, 2, 3A, 3B, 3C, and LP; LMPs 1, 2A, and 2B; and untranslated RNAs such as EBERs, BARTs, and microRNAs) in rabbits treated with cyclosporine A (Khan et al. 2015).

19.3 Lymphocryptoviruses and Rhadinoviruses of Non-human Primates

Gammaherpesviruses, similar to other herpesviruses, are thought to have coevolved with their respective hosts, and most old-world non-human primates (NHPs) and at least a fraction of new-world NHPs carry their own distinctive LCVs and RVs closely related to EBV and KSHV, respectively. Since some of these viruses share remarkably similar genomic contents, life cycle, and pathogenesis with either EBV or KSHV, the use of these viruses as surrogate models for the human viruses has been explored. Meanwhile, recent studies identified two species of LCVs in a number of both old-world and new-world NHPs such as gorillas, baboons, and squirrel monkeys, suggesting the possibility that humans might harbor a yet to be identified species of LCV (Damania 2007). Similarly, both RV1 and RV2 species have been identified in most old-world NHPs so far examined, suggesting that humans might also harbor an RV2 species that have not been discovered (Damania 2007). Gammaherpesviruses of NHPs are thus valuable not only as practical models of EBV and KSHV but also as source of insight into the evolution of human gammaherpesviruses.

19.3.1 Rhesus LCV as a Model for EBV Infection

Among the NHP LCVs closely related to EBV, the rhesus LCV (rhLCV, *Macacine gammaherpesvirus 4* by ICTV) has been most extensively studied (Moghaddam et al. 1997) [reviewed in Mühe and Wang (2015)]. Sequence analysis identified in the rhLCV genome a homolog for each known EBV gene, with better conservation in lytic-cycle genes (49–98% amino acid identity) than in latent-cycle genes (28–60%) (Rivailler et al. 2002b). Interestingly, two strains of rhLCV, corresponding to the type 1 and type 2 EBV strains, distinguished by sequence variation in the rhesus EBNA2 homolog, have been identified, suggesting the presence of a same pressure for the evolution of the two types of LCV in humans and rhesus monkeys.

RhLCV, just like EBV, is ubiquitous among both wild and captive rhesus macaques; virtually all animals get seropositive by the age of 1 year (Mühe and Wang 2015), resulting in asymptomatic persistent infection in memory B cells. Upon primary infection of naïve macaques, rhLCV induces signs and symptoms reminiscent of human symptomatic primary EBV infection, including lymphadenopathy, hepatosplenomegaly, and atypical lymphocytosis. The acute phase of primary infection, a stage difficult to investigate with EBV, can be analyzed in detail with rhLCV (Rivailler et al. 2004). In the peripheral blood of macaques inoculated orally with rhLCV, EBER RT-PCR turns positive at 7 days post-infection (p.i.); DNA PCR turns positive at 2–3 weeks p.i.; and atypical lymphocytosis and lymphadenopathy are recognized at 3–15 weeks p.i. This acute phase is followed by asymptomatic persistent infection similar to EBV latency in humans that is maintained by immunosurveillance. RhLCV immortalizes B lymphocytes of rhesus macaques and transplantation of autologous rhLCV-immortalized B cells in a rhesus macaque infected with the chimeric simian/human immunodeficiency virus SHIV 89.6P resulted in the development of monoclonal B-cell lymphoma (Rivailler et al. 2004). Oral hairy leukoplakia (OHL), an immunodeficiency-associated epithelial lesion caused by local EBV replication, is also reproduced in rhesus macaques coinfected with rhLCV and SIV (Kutok et al. 2004).

RhLCV-induced humoral and cell-mediated immune responses have been characterized in detail and are considered as a suitable model to test hypotheses and strategies for the development of EBV vaccines. So far, the effect of vaccination with soluble rhgp350 (the rhLCV homolog of the EBV gp350) has been demonstrated; upon challenge with rhLCV, immunized animals had reduced rates of seroconversion and reduced level of viral DNA (viral set point) when seroconverted (Sashihara et al. 2011). Similarly, an adenovirus-based vector encoding rhEBNA1, given to rhesus macaques persistently infected with rhLCV, was shown to expand both CD4+ and CD8+ T-cells specific to the viral protein (Leskowitz et al. 2014).

Reverse genetics studies are feasible with rhLCV, and a bacterial artificial chromosome (BAC)-based rhLCV recombinant with defective rhBARF1 has been characterized. RhBARF1 encodes a soluble form of the colony stimulating factor 1 (CSF-1) receptor that inhibits CSF-1 signaling. Animals infected with this RhBARF1-defective rhLCV had decreased viral load in acute infection and low numbers of infected B cells in persistent infection, giving evidence for the putative immunoevasion function of BARF1 (Ohashi et al. 2012).

Although rhLCV is a remarkable experimental model to recapitulate major features of EBV infection in an NHP species, it should be kept in mind that there are significant differences in the nature of the two viruses (EBV vs. rhLCV) and the two host species (human vs. rhesus macaque), as evidenced by the findings that EBV does not transform rhesus B cells and rhLCV does not transform human B cells. Experimental infection of rhesus macaques with EBV has not been feasible even when the animals were rhLCV-seronegative. High cost, limited availability of rhLCV-naïve host animals, and ethical considerations tend to restrict the use of this model only in selected studies.

19.3.2 Rhesus RVs as Models for KSHV Infection

Retroperitoneal fibromatosis (RF) is a vascular fibroproliferative disease originally found in macaque populations housed in primate research facilities, and its histological similarity with KS, including proliferation of spindle-shaped cells and neoangiogenesis, had been recognized even before the discovery of KSHV [reviewed in Westmoreland and Mansfield (2008)]. Persistent infection observed in animals affected with RF with an immunodeficiency-inducing retrovirus [simian retrovirus 2 (SRV-2)] was also reminiscent of KS. Partial genomic DNA of a rhadinovirus species of the RV1 group, later termed RF herpesvirus (RFHV), was identified in RF lesion by degenerative PCR (Rose et al. 1997). An RFHV-encoded nuclear protein homologous to the KSHV latency-associated nuclear antigen (LANA) protein was identified in the spindle-shaped cells in RF lesions, and RFHV is now considered as the etiologic agent of RF. The complete genomic sequence of an RFHV strain isolated from *Macaca nemestrina* (RFHVMn) has been recently obtained (Bruce et al. 2013). The RFHVMn genome is collinear with that of KSHV and has a homolog for every known KSHV gene, except for ORF11, K5, and K6. RFHV is thus the virus most closely related to KSHV so far sequenced (Bruce et al. 2013). However, RFHV has not been isolated and grown in vitro, and no information on its biology has been obtained, making it difficult at present to use the virus as a model of KSHV.

The rhesus rhadinovirus (RRV, *Macacine gammaherpesvirus 5* by ICTV), an RV2 virus, naturally infects rhesus macaques and is closely related to KSHV (Wong et al. 1999) [reviewed in Estep and Wong (2013)]. The genome of a strain (17577) of RRV has been shown to be collinear with the KSHV genome, and 67 of its total 79 ORFs have a homologous counterpart in KSHV (Searles et al. 1999). Among these 67 ORFs, 48 show >50% amino acid similarity, and 27 show >60% similarity (Searles et al. 1999). RRV encodes viral homologs of cellular proteins, such as IL-6, a G-protein coupled receptor (GPCR), interferon-regulatory factors (IRFs), and CD200. KSHV also encodes homologs for these cellular proteins that are assumed to play critical roles in KSHV pathogenesis. BAC-based viral recombinants have been available for in vitro and in vivo functional analyses of certain RRV genes, including vCD200 and vIRFs (Estep et al. 2014; Robinson et al. 2012). Although RRV is highly prevalent in captive rhesus macaques in primate facilities, it rarely induces disease in them. In immunocompromised conditions caused by the simian immunodeficiency virus mac239 (SIV_{mac239}), however, it induces B-cell LPD resembling the plasma cell variant of MCD (Wong et al. 1999; Mansfield et al. 1999). Interestingly, KS-like lesions with proliferating spindle-like cells and angiogenesis were induced in a rhesus macaque that was coinfected with RRV and SIV_{mac239} (Orzechowska et al. 2008). Although RRV belongs to the RV2 group and its evolutionary relationship with KSHV is more distant as compared with the RV1 virus RFHV, it is widely used as a model of KSHV, because RRV can be grown to high titers in cell culture.

19.3.3 Lymphocryptoviruses and Rhadinoviruses of New-World NHPs

Although earlier serological studies revealed no evidence for LCVs in new-world NHPs, a novel LCV (termed marmoset LCV (maLCV), *Callitrichine gammaherpesvirus 3* by ICTV) was identified in fatal B-cell LPD of common marmosets (Ramer et al. 2000). Similar to LCVs of old-world primates, maLCV transforms B cells and encodes distant homologs for certain well-characterized EBV proteins such as EBNA1, BZLF1, and gp350, but not EBERs and BCRF1 (vIL-10) (Rivailler et al. 2002a). Positional homologs for EBV EBNA2 and LMP1 have been identified in the maLCV genome, but they have no sequence similarity with the EBV counterparts. Interestingly, maLCV encodes only one copy of the positional EBNA3 homolog, supporting the hypothesis that EBNAs 3A, 3B, and 3C of EBV have been generated through gene duplication during evolution of the virus (Rivailler et al. 2002a). Later studies suggested that a wide variety of new-world primates (not only the common marmoset) harbor LCVs [reviewed in Mühe and Wang (2015)]. Given the distant relationship with EBV, these new-world primates LCVs seem more interesting from the standpoint of LCV evolution rather than utility as an experimental model.

Herpesvirus saimiri (HVS, *Saimiriine gammaherpesvirus 2* by ICTV) is an RV2 species that is widespread in squirrel monkeys [reviewed in Ensser and Fleckenstein (2007)]. HVS does not cause any disease in the natural host, but cross-species infection to other primate species, including old-world primates, results in aggressive lymphomas. In contrast to most other RVs, HVS transforms T-cells rather than B cells and represents a reliable tool to establish a human T-cell line. Although HVS shares a collinear genome with KSHV, its homology to KSHV is not as remarkable as that of RFHV and RRV.

19.4 Murine Gammaherpesvirus 68

The murine gammaherpesvirus 68 (MHV-68, *Murid gammaherpesvirus 4* in ICTV) shares some common properties with KSHV and EBV in life cycle (e.g., B-cell tropism and persistence in memory B cells) and pathogenesis (e.g., B-cell LPD) and has been characterized as a model for human gammaherpesvirus infections. MHV-68 was originally isolated from the bank vole (*Myodes glareolus*) but infects laboratory mouse strains as well [reviewed in Barton et al. (2011)]. MHV-68 is a rhadinovirus and its genome exhibits higher homology with that of KSHV than that of EBV. The MHV-68 genome contains at least 80 genes, among which 63 are homologs of KSHV genes and many of these 63 genes have homologs also in the EBV genome (Virgin et al. 1997). MHV-68 encodes several proteins that are homologous to KSHV proteins critically involved in its pathogenesis, including LANA, the complement regulatory protein KCP, a D-type cyclin (v-cyclin), a bcl-2 homolog with anti-apoptotic function (v-bcl-2), and viral GPCR (vGPCR). Although EBV genes involved in latent infection have no homologs in MHV-68, there is a functional similarity among EBV EBNA1, KSHV LANA, and MHV-68 LANA, all of which play essential roles in the replication and maintenance of viral episomes.

Similar to EBV and KSHV, MHV-68 does not usually induce any disease in immunocompetent hosts, but in immunocompromised mice, it causes B-cell lymphomas (Sunil-Chandra et al. 1994). In addition, MHV-68 causes arteritis and multi-organ fibrosis that are considered to be caused by viral replication. MHV-68 however does not induce vascular endothelial tumor like KS. Intranasal or intraperitoneal inoculation of MHV-68 is followed first by viral replication in the alveolar epithelium in the lungs and then by latent infection in the spleen and the peritoneum. The main reservoir of MHV-68 in latent infection is B cells (mainly class-switched memory B cells), but macrophages and splenic dendritic cells also harbor the virus. In contrast to EBV, but similar to KSHV, MHV-68 lacks the ability to transform mature B lymphocytes, although a recent report showed that it can transform B cells derived from murine fetal liver (Liang et al. 2011). Differentiation of B cells latently infected with MHV-68 toward plasma cells leads to the reactivation of the viral replicative cycle, a property shared by both EBV and KSHV (Liang et al. 2009).

Immune responses to MHV-68 have been well characterized, and some parallels with human gammaherpesviruses have been revealed [reviewed in Barton et al. (2011)]. IFN-α and IFN-β play a critical role in the control of acute MHV-68 infection since mice with IFN-α/IFN-β knockout succumb to even low doses of MHV-68 infection. IFN-γ, CD4+ T-cells, and CD8+ T-cells all play critical roles in the control of MHV-68 infection, especially in its latent phase. Mice deficient for IFN-γ suffer from arteritis and/or fibrosis due to increased MHV-68 replication; mice deficient for CD4+ T-cells allow persistent high-level viral replication resulting in tissue damages; and those deficient for CD8+ T-cells allow increase in the number of latently infected cells. Basic research on gammaherpesvirus vaccine development has been performed with MHV-68; immunization with the MHV-68 gp150/M7 (distant homolog of the EBV gp350) protected mice from IM-like syndrome caused by MHV-68 challenges but not from seroconversion [reviewed in Wu et al. (2012)]. Accumulated knowledge and resources of mouse genetics provide an opportunity to investigate the role of host genetic background in the immune control of MHV-68 infection. MHV-68 infection of mice knocked-out for various genes involved in innate immunity revealed that deficiency of genes involved in DNA sensing facilitates the establishment of latent infection (Sun et al. 2015). Research in this direction might give insights into the host genetics of human gammaherpesvirus infections.

Mice that are latently infected with MHV-68 were shown to have increased resistance to various pathogens such as *Listeria monocytogenes* and *Yersinia pestis*, and this was explained by general upregulation of innate immunity by the virus (Barton et al. 2007). Similarly, the "armed" state of NK cells, characterized by abundant expression of perforin and granzyme B and readiness to kill target cells, is induced following the establishment of latent infection with MHV-68 (White et al. 2010). These findings suggest an interesting possibility that herpesvirus latency remodulates host immune system and may have some symbiotic effects to protect the host from other infections, although it is not known whether human gammaherpesvirus infections also have a similar effect.

19.5 Humanized Mice

19.5.1 Early-Generation Humanized Mouse Models of Gammaherpesvirus Infections

Studies on gammaherpesvirus pathogenesis and therapeutics had been hampered by the absence of suitable small animal models of infection, but the generation of the highly immunodeficient mouse strain C.B-17 *scid* opened the way for the development of the first mouse model of human gammaherpesvirus infection. Due to mutation of the gene coding for a subunit of DNA-dependent protein kinase, mice of this strain lack both B and T lymphocytes, resulting in the severe combined immunodeficiency (*scid*) phenotype (Bosma et al. 1983). Two types of early-generation humanized mice were generated from C.B-17 *scid* mice, namely, *scid*-hu PBL and *scid*-hu thy/liv. *Scid*-hu PBL mice were prepared by transplanting human peripheral blood mononuclear cells (PBMCs) into the abdominal cavity of C.B-17 *scid* mice, resulting in the engraftment of various human blood components including T and B cells (Mosier et al. 1988). Transplantation of PBMCs from EBV-seropositive donors but not from seronegative ones resulted in the development of EBV-positive B-cell LPD (Mosier et al. 1988). Inoculation of EBV to mice that had been transplanted with PBMCs from seronegative donors also resulted in similar EBV-positive LPD (Cannon et al. 1990). Later studies showed that EBV-positive LPD in *scid*-hu PBL mice thus generated was similar to the typical type of EBV-associated LPD in immunocompromised patients, regarding histology [mainly diffuse large B-cell lymphoma (DLBCL)], surface marker expression (activated B-cell phenotype), and EBV gene expression (latency III) [Rowe et al. (1991) and reviewed in Johannessen and Crawford (1999)]. Studies with *scid*-hu PBL mice revealed critical roles for CD4+ T-cells, IL-10 signaling, and CXCL12/CXCR4 signaling in the pathogenesis of EBV-associated LPD [reviewed in Johannessen and Crawford (1999)]. *Scid*-hu thy/liv mice, on the other hand, were prepared by transplanting human fetal thymus and liver tissues under the renal capsule of C.B-17 *scid* mice, resulting in the development of human T and B cells (McCune et al. 1988). Although *scid*-hu thy/liv mice were primarily used as a model of HIV-1 infection, infection models for a variety of other viruses, including KSHV, were prepared with them (Dittmer et al. 1999). Direct inoculation of KSHV into the thymus/liver implant resulted in both latent and lytic infections, with viral replication predominantly observed in B cells. Coinfection of *scid*-hu thy/liv mice with KSHV and HIV-1 however did not result in any significant differences compared with single KSHV infection. In one study, C.B-17 *scid* mice were transplanted with human skin graft, and inoculation of KSHV to this transplant resulted in the development of KS-like lesions characterized by angiogenesis and proliferation of spindle-shaped cells expressing LANA (Foreman et al. 2001).

19.5.2 New-Generation Humanized Mouse Models of EBV Infection

There were certain limitations in the early-generation humanized mice; the efficiency of engraftment of human cells was very low and their normal functions were not clearly recognized. Immune responses to viruses could not be reproduced in these models. Means to overcome these limitations were provided by the development of the so-called new-generation humanized mice that became possible when new strains of more severely immunodeficient mice were generated. The NOD/*scid* strain was generated by introducing the *scid* mutation into the nonobese diabetic (NOD) mouse strain. Chimeric mice prepared by implanting NOD/*scid* mice with human fetal bone, thymus, and skin grafts were successfully infected with KSHV and produced human antibodies specific to the virus (Parsons et al. 2006). More recently three strains of severely immunodeficient mice, namely, NOD/Shi-*scid* *Il2rg*^{null} (NOG), Balb/c *Rag2*$^{-/-}$*Il2rg*$^{-/-}$ (BRG), and NOD/LtSz-*scid Il2rg*$^{-/-}$ (NSG), have been mainly used to prepare new-generation humanized mice (Ito et al. 2002; Traggiai et al. 2004; Shultz et al. 2005). Reconstitution of human B, T, and NK lymphocytes, macrophages, and dendritic cells was observed in these mice following transplantation of human hematopoietic stem cells (HSCs). A specific protocol for preparing humanized mice involved transplantation of human fetal liver and thymus tissues as well as HSCs isolated from the same liver (BLT mice), enabling proper intrathymic education of human T-cell progenitors (Melkus et al. 2006). Traggiai and others demonstrated EBV-induced B-cell proliferation and suggested the induction of T-cell responses to the virus in humanized BRG mice (Traggiai et al. 2004). EBV-specific T-cell responses restricted by human MHC class I were induced in EBV-infected BLT-NOD/*scid* mice (Melkus et al. 2006). Following these pioneering works, increasing efforts have been made to reproduce various aspects of human EBV infection in new-generation humanized mice [reviewed in Fujiwara et al. (2015)].

19.5.3 EBV Pathogenesis in Humanized Mice

Humanized NOG mice inoculated intravenously with EBV were shown to reproduce cardinal features of EBV-associated B-cell LPD (Yajima et al. 2008) (Fig. 19.2). The development of LPD was dependent on the dose of the virus inoculated; EBV doses >10^2 TD$_{50}$ induced LPD in most mice. Histology of this LPD was mainly the DLBCL type with the expression of B-cell activation markers such as CD23 and the germinal center marker Mum-1, and EBV gene expression was consistent with latency III. It was thus evident that the typical form of EBV-associated B-cell LPD in immunocompromised hosts was reproduced in humanized NOG mice. In some LPD tissues, Reed-Sternberg-like cells with multiple nuclei and Hodgkin-like cells with marked nucleoli were observed (Yajima et al. 2008). Following low-dose EBV

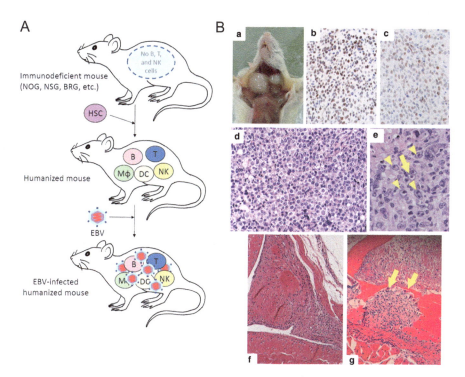

Fig. 19.2 Humanized mouse model of EBV infection. (**A**) Schematic illustration showing the process of generating a humanized mouse. (**B**) Representative findings obtained from EBV-infected humanized NOG mice. (**a–e**) B-cell LPD. Macroscopic view (**a**), in situ hybridization for EBER (**b**), immunostaining for EBNA2 (**c**), and HE staining (**d, e**) are shown. Hodgkin-like cells (arrowheads) and a Reed-Sternberg-like cell (arrow) are shown in (**e**). (**f–g**) Erosive arthritis. Massive synovial proliferation (**f**) and bone destruction with pannus formation (**g**) are shown. Arrows in (**g**) indicate multinucleated osteoclast-like cells. Reproduced from Yajima et al. (2008) with permission from Oxford University Press and from Fujiwara et al. (2015) and Kuwana et al. (2011) under Creative Commons Attribution (CC-BY) license

infection (<10^1 TD$_{50}$), most mice did not develop LPD and survived for up to 250 days without any signs of disease (Yajima et al. 2008). Histological analysis of these mice revealed a small number of EBV-infected B cells in the liver and kidney, indicating the establishment of persistent infection that may be similar to EBV latency in humans. The development of similar EBV-induced B-cell lymphoma with the latency III type viral gene expression was demonstrated subsequently in humanized NSG mice (Strowig et al. 2009). Recently, humanized NSG mice with predominant T-cell development were shown to develop mainly Hodgkin-like lymphomas, whereas those with B-cell predominance tended to develop non-Hodgkin-type DLBCL (Lee et al. 2015).

Features of hemophagocytic lymphohistiocytosis (HLH) were recapitulated in EBV-infected humanized NOG mice (Sato et al. 2011). In this study, EBV-infected mice exhibited persistent viremia, leukocytosis, IFN-γ hypercytokinemia, marked

CD8+ T-cell proliferation, and prominent hemophagocytosis in the bone marrow, spleen, and liver. These results appear significantly different from those by Yajima and others, who used the same NOG mouse strain and observed mainly B-cell LPD (Yajima et al. 2008). However, there were a number of differences in the protocol for preparation of humanized mice between the two groups, including the route of HSC transplantation, the timing of transplantation, and the sex of mice. These differences may have resulted in variance in the nature of T-cell responses to EBV infection.

EBV has been associated with the etiology of a variety of autoimmune diseases, including rheumatoid arthritis (RA), multiple sclerosis, and systemic lupus erythematosus [reviewed in Niller et al. (2011)]. Patients with RA have higher anti-EBV antibody titer and higher peripheral blood EBV DNA load as compared with normal controls and patients with other inflammatory joint diseases. They have impaired T-cell responses to the virus, and EBV-infected cells with the expression of EBERs and various viral proteins have been identified in RA lesions of a fraction of patients (Takei et al. 1997). When major joints of EBV-infected humanized NOG mice were examined, histological changes characteristic to RA were demonstrated, including massive synovial proliferation and edema in the bone marrow adjacent to affected joint (Kuwana et al. 2011). Moreover, a histological structure termed pannus, a pathognomonic finding in RA, representing inflammatory granulation tissue containing osteoclasts, was clearly demonstrated in these mice. Although only a few EBER-positive cells were found in the affected synovium, many such cells were demonstrated in the bone marrow adjacent to the affected joint. Human B cells, T-cells, and macrophages are seen infiltrating the synovium and the adjacent bone marrow of these mice. These findings indicated that EBV can induce RA-like arthritis in humanized mice. Further studies, however, especially those on the molecular mechanisms of this arthritis, are required to verify that these mice can be considered as a proper model mouse of RA.

19.5.4 Immune Responses to EBV in Humanized Mice

EBV-specific T-cell responses restricted by human MHC class I were demonstrated in humanized NOG mice (Melkus et al. 2006; Yajima et al. 2008). These responses had a protective effect in that antibody-mediated depletion of either CD8-positive or CD3-positive T-cells in EBV-infected mice reduced their lifespan (Yajima et al. 2009). IgM antibody specific to a major component ($p18^{BFRF3}$) of the viral capsid antigen (VCA) was also identified (Yajima et al. 2008). EBV-specific T-cell responses were also characterized in humanized NSG mice, and protective roles for both CD4+ and CD8+ T-cells against EBV-induced lymphomagenesis were elucidated (Strowig et al. 2009). These cytotoxic T-cell responses were shown to be directed predominantly to lytic-cycle EBV proteins (Strowig et al. 2009). Humanized mice prepared from an NSG substrain carrying human HLA-A2 transgene were shown to mount efficient EBV-specific T-cell responses restricted by the particular type of HLA (Shultz et al. 2010).

Innate immune responses to EBV have been also characterized in humanized mice. In humanized NSG mice, NK cells, especially those of the early differentiation phenotype (NKG2A+ KIR-), were shown to play a critical role in the control of primary EBV infection (Chijioke et al. 2013). Depletion of NK cells resulted in increased level of EBV DNA load in the spleen, exaggerated CD8+ T-cell response, and increased risk of EBV-positive lymphoproliferation (Chijioke et al. 2013). Expansion of NK cells of the early differentiation phenotype was subsequently confirmed in human patients with IM (Azzi et al. 2014). Since NK cells of the early differentiation phenotype are most abundant in newborns and decrease thereafter with age, it is speculated that preferential development of IM in adolescents and young adults rather than in small children might be due to relative deficiency of this particular fraction of NK cells (Azzi et al. 2014).

19.5.5 EBV Reverse Genetics in Humanized Mice

Loss-of-function mutations of some EBV genes do not exhibit any phenotypic changes in in vitro experiments with cultured cells, and their functions had remained unknown. EBNA3B had been presumed to have an important function given its stable conservation in virtually every wild-type EBV isolate; however an EBNA3B knockout (KO) EBV recombinant was able to immortalize human B cells as efficiently as the wild-type virus. Unexpectedly, characterization of EBNA3B KO EBV in humanized NSG mice indicated that tumorigenesis by the virus was enhanced by this mutation (White et al. 2012). This was explained by upregulation of the T-cell attracting chemokine CXCL10 by EBNA3B, resulting in enhanced elimination of EBV-infected cells by T-cells. EBNA3B is thus the first example of a virus-encoded tumor suppressor gene and is speculated to have evolved to minimize oncogenic risk to the host. Importantly, these novel findings obtained from humanized mice were substantiated by human studies; various EBNA3B mutations have been identified in a number of EBV-positive DLBCL cases, implying that EBNA3B mutation gives some advantages in actual human lymphomagenesis.

LMP1 KO EBV recombinants had been shown to lose its B-cell transforming activity in vitro (Dirmeier et al. 2003). To examine the in vivo role of LMP1 in EBV-associated LPD, a modified version of humanized NSG mice, prepared by intraperitoneal injection of EBV-infected cord blood mononuclear cells (CBMCs), was used (Ma et al. 2015). Surprisingly, transplantation of CBMCs infected with an LMP1 KO EBV recombinant induced EBV-positive B-cell lymphoma almost as efficiently as that of the same cell population infected with the wild-type EBV (Ma et al. 2015). Interestingly, depletion of CD4+ T-cells following transplantation abolished tumor formation of LMP1 KO EBV infected cells but not those infected with wild-type virus. These results suggested that in vivo EBV can induce B-cell proliferation without LMP1, if CD4+ T-cells provide CD40 signals. Besides the EBV mutants described above, mutants of the BART miRNA cluster (Lin et al. 2015; Qiu et al. 2015), BPLF1 (Whitehurst et al. 2015), EBERs (Gregorovic et al. 2015), the BHRF1 miRNA cluster (Wahl et al. 2013), BZLF1 (Ma et al. 2011), and LMP2A (Ma et al. 2017) were characterized in humanized mice, and the results elucidated in vivo roles for these EBV genes that had not been revealed in previous in vitro studies.

19.5.6 Experimental Treatment of EBV-Associated Diseases in Humanized Mice

Humanized mouse models of EBV infection have been used to evaluate novel experimental therapies for EBV-associated diseases. The modified version of humanized NSG mice described above was used to examine the effect of PD-1 and CTLA-4 blockade in the treatment of EBV-associated B-cell LPD (Ma et al. 2016). Double blockade of the T-cell inhibitory receptors PD-1 and CTLA4 enhanced T-cell responses to EBV and suppressed the outgrowth of EBV-positive lymphoma in these mice. Humanized BRG mice were used to show that targeted activation of Vγ9Vδ2 T-cells can enhance the control of EBV-induced LPD (Xiang et al. 2014). Transfer of autologous T-cells that had been activated and expanded ex vivo by IL-2 and anti-CD3 antibody to EBV-infected humanized NOG mice resulted in the extension of their lifespan (Matsuda et al. 2015). Transfer of CD8+ T-cells specific to the EBV lytic-cycle protein BMLF1 prior to EBV inoculation to humanized NSG mice revealed a transient virus-controlling effect of this T-cell population (Antsiferova et al. 2014).

19.5.7 New-Generation Humanized Mouse Models of KSHV Infection

BLT-NSG mice were used to reproduce various aspects of KSHV infection (Wang et al. 2014). KSHV DNA and transcripts and proteins (LANA and K8.1) encoded by both latent- and lytic-cycle KSHV genes were demonstrated in various tissues including the spleen and the skin of the mice following either intraperitoneal, oral, or intravaginal inoculations. Infection of B cells and macrophages was demonstrated but that in endothelial cells was not clearly shown. It is noteworthy that the normal oral route of human KSHV infection was reproduced in this model, although KSHV-related diseases and immune responses were not described. In a related study, human CD34+ hematopoietic progenitor cells (HPCs) were inoculated with KSHV in vitro and transplanted into NOD/*scid* mice. Human CD14+ and CD19+ cells reconstituted from these infected HPCs contained KSHV, suggesting that this system might be used as an in vivo model of persistent KSHV infection (Wu et al. 2006). When cells of the endothelial lineage isolated from mouse bone marrow were transfected with a BAC plasmid containing the KSHV genome and transplanted to nude mice or NOD/*scid* mice, they developed KS-like tumor with spindle-shaped cells and angiogenesis (Mutlu et al. 2007).

19.6 Mouse Xenograft Models

EBV infects not only B cells but also T and NK cells and in rare occasions induces their unlimited proliferation in apparently immunocompetent hosts (EBV-associated T-/NK-cell LPD). No EBV infection of human T or NK cells however has been demonstrated in humanized mice so far. To model EBV-induced T- and NK-cell lymphoproliferation in mice, PBMCs isolated from patients with CAEBV, a representative EBV-associated T-/NK-cell LPD, were transplanted intravenously to NOG mice (Imadome et al. 2011). EBV-infected cells of CD4+, CD8+, γδT, or NK lineages, depending on which cell type is infected in the donor patient, were successfully engrafted. Importantly, CD4+ T-cells were required for the engraftment of EBV-infected cells, and depletion of CD4+ cells by administration of anti-CD4 antibody following transplantation of patient-derived PBMCs blocked engraftment. These results point to the possibility of a novel anti-CAEBV therapy targeting CD4+ T-cells. Similar transplantation of PBMCs isolated from patients with EBV-associated HLH also resulted in the engraftment of EBV-infected T-cells (Imadome et al. 2011). NOG mice engrafted subcutaneously with cells of the SNK6 line derived from EBV-positive NK-cell lymphoma were used in some other studies as an in vivo model of EBV-associated T-/NK-cell LPD to evaluate drug candidates for the disease (Murata et al. 2013).

An orthotopic xenograft model of EBV-positive nasopharyngeal carcinoma was developed by transplanting cells of an EBV-positive NPC line in the nasopharyngeal epithelium of NSG mice (Smith et al. 2011). This model was subsequently used to evaluate the in vivo functions of EBV microRNAs of the BART cluster (Qiu et al. 2015).

Xenotransplantation of the KSHV-infected, PEL-derived cell line BCBL-1 to C.B-17 *scid* mice resulted in immunoblastic lymphoma at the site of injection (Picchio et al. 1997). A marked murine angiogenesis was observed in this lymphoma, suggesting that this system may be also used as a model of KSHV-induced angiogenesis. More recently, a direct xenograft model of PEL was developed by transplanting cells freshly isolated from pleural fluid of a PEL patient to the peritoneal cavity of NOD/*scid* mice (Sarosiek et al. 2010).

19.7 Concluding Remarks

Humanized mice are a unique small animal model in which interactions between a virus and human cells, as well as those between the virus and the human immune system, can be analyzed in mice. Novel findings obtained first in humanized mice have been confirmed later in human studies, and they are now becoming a standard in vivo model of various human viruses including EBV. It is expected that they will be used more extensively to model KSHV infection in the near future. Humanized mouse models of EBV and KSHV infections will facilitate our understanding of their pathogenesis and the development of therapeutic and prophylactic approaches against these viruses. NHP models, on the other hand, are important as animal models closely related to humans. They will remain essential as models that allow the evaluation of novel vaccines and chemotherapeutic interventions in primate hosts that closely approximate the human condition.

Acknowledgments Work in my laboratory has been carried out in collaboration with many colleagues and students to whom I express my utmost gratitude. The work was supported by grants from the Ministry of Health, Labour and Welfare of Japan for the Research on Measures for Intractable Diseases (H21-Nanchi-094, H22-Nanchi-080, H24-Nanchi-046) and by the Practical Research Project for Rare/Intractable Diseases (16ek0109098) from Japan Agency for Medical Research and Development, AMED.

References

Antsiferova O, Muller A, Ramer PC, Chijioke O, Chatterjee B, Raykova A, Planas R, Sospedra M, Shumilov A, Tsai MH, Delecluse HJ, Münz C (2014) Adoptive transfer of EBV specific CD8+ T cell clones can transiently control EBV infection in humanized mice. PLoS Pathog 10(8):e1004333. https://doi.org/10.1371/journal.ppat.1004333 PPATHOGENS-D-14-00808 [pii]

Azzi T, Lunemann A, Murer A, Ueda S, Beziat V, Malmberg KJ, Staubli G, Gysin C, Berger C, Münz C, Chijioke O, Nadal D (2014) Role for early-differentiated natural killer cells in infectious mononucleosis. Blood 124(16):2533–2543. https://doi.org/10.1182/blood-2014-01-553024 blood-2014-01-553024 [pii]

Barton ES, White DW, Cathelyn JS, Brett-McClellan KA, Engle M, Diamond MS, Miller VL, Virgin HWT (2007) Herpesvirus latency confers symbiotic protection from bacterial infection. Nature 447 (7142):326–329. nature05762 [pii] https://doi.org/10.1038/nature05762

Barton E, Mandal P, Speck SH (2011) Pathogenesis and host control of gammaherpesviruses: lessons from the mouse. Annu Rev Immunol 29:351–397. https://doi.org/10.1146/annurev-immunol-072710-081639

Bosma GC, Custer RP, Bosma MJ (1983) A severe combined immunodeficiency mutation in the mouse. Nature 301(5900):527–530

Bruce AG, Ryan JT, Thomas MJ, Peng X, Grundhoff A, Tsai CC, Rose TM (2013) Next-generation sequence analysis of the genome of RFHVMn, the macaque homolog of Kaposi's sarcoma (KS)-associated herpesvirus, from a KS-like tumor of a pig-tailed macaque. J Virol 87 (24):13676–13693. https://doi.org/10.1128/JVI.02331-13 JVI.02331-13 [pii]

Cannon MJ, Pisa P, Fox RI, Cooper NR (1990) Epstein-Barr virus induces aggressive lymphoproliferative disorders of human B cell origin in SCID/hu chimeric mice. J Clin Invest 85(4):1333–1337

Chang H, Wachtman LM, Pearson CB, Lee JS, Lee HR, Lee SH, Vieira J, Mansfield KG, Jung JU (2009) Non-human primate model of Kaposi's sarcoma-associated herpesvirus infection. PLoS Pathog 5(10):e1000606. https://doi.org/10.1371/journal.ppat.1000606

Chijioke O, Muller A, Feederle R, Barros MH, Krieg C, Emmel V, Marcenaro E, Leung CS, Antsiferova O, Landtwing V, Bossart W, Moretta A, Hassan R, Boyman O, Niedobitek G, Delecluse HJ, Capaul R, Münz C (2013) Human natural killer cells prevent infectious mononucleosis features by targeting lytic Epstein-Barr virus infection. Cell Rep 5(6):1489–1498. https://doi.org/10.1016/j.celrep.2013.11.041 S2211-1247(13)00725-0 [pii]

Damania B (2007) EBV and KSHV-related herpesviruses in non-human primates. In: Arvin A, Campadelli-Fiume G, Mocarski E, Moore PS, Roizman B, Whitley R, Yamanishi K (eds) Human herpesviruses: biology, therapy, and immunoprophylaxis. Cambridge University Press, New York, pp 1013–1114

Damania BA, Cesarman E (2013) Kaposi's sarcoma-associated herpesvirus. In: Knipe DM, Howley PM, Cohen JI, Griffin DE, Lamb RA, Martin MA, Racaniello VR, Roizman B (eds) Fields virology, vol II. Wolters Kluwer/Lippincott Williams & Wilkins, Philadelphia, pp 2080–2128

Dirmeier U, Neuhierl B, Kilger E, Reisbach G, Sandberg ML, Hammerschmidt W (2003) Latent membrane protein 1 is critical for efficient growth transformation of human B cells by epstein-barr virus. Cancer Res 63(11):2982–2989

Dittmer D, Stoddart C, Renne R, Linquist-Stepps V, Moreno ME, Bare C, McCune JM, Ganem D (1999) Experimental transmission of Kaposi's sarcoma-associated herpesvirus (KSHV/HHV-8) to SCID-hu Thy/Liv mice. J Exp Med 190(12):1857–1868

Ensser A, Fleckenstein B (2007) Gammaherpesviruses of new world primates. In: Arvin A, Campadelli-Fiume G, Mocarski E, Moore PS, Roizman B, Whitley R, Yamanishi K (eds) Human herpesviruses: biology, therapy, and immunoprophylaxis. Cambridge University Press, New York, pp 1076–1092

Epstein MA, Morgan AJ, Finerty S, Randle BJ, Kirkwood JK (1985) Protection of cottontop tamarins against Epstein-Barr virus-induced malignant lymphoma by a prototype subunit vaccine. Nature 318(6043):287–289

Estep RD, Wong SW (2013) Rhesus macaque rhadinovirus-associated disease. Curr Opin Virol 3(3):245–250. https://doi.org/10.1016/j.coviro.2013.05.016 S1879-6257(13)00076-X [pii]

Estep RD, Rawlings SD, Li H, Manoharan M, Blaine ET, O'Connor MA, Messaoudi I, Axthelm MK, Wong SW (2014) The rhesus rhadinovirus CD200 homologue affects immune responses and viral loads during in vivo infection. J Virol 88 (18):10635–10654. https://doi.org/10.1128/JVI.01276-14 JVI.01276-14 [pii]

Falk L, Deinhardt F, Wolfe L, Johnson D, Hilgers J, de The G (1976) Epstein-Barr virus: experimental infection of Callithrix Jacchus marmosets. Int J Cancer 17(6):785–788

Foreman KE, Friborg J, Chandran B, Katano H, Sata T, Mercader M, Nabel GJ, Nickoloff BJ (2001) Injection of human herpesvirus-8 in human skin engrafted on SCID mice induces Kaposi's sarcoma-like lesions. J Dermatol Sci 26(3):182–193 doi:S0923181101000871 [pii]

Fujiwara S, Kimura H, Imadome K, Arai A, Kodama E, Morio T, Shimizu N, Wakiguchi H (2014) Current research on chronic active Epstein-Barr virus infection in Japan. Pediatr Int 56(2):159–166. https://doi.org/10.1111/ped.12314

Fujiwara S, Imadome K, Takei M (2015) Modeling EBV infection and pathogenesis in new-generation humanized mice. Exp Mol Med 47:e135. https://doi.org/10.1038/emm.2014.88 emm201488 [pii]

Gregorovic G, Boulden EA, Bosshard R, Elgueta Karstegl C, Skalsky R, Cullen BR, Gujer C, Ramer P, Münz C, Farrell PJ (2015) Epstein-Barr Viruses (EBVs) deficient in EBV-encoded RNAs have higher levels of latent membrane protein 2 RNA expression in Lymphoblastoid cell lines and efficiently establish persistent infections in humanized mice. J Virol 89(22):11711–11714. https://doi.org/10.1128/JVI.01873-15 JVI.01873-15 [pii]

Hislop AD, Taylor GS, Sauce D, Rickinson AB (2007) Cellular responses to viral infection in humans: lessons from Epstein-Barr virus. Annu Rev Immunol 25:587–617. https://doi.org/10.1146/annurev.immunol.25.022106.141553

Imadome K, Yajima M, Arai A, Nakazawa A, Kawano F, Ichikawa S, Shimizu N, Yamamoto N, Morio T, Ohga S, Nakamura H, Ito M, Miura O, Komano J, Fujiwara S (2011) Novel mouse xenograft models reveal a critical role of CD4+ T cells in the proliferation of EBV-infected T and NK cells. PLoS Pathog 7(10):e1002326. https://doi.org/10.1371/journal.ppat.1002326 PPATHOGENS-D-11-00208 [pii]

Ito M, Hiramatsu H, Kobayashi K, Suzue K, Kawahata M, Hioki K, Ueyama Y, Koyanagi Y, Sugamura K, Tsuji K, Heike T, Nakahata T (2002) NOD/SCID/gamma(c)(null) mouse: an excellent recipient mouse model for engraftment of human cells. Blood 100(9):3175–3182

Johannessen I, Crawford DH (1999) In vivo models for Epstein-Barr virus (EBV)-associated B cell lymphoproliferative disease (BLPD). Rev Med Virol 9(4):263–277. https://doi.org/10.1002/(SICI)1099-1654(199910/12)9:4<263::AID-RMV256>3.0.CO;2-D [pii]

Khan G, Ahmed W, Philip PS, Ali MH, Adem A (2015) Healthy rabbits are susceptible to Epstein-Barr virus infection and infected cells proliferate in immunosuppressed animals. Virol J 12:28. https://doi.org/10.1186/s12985-015-0260-1 s12985-015-0260-1 [pii]

Kutok JL, Klumpp S, Simon M, MacKey JJ, Nguyen V, Middeldorp JM, Aster JC, Wang F (2004) Molecular evidence for rhesus lymphocryptovirus infection of epithelial cells in immunosuppressed rhesus macaques. J Virol 78(7):3455–3461

Kuwana Y, Takei M, Yajima M, Imadome K, Inomata H, Shiozaki M, Ikumi N, Nozaki T, Shiraiwa H, Kitamura N, Takeuchi J, Sawada S, Yamamoto N, Shimizu N, Ito M, Fujiwara S (2011) Epstein-Barr virus induces erosive arthritis in humanized mice. PLoS One 6(10):e26630. https://doi.org/10.1371/journal.pone.0026630 PONE-D-11-16695 [pii]

Lee EK, Joo EH, Song KA, Choi B, Kim M, Kim SH, Kim SJ, Kang MS (2015) Effects of lymphocyte profile on development of EBV-induced lymphoma subtypes in humanized mice. Proc Natl Acad Sci USA 112(42):13081–13086. https://doi.org/10.1073/pnas.1407075112 1407075112 [pii]

Leskowitz R, Fogg MH, Zhou XY, Kaur A, Silveira EL, Villinger F, Lieberman PM, Wang F, Ertl HC (2014) Adenovirus-based vaccines against rhesus lymphocryptovirus EBNA-1 induce expansion of specific CD8+ and CD4+ T cells in persistently infected rhesus macaques. J Virol 88 9:4721–4735. https://doi.org/10.1128/JVI.03744-13 JVI.03744-13 [pii]

Liang X, Collins CM, Mendel JB, Iwakoshi NN, Speck SH (2009) Gammaherpesvirus-driven plasma cell differentiation regulates virus reactivation from latently infected B lymphocytes. PLoS Pathog 5(11):e1000677. https://doi.org/10.1371/journal.ppat.1000677

Liang X, Paden CR, Morales FM, Powers RP, Jacob J, Speck SH (2011) Murine gamma-herpesvirus immortalization of fetal liver-derived B cells requires both the viral cyclin D homolog and latency-associated nuclear antigen. PLoS Pathog 7(9):e1002220. https://doi.org/10.1371/journal.ppat.1002220 PPATHOGENS-D-11-00810 [pii]

Lin X, Tsai MH, Shumilov A, Poirey R, Bannert H, Middeldorp JM, Feederle R, Delecluse HJ (2015) The Epstein-Barr virus BART miRNA cluster of the M81 strain modulates multiple functions in primary B cells. PLoS Pathog 11(12):e1005344. https://doi.org/10.1371/journal.ppat.1005344 PPATHOGENS-D-15-01749 [pii]

Longnecker RM, Kieff E, Cohen JI (2013) Epstein-Barr virus. In: Knipe DM, Howley PM, Cohen JI, Griffin DE, Lamb RA, Martin MA, Racaniello VR, Roizman B (eds) Fields virology, vol II, 6th edn. Wolters Kluwer/Lippincott Williams & Wilkins, Philadelphia, pp 1898–1959

Ma SD, Hegde S, Young KH, Sullivan R, Rajesh D, Zhou Y, Jankowska-Gan E, Burlingham WJ, Sun X, Gulley ML, Tang W, Gumperz JE, Kenney SC (2011) A new model of Epstein-Barr virus infection reveals an important role for early lytic viral protein expression in the development of lymphomas. J Virol 85(1):165–177. https://doi.org/10.1128/JVI.01512-10 JVI.01512-10 [pii]

Ma SD, Xu X, Plowshay J, Ranheim EA, Burlingham WJ, Jensen JL, Asimakopoulos F, Tang W, Gulley ML, Cesarman E, Gumperz JE, Kenney SC (2015) LMP1-deficient Epstein-Barr virus mutant requires T cells for lymphomagenesis. J Clin Invest 125(1):304–315. https://doi.org/10.1172/JCI76357 76357 [pii]

Ma SD, Xu X, Jones R, Delecluse HJ, Zumwalde NA, Sharma A, Gumperz JE, Kenney SC (2016) PD-1/CTLA-4 blockade inhibits Epstein-Barr virus-induced lymphoma growth in a cord blood humanized-mouse model. PLoS Pathog 12(5):e1005642. https://doi.org/10.1371/journal.ppat.1005642 PPATHOGENS-D-15-02021 [pii]

Ma SD, Tsai MH, Romero-Masters JC, Ranheim EA, Huebner SM, Bristol J, Delecluse HJ, Kenney SC (2017) LMP1 and LMP2A collaborate to promote Epstein-Barr virus (EBV)-induced B cell lymphomas in a cord blood-humanized mouse model but are not essential. J Virol. JVI.01928-16 [pii] https://doi.org/10.1128/JVI.01928-16

Mansfield KG, Westmoreland SV, DeBakker CD, Czajak S, Lackner AA, Desrosiers RC (1999) Experimental infection of rhesus and pig-tailed macaques with macaque rhadinoviruses. J Virol 73(12):10320–10328

Matsuda G, Imadome K, Kawano F, Mochizuki M, Ochiai N, Morio T, Shimizu N, Fujiwara S (2015) Cellular immunotherapy with ex vivo expanded cord blood T cells in a humanized mouse model of EBV-associated lymphoproliferative disease. Immunotherapy 7(4):335–341. https://doi.org/10.2217/imt.15.2

McCune JM, Namikawa R, Kaneshima H, Shultz LD, Lieberman M, Weissman IL (1988) The SCID-hu mouse: murine model for the analysis of human hematolymphoid differentiation and function. Science 241(4873):1632–1639

Melkus MW, Estes JD, Padgett-Thomas A, Gatlin J, Denton PW, Othieno FA, Wege AK, Haase AT, Garcia JV (2006) Humanized mice mount specific adaptive and innate immune responses to EBV and TSST-1. Nat Med 12(11):1316–1322

Moghaddam A, Rosenzweig M, Lee-Parritz D, Annis B, Johnson RP, Wang F (1997) An animal model for acute and persistent Epstein-Barr virus infection. Science 276(5321):2030–2033

Mosier DE, Gulizia RJ, Baird SM, Wilson DB (1988) Transfer of a functional human immune system to mice with severe combined immunodeficiency. Nature 335(6187):256–259

Mühe J, Wang F (2015) Non-human primate Lymphocryptoviruses: past, present, and future. Curr Top Microbiol Immunol 391:385–405. https://doi.org/10.1007/978-3-319-22834-1_13

Murata T, Iwata S, Siddiquey MN, Kanazawa T, Goshima F, Kawashima D, Kimura H, Tsurumi T (2013) Heat shock protein 90 inhibitors repress latent membrane protein 1 (LMP1) expression and proliferation of Epstein-Barr virus-positive natural killer cell lymphoma. PLoS One 8(5):e63566. https://doi.org/10.1371/journal.pone.0063566 PONE-D-12-34778 [pii]

Mutlu AD, Cavallin LE, Vincent L, Chiozzini C, Eroles P, Duran EM, Asgari Z, Hooper AT, La Perle KM, Hilsher C, Gao SJ, Dittmer DP, Rafii S, Mesri EA (2007) In vivo-restricted and reversible malignancy induced by human herpesvirus-8 KSHV: a cell and animal model of virally induced Kaposi's sarcoma. Cancer Cell 11(3):245–258. S1535-6108(07)00031-1 [pii] https://doi.org/10.1016/j.ccr.2007.01.015

Niller HH, Wolf H, Ay E, Minarovits J (2011) Epigenetic dysregulation of Epstein-Barr virus latency and development of autoimmune disease. Adv Exp Med Biol 711:82–102

Ohashi M, Fogg MH, Orlova N, Quink C, Wang F (2012) An Epstein-Barr virus encoded inhibitor of Colony stimulating Factor-1 signaling is an important determinant for acute and persistent EBV infection. PLoS Pathog 8(12):e1003095. https://doi.org/10.1371/journal.ppat.1003095 PPATHOGENS-D-12-02098 [pii]

Okuno K, Takashima K, Kanai K, Ohashi M, Hyuga R, Sugihara H, Kuwamoto S, Kato M, Sano H, Sairenji T, Kanzaki S, Hayashi K (2010) Epstein-Barr virus can infect rabbits by the intranasal or peroral route: an animal model for natural primary EBV infection in humans. J Med Virol 82(6):977–986. https://doi.org/10.1002/jmv.21597

Orzechowska BU, Powers MF, Sprague J, Li H, Yen B, Searles RP, Axthelm MK, Wong SW (2008) Rhesus macaque rhadinovirus-associated non-Hodgkin lymphoma: animal model for KSHV-associated malignancies. Blood 112(10):4227–4234. https://doi.org/10.1182/blood-2008-04-151498 blood-2008-04-151498 [pii]

Parsons CH, Adang LA, Overdevest J, O'Connor CM, Taylor JR Jr, Camerini D, Kedes DH (2006) KSHV targets multiple leukocyte lineages during long-term productive infection in NOD/SCID mice. J Clin Invest 116(7):1963–1973. https://doi.org/10.1172/JCI27249

Picchio GR, Sabbe RE, Gulizia RJ, McGrath M, Herndier BG, Mosier DE (1997) The KSHV/HHV8-infected BCBL-1 lymphoma line causes tumors in SCID mice but fails to transmit virus to a human peripheral blood mononuclear cell graft. Virology 238(1):22–29. S0042-6822(97)98822-X [pii] https://doi.org/10.1006/viro.1997.8822

Plummer M, de Martel C, Vignat J, Ferlay J, Bray F, Franceschi S (2016) Global burden of cancers attributable to infections in 2012: a synthetic analysis. Lancet Glob Health 4(9):e609–e616. https://doi.org/10.1016/S2214-109X(16)30143-7 S2214-109X(16)30143-7 [pii]

Qiu J, Smith P, Leahy L, Thorley-Lawson DA (2015) The Epstein-Barr virus encoded BART miRNAs potentiate tumor growth in vivo. PLoS Pathog 11(1):e1004561. https://doi.org/10.1371/journal.ppat.1004561 PPATHOGENS-D-14-01638 [pii]

Ramer JC, Garber RL, Steele KE, Boyson JF, O'Rourke C, Thomson JA (2000) Fatal lymphoproliferative disease associated with a novel gammaherpesvirus in a captive population of common marmosets. Comp Med 50(1):59–68

Renne R, Dittmer D, Kedes D, Schmidt K, Desrosiers RC, Luciw PA, Ganem D (2004) Experimental transmission of Kaposi's sarcoma-associated herpesvirus (KSHV/HHV-8) to SIV-positive and SIV-negative rhesus macaques. J Med Primatol 33(1):1–9. https://doi.org/10.1046/j.1600-0684.2003.00043.x JMP043 [pii]

Rivailler P, Cho YG, Wang F (2002a) Complete genomic sequence of an Epstein-Barr virus-related herpesvirus naturally infecting a new world primate: a defining point in the evolution of oncogenic lymphocryptoviruses. J Virol 76(23):12055–12068

Rivailler P, Jiang H, Cho YG, Quink C, Wang F (2002b) Complete nucleotide sequence of the rhesus lymphocryptovirus: genetic validation for an Epstein-Barr virus animal model. J Virol 76(1):421–426

Rivailler P, Carville A, Kaur A, Rao P, Quink C, Kutok JL, Westmoreland S, Klumpp S, Simon M, Aster JC, Wang F (2004) Experimental rhesus lymphocryptovirus infection in immunosuppressed macaques: an animal model for Epstein-Barr virus pathogenesis in the immunosuppressed host. Blood 104(5):1482–1489. https://doi.org/10.1182/blood-2004-01-0342 2004-01-0342 [pii]

Robinson BA, Estep RD, Messaoudi I, Rogers KS, Wong SW (2012) Viral interferon regulatory factors decrease the induction of type I and type II interferon during rhesus macaque rhadinovirus infection. J Virol 86(4):2197–2211. https://doi.org/10.1128/JVI.05047-11 JVI.05047-11 [pii]

Rohner E, Wyss N, Trelle S, Mbulaiteye SM, Egger M, Novak U, Zwahlen M, Bohlius J (2014) HHV-8 seroprevalence: a global view. Syst Rev 3:11. https://doi.org/10.1186/2046-4053-3-11 2046-4053-3-11 [pii]

Rose TM, Strand KB, Schultz ER, Schaefer G, Rankin GW Jr, Thouless ME, Tsai CC, Bosch ML (1997) Identification of two homologs of the Kaposi's sarcoma-associated herpesvirus (human herpesvirus 8) in retroperitoneal fibromatosis of different macaque species. J Virol 71(5):4138–4144

Rowe M, Young LS, Crocker J, Stokes H, Henderson S, Rickinson AB (1991) Epstein-Barr virus (EBV)-associated lymphoproliferative disease in the SCID mouse model: implications for the pathogenesis of EBV-positive lymphomas in man. J Exp Med 173(1):147–158

Sarosiek KA, Cavallin LE, Bhatt S, Toomey NL, Natkunam Y, Blasini W, Gentles AJ, Ramos JC, Mesri EA, Lossos IS (2010) Efficacy of bortezomib in a direct xenograft model of primary effusion lymphoma. Proc Natl Acad Sci USA 107(29):13069–13074. https://doi.org/10.1073/pnas.1002985107 1002985107 [pii]

Sashihara J, Hoshino Y, Bowman JJ, Krogmann T, Burbelo PD, Coffield VM, Kamrud K, Cohen JI (2011) Soluble rhesus lymphocryptovirus gp350 protects against infection and reduces viral loads in animals that become infected with virus after challenge. PLoS Pathog 7(10):e1002308. https://doi.org/10.1371/journal.ppat.1002308 PPATHOGENS-D-11-01479 [pii]

Sato K, Misawa N, Nie C, Satou Y, Iwakiri D, Matsuoka M, Takahashi R, Kuzushima K, Ito M, Takada K, Koyanagi Y (2011) A novel animal model of Epstein-Barr virus-associated hemophagocytic lymphohistiocytosis in humanized mice. Blood 117(21):5663–5673. https://doi.org/10.1182/blood-2010-09-305979 blood-2010-09-305979 [pii]

Searles RP, Bergquam EP, Axthelm MK, Wong SW (1999) Sequence and genomic analysis of a Rhesus macaque rhadinovirus with similarity to Kaposi's sarcoma-associated herpesvirus/human herpesvirus 8. J Virol 73(4):3040–3053

Shope T, Dechairo D, Miller G (1973) Malignant lymphoma in cottontop marmosets after inoculation with Epstein-Barr virus. Proc Natl Acad Sci USA 70(9):2487–2491

Shultz LD, Lyons BL, Burzenski LM, Gott B, Chen X, Chaleff S, Kotb M, Gillies SD, King M, Mangada J, Greiner DL, Handgretinger R (2005) Human lymphoid and myeloid cell development in NOD/LtSz-scid IL2R gamma null mice engrafted with mobilized human hemopoietic stem cells. J Immunol 174(10):6477–6489. doi:174/10/6477 [pii]

Shultz LD, Saito Y, Najima Y, Tanaka S, Ochi T, Tomizawa M, Doi T, Sone A, Suzuki N, Fujiwara H, Yasukawa M, Ishikawa F (2010) Generation of functional human T-cell subsets with HLA-restricted immune responses in HLA class I expressing NOD/SCID/IL2r gamma(null) humanized mice. Proc Natl Acad Sci USA 107(29):13022–13027. https://doi.org/10.1073/pnas.1000475107 1000475107 [pii]

Smith PA, Merritt D, Barr L, Thorley-Lawson DA (2011) An orthotopic model of metastatic nasopharyngeal carcinoma and its application in elucidating a therapeutic target that inhibits metastasis. Genes Cancer 2(11):1023–1033. https://doi.org/10.1177/1947601912440878 10.1177_1947601912440878 [pii]

Strowig T, Gurer C, Ploss A, Liu YF, Arrey F, Sashihara J, Koo G, Rice CM, Young JW, Chadburn A, Cohen JI, Münz C (2009) Priming of protective T cell responses against virus-induced tumors in mice with human immune system components. J Exp Med 206(6):1423–1434. https://doi.org/10.1084/jem.20081720 jem.20081720 [pii]

Sun C, Schattgen SA, Pisitkun P, Jorgensen JP, Hilterbrand AT, Wang LJ, West JA, Hansen K, Horan KA, Jakobsen MR, O'Hare P, Adler H, Sun R, Ploegh HL, Damania B, Upton JW, Fitzgerald KA, Paludan SR (2015) Evasion of innate cytosolic DNA sensing by a gammaherpesvirus facilitates establishment of latent infection. J Immunol 194 (4):1819–1831. https://doi.org/10.4049/jimmunol.1402495 jimmunol.1402495 [pii]

Sunil-Chandra NP, Arno J, Fazakerley J, Nash AA (1994) Lymphoproliferative disease in mice infected with murine gammaherpesvirus 68. Am J Pathol 145(4):818–826

Takashima K, Ohashi M, Kitamura Y, Ando K, Nagashima K, Sugihara H, Okuno K, Sairenji T, Hayashi K (2008) A new animal model for primary and persistent Epstein-Barr virus infection: human EBV-infected rabbit characteristics determined using sequential imaging and pathological analysis. J Med Virol 80(3):455–466. https://doi.org/10.1002/jmv.21102

Takei M, Mitamura K, Fujiwara S, Horie T, Ryu J, Osaka S, Yoshino S, Sawada S (1997) Detection of Epstein-Barr virus-encoded small RNA 1 and latent membrane protein 1 in synovial lining cells from rheumatoid arthritis patients. Int Immunol 9(5):739–743

Traggiai E, Chicha L, Mazzucchelli L, Bronz L, Piffaretti JC, Lanzavecchia A, Manz MG (2004) Development of a human adaptive immune system in cord blood cell-transplanted mice. Science 304(5667):104–107

Virgin HW, Latreille P, Wamsley P, Hallsworth K, Weck KE, Dal Canto AJ, Speck SH (1997) Complete sequence and genomic analysis of murine gammaherpesvirus 68. J Virol 71(8):5894–5904

Wahl A, Linnstaedt SD, Esoda C, Krisko JF, Martinez-Torres F, Delecluse HJ, Cullen BR, Garcia JV (2013) A cluster of virus-encoded microRNAs accelerates acute systemic Epstein-Barr virus infection but does not significantly enhance virus-induced oncogenesis in vivo. J Virol 87 (10):5437–5446. https://doi.org/10.1128/JVI.00281-13 JVI.00281-13 [pii]

Wang LX, Kang G, Kumar P, Lu W, Li Y, Zhou Y, Li Q, Wood C (2014) Humanized-BLT mouse model of Kaposi's sarcoma-associated herpesvirus infection. Proc Natl Acad Sci USA 111(8):3146–3151. https://doi.org/10.1073/pnas.1318175111 1318175111 [pii]

Westmoreland SV, Mansfield KG (2008) Comparative pathobiology of Kaposi sarcoma-associated herpesvirus and related primate rhadinoviruses. Comp Med 58(1):31–42

White DW, Keppel CR, Schneider SE, Reese TA, Coder J, Payton JE, Ley TJ, Virgin HW, Fehniger TA (2010) Latent herpesvirus infection arms NK cells. Blood 115(22):4377–4383. https://doi.org/10.1182/blood-2009-09-245464 blood-2009-09-245464 [pii]

White RE, Ramer PC, Naresh KN, Meixlsperger S, Pinaud L, Rooney C, Savoldo B, Coutinho R, Bodor C, Gribben J, Ibrahim HA, Bower M, Nourse JP, Gandhi MK, Middeldorp J, Cader FZ, Murray P, Münz C, Allday MJ (2012) EBNA3B-deficient EBV promotes B cell lymphomagenesis in humanized mice and is found in human tumors. J Clin Invest 122(4):1487–1502. https://doi.org/10.1172/JCI58092 58092 [pii]

Whitehurst CB, Li G, Montgomery SA, Montgomery ND, Su L, Pagano JS (2015) Knockout of Epstein-Barr virus BPLF1 retards B-cell transformation and lymphoma formation in humanized mice. MBio 6(5):e01574–e01515. https://doi.org/10.1128/mBio.01574-15 e01574-15 [pii] mBio.01574-15 [pii]

Wong SW, Bergquam EP, Swanson RM, Lee FW, Shiigi SM, Avery NA, Fanton JW, Axthelm MK (1999) Induction of B cell hyperplasia in simian immunodeficiency virus-infected rhesus macaques with the simian homologue of Kaposi's sarcoma-associated herpesvirus. J Exp Med 190(6):827–840

Wu W, Vieira J, Fiore N, Banerjee P, Sieburg M, Rochford R, Harrington W, Jr., Feuer G (2006) KSHV/HHV-8 infection of human hematopoietic progenitor (CD34+) cells: persistence of infection during hematopoiesis in vitro and in vivo. Blood 108(1):141–151. 2005-04-1697 [pii] https://doi.org/10.1182/blood-2005-04-1697

Wu TT, Qian J, Ang J, Sun R (2012) Vaccine prospect of Kaposi sarcoma-associated herpesvirus. Curr Opin Virol 2(4):482–488. https://doi.org/10.1016/j.coviro.2012.06.005 S1879-6257(12)00100-9 [pii]

Xiang Z, Liu Y, Zheng J, Liu M, Lv A, Gao Y, Hu H, Lam KT, Chan GC, Yang Y, Chen H, Tsao GS, Bonneville M, Lau YL, Tu W (2014) Targeted activation of human Vgamma9Vdelta2-T cells controls epstein-barr virus-induced B cell lymphoproliferative disease. Cancer Cell 26(4):565–576. https://doi.org/10.1016/j.ccr.2014.07.026 S1535-6108(14)00314-6 [pii]

Yajima M, Imadome K, Nakagawa A, Watanabe S, Terashima K, Nakamura H, Ito M, Shimizu N, Honda M, Yamamoto N, Fujiwara S (2008) A new humanized mouse model of Epstein-Barr virus infection that reproduces persistent infection, lymphoproliferative disorder, and cell-mediated and humoral immune responses. J Infect Dis 198(5):673–682. https://doi.org/10.1086/590502

Yajima M, Imadome K, Nakagawa A, Watanabe S, Terashima K, Nakamura H, Ito M, Shimizu N, Yamamoto N, Fujiwara S (2009) T cell-mediated control of Epstein-Barr virus infection in humanized mice. J Infect Dis 200(10):1611–1615. https://doi.org/10.1086/644644

Chapter 20
Gastritis-Infection-Cancer Sequence of Epstein-Barr Virus-Associated Gastric Cancer

Masashi Fukayama, Akiko Kunita, and Atsushi Kaneda

Abstract Epstein-Barr virus-associated gastric cancer (EBVaGC) is a representative EBV-infected epithelial neoplasm, which is now included as one of the four subtypes of The Cancer Genome Atlas molecular classification of gastric cancer. In this review, we portray a gastritis-infection-cancer sequence of EBVaGC. This virus-associated type of gastric cancer demonstrates clonal growth of EBV-infected epithelial cells within the mucosa of atrophic gastritis. Its core molecular abnormality is the EBV-specific hyper-epigenotype of CpG island promoter methylation, which induces silencing of tumor suppressor genes. This is due to the infection-induced disruption of the balance between DNA methylation and DNA demethylation activities. Abnormalities in the host cell genome, including phosphatidylinositol-4,5-biphosphate 3-kinase catalytic subunit α (*PIK3CA*), AT-rich interaction domain 1A (*ARID1A*), and programmed death-ligand 1 (*PD-L1*), are associated with the development and progression of EBVaGC. Furthermore, post-transcriptional modulation affects the transformation processes of EBV-infected cells, such as epithelial mesenchymal transition and anti-apoptosis, via cellular and viral microRNAs (miRNAs). Once established, cancer cells of EBVaGC remodel their microenvironment, at least partly, via the delivery of exosomes containing cellular and viral miRNAs. After exosomes are incorporated, these molecules change the functions of stromal cells, tuning the microenvironment for EBVaGC. During this series of events, EBV hijacks and uses cellular machineries, such as DNA methylation and the miRNA delivery system. This portrait of gastritis-infection-cancer sequences highlights the survival strategies of EBV in the stomach epithelial cells and may be useful for the integration of therapeutic modalities against EBV-driven gastric cancer.

M. Fukayama (✉) · A. Kunita
Department of Pathology, Graduate School of Medicine, The University of Tokyo, Tokyo, Japan
e-mail: mfukayama-tky@umin.ac.jp

A. Kaneda
Department of Molecular Oncology, Graduate School of Medicine, Chiba University, Chiba, Japan

Keywords Gastric cancer · EBVaGC · EBER-ISH · TCGA · CpG methylation · PIK3CA · ARID1A · PD-L1 · microRNA

20.1 Introduction

Epstein-Barr virus-associated gastric cancer (EBVaGC) is a distinct subtype of gastric cancer, consisting of clonal growth of EBV-infected stomach epithelial cells (Fukayama and Ushiku 2011, Shinozaki-Ushiku et al. 2015a). Previous studies investigating this unique type of gastric cancer have revealed its characteristic molecular abnormalities, and The Cancer Genome Atlas (TCGA) Research Network recently included EBVaGC as one of the four subtypes in a novel molecular classification of gastric cancer, in addition to the microsatellite instability (MSI), chromosomal instability (CI), and genomically stable (GS) subtypes (The Cancer Genome Atlas Research Network 2014).

In this chapter, we portray a gastritis-infection-cancer sequence of EBVaGC (Fig. 20.1) following a brief review of the epidemiological and pathological aspects of EBVaGC (Table 20.1). A complete understanding of the cellular and microenvironment abnormalities of this subtype of gastric cancer remains to be elucidated; thus, we will highlight important questions that should be addressed by future studies in the following sections.

20.2 Brief Review of Epidemiology and Pathology of EBVaGC

The presence of EBV-DNA was first identified by PCR analysis of gastric cancer tissues demonstrating lymphoepithelioma-like histology in 1990 (Burke et al. 1990). Subsequent studies have adopted in situ hybridization (ISH) targeting EBV-encoded small RNAs (EBERs) for the detection of EBV infection, since it can easily be performed with formalin-fixed and paraffin-embedded tumor tissue sections (Shibata et al. 1991, Shibata and Weiss 1992, Fukayama et al. 1994). Thus, from the beginning, the research was in combination with histopathological observations, and it has been suggested that EBV is associated with cancer. Almost all the cancer cells demonstrated positive signals of EBERs, whereas infiltrating immune cells in and around the tumor were negative, with the exception of a small number of positive lymphocytes.

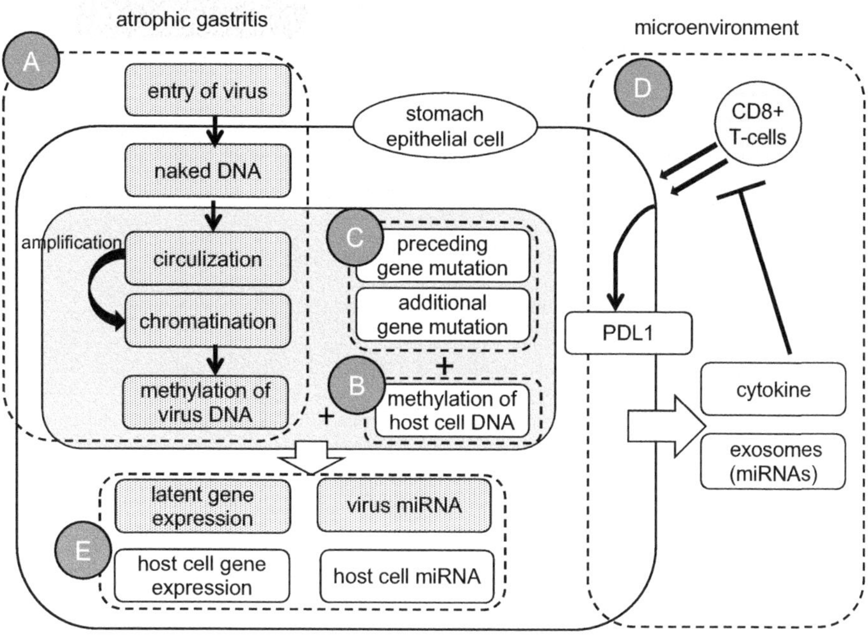

Fig. 20.1 Gastritis-infection-cancer sequence of EBV-associated gastric cancer. (**a**) EBV infection is established in a stem cell of stomach epithelia with a background of atrophic gastritis. Virus DNA in a host cell nucleus takes a circular form and is tethered to cell chromatin, and then its CpG motifs are widely methylated toward latent infection. (**b**) Hypermethylation of CpG DNA of host genome is induced after viral DNA methylation. Then clonal growth begins. (**c**) Precursor and additional gene mutations in the infected cell affect the development and progression of cancer. (**d**) Microenvironment is altered by cytokines and exosomes from cancer cells. Expression of PD-L1 is increased in response to cytokines from CD8+ T-cells. (**e**) Abnormal gene expressions in a host cell, latent gene expression of EBV, and miRNAs derived from host cell and virus are associated with cancer development and modulation of microenvironment

20.2.1 Epidemiology

Owing to high sensitivity and specificity, positive signals with EBER-ISH in cancer cell nuclei have been regarded as a gold standard to define EBVaGC. Using EBER-ISH, the following epidemiological data were obtained. The frequency of EBVaGC was 7–10% of gastric cancer; therefore, the annual worldwide EBVaGC-associated mortalities were estimated to be 70,000 cases (Khan and Hashim 2014). This frequency is comparable to that of nasopharyngeal carcinoma (NPC) and is much higher compared with the incidence of EBV-associated lymphoid malignancies

Table 20.1 Clinicopathological features of EBV-associated gastric cancer

Patient	
Clinical	Nearly 10% of total gastric carcinoma Male predominance Relatively younger age[a] Previous stomach resection
Cancer behavior	Relatively favorable prognosis
Pathology	
Macroscopic	Location at gastric cardia/body, remnant stomach Ulcerated or saucerlike tumor Marked thickening of the gastric wall Multiple lesions[a]
Microscopic	Moderately to poorly differentiated adenocarcinoma Lymphocytic infiltration in various degrees Lymphoepithelioma-like, Crohn's disease-like, conventional Lace pattern within the mucosa
Extension	Lower rate of lymph node involvement[a]
Background	Atrophic gastritis (moderate to severe in degree) Gastritis cystica profunda

[a]The findings have been suggested, but more evidence is necessary for confirmation

(~8,000 cases of Hodgkin's lymphoma and ~2, 000 of Burkitt's lymphoma) (Khan and Hashim 2014). In regard to the regional occurrences, an inverse association has been suggested between the background incidence of gastric cancer and the proportion of EBVaGC (Tashiro et al. 1998, Murphy et al. 2009). This phenomenon contrasts to the regional preference of other types of EBV-positive neoplasms, including cases of NPC in China and Southeast Asia and endemic Burkitt's lymphoma in Africa and New Guinea. As a major carcinogen of gastric cancer, *Helicobacter pylori* (*H. pylori*) infection may serve a fundamental role in the developmental process of EBVaGC. The infection rates of *H. pylori* do not vary between EBVaGC and EBV-negative gastric cancer (Camargo et al. 2016); however, a severe degree of atrophic gastritis is characteristic of EBVaGC (Kaizaki et al. 1999).

20.2.2 Pathology

Clinicopathological features of EBVaGC include male predominance and a predominant location in the proximal and middle regions of the stomach. The histological features demonstrate moderately to poorly differentiated adenocarcinoma and relatively rich lymphocytic infiltration (Table 20.1) (Fukayama and Ushiku 2011, Shinozaki-Ushiku et al. 2015a). Song et al. (2010) divided EBVaGC into three subgroups based on the cellular immune response: lymphoepithelioma-like carcinoma (LELC), Crohn's disease-like reaction (CLR), and conventional adenocarcinoma (CA). The prognosis of CA-EBVaGC was the poorest, followed by CLR and LELC. When cases of CA were excluded, the prognosis of LELC and CRL cases combined was significantly improved compared with that of EBV-negative GC. However, a recent meta-analysis (Murphy et al. 2009) demonstrated that the prognosis of EBVaGC was better than that of EBV-negative GC, particularly in Asian individuals.

Virologic aspects of EBVaGC were first reported by the research group of Prof. Takada (Imai et al. 1994, Sugiura et al. 1996), including the episomal form of EBV-DNA and the expression of a limited number of EBV latent genes (EBNA1, LMP2A, and EBER). A previous study investigating the molecular abnormalities in EBVaGC consecutively demonstrated a paucity of apoptosis, rare abnormalities in p53 expression level, and a low frequency of allelic imbalance (loss of heterozygosity and MSI) (Chong et al. 1994). Subsequently, CpG island DNA methylation was identified in the promoter regions of numerous cancer-associated genes in EBVaGC (Kang et al. 2002, Osawa et al. 2002), which led to recognition as the core abnormality of EBVaGC, i.e., EBV-specific hyper-epigenotype or EBV-CpG island methylator phenotype (CIMP) (Matsusaka et al. 2011).

20.3 The Gastritis-Infection-Cancer Sequence

Figure 20.1 presents the developmental processes of EBVaGC (Fukayama and Ushiku 2011, Fukayama et al. 2008). EBV infection is established in a stem cell at the mucosa of atrophic gastritis (**A**). The latent infection simultaneously induces hypermethylation of CpG DNA of the host genome, which is partly due to a loss of balance between DNA methyltransferase1 (DNMT1)-mediated methylation and ten-eleven translocation (TET) enzyme-mediated demethylation by overdriving the methylation machinery in the founder cell, which is followed by clonal cell growth (**B**). In this process, inflammation-associated signals serve important roles in infected cells. Genomic mutations in precursor or cancer cells affect the development and progression of EBVaGC (**C**). Against the host immune response, cancer cells with EBV adapt and alter the constituent cells of their microenvironment (**D**). In these processes, abnormal cellular and viral microRNAs (miRNAs) serve important roles in epithelial-mesenchymal transition, resistance against apoptosis of founder clones, and delivery of exosomes as a tool for modulating the microenvironment (**E**). As a result, EBV achieves coexistence of infected cells with immune cells.

20.4 From Infection to Generation of the Founder Clone

EBV is a DNA virus that takes a double-stranded and linear form within viral particles. It becomes circular and episomal within the host nuclei following infection. Varying numbers of terminal repeats (TRs) are excised from both ends of EBV-DNA at fusion of the linear DNA, and therefore, the number of TRs varies in each infected cell. Southern blot analysis of EBV-TRs, when applied to EBVaGC, demonstrated that EBV was monoclonal in cancers at advanced stages (Imai et al. 1994, Fukayama et al. 1994, Song et al. 2010), whereas it was monoclonal or oligoclonal in intramucosal cancers (Uozaki and Fukayama 2008). In addition, positive EBER signals were present in the nuclei of all cancer cells, indicating that EBVaGC consists of clonal growth of EBV-infected cells. Therefore, EBV infection is an early event in carcinogenesis of EBVaGC, which means that only one of the EBV-infected founder clones becomes predominant and progresses to the monoclonal invasive cancer.

EBV remains in an episomal form in the host nucleus. Recent whole genome sequencing studies (Strong et al. 2013, Liu et al. 2016) confirmed that chromosomal integration of viral DNA into host genome is an extremely rare occurrence. The copy number of EBV-DNA per cell is variable in each case of EBVaGC from several to <100 copies, according to an estimate revealed by Southern blotting (Imai et al. 1994, Fukayama et al. 1994, Uozaki and Fukayama 2008). It has been suggested that copies of the same number are transmitted to the daughter cells, and thus the number of EBV genomes is maintained during the clonal growth of EBVaGC (Nanbo et al. 2007). Viral nuclear protein, EBNA1, serves a role in the maintenance of infection in the host cells. EBNA1 tethers the viral episomes to the host chromosomes and recruits a cellular origin-recognition complex to the origin of DNA replication (*oriP*) of the viral DNA (Kanda et al. 2013). Subsequently, using the cellular machinery, viral replication and segregation are synchronized with the host cell replication and mitosis (Nanbo et al. 2007).

When we describe the process from viral entry to establishment of clonal founder cells (**A** in Fig. 20.1), several questions arise that are subsequently addressed in the following sections.

20.4.1 Entry of EBV to Stomach Epithelial Cells

Based on the current view regarding the mechanism underlying EBV infection of epithelial cells (Hutt-Fletcher 2016), the infection processes in the stomach mucosa are as follows; EBV is reactivated (from latent to replicative infection) in B lymphocytes of the stomach mucosa, and the virus particles are transferred from EBV-infected B lymphocytes to gastric epithelial cells. Gastric epithelial cells lack expressions of major histocompatibility complex class II and CD21, which function as receptors for EBV particles in B lymphocytes. Therefore, it was proposed that the entry of EBV virions is dependent on fusion with the membrane of epithelial cells, which is triggered by direct interaction between viral glycoproteins gH-gL complex and the $\alpha v \beta 5$, $\alpha v \beta 6$, or $\alpha v \beta 8$ integrin.

In experimental conditions (Imai et al. 1998), EBV infection in gastric cancer cell lines can be established with approximately 800-fold-higher efficiency by cell-to-cell contact using an EBV-positive lymphocyte cell line, Akata, compared with direct incubation with viral particles. Nanbo et al. (2016) proposed the cell-to-cell transmission mechanism; EBV latently infected lymphocytes contact with epithelial cells via integrin β1/β2 and other molecules, which induced the lytic cycle of EBV and intracellular adhesion molecule-1 (ICAM-1) translocation to the cell surface, stabilizing cell-to-cell contact on the side of the lymphocytes. The viral particle is then transmitted by clathrin-mediated endocytosis to epithelial cells. Further series of events include absorption of viral particles, transportation of EBV-DNA to the nucleus, and subsequent replication of viral DNA for establishing latent gene expression.

20.4.2 Amplification of EBV-DNA

It has not been clarified how the copy number of EBV-DNA is determined in the nucleus of a founder clone (Lieberman 2013). As an initial step, the following events proceed in addition to the establishment of viral latent gene expressions: circularization and chromatinization of the viral genome including assembly of viral chromatin, patterning of histone posttranslational modifications and DNA methylation, and formation of higher-order chromosome conformations. Shannon-Lowe et al. (2009) demonstrated that early gene expression of the lytic cycle was required for amplification of the EBV genome in a semi-permissive epithelial cell line, AGS. In the same experimental conditions, infection of B cells with one or two copies of EBV induced rapid amplification of the viral genome to >20 copies per cell. Conversely, this amplification was not normally observed following infection of primary epithelial cells or undifferentiated epithelial cell lines, and the viral DNA was eventually lost. Thus, there may be certain prerequisites for the founder cells to accomplish amplification of EBV genome in the initial step of latent infection. Mutations of the genes, phosphatidylinositol-4,5-biphosphate 3-kinase catalytic subunit α (*PIK3CA*) and AT-rich interaction domain 1A (*ARID1A*), are possible candidates as cellular prerequisites (Abe et al. 2015), which will be discussed in Sect. 20.6.

20.4.3 Survival Advantage of EBV-Infected Cells

The important question is whether EBV provides a survival advantage to the non-neoplastic stomach epithelial cells. Recently, Kanda et al. (2016) cloned EBV virus from gastric cancer cell lines (SNU719 and YCCEL1) and established stable latent infection with these viruses in immortalized keratinocytes (using the expression of a mutant form of cyclin-dependent kinase 4/cyclin D1/human telomerase reverse transcriptase). Viral gene expression was restricted to a low level of EBNA1 and BamHI-A region rightward transcripts (BART) miRNAs. Of note, EBV-infected keratinocytes demonstrated resistance to mutant Ras-induced vacuolar degeneration and cell death, which are inevitable in uninfected keratinocytes. However, EBV-infected keratinocytes did not proliferate following survival of oncogene-inducing

:d a survival advantage to the
rization. These results resem-
immortalized nasopharyngeal
n of cyclin D1 or knockdown
and senescence (Tsang et al.

necessary to estimate the fre-
n vivo. The majority of infec-
hing the latent infection in an
 epithelial cells on the surface
kayama et al. 1994). One or a
lls are far infrequent, but they
iiku 2011).

20.4.4 Role of Background Gastritis and **H. pylori** *Infection*

As aforementioned in Sect. 20.2.1, a severe degree of atrophic gastritis is frequently observed in the surrounding mucosa of EBVaGC. Lytic infection of EBV in the stomach mucosa has been revealed to be associated with gastritis and *H. pylori* infection (Shukla et al. 2012). Recently, the cooperation of *H. pylori* in EBVaGC was demonstrated (Saju et al. 2016). SHP1, a host cell phosphatase, interacts with *H. pylori* CagA and dephosphorylates tyrosine-phosphorylated CagA, which induces repression of the CagA-SHP2 signaling pathway. Of note, SHP1 was epigenetically downregulated in EBV-infected cells, resulting in acceleration of oncogenic activity of CagA. Thus, *H. pylori* CagA directly contributed to the development of EBVaGC by constitutively activating SHP2.

20.5 DNA Methylation and Virus-Host Interactions

Epigenetic regulation is essential for virus and host cells of EBV-associated neoplasms (Kaneda et al. 2012). For the virus, methylation of viral DNA serves an important role for latent infection (Hammerschmidt 2015). The transcription initiation sites for lytic genes of EBV are densely methylated in latently infected cells. In response to reactivation signals, viral immediate early transcription factors, Rta (BRLF1, IE2) and Zta (ZEBRA, BZLF1), are expressed, and Zta selectively binds to the methylated CpG motifs of viral promotors in early genes, which induces the cells to transition from latency to lytic infection. Thus EBV utilizes the methylation state as a marker of latency.

Unexpectedly, DNA methylation was revealed to occur at high frequencies in CpG islands of the promotor region of host genes in EBVaGC (Kang et al. 2002, Osawa et al. 2002) (**B** in Fig. 20.1). Various tumor suppressor genes, including *p14*, *p16*, and *E-cadherin*, are suppressed by promoter methylation in EBVaGC. This process is not random, as mutL homolog 1 (*MLH1*) in particular, which is methylated in gastric cancers of MSI subtype, is not methylated in EBVaGC (Fukayama et al. 2008). We performed genome-wide comprehensive methylation analysis of gastric cancers (Matsusaka et al. 2011) and identified three distinct epigenotypes [EBV(−)/low methylation, EBV(−)/high methylation, and EBV(+)/high methylation] (Fig. 20.2.a). EBV epigenotype consisted of three sets of methylated genes,

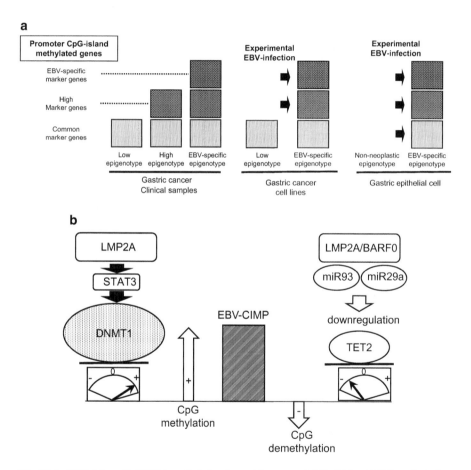

Fig. 20.2 EBV-driven CpG DNA methylation in gastric cancer. (**a**) CpG island DNA methylation in EBV-associated gastric cancer. (**b**) Infection-induced disruption of the balance between DNA methylation and DNA demethylation activities. Viral latent protein, LMP2A, induces overexpression of host cell DNMT1 through phosphorylation of STAT3, which leads to CpG methylation in promotor region of cancer-repressive genes. TET2, a member of TET family enzymes which promotes reversal of DNA methylation, is repressed by EBV infection through LMP2A, BARF0, and increased miR93 and miR29a. Consequently, methylation/demethylation balance inclines toward hypermethylation conditions

common marker genes, high marker genes, and EBVaGC-specific genes. Furthermore, using MKN7, a gastric cancer cell line with low CpG methylation, we revealed that global methylation was induced by EBV infection in both sets of high marker and EBV-specific marker genes.

20.5.1 Mechanisms Underlying EBV-Driven Hypermethylation

The mechanisms of EBV-driven hypermethylation (Fig. 20.2) have not yet been fully elucidated. We observed that viral latent protein, LMP2A, induced overexpression of DNMT1 via phosphorylation of signal transducer and activator of transcription 3 (STAT3), leading to methylation of the phosphatase and tensin homolog (PTEN) promotor in infected gastric cancer cell lines (Hino et al. 2009); however, this pathway only partially explains the global CpG island methylation of EBVaGC. DNA methylation is a dynamic process, which is balanced between driving force and resistance factors. Cytosine methylation is due to DNMTs, whereas its demethylation is mediated via oxidation of 5-methylcytosine by TET family of dioxygenases. We recently conducted an RNA-sequencing analysis of gastric epithelial cells with and without EBV infection and revealed that *TET* family genes, in particular *TET2*, were repressed by EBV infection at the mRNA and protein levels (Namba-Fukuyo et al. 2016). Hydroxymethylation target genes for TET2 enzyme overlapped significantly with methylation target genes in EBV-infected cells. When TET2 was knocked down by short hairpin RNA, EBV infection induced higher methylation in methylation-acquired promoters or de novo methylation acquisition in methylation-protected promoters, which induced gene repression. However, the role of TET2 is complementary as a resistance factor against de novo methylation, since TET2 knockdown alone did not induce de novo DNA methylation without EBV infection.

20.5.2 Chronological Sequence of DNA Methylation

Chromatin remodeling, including histone modifications, constitutes EBV-hyper-epigenotype and has been intensively investigated (Funata et al. 2017). But for the time being, we speculate that hypermethylation in EBVaGC may be due to overdriving of methylation mechanisms inherent to host cells (Fukayama et al. 2008, Kaneda et al. 2012). Using an experimental EBV infection system (Matsusaka et al. 2017), viral DNA methylation was revealed to precede the host DNA methylation by 1 week, and methylation of the host genome commences following completion of viral DNA methylation. The methylation of viral genes may be due to a host defense mechanism against foreign DNA, and the activated methylation machinery induces repression of tumor suppressor genes, which then promote the survival of infected cells. Furthermore, suppression of viral latent proteins benefits the virus by allowing it to evade host immune responses. Thus, overdriving of the cellular methylation mechanisms may be one of the survival strategies of EBV.

20.6 Molecular Abnormalities of Host Cells

As mentioned in the introduction (Sect. 20.1), the TCGA molecular classification of gastric cancer consists of four subtypes (The Cancer Genome Atlas Research Network 2014). Among them, EBVaGC is characterized by EBV-specific hyper-epigenotype or EBV-CIMP. The study further identified recurrent mutations of *ARID1A* (55%) and *PIK3CA* (80%). Other frequent mutations included BCL-6 corepressor (*BCOR*, 23%); however, mutation of tumor protein p53 (*TP53*) was extremely rare. Chromosomal abnormality was demonstrated to be uncommon in EBVaGC, but amplification of *JAK2*, programmed death-ligand 1 (*PD-L*), and *PD-L2* were frequent in EBVaGC.

When we consider the frequencies of these mutations, there are two possible explanations (Abe et al. 2015) (**C** in Fig. 20.1), although both are not exclusive: (i) higher frequencies suggest that these mutations occur during an earlier phase of cancer development, but lower frequencies require later timing of mutation, and (ii) there are several specific subgroups harboring specific sets of mutations.

20.6.1 PIK3CA

PIK3CA is an important regulator of the PI3K/protein kinase B (Akt) signaling pathway. Mutations in *PIK3CA* are frequent in cancer of various organs, including colon, breast, endometrium, and ovary cancers. These mutations are mostly hotspot mutations located on exon 9 (E542K and E545K) and exon 20 (H1047R). Referring to other sequencing studies of gastric cancer, the frequencies of *PIK3CA* mutation vary considerably in EBVaGC (Wang et al. 2014, Fang et al. 2016). The variable frequencies may be due to the differences in the detection methods or the patients' characteristics, including ethnicity and location of cancer. The mutation sites in *PIK3CA* are distributed widely in EBVaGC with only 28% of mutations observed in hotspots (The Cancer Genome Atlas Research Network 2014). A considerable number of the non-hotspot mutants were demonstrated to possess moderate-to-weak transformation activity and/or lipid phosphorylation activity (Burke et al. 2012). It is possible that *PIK3CA* mutation precedes EBV infection, which subsequently augments the functions of mutated *PIK3CA* in order to activate the PI3K/AKT signaling pathway. Coexistence of *PIK3CA* mutations with *ARID1A* mutation was frequently observed in ovarian clear cell cancer and in specific foci of endometriosis, in particular atypical endometriosis (Yamamoto et al. 2012). Further studies investigating *PIK3CA* mutations in nonneoplastic mucosa are required in order to clarify its association with EBV infection.

20.6.2 ARID1A

ARID1A is a subunit of the SWItch/sucrose non-fermentable (SWI/SNF) chromatin remodeling complex and is thought to have a tumor suppression role by regulating chromatin structure and gene expression. There are no hotspots in *ARID1A* mutations, but most are missense or truncation mutations, which result in loss or low expression levels of ARID1A protein (Wang et al. 2011). In order to identify its abnormalities, we used immunohistochemistry to analyze a large series of gastric cancer tissue microarrays, and it was revealed that expression of ARID1A was frequently lost in EBVaGC (Abe et al. 2012). Homogeneous loss within the tumor was characteristic to EBVaGC, and this finding was independent of invasion depth, which indicates that abnormality of ARID1A is an early event in EBVaGC carcinogenesis. In a deep sequencing study of nonneoplastic gastric mucosa (Shimizu et al. 2014), *ARID1A* mutations were present in chronically inflamed and *H. pylori*-infected gastric mucosa. These observations suggest that the mutation event precedes EBV infection. Mutations in *ARID1A*, leading to altered chromatin structures, may facilitate viral entry to the nucleus, subsequent CpG methylation, or activation of the PIK3CA/AKT signaling pathway (Zhang et al. 2016).

20.6.3 Amplification of PD-L1

A comprehensive analysis of gastric cancer revealed that *PD-L1* was amplified in EBVaGC (The Cancer Genome Atlas Research Network 2014). PD-L1 is a ligand of programmed death-1 (PD-1), which is a co-receptor of mature CD4+ and CD8+ T-cells. PD-L1 is mainly expressed on the surface of antigen-presenting cells and various tumor cells. Following ligation with PD-L1, the PD-1 signaling pathway is activated and delivers inhibitory signals for T-cell activation, which results in immune tolerance or evasion. The expression of PD-L1 was frequently detected in cancer cells of EBVaGC (34%), and fluorescence in situ hybridization analysis demonstrated *PD-L1* gene amplification in 11% of cases (Saito et al. 2017). *PD-L1*-amplified cells corresponded with PD-L1-positive cells demonstrating high intensities of staining, whereas other cancer cells revealed weak or moderate intensities of staining. Thus, *PD-L1* amplification occurs as a clonal evolution in the PD-L1-expressing cancer cells in EBVaGC. Kataoka et al. (2016) recently reported structural disruption of *PD-L1* 3′-UTR, which leads marked elevation of aberrant PD-L1 transcripts in adult T-cell lymphomas and a subset of gastric cancers including EBVaGC.

20.7 Tumor Microenvironment

The mechanisms underlying EBV-mediated alterations of the microenvironment of the infected cells require further investigation (**D** in Fig. 20.1). The microenvironment and cancer cells are postulated to cooperate for the development and progression of cancer. The microenvironment of EBVaGC appears characteristic, including the prominent infiltration of CD8+ cytotoxic T-cells (CTLs). Based on gene expression profiling in EBVaGC (Kim et al. 2015b), the majority of alterations occur in immune response genes, which may contribute to the recruitment of reactive immune cells. Integrated pathway analysis of TCGA (The Cancer Genome Atlas Research Network, 2014) demonstrated that IL-12-mediated signaling was characteristic of EBVaGC, suggesting a robust immune cell presence (antitumor response).

20.7.1 How to Survive in the Milieu of Antitumor Response

IL-12 exerts antitumor responses via interferon-γ. How do cancer cells of EBVaGC evade the immune response? In our study by immunohistochemistry, PD-L1 expression in cancer cells was associated with CD8+ immune cell density (Saito et al. 2017), suggesting that cancer cells increased PD-L1 expression level in response to cytokines from CD8 + CTL (adaptive expression). Furthermore, PD-L1+ immune cells infiltrated into the cancer stroma of EBVaGC more frequently compared with EBV-negative GC, particularly diffuse and intestinal subtypes (corresponding to GS and CI subtypes of gastric cancer, respectively). PD-L1+ immune cells were also frequently observed in MLH1-negative cases (MSI subtype), suggesting that EBVaGC and MSI subtypes acquire the ability to recruit PD-L1+ immune cells or induce PD-L1 expression in infiltrating cells against a burden of lymphocyte infiltration.

In parallel with the case of PD-L1 in immune cells, the microenvironment of EBVaGC is remodeled by accumulation of forkhead box P3+ cells (regulatory T-cells, Treg) (Zhang et al. 2015) and CD204+ macrophages (tumor-associated macrophages, TAM) (Ichimura et al. 2016), although the latter are relatively less frequent. Due to the finely balanced milieu, a small population of CD66b + neutrophils (anticancer tumor-associated neutrophils) may affect the behavior of EBVaGC (Abe et al. 2016).

20.7.2 Tools for Immune Evasion

In infiltrating cells of EBVaGC, the expression level of indoleamine 2,3-dioxygenase (IDO1) is upregulated (Strong et al. 2013, Kim et al. 2015b). IDO1 is a potent immune-inhibitory molecule that inhibits the immune responses via the local depletion of tryptophan, which is essential for anabolic functions in T-cells. Cytokines and chemokines produced by EBVaGC may affect the profile of infiltrating immune cells (Chong et al. 2002). Recently, exosomes have attracted attention as a modality of intercellular communications (Pegtel et al. 2010). Exosomes are small vesicles (30–150 nm in diameter) with lipid bilayer membranes and are derived from endoplasmic multivesicular bodies containing proteins, lipids, and RNAs from their cell origin. Exosomes are released into the extracellular space and deliver their contents from cell to cell. The molecular profile of exosomes has been vigorously investigated in NPC, including LMP1-positive exosomes (Meckes 2015). These exosomes contain other EBV-specific molecules, including viral miRNAs and cellular miRNAs of infected cells. When incorporated with recipient cells, contents of exosomes alter the recipient cellular functions (Meckes et al. 2010), and these changes contribute to the recruitment and expansion of Treg (Mrizak et al. 2014). Furthermore, increased circulating exosome concentrations were associated with advanced lymph node stage and poor prognosis in patients with NPC (Ye et al. 2014). Thus, exosomes from the tumor tissues and the patients' plasma should also be investigated in EBVaGC.

20.8 Host Cell and Virus miRNAs

miRNAs are small, noncoding RNA molecules ~22 nucleotides in length and are novel, posttranscriptional regulators of gene expression. Cancer cells of EBVaGC express both cellular and viral miRNAs, which provide characteristic features to themselves (**E** in Fig. 20.1).

20.8.1 Host Cell miRNAs

We previously reported that two cellular miRNAs, hsa-miR-200a and hsa-miR-200b, were downregulated in EBVaGC tissue samples and cell lines (Shinozaki et al. 2010). These miRNAs target the transcription repressors, zinc finger E-box-binding homeobox (ZEB)1 and ZEB2, which regulate E-cadherin expression levels. Downregulation of these miRNAs ultimately reduced E-cadherin expression level and induced epithelial-to-mesenchymal transition. EBV latent gene products, BARF0, EBNA1, LMP2A, and EBERs, cooperatively suppressed hsa-miR-200a and hsa-miR-200b expression in cell lines, whereas decrease of E-cadherin expression was significant only in EBER-transfected cells. Marquiz et al. (2014) demonstrated that EBV infection to AGS cells induced a decrease in the expression levels of tumor suppressor miRNAs (anti-oncomirs), including the let-7 and miR-200 families.

20.8.2 Virus miRNAs

A total of 25 EBV miRNA precursors and 44 mature EBV miRNAs have been registered (Shinozaki-Ushiku et al. 2015b). EBV miRNAs are produced from these 25 precursors, 3 of which are transcribed from *BHRF1* and the remaining 22 are from two regions of *BARTs* of the EBV genome. *BHRF*-derived miRNAs are only transcribed in EBV-infected lymphoblastic cells, whereas *BART*-derived miRNAs are expressed in EBV-infected neoplastic cells. Recent RNA-sequencing studies of EBVaGC demonstrated that the majority of the transcripts of EBV consist of BARTs. Numerous spliced transcripts, other than pre-miRNAs, remain in the nucleus of infected cells without definite evidence of translated proteins. Marquiz et al. (2015) demonstrated that considerable components of *BARTs* may act as long noncoding RNAs, which are involved in repressive complexes with cellular genes. Conversely, miRNAs are now intensively studied in cancer and viruses. Based on the study analyzing tissue samples from patients with EBVaGC (Shinozaki-Ushiku et al. 2015b), we revealed that ebv-miR-BART1-3p, 3, 4, 5, and 17-5p (cluster 1), 7, 9, 10, and 18-5p (cluster 2) and 2-5p were expressed at relatively high levels.

Functions of viral miRNAs, in particular those involved in the maintenance of viral latency and transformation of host cells, have been intensively investigated in NPC (Lo et al. 2012); however, they remain largely undetermined in EBVaGC (Table 20.2) (Shinozaki-Ushiku et al. 2015b, Choy et al. 2008, He et al. 2016, Tsai et al. 2017, Kang et al. 2017, Kim et al. 2015a, Jung et al. 2014, Choi and Lee 2017, Choi et al. 2013, Kang et al. 2017, Kanda et al. 2015). We demonstrated that ebv-miR-BART4-5p regulates the expression of a pro-apoptotic protein, BH3-interacting domain death (BID) agonist in EBVaGC (Shinozaki-Ushiku et al. 2015b). The rate of apoptosis revealed to be reduced in tissue samples of EBVaGC, and the gastric cancer cell lines infected with EBV exhibited downregulation of BID and resistance to apoptosis. However, it remains unknown whether the functions of each miRNA can be precisely determined. Marquitz et al. (2011) demonstrated that BART cluster 1 and a number of miRNAs targeted a pro-apoptotic molecule, BCL2-like 11 (Bim). In the reporter assay, the *Bim* 3′ untranslated region was inhibited by both clusters but not by an individual miRNA. The complexity of functional effects of viral miRNAs was also revealed by Kanda et al. (2015). They identified BART22 as being responsible for NDRG1 downregulation in epithelial cells by performing in vitro reporter assays. Conversely, EBV genetic analyses by deletion of fragments of cluster 2 (encompassing BART22) demonstrated that the entire BART cluster 2 was responsible for downregulation. Thus, functions of ebv-miR-BARTs are redundant, and deregulation of cellular miRNAs may compensate the results. Furthermore, in vitro experiments may not reflect in vivo functions, as highlighted by Qiu et al. (2015).

Table 20.2 BART microRNAs in EBV-associated gastric cancer

BART-miRNA	Amount[a]	Target	Subject (gastric cancer)	Comment	Reference
4-5p	High	Bid	GC tissues of FFPE (n = 10) KT, SNU719, MKN1-EBV, NUGC3-EBV, AGS-EBV	Profiling of BARTs in EBVaGC Anti-apoptosis	Shinozaki-Ushiku et al. (2015b)
5-5p	High	PUMA	AGS-EBV (AGS/Bx1)	Anti-apoptosis	Choy et al. (2008)
6-3p	Intermediate	lncRNA LOC553103	(AGS)	Reverse of EMT Inhibition of cell metastasis and invasion	He et al. (2016)
9	High	pre-miR-200a, pre-miR-141	GC tissues of FFPE (n = 52) SNU719	Profiling of BARTs in EBVaGC Downregulation of miR-200, miR-141	Tsai et al. (2017)
15-3p	Intermediate	Bruce	AGS-EBV, SNU719 Exosome	Induction of apoptosis Enriched in exosomes, suggesting transmission to the neighboring cells	Choi et al. (2013)
15-3p	Intermediate	TAX1BP1	AGS-EBV	Activation of NFkB Enhancement of chemosensitivity to 5-FU	Choi and Lee (2017)
20-5p	Low	Prognosis of the patients with EBVaGC	GC tissues of FFAP (n = 59)	Expression of miR-BART1-5p, 4-5p, and 20-5p and prognosis of the patients BART20-5p with worse recurrence-free survival	Kang et al. (2017)
20-5p	Low	BAD	AGS-EBV	Anti-apoptosis, growth promotion, resistance to 5FU and docetaxel	Kim et al. (2015a, b)
20-5p	Low	BZLF1, BRLF1	AGS-EBV	Maintenance of latent infection	Jung et al. (2014)
Cluster 1	Low-high	Bim	AGS-EBV	Capability of apoptosis Reporter assay of Bim 3′ UTR could not determine responsible miRNAs	Marquitz et al. (2011)
Cluster 1 and 2	Low-high	Cell growth in NOD mice	Xenograft model AGS-EBV (AGS/Bx1)	Increase of tumor growth in vivo Not associated with expression of specific BART miRNA, but with upregulation of all	Qiu et al. (2015)

GC gastric cancer, *EBVaGC* EBV-associated gastric cancer, *FFPE* formalin-fixed paraffin-embedded specimens

[a]Amount is described according to Shinozaki-Ushiku et al. (2015b)

20.9 Future Perspective

This review is a portrait of the gastritis-infection-cancer sequence of EBVaGC. A problem facing investigations of this virus-driven stomach cancer is a lack of a suitable model system to reconstitute the process from virus entry to generation of the founder clone. To clarify these initial steps, organoid culture of the stomach epithelial cells (McCracken et al. 2014, Bartfeld et al. 2015) should be applied to this issue in combination with live imaging of EBV and *H. pylori* infection. Application of single-cell sequencing and stem cell biology will be required for further analyses.

In order to concur advanced and metastatic cancers of this virus-induced and virus-driven cancer, multiple counteractions to the viral strategy should be developed including (i) augmentation of the immune response, for example, therapeutic vaccine (Taylor and Steven 2016, Taylor et al. 2014), adoptive T-cell transfer (Manzo et al. 2015, Su et al. 2016), and immune checkpoint therapy; (ii) prevention of exosome delivery; (iii) damping inflammatory signals (Ho et al. 2013); (iv) reversing CpG methylation of viral and cellular genomes (Yau et al. 2014, Hui et al. 2016), which also induces transition to lytic infection; and (v) other modalities of directly inducting lytic infection (including radiotherapy and chemotherapy) (Fu et al. 2008, Lee et al. 2015). A number of clinical trials for EBVaGC or virus-associated cancers are currently ongoing (four trials are now recruiting as of April 2017; EBV and gastric cancer at https://clinicaltrials.gov/).

Acknowledgments This work was supported by Grants-in-Aid for Scientific Research (KAKENHI) from the Japan Society for the Promotion of Science (26253021 to M.F.) and from the Core Research for Evolutionary Science and Technology program from the Japan Science and Technology Agency to A.K. and M.F.

References

Abe H, Morikawa T, Saito R et al (2016) In Epstein-Barr virus-associated gastric carcinoma a high density of CD66b-positive tumor-associated neutrophils is associated with intestinal-type histology and low frequency of lymph node metastasis. Virchows Arch 468:539–548

Abe H, Kaneda A, Fukayama M (2015) Epstein-Barr virus associated gastric carcinoma: use of host cell machineries and somatic gene mutations. Pathobiology 82:212–223

Abe H, Maeda D, Hino R et al (2012) ARID1A expression loss in gastric cancer: pathway-dependent roles with and without Epstein-Barr virus infection and microsatellite instability. Virchows Arch 461:367–377

Bartfeld S, Bayram T, van de Wetering M et al (2015) In vitro expansion of human gastric epithelial stem cells and their responses to bacterial infection. Gastroenterology 148:126–136.e6

Burke AP, Yen TS, Shekitka KM et al (1990) Lymphoepithelial carcinoma of the stomach with Epstein-Barr virus demonstrated by polymerase chain reaction. Mod Pathol 3:377–380

Burke JE, Perisic O, Masson GR et al (2012) Oncogenic mutations mimic and enhance dynamic events in the natural activation of phosphoinositide 3-kinase p110α (PIK3CA). Proc Natl Acad Sci U S A 109:15259–15264

Camargo MC, Kim KM, Matsuo K et al (2016) Anti-*Helicobacter pylori* antibody profiles in Epstein-Barr virus (EBV)-positive and EBV-negative gastric cancer. Helicobacter 21:153–157

Choi H, Lee SK (2017) TAX1BP1 downregulation by EBV-miR-BART15-3p enhances chemosensitivity of gastric cancer cells to 5-FU. Arch Virol 162:369–377

Choi H, Lee H, Kim SR et al (2013) Epstein-Barr virus-encoded microRNA BART15-3p promotes cell apoptosis partially by targeting BRUCE. J Virol 87:8135–8144

Chong JM, Sakuma K, Sudo M et al (2002) Interleukin-1beta expression in human gastric carcinoma with Epstein-Barr virus infection. J Virol 76:6825–6831

Chong JM, Fukayama M, Hayashi Y et al (1994) Microsatellite instability in the progression of gastric carcinoma. Cancer Res 54:4595–4597

Choy EY, Siu KL, Kok KH et al (2008) An Epstein-Barr virus-encoded microRNA targets PUMA to promote host cell survival. J Exp Med 205:2551–2560

Fang WL, Huang KH, Lan YT et al (2016) Mutations in PI3K/AKT pathway genes and amplifications of PIK3CA are associated with patterns of recurrence in gastric cancers. Oncotarget 7:6201–6220

Fu DX, Tanhehco Y, Chen J et al (2008) Bortezomib-induced enzyme-targeted radiation therapy in herpesvirus-associated tumors. Nat Med 14:1118–1122

Fukayama M, Ushiku T (2011) Epstein-Barr virus-associated gastric carcinoma. Pathol Res Pract 207:529–537

Fukayama M, Hino R, Uozaki H (2008) Epstein-Barr virus and gastric carcinoma: virus-host interactions leading to carcinoma. Cancer Sci 99:1726–1733

Fukayama M, Hayashi Y, Iwasaki Y et al (1994) Epstein-Barr virus-associated gastric carcinoma and Epstein-Barr virus infection of the stomach. Lab Investig 71:73–81

Funata S, Matsusaka K, Yamanaka R, Yamamoto S, Okabe A, Fukuyo M, Aburatani H, Fukayama M, Kaneda A (2017) Histone modification alteration coordinated with acquisition of promoter DNA methylation during Epstein-Barr virus infection. Oncotarget 8(33)

Hammerschmidt W (2015) The epigenetic life cycle of Epstein-Barr virus. Curr Top Microbiol Immunol 390(Pt 1):103–117

He B, Li W, Wu Y et al (2016) Epstein-Barr virus-encoded miR-BART6-3p inhibits cancer cell metastasis and invasion by targeting long non-coding RNA LOC553103. Cell Death Dis 7(9):e2353

Hino R, Uozaki H, Murakami N et al (2009) Activation of DNA methyltransferase 1 by EBV latent membrane protein 2A leads to promoter hypermethylation of PTEN gene in gastric carcinoma. Cancer Res 69:2766–2774

Ho Y, Tsao SW, Zeng M et al (2013) STAT3 as a therapeutic target for Epstein-Barr virus (EBV): associated nasopharyngeal carcinoma. Cancer Lett 330(2):141–149

Hui KF, Cheung AK, Choi CK et al (2016) Inhibition of class I histone deacetylases by romidepsin potently induces Epstein-Barr virus lytic cycle and mediates enhanced cell death with ganciclovir. Int J Cancer 138:125–136

Hutt-Fletcher LM (2016) The long and complicated relationship between Epstein-Barr Virus and epithelial cells. J Virol 91(1):pii: e01677-16

Ichimura T, Abe H, Morikawa T et al (2016) Low density of CD204-positive M2-type tumor-associated macrophages in Epstein-Barr virus-associated gastric cancer: a clinicopathologic study with digital image analysis. Hum Pathol 56:74–80

Imai S, Nishikawa J, Takada K (1998) Cell-to-cell contact as an efficient mode of Epstein-Barr virus infection of diverse human epithelial cells. J Virol 72:4371–4378

Imai S, Koizumi S, Sugiura M et al (1994) Gastric carcinoma: monoclonal epithelial malignant cells expressing Epstein-Barr virus latent infection protein. Proc Natl Acad Sci U S A 91:9131–9135

Kaizaki Y, Sakurai S, Chong JM et al (1999) Atrophic gastritis, Epstein-Barr virus infection, and Epstein-Barr virus-associated gastric carcinoma. Gastric Cancer 2:101–108

Kanda T, Furuse Y, Oshitani H et al (2016) Highly efficient CRISPR/Cas9-mediated cloning and functional characterization of gastric cancer-derived Epstein-Barr virus strains. J Virol 90:4383–4393

Kanda T, Miyata M, Kano M et al (2015) Clustered microRNAs of the Epstein-Barr virus cooperatively downregulate an epithelial cell-specific metastasis suppressor. J Virol 89:2684–2697

Kanda T, Horikoshi N, Murata T et al (2013) Interaction between basic residues of Epstein-Barr virus EBNA1 protein and cellular chromatin mediates viral plasmid maintenance. J Biol Chem 2013(288):24189–24199

Kaneda A, Matsusaka K, Aburatani H et al (2012) Epstein-Barr virus infection as an epigenetic driver of tumorigenesis. Cancer Res 72:3445–3450

Kang BW, Choi Y, Kwon OK et al (2017) High level of viral microRNA-BART20-5p expression is associated with worse survival of patients with Epstein-Barr virus-associated gastric cancer. Oncotarget 8:14988–14994

Kang GH, Lee S, Kim WH et al (2002) Epstein-Barr virus-positive gastric carcinoma demonstrates frequent aberrant methylation of multiple genes and constitutes CpG island methylator phenotype-positive gastric carcinoma. Am J Pathol 160:787–794

Kataoka K, Shiraishi Y, Takeda Y et al (2016) Aberrant PD-L1 expression through 3′-UTR disruption in multiple cancers. Nature 534(7607):402–406

Khan G, Hashim MJ (2014) Global burden of deaths from Epstein-Barr virus attributable malignancies 1990–2010. Infect Agent Cancer 9:38

Kim H, Choi H, Lee SK (2015a) Epstein-Barr virus miR-BART20-5p regulates cell proliferation and apoptosis by targeting BAD. Cancer Lett 356(2 Pt B):733–742

Kim SY, Park C, Kim HJ et al (2015b) Deregulation of immune response genes in patients with Epstein-Barr virus-associated gastric cancer and outcomes. Gastroenterology 148:137–147.e9

Jung YJ, Choi H, Kim H et al (2014) MicroRNA miR-BART20-5p stabilizes Epstein-Barr virus latency by directly targeting BZLF1 and BRLF1. J Virol 88:9027–9037

Lee HG, Kim H, Kim EJ et al (2015) Targeted therapy for Epstein-Barr virus-associated gastric carcinoma using low-dose gemcitabine-induced lytic activation. Oncotarget 6:31018–31029

Lieberman PM (2013) Keeping it quiet: chromatin control of gammaherpesvirus latency. Nat Rev Microbiol 11:863–875

Liu Y, Yang W, Pan Y et al (2016) Genome-wide analysis of Epstein-Barr virus (EBV) isolated from EBV-associated gastric carcinoma (EBVaGC). Oncotarget 7:4903–4914

Lo AK, Dawson CW, Jin DY et al (2012) The pathological roles of BART miRNAs in nasopharyngeal carcinoma. J Pathol 227:392–403

McCracken KW, Cata EM, Crawford CM et al (2014) Modelling human development and disease in pluripotent stem-cell-derived gastric organoids. Nature 516:400–404

Manzo T, Heslop HE, Rooney CM (2015) Antigen-specific T cell therapies for cancer. Hum Mol Genet 24(R1):R67–R73

Marquitz AR, Mathur A, Edwards RH et al (2015) Host gene expression is regulated by two types of noncoding RNAs transcribed from the Epstein-Barr virus BamHI A rightward transcript region. J Virol 89:11256–11268

Marquitz AR, Mathur A, Chugh PE et al (2014) Expression profile of microRNAs in Epstein-Barr virus-infected AGS gastric carcinoma cells. J Virol 88:1389–1393

Marquitz AR, Mathur A, Nam CS et al (2011) The Epstein-Barr virus BART microRNAs target the pro-apoptotic protein Bim. Virology 412:392–400

Matsusaka K, Funata S, Fukuyo M et al (2017) Epstein-Barr virus infection induces genome-wide *de novo* DNA methylation in non-neoplastic gastric epithelial cells. J Pathol 242(4):391–399

Matsusaka K, Kaneda A, Nagae G et al (2011) Classification of Epstein-Barr virus-positive gastric cancers by definition of DNA methylation epigenotypes. Cancer Res 71:7187–7197

Meckes DG Jr (2015) Exosomal communication goes viral. J Virol 89:5200–5203

Meckes DG Jr, Shair KH, Marquitz AR et al (2010) Human tumor virus utilizes exosomes for intercellular communication. Proc Natl Acad Sci U S A 107:20370–20375

Mrizak D, Martin N, Barjon C et al (2014) Effect of nasopharyngeal carcinoma-derived exosomes on human regulatory T cells. J Natl Cancer Inst 107:363

Murphy G, Pfeiffer R, Camargo MC et al (2009) Meta-analysis shows that prevalence of Epstein-Barr virus-positive gastric cancer differs based on sex and anatomic location. Gastroenterol 137:824–833

Namba-Fukuyo H, Funata S, Matsusaka K et al (2016) TET2 functions as a resistance factor against DNA methylation acquisition during Epstein-Barr virus infection. Oncotarget 7:81512–81526

Nanbo A, Kachi K, Yoshiyama H et al (2016) Epstein-Barr virus exploits host endocytic machinery for cell-to-cell viral transmission rather than a virological synapse. J Gen Virol 97:2989–3006

Nanbo A, Sugden A, Sugden B (2007) The coupling of synthesis and partitioning of EBV's plasmid replicon is revealed in live cells. EMBO J 26:4252–4262

Osawa T, Chong JM, Sudo M et al (2002) Reduced expression and promoter methylation of p16 gene in Epstein-Barr virus-associated gastric carcinoma. Jpn J Cancer Res 93:1195–1200

Pegtel DM, Cosmopoulos K, Thorley-Lawson DA et al (2010) Functional delivery of viral miRNAs via exosomes. Proc Natl Acad Sci U S A 107:6328–6333

Qiu J, Smith P, Leahy L et al (2015) The Epstein-Barr virus encoded BART miRNAs potentiate tumor growth in vivo. PLoS Pathog 11:e1004561

Saito R, Abe H, Kunita A et al (2017) Overexpression and gene amplification of PD-L1 in cancer cells and PD-L1(+) immune cells in Epstein-Barr virus-associated gastric cancer: the prognostic implications. Mod Pathol 30:427–439

Saju P, Murata-Kamiya N, Hayashi T et al (2016) Host SHP1 phosphatase antagonizes helicobacter pylori CagA and can be downregulated by Epstein-Barr virus. Nat Microbiol 1:16026

Shannon-Lowe C, Adland E, Bell AI et al (2009) Features distinguishing Epstein-Barr virus infections of epithelial cells and B cells: viral genome expression, genome maintenance, and genome amplification. J Virol 83:7749–7760

Shibata D, Tokunaga M, Uemura Y et al (1991) Association of Epstein-Barr virus with undifferentiated gastric carcinomas with intense lymphoid infiltration. Lymphoepithelioma-like carcinoma. Am J Pathol 139:469–474

Shibata D, Weiss LM (1992) Epstein-Barr virus-associated gastric adenocarcinoma. Am J Pathol 140:769–774

Shimizu T, Marusawa H, Matsumoto Y et al (2014) Accumulation of somatic mutations in TP53 in gastric epithelium with Helicobacter pylori infection. Gastroenterology 147:407–417.e3

Shinozaki A, Sakatani T, Ushiku T et al (2010) Downregulation of microRNA-200 in EBV-associated gastric carcinoma. Cancer Res 70:4719–4727

Shinozaki-Ushiku A, Kunita A, Fukayama M (2015a) Update on Epstein-Barr virus and gastric cancer (review). Int J Oncol 46:1421–1434

Shinozaki-Ushiku A, Kunita A, Isogai M et al (2015b) Profiling of virus-encoded microRNAs in Epstein-Barr virus-associated gastric carcinoma and their roles in gastric carcinogenesis. J Virol 89:5581–5591

Shukla SK, Prasad KN, Tripathi A et al (2012) Expression profile of latent and lytic transcripts of Epstein-Barr virus in patients with gastroduodenal diseases: a study from northern India. J Med Virol 84:1289–1297

Song HJ, Srivastava A, Lee J et al (2010) Host inflammatory response predicts survival of patients with Epstein-Barr virus-associated gastric carcinoma. Gastroenterology 139:84–92.e82

Strong MJ, Xu G, Coco J et al (2013) Differences in gastric carcinoma microenvironment stratify according to EBV infection intensity: implications for possible immune adjuvant therapy. PLoS Pathog 9:e1003341

Su S, Zou Z, Chen F et al (2016) CRISPR-Cas9-mediated disruption of PD-1 on human T cells for adoptive cellular therapies of EBV positive gastric cancer. Oncoimmunology 6:e1249558

Sugiura M, Imai S, Tokunaga M et al (1996) Transcriptional analysis of Epstein-Barr virus gene expression in EBV-positive gastric carcinoma: unique viral latency in the tumour cells. Br J Cancer 74:625–631

Tashiro Y, Arikawa J, Itho T et al (1998) Clinico–pathological findings of Epstein–Barr virus-related gastric cancer. In: Osato T, Takada K, Tokunaga M (eds) Epstein–Barr Virus and Human Cancer (Gann Monograph on Cancer Research, No 45). S. Karger Ag, Barsel, pp 87–97

Taylor GS, Steven NM (2016) Therapeutic vaccination strategies to treat nasopharyngeal carcinoma. Chin Clin Oncol 5:23

Taylor GS, Jia H, Harrington K et al (2014) A recombinant modified vaccinia ankara vaccine encoding Epstein-Barr Virus (EBV) target antigens: a phase I trial in UK patients with EBV-positive cancer. Clin Cancer Res 20:5009–5022

The Cancer Genome Atlas Research Network (2014) Comprehensive molecular characterization of gastric adenocarcinoma. Nature 513:202–209

Tsai CY, Liu YY, Liu KH et al (2017) Comprehensive profiling of virus microRNAs of Epstein-Barr virus-associated gastric carcinoma: highlighting the interactions of ebv-Bart9 and host tumor cells. J Gastroenterol Hepatol 32:82–91

Tsang CM, Yip YL, Lo KW et al (2012) Cyclin D1 overexpression supports stable EBV infection in nasopharyngeal epithelial cells. Proc Natl Acad Sci U S A 2012(109):E3473–E3482

Uozaki H, Fukayama M (2008) Epstein-Barr virus and gastric carcinoma-viral carcinogenesis through epigenetic mechanisms. Int J Clin Exp Pathol 1:198–216

Wang K, Yuen ST, Xu J et al (2014) Whole-genome sequencing and comprehensive molecular profiling identify new driver mutations in gastric cancer. Nat Genet 46:573–582

Wang K, Kan J, Yuen ST et al (2011) Exome sequencing identifies frequent mutation of ARID1A in molecular subtypes of gastric cancer. Nat Genet 43:1219–1223

Yamamoto S, Tsuda H, Takano M et al (2012) Loss of ARID1A protein expression occurs as an early event in ovarian clear-cell carcinoma development and frequently coexists with PIK3CA mutations. Mod Pathol 2012(25):615–624

Yau TO, Tang CM, Yu J (2014) Epigenetic dysregulation in Epstein-Barr virus-associated gastric carcinoma: disease and treatments. World J Gastroenterol 20:6448–6456

Ye SB, Li ZL, Luo DH et al (2014) Tumor-derived exosomes promote tumor progression and T-cell dysfunction through the regulation of enriched exosomal microRNAs in human nasopharyngeal carcinoma. Oncotarget 5:5439–5452

Zhang NN, Chen JN, Xiao L et al (2015) Accumulation mechanisms of CD4(+)CD25(+)FOXP3(+) regulatory T cells in EBV-associated gastric carcinoma. Sci Rep 5:18057

Zhang Q, Yan HB, Wang J et al (2016) Chromatin remodeling gene AT-rich interactive domain-containing protein 1A suppresses gastric cancer cell proliferation by targeting PIK3CA and PDK1. Oncotarget 7:46127–46141

Chapter 21
EBV in T-/NK-Cell Tumorigenesis

Hiroshi Kimura

Abstract Epstein–Barr virus (EBV), which is associated with B-cell proliferative disorders, also transforms T- or natural killer (NK)-lineage cells and has been connected with various T- or NK (T/NK)-cell malignancies, such as extranodal NK/T-cell lymphoma-nasal type and aggressive NK-cell leukemia. Chronic active EBV (CAEBV) disease, which occurs most often in children and young adults in East Asia, is an EBV-associated T-/NK-cell lymphoproliferative disease. Patients with CAEBV often progress to overt lymphoma or leukemia over a long-term clinical course. EBV's transforming capacity in B cells is well characterized, but the molecular pathogenesis of clonal expansion caused by EBV in T/NK cells has not yet been clarified. In the primary infection, EBV infects B cells and epithelial cells and may also infect some T/NK cells. In some individuals, because of poor presentation by specific human leukocyte antigens or the genetic background, EBV-infected T/NK cells evade host immunity and survive. Occasionally, with the help of viral oncogenes, EBV-associated T/NK lymphoproliferative diseases, such as CAEBV, may develop. The subsequent accumulation of genetic mutations and/or epigenetic modifications in driver genes, such as DDX3X and TP53, may lead to overt lymphoma and leukemia. Activation-induced cytidine deaminase and the APOBEC3 family, driven by EBV infection, may induce chromosomal recombination and somatic mutations.

Keywords AID · CAEBV · Chronic active EBV disease · DDX3X · ENKL · Extranodal NK/T-cell lymphoma-nasal type · EBV-T/NK LPD · Lymphomagenesis · Lymphoproliferative disease

---

H. Kimura (✉)
Department of Virology, Nagoya University Graduate School of Medicine,
Nagoya, Aichi, Japan
e-mail: hkimura@med.nagoya-u.ac.jp

21.1 Introduction

Epstein–Barr virus (EBV) infects only human beings, so its host range is very limited. Host-cell ranges are also narrow, primarily because of receptor tropism. EBV preferentially infects B cells and is associated with various B-cell-origin diseases (Longnecker et al. 2013; Grywalska and Rolinski 2015), such as infectious mononucleosis, Burkitt lymphoma, and posttransplant lymphoproliferative disorders (Table 21.1). EBV also infects epithelial cells and has been linked to several epithelial-cell-origin diseases, including nasopharyngeal carcinoma and gastric cancer (Table 21.1). However, EBV can infect non-B lymphocytes, such as T-cells and natural killer (NK) cells, and its association with T- and NK (T/NK)-cell-origin diseases has been demonstrated (Cohen et al. 2009).

Table 21.1 EBV-associated diseases (non-T/NK cells)

Disease entity	Association to EBV (%)	Infected cells	Latency type	Population at high risk
Infectious mononucleosis	100	B	III	Adolescents
Burkitt lymphoma, endemic	100	B	I	Equatorial Africa, New Guinea
Burkitt lymphoma, sporadic	30	B	I	
Hodgkin lymphoma, mixed cellularity	60–80	B	II	
Hodgkin lymphoma, nodular sclerosis	20–40	B	II	
Lymphomatoid granulomatosis	100	B	II	Western countries
EBV+ diffuse large B-cell lymphoma of elderly	100	B	III?	
Posttransplant lymphoproliferative disorders	>90	B	III	Recipients with heart, lung, or intestine transplantation
Lymphoma associated with HIV infection	40	B	I-III	
Primary effusion lymphoma[a]	70–80	B	III	HIV-infected individuals
Plasmablastic lymphoma	70	Plasma blasts	I?	HIV-infected individuals
Hairy leukoplakia	100	Epithelial cells	Lytic infection	HIV-infected individuals
Nasopharyngeal carcinoma	100	Epithelial cells	II	Southern China
Gastric cancer	9	Epithelial cells	I	

HIV human immunodeficiency virus
[a]Universally associated with human herpesvirus 8

EBV, which was originally isolated from Burkitt lymphoma in 1964 (Epstein et al. 1964), has been studied extensively in the context of B-cell lymphomagenesis for more than 50 years. Because EBV-associated T-/NK-cell tumors are rare and the generation and handling of EBV-positive T/NK cells are more difficult than B or epithelial cells, the precise mechanism of T-/NK-cell tumorigenesis has not yet been clarified. In this chapter, I will summarize the pathogenesis of EBV-associated T-/NK-cell neoplasms. In particular, oncogenicity in EBV-associated T-/NK-cell lymphoma will be the focus.

21.2 A Historical Perspective on T-/NK-Cell Tumorigenesis

An association between EBV, Burkitt lymphoma, and nasopharyngeal carcinoma was demonstrated soon after the first isolation of EBV (Epstein et al. 1964; Zur Hausen et al. 1970). However, a linkage to T-/NK-cell tumors was observed for the first time in the late 1980s. In 1988, Jones et al. reported T-cell lymphomas containing EBV DNA (Jones et al. 1988), nearly 25 years after the discovery of EBV (Table 21.2). This delay was partly due to the rarity of T-/NK-cell tumors, but the main reason was that there had been no reliable way to detect EBV-infected cells. The development of EBV-encoded small RNA (EBER) in situ hybridization boosted the discovery of a variety of EBV-associated diseases (Weiss et al. 1989). In 1989, the first case of EBV-associated NK-cell lymphoproliferative disease was reported (Kawa-Ha et al. 1989). Since then, linkages between previously known T-/NK-cell tumors and EBV have been found (Table 21.2) (Kikuta et al. 1988; Harabuchi et al. 1990; Hart et al. 1992; Kawaguchi et al. 1993; Ishihara et al. 1997; Iwatsuki et al. 1999).

In the 1990s, several EBV-positive T-/NK-cell lines were established. Most of them were derived from patient specimens (Imai et al. 1996; Tsuchiyama et al. 1998; Tsuge et al. 1999; Zhang et al. 2003), but some were established by in vitro infection of EBV-negative cell lines (Table 21.2) (Paterson et al. 1995; Fujiwara and Ono 1995; Isobe et al. 2004). These cell lines have helped greatly in studying the pathogenesis of EBV in T-/NK-cell tumors, combined with the development of mouse xenograft models (Imadome et al. 2011; Fujiwara et al. 2014).

In the twenty-first century, novel techniques, such as DNA microarrays, array comparative genomic hybridization, and "next-generation" sequencing, have been applied to genome-wide association studies and whole-genome/exome sequencing. With these techniques, the tumorigenesis of EBV-associated T-/NK-cell tumors is now being clarified to some extent (Iqbal et al. 2009; Karube et al. 2011; Jiang et al. 2015; Li et al. 2016), although the precise mechanism(s) remain unresolved (Table 21.2).

Table 21.2 Discoveries and events associated with T-/NK-cell tumorigenesis and EBV

Year	Investigators	Discovery and event	Reference
1988	Jones et al.	Linkage to T-cell lymphoma	Jones et al. (1988)
1988	Kikuta et al.	T-cell infection in chronic active EBV disease	Kikuta et al. (1988)
1989	Kawa-Ha et al.	Linkage to NK-cell lymphoproliferative disease	Kawa-Ha et al. (1989)
1990	Harabuchi et al.	Linkage to extranodal NK/T-cell lymphoma	Harabuchi et al. (1990)
1992	Hart et al.	Linkage to aggressive NK-cell leukemia	Hart et al. (1992)
1993	Kawaguchi et al.	T-cell infection in EBV-associated hemophagocytic lymphohistiocytosis	Kawaguchi et al. (1993)
1995	Paterson et al.	In vitro infection of T-cell line	Paterson et al. (1995)
1995	Fujiwara et al.	Establishment of T-cell lines by in vitro infection of HTLV-1-positive T-cell line	Fujiwara and Ono (1995)
1996	Imai et al.	Establishment of T-cell lines from clinical specimens	Imai et al. (1996)
1997	Ishihara et al.	Linkage to severe mosquito bite allergy	Ishihara et al. (1997)
1998	Tsuchiyama et al.	Establishment of an NK-cell line	Tsuchiyama et al. (1998)
1999	Iwatsuki et al.	Linkage to hydroa vacciniforme	Iwatsuki et al. (1999)
2004	Isobe et al.	Establishment of an NK-cell line by in vitro infection to an NK-cell leukemia line	Isobe et al. (2004)
2009	Iqbal et al.	Mutational analysis by array comparative genomic hybridization in NK-cell malignancies	Iqbal et al. (2009)
2011	Imadome et al.	Mouse xenograft model of chronic active EBV disease	Imadome et al. (2011)
2015	Jiang et al.	Whole exome sequencing in extranodal NK/T-cell lymphoma	Jiang et al. (2015)
2016	Li et al.	Genome-wide association study in extranodal NK/T-cell lymphoma	Li et al. (2016)

21.3 EBV-Associated T-/NK-Cell Tumors

There are various EBV-associated T-/NK-cell tumors, from lymphoproliferative disease to overt leukemia/lymphoma (Table 21.3). Although some have names suggesting seemingly benign diseases (e.g., mosquito bite allergy), all are basically neoplasms where EBV-infected T/NK cells proliferate with clonality and infiltrate organs.

In the 2008 WHO classification of tumors of hematopoietic and lymphoid tissues, the following diseases are listed as mature T-cell and NK-cell neoplasms associated with EBV: aggressive NK-cell leukemia, extranodal NK/T-cell lymphoma-nasal type (ENKL), systemic EBV-positive T-cell lymphoproliferative disease of childhood, hydroa vacciniforme-like lymphoma, and EBV$^+$ peripheral T-cell lymphoma, not otherwise specified (Chan et al. 2008a, b; Quintanilla-Martinez

Table 21.3 EBV-associated T- or NK-cell lymphoproliferative diseases/lymphoma

Disease entity	Association with EBV (%)	Infected cells	Latency type	Population at high risk
Angioimmunoblastic T-cell lymphoma	>90	B[a]	II	
Aggressive NK-cell leukemia	>90	NK	II	Asians
Extranodal NK/T-cell lymphoma, nasal type	100	NK, T	II	East Asians
Peripheral T-cell lymphoma, not otherwise specified	30	T	II	
Chronic active EBV disease of T/NK type	100	T, NK (B)	II	East Asians
Severe mosquito bite allergy	100	NK (T)	II	East Asians
EBV-associated hemophagocytic lymphohistiocytosis	100	$CD8^+$ T, NK	II	
Systemic EBV^+ T-cell lymphoma of childhood	100	T	II	East Asians
Hydroa vacciniforme-like lymphoproliferative disorder	100	$\gamma\delta$T, NK	II	Asians, Native Americans

[a]Neoplastic T-cells are EBV-negative

et al. 2008; Pileri et al. 2008). Except for peripheral T-cell lymphoma, EBV is associated strongly with each disease (Table 21.3). In the 2016 revision of the WHO classification, EBV^+ T-cell lymphoproliferative disease of childhood and hydroa vacciniforme-like lymphoma have been renamed EBV^+ T-cell lymphoma of childhood and hydroa vacciniforme-like lymphoproliferative disorder, respectively (Swerdlow et al. 2016; Quintanilla-Martinez et al. 2017). Chronic active EBV disease (CAEBV) of the T-/NK-cell type and severe mosquito bite allergy are also described in the text of the classification under the umbrella term of EBV-associated T-/NK-cell lymphoproliferative disorders (EBV-T/NK LPD) in the pediatric age group (Quintanilla-Martinez et al. 2017). Because the etiology of each disease is not yet fully understood, the classification and terminology are incomplete and need further improvement.

ENKL, which is also called nasal NK-/T-cell lymphoma, is a predominantly extranodal lymphoma, characterized by vascular damage, necrosis, and a cytotoxic phenotype (Elenitoba-Johnson et al. 1998; Chan et al. 2008b). The upper aerodigestive tract, including the nasal cavity and paranasal sinuses, is most commonly involved. This disease often progresses, with extensive midfacial destructive lesions (previously called lethal midline granuloma), and sometimes disseminates to other sites, including the skin and gastrointestinal tract. As it has a poor response to chemotherapy, the prognosis of ENKL is poor (Chan et al. 2008b). EBV is invariantly associated with this lymphoma, so an etiological link with EBV infection has generally been assumed. ENKL is more prevalent in East Asians and Native Americans in Mexico, Central America, and South America. The most typical immunophenotype of ENKL is $CD2^+$, $CD3^-$, and $CD56^+$, indicating NK cells.

CAEBV is a potentially life-threatening illness of children and young adults, characterized by the clonal proliferation of EBV-infected lymphocytes (Okano et al. 2005; Cohen et al. 2009; Kimura et al. 2012). The T-/NK-cell type of this disease has a strong racial predisposition, with most cases occurring in East Asians and some cases in Native American populations in the Western Hemisphere (Cohen et al. 2011; Quintanilla-Martinez et al. 2017); this distribution is analogous to that of ENKL. Typically, patients with CAEBV develop fever, hepatosplenomegaly, and lymphadenopathy; other common symptoms are thrombocytopenia, anemia, skin rash, diarrhea, and uveitis (Kimura et al. 2003). Patients often have abnormal liver function tests and an abnormal EBV serology, with high antiviral capsid antigen IgG antibodies (Straus et al. 1985). This disease is refractory to antiviral agents and conventional chemotherapies and thus has a poor prognosis. Stem-cell transplantation alone is a curative treatment for the disease, although the incidence of transplantation-related complications is high (Gotoh et al. 2008; Kawa et al. 2011).

21.4 Etiological Factors in the Development of T-/NK-Cell Tumors

Although EBV is ubiquitous, it remains unclear why the virus causes T-/NK-cell tumors in only some individuals (Kimura 2006). Some of them have risk factors, many of which are geographical (Table 21.3). This contrasts with B-cell diseases, where immunodeficiency is the main risk factor (Table 21.1). There are several hypotheses to account for the regional or racial deviation.

First, the genetic background may be related to the development of T-/NK-cell tumors. A positive association with human leukocyte antigen (HLA) A26 and a negative association with B52 were observed in EBV-T/NK LPD (Ito et al. 2013b). Interestingly, both the A26 and B52 alleles are frequently seen in East Asia and Mexico, where the prevalence of EBV-T/NK LPD is high. There are at least two possible explanations for these associations. One is that HLA-A26 does not effectively present epitopes from EBV latent antigens. Another is that genetic traits related to lymphomagenesis, which are codominantly expressed with HLA-A26, play important roles. Associations with HLA loci have been reported in other EBV-associated malignancies that show geographically distinct distributions. HLA-A02:07, which is common among Chinese people but not among Caucasians, is associated with nasopharyngeal carcinoma (Hildesheim et al. 2002). However, HLA-A1 is associated with an increased risk of developing EBV+ Hodgkin lymphoma (Niens et al. 2007).

Second, specific EBV strains or variants that can efficiently infect T/NK cells or can evade innate or acquired immunity may have a higher tendency to develop T-/NK-cell tumors. For example, nasopharyngeal carcinoma has a geographic bias to East Asia. Associations between nasopharyngeal carcinoma and specific strains or variants of EBV have been extensively studied, although definitive conclusions have not been reached (Neves et al. 2017). There are anecdotal papers on defective EBV or specific variants among T-/NK-cell tumors (Alfieri and Joncas 1987; Itakura et al. 1996). However, so far, no direct evidence has demonstrated a relationship with specific strain variants or mutants (Shibata et al. 2006; Kimura 2006). Recently, Coleman et al. showed that type 2 EBV can more efficiently infect T-cells than type 1 EBV (Coleman et al. 2015), although type 2 is not common in East Asia, where T-/NK-cell tumors are more common. Indeed, all of the established EBV+ T-/NK-cell lines involve type 1 EBV (unpublished data).

Finally, environmental factors may be associated with the progression or development of T-/NK-cell tumors. In Burkitt lymphoma, which is endemic in central Africa, co-infections with malaria or plant exposure (*Euphorbia tirucalli*) may enhance the tumorigenesis of EBV (Osato et al. 1987; Longnecker et al. 2013). The consumption of salted fish is believed to be a risk factor for nasopharyngeal carcinoma (Louie et al. 1977). On the other hand, case-control studies have shown that exposure to pesticides and chemical solvents could cause ENKL (Xu et al. 2007; Aozasa et al. 2008). Agricultural pesticide use is associated with chromosomal translocation in non-Hodgkin lymphoma (Xu et al. 2007). Similar environmental and lifestyle factors may be associated with the development of ENKL.

21.5 Infection of T or NK Cells

In primary infection, EBV in saliva infects naïve B cells, directly or indirectly via epithelial cells, and EBV-infected B cells become transformed and proliferate (Cohen 2000; Thorley-Lawson and Gross 2004). EBV attaches to B cells through binding gp350/220 to CD21, which is the receptor for the C3d component of complement, and gH/gL/gp42 to HLA class II molecules (Hutt-Fletcher 2007; Longnecker et al. 2013). Both CD21 and HLA class II are expressed on the surface of B cells, so B cells are natural hosts of EBV. During primary infection, epithelial cells of Waldeyer's ring also become infected with EBV (Borza and Hutt-Fletcher 2002). EBV attaches to epithelial cells through the binding of BMRF2 to integrins ($\alpha 1$, $\alpha 3$, $\alpha 5$, and αv integrins) and gH/gL to $\alpha v \beta 6$ and $\alpha v \beta 8$ integrins (Tugizov et al. 2003; Longnecker et al. 2013).

Although it is limited, there is some evidence that EBV infects T/NK cells in the primary infection. EBER-positive T/NK cells are seen in both tonsils and peripheral blood from patients with acute infectious mononucleosis (Anagnostopoulos et al. 1995; Hudnall et al. 2005; Kasahara et al. 2001). However, the mechanism by which EBV attaches and enters T/NK cells remains unknown. Basically, T/NK cells express neither CD21 nor HLA class II. T/NK cells express some integrins, and their expression is increased when stimulated. Thus, integrins may function as the receptors in T/NK cells as well as in epithelial cells. It is also possible that EBV may attach to T-cells via CD21, which is expressed in premature T-cells and common lymphoid progenitors (Fischer et al. 1991). It has been shown that EBV can infect premature T-cells and common lymphoid progenitors (Panzer-Grumayer et al.

1993; Ichigi et al. 1993; Paterson et al. 1995; Fischer et al. 1999). Interestingly, there have been several reports that show dual infections in both T-cell and NK-cell lineages in patients with CAEBV (Endo et al. 2004; Ohga et al. 2011; Sugimoto et al. 2014), supporting the hypothesis that EBV may infect common progenitor cells. Regarding cell phenotypes, EBV-infected T-cells are variable: $CD4^+$ T, $CD8^+$ T, $CD4^+$ $CD8^+$ T, $CD4^-$ $CD8^-$ T, and γδ T-cells have been reported (Kimura et al. 2012).

Another possibility is that EBV may infect from EBV-infected B cells or epithelial cells to T/NK cells by cell-to-cell infection. It has been reported that NK cells activated by EBV-infected B cells acquire CD21 molecules by synaptic transfer, and these ectopic receptors allow EBV binding to NK cells (Tabiasco et al. 2003). Such cell-to-cell infection through an immunological synapse has been observed in HTLV-1 infection between T-cells (Van Prooyen et al. 2010). EBV-infected T/NK cells usually express cytotoxic molecules, such as perforin, granzyme, and T-cell intracytoplasmic antigen (TIA)-1 (Ohshima et al. 1997b, 1999; Quintanilla-Martinez et al. 2000), indicating that they have a killer cell phenotype. Indeed, NK cells, $CD8^+$ T-cells, and γδ T-cells, which are typical EBV-infected cell types seen in EBV-associated T-/NK-cell tumors, are basically killer cells. These results suggest that T/NK cells that attempt to kill EBV-infected B or epithelial cells may become infected with EBV through close contact through an immunological synapse (Kimura et al. 2013).

21.6 EBV Gene Expression in T-/NK-Cell Tumors

Based on the pattern in normal B cells, EBV latency programs are classified as 0, I, II, or III (Longnecker et al. 2013). Similar to Hodgkin lymphoma and nasopharyngeal carcinoma, EBV-infected T/NK cells belong to latency type II, where only three viral proteins, EBNA1, LMP1, and LMP2A, are expressed (Chen et al. 1993; Kimura et al. 2005; Ito et al. 2013a). Noncoding RNAs, such as EBERs and BARTs, are also expressed in this latency type. In latency type II, immunodominant antigens, EBNA2 and EBNA3s, are not expressed. Thus, EBV-infected T/NK cells do not express major antigens against cytotoxic T lymphocytes and thus have an advantage in evading host immunity.

In CAEBV, which also belongs to type II latency, the frequency of LMP1 expression is variable among patients (Iwata et al. 2010; Kanemitsu et al. 2012). It seems that EBV latent gene expression in T/NK cells is heterogeneous, including variant transcripts (Yoshioka et al. 2001; Fox et al. 2010). LMP1 expression has a favorable impact on clinical outcomes in extranodal NK/T-cell lymphoma (Kanemitsu et al. 2012; Yamaguchi et al. 2014). This would seem to contradict the notion that LMP-1 is a potent oncoprotein that promotes proliferation and inhibit apoptosis. More recently, mutual exclusivity among tumors with somatic NF-κB pathway aberrations and LMP1 overexpression has been reported in nasopharyngeal carcinoma, suggesting that NF-κB activation is selected for by both somatic and viral events (Li et al. 2017). Loss of EBV was reported in a patient with cutaneous ENKL (Teo and Tan 2011). In vitro disappearance of EBV has been reported in EBV-infected B-cell lines (Shimizu et al. 1994). These results support the idea that subclones of originally EBV-positive lymphoma cells may have evolved alternative, virus-independent mechanisms to sustain oncogenic signaling pathways (Gru et al. 2015).

21.7 Tumorigenesis

Tumorigenesis in EBV-associated lymphoma/leukemia is a multistage process. There are two potential models of the multistage process. One is that EBV infects T/NK cells, followed by host-gene mutations and/or epigenetic modifications (Fig. 21.1a). In this model, the EBV-infected cells proliferate and evade apoptosis with the help of viral oncogenes. In the long run, mutations/modifications accumulate, leading to the development of overt lymphoma/leukemia. EBV-T/NK LPD may be an intermediate phase in this process (Fig. 21.1a). CAEBV, which is a representative EBV-T/NK LPD, usually develops in children or adolescents. Onset at an early age matches the infection's first mode. Another model is that

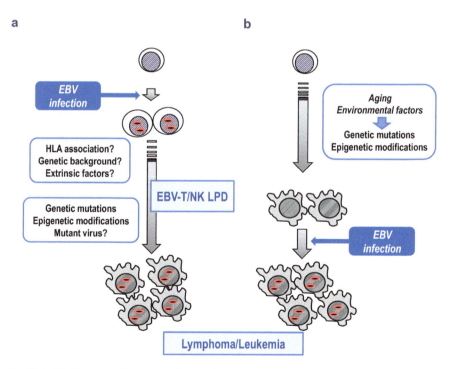

Fig. 21.1 Multistage models of tumorigenesis in EBV-associated T-/NK-cell lymphoma or leukemia (**a**) infection-genetic alteration sequence (**b**) genetic alteration-infection sequence

EBV infects cells where mutations/modifications have already accumulated (Fig. 21.1b). This second model corresponds to ENKL, which develops in older people (Chan et al. 2008b). Aging and long-term exposure to environmental factors could induce genetic mutations in T/NK cells of the nasal cavity. A bimodal age distribution has been noted in patients with ENKL: adolescents and the elderly (Takahashi et al. 2011). Patients with different peaks may have different processes of lymphomagenesis.

In the process of lymphomagenesis, mutations in host cell genes are necessary. What causes such genetic changes? Activation-induced cytidine deaminase (AID), which belongs to the APOBEC3 protein family, is one candidate. AID is expressed in germinal center B cells and induces somatic hypermutations and class switch recombination (Honjo et al. 2004). This enzyme is necessary for the chromosomal breaks in *c-myc* and its translocations (Robbiani et al. 2008). EBV-oncoprotein LMP-1 increases genomic instability through upregulation of AID in B-cell lymphomas (He et al. 2003; Kim et al. 2013). More recently, it has been shown that viral tegument protein BNRF1 can induce centrosome amplification and chromosomal instability, thereby conferring a risk of the development of tumors (Shumilov et al. 2017). By contrast, APOBEC3 cytidine deaminases can target and edit the EBV genome (Suspene et al. 2011). Thus, EBV infection per se may play an important role in *c-myc* translocation and lymphomagenesis of Burkitt lymphoma and other B-cell lymphomas. In AID transgenic mice, B-cell lymphoma develops (Okazaki et al. 2003). Interestingly, not only B-cell but also T-cell lymphoma develops in these mice. Furthermore, AID is expressed in HTLV-1⁺ cell lines and HTLV-1⁺ T-cells in peripheral blood (Ishikawa et al. 2011; Nakamura et al. 2011b). AID expression is high not only in EBV-positive T-/NK-cell lines but also in EBV-infected NK cells from patients with EBV-T/NK LPD (Nakamura et al. 2011a). In patients with ENKL, cytogenetic abnormalities are seen on the sixth chromosome (Ohshima et al. 1997a), although currently it is unclear whether this is a primary or progression-associated event (Chan et al. 2008b).

The next question is which are the key genes for development into T-/NK-cell lymphoma. Recent genetic studies have revealed that some of the driver gene mutations are related to the development of ENKL. Using a CGH array, PRDM1, HACE1, and FOXO3 were identified as driver gene candidates (Iqbal et al. 2009; Huang et al. 2010; Karube et al. 2011). More recently, JAK3 mutations were identified by exome sequencing (Koo et al. 2012), although the frequency is apparently not as high as first reported (Kimura et al. 2014; Guo et al. 2014). By whole exome sequencing, Jiang et al. reported that these driver gene mutations, such as DDX3X, TP53, and STAT3, were found at higher frequencies in Chinese patients with ENKL (Jiang et al. 2015) (Fig. 21.2). However, BCOR1 was the leading mutated gene in Japanese patients, followed by TP53 and DDX3X (Dobashi et al. 2016). DDX3X; DEAD-box helicase 3, X-linked, is an ATP-dependent RNA helicase and plays roles in transcriptional regulation, pre-mRNA splicing, and mRNA export. Its mutations are seen frequently not only in ENKL but also in Burkitt lymphoma (Schmitz et al. 2012). Although DDX3X mutations are also seen in EBV-unrelated

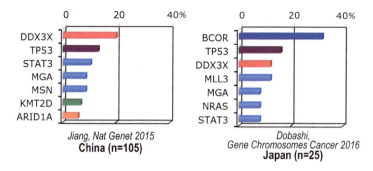

Fig. 21.2 Frequency of driver gene mutations in extranodal NK/T-cell lymphoma-nasal type (ENKL)

diseases, such as medulloblastoma (Pugh et al. 2012), its mutation may be associated with the development of EBV-associated tumors.

References

Alfieri C, Joncas JH (1987) Biomolecular analysis of a defective nontransforming Epstein-Barr virus (EBV) from a patient with chronic active EBV infection. J Virol 61:3306–3309

Anagnostopoulos I, Hummel M, Kreschel C, Stein H (1995) Morphology, immunophenotype, and distribution of latently and/or productively Epstein-Barr virus-infected cells in acute infectious mononucleosis: implications for the interindividual infection route of Epstein-Barr virus. Blood 85:744–750

Aozasa K, Takakuwa T, Hongyo T, Yang WI (2008) Nasal NK/T-cell lymphoma: epidemiology and pathogenesis. Int J Hematol 87:110–117

Borza CM, Hutt-Fletcher LM (2002) Alternate replication in B cells and epithelial cells switches tropism of Epstein-Barr virus. Nat Med 8:594–599

Chan JKC, Jaffe ES, Ralfkiaer E, Ko YH (2008a) Aggressive NK-cell leukaemia. In: Swerdlow SH, Campo E, Harris NL, Jaffe ES, Pileri SA, Stein H, Thiele J (eds) WHO classification of tumours of haematopoietic and lymphoid tissues. WHO Press, Lyon

Chan JKC, Quintanilla-Martinez L, Ferry JA, Peh S-C (2008b) Extranodal NK/T-cell lymphoma, nasal type. In: Swerdlow SH, Campo E, Harris NL, Jaffe ES, Pileri SA, Stein H, Thiele J (eds) WHO classification of tumours of haematopoietic and lymphoid tissues, 4th edn. WHO Press, Lyon

Chen CL, Sadler RH, Walling DM, Su IJ, Hsieh HC, Raab-Traub N (1993) Epstein-Barr virus (EBV) gene expression in EBV-positive peripheral T-cell lymphomas. J Virol 67:6303–6308

Cohen JI (2000) Epstein-Barr virus infection. N Engl J Med 343:481–492

Cohen JI, Kimura H, Nakamura S, Ko YH, Jaffe ES (2009) Epstein-Barr virus-associated lymphoproliferative disease in non-immunocompromised hosts: a status report and summary of an international meeting, 8–9 September 2008. Ann Oncol 20:1472–1482

Cohen JI, Jaffe ES, Dale JK, Pittaluga S, Heslop HE, Rooney CM, Gottschalk S, Bollard CM, Rao VK, Marques A, Burbelo PD, Turk SP, Fulton R, Wayne AS, Little RF, Cairo MS, El-Mallawany NK, Fowler D, Sportes C, Bishop MR, Wilson W, Straus SE (2011) Characterization and treatment of chronic active Epstein-Barr virus disease: a 28-year experience in the United States. Blood 117:5835–5849

Coleman CB, Wohlford EM, Smith NA, King CA, Ritchie JA, Baresel PC, Kimura H, Rochford R (2015) Epstein-Barr virus type 2 latently infects T cells, inducing an atypical activation characterized by expression of lymphotactic cytokines. J Virol 89:2301–2312

Dobashi A, Tsuyama N, Asaka R, Togashi Y, Ueda K, Sakata S, Baba S, Sakamoto K, Hatake K, Takeuchi K (2016) Frequent BCOR aberrations in extranodal NK/T-Cell lymphoma, nasal type. Genes Chromosomes Cancer 55:460–471

Elenitoba-Johnson KS, Zarate-Osorno A, Meneses A, Krenacs L, Kingma DW, Raffeld M, Jaffe ES (1998) Cytotoxic granular protein expression, Epstein-Barr virus strain type, and latent membrane protein-1 oncogene deletions in nasal T-lymphocyte/natural killer cell lymphomas from Mexico. Mod Pathol 11:754–761

Endo R, Yoshioka M, Ebihara T, Ishiguro N, Kikuta H, Kobayashi K (2004) Clonal expansion of multiphenotypic Epstein-Barr virus-infected lymphocytes in chronic active Epstein-Barr virus infection. Med Hypotheses 63:582–587

Epstein MA, Achong BG, Barr YM (1964) Virus particles in cultured lymphoblasts from Burkitt's lymphoma. Lancet 1:702–703

Fischer E, Delibrias C, Kazatchkine MD (1991) Expression of CR2 (the C3dg/EBV receptor, CD21) on normal human peripheral blood T lymphocytes. J Immunol 146:865–869

Fischer EM, Mouhoub A, Maillet F, Fremeaux-Bacchi V, Krief C, Gould H, Berrih-Aknin S, Kazatchkine MD (1999) Expression of CD21 is developmentally regulated during thymic maturation of human T lymphocytes. Int Immunol 11:1841–1849

Fox CP, Haigh TA, Taylor GS, Long HM, Lee SP, Shannon-Lowe C, O'connor S, Bollard CM, Iqbal J, Chan WC, Rickinson AB, Bell AI, Rowe M (2010) A novel latent membrane 2 transcript expressed in Epstein-Barr virus-positive NK- and T-cell lymphoproliferative disease encodes a target for cellular immunotherapy. Blood 116:3695–3704

Fujiwara S, Ono Y (1995) Isolation of Epstein-Barr virus-infected clones of the human T-cell line MT-2: use of recombinant viruses with a positive selection marker. J Virol 69:3900–3903

Fujiwara S, Kimura H, Imadome K, Arai A, Kodama E, Morio T, Shimizu N, Wakiguchi H (2014) Current research on chronic active Epstein-Barr virus infection in Japan. Pediatr Int 56:159–166

Gotoh K, Ito Y, Shibata-Watanabe Y, Kawada J, Takahashi Y, Yagasaki H, Kojima S, Nishiyama Y, Kimura H (2008) Clinical and virological characteristics of 15 patients with chronic active Epstein-Barr virus infection treated with hematopoietic stem cell transplantation. Clin Infect Dis 46:1525–1534

Gru AA, Haverkos BH, Freud AG, Hastings J, Nowacki NB, Barrionuevo C, Vigil CE, Rochford R, Natkunam Y, Baiocchi RA, Porcu P (2015) The Epstein-Barr virus (EBV) in T cell and NK cell lymphomas: time for a reassessment. Curr Hematol Malig Rep 10:456–467

Grywalska E, Rolinski J (2015) Epstein-Barr virus-associated lymphomas. Semin Oncol 42:291–303

Guo Y, Arakawa F, Miyoshi H, Niino D, Kawano R, Ohshima K (2014) Activated janus kinase 3 expression not by activating mutations identified in natural killer/T-cell lymphoma. Pathol Int 64(6):263

Harabuchi Y, Yamanaka N, Kataura A, Imai S, Kinoshita T, Mizuno F, Osato T (1990) Epstein-Barr virus in nasal T-cell lymphomas in patients with lethal midline granuloma. Lancet 335:128–130

Hart DN, Baker BW, Inglis MJ, Nimmo JC, Starling GC, Deacon E, Rowe M, Beard ME (1992) Epstein-Barr viral DNA in acute large granular lymphocyte (natural killer) leukemic cells. Blood 79:2116–2123

He B, Raab-Traub N, Casali P, Cerutti A (2003) EBV-encoded latent membrane protein 1 cooperates with BAFF/BLyS and APRIL to induce T cell-independent Ig heavy chain class switching. J Immunol 171:5215–5224

Hildesheim A, Apple RJ, Chen CJ, Wang SS, Cheng YJ, Klitz W, Mack SJ, Chen IH, Hsu MM, Yang CS, Brinton LA, Levine PH, Erlich HA (2002) Association of HLA class I and II alleles and extended haplotypes with nasopharyngeal carcinoma in Taiwan. J Natl Cancer Inst 94:1780–1789

Honjo T, Muramatsu M, Fagarasan S (2004) AID: how does it aid antibody diversity? Immunity 20:659–668

Huang Y, De Reynies A, De Leval L, Ghazi B, Martin-Garcia N, Travert M, Bosq J, Briere J, Petit B, Thomas E, Coppo P, Marafioti T, Emile JF, Delfau-Larue MH, Schmitt C, Gaulard P (2010) Gene expression profiling identifies emerging oncogenic pathways operating in extranodal NK/T-cell lymphoma, nasal type. Blood 115:1226–1237

Hudnall SD, Ge Y, Wei L, Yang NP, Wang HQ, Chen T (2005) Distribution and phenotype of Epstein-Barr virus-infected cells in human pharyngeal tonsils. Mod Pathol 18:519–527

Hutt-Fletcher LM (2007) Epstein-Barr virus entry. J Virol 81:7825–7832

Ichigi Y, Naitoh K, Tokushima M, Haraoka S, Tagoh H, Kimoto M, Muraguchi A (1993) Generation of cells with morphological and antigenic properties of microglia from cloned EBV-transformed lymphoid progenitor cells derived from human fetal liver. Cell Immunol 149:193–207

Imadome K, Yajima M, Arai A, Nakazawa A, Kawano F, Ichikawa S, Shimizu N, Yamamoto N, Morio T, Ohga S, Nakamura H, Ito M, Miura O, Komano J, Fujiwara S (2011) Novel mouse xenograft models reveal a critical role of CD4+ T cells in the proliferation of EBV-infected T and NK cells. PLoS Pathog 7:e1002326

Imai S, Sugiura M, Oikawa O, Koizumi S, Hirao M, Kimura H, Hayashibara H, Terai N, Tsutsumi H, Oda T, Chiba S, Osato T (1996) Epstein-Barr virus (EBV)-carrying and -expressing T-cell lines established from severe chronic active EBV infection. Blood 87:1446–1457

Iqbal J, Kucuk C, Deleeuw RJ, Srivastava G, Tam W, Geng H, Klinkebiel D, Christman JK, Patel K, Cao K, Shen L, Dybkaer K, Tsui IF, Ali H, Shimizu N, Au WY, Lam WL, Chan WC (2009) Genomic analyses reveal global functional alterations that promote tumor growth and novel tumor suppressor genes in natural killer-cell malignancies. Leukemia 23:1139–1151

Ishihara S, Okada S, Wakiguchi H, Kurashige T, Hirai K, Kawa-Ha K (1997) Clonal lymphoproliferation following chronic active Epstein-Barr virus infection and hypersensitivity to mosquito bites. Am J Hematol 54:276–281

Ishikawa C, Nakachi S, Senba M, Sugai M, Mori N (2011) Activation of AID by human T-cell leukemia virus Tax oncoprotein and the possible role of its constitutive expression in ATL genesis. Carcinogenesis 32:110–119

Isobe Y, Sugimoto K, Yang L, Tamayose K, Egashira M, Kaneko T, Takada K, Oshimi K (2004) Epstein-Barr virus infection of human natural killer cell lines and peripheral blood natural killer cells. Cancer Res 64:2167–2174

Itakura O, Yamada S, Narita M, Kikuta H (1996) High prevalence of a 30-base pair deletion and single-base mutations within the carboxy terminal end of the LMP-1 oncogene of Epstein-Barr virus in the Japanese population. Oncogene 13:1549–1553

Ito Y, Kawamura Y, Iwata S, Kawada J, Yoshikawa T, Kimura H (2013a) Demonstration of type II latency in T lymphocytes of Epstein-Barr virus-associated hemophagocytic lymphohistiocytosis. Pediatr Blood Cancer 60:326–328

Ito Y, Suzuki R, Torii Y, Kawa K, Kikuta A, Kojima S, Kimura H (2013b) HLA-A*26 and HLA-B*52 are associated with a risk of developing EBV-associated T/NK lymphoproliferative disease. Blood e-Letter, bloodjournal_el; 8085

Iwata S, Wada K, Tobita S, Gotoh K, Ito Y, Demachi-Okamura A, Shimizu N, Nishiyama Y, Kimura H (2010) Quantitative analysis of Epstein-Barr virus (EBV)-related gene expression in patients with chronic active EBV infection. J Gen Virol 91:42–50

Iwatsuki K, Xu Z, Takata M, Iguchi M, Ohtsuka M, Akiba H, Mitsuhashi Y, Takenoshita H, Sugiuchi R, Tagami H, Kaneko F (1999) The association of latent Epstein-Barr virus infection with hydroa vacciniforme. Br J Dermatol 140:715–721

Jiang L, Gu ZH, Yan ZX, Zhao X, Xie YY, Zhang ZG, Pan CM, Hu Y, Cai CP, Dong Y, Huang JY, Wang L, Shen Y, Meng G, Zhou JF, Hu JD, Wang JF, Liu YH, Yang LH, Zhang F, Wang JM, Wang Z, Peng ZG, Chen FY, Sun ZM, Ding H, Shi JM, Hou J, Yan JS, Shi JY, Xu L, Li Y, Lu J, Zheng Z, Xue W, Zhao WL, Chen Z, Chen SJ (2015) Exome sequencing identifies somatic mutations of DDX3X in natural killer/T-cell lymphoma. Nat Genet 47:1061–1066

Jones J, Shurin S, Abramowsky C, Tubbs R, Sciotto C, Wahl R, Sands J, Gottman D, Katz B, Sklar J (1988) T-cell lymphomas containing Epstein-Barr viral DNA in patients with chronic Epstein-Barr virus infections. N Engl J Med 318:733–741

Kanemitsu N, Isobe Y, Masuda A, Momose S, Higashi M, Tamaru JI, Sugimoto K, Komatsu N (2012) Expression of Epstein-Barr virus-encoded proteins in extranodal NK/T-cell lymphoma, nasal type (ENKL): differences in biological and clinical behaviors of LMP1-positive and -negative ENKL. Clin Cancer Res 18:2164–2172

Karube K, Nakagawa M, Tsuzuki S, Takeuchi I, Honma K, Nakashima Y, Shimizu N, Ko YH, Morishima Y, Ohshima K, Nakamura S, Seto M (2011) Identification of FOXO3 and PRDM1 as tumor-suppressor gene candidates in NK-cell neoplasms by genomic and functional analyses. Blood 118:3195–3204

Kasahara Y, Yachie A, Takei K, Kanegane C, Okada K, Ohta K, Seki H, Igarashi N, Maruhashi K, Katayama K, Katoh E, Terao G, Sakiyama Y, Koizumi S (2001) Differential cellular targets of Epstein-Barr virus (EBV) infection between acute EBV-associated hemophagocytic lymphohistiocytosis and chronic active EBV infection. Blood 98:1882–1888

Kawa K, Sawada A, Sato M, Okamura T, Sakata N, Kondo O, Kimoto T, Yamada K, Tokimasa S, Yasui M, Inoue M (2011) Excellent outcome of allogeneic hematopoietic SCT with reduced-intensity conditioning for the treatment of chronic active EBV infection. Bone Marrow Transplant 46:77–83

Kawaguchi H, Miyashita T, Herbst H, Niedobitek G, Asada M, Tsuchida M, Hanada R, Kinoshita A, Sakurai M, Kobayashi N, Et A (1993) Epstein-Barr virus-infected T lymphocytes in Epstein-Barr virus-associated hemophagocytic syndrome. J Clin Invest 92:1444–1450

Kawa-Ha K, Ishihara S, Ninomiya T, Yumura-Yagi K, Hara J, Murayama F, Tawa A, Hirai K (1989) CD3-negative lymphoproliferative disease of granular lymphocytes containing Epstein-Barr viral DNA. J Clin Invest 84:51–55

Kikuta H, Taguchi Y, Tomizawa K, Kojima K, Kawamura N, Ishizaka A, Sakiyama Y, Matsumoto S, Imai S, Kinoshita T, Et A (1988) Epstein-Barr virus genome-positive T lymphocytes in a boy with chronic active EBV infection associated with Kawasaki-like disease. Nature 333:455–457

Kim JH, Kim WS, Park C (2013) Epstein-Barr virus latent membrane protein 1 increases genomic instability through Egr-1-mediated up-regulation of activation-induced cytidine deaminase in B-cell lymphoma. Leuk Lymphoma 54:2035–2040

Kimura H (2006) Pathogenesis of chronic active Epstein-Barr virus infection: is this an infectious disease, lymphoproliferative disorder, or immunodeficiency? Rev Med Virol 16:251–261

Kimura H, Morishima T, Kanegane H, Ohga S, Hoshino Y, Maeda A, Imai S, Okano M, Morio T, Yokota S, Tsuchiya S, Yachie A, Imashuku S, Kawa K, Wakiguchi H (2003) Prognostic factors for chronic active Epstein-Barr virus infection. J Infect Dis 187:527–533

Kimura H, Hoshino Y, Hara S, Sugaya N, Kawada J, Shibata Y, Kojima S, Nagasaka T, Kuzushima K, Morishima T (2005) Differences between T cell-type and natural killer cell-type chronic active Epstein-Barr virus infection. J Infect Dis 191:531–539

Kimura H, Ito Y, Kawabe S, Gotoh K, Takahashi Y, Kojima S, Naoe T, Esaki S, Kikuta A, Sawada A, Kawa K, Ohshima K, Nakamura S (2012) EBV-associated T/NK-cell lymphoproliferative diseases in nonimmunocompromised hosts: prospective analysis of 108 cases. Blood 119:673–686

Kimura H, Kawada J, Ito Y (2013) Epstein-Barr virus-associated lymphoid malignancies: the expanding spectrum of hematopoietic neoplasms. Nagoya J Med Sci 75:169–179

Kimura H, Karube K, Ito Y, Hirano K, Suzuki M, Iwata S, Seto M (2014) Rare occurrence of JAK3 mutations in natural killer cell neoplasms in Japan. Leuk Lymphoma 55:962–963

Koo GC, Tan SY, Tang T, Poon SL, Allen GE, Tan L, Chong SC, Ong WS, Tay K, Tao M, Quek R, Loong S, Yeoh KW, Yap SP, Lee KA, Lim LC, Tan D, Goh C, Cutcutache I, Yu W, Ng CC, Rajasegaran V, Heng HL, Gan A, Ong CK, Rozen S, Tan P, Teh BT, Lim ST (2012) Janus kinase 3-activating mutations identified in natural killer/T-cell lymphoma. Cancer Discov 2(7):591

Li Z, Xia Y, Feng LN, Chen JR, Li HM, Cui J, Cai QQ, Sim KS, Nairismagi ML, Laurensia Y, Meah WY, Liu WS, Guo YM, Chen LZ, Feng QS, Pang CP, Chen LJ, Chew SH, Ebstein RP, Foo JN, Liu J, Ha J, Khoo LP, Chin ST, Zeng YX, Aung T, Chowbay B, Diong CP, Zhang F, Liu YH, Tang T, Tao M, Quek R, Mohamad F, Tan SY, Teh BT, Ng SB, Chng WJ, Ong CK, Okada Y, Raychaudhuri S, Lim ST, Tan W, Peng RJ, Khor CC, Bei JX (2016) Genetic risk of extranodal natural killer T-cell lymphoma: a genome-wide association study. Lancet Oncol 17:1240–1247

Li YY, Chung GT, Lui VW, To KF, Ma BB, Chow C, Woo JK, Yip KY, Seo J, Hui EP, Mak MK, Rusan M, Chau NG, Or YY, Law MH, Law PP, Liu ZW, Ngan HL, Hau PM, Verhoeft KR, Poon PH, Yoo SK, Shin JY, Lee SD, Lun SW, Jia L, Chan AW, Chan JY, Lai PB, Fung CY, Hung ST, Wang L, Chang AM, Chiosea SI, Hedberg ML, Tsao SW, Van Hasselt AC, Chan AT, Grandis JR, Hammerman PS, Lo KW (2017) Exome and genome sequencing of nasopharynx cancer identifies NF-kappaB pathway activating mutations. Nat Commun 8:14121

Longnecker R, Kieff E, Cohen JI (2013) Epstein-Barr virus. In: Knipe DM, Howley PM (eds) Fields virology, 6th edn. Lippincott Williams & Willkins, Philadelphia

Louie E, Henderson BE, Buell P, Jing JS, Menck HR, Pike MC (1977) Selected observations on the epidemiology of pharyngeal cancers. Natl Cancer Inst Monogr 47:129–133

Nakamura M, Iwata S, Kimura H, Tokura Y (2011a) Elevated expression of activation-induced cytidine deaminase in T and NK cells from patients with chronic active Epstein-Barr virus infection. Eur J Dermatol 21:780–782

Nakamura M, Sugita K, Sawada Y, Yoshiki R, Hino R, Tokura Y (2011b) High levels of activation-induced cytidine deaminase expression in adult T-cell leukaemia/lymphoma. Br J Dermatol 165:437–439

Neves M, Marinho-Dias J, Ribeiro J, Sousa H (2017) Epstein-Barr virus strains and variations: geographic or disease-specific variants? J Med Virol 89:373–387

Niens M, Jarrett RF, Hepkema B, Nolte IM, Diepstra A, Platteel M, Kouprie N, Delury CP, Gallagher A, Visser L, Poppema S, Te Meerman GJ, Van Den Berg A (2007) HLA-A*02 is associated with a reduced risk and HLA-A*01 with an increased risk of developing EBV+ Hodgkin lymphoma. Blood 110:3310–3315

Ohga S, Ishimura M, Yoshimoto G, Miyamoto T, Takada H, Tanaka T, Ohshima K, Ogawa Y, Imadome K, Abe Y, Akashi K, Hara T (2011) Clonal origin of Epstein-Barr virus (EBV)-infected T/NK-cell subpopulations in EBV-positive T/NK-cell lymphoproliferative disorders of childhood. J Clin Virol 51:31–37

Ohshima K, Suzumiya J, Ohga S, Ohgami A, Kikuchi M (1997a) Integrated Epstein-Barr virus (EBV) and chromosomal abnormality in chronic active EBV infection. Int J Cancer 71:943–947

Ohshima K, Suzumiya J, Shimazaki K, Kato A, Tanaka T, Kanda M, Kikuchi M (1997b) Nasal T/NK cell lymphomas commonly express perforin and Fas ligand: important mediators of tissue damage. Histopathology 31:444–450

Ohshima K, Suzumiya J, Sugihara M, Nagafuchi S, Ohga S, Kikuchi M (1999) CD95 (Fas) ligand expression of Epstein-Barr virus (EBV)-infected lymphocytes: a possible mechanism of immune evasion in chronic active EBV infection. Pathol Int 49:9–13

Okano M, Kawa K, Kimura H, Yachie A, Wakiguchi H, Maeda A, Imai S, Ohga S, Kanegane H, Tsuchiya S, Morio T, Mori M, Yokota S, Imashuku S (2005) Proposed guidelines for diagnosing chronic active Epstein-Barr virus infection. Am J Hematol 80:64–69

Okazaki IM, Hiai H, Kakazu N, Yamada S, Muramatsu M, Kinoshita K, Honjo T (2003) Constitutive expression of AID leads to tumorigenesis. J Exp Med 197:1173–1181

Osato T, Mizuno F, Imai S, Aya T, Koizumi S, Kinoshita T, Tokuda H, Ito Y, Hirai N, Hirota M et al (1987) African Burkitt's lymphoma and an Epstein-Barr virus-enhancing plant Euphorbia tirucalli. Lancet 1:1257–1258

Panzer-Grumayer ER, Panzer S, Wolf M, Majdic O, Haas OA, Kersey JH (1993) Characterization of CD7+CD19+ lymphoid cells after Epstein-Barr virus transformation. J Immunol 151:92–99

Paterson RL, Kelleher C, Amankonah TD, Streib JE, Xu JW, Jones JF, Gelfand EW (1995) Model of Epstein-Barr virus infection of human thymocytes: expression of viral genome and impact on cellular receptor expression in the T-lymphoblastic cell line, HPB-ALL. Blood 85:456–464

Pileri SA, Weisenburger DD, Sng I, Jaffe ES (2008) Peripheral T-cell lymphoma, not otherwise specified. In: Swerdlow SH, Campo E, Harris NL, Jaffe ES, Pileri SA, Stein H, Thiele J (eds) WHO classification of tumours of haematopoietic and lymphoid tissues, 4th edn. WHO Press, Lyon

Pugh TJ, Weeraratne SD, Archer TC, Pomeranz Krummel DA, Auclair D, Bochicchio J, Carneiro MO, Carter SL, Cibulskis K, Erlich RL, Greulich H, Lawrence MS, Lennon NJ, McKenna A, Meldrim J, Ramos AH, Ross MG, Russ C, Shefler E, Sivachenko A, Sogoloff B, Stojanov P, Tamayo P, Mesirov JP, Amani V, Teider N, Sengupta S, Francois JP, Northcott PA, Taylor MD, Yu F, Crabtree GR, Kautzman AG, Gabriel SB, Getz G, Jager N, Jones DT, Lichter P, Pfister SM, Roberts TM, Meyerson M, Pomeroy SL, Cho YJ (2012) Medulloblastoma exome sequencing uncovers subtype-specific somatic mutations. Nature 488:106–110

Quintanilla-Martinez L, Kumar S, Fend F, Reyes E, Teruya-Feldstein J, Kingma DW, Sorbara L, Raffeld M, Straus SE, Jaffe ES (2000) Fulminant EBV(+) T-cell lymphoproliferative disorder following acute/chronic EBV infection: a distinct clinicopathologic syndrome. Blood 96:443–451

Quintanilla-Martinez L, Kimura H, Jaffe ES (2008) EBV+ T-cell lymphoma of childhood. In: Swerdlow SH, Campo E, Harris NL, Jaffe ES, Pileri SA, Stein H, Thiele J (eds) WHO classification of tumours of haematopoietic and lymphoid tissues, 4th edn. WHO Press, Lyon

Quintanilla-Martinez L, Ko YH, Kimura H, Jaffe ES (2017) EBV–positive T-cell and NK-cell lymphoproliferative diseases of childhood. In: Swerdlow SH, Campo E, Harris NL, Jaffe ES, Pileri SA, Stein H, Thiele J (eds) WHO classification of tumours of haematopoietic and lymphoid tissues, revised 4th edn. WHO Press, Lyon

Robbiani DF, Bothmer A, Callen E, Reina-San-Martin B, Dorsett Y, Difilippantonio S, Bolland DJ, Chen HT, Corcoran AE, Nussenzweig A, Nussenzweig MC (2008) AID is required for the chromosomal breaks in c-myc that lead to c-myc/IgH translocations. Cell 135:1028–1038

Schmitz R, Young RM, Ceribelli M, Jhavar S, Xiao W, Zhang M, Wright G, Shaffer AL, Hodson DJ, Buras E, Liu X, Powell J, Yang Y, Xu W, Zhao H, Kohlhammer H, Rosenwald A, Kluin P, Muller-Hermelink HK, Ott G, Gascoyne RD, Connors JM, Rimsza LM, Campo E, Jaffe ES, Delabie J, Smeland EB, Ogwang MD, Reynolds SJ, Fisher RI, Braziel RM, Tubbs RR, Cook JR, Weisenburger DD, Chan WC, Pittaluga S, Wilson W, Waldmann TA, Rowe M, Mbulaiteye SM, Rickinson AB, Staudt LM (2012) Burkitt lymphoma pathogenesis and therapeutic targets from structural and functional genomics. Nature 490:116–120

Shibata Y, Hoshino Y, Hara S, Yagasaki H, Kojima S, Nishiyama Y, Morishima T, Kimura H (2006) Clonality analysis by sequence variation of the latent membrane protein 1 gene in patients with chronic active Epstein-Barr virus infection. J Med Virol 78:770–779

Shimizu N, Tanabe-Tochikura A, Kuroiwa Y, Takada K (1994) Isolation of Epstein-Barr virus (EBV)-negative cell clones from the EBV-positive Burkitt's lymphoma (BL) line Akata: malignant phenotypes of BL cells are dependent on EBV. J Virol 68:6069–6073

Shumilov A, Tsai MH, Schlosser YT, Kratz AS, Bernhardt K, Fink S, Mizani T, Lin X, Jauch A, Mautner J, Kopp-Schneider A, Feederle R, Hoffmann I, Delecluse HJ (2017) Epstein-Barr virus particles induce centrosome amplification and chromosomal instability. Nat Commun 8:14257

Straus SE, Tosato G, Armstrong G, Lawley T, Preble OT, Henle W, Davey R, Pearson G, Epstein J, Brus I, Et A (1985) Persisting illness and fatigue in adults with evidence of Epstein-Barr virus infection. Ann Intern Med 102:7–16

Sugimoto KJ, Shimada A, Wakabayashi M, Imai H, Sekiguchi Y, Nakamura N, Sawada T, Ota Y, Takeuchi K, Ito Y, Kimura H, Komatsu N, Noguchi M (2014) A probable identical Epstein-Barr virus clone-positive composite lymphoma with aggressive natural killer-cell leukemia and cytotoxic T-cell lymphoma. Int J Clin Exp Pathol 7:411–417

Suspene R, Aynaud MM, Koch S, Pasdeloup D, Labetoulle M, Gaertner B, Vartanian JP, Meyerhans A, Wain-Hobson S (2011) Genetic editing of herpes simplex virus 1 and Epstein-Barr her-

pesvirus genomes by human APOBEC3 cytidine deaminases in culture and in vivo. J Virol 85:7594–7602
Swerdlow SH, Campo E, Pileri SA, Harris NL, Stein H, Siebert R, Advani R, Ghielmini M, Salles GA, Zelenetz AD, Jaffe ES (2016) The 2016 revision of the World Health Organization classification of lymphoid neoplasms. Blood 127:2375–2390
Tabiasco J, Vercellone A, Meggetto F, Hudrisier D, Brousset P, Fournie JJ (2003) Acquisition of viral receptor by NK cells through immunological synapse. J Immunol 170:5993–5998
Takahashi E, Ohshima K, Kimura H, Hara K, Suzuki R, Kawa K, Eimoto T, Nakamura S (2011) Clinicopathological analysis of the age-related differences in patients with Epstein-Barr virus (EBV)-associated extranasal natural killer (NK)/T-cell lymphoma with reference to the relationship with aggressive NK cell leukaemia and chronic active EBV infection-associated lymphoproliferative disorders. Histopathology 59:660–671
Teo WL, Tan SY (2011) Loss of Epstein-Barr virus-encoded RNA expression in cutaneous dissemination of natural killer/T-cell lymphoma. J Clin Oncol 29:e342–e343
Thorley-Lawson DA, Gross A (2004) Persistence of the Epstein-Barr virus and the origins of associated lymphomas. N Engl J Med 350:1328–1337
Tsuchiyama J, Yoshino T, Mori M, Kondoh E, Oka T, Akagi T, Hiraki A, Nakayama H, Shibuya A, Ma Y, Kawabata T, Okada S, Harada M (1998) Characterization of a novel human natural killer-cell line (NK-YS) established from natural killer cell lymphoma/leukemia associated with Epstein-Barr virus infection. Blood 92:1374–1383
Tsuge I, Morishima T, Morita M, Kimura H, Kuzushima K, Matsuoka H (1999) Characterization of Epstein-Barr virus (EBV)-infected natural killer (NK) cell proliferation in patients with severe mosquito allergy; establishment of an IL-2-dependent NK-like cell line. Clin Exp Immunol 115:385–392
Tugizov SM, Berline JW, Palefsky JM (2003) Epstein-Barr virus infection of polarized tongue and nasopharyngeal epithelial cells. Nat Med 9:307–314
Van Prooyen N, Gold H, Andresen V, Schwartz O, Jones K, Ruscetti F, Lockett S, Gudla P, Venzon D, Franchini G (2010) Human T-cell leukemia virus type 1 p8 protein increases cellular conduits and virus transmission. Proc Natl Acad Sci U S A 107:20738–20743
Weiss LM, Movahed LA, Warnke RA, Sklar J (1989) Detection of Epstein-Barr viral genomes in reed-Sternberg cells of Hodgkin's disease. N Engl J Med 320:502–506
Xu JX, Hoshida Y, Yang WI, Inohara H, Kubo T, Kim GE, Yoon JH, Kojya S, Bandoh N, Harabuchi Y, Tsutsumi K, Koizuka I, Jia XS, Kirihata M, Tsukuma H, Aozasa K (2007) Life-style and environmental factors in the development of nasal NK/T-cell lymphoma: a case-control study in East Asia. Int J Cancer 120:406–410
Yamaguchi M, Takata K, Yoshino T, Ishizuka N, Oguchi M, Kobayashi Y, Isobe Y, Ishizawa K, Kubota N, Itoh K, Usui N, Miyazaki K, Wasada I, Nakamura S, Matsuno Y, Oshimi K, Kinoshita T, Tsukasaki K, Tobinai K (2014) Prognostic biomarkers in patients with localized natural killer/T-cell lymphoma treated with concurrent chemoradiotherapy. Cancer Sci 105:1435–1441
Yoshioka M, Ishiguro N, Ishiko H, Ma X, Kikuta H, Kobayashi K (2001) Heterogeneous, restricted patterns of Epstein-Barr virus (EBV) latent gene expression in patients with chronic active EBV infection. J Gen Virol 82:2385–2392
Zhang Y, Nagata H, Ikeuchi T, Mukai H, Oyoshi MK, Demachi A, Morio T, Wakiguchi H, Kimura N, Shimizu N, Yamamoto K (2003) Common cytological and cytogenetic features of Epstein-Barr virus (EBV)-positive natural killer (NK) cells and cell lines derived from patients with nasal T/NK-cell lymphomas, chronic active EBV infection and hydroa vacciniforme-like eruptions. Br J Haematol 121:805–814
Zur Hausen H, Schulte-Holthausen H, Klein G, Henle W, Henle G, Clifford P, Santesson L (1970) EBV DNA in biopsies of Burkitt tumours and anaplastic carcinomas of the nasopharynx. Nature 228:1056–1058

Chapter 22
Vaccine Development for Epstein-Barr Virus

Jeffrey I. Cohen

Abstract Epstein-Barr virus (EBV) is the primary cause of infectious mononucleosis and is associated with several malignancies, including nasopharyngeal carcinoma, gastric carcinoma, Hodgkin lymphoma, Burkitt lymphoma, and lymphomas in immunocompromised persons, as well as multiple sclerosis. A vaccine is currently unavailable. While monomeric EBV gp350 was shown in a phase 2 trial to reduce the incidence of infectious mononucleosis, but not the rate of EBV infection, newer formulations of gp350 including multimeric forms, viruslike particles, and nanoparticles may be more effective. A vaccine that also includes additional viral glycoproteins, lytic proteins, or latency proteins might improve the effectiveness of an EBV gp350 vaccine. Clinical trials to determine if an EBV vaccine can reduce the rate of infectious mononucleosis or posttransplant lymphoproliferative disease should be performed. The former is important since infectious mononucleosis can be associated with debilitating fatigue as well as other complications, and EBV infectious mononucleosis is associated with increased rates of Hodgkin lymphoma and multiple sclerosis. A vaccine to reduce EBV posttransplant lymphoproliferative disease would be an important proof of principle to prevent an EBV-associated malignancy. Trials of an EBV vaccine to reduce the incidence of Hodgkin lymphoma, multiple sclerosis, or Burkitt lymphoma would be difficult but feasible.

Keywords Epstein-Barr virus · Infectious mononucleosis · Nasopharyngeal carcinoma · Burkitt lymphoma · Hodgkin lymphoma · Gastric carcinoma

Most primary infections with Epstein-Barr virus (EBV) occur in infants or young children, and these infections are asymptomatic or result in nonspecific symptoms (Cohen 2000). The virus infects epithelial cells in the oropharynx where it replicates

---

J. I. Cohen (✉)
Laboratory of Infectious Diseases, National Institutes of Health, Bethesda, MD, USA
e-mail: jcohen@niaid.nih.gov

and subsequently infects B lymphocytes, or it may infect B cells in the tonsillar crypts directly. These B cells circulate throughout the body and may undergo lytic infection with production of progeny virus or, more often, undergo latent infection with very limited viral gene expression.

22.1 The Burden of EBV

Epstein-Barr virus is the principal cause of infectious mononucleosis and is a cofactor for several epithelial and lymphoid cell malignancies. The incidence of infectious mononucleosis in the United States is about 500 cases per 100,000 persons each year (Luzuriaga and Sullivan 2010). While infectious mononucleosis is often thought of as a mild disease, about 20% of patients will have persistent fatigue at 2 months and 13% at 6 months (Rea et al. 2001). About 1% of patients will have severe neurologic, hematologic, or liver complications from the disease. Infectious mononucleosis is the most common cause of lost time for new Army recruits.

EBV is associated with several malignancies; the criteria for association of EBV with cancer include finding the viral genome in every tumor cell, the presence of viral gene expression, and evidence that EBV is clonal (or oligoclonal) in the tumor cells. Each year worldwide there are about 84,000 cases of gastric carcinoma, 78,000 cases of nasopharyngeal carcinoma, 29,000 cases of Hodgkin lymphoma, 7000 cases of Burkitt lymphoma, and 2000 cases of lymphoma in transplant recipients associated with EBV (reviewed in Cohen et al. 2011). About 9% of gastric carcinomas are associated with EBV; 90% of gastric lymphoepitheliomas, 7% of moderately to well-differentiated adenocarcinomas, and 6% of poorly differentiated gastric adenocarcinomas are EBV-positive. Virtually all anaplastic nasopharyngeal carcinomas contain EBV genomes. The incidence of nasopharyngeal carcinoma is particularly high in southern China with a rate of 80 per 100,000 in men >40 years old. About 30–40% of Hodgkin lymphomas in developed countries are EBV-positive, while 80–90% of these lymphomas are EBV-positive in developing countries. About 85% of Burkitt lymphomas in Africa are EBV-positive, while about 15% of these tumors in the United States are virus-positive. In sub-Sahara Africa, the incidence of Burkitt lymphoma is 20 per 100,000 in children between the ages of 5 and 9 years old. The rate of EBV posttransplant lymphomas varies among the type of transplant ranging from about 1% in renal and hematopoietic stem cell transplant recipients to about 10% in intestinal transplant recipients. Up to 10% of seronegative children receiving a solid organ transplant may develop EBV posttransplant lymphoproliferative disease. Patients with HIV are at increased risk for EBV-associated malignancies including Burkitt lymphoma, diffuse large B cell lymphoma, Hodgkin lymphoma, immunoblastic lymphoma, primary central nervous system lymphoma, and smooth muscle tumors. EBV is also associated with other tumors in otherwise healthy persons including angioimmunoblastic T-cell lymphoma, extranodal NK/T-cell nasal lymphoma, diffuse large B cell lymphoma, and peripheral T-cell lymphoma.

EBV has also been associated with multiple sclerosis. A meta-analysis of 14 studies showed a relative risk of 2.3-fold for multiple sclerosis after EBV infectious mononucleosis (Thacker et al. 2006). A case-control study of persons who developed multiple sclerosis showed that 100% of EBV-seronegative persons became EBV-seropositive before the onset of multiple sclerosis, while only 36% of persons without multiple sclerosis became EBV-seropositive during the same time frame (Levin et al. 2010). A prospective study of military personnel showed that the risk of multiple sclerosis increased as serum titers to the anti-EBV nuclear antigen complex increased; the risk was 36-fold higher in persons with titers ≥320 compared to those with titers <20 (Munger et al. 2011).

EBV is associated with lymphoproliferative disease in immunodeficient patients (reviewed in Cohen 2015b). Boys with X-linked lymphoproliferative disease type 1, who have mutations in *SH2D1A*, can develop fatal infectious mononucleosis with infiltration of multiple organs by lymphocytes and histiocytes. Mutations in other genes including *BIRC4*, *CD27*, *CD70*, *CORO1A*, *FAAP24*, *LRBA*, and *MAGT1* predispose patients to severe EBV disease, usually in the absence of increased susceptibility to other pathogens. Mutations in *STXBP2*, *PRF1*, or *UNC13D* predispose to severe EBV infections and hemophagocytic lymphohistiocytosis. Finally, mutations in other genes including *ATM*, *CARD11*, *CTPS1*, *FCGR3A*, *GATA2*, *MCM4*, *PIK3CD*, *PIK3R1*, and *STK4*, as well as genes associated with severe combined immunodeficiency, increase the risk of severe EBV disease as well as infections due to other pathogens.

22.2 EBV Glycoproteins as Vaccine Candidates

Glycoproteins, present on the surface of viruses and virus-infected cells, have typically been primary candidates for development of vaccines to prevent infection and/or disease. EBV infection of B cells requires the function of several glycoproteins (reviewed in Longnecker et al. 2013). EBV glycoprotein gp350 is important for attachment of the virus to B cells. EBV gp350 binds to its receptor, CD21 (also known as complement receptor CR2) or CD35 (also termed complement receptor 1). This results in attachment of the virus to the B cell, and the virus is then taken up by endocytosis with fusion of the viral envelope to the host cell membrane mediated by EBV gp42 binding to MHC class II. Thereafter, gH/gL are thought to activate gB for fusion of the viral membrane to the plasma membrane of B cells. gH/gL and gB are essential for herpesvirus infection of cells and gp42 is required for EBV entry into B cells.

gp350 is a type I membrane protein and is the most abundant glycoprotein on the surface of virus-infected cells and on virions. gp350 is not strictly essential for virus infection but is important for efficient infection of B cells in vitro (Janz et al. 2000). The amino acid sequence of gp350 is highly conserved among different isolates especially in the amino terminal region; however, there are differences in the amino terminal region between EBV types 1 and 2 (Lees et al. 1993; Kawaguchi et al.

2009). Recent sequencing of clinical isolates indicates that gp350 is less conserved than other glycoproteins important for infection (Palser et al. 2015; Santpere et al. 2014), likely due to pressure to evolve in response to its role as a target for cytotoxic T-cells. There is little change in gp350 amino acid sequence in individuals between the time of acute infectious mononucleosis and convalescence; the few changes that do occur are located outside the CR2-binding domain (Weiss et al. 2016). In general, the CR2-binding site on gp350 is highly conserved.

EBV infection of epithelial cells involves EBV BMRF2 binding to integrins, followed by gH/gL binding to integrins and ephrin receptor A2, triggering activation of gB and fusion of the viral envelope to the plasma membrane of the epithelial cell. EBV infection of epithelial cells occurs at the cell surface, not through endocytosis.

22.3 EBV Lytic Proteins as Vaccine Candidates

The symptoms of infectious mononucleosis are thought to be due to the T-cell response to the virus (Silins et al. 2001). T-cell responses are important for controlling reactivation of the virus and the level of virus in the blood (reviewed in Taylor et al. 2015). Therefore, a vaccine that induces an effective T-cell response to EBV might reduce symptomatic disease and/or lower the viral load. The level of virus in the blood has shown to be a risk factor for development of lymphoproliferative disease after hematopoietic cell transplantation (van Esser et al. 2001; Aalto et al. 2007). Since EBV-seronegative recipients of solid organ transplants typically become infected from EBV in the transplanted organ, a vaccine to control proliferation of virus-infected cells in the organ may require T-cells in addition to antibody.

Different types of immunogens to induce T-cell responses have been suggested for a prophylactic EBV vaccine (Brooks et al. 2016). EBV immediate-early proteins Zta (encoded by BZLF1) and Rta (encoded by BRLF1) are the first genes expressed during infection, and these are produced before most of the immune evasion genes are expressed which dampen T-cell responses (reviewed in Longnecker et al. 2013). Destruction of EBV-infected cells expressing Zta or Rta would reduce the likelihood of the cells producing late proteins and virions. Zta and Rta are important T-cell targets in patients with infectious mononucleosis (Callan et al. 1998, Steven et al. 1997; Precopio et al. 2003). Patients with EBV posttransplant lymphoproliferative disease who resolve their disease after a reduction in immunosuppression have an increase in CD8 T-cells to Zta (Porcu et al. 2002). Zta and Rta are recognized by CD8 T-cells more often than early or late proteins (Pudney et al. 2005). However, a recent study suggests that CD8 T-cells do not recognize Zta or Rta within the first day after EBV infection in vitro (Brooks et al. 2016).

Another approach is to induce T-cell responses to early or late lytic EBV proteins. While initial studies showed that CD4 T-cell responses that lyse virus-infected B cells are directed against structural proteins including EBV gp350 and glycoprotein B (Adhikary et al. 2006, 2007), more recent studies show that multiple lytic proteins including BMLF1 (a posttranscriptional regulatory protein), BMRF1 (polymerase-associated processivity factor), BNRF1 (the major tegument protein), BORF1 (DNA packaging protein), BcLF1 (major capsid protein), and BXLF1 (thymidine kinase) are targets of CD4 cells (Long et al. 2011; reviewed in Taylor et al. 2015). BMLF1 and BMRF1 are expressed early in infection, before virus structural proteins are made, and are targets of both CD4 and CD8 cells. EBV structural proteins in the nucleocapsid or envelope are presented directly to newly infected cells and can be processed and recognized by CD4 T-cells. These proteins can be detected by CD4 T-cells early after infection (Adhikary et al. 2006), and gp350, gH, and gB are recognized by CD4 T-cells within the first day after infection in vitro (Brooks et al. 2016).

22.4 EBV Latent Proteins as Vaccine Candidates

Another approach for a prophylactic EBV vaccine is to induce T-cell responses to EBV latency proteins, such those initially expressed during EBV infection of B cells. By 12 h after infection, EBV nuclear antigen 2 (EBNA-2) and EBNA leader protein (EBNA-LP) are detected (Alfieri et al. 1991). A recent in vitro study showed that several epitopes within EBNA2 induce immunodominant CD8 T-cell responses and that EBNA-2 CD8 T-cells recognize EBV-infected B cells within 1 day after virus infection of B cells before CD8 T-cells that recognize other latent proteins (Brooks et al. 2016). These EBNA-2-specific T-cell responses inhibit outgrowth of EBV-transformed B cell lines. EBNA-2 and EBNA-LP are also targets of CD4 T-cells (reviewed in Taylor et al. 2015). Thus, a prophylactic vaccine that induces T-cell responses to the first viral proteins expressed after infection in B cells, such as EBNA-2 or EBNA-LP, might destroy any newly infected cells.

22.5 Adaptive Immunity to EBV

Infection with EBV induces antibodies and T-cells specific for viral proteins. Glycoprotein gp350 is the principal target of neutralizing antibody for EBV infection of B cells (North et al. 1980; Thorley-Lawson and Poodry 1982). Injection of antibody to gp350 prevents lymphoproliferative disease in an immunocompromised mice model (Haque et al. 2006). Antibody to gp42 also neutralizes infection of B cells, while antibody to gH/gL (Li et al. 1995) and BMRF1 (Tugizov et al. 2003) neutralizes EBV infection of epithelial cells. EBV neutralizing antibody in human sera correlates better with levels of antibody to gp350 than gp42 (Sashihara et al. 2009). B cell neutralizing antibody reaches peak levels at a median of about 180 days after the onset of infectious mononucleosis (Bu et al. 2016). Immunoprecipitating antibody to EBV gp350 and gp42 achieves peak levels only at a median of about 900 and 400 days, respectively after onset of symptoms. Thus, antibody maturation can take over a year to occur after primary EBV infection. These findings are consistent with other activities mediated by antibody to EBV proteins which also require time to develop. Antibody-dependent cellular cytotoxicity (ADCC) directed against cells expressing gp350 was not detected in sera from persons at the onset of infectious mononucleosis but was detected in healthy EBV-seropositive persons (Xu et al. 1998). Similarly, antibody-dependent cell-mediated phagocytosis (ADCP) was rarely detected during the initial phase of infectious mononucleosis but was frequently present 6 months later (Weiss et al. 2016). At present it is unknown which activities mediated by antibodies are most important for protection against EBV infection or disease.

CD8 T-cell responses during infectious mononucleosis are targeted to EBV lytic antigens including gp350, gH, gL, and gB (reviewed in Taylor et al. 2015); over time the number of CD8 T-cells recognizing lytic antigens declines, and T-cells recognizing EBV latency proteins increase (Hislop et al. 2002). CD8 T-cells during infectious mononucleosis recognize immediate-early proteins most often and late proteins least often (Pudney et al. 2005); the EBNA-3 proteins are the predominant latency proteins targeted by CD8 T-cell targets (Steven et al. 1997). In contrast, during infectious mononucleosis CD4 T-cells are directed more toward latent antigens, especially EBNA-3 proteins (Woodberry et al. 2005). CD4 T-cells recognize immediate-early, early, and late proteins without a preference for the kinetic class of gene expression in EBV-seropositive persons (Long et al. 2011). Glycoproteins including gp350, gH, gL, gp42, and gB are also recognized by CD4 T-cells (reviewed in Taylor et al. 2015). These findings suggest that a vaccine targeting CD8 T-cells might focus more on lytic antigens especially immediate-early proteins, while a vaccine targeting CD4 T-cells might focus on EBNA-3.

22.6 EBV Glycoprotein Vaccines: Immunogenicity in Animals

EBV gp350 has been shown to induce EBV neutralizing antibodies, ADCC, or T-cell responses in animals using a number of different platforms. EBV neutralizing antibodies were first reported in rabbits (Thorley-Lawson 1979) and cottontop tamarins (Morgan et al. 1984) immunized with gp350 purified from virus-infected cells. Owl monkeys immunized with gp350 purified from EBV-infected cells developed EBV neutralizing and ADCC antibodies (Qualtiere et al. 1982). Subsequently recombinant gp350 purified from mammalian cells was shown to induce neutralizing EBV antibody in rabbits (Emini et al. 1988; Jackman et al. 1999) and cottontop tamarins (Finerty et al. 1992). Mice vaccinated with recombinant gp350 adjuvanted with a TLR4 agonist (glucopyranosyl lipid A) in emulsion developed EBV neutralizing antibodies and gp350-specific CD4 T-cell responses (Heeke et al. 2016). Immunization of HLA-A2 transgenic mice with a gp350 peptide induced cytotoxic T-cell responses and protected the animals against vaccinia virus expressing gp350 (Khanna et al. 1999). Immunization of mice with a plasmid expressing gp350 induced antibodies that mediated ADCC and gp350-specific cytotoxic T-cells (Jung et al. 2001). In another approach, neutralizing antibodies were detected in mice immunized with vaccinia virus expressing gp350 or a combination of four vaccinia viruses expressing gp350, gB, EBNA-2, or EBNA-3C mixed together (Lockey et al. 2008). CD4 T-cell responses to EBNA-2 were detected in mice vaccinated with the combination of the four vaccinia viruses.

New approaches have recently been developed to express gp350 in a multimeric configuration. First, a tetrameric gp350 construct was expressed in Chinese hamster ovary cells that induced 19-fold higher levels of neutralizing antibodies in mice than soluble gp350; however, the neutralization assay did not use EBV, but instead the ability to block binding of gp350 to a cell line expressing CD21 (Cui et al. 2013). In a follow-up paper from the same group, mice immunized with the tetrameric gp350 showed fourfold higher titers compared with animals immunized with monomeric gp350 using a virus neutralizing assay (Cui et al. 2016). Second, the ectodomain of gp350 was fused to the F protein of Newcastle disease virus, and the chimeric protein was incorporated into the membrane of viruslike particles (VLPs) composed of the Newcastle disease virus matrix and nucleoprotein. Mice immunized with these gp350 VLPs produced higher levels of EBV neutralizing antibodies than those immunized with soluble gp350, although the differences were not statistically significant (Ogembo et al. 2015). Third, a portion of the ectodomain of gp350 (containing the CR2-binding domain of gp350) was fused to ferritin or encapsulin, and nanoparticles were produced that contain 24 or 60 copies of gp350, respectively (Kanekiyo et al. 2015). Immunization of mice with the nanoparticles induced neutralizing titers that were about 1000-fold higher than that obtained with soluble gp350; immunization of cynomolgus monkeys that were seropositive for cynomolgus monkey lymphocryptovirus (an ortholog of EBV) with the nanoparticles induced three- to tenfold higher neutralizing titers than that obtained with soluble gp350. Vaccination of mice with ferritin-gp350 nanoparticles protected the animals from challenge with vaccinia virus expressing gp350; vaccination with encapsulin-gp350 nanoparticles did not protect the mice.

Other EBV glycoproteins have also been used to induce neutralizing antibody to EBV or T-cell responses in mice and rabbits. Vaccination of rabbits with trimeric gB, monomeric gH/gL, or trimeric gH/gL induced 18-fold, 20-fold, or >100-fold higher levels of EBV neutralizing antibody than monomeric gp350 (Cui et al. 2016). In another approach, two Newcastle disease virus VLPs were constructed; one containing the ectodomain of EBV gH fused to the Newcastle disease virus F protein, the EBV gL ectodomain fused to the Newcastle disease virus HN protein, and the carboxyl half of EBV EBNA1 fused to the Newcastle disease virus NP protein (Perez et al. 2016). The second VLP contained the ectodomain EBV gB fused to the Newcastle disease virus F protein, and EBV LMP2 fused to the Newcastle disease virus NP protein. Mice immunized with either of the two EBV Newcastle disease virus VLPs produced high levels of EBV neutralizing antibodies and EBV-specific T-cell responses in mice. Immunization of HLA-A2 transgenic mice with gH peptides induced cytotoxic T-cell responses and protected the animals against vaccinia virus expressing gH (Khanna et al. 1999).

Another approach to an EBV vaccine used EBV VLPs. A producer cell line containing an EBV genome deleted for EBNA2, LMP1, EBNA3A, and EBNA3C, but still containing viral proteins needed for assembly and release of virions, was used to produce EBV VLPs (Hettich et al. 2006). Mouse immunized with these EBV VLPs produced neutralizing antibody and T-cell responses to viral proteins (Ruiss et al. 2011).

22.7 EBV Lytic Proteins: Immunogenicity in Animals

EBV BZLF1 encodes the immediate-early protein (Zta). SCID mice injected with human peripheral blood mononuclear cells from an EBV-seropositive donor were vaccinated with dendritic cells transduced either by an adenovirus expressing Zta or adenovirus with an empty vector (Hartlage et al. 2015). Mice receiving dendritic cells expressing Zta developed Zta-specific T-cell responses and had delayed development of EBV lymphoproliferative disease compared with animals receiving dendritic cells not expressing Zta.

22.8 Animal Studies Using EBV Challenge Models

Most prophylactic EBV vaccines have used gp350. Initial experiments focused on cottontop tamarins, which develop EBV-positive mono- or oligoclonal large B cell lymphomas after parental inoculation with high titers of virus (Cleary et al. 1985). The first proof of principle for an EBV vaccine was reported by Epstein et al. (1985). Tamarins vaccinated intraperitoneally with purified cell membranes containing gp350, isolated from virus-infected (B95-8) cells, developed neutralizing antibody to EBV and were protected from EBV tumors after challenge with virus. In additional experiments, animals vaccinated with gp350 incorporated into liposomes developed neutralizing antibody and were also protected against challenge with EBV. Subsequent experiments performed in cottontop tamarins showed that purified gp350 in immunostimulating complexes (ISCOMs) or muramyl dipeptide in squalene, recombinant gp350 in alum or muramyl dipeptide in squalene, or adenovirus or vaccinia virus expressing gp350 protected animals from lymphoma after challenge with EBV (reviewed in Cohen 2015a). Infection of common marmosets with EBV results in lymphocytosis and development of heterophile antibodies similar to those seen with infectious mononucleosis (Wedderburn et al. 1984). Vaccination of common marmosets with vaccinia or adenovirus expressing gp350 reduced EBV replication after challenge with the virus (reviewed in Cohen 2015a).

Analysis of these studies showed that neutralizing antibody did not always correlate with protection from disease. In two studies, cottontop tamarins immunized with adenovirus or vaccinia virus expressing gp350 did not develop detectable levels of EBV neutralizing antibody but were protected from challenge with EBV (Morgan et al. 1988; Ragot et al. 1993). In another study, not all cottontop tamarins that developed high neutralizing titers to EBV after vaccine were protected against challenge with the virus (Epstein et al. 1986). The reasons for these findings are not clear at present and may due to insufficient time for maturation of neutralizing antibody responses or to other activities of antibodies including ADCC, ADCP, or complement-dependent cytotoxicity. Alternatively these vaccines could induce protective CD4 or CD8 T-cell immunity.

Another model that has been used to test gp350 vaccines is rhesus lymphocryptovirus (LCV) infection in rhesus macaques. Virtually all adult rhesus macaques are naturally infected with rhesus LCV. Oral inoculation of seronegative animals results in atypical lymphocytes in the blood, lymphadenopathy, latent infection in circulating B cells, virus shedding from the saliva, and antibody responses to lytic and latent EBV antigens (Moghaddam et al. 1997). Rhesus LCV was used as a model to test different types of vaccines (Sashihara et al. 2011). Rhesus macaques were vaccinated with rhesus LCV gp350, viruslike replicon particles expressing rhesus LCV gp350, a mixture of replicon particles expressing rhesus LCV gp350, EBNA-3A, and EBNA-3B, or saline. The highest levels of antibodies to gp350 were observed in animals vaccinated with soluble gp350. Rhesus LCV-specific CD4 and CD8 T-cell responses were observed in animals vaccinated with viruslike replicon particles expressing EBNA-3A and EBNA-3B, but not with particles expressing gp350. Rhesus macaques vaccinated with rhesus LCV gp350 had the best level of protection after challenge; animals that still became infected after challenge had the lowest level of rhesus LCV DNA in the blood nearly 3 years after infection. These results emphasize the important role of immune responses to gp350 for protecting animals from infection and for reducing the level of EBV in the blood in animals that still become infected after challenge.

22.9 EBV Vaccine Trials in Humans

The first EBV vaccine trial in humans used live recombinant vaccinia virus expressing gp350 (Gu et al. 1995). Vaccination of adults that were seropositive for both EBV and vaccinia virus did not induce increased titers to EBV. Vaccination of children that were both EBV-seropositive and vaccinia virus-seronegative boosted EBV neutralizing antibody titers. Vaccination of infants that were seronegative for both EBV and vaccinia virus induced neutralizing antibodies in all 9 infants; one-third became infected with EBV within 16 months after vaccination, while 10 of 10 unvaccinated control infants became infected. The numbers were too small to prove efficacy. Live vaccinia virus is no longer a practical platform for a vaccine in the general population.

A recombinant gp350 vaccine produced in Chinese hamster ovary cells was tested in two double-blind randomized controlled studies (Moutschen et al. 2007). In a phase 1 study that included EBV-seropositive and EBV-seronegative adults, all seronegative vaccine recipients produced ELISA antibody to gp350; more subjects who received the vaccine in alum/monophosphoryl lipid A (MPL) adjuvant developed neutralizing antibodies than those receiving the vaccine in alum alone. In a phase 1/2 study, EBV-seronegative adults received gp350 in alum, alum/MPL, or no adjuvant. ELISA antibody titers to gp350 were induced in all the subjects in this study; neutralizing titers developed in 50–60% of persons, and more persons receiving the vaccine in alum adjuvant developed neutralizing titers than those receiving vaccine in alum/MPL or no adjuvant. One serious adverse event occurred that was suspected to be related to the vaccine; a subject who received gp350 in alum/MPL developed headache, meningismus, and oligoarthritis which resolved after 2 months.

A phase 2 double-blind placebo-controlled trial was then performed in EBV-seronegative adults using 50 ug of gp350 in alum/MPL (Sokal et al. 2007). Eighty-eight of 90 persons in the vaccine group and 90 of 91 in the placebo completed the study. Subjects received three doses of vaccine or placebo at 0, 1, and 5 months and were followed for symptoms of infectious mononucleosis for 18 months after the second dose of vaccine. Using an according to protocol analysis, there were fewer cases of infectious mononucleosis in the vaccine group, but the difference did not reach statistical significance ($p = 0.06$); in the intention to treat analysis, the difference was significant ($p = 0.03$) with a vaccine efficacy to prevent infectious mononucleosis of 78%. The incidence of asymptomatic EBV infection was similar in both groups. One month after the third dose of vaccine, 99% of subjects had gp350 antibodies, and these antibodies persisted for 18 months; 70% of vaccinated subjects developed competition ELISA antibodies (a surrogate for neutralizing antibody); EBV DNA levels in the blood were not measured. No serious adverse events were reported.

A phase 1 trial using 12.5 µg or 25 µg of EBV gp350 vaccine in alum adjuvant was performed in EBV-seronegative children with chronic renal insufficiency while waiting for kidney transplants (Rees et al. 2009). All 13 children who could be evaluated developed antibody to EBV, but only 4 developed EBV neutralizing antibody. Antibody levels fell rapidly, and 2 of 13 children became infected with EBV. The authors concluded that additional doses of vaccine and/or a more potent adjuvant would be needed for such a vaccine to reduce EBV disease.

A randomized single-blind, placebo-controlled phase 1 study of two doses of an EBV peptide vaccine was tested in HLA B*801 EBV-seronegative young adults (Elliott et al. 2008). Subjects received 5 ug or 50 ug of an EBNA-3A peptide with tetanus toxoid in a water in oil emulsion (Montanide ISA 720) or placebo. None of the 10 subjects who received the vaccine developed infectious mononucleosis, while 1 of the 4 placebo recipients developed the disease; 4 of the 10 vaccine recipients developed asymptomatic EBV infection, while 1 of the 4 placebo recipients had an asymptomatic infection. EBV-specific T-cell responses were detected in 8 of 9 subjects who received the vaccine. No serious adverse events were noted. While the numbers were small, the results resembled the phase 2 gp350 trial with a possible reduction in infectious mononucleosis with the vaccine without a reduction in asymptomatic EBV infection.

22.10 Future Vaccine Trials

The phase 2 gp350 trial showed that gp350 in alum/MPL reduced the rate of infectious mononucleosis by 78% (Sokal et al. 2007). A phase 3 trial of a vaccine that includes gp350 should be tested in young seronegative adults to determine if the vaccine definitely can reduce the rate of infectious mononucleosis. Infectious mononucleosis is associated with increased rates of Hodgkin lymphoma and multiple sclerosis. About 1 of 800 persons in Sweden and Denmark with EBV-positive infectious mononucleosis developed Hodgkin lymphoma (Hjalgrim et al. 2003). The risk increased within 1 year after onset of mononucleosis and declined back to the rate seen in controls at about 12 years. The median time from infectious mononucleosis to EBV-positive Hodgkin lymphoma was 4 years, and the relative risk of EBV-positive Hodgkin lymphoma was 4.0 after EBV infectious mononucleosis. Therefore, a vaccine that reduces infectious mononucleosis might reduce Hodgkin lymphoma, although the relatively low rate of lymphoma would require a very large study.

The prevalence of multiple sclerosis in the United States is about 1 to 1.5 per 1000. A meta-analysis of 18 studies showed that the relative risk of multiple sclerosis after EBV infectious mononucleosis was 2.2 (Handel et al. 2010). In a nested, case-controlled study of persons who developed multiple sclerosis who had serial serum samples stored, the mean interval between primary EBV infection and onset of multiple sclerosis was estimated to be 5.6 years (with a range of 2.3–9.4 years) (Levin et al. 2010). These data suggest that a vaccine that prevents infectious mononucleosis might reduce the rate of multiple sclerosis. Importantly, such a vaccine might definitively demonstrate (or refute) a causative role of EBV in multiple sclerosis.

EBV-positive posttransplant lymphoproliferative disease usually occurs within 1 year of hematopoietic stem cell transplantation and within 3 years after solid organ transplantation. The rate of EBV posttransplant lymphoproliferative disease is 24–33-fold higher in persons with primary infection after transplant (Preiksaitis and Cockfield, 1998). About 6% of seronegative persons who receive solid organ transplants develop posttransplant lymphoproliferative disease (Sarabu and Hricik 2015). The EBV level in the blood is predictive of posttransplant lymphoproliferative disease (van Esser et al. 2001; Aalto et al. 2007), and rituximab (monoclonal anti-CD20 antibody), given when the viral load is increasing in the blood, usually reduces the viral load to undetectable levels and may reduce posttransplant lymphoproliferative disease (van Esser et al. 2002). Therefore, a vaccine that prevents EBV infection or reduces the viral load during primary infection might reduce the rate of EBV posttransplant lymphoproliferative disease. Similarly, an elevated EBV load in the blood of patients with HIV was associated with a 2.5-fold increased risk of developing systemic B cell lymphoma a median of 10 months after blood was drawn (Leruez-Ville et al. 2012). A prior study in which rhesus macaques were vaccinated with a rhesus LCV gp350 vaccine and challenged with wild-type rhesus LCV showed that animals that became infected after challenge had a lower viral load nearly 3 years after challenge compared with control animals (Sashihara et al. 2011). Unfortunately, viral loads were not measured in the phase 1, 1/2, and 2 trials of the recombinant gp350 vaccine in humans (Moutschen et al. 2007, Sokal et al. 2007), but sensitive assays can reliably measure EBV DNA levels in healthy seropositive persons (Hoshino et al. 2009). This may be important since there is a significant correlation between the level of EBV DNA in the blood and the severity of symptoms with infectious mononucleosis (Balfour et al. 2013). Thus, an EBV vaccine that does not prevent infection might still reduce the viral load in the blood after infection and decrease the risk of severe infectious mononucleosis and EBV-associated malignancies.

An EBV vaccine might also be used to prevent X-linked lymphoproliferative disease type 1 in EBV-seronegative boys with mutations in *SH2D1A* or in other patients with genetic disorders that predispose to EBV malignancies (reviewed in Cohen 2015b) who have not yet become infected. A potential concern is that patients who have X-linked lymphoproliferative disease type 1 might not have a normal response to the EBV vaccine. Mice with mutations in *SH2D1A* have acute IgG antibody responses but a near complete absence of antigen-specific long-lived plasma cells and memory B cells (Crotty et al. 2003).

Burkitt lymphoma is a common malignancy in children in sub-Sahara Africa. About 50% of these children are infected with EBV before 1 year of age, so a vaccine would have to be given at a very young age, and recipients would need to be followed for 5–10 years to determine if the vaccine reduces the rate of disease.

A vaccine trial to reduce rates of nasopharyngeal carcinoma or EBV-positive gastric carcinoma would be difficult to perform due to long latency period between primary infection and development of these carcinomas. Even if such a vaccine does not prevent infection, it still might reduce the rate of these malignancies. Higher EBV gp350 and B cell neutralizing antibody levels correlated with a reduced risk of nasopharyngeal carcinoma in one study (Coghill et al. 2016). This suggests that a vaccine that induces persistently elevated levels of EBV B cell neutralizing antibodies might reduce the rate of nasopharyngeal carcinoma.

At a meeting held at the National Institutes of Health in 2011 (Cohen et al. 2011), there was a strong consensus that clinical trials be performed with vaccine candidates with a goal to prevent infectious mononucleosis and EBV-associated cancers. In addition, priorities for future research that were identified included determining disease-predictive surrogate markers of EBV malignancies to use as endpoints for EBV vaccine trials, identifying immune correlates of protection from EBV infection and disease, establishing epidemiologic studies to determine the benefit of an EBV vaccine, developing a plan to determine vaccine efficacy for preventing malignancies, and establishing a strategy to facilitate collaborations between academic, industry, and government organizations to accelerate EBV vaccine development.

Acknowledgment This work was supported by the intramural research program of the National Institute of Allergy and Infectious Diseases.

References

Aalto SM, Juvonen E, Tarkkanen J, Volin L, Haario H, Ruutu T, Hedman K (2007) Epstein-Barr viral load and disease prediction in a large cohort of allogeneic stem cell transplant recipients. Clin Infect Dis 45:1305–1309

Adhikary D, Behrends U, Moosmann A, Witter K, Bornkamm GW, Mautner J (2006) Control of Epstein-Barr virus infection in vitro by T helper cells specific for virion glycoproteins. J Exp Med 203:995–1006

Adhikary D, Behrends U, Boerschmann H, Pfunder A, Burdach S, Moosmann A, Witter K, Bornkamm GW, Mautner J (2007) Immunodominance of lytic cycle antigens in Epstein-Barr virus-specific CD4+ T cell preparations for therapy. PLoS One 2:e583

Alfieri C, Birkenbach M, Kieff E (1991) Early events in Epstein-Barr virus infection of human B lymphocytes. Virology 181:595–608

Balfour HH Jr, Odumade OA, Schmeling DO, Mullan BD, Ed JA, Knight JA, Vezina HE, Thomas W, Hogquist KA (2013) Behavioral, virologic, and immunologic factors associated with acquisition and severity of primary Epstein-Barr virus infection in university students. J Infect Dis 207:80–88

Brooks JM, Long HM, Tierney RJ, Shannon-Lowe C, Leese AM, Fitzpatrick M, Taylor GS, Rickinson AB (2016) Early T cell recognition of B cells following Epstein-Barr virus infection: identifying potential targets for prophylactic vaccination. PLoS Pathog 12:e1005549

Bu W, Hayes GM, Liu H, Gemmell L, Schmeling DO, Radecki P, Aguilar F, Burbelo PD, Woo J, Balfour HH Jr, Cohen JI (2016) Kinetics of Epstein-Barr virus (EBV) neutralizing and virus-specific antibodies after primary infection with EBV. Clin Vaccine Immunol 23:363–369

Callan MF, Tan L, Annels N, Ogg GS, Wilson JD (1998) Direct visualization of antigen-specific CD8 T cells during the primary immune response to Epstein-Barr virus in vivo. J Exp Med 187:1395–1402

Cleary ML, Epstein MA, Finerty S, Dorfman RF, Bornkamm GW, Kirkwood JK, Morgan AJ, Sklar J (1985) Individual tumors of multifocal EB virus-induced malignant lymphomas in tamarins arise from different B-cell clones. Science 228:722–724

Coghill AE, Bu W, Nguyen H, Hsu WL, Yu KJ, Lou PJ, Wang CP, Chen CJ, Hildesheim A, Cohen JI (2016) High levels of antibody that neutralize B-cell infection of Epstein-Barr virus and that bind EBV gp350 are associated with a lower risk of nasopharyngeal carcinoma. Clin Cancer Res 22:3451–3457

Cohen JI (2000) Epstein-Barr virus infection. N Engl J Med 343:481–492

Cohen JI (2015a) Epstein-Barr virus vaccines. Clin Transl Immunol 4:e32

Cohen JI (2015b) Primary immunodeficiencies associated with EBV disease. Curr Top Microbiol Immunol 390(Pt 1):241–265

Cohen JI, Fauci AS, Varmus H, Nabel GJ (2011) Epstein-Barr virus: an important vaccine target for cancer prevention. Sci Transl Med 3:107fs7

Crotty S, Kersh EN, Cannons J, Schwartzberg PL, Ahmed R (2003) SAP is required for generating long-term humoral immunity. Nature 421:282–287

Cui X, Cao Z, Sen G, Chattopadhyay G, Fuller DH, Fuller JT, Snapper DM, Snow AL, Mond JJ, Snapper CM (2013) A novel tetrameric gp350 1-470 as a potential Epstein-Barr virus vaccine. Vaccine 31:3039–3045

Cui X, Cao Z, Chen Q, Arjunaraja S, Snow AL, Snapper CM (2016) Rabbits immunized with Epstein-Barr virus gH/gL or gB recombinant proteins elicit higher serum virus neutralizing activity than gp350. Vaccine 34:4050–4055

Elliott SL, Suhrbier A, Miles JJ, Lawrence G, Pye SJ, Le TT, Rosenstengel A, Nguyen T, Allworth A, Burrows SR, Cox J, Pye D, Moss DJ, Bharadwaj M (2008) Phase I trial of a CD8+ T-cell peptide epitope-based vaccine for infectious mononucleosis. J Virol 82:1448–1457

Emini EA, Schleif WA, Armstrong ME, Silberklang M, Schultz LD, Lehman D, Maigetter RZ, Qualtiere LF, Pearson GR, Ellis RW (1988) Antigenic analysis of the Epstein-Barr virus major membrane antigen (gp350/220) expressed in yeast and mammalian cells: implications for the development of a subunit vaccine. Virology 166:387–393

Epstein MA, Morgan AJ, Finerty S, Randle BJ, Kirkwood JK (1985) Protection of cottontop tamarins against Epstein-Barr virus-induced malignant lymphoma by a prototype subunit vaccine. Nature 318:287–289

Epstein MA, Randle BJ, Finerty S, Kirkwood JK (1986) Not all potently neutralizing, vaccine-induced antibodies to Epstein-Barr virus ensure protection of susceptible experimental animals. Clin Exp Immunol 63:485–490

Finerty S, Tarlton J, Mackett M, Conway M, Arrand JR, Watkins PE, Morgan AJ (1992) Protective immunization against Epstein-Barr virus-induced disease in cottontop tamarins using the virus envelope glycoprotein gp340 produced from a bovine papillomavirus expression vector. J Gen Virol 73:449–453

Gu SY, Huang TM, Ruan L, Miao YH, Lu H, Chu CM, Motz M, Wolf H (1995) First EBV vaccine trial in humans using recombinant vaccinia virus expressing the major membrane antigen. Dev Biol Stand 84:171–177

Handel AE, Williamson AJ, Disanto G, Handunnetthi L, Giovannoni G, Ramagopalan SV (2010) An updated meta-analysis of risk of multiple sclerosis following infectious mononucleosis. PLoS One. 5. pii: e12496

Haque T, Johannessen I, Dombagoda D, Sengupta C, Burns DM, Bird P, Hale G, Mieli-Vergani G, Crawford DH (2006) A mouse monoclonal antibody against Epstein-Barr virus envelope glycoprotein 350 prevents infection both in vitro and in vivo. J Infect Dis 194:584–587

Hartlage AS, Liu T, Patton JT, Garman SL, Zhang X, Kurt H, Lozanski G, Lustberg ME, Caligiuri MA, Baiocchi RA (2015) The Epstein-Barr virus lytic protein BZLF1 as a candidate target antigen for vaccine development. Cancer Immunol Res 3:787–794

Heeke DS, Lin R, Rao E, Woo JC, McCarthy MP, Marshall JD (2016) Identification of GLA/SE as an effective adjuvant for the induction of robust humoral and cell-mediated immune responses to EBV-gp350 in mice and rabbits. Vaccine 34:2562–2569

Hettich E, Janz A, Zeidler R, Pich D, Hellebrand E, Weissflog B, Moosmann A, Hammerschmidt W (2006) Genetic design of an optimized packaging cell line for gene vectors transducing human B cells. Gene Ther 13:844–856

Hislop AD, Annels NE, Gudgeon NH, Leese AM, Rickinson AB (2002) Epitope-specific evolution of human CD8(+) T cell responses from primary to persistent phases of Epstein-Barr virus infection. J Exp Med 195:893–905

Hjalgrim H, Askling J, Rostgaard K, Hamilton-Dutoit S, Frisch M, Zhang JS, Madsen M, Rosdahl N, Konradsen HB, Storm HH, Melbye M (2003) Characteristics of Hodgkin's lymphoma after infectious mononucleosis. N Engl J Med 349:1324–1332

Hoshino Y, Katano H, Zou P, Hohman P, Marques A, Tyring SK, Follmann D, Cohen JI (2009) Long-term administration of valacyclovir reduces the number of Epstein-Barr virus (EBV)-infected B cells but not the number of EBV DNA copies per B cell in healthy volunteers. J Virol 83:11857–11861

Jackman WT, Mann KA, Hoffmann HJ, Spaete RR (1999) Expression of Epstein-Barr virus gp350 as a single chain glycoprotein for an EBV subunit vaccine. Vaccine 17:660–668

Janz A, Oezel M, Kurzeder C, Mautner J, Pich D, Kost M, Hammerschmidt W, Delecluse HJ (2000) Infectious Epstein-Barr virus lacking major glycoprotein BLLF1 (gp350/220) demonstrates the existence of additional viral ligands. J Virol 74:10142–10152

Jung S, Chung YK, Chang SH, Kim J, Kim HR, Jang HS, Lee JC, Chung GH, Jang YS (2001) DNA-mediated immunization of glycoprotein 350 of Epstein-Barr virus induces the effective humoral and cellular immune responses against the antigen. Mol Cells 12:41–49

Kanekiyo M, Bu W, Joyce MG, Meng G, Whittle JR, Baxa U, Yamamoto T, Narpala S, Todd JP, Rao SS, McDermott AB, Koup RA, Rossmann MG, Mascola JR, Graham BS, Cohen JI, Nabel GJ (2015) Rational design of an Epstein-Barr virus vaccine targeting the receptor-binding site. Cell 162:1090–1100

Kawaguchi A, Kanai K, Satoh Y, Touge C, Nagata K, Sairenji T, Inoue Y (2009) The evolution of Epstein-Barr virus inferred from the conservation and mutation of the virus glycoprotein gp350/220 gene. Virus Genes 38:215–223

Khanna R, Sherritt M, Burrows SR (1999) EBV structural antigens, gp350 and gp85, as targets for ex vivo virus-specific CTL during acute infectious mononucleosis: potential use of gp350/gp85 CTL epitopes for vaccine design. J Immunol 162:3063–3069

Lees JF, Arrand JE, Pepper SD, Stewart JP, Mackett M, Arrand JR (1993) The Epstein-Barr virus candidate vaccine antigen gp340/220 is highly conserved between virus types A and B. Virology 195:578–586

Leruez-Ville M, Seng R, Morand P, Boufassa F, Boue F, Deveau C, Rouzioux C, Goujard C, Seigneurin JM, Meyer L (2012) Blood Epstein-Barr virus DNA load and risk of progression to AIDS-related systemic B lymphoma. HIV Med 13:479–487

Levin LI, Munger KL, O'Reilly EJ, Falk KI, Ascherio A (2010) Primary infection with the Epstein-Barr virus and risk of multiple sclerosis. Ann Neurol 67:824–830

Li Q, Turk SM, Hutt-Fletcher LM (1995) The Epstein-Barr virus (EBV) BZLF2 gene product associates with the gH and gL homologs of EBV and carries an epitope critical to infection of B cells but not of epithelial cells. J Virol 69:3987–3994

Lockey TD, Zhan X, Surman S, Sample CE, Hurwitz JL (2008) Epstein-Barr virus vaccine development: a lytic and latent protein cocktail. Front Biosci 13:5916–5927

Long HM, Leese AM, Chagoury OL, Connerty SR, Quarcoopome J, Quin LL, Shannon-Lowe C, Rickinson AB (2011) Cytotoxic CD4+ T cell responses to EBV contrast with CD8 responses in breadth of lytic cycle antigen choice and in lytic cycle recognition. J Immunol 187:92–101

Longnecker L, Kieff E, Cohen JI (2013) Epstein-Barr virus. In: Knipe DM, Howley PM, Cohen JI, Griffith DE, Lamb RA, Martin MA, Racaniello V, Roizman B (eds) Fields virology, 6th edn. Lippincott Williams & Wilkins, Philadelphia, pp 1898–1959

Luzuriaga K, Sullivan JL (2010) Infectious mononucleosis. N Engl J Med 362:1993–2000

Moghaddam A, Rosenzweig M, Lee-Parritz D, Annis B, Johnson RP, Wang F (1997) An animal model for acute and persistent Epstein-Barr virus infection. Science 276:2030–2033

Morgan AJ, Epstein MA, North JR (1984) Comparative immunogenicity studies on Epstein-Barr virus membrane antigen (MA) gp340 with novel adjuvants in mice, rabbits, and cotton-top tamarins. J Med Virol 13:281–292

Morgan AJ, Mackett M, Finerty S, Arrand JR, Scullion FT, Epstein MA (1988) Recombinant vaccinia virus expressing Epstein-Barr virus glycoprotein gp340 protects cottontop tamarins against EB virus-induced malignant lymphomas. J Med Virol 25:189–195

Moutschen M, Léonard P, Sokal EM, Smets F, Haumont M, Mazzu P, Bollen A, Denamur F, Peeters P, Dubin G, Denis M (2007) Phase I/II studies to evaluate safety and immunogenicity of a recombinant gp350 Epstein-Barr virus vaccine in healthy adults. Vaccine 25:4697–4705

Munger KL, Levin LI, O'Reilly EJ, Falk KI, Ascherio A (2011) Anti-Epstein-Barr virus antibodies as serological markers of multiple sclerosis: a prospective study among United States military personnel. Mult Scler 17:1185–1193

North JR, Morgan AJ, Epstein MA (1980) Observations on the EB virus envelope and virus-determined membrane antigen (MA) polypeptides. Int J Cancer 26:231–240

Ogembo JG, Muraswki MR, McGinnes LW, Parcharidou A, Sutiwisesak R, Tison T, Avendano J, Agnani D, Finberg RW, Morrison TG, Fingeroth JD (2015) A chimeric EBV gp350/220-based VLP replicates the virion B-cell attachment mechanism and elicits long-lasting neutralizing antibodies in mice. J Transl Med 13:50

Palser AL, Grayson NE, White RE, Corton C, Correia S, Ba Abdullah MM, Watson SJ, Cotten M, Arrand JR, Murray PG, Allday MJ, Rickinson AB, Young LS, Farrell PJ, Kellam P (2015) Genome diversity of Epstein-Barr virus from multiple tumor types and normal infection. J Virol 89:5222–5237

Perez EM, Foley J, Tison T, Silva R, Ogembo JG (2016) Novel Epstein-Barr virus-like particles incorporating gH/gL-EBNA1 or gB-LMP2 induce high neutralizing antibody titers and EBV-specific T-cell responses in immunized mice. Oncotarget. Dec 1, online

Porcu P, Eisenbeis CF, Pelletier RP, Davies EA, Baiocchi RA, Roychowdhury S, Vourganti S, Nuovo GJ, Marsh WL, Ferketich AK, Henry ML, Ferguson RM, Caligiuri MA (2002) Successful treatment of posttransplantation lymphoproliferative disorder (PTLD) following renal allografting is associated with sustained CD8(+) T-cell restoration. Blood 100:2341–2348

Precopio ML, Sullivan JL, Willard C, Somasundaran M, Luzuriaga K (2003) Differential kinetics and specificity of EBV-specific CD4+ and CD8+ T cells during primary infection. J Immunol 170:2590–2598

Preiksaitis JK, Cockfield SM (1998) Epstein-Barr virus and lymphoproliferative disease after transplantation. In: Bowden RA, Ljungman P, Paya CV (eds) Transplant infections. Lippincott-Raven, Philadelphia, pp 245–263

Pudney VA, Leese AM, Rickinson AB, Hisop AD (2005) CD8+ immunodominance among Epstein-Barr virus lytic cycle antigens directly reflects the efficiency of antigen presentation in lytically infected cells. J Exp Med 201:349–360

Qualtiere LF, Chase R, Pearson GR (1982) Purification and biologic characterization of a major Epstein Barr virus-induced membrane glycoprotein. J Immunol 129:814–818

Ragot T, Finerty S, Watkins PE, Perricaudet M, Morgan AJ (1993) Replication-defective recombinant adenovirus expressing the Epstein-Barr virus (EBV) envelope glycoprotein gp340/220 induces protective immunity against EBV-induced lymphomas in the cottontop tamarin. J Gen Virol 74:501–507

Rea TD, Russo JE, Katon W, Ashley RL, Buchwald DS (2001) Prospective study of the natural history of infectious mononucleosis caused by Epstein-Barr virus. J Am Board Fam Pract 14:234–242

Rees L, Tizard EJ, Morgan AJ, Cubitt WD, Finerty S, Oyewole-Eletu TA, Owen K, Royed C, Stevens SJ, Shroff RC, Tanday MK, Wilson AD, Middeldorp JM, Amlot PL, Steven NM (2009) A phase I trial of Epstein-Barr virus gp350 vaccine for children with chronic kidney disease awaiting transplantation. Transplantation 88:1025–1029

Ruiss R, Jochum S, Wanner G, Reisbach G, Hammerschmidt W, Zeidler R (2011) A virus-like particle-based Epstein-Barr virus vaccine. J Virol 85:13105–13113

Santpere G, Darre F, Blanco S, Alcami A, Villoslada P, Mar Albà M, Navarro A (2014) Genome-wide analysis of wild-type Epstein-Barr virus genomes derived from healthy individuals of the 1,000 Genomes Project. Genome Biol Evol 6:846–860

Sarabu N, Hricik DE (2015) Introduction to solid organ transplantation. In: Ljungman P, Snydman D, Boeckh M (eds) Transplant infections. Springer, Basel, pp 19–30

Sashihara J, Burbelo PD, Savoldo B, Pierson TC, Cohen JI (2009) Human antibody titers to Epstein-Barr Virus (EBV) gp350 correlate with neutralization of infectivity better than antibody titers to EBV gp42 using a rapid flow cytometry-based EBV neutralization assay. Virology 391:249–256

Sashihara J, Hoshino Y, Bowman JJ, Krogmann T, Burbelo PD, Coffield VM, Kamrud K, Cohen JI (2011) Soluble rhesus lymphocryptovirus gp350 protects against infection and reduces viral loads in animals that become infected with virus after challenge. PLoS Pathog 7:e1002308

Silins SL, Sherritt MA, Silleri JM, Cross SM, Elliott SL, Bharadwaj M, Le TT, Morrison LE, Khanna R, Moss DJ, Suhrbier A, Misko IS (2001) Asymptomatic primary Epstein-Barr virus infection occurs in the absence of blood T-cell repertoire perturbations despite high levels of systemic viral load. Blood 98:3739–3744

Sokal EM, Hoppenbrouwers K, Vandermeulen C, Moutschen M, Léonard P, Moreels A, Haumont M, Bollen A, Smets F, Denis M (2007) Recombinant gp350 vaccine for infectious mononucleosis: a phase 2, randomized, double-blind, placebo-controlled trial to evaluate the safety, immunogenicity, and efficacy of an Epstein-Barr virus vaccine in healthy young adults. J Infect Dis 196:1749–1753

Steven NM, Annels NE, Kumar A, Leese AM, Kurilla MG (1997) Immediate early and early lytic cycle proteins are frequent targets of the Epstein-Barr virus induced cytotoxic T cell response. J Exp Med 185:1605–1617

Taylor GS, Long HM, Brooks JM, Rickinson AB, Hislop AD (2015) The immunology of Epstein-Barr virus-induced disease. Annu Rev Immunol 33:787–821

Thacker EL, Mirzaei F, Ascherio A (2006) Infectious mononucleosis and risk for multiple sclerosis: a meta-analysis. Ann Neurol 59:499–503

Thorley-Lawson DA (1979) A virus-free immunogen effective against Epstein-Barr virus. Nature 281:486–488

Thorley-Lawson DA, Poodry CA (1982) Identification and isolation of the main component (gp350-gp220) of Epstein-Barr virus responsible for generating neutralizing antibodies in vivo. J Virol 43:730–736

Tugizov SM, Berline JW, Palefsky JM (2003) Epstein-Barr virus infection of polarized tongue and nasopharyngeal epithelial cells. Nat Med 9:307–314

van Esser JW, van der Holt B, Meijer E, Niesters HG, Trenschel R, Thijsen SF, van Loon AM, Frassoni F, Bacigalupo A, Schaefer UW, Osterhaus AD, Gratama JW, Löwenberg B, Verdonck LF, Cornelissen JJ (2001) Epstein-Barr virus (EBV) reactivation is a frequent event after allogeneic stem cell transplantation (SCT) and quantitatively predicts EBV-lymphoproliferative disease following T-cell–depleted SCT. Blood 98:972–978

van Esser JW, Niesters HG, van der Holt B, Meijer E, Osterhaus AD, Gratama JW, Verdonck LF, Löwenberg B, Cornelissen JJ (2002) Prevention of Epstein-Barr virus-lymphoproliferative disease by molecular monitoring and preemptive rituximab in high-risk patients after allogeneic stem cell transplantation. Blood 99:4364–4369

Wedderburn N, Edwards JM, Desgranges C, Fontaine C, Cohen B, de Thé G (1984) Infectious mononucleosis-like response in common marmosets infected with Epstein-Barr virus. J Infect Dis 150:878–882

Weiss ER, Alter G, Ogembo JG, Henderson JL, Tabak B, Bakiş Y, Somasundaran M, Garber M, Selin L, Luzuriaga K (2016) High Epstein-Barr virus load and genomic diversity are associated with generation of gp350-specific neutralizing antibodies following acute infectious mononucleosis. J Virol. 91. pii: e01562-16

Woodberry T, Suscovich TJ, Henry LM, Davis JK, Frahm N, Walker BD, Scadden DT, Wang F, Brander C (2005) Differential targeting and shifts in the immunodominance of Epstein-Barr virus--specific CD8 and CD4 T cell responses during acute and persistent infection. J Infect Dis 192:1513–1524

Xu J, Ahmad A, Blagdon M, D'Addario M, Jones JF, Dolcetti R, Vaccher E, Prasad U, Menezes J (1998) The Epstein-Barr virus (EBV) major envelope glycoprotein gp350/220-specific antibody reactivities in the sera of patients with different EBV-associated diseases. Int J Cancer 79:481–486

Index

A
A-capsids, 171
Accessory genes, 68
Activation-induced cytidine deaminase (AID), 468
Acute GVHD, 256
ACV suppressive therapy, 94
Acyclovir (ACV), 68, 90, 94–96, 103
Adipogenesis, 190
Adverse events, 136
AIDS-associated lymphoma, 364
Akt, 329–331
Allele frequencies, 128
Amenamevir, 114
Amplicon, 79
Animal models, 150
Anti-apoptotic, 403
Antigenemia, 258
Antiviral, 171
Antiviral drugs, 262
Apaf-1, 339
APC, 337
Apoptosis, 339–342
$AS01_B$, 135
ASP2151 (amenamevir), 104
A specific inhibitor of MLCK, ML-7, 15
Assembly protein, 172
AT-rich interaction domain 1A (*ARID1A*), 448
Atrophic gastritis, 444
Attenuated canary pox (ALVAC), 281, 285
Autophosphorylation, 50, 51
Axin, 337

B
Bacterial artificial chromosome (BAC), 72
BamHI-A region rightward transcripts (BART), 384–386, 388, 451
BAY 57-1293 (pritelivir), 104
B capsids, 171
Bcl-2, 341
Bcl-2-associated agonist of cell death (BAD), 54–55
Betaherpesvirinae, 168
Betaherpesvirus subfamily, 168
BHRF1, 384, 387–388
BIR model, 214
BLT mice, 424
BMRF2, 465, 480
Breakthrough varicella, 126
BRG, 424
Brivudin, 112
BRLF (Rta), 313
Bromovinyluracil, 111
Burkitt lymphomas, 386, 460, 478
BZLF1, 396

C
Cancer, 63
Capsid assembly, 23–24
Capsid proteins, 172, 228–231
Capsids, 168, 171, 405–406
Capsid types, 171–172
Capsid-vertex specific component, 24
Caspase, 339

β-catenin, 337
CBF1, 325, 336
C.B-17 *scid*, 423
CD134, 153
CD46, 152
Cdk, 342
CD4+ T-cells, 429
C/EBPβ, 338
Cell cycle, 342–343
Cell-to-cell spread, 35, 36
Cell tropisms, 129, 149–150
Cerebrospinal fluid (CSF), 88, 90, 92, 95
Chain termination, 104
Chemokine receptor US28, 240
Chromosomally integrated HHV-6 (ciHHV-6), 259
Chronic active EBV (CAEBV), 464
Cidofovir, 113, 261
CIMP, 441
Clinical trials, 73
Clonal growth, 442
CMV gastroenteritis, 253
CMV pneumonia, 253
CMV retinitis, 254
c-My, 338
Combination therapy, 79–80
Common marmosets, 417
Computed tomography (CT), 88
Concluding remarks, 15
Conclusion, 220–221
Conserved tegument proteins, 31
Consensus phosphorylation target sequence of Us3, The, 57
Copy number, 443
Core SNPs, 130
Corneal infection, 11
Cortical actin, 36
Corticosteroid, 91
Cotton-top tamarins, 417
Cranial model, 11
CSL, 325, 336
cVAC, 178–180
Cytokine mimic cmvIL-10, 241–242
Cytoplasmic virion assembly, 31
Cytoplasmic virion assembly compartment (cVAC), 178

D
DDX3X, 468
De-envelopment, 27
Deep sequencing technology, 130–131
Delirium, 256
Dense bodies, 170, 286

Detection active infection versus iciHHV-6, 216–219
Diffuse large B-cell lymphoma (DLBCL), 423
Direct repeat regions (DR1 and DR2), 314
Disabled infectious single cycle (DISC) vaccine, 279–282
DNA methylation, 444
DNA packaging, 24
DNA polymerase, 238
DNA polymerase (UL30/UL42), 113
DNMT1, 446
Droplet digital PCR, 216
Dynamin, 191

E
E3 ubiquitin ligase, 325, 333, 336, 342
EBER-ISH, 439
EBNA1, 381–382
EBNA2, 336, 379, 481
EBNA3, 380–381, 389
EBNA-LP, 379
EBVaGC, 438
EBV-encoded small RNA (EBER), 384
EBV epigenotype, 445
EBV glycoproteins, 479–480
EBV gp350 vaccine, 486
EBV-positive gastric cancer, 388
EBV-positive T-/NK-cell lines, 461
EBV-T/NK LPD, 463
Egress, 178
Electroencephalogram (EEG), 89
Ellen, 128
Emerin, 56
Encapsidation, 176–177
Endogenous PILRα, 8
Endosomal sorting complex required for transport (ESCRT), 34, 191
Endosomes, 32
Entry receptors, 4
Envelope, 170
Enveloped virus-like particle (eVLP), 280, 283
Envelope glycoproteins, 234–238
Envelope proteins, 26, 30–36
Envelopment, 178, 183–186
Episome, 301
Epstein-Barr virus
Epstein-Barr virus (EBV), 10, 182, 358
ESCRT, 192
Essential genes, 68
Euphorbia tirucalli, 465
Exosomes, 450
Extranodal NK/T-cell lymphoma-nasal type (ENKL), 463, 468

Index

F
Famciclovir, 68, 104
Fatty acid metabolism, 189–191
Figure 1, 210
Figure 2, 212
Figure 3, 217
FLICE inhibitory protein (v-FLIP), 303
5-fluorouracil (5FU), 111
Foscarnet, 104, 112, 261

G
G207, 73
Ganciclovir, 261
gB, 7, 52, 235, 331, 483
gC, 8
gD, 6
Genital herpes, 118
Genomes, 168
gE-specific, 135
gE-specific cell-mediated immune response, 136
gH/gL, 7, 236, 483
gH/gL/gO, 237
gH/gL/gO complex, 153
gH/gL/gp42, 465
gH/gL/gQ1/gQ2 complex, 150–153
gH/gL/UL128/UL130/UL131A, 237
gK/UL20, 8
Glioma, 64
Glycoprotein, 407
Glycoprotein B (gB), 273–275, 277, 280–288
Glycoprotein E (gE), 135
gM/gN complex, 154–155
gp350, 479, 481
gp350 peptide, 482
gp350/220, 465
GPCMV, 186
gpK8.1A, 332
GSK-3β, 337
G-string, 110
Guinea pig CMV (GPCMV), 185, 274, 287–288

H
Helicase-primase (HP), 104, 113
Helicase-primase inhibitors (HPIs), 104
Hematopoietic stem cell transplant (HSCT), 272
Heparan, 147
HER-2, 78
Herpes simplex encephalitis (HSE), 86–91
Herpes simplex virus (HSV), 64, 86–94, 103
Herpesviridae family, 4
Herpesvirus saimiri, 421
γ2-herpesvirus, 300
Herpes zoster (HZ), 124
hESC-derived neurons, 132
Hexon, 230
HF10, 76–77
HHV-6 encephalitis, 262
HHV-6 integration, 210–216
HHV-6 integration models, 214
HHV-6 life cycle, 209–210
High mortality, 253
High-risk, 252
Host glycoproteins, 156
HPI-resistant virus, 117
H. pylori, 444
HSV-1, 86
HSV1716, 74
HSV-2, 86
HSV DNA, 90, 92
HSV entry pathway, 8
HSV fusion, 4
Human betaherpesvirus 5, 168
Human betaherpesvirus 6A, 168
Human betaherpesvirus 6B, 168
Human cytomegalovirus, 168
Human embryonic stem cell (hESC)-derived neurons, 132
Human herpesvirus 6A, 168
Human herpesvirus 6B, 168
Human herpesvirus 8 (HHV-8), 358
Human immunodeficiency virus (HIV), 300
Human leukocyte antigen (HLA), 464
Humanized mice, 423–428
HVEM-1-restricted or nectin-1-restricted mutants, 14
Hyperimmune globulin (HIG), 274–276
Hypersensitivity, 109
HZ/su, 135
HZ subunit vaccine, 124

I
iciHHV-6, 259
iciHHV-6 and transplantation, 220
ICP34.5, 72
ICP4, 71
ICP47, 71
IFN, 323–326
IKK, 333
IKKγ, 313
IL-13, 78

Immediate early protein IE1, 240–241
Immune checkpoint inhibitors, 80
Immune correlates, 272, 279
Immune evasion, 274, 279, 287, 288, 403
Immunoevasin US2, 242–243
Impacts of ciHHV-6 on cell biology, 219–220
Inhibitor of apoptosis protein (IAP), 341
In vitro latency system, 132
In vitro VZV latency system, 131–133
Infected cell polypeptide 0 (ICP0), 71
Infectious mononucleosis (IM), 414, 478
Inherited chromosomally integrated HHV-6, 210
Inner nuclear membrane, 177
Inner tegument, 27, 29, 30
Integrin α3β1, 332
Interaction between gB and PILRα, 9
Interferon-stimulated genes (ISG), 323
Ipilimumab, 80
IRF3, 57, 87
IκB, 333

J
JAK, 324
JAK3, 468
Jun N-terminal kinases (JNK), 331

K
K1, 316, 329, 335
K15, 316, 335
K2 (v-IL6), 316
K7, 335, 341
Kaposin, 314–315, 322, 338–339
Kaposi's sarcoma (KS), 301, 362–364, 414
Kaposi's sarcoma-associated herpesvirus (KSHV), 182, 300, 321, 358
Kaposin (K12), 303
K-bZIP, 325, 332
KICS, 369–370
KIF3A, 54
Kinesin, 35, 36
KSHV-DLBCL, NOS, 366
KSHV inflammatory cytokine syndrome, 369–370
KSHV MicroRNAs (K-miRNAs), 315

L
Lamin A/C, 56–57
LANA-1, 359, 364
LANA- binding sites (LBS), 304

Latency, 123, 209, 300
Latency-associated nuclear antigen (LANA), 303, 322, 328, 336, 338, 342
Latency type, 378
Latency type II, 466
Latent infection, 303
Live-attenuated vaccine, 279
LMP1, 383, 389, 466
LMP2, 383
lncRNA, 328
LOGO algorithm, 57
Lymphoblastoid cell line (LCL), 377, 379–381, 383–386
Lymphocryptovirus (LCV), 485
Lymphocytic choriomeningitis virus (LCMV), 281, 286, 288
Lymphoproliferative disease (LPD), 414

M
Macaque
  cynomolgus macaque, 288
  rhesus macaque, 274, 285, 288
Magnetic resonance imaging (MRI), 88, 93, 256
Marmoset LCV (maLCV), 420
MDM2, 343
Mechanisms of chromosomal integration, 212–215
Membrane fusion, 26
Methylation mechanisms, 446
MF59, 275, 280, 282
MHC class I, 426
Microenvironment, 449
Microfluidic device, 132
microRNAs, 303, 384–386
Microtubules, 35, 182
miR-200, 450
miR-K1, 326, 335
miR-K10a, 327
miR-K3, 326
miR-K4-5p, 326
miR-K5, 327
miR-K7, 327
miRNA, 360, 384–386, 388
Mitogen-activated protein kinase (MAPK), 331–333
Modified vaccinia Ankara (MVA), 281, 285, 286, 288
Mouse, 421
Mouse cytomegalovirus, 168
Mouse herpesvirus 68, 185
Monophosphoryl lipid A (MPL), 486

Index 499

mTOR, 329
Multicentric Castleman disease (MCD), 301, 322, 368–369, 414
Multiple sclerosis, 479, 487
Multivesicular bodies, 185, 186, 192
Murid betaherpesvirus 1, 168
Murine CMV (MCMV), 274, 287
Murine gammaherpesvirus 68, 421
Mutant viruses, 13–14
Myc, 386
Myelin-associated glycoprotein (MAG), 9

N

Nasopharyngeal carcinoma, 388, 389, 478
Natural killer cells
    decidual NK, 273
Natural killer cells (dNK cells), 273
Natural killer cells (NK cells), 273, 427
NEMO, 87
Neonatal herpes, 91
Neonatal HSV infection, 91
Neonatal HSV meningoencephalitis, 91–94
Neurotropism, 129
Next-generation sequencing, 127–128
NFκB, 313
NF-κB, 333–336
Nivolumab, 64
NKG2D, 274
N-methyl-D-aspartate receptor (NMDAR), 88
NMHC-IIA, 9
NOD/scid, 424
NOG, 424
Noncanonical TATA box, 403
Non-human primates (NHPs), 418
Noninfectious, enveloped particles (NIEPs), 170
Non-muscle myosin heavy chain IIA (NMHC-IIA), 10
Notch, 335–336
NSG, 424
Nuclear egress, 25–27, 177–178
Nuclear egress complex, 25, 177, 234
Nuclear lamina, 177
Nuclear matrix, 311
NV1020, 74

O

Oncolytic virotherapy, 64
OncoVEXGM-CSF, 74
Oral hairy leukoplakia (OHL), 419
ORF0, 128
ORF16 (v-BCL2), 316

ORF45, 333
ORF50, 359
ORF62, 127
Organoid culture, 453
OriLyt, 401
Ori-P, 305
Outer nuclear membrane, 177
Outer tegument, 27

P

p38MAPK, 339
p53, 342
Patient mortality, 257
PD-L1, 448
PD-L1+ immune cells, 449
PEL, 364–366
Pembrolizumab, 80
Penciclovir, 104
Pentameric complex
    Pentamer, 275
Penton, 172
Perinuclear virions, 25, 26, 30
Periodic lateralizing epileptiform discharges (PLEDs), 89
Persistent infection, 425
Phase 1, 279, 282–286
Phase 2, 274, 279, 282, 284, 286
Phase 3, 274, 289
Phosphorylation, 46
PI3K, 329
PIK3CA, 447
PILRα, 8–9
PKA, 55
Placenta, 272–276, 278, 286–288
Pneumonitis, 255
pOka, 125
Polyadenylated nuclear (PAN) RNA, 328
Polymerase chain reaction (PCR), 90, 92, 95
Population diversity, 130
Porcine reproductive and respiratory syndrome virus (PRRSV), 10
Post transplant acute limbic encephalitis (PALE), 255
Postherpetic neuralgia (PHN), 112, 124
Post-nuclear Assembly, 178–194
Posttransplant lymphoproliferative disease, 487
pp65, 277, 280, 281, 284–288
pRb, 342
Preemptive treatment, 261, 262
Pre-latent abortive lytic, 403
Primary effusion lymphoma (PEL), 301, 322, 364–366, 414

Primary enveloped particles, 25
Primary envelopment, 25, 30, 177
Pritelivir, 114
Proofreading, 111
Prophylactic treatments, 262
Protease, 238
Protein kinase, 46
Pseudorabies virus, 185

Q
Quantitative PCR, 216
Quiescent state, 129

R
Rab, 36
Rabbits, 417
RBP-Jκ, 325, 336
Reactivation, 124, 252
Reactivation from integration, 215–216
Real-time PCR, 90, 258
Real-time polymerase chain reaction, 258
Recognition sites, 109
Recombinant gp350, 482
Recombinant gp350 vaccine, 485
Recombinant HSV-1 carrying gB-T53A/T480A, The, 14
Reinfection, 252
Relapse, 254
Replication compartments, 401
Resistant mutants, 109
Retrograde axonal transport, 131
Retroperitoneal fibromatosis (RF), 419
Reverse Genetics, 427–428
RF herpesvirus (RFHV), 419
Rhesus CMV (RhCMV), 186, 274, 285, 288, 289
Rhesus LCV, 418
Rhesus macaques, 417, 418
Rhesus rhadinovirus, 420
Rheumatoid arthritis (RA), 426
Ribonucleotide reductase (RR), 72, 106
Replication and transcription activator (RTA), 322, 325, 332, 336, 359, 480
RNA-DNA hybrid, 402
RT-PCR assays, 259

S
Scaffold, 24, 25, 172
Scid-hu, 423
SCID-hu mouse model, 128

Secondary envelopment, 26, 27, 30, 32–36, 178, 182–183
Secretory vesicles, 193
Seroprevalence, 361
Serostatus, 260
Severe fever with thrombocytopenia syndrome virus (SFTSV), 10
Severely combined immunodeficiency (SCID), 129
Signal-regulated protein kinases (ERK), 331
Significance of gB receptors in vivo, 14–15
Single nucleotide polymorphisms (SNPs), 127
Significance of gD receptors in vivo using knockout mice, The, 11–13
Skin tropism, 131
SNARE, 36
Solid organ transplant (SOT), 272
Sorivudine, 104, 106, 111
SSA model, 215
STAT, 324
STAT1, 87
Summarized structural information, 229–230
Survival advantage, 443
Synergism, 118

T
Talimogene laherparepvec (T-VEC), 74–76
Targeted enrichment methodology, 130
TBK1, 87
T-cell immunity, 135
T-cell intracytoplasmic antigen (TIA)-1, 466
Tegument, 27, 169, 406–407
Tegument assembly, 27–32
Tegument structure, 31
Tegumentation, 178, 180–186
Temporal lobe, 87, 88
Terminal repeats (TR), 305
Terminase, 24, 176, 239–240
TET2, 446
TGF-β-activated kinase 1 (TAK1), 333
Thymidine kinase (TK), 104
Thymidylate synthase (TS), 106
TK altered, 109
TK deficient, 109
TLR 3, 87
TNF, 339
Toll-like receptors (TLRs), 87, 273, 274
TRAF3, 87
TRAF6, 333
Transcytosis, 276
Trans-Golgi network, 32
TRIF, 87

Triplex, 230
Trophoblasts, 273, 275
TSC2, 55–56
Tumor suppressor gene, 427
Two HSV entry pathways, 5–6
Type 2 EBV, 465

U
U14, 232
UL141, 242
UL16, 244
UL18, 243
UL31, 51
UL32, 232
UL34, 52
UL47, 53
UL8 (cofactor, VZVORF52), 113
UL5 (helicase, VZVORF55), 113
UL52 (primase, VZVORF6), 113
UNC93B1, 87
Us3, 46–50

V
Vaccine, 263
Vaginal epithelium, 11
Valaciclovir, 91
Valacyclovir, 104
Varicella, 123
Varicella-zoster (VZV) infection, 103
Varicella-zoster virus (VZV), 94–96
vBcl-2, 322, 341
vCyclin, 322, 341, 342
v-cyclin (*v-CYC, ORF72*), 303
vdUTPase, 53–54
Vesicle fusion, 36
vFLIP, 322, 335, 341
v-FLIP, 313
vGPCR, 322, 330, 335, 336
v-GPCR, 316
Vidarabine, 104
vIL6, 322, 331, 336
vIL-6, 360, 364
ViPK, 46

Viral isolation, 257
Viral Long Non-coding RNA, 328–329
Viral microRNA, 322
Viral miRNA, 326–328, 335
vIRF, 322–326
vIRF1, 322, 324, 343
vIRF2, 322, 325, 332
vIRF3, 322, 324
vIRF4, 322, 325, 343
Virion egress, 192–194
Virion envelopment and egress vesicles (VEEVs), 192
Virion structure, 146
Virion-containing vesicles, 35, 36
Virus egress, 35
Virus entry, 147–149
Viruslike particles (VLPs), 483
Virus-like replicon particle (VRP), 286
vOka virulence, 127
VP24 protease, 24
VZV DNA, 95
VZV encephalitis, 95
VZV vasculopathy, 96
VZV-specific CD4 T-cells, 134
VZV-specific cellular immunity, 124
VZV-specific long-lived memory T-cells, 133
VZV-specific T-cells, 134
VZV-specific T-cell immunity, 124

W
Whole-genome sequencing, 128
Wnt, 337
Wp-restricted latency, 387–388

X
Xenograft, 429

Z
ZOE-50, 135
ZOE-70, 136
Zta, 480, 484

Printed by Printforce, the Netherlands